AF410868

Synthesis and Characterization of Biomedical Materials

Synthesis and Characterization of Biomedical Materials

Editors

Leszek Adam Dobrzański
Anna D. Dobrzańska-Danikiewicz
Lech Bolesław Dobrzański
Joanna Dobrzańska

MDPI • Basel • Beijing • Wuhan • Barcelona • Belgrade • Manchester • Tokyo • Cluj • Tianjin

Editors
Leszek Adam Dobrzański
Medical and Dental Engineering Centre for
Research, Design and Production ASKLEPIOS
&
Koszalin University of Technology
Poland

Anna D. Dobrzańska-Danikiewicz
University of Zielona Góra
Poland

Lech Bolesław Dobrzański
Medical and Dental Engineering Centre for
Research, Design and Production ASKLEPIOS
Poland

Joanna Dobrzańska
Medical and Dental Engineering Centre for
Research, Design and Production ASKLEPIOS
Poland

Editorial Office
MDPI
St. Alban-Anlage 66
4052 Basel, Switzerland

This is a reprint of articles from the Special Issue published online in the open access journal *Processes* (ISSN 2227-9717) (available at: https://www.mdpi.com/journal/processes/special_issues/Biomedical_Materials).

For citation purposes, cite each article independently as indicated on the article page online and as indicated below:

LastName, A.A.; LastName, B.B.; LastName, C.C. Article Title. *Journal Name* **Year**, *Volume Number*, Page Range.

ISBN 978-3-0365-2522-8 (Hbk)
ISBN 978-3-0365-2523-5 (PDF)

Cover image courtesy of L.A. Dobrzański, L.B. Dobrzański, A.D. Dobrzańska-Danikiewicz, J. Dobrzańska and A. Achtelik-Franczak

Contents

About the Editors

Leszek Adam Dobrzański Prof. DSc, PhD, MSc, Eng, Hon. Prof., M Dr HC is working as a Full Professor of Materials Engineering, Nanotechnology, Biomaterials, Medical, Dental, and Manufacturing Engineering, Management and Organization, and the Director of the Science Centre ASKLEPIOS and the Supervisory Council Chairman of the Medical and Dental Engineering Centre for Research, Design and Production ASKLEPIOS Ltd in Gliwice, Poland. He also cooperates with the Koszalin University of Technology, where he works as a Full Professor. He is an Emeritus Professor and a former long-time Dean of the Faculty of Mechanical Engineering at the Silesian University of Technology in Gliwice, Poland, and a former Vice-rector of the Academy of Informatics and Management in Bielsko Biała, Poland, and then of the Silesian University of Technology in Gliwice, Poland. He is Vice President and a Fellow of the Academy of Engineers in Poland, a Foreign Fellow of the Ukrainian Academy of Engineering Sciences, and a Foreign Fellow of the Slovak Academy of Engineering Sciences, as well as being the President and Fellow of the World Academy of Materials and Manufacturing Engineering, and President of the International Association of Computational Materials Science and Surface Engineering. He is the Editor-in-Chief of the Journal of Achievements in Materials and Manufacturing Engineering, the Archives of Materials Science and Engineering and the Open Access Library, and a member of the editorial boards of many scientific journals. He is a creator and leader of the Scientific School, which personally promoted a group of 62 completed PhD theses and promoted ca. 1000 MSc and BSc theses. He headed more than 35 research and development projects, a few high-value investment projects to create new laboratory buildings perfectly equipped with super modern research and technological equipment, and about 20 international projects for research and academic staff mobility. He was the main organizer of almost 100 worldwide scientific conferences, and has given more than 100 openings and invited lectures during the international conferences in many world countries. He was a Visiting Professor in about 100 universities in 50 countries. His scientific output includes nearly 2500 scientific publications, including 50 books and monographs and 50 chapters in books and monographs, 250 papers in English in journals covered in the Web of Science Core Collection, and more than 60 patents. He has been cited ca. 16,000 times in worldwide scientific journals; h-index—52(GS), 32 (SC), 26 (WS). ORCID ID: https://orcid.org/0000-0002-7584-8917.

Anna D. Dobrzańska-Danikiewicz Prof., DSc, PhD, MSc Eng. represents the following scientific disciplines: Materials Science, Mechanical Engineering, and Biomedical Engineering. She is a specialist in nanotechnology, materials surface engineering, biomaterials, production engineering, and knowledge management. She is working as a Full Professor at the University of Zielona Góra, Poland. Her published scientific accomplishments comprise 250 items, including 12 scientific books. She participated in 25 foreign scientific conferences in countries including Mexico, Canada, USA, India, Turkey, Austria, France, Portugal, Spain, Switzerland, Germany, Italy, and Russia. She has participated in foreign internships, given guest lectures and taken part in European programs as an independent expert scoring proposals for co-financing, a lecturer, a scientist, a PhD-student, and a student staying at 12 foreign internships. She participated in 35 research, implementation, and education projects, including four times as a Principal Investigator in a managerial position. She is an expert who evaluated ca. 500 research, implementation, and education projects commissioned by public institutions, entrepreneurs, and the EACEA (EC). She also appraised 50 articles submitted to print in specialistic scientific journals (WoS, Scopus). From 2018, she has worked as the co-editor of Special Issues such as "Advanced Nanoengineering Materials" (Hindawi), "Carbon Nanomaterials and Nanocomposites" (ASTM International), and "Synthesis and Characterization of Biomedical Materials" (MDPI). She is the Deputy Editor-in-Chief of the Journal of Achievements in Materials and Manufacturing Engineering and the Archives of Materials Science and Engineering. She was the supervisor of 4 PhD Theses and ca. 70 MSc and Eng. theses. She is the co-creator of 5 granted patents and winner of 25 common awards for inventive activity. According to the referring journals, her publications have been referred to 687 times (SC), 237 times (WS), 1471 times (GS), as well as the h Hirsch index, which equals, respectively, 15 (SC), 10 (WS) and 20 (GS). ORCID ID: https://orcid.org/0000-0001-7335-5759.

Lech Bolesław Dobrzański PhD, MSc, Eng is a specialist in the following scientific disciplines: dental engineering, dentistry, materials engineering, and biomedical electronics. He is a medical manager working as Chief Executive Officer and President of the Medical and Dental Engineering Center for Research, Design, and Production ASKLEPIOS Ltd in Gliwice, Poland (since 2016) and CEO and President of the Center of Medicine and Dentistry SOBIESKI (since 2011). He is Deputy R&D Project Manager, was the coordinator of 15 projects financed by the European Union, and the coordinator of 10 research and development projects (PLN 22.5 million), including the Deputy Manager of the Research and Development Project (PLN 14.5 million). Since 2004, he has been continuously holding managerial positions in nongovernmental organizations and private companies. Since 2004, he has been the President of the Management Board of the SIGNUM Association. As part of his nongovernmental activity, he coordinated about 50 projects on various topics and promotional projects. As part of his commercial activities, as CEO of the Center SOBIESKI, he has carried out four already completed research and development and investment projects related to expanding the company's offer, increasing its competitiveness, and creating a unique offer of dental services and implementation of innovative prosthetic products on a national scale. He is the author of 50 articles and chapters in scientific books of international scope. He has received around 20 awards at inventions fairs around the world. H index: 13 (Google Scholar), 9 (Scopus), 5 (Web of Science) ORCID ID: https://orcid.org/0000-0002-4671-3060.

Joanna Dobrzańska PhD, MD is Vice President of the Center of Medicine and Dentistry SOBIESKI in Gliwice, Poland, and Member of the Supervisory Board of the Medical and Dental Engineering Center for Research, Design and Production ASKLEPIOS Ltd in Gliwice, Poland. She is a dentist practicing for 15 years, with experience performing medical experiments related to implementing innovative prosthetic restorations and implants. She practices prosthetics, implantology, implantology, endodontics, and cosmetic dentistry. From 2008 to 2012 she was a researcher at the Medical University of Silesia in Katowice. She is the head of the Microscopy Laboratory at the Asklepios Center. Since 2011, she has continuously participated in research and development projects as an R&D employee and co-manager. She participated in projects related to the introduction of new and improved prosthetic products to the market, in projects expanding the offer of her companies with unique services, as well as currently being a key member of the R&D team of the IMSKA-MAT project, where she serves as the main medical specialist. Since 2010, she has been authorized as the Radiological Protection Inspector. She is the author of 30 articles and chapters in books of national and international scope. She has received 12 awards at invention fairs around the world. H index: 6 (Google Scholar), 5 (Scopus), ORCID ID: https://orcid.org/0000-0002-0934-5991.

Preface to "Synthesis and Characterization of Biomedical Materials"

A balance between numerous factors, such as the economic level, sound condition of the natural and man-made environment, quality of life, health protection potential, scope of education, and level of information flow, when sufficiently high, can be considered a sign that well-being has been achieved. The United Nations has set 17 Sustainable Development Goals that guarantee this high level of prosperity, which all countries and societies in the world aim for. They include good health and well-being, and the greatest concern is extending human life while maintaining good health among aging societies. It is worth mentioning that the issue of ensuring health has been a constant concern of mankind since ancient times. It is enough to refer to the works of Hippocrates, dated more than 2.5 thousand years ago, which are quoted in this book several times, and even to older accounts, such as the ancient Greek myth of Asclepius. Health is also an age-old philosophical problem and remains imprecisely defined to this day. It can be imagined as a struggle between two opposing elements. A balance between these elements ensures health and well-being, while the loss of this balance plunges a person into illness. In this context, the current and earlier crises caused by the COVID-19 pandemic, human immunodeficiency virus (HIV), or hepatitis C virus (HCV) require greater attention than ever before regarding aspects of health safety and the organization and modernization of health systems, as well as the provision of appropriate specialized devices that can improve this safety.

Providing the highest possible level of protection of life and health is one of the basic human rights in the 21st century era of civilization. It is both a privilege of societies and must be universally available, and an inalienable duty of the rulers, especially in developed countries of democratic systems. Such activities are and must be addressed to every citizen because only the superposition of good health and well-being for all can ensure a positive result for the general public. This statement includes the individualization and personalization of all activities. This is the greatest difficulty in the holistic solution of this problem, affecting almost 8 billion people on Earth at present, without exception. One the most important aspects is protection against serious health threats, particularly civilization diseases, pandemics, and bioterrorism, to improve the health of the general public by ensuring medical care and health safety. Hence, the measures taken include the prevention of the risk of health loss and disease, which requires early diagnosis of the disease and the efficient and effective implementation of medical procedures. The necessary condition for achieving these goals is an improvement in the effectiveness of healthcare systems, modernization of the healthcare infrastructure, and the development and implementation of increasingly perfect, comprehensive, and innovative medical therapies.

Taking responsibility for the achievement of these goals requires continuous action to minimize, on a universal scale, any factors involved in loss of the dynamic balance between health and disease. Therefore, ensuring health and well-being requires universal activity and work toward this goal by the general public. It should be realized that many entities and areas of social life are responsible for achieving the set goals in terms of ensuring health and well-being. It cannot be assumed that the problem is only with medicine. and with doctors and other healthcare professionals. This is not the place to exhaustively list the social and professional groups affected by this statement. It is enough to mention that modern medicine makes extensive use of the achievements of basic, technical, and physicochemical sciences. PT readers can find statements in the scientific literature as well as in the daily press that medicine's many successes have their source in the implementation of solutions, devices, and technologies developed by engineers, based on the results of research in many scientific disciplines, sometimes, but only seemingly, very distant from medicine. Generally, these can be described as bioengineering.

The manufacturing of any product, including medical devices or other products that directly or indirectly serve medicine, must now follow the rules of the stage of the industrial revolution referred to as Industry 4.0 and the accompanying stage of development of the advanced information society, which could be consistently referred to as Society 4.0 (although, in Japan, the same stage of development was recognized as Society 5.0). Similarly, the currently achieved development stages can be defined as Bioengineering 4.0, Dentistry 4.0, Economy 4.0, and, therefore, also Medicine 4.0, and this is because, in each case, it is necessary to use products, materials, technologies, and design methodologies that require advanced cyber-physical systems that are appropriate for the Industry 4.0 stage. It should be realized that no product will be created without an appropriate, highly developed engineering design, responding to very specific needs in order to provide the required functional functions, which now require the use of very advanced information technologies and modern systems that collect and transfer information, including cloud computing. However, regardless of the design method, no product can be created without the use of engineering materials and, in some cases, natural technical materials. Detailed references to this state can be found in the individual articles in this book. The material engineering paradigm, to ensure the expected utility function, requires the optimal design of the expected material from among more than 100,000 already available or completely new materials, which, shaped in a properly designed expected technological process, will obtain the appropriate expected geometric form and the expected optimal structure, including surface structure. This will provide this product with the expected physicochemical properties which, in turn, will enable the achievement of the expected functional properties, ensuring the expected utility functions of the product. As stated in one of the articles in the book, this is the 6xE (expected) rule. For this reason, one of the main determinants of technological development at the Industry 4.0 stage is the development and implementation of new engineering materials according to the Materials 4.0 standards. Regarding the stages of Medicine 4.0 and Dentistry 4.0, the most important is the equally advanced stage of development of Biomaterials 4.0 biomedical materials.

The book focuses on the discussion of recent issues related to the production and implementation of biomedical materials for applications in medicine and dentistry. Frequent neoplastic diseases, traffic and sports accidents, as well as often neglected inflammations, often force the replacement of lost parts of the human body with artificial substitutes, for both organs and tissues. After application to the organism, biomedical materials remain in constant contact with living human tissues and body fluids and are exposed to the influence of microorganisms. The basic requirement is the biocompatibility of biomaterials, which should be established and tested before clinical application. There are many requirements for biomedical materials, resulting from extensive interdisciplinary knowledge not only in the field of medicine, materials science and engineering, but also tissue engineering, chemistry, physics and biology. An extensive knowledge of the material design and selection of materials is also required, as well as the use of various technologies for the manufacturing and synthesis of biomedical materials, and advanced technologies for their processing. The necessity of adjusting the geometrical features and properties of a specific restoration, prosthesis or other implants to the anatomical features and the disease case of a particular patient often require an individual approach. Then, design and additive manufacturing methods are applied. This means that, in this area, Biomaterial 4.0 stage was achieved by analogy to the Industry 4.0 stage of the industrial and social revolution. The chapters of this book provide detailed information on the development of both modern biomedical materials for use in modern medicine and dentistry, and tissue engineering. We sincerely thank all PT authors for their efforts in preparing their chapters. Our sincere thanks go to MDPI for the decision to publish this book, and especially to Ms. Tami Hu Section Managing Editor, as, without her activity, inspiration, and care, this book could not have been created.

We strongly encourage the wide group of PT readers to explore the book that we are presenting. The book will surely find an audience among practitioners, doctors, and engineers, who can use the experiences in their own daily clinical and technological practice, and among scientists from both fields, as an inspiration for their own research, and as a valuable source of knowledge among Ph.D. students and students of medical and dental studies, as well as engineering. We wish hope you enjoy exploring the interesting knowledge contained in individual chapters, and that this reading will be useful to you in your studies and/or professional activities, research and in productive and clinical work, regardless of your research activities and profession.

Leszek Adam Dobrzański, Anna D. Dobrzańska-Danikiewicz, Lech Bolesław Dobrzański ,
Joanna Dobrzańska
Editors

MDPI

Editorial

The Importance of Synthesis and Characterization of Biomedical Materials for the Current State of Medicine and Dentistry

Leszek A. Dobrzański [1],*, Anna D. Dobrzańska-Danikiewicz [2], Lech B. Dobrzański [1] and Joanna Dobrzańska [1]

[1]   Medical and Dental Engineering Centre for Research, Design and Production ASKLEPIOS, 13 D Krolowej Bony St., 44-100 Gliwice, Poland; dobrzanski@centrumasklepios.pl (L.B.D.); Joanna.dobrzanska@centrumasklepios.pl (J.D.)

[2]   Faculty of Mechanical Engineering, University of Zielona Góra, 4 Prof. Z. Szafrana St., 65-516 Zielona Góra, Poland; anna.dobrzanska.danikiewicz@gmail.com

*   Correspondence: leszek.dobrzanski@centrumasklepios.pl

> *"Noble health,*
> *no one will know*
> *what you taste*
> *like until you break down"*
>
> *"Ślachetne zdrowie,*
> *nikt się nie dowie,*
> *jako smakujesz,*
> *aż się zepsujesz"*
> *(in Polish)*
>
> Jan Kochanowski (1530–1584)

Citation: Dobrzański, L.A.; Dobrzańska-Danikiewicz, A.D.; Dobrzański, L.B.; Dobrzańska, J. The Importance of Synthesis and Characterization of Biomedical Materials for the Current State of Medicine and Dentistry. *Processes* **2021**, *9*, 978. https://doi.org/10.3390/pr9060978

Received: 27 May 2021
Accepted: 28 May 2021
Published: 1 June 2021

Publisher's Note: MDPI stays neutral with regard to jurisdictional claims in published maps and institutional affiliations.

From time immemorial to the present day, health has been considered to be of the highest value. In the writings of the father of medicine, Hippocrates (460–377 BC), known as *"The Hippocratic Corpus"* (Latin: "Corpus Hippocraticum"), or *Hippocratic Collection* [1,2], collected 100 years after his death and first translated and printed in 1525 in Venice, Italy, we can find the sentence "A wise man should know that health is his most precious possession and he should learn how to heal his diseases himself." Aulus Cornelius Celsus (53 BC to around 7 AD), a Roman scholar and encyclopedist, in his work *"De medicina libri VIII"* [3,4], published first in Florence in 1478, in turn, wrote that "It is not important what causes disease, but what removes it". In the Middle Ages in Poland, Jan Kochanowski (1530–1584), a poet and one of the greatest authors of the Renaissance in Europe, who contributed most to the development of the Polish literary language, wrote in his trifle poem the words we have chosen as the motto of this book [5]. We personally translated them into English, but unfortunately, we were unable to recreate either the rhythm or the rhyme of this fragment of the poem. Undoubtedly, the poet was already writing about the extraordinary value of human health. Today, the right to healthcare and maintaining good health is one of the primary and inalienable human rights. It is of particularly special value in the era of the global COVID-19 pandemic.

Modern medicine uses progress not only in basic sciences but also in technical and physicochemical sciences [6,7]. One of the best examples of such a combination of competencies is in biomaterials [8–19]. A biomaterial is a substance that is designed to interact with biological systems for therapeutic purposes. Biomaterials are used in various fields of medicine, especially in regenerative medicine, surgery, and dentistry, for the treatment, repair, or replacement of body tissues, for the improvement or restoration of tissue function, and for diagnostic purposes. Therefore, biomaterials used in contact with living tissues, organisms, or microorganisms must consider various issues related not only to medicine but also biology, chemistry, tissue engineering, and materials engineering. Biomaterials can be derived from nature or synthesized in a laboratory using a variety of methods involving metals, polymers, ceramics, or composites, as well as porous materials and biological-engineering materials [18]. Often the expected properties of biomaterials are obtained by surface engineering methods [20]. Biomaterials are often used or adapted

1

for medical applications, and thus, include all or part of a living biomedical structure or device that performs, enhances, or replaces a natural function. Such functions can be relatively passive, such as in the case of a heart valve, or bioactive and more interactive, as in the case of hip implants. Biomaterials are also widely used in dental, surgery, and drug delivery applications, e.g., in the form of pharmaceutical products that are placed in the body to release the drug over long periods. The biomaterials can be autografts, allografts, or xenografts used as transplant materials. Biomaterials are used, among others, for the replacement of joints, bone plates, bone cement, surgical sutures, clips and staples for closing wounds, pins and screws for stabilizing fractures, surgical meshes, breast implants, artificial ligaments and tendons, dental implants, stabilizers, vascular prostheses blood vessels, heart valves, vascular grafts, stents, nerve cables, skin repair devices, intraocular lenses for eye surgery, contact lenses, and drug delivery systems. Biomaterials must be compatible with the organism, and biocompatibility issues must be resolved before the product can be used in a clinical setting. For this reason, biomaterials are usually subjected to numerous tests, and the technologies for their evaluation must be certified.

Currently, the manufacturing and synthesis of biomaterials require the use of various technologies and methods. Often, these are methods that allow obtaining the correct material, which is then processed using advanced material processing technologies to manufacture a specific prosthesis or other types of implant. Often, however, it is necessary to directly manufacture a specific product with individualized geometric features and properties tailored to the requirements of a particular patient. Special technologies are required for diagnostic materials as well as for long-term drug release systems. Technologies in the current stage of the industrial revolution, Industry 4.0, are increasingly used in the biomaterial production cycle [21–35].

This book, including this Editorial on "The Importance of Synthesis and Characterization of Biomedical Materials for the Current State of Medicine and Dentistry" by Leszek A. Dobrzański, Anna D. Dobrzańska-Danikiewicz, Lech B. Dobrzański, and Joanna Dobrzańska as Editors [36], contains 20 chapters. The book aims to summarize the latest achievements in the development and manufacturing of modern biomaterials used in modern medicine and dentistry; for example, in cases where as a result of a traffic or sports accident, aging, resection of organs after oncological surgery, or dangerous inflammation, there is a need to replace lost organs, tissues, and parts of the human body. The book also includes chapters on the design of biomaterials and related technologies, including computer-aided design and manufacturing CAD/CAM methods, and research on their structure and properties, including biological research characterizing the reactions of the human body to the implantation or introduction of various types of biomaterials, both in the field of regenerative medicine and regenerative dentistry. A large part of the content contained in the next few chapters concerns dentistry and the issues mentioned above related to this particular branch of medicine, as well as implantation and prosthetic restorations of teeth.

After this introductory chapter, the first is a chapter by Leszek A. Dobrzański, Anna D. Dobrzańska-Danikiewicz, and Lech B. Dobrzański from Poland entitled "Effect of Biomedical Materials in the Implementation of a Long and Healthy Life Policy" [37]. Its purpose is a literature review showing the importance of bioengineering and the global production of biomaterials required for the level of health care in the world. All the considerations in this chapter are preceded by a discussion of health and diseases and the general well-being of societies. It turns out that the views of neither doctors nor philosophers as to the definition of these concepts are still not fully consistent; therefore, the intuitive definition contained in the Constitution of the World Health Organization decided at the founding meeting of this organization in 1948 should be adopted. Particular attention is paid to the relationship of health with the highly developed industry of the Industry 4.0 stage and the role assigned to biomaterials in this process. The need to support medicine on an ongoing basis through engineering activities is emphasized; largely applying to a wide group of medical devices including implants, implant-scaffolds, scaffolds, and prostheses, as well as the synthesis of entire systems by engineers and various methods of drug delivery, which by definition are not medical devices. There are strict rules on the use of medical devices, and adherence to strict moral imperatives is the responsibility of doctors and engineers working together. A far-reaching simplification is a commonly known principle of "do no harm first" because the ethical principles followed by engineers and doctors resulting from the so-called "Hippocratic oath" come down to a total of nine main findings discussed in the chapter. The tenth principle concerns the obligation to present to the patient the truth about his/her health condition and the therapeutic activities undertaken. It is crucial to consider the COVID-19 coronavirus pandemic, which is currently determining the health situation in the world. Hence, the need to introduce systemic measures to protect the health of both patients and medical staff against the spread of the lethal SARS-CoV-2 virus is also discussed, which has greatly expanded the scope of cooperation between medical and engineering activities. The next part of the chapter concentrates on the issues of bioengineering, mainly medical and dental engineering, taking into account the current stage of the Industry 4.0 industrial revolution implemented in these areas. Analogous to the concept of Dentistry 4.0, a general approach to Bioengineering 4.0 has been proposed. The basics of production management and the quality loop in the product life cycle are analyzed. The chapter contains an analysis of the synthesis and characteristics of biomedical materials supporting medicine and dentistry, with particular emphasis on the methods of additive manufacturing. Numerous examples of clinical applications are presented to support the consideration of biomedical materials. The economic conditions for implementing various groups of biomedical materials are supported by forecasts for the development of global markets for biomaterials, regenerative medicine, and tissue engineering. In Part Seven, the summary and concluding remarks are made against the background of a historical retrospective, which emphasizes that the technological processes of manufacturing and processing show a systematic increase in the global production of biomedical materials. The increase is a strong determinant for the implementation of policy for a long and healthy life around the world—one of the 17 sustainable development goals selected by the United Nations. The chapter is an extensive substantive introduction to the broad issues of biomedical materials and related technological processes presented in the book.

Looking at the 18 articles of this book, the first group concerns biomaterials used in dentistry and their relevant technological processes.

"The Concept of Sustainable Development of Modern Dentistry" [38] is presented in the next chapter by Leszek A. Dobrzański, Lech B. Dobrzański, Anna D. Dobrzańska-Danikiewicz, and Joanna Dobrzańska, the editors of this book. The paper concerns an assessment of the current state of dentistry in the world and the prospects of its sustainable development. A traditional Chinese censer was adopted as the pattern, characterized by strong and stable support on three legs. The dominant diseases of the oral cavity are caries and periodontal diseases, with the inevitable consequence of toothlessness. An estimated 3.5–5 billion people suffer from caries. Moreover, each of these diseases has a wide effect on the development of systemic complications. The territorial range of these diseases and their significant differentiation in severity in different countries and their impact on disability-adjusted life years index (DALY) are presented. Edentulousness has a significant impact on the oral health-related quality of life (OHRQoL). The etiology of these diseases is presented, as well as the preventive and therapeutic strategies undertaken as a result of modifying the Deming circle through the "fives' rules" idea. The state of development of Dentistry 4.0 is an element of the current stage of the industrial revolution Industry 4.0 and the great achievements of modern dental engineering. Dental treatment examples from the authors' own clinical practice are given. The systemic safety of a huge number of dentists in the world is discussed, in place of the passive strategy of using more and more advanced personal protective equipment (PPE), introducing the authors' own strategy for the active prevention of the spread of pathogenic microorganisms, including SARS-CoV-2. The ethical aspects of dentists' activity towards their own patients and the ethical obligations of the dentist community towards society are discussed in detail. This paper is a polemic, arguing against the view presented by a group of eminent specialists in the middle of the previous year in *The Lancet*. It is impossible to disagree with these views when it comes to waiting for egalitarianism in dental care, increasing the scope of prevention, and eliminating discrimination in this area on the basis of scarcity and poverty. The views on the discrimination of dentistry in relation to other branches of medicine are far more debatable. Therefore, relevant world statistics for other branches of medicine are presented. The authors of this paper do not agree with the thesis that interventional dental treatment can be replaced with properly implemented prophylaxis. The final remarks, therefore, present a discussion on the prospects for the development of dentistry based on three pillars, analogous to the traditional Chinese censer obtaining a stable balance thanks to its three legs. The Dentistry Sustainable Development (DSD) > 2020 model, consisting of Global Dental Prevention (GDP), Advanced Interventionist Dentistry 4.0 (AID 4.0), and Dentistry Safety System (DSS), is presented.

The next chapter deals with a very special aspect of dentistry—the safety of medical staff during dental procedures. The chapter entitled "Non-Antagonistic Contradictoriness of the Progress of Advanced Digitized Production with SARS-CoV-2 Virus Transmission in the Area of Dental Engineering" [39] was written by a team of six, composed of Leszek A. Dobrzański, Lech B. Dobrzański, Anna D. Dobrzańska-Danikiewicz, Joanna Dobrzańska, Karolina Rudziarczyk, and Anna Achtelik-Franczak from the ASKLEPIOS Center in Gliwice, Poland. The general goals of advanced digitized production in the Industry 4.0 stage of the industrial revolution are presented along with the extended holistic model of Industry 4.0, introduced by the authors, indicating the importance of material design and the selection of appropriate manufacturing technology. The effect of the global lockdown caused by the SARS-CoV-2 virus transmission pandemic was a drastic decrease in production, resulting in a significant decrease in the gross domestic product (GDP) in all countries, and huge problems in health care, including dentistry. Dentists belong to the highest risk group because the doctor works in the patient's respiratory tract. This chapter presents a breakthrough authors' solution, implemented by the active SPEC strategy (System-Prevention-Efficiency-Cause), and aims to eliminate clinical aerosol at the source by negative pressure aspirating bioaerosol at the patient's mouth line. The comparative benchmarking analysis and its results show that only the proprietary solution with a set of devices eliminates the threat at the source, while the remaining known methods do not meet the expectations. The details of this solution are described. Photopolymer materials and additive digital light printing (DLP) technology were used.

The next chapter covers "Dentistry 4.0 Concept in the Design and Manufacturing of Prosthetic Dental Restorations" [40]. The chapter was prepared by Leszek A. Dobrzanski and Lech B. Dobrzanski from Poland. It presents an overview of work on the current trends in the development of technical support for dental prosthetics under the name Dentistry 4.0, similar to the concept of the highest stage of Industry 4.0 of the industrial revolution. A new model of cooperation between dentists and dental engineers was formed. The research on the importance of cone-beam computed tomography CBCT in the planning of prosthetic treatment as well as in the design and manufacturing of prosthetic restorations is described. Procedures and technologies of implant preparation with the use of computer-aided design and manufacturing (CAD/CAM) methods and additive manufacturing technologies (AM), including selective laser sintering (SLS). Examples of this approach for several types of prosthetic restorations have been described.

Another chapter on "Virtual Approach to the Comparative Analysis of Biomaterials Used in Endodontic Treatment" [41] was developed by a team consisting of Joanna Dobrzańska, Lech B. Dobrzański, Klaudiusz Gołombek, Leszek A. Dobrzański, and Anna D. Dobrzańska-Danikiewicz from Poland. The importance of endodontics is presented within the framework of the own concept of Sustainable Development of Dentistry (DSD), presented previously. The materials and methods of clinical management in endodontic procedures are characterized, as well as forecasts for the development of the global endodontic market. A detailed methodological approach is presented to objectify the assessment of endodontic treatment with the use of procedural benchmarking methods and contextual matrices. In practice, the so-called "digital twins" approach to virtual benchmarking in endodontic treatment. The selection of materials, techniques for developing and filling root canals, and methods for assessing the effectiveness of fillings are virtually optimized. The full usefulness of the research on the effectiveness and tightness of root canal filling with the use of a scanning electron microscope is indicated.

The next chapter, entitled "Application Solid Laser-Sintered or Machined Ti6Al4V Alloy in Manufacturing of Dental Implants and Dental Prosthetic Restorations According to Dentistry 4.0 Concept" [42] was developed by a team consisting of Leszek A. Dobrzański, Lech B. Dobrzański, Anna Achtelik- Franczak, and Joanna Dobrzańska from Poland. The influence of milling technology in a numerically controlled computer machining center (CNC) and selective laser sintering (SLS) on the properties of the Ti6Al4V solid titanium alloy are compared. Even slight changes in the technological conditions in the SLS manufacturing can cause a difference of almost 2.5 times in the properties of the obtained material. The strength after milling is approx. 30% lower than in the case of additive manufacturing. At the same time, plug-and-play factory conditions can only provide approx. 60% of the actual and achievable strength properties, and therefore such a procedure cannot be approved. Biological tests with osteoblasts confirm a good tendency for the proliferation of living cells on the medium produced under optimal conditions. This chapter shows, based on the example of a specific biomaterial, how to practically implement the assumptions of the concept of Dentistry 4.0 in the production of dental prosthetics and implanted devices.

The chapter entitled "Corrosion Resistance of Cr-Co Alloys Subjected to Porcelain Firing Heat Treatment—In Vitro Study" [43] was prepared by the team of authors Dorota Rylska, Bartłomiej Januszewicz, Grzegorz Sokołowski, and Jerzy Sokołowski from Poland. This chapter describes the results of a comparative assessment of the effect of heat treatment used for firing dental ceramics on the corrosion properties of Co-Cr alloys manufactured by various methods: casting, milling of the compact, and its secondary sintering and selective laser sintering, which the authors call laser melting. Heat treatment under ceramic firing conditions was applied in all cases. Both the structure and the corrosion properties were investigated by means of electrochemical, potentiodynamic polarization tests. Among the heat-treated alloys, the material that was selectively laser sintered, then conventionally sintered after machining, showed the highest corrosion resistance, while the cast material turned out to be the least resistant. It is explained that this is related to the uniform distribution of the alloying elements in a homogeneous structure and the reduced porosity. The non-porosity increases the corrosion resistance and reduces the risk of crevice corrosion. The segregation of chemical elements, directly dependent on the manufacturing technology, has a fundamental impact on the corrosion behavior of the tested materials.

Chapter titled "Fretting Wear in Orthodontic and Prosthetic Alloys with Ti (C, N) Coatings" [44] was developed by a team consisting of Katarzyna Banaszek, Leszek Klimek, Jan. R. Dąbrowski, and Wojciech Jastrzębski. The aim of the presented research was to compare the fretting wear resistance of orthodontic and prosthetic Ni-Cr-Mo alloys coated with Ti (C, N) and without the coating. Samples with Ti (C, N) coatings showed higher resistance to fretting wear, and the wear in each case is more than twice lower. The use of coatings reduces the adverse effects of this type of wear by reducing the number of ions released during orthodontic treatment or wearing dentures.

The next chapter was developed by an eight-person team consisting of Katarzyna Mydłowska, Ewa Czerwińska, Adam Gilewicz, Ewa Dobruchowska, Ewa Jakubczyk, Łukasz Szparaga, Przemysław Ceynowa, and Jerzy Ratajski from Poland. The chapter is titled "Evolution of Phase Composition and Antibacterial Activity of Zr-C Thin Films" [45]. The results of extensive material science research, including electron microscopy, X-ray diffraction, and Raman spectroscopy, are presented. Zr-C coatings deposited on 304 L steel were tested by reactive magnetron sputtering from a Zr target in an Ar-C2H2 atmosphere at various acetylene flow rates. The concentration of C in the coatings ranged from 21 to 79%. Favorable mechanical properties of the coatings and relatively good antibacterial and anti-corrosive properties were obtained in the environment of artificial saliva where the concentration of C is greater than 50%. The Zr-C coatings obtained in this way can be used for medical purposes, especially in dentistry.

A large international team consisting of Sadaqat Ali, Ahmad Majdi Abdul Rani, Riaz Ahmad Mufti, Farooq I. Azam, Sri Hastuty, Zeeshan Baig, Murid Hussain, and Nasir Shehzad from Malaysia, Pakistan, and Indonesia prepared a chapter entitled "The Influence of Nitrogen Absorption on Microstructure, Properties and Cytotoxicity Assessment of 316 L Stainless Steel Alloy Reinforced with Boron and Niobium" [46]. In the past, 316 L stainless steel (SS) has been used in the manufacture of implants, although leaching of nickel ions limits its suitability for this purpose. The composition of the SS was modified by the addition of boron and niobium and then sintered under nitrogen for 8 h. X-ray diffraction (XRD) results showed the formation of strong nitrides, indicating nitrogen diffusion into the SS matrix and deposition of a nitride layer on the surface, making it difficult to wash out the nickel ions. Corrosion resistance in artificial saliva is improved and the developed upgraded stainless steels are compatible with living cells and can be used as implant materials.

"Optimization of Sintering Parameters of 316 L Stainless Steel for In-Situ Nitrogen Absorption and Surface Nitriding Using Response Surface Methodology" [47] is the title of a chapter prepared by a team from Malaysia and Indonesia composed of Sadaqat Ali, Ahmad Majdi Abdul Rani, Riaz Ahmad Mufti, Syed Waqar Ahmed, Zeeshan Baig, Sri Hastuty, Muhammad Al'Hapis Abdul Razak, and Abdul Azeez Abdu Aliyu. It is impossible not to notice that due to the nickel content, the possibilities of using this steel for medical purposes are limited. The simultaneous sintering and surface nitriding of 316 L stainless steel by powder metallurgy is discussed in this chapter. The developed technique of simultaneous sintering and surface nitriding could help improve the corrosion resistance of this material and control the leaching of metal ions for the potential use of this steel in biomedical applications.

The team consisting of Tsanka Dikova, Jordan Maximov, Vladimir Todorov, Georgi Georgiev, and Vladimir Panov from Bulgaria developed the chapter on "Optimization of the Photo-Polymerization Process of Dental Composites" [48]. The purpose of this chapter is to optimize the parameters of the photo-polymerization process of dental composites in order to obtain maximum hardness. Several alternative materials were considered, including a versatile light-curing composite, a bulk composite fill, and a flowable composite. The photo-polymerization of the samples was carried out under various conditions. For all composites, hardness regression models were determined on the upper and lower surfaces of the composite layer, for a total of 21 photo-polymerization modes that changed depending on the light intensity and exposure time. It turned out that the recommendations of the manufacturers of these materials were not confirmed by the test results. Therefore, Tables with recommended photo-polymerization regimes for the tested types of composites have been developed, which guarantee the high hardness of the composite filling. The research results developed in this way are of great practical importance as they can facilitate the daily work of dentists in dental clinics.

"Finite Element Analysis in Setting of Fillings of V-Shaped Tooth Defects Made with Glass-Ionomer Cement and Flowable Composite" [49] is the second chapter developed in Bulgaria by a team of Tsanka Dikova, Tihomir Vasilev, Vesela Hristova, and Vladimir Panov. Two different materials are used—resin reinforced self-curing glass ionomer cement (GIC) and flowable photocurable composite (FPC). There is an analogous non-uniform distribution of von Mises equivalent stresses on V-shaped cavity fillings made with GIC and FPC. Maximum stresses arise along the boundaries of the filling on the vestibular surface of the tooth and on the bottom of the filling. The von Mises equivalent stress values of the GIC fillings are higher than the FPC values. An experimental study of micro-leaks confirmed the adequacy of the models used in FEM.

The fourteenth chapter, does not concern dentistry anymore and was developed in Poland by the team of Anna Ślósarczyk, Joanna Czechowska, Ewelina Cichoń, and

Aneta Zima. The title of the chapter is "New Hybrid Bioactive Composites for Bone Substitution" [50]. The chapter describes the results of optimization of the composition of hybrid inorganic–organic materials and covers the study of bio-micro-concrete containing hydroxyapatite (HAp)-chitosan (CTS) granules dispersed in a matrix of α-tricalcium phosphate (αTCP). The resulting materials proved to be promising bone substitutes for non-load-bearing applications.

A team from Portugal consisting of Basam A.E. Ben-Arfa and Robert C. Pullar prepared a chapter entitled "A Comparison of Bioactive Glass Scaffolds Fabricated by Robocasting from Powders Made by Sol-Gel and Melt-Quenching Methods" [51]. Bioactive glass scaffold is used in bone and tissue biomedical implants and their fabrication using additive manufacturing/3D printing techniques such as robocasting is of great interest. Most bioactive glasses are based on silica forming a glass network, with calcium and phosphorus content for new bone growth, and a glass modifier such as sodium. In relation to 4555 Bioglass, robocasting was used for the first time in 2013, and in 2019, sol-gel with high silica content (HSSGG) was used in 2019.

The chapter titled "Injectable Chitosan Scaffolds with Calcium β-Glycerophosphate as the Only Neutralizing Agent" [52] was developed by the team of Piotr Owczarz, Anna Ryl, Marek Dziubinski, and Jan Sielski from Poland. A method of preparing thermosensitive hydrogel chitosan is described, using calcium β-glycerophosphate as the only pH neutralizing agent and supporting the cross-linking process. The presence of calcium ions instead of sodium ions is especially important for scaffolds in bone engineering. It is possible to enrich the obtained cell grids with calcium ions with the addition of calcium carbonate with a physiological ratio of calcium to phosphorus (1.6–1.8): 1.

"Green and Facile Synthesis of Dendritic and Branched Gold Nanoparticles by Gelatin and Investigation of Their Biocompatibility on Fibroblast Cells" [53] is another chapter prepared by a team from Vietnam consisting of Quoc Khuong Vo, My Nuong Nguyen Thi, Phuong Phong Nguyen Thi, and Duy Trinh Nguyen. Gold nanostar (AuNPs) and gold nano dendrites were synthesized by a one-pot and environmentally friendly method in the presence of gelatin. Subsequently, the influence of gelatin concentrations and reaction conditions on branched (AuNP) growth was investigated. Interestingly, the conversion of morphology between the dendritic and branched nanostructures can be achieved by changing the pH value of the gelatin solution. The role of gelatin as a protective agent through electrostatic and steric interaction is also disclosed. On the basis of the performed characteristics, it is possible to propose a growth mechanism explaining the evolution of the branched morphology of AuNP. Moreover, branched AuNPs are highly reliable at a concentration of 100 μg/mL when performed in the SRB test with human foreskin fibroblast cells.

Another team from Poland, consisting of Jagoda Kurowiak, Agnieszka Kaczmarek-Pawelska, Agnieszka G. Mackiewicz, and Romuald Bedzinski prepared the chapter titled "Analysis of the Degradation Process of Alginate-Based Hydrogels in Artificial Urine for Use as a Bioresorbable Material in the Treatment of Urethral Injuries" [54]. Natural polymer hydrogels such as sodium alginate have great potential in regenerative medicine due to their biocompatibility, biodegradability, mechanical properties, bio-resorbability, and relatively low cost. Sodium alginate, a polysaccharide derived from brown algae, is the most widely researched and used biomaterial in biomedical applications. Alginate dressings are also useful as a delivery platform to provide a controlled release of therapeutic substances (e.g., analgesics, antibacterials, and anti-inflammatory agents). The process of alginate hydrogel degradation was analyzed. Hydrogels were prepared by the dip-coating method. An original hybrid cross-linking process using not one, but a mixture of two cross-linkers (calcium chloride and barium chloride) was also described. Various cross-linking agents allow the production of hydrogels with a spectrum of mechanical properties similar to that of the urethral tissue. The obtained hydrogels have a different degree of degradation in artificial urine and can be used as a material for healing urethral injuries, especially urethral strictures, which have a significant impact on the quality of life of patients.

An international team from China and Canada consisting of Jiang Xu, Yuyan Chen, Xizhi Jiang, Zhongzheng Gui, and Lei Zhang presented a chapter titled "Development of Hydrophilic Drug Encapsulation and Controlled Release Using a Modified Nanoprecipitation Method" [55]. A novel drip nanoprecipitation method was developed that separates hydrophilic drugs and polymers into an aqueous phase (continuous phase) and an organic phase (dispersed phase), both individually and in a mixing process. This method successfully produced NPs loaded with ciprofloxacin-loaded NPs by Poly (d, l-lactic acid)-Dextran (PLA-DEX) and Polylactic acid-co-glycolic acid-Polyethylene glycol (PLGA-PEG) ensuring the ability to charge to 27.2 wt.% and sustained release in vitro for up to six days. The drug content in NP can be precisely adjusted by changing the initial concentration of ciprofloxacin. This modified nanoprecipitation method is a rapid, easy and reproducible technique to produce nanoscale drug delivery vehicles with high drug delivery capacity.

Twentieth in this book, so last but not least, but certainly compiled by the most numerous 11-person team from Vietnam: Ngoc-Tram Nguyen-Thi, Linh Phuong Pham Tran, Ngoc Thuy Trang Le, Minh-Tri Cao, The -Nam Tran, Ngoc Tung Nguyen, Cong Hao Nguyen, Dai-Hai Nguyen, Van Thai Than, Quang Tri Le, and Nguyen Quang Trung, is the chapter is entitled "The Engineering of Porous Silica and Hollow Silica Nanoparticles to Enhance Drug-loading Capacity" [56]. The hollow mesoporous silica nanoparticles (HMSN) are used to increase drug loading capacity. Porous nanosilica (PNS) and HMSN were prepared by sol-gel and matrix-assisted methods and then used for rhodamine (RhB) loading. Based on the results of the conducted research, it is suggested that the prepared HMSN nanocarriers can act as high-capacity carriers in drug delivery applications.

It is with personal satisfaction that we present this book to P.T (Pleno Titulo). Readers. We are deeply convinced that it will contribute to the popularization and the level of knowledge about biomaterials, which more and more often occupy an important place in modern medicine. As the Hippocratic Oath dictates, despite the fact that its text has been appropriately modified today, it is not enough to simply not harm the patient—that would be far too little. Maximum effort and knowledge should be made to actively help patients to ensure their long health and well-being, both mentally and physically. As we wrote at the beginning of this Editorial, engineers and their highly specialized knowledge of the possibility of using various biomaterials and the advanced technologies appropriate for them have a huge role to play in these activities. Just because in this area, as in many other areas of contemporary design and production of numerous products, the assumptions and achievements of the Industry 4.0 stage are increasingly used, more and more patients in the world after a serious illness or accident can live peacefully and healthily for subsequent and often many years, because doctors applied the required medical devices, most often implanted into the body, entirely conceived and manufactured by engineers, and precisely with the use of biomaterials.

The book will surely find an audience among practitioners, doctors, and engineers who can use the described experiences in their own daily clinical and technological practice, among scientists from both fields, inspiring them to their own research and scientific research, and as a valuable source of knowledge among Ph.D. students and students, both in medical and dental studies, as well as in engineering.

Both the authors and we, as editors, put a lot of effort and effort to make it materialize. We would like to thank all the authors for their cooperation, and special thanks to Ms. Tami Hu who supported us in terms of organization on behalf of the MDPI Branch in Beijing, China, without the inspiration and help of who this book could never have been written.

It, therefore, remains to wish PT Readers a pleasant and fruitful reading.

Conflicts of Interest: The authors declare no conflict of interest.

References

1. Hippocrates, C.J. *Coi Medicorum Omnium Longe Principis, Opera Quae Apud Nos Extant Omnia*; Apud Ant. Vincentium: New York, NY, USA, 1555. Available online: https://search.lib.virginia.edu/sources/uva_library/items/u3623846 (accessed on 26 March 2021).
2. Iniesta, I. Hippocratic Corpus. *BMJ* **2011**, *342*, d688. [CrossRef]

3. Celsi, A.C. De Medicina Libri Octo. Available online: https://archive.org/details/aurelcornelcelsi00cels?ref=ol&view=theater (accessed on 26 March 2021).
4. Celsi, A.C. In Hoc Uolumine Haec Continentur Aurelij Cornelij Celsi Medicinae libri 8. In Aedibus Aldi, et Andreae Asulani Soceri. 1528. Available online: https://openlibrary.org/works/OL16537329W/In_hoc_uolumine_haec_continentur_Aurelij_Cornelij_Celsi_Medicinae_libri_8?edition=inhocuoluminehae00cels (accessed on 26 March 2021).
5. Kochanowski, J. *Dzieła Polskie*, 8th ed.; Państwowy Instytut Wydawniczy: Warsaw, Poland, 1976; Volume 1, Available online: https://wolnelektury.pl/media/book/txt/fraszki-ksiegi-trzecie-na-zdrowie.txt (accessed on 26 March 2021).
6. Herold, K.E.; Vossoughi, J.; Bentley, W.E. (Eds.) *26th Southern Biomedical Engineering Conference SBEC 2010, College Park, MD, USA, 30 April–2 May 2010*; Springer Science & Business Media, 2010. Available online: https://archive.org/details/thsouthernbiomed00hero/page/n3/mode/2up (accessed on 26 March 2021).
7. Enderle, J.D.; Bronzino, J.D. (Eds.) *Introduction to Biomedical Engineering*, 3rd ed.; Academic Press: Oxford, UK, 2012. [CrossRef]
8. Bronzino, J.D.; Peterson, D.R. *The Biomedical Engineering Handbook: Biomedical Engineering Fundamentals*, 4th ed.; CRC Press: Boca Raton, FL, USA, 2019; Volume 1.
9. Wiliams, D.F. (Ed.) *Definitions in Biomaterials*; Elsevier: Amsterdam, The Netherlands; Oxford, UK; New York, NY, USA; Tokyo, Japan, 1987.
10. Dobrzański, L.A. *Materiały Inżynierskie i Projektowanie Materiałowe: Podstawy Nauki o Materiałach i Metaloznawstwo*, 2nd ed.; WNT: Warsaw, Poland, 2006.
11. Ratner, B.D.; Hoffman, A.S.; Schoen, F.J.; Lemons, J.E. (Eds.) *Biomaterials Science: An Introduction to Materials in Medicine*, 3rd ed.; Academic Press: Oxford, UK, 2013.
12. Dobrzański, L.A. Metale i ich stopy. In *Open Access Library*; Dobrzański, L.A., Ed.; International OCSCO World Press: Gliwice, Poland, 2017; Volume VII, pp. 1–982.
13. Hin, T.S. (Ed.) *Engineering Materials for Biomedical Applications*; World Scientific: Singapore, 2004. [CrossRef]
14. Dobrzański, L.A. (Ed.) *Biomaterials in Regenerative Medicine*; IntechOpen: Rijeka, Croatia, 2018.
15. Pignatello, R. (Ed.) *Advances in Biomaterials Science and Biomedical Applications*; IntechOpen: Rijeka, Croatia, 2013.
16. Jenkins, M. (Ed.) *Biomedical Polymers*; Woodhead Publishing: Cambridge, UK, 2007; pp. 1–236.
17. Kokubo, T. (Ed.) *Biocermics and Their Clinical Applications*; Woodhead Publishing: Cambridge, UK, 2008; pp. 1–784.
18. Dobrzański, L.A.; Dobrzańska-Danikiewicz, A.D. (Eds.) Metalowe materiały mikroporowate i lite do zastosowań medycznych i stomatologicznych. In *Open Access Library*; International OCSCO World Press: Gliwice, Poland, 2017; Volume VII, pp. 1–580.
19. Curtis, R.V.; Watson, T.F. (Eds.) *Dental Biomaterials*; Woodhead Publishing: Cambridge, UK, 2008; pp. 1–528.
20. Dobrzański, L.A.; Dobrzańska-Danikiewicz, A.D. Inżynieria powierzchni materiałów: Kompendium wiedzy i podręcznik akademicki. In *Open Access Library*; Dobrzański, L.A., Ed.; International OCSCO World Press: Gliwice, Poland, 2018; Volume VIII, pp. 1–1138.
21. Kagermann, H.; Wahlster, W.; Helbig, J. *Recommendations for Implementing the Strategic Initiative INDUSTRIE 4.0: Final Report of the Industrie 4.0 Working Group*; Federal Ministry of Education and Research: Bonn, Germany, 2013.
22. Kagermann, H. Chancen von Industrie 4.0 Nutzen. In *Industrie 4.0 in Produktion, Automatisierung und Logistik*; Springer Fachmedien Wiesbaden: Wiesbaden, Germany, 2014; pp. 603–614.
23. Hermann, M.; Pentek, T.; Otto, B. *Design Principles for Industrie 4.0 Scenarios: A Literature Review*; Technische Universität Dortmund: Dortmund, Germany, 2015.
24. Rüßmann, M.; Lorenz, M.; Gerbert, P.; Waldner, M.; Justus, J.; Engel, P.; Harnisch, M. *Industry 4.0: The Future of Productivity and Growth in Manufacturing Industries*; Boston Consulting Group: Boston, MA, USA, 2015.
25. Jose, R.; Ramakrishna, S. Materials 4.0: Materials Big Data Enabled Materials Discovery. *Appl. Mater. Today* **2018**, *10*, 127–132. [CrossRef]
26. Dobrzański, L.A.; Dobrzański, L.B. Approach to the Design and Manufacturing of Prosthetic Dental Restorations According to the Rules of Industry 4.0. *Mater. Perform. Charact.* **2020**, *9*, 394–476. [CrossRef]
27. Dobrzański, L.A.; Dobrzańska-Danikiewicz, A.D.; Achtelik-Franczak, A.; Dobrzański, L.B.; Szindler, M.; Gaweł, T.G. Porous selective laser melted Ti and Ti6Al4V materials for medical applications. In *Powder Metallurgy—Fundamentals and Case Studies*; Dobrza'nski, L.A., Ed.; IntechOpen: Rijeka, Croatia, 2017; pp. 161–181.
28. Dobrzański, L.A. Role of materials design in maintenance engineering in the context of industry 4.0 idea. *JAMME* **2019**, *96*, 12–49. [CrossRef]
29. Dobrzański, L.A.; Dobrzańska-Danikiewicz, A.D. Why Are Carbon-Based Materials Important in Civilization Progress and Especially in the Industry 4.0 Stage of the Industrial Revolution. *Mater. Perform. Charact.* **2019**, *8*, 337–370. [CrossRef]
30. Dobrzański, L.A.; Dobrzański, L.B.; Dobrzańska-Danikiewicz, A.D. Overview of conventional technologies using the powders of metals, their alloys and ceramics in Industry 4.0 stage. *JAMME* **2020**, *98*, 56–85. [CrossRef]
31. Dobrzański, L.A.; Dobrzański, L.B.; Dobrzańska-Danikiewicz, A.D.; Kraszewska, M. Manufacturing powders of metals, their alloys and ceramics and the importance of conventional and additive technologies for products manufacturing in Industry 4.0 stage. *Arch. Mater. Sci. Eng.* **2020**, *102*, 13–41. [CrossRef]
32. Dobrzański, L.B.; Achtelik-Franczak, A.; Dobrzańska, J.; Dobrzański, L.A. The digitisation for the immediate dental implantation of incisors with immediate individual prosthetic restoration. *JAMME* **2019**, *97*, 57–68. [CrossRef]
33. Dobrzański, L.A.; Dobrzańska-Danikiewicz, A.D. Applications of Laser Processing of Materials in Surface Engineering in the Industry 4.0 Stage of the Industrial Revolution. *Mater. Perform. Charact.* **2019**, *8*, 1091–1129. [CrossRef]

34. Dobrzański, L.A. Effect of heat and surface treatment on the structure and properties of the Mg-Al-Zn-Mn casting alloys. In *Magnesium and Its Alloys*; Dobrzański, L.A., Totten, G.E., Bamberger, M., Eds.; CRC Press: Boca Raton, FL, USA, 2019; pp. 91–202.

35. Dobrzański, L.A. Introductory Chapter: Multi-Aspect Bibliographic Analysis of the Synergy of Technical, Biological and Medical Sciences Concerning Materials and Technologies Used for Medical and Dental Implantable Devices. In *Biomaterials in Regenerative Medicine*; Dobrzański, L.A., Ed.; IntechOpeen: Rijeka, Croatia, 2018; pp. 1–43. [CrossRef]

36. Dobrzański, L.A.; Dobrzańska-Danikiewicz, A.D.; Dobrzański, L.B.; Dobrzańska, J. The importance of synthesis and characterization of biomedical materials for the current state of medicine and dentistry. *Processes* **2021**, in press.

37. Dobrzański, L.A.; Dobrzańska-Danikiewicz, A.D.; Dobrzański, L.B. Effect of Biomedical Materials in the implementation of a long and healthy life policy. *Processes* **2021**, *9*, 865. [CrossRef]

38. Dobrzański, L.A.; Dobrzański, L.B.; Dobrzańska-Danikiewicz, A.D.; Dobrzańska, J. The Concept of Sustainable Development of Modern Dentistry. *Processes* **2020**, *8*, 1605. [CrossRef]

39. Dobrzański, L.A.; Dobrzański, L.B.; Dobrzańska-Danikiewicz, A.D.; Dobrzańska, J.; Rudziarczyk, K.; Achtelik-Franczak, A. Non-Antagonistic Contradictoriness of the Progress of Advanced Digitized Production with SARS-CoV-2 Virus Transmission in the Area of Dental Engineering. *Processes* **2020**, *8*, 1097. [CrossRef]

40. Dobrzański, L.A.; Dobrzański, L.B. Dentistry 4.0 Concept in the Design and Manufacturing of Prosthetic Dental Restorations. *Processes* **2020**, *8*, 525. [CrossRef]

41. Dobrzańska, J.; Dobrzański, L.B.; Gołombek, K.; Dobrzański, L.A.; Dobrzańska-Danikiewicz, A.D. Virtual Approach to the Comparative Analysis of Biomaterials Used in Endodontic Treatment. *Processes* **2021**, *9*, 926. [CrossRef]

42. Dobrzański, L.A.; Dobrzański, L.B.; Achtelik-Franczak, A.; Dobrzańska, J. Application Solid Laser-Sintered or Machined Ti6Al4V Alloy in Manufacturing of Dental Implants and Dental Prosthetic Restorations According to Dentistry 4.0 Concept. *Processes* **2020**, *8*, 664. [CrossRef]

43. Rylska, D.; Januszewicz, B.; Sokołowski, G.; Sokołowski, J. Corrosion Resistance of Cr–Co Alloys Subjected to Porcelain Firing Heat Treatment—In Vitro Study. *Processes* **2021**, *9*, 636. [CrossRef]

44. Banaszek, K.; Klimek, L.; Dąbrowski, J.R.; Jastrzębski, W. Fretting Wear in Orthodontic and Prosthetic Alloys with Ti(C., N) Coatings. *Processes* **2019**, *7*, 874. [CrossRef]

45. Mydłowska, K.; Czerwińska, E.; Gilewicz, A.; Dobruchowska, E.; Jakubczyk, E.; Szparaga, Ł.; Ceynowa, P.; Ratajski, J. Evolution of Phase Composition and Antibacterial Activity of Zr–C Thin Films. *Processes* **2020**, *8*, 260. [CrossRef]

46. Ali, S.; Abdul Rani, A.M.; Mufti, R.A.; Azam, F.I.; Hastuty, S.; Baig, Z.; Hussain, M.; Shehzad, N. The Influence of Nitrogen Absorption on Microstructure, Properties and Cytotoxicity Assessment of 316L Stainless Steel Alloy Reinforced with Boron and Niobium. *Processes* **2019**, *7*, 506. [CrossRef]

47. Ali, S.; Abdul Rani, A.M.; Ahmad Mufti, R.; Ahmed, S.W.; Baig, Z.; Hastuty, S.; Razak, M.A.A.; Abdu Aliyu, A.A. Optimization of Sintering Parameters of 316L Stainless Steel for In-Situ Nitrogen Absorption and Surface Nitriding Using Response Surface Methodology. *Processes* **2020**, *8*, 297. [CrossRef]

48. Dikova, T.; Maximov, J.; Todorov, V.; Georgiev, G.; Panov, V. Optimization of Photopolymerization Process of Dental Composites. *Processes* **2021**, *9*, 779. [CrossRef]

49. Dikova, T.; Vasilev, T.; Hristova, V.; Panov, V. Finite Element Analysis in Setting of Fillings of V-Shaped Tooth Defects Made with Glass-Ionomer Cement and Flowable Composite. *Processes* **2020**, *8*, 363. [CrossRef]

50. Ślósarczyk, A.; Czechowska, J.; Cichoń, E.; Zima, A. New Hybrid Bioactive Composites for Bone Substitution. *Processes* **2020**, *8*, 335. [CrossRef]

51. Ben-Arfa, B.A.E.; Pullar, R.C. A Comparison of Bioactive Glass Scaffolds Fabricated by Robocasting from Powders Made by Sol–Gel and Melt-Quenching Methods. *Processes* **2020**, *8*, 615. [CrossRef]

52. Owczarz, P.; Rył, A.; Dziubiński, M.; Sielski, J. Injectable Chitosan Scaffolds with Calcium β-Glycerophosphate as the Only Neutralizing Agent. *Processes* **2019**, *7*, 297. [CrossRef]

53. Vo, Q.K.; Nguyen Thi, M.N.; Nguyen Thi, P.P.; Nguyen, D.T. Green and Facile Synthesis of Dendritic and Branched Gold Nanoparticles by Gelatin and Investigation of Their Biocompatibility on Fibroblast Cells. *Processes* **2019**, *7*, 631. [CrossRef]

54. Kurowiak, J.; Kaczmarek-Pawelska, A.; Mackiewicz, A.G.; Bedzinski, R. Analysis of the Degradation Process of Alginate-Based Hydrogels in Artificial Urine for Use as a Bioresorbable Material in the Treatment of Urethral Injuries. *Processes* **2020**, *8*, 304. [CrossRef]

55. Xu, J.; Chen, Y.; Jiang, X.; Gui, Z.; Zhang, L. Development of Hydrophilic Drug Encapsulation and Controlled Release Using a Modified Nanoprecipitation Method. *Processes* **2019**, *7*, 331. [CrossRef]

56. Nguyen-Thi, N.-T.; Pham Tran, L.P.; Le, N.T.T.; Cao, M.-T.; Tran, T.-N.; Nguyen, N.T.; Nguyen, C.H.; Nguyen, D.-H.; Than, V.T.; Le, Q.T.; et al. The Engineering of Porous Silica and Hollow Silica Nanoparticles to Enhance Drug-loading Capacity. *Processes* **2019**, *7*, 805. [CrossRef]

Review

Effect of Biomedical Materials in the Implementation of a Long and Healthy Life Policy

Leszek A. Dobrzański [1,*], Anna D. Dobrzańska-Danikiewicz [2] and Lech B. Dobrzański [1]

[1] Medical and Dental Engineering Centre for Research, Design and Production ASKLEPIOS, 12/1 Jan III Sobieski Street, 44-100 Gliwice, Poland; dobrzanski@centrumasklepios.pl

[2] Department of Mechanical Engineering, University of Zielona Góra, 4 Prof. Z. Szafrana St., 65-516 Zielona Góra, Poland; anna.dobrzanska.danikiewicz@gmail.com

* Correspondence: leszek.dobrzanski@centrumasklepios.pl

Abstract: This paper is divided into seven main parts. Its purpose is to review the literature to demonstrate the importance of developing bioengineering and global production of biomaterials to care for the level of healthcare in the world. First, the general description of health as a universal human value and assumptions of a long and healthy life policy is presented. The ethical aspects of the mission of medical doctors and dentists were emphasized. The coronavirus, COVID-19, pandemic has had a significant impact on health issues, determining the world's health situation. The scope of the diseases is given, and specific methods of their prevention are discussed. The next part focuses on bioengineering issues, mainly medical engineering and dental engineering, and the need for doctors to use technical solutions supporting medicine and dentistry, taking into account the current stage Industry 4.0 of the industrial revolution. The concept of Dentistry 4.0 was generally presented, and a general Bioengineering 4.0 approach was suggested. The basics of production management and the quality loop of the product life cycle were analyzed. The general classification of medical devices and biomedical materials necessary for their production was presented. The paper contains an analysis of the synthesis and characterization of biomedical materials supporting medicine and dentistry, emphasizing additive manufacturing methods. Numerous examples of clinical applications supported considerations regarding biomedical materials. The economic conditions for implementing various biomedical materials groups were supported by forecasts for developing global markets for biomaterials, regenerative medicine, and tissue engineering. In the seventh part, recapitulation and final remarks against the background of historical retrospection, it was emphasized that the technological processes of production and processing of biomedical materials and the systematic increase in their global production are a determinant of the implementation of a long and healthy policy.

Keywords: health; well-being; long and healthy life policy; medicine; dentistry; medical ethics; COVID-19 pandemic; bioengineering; medical engineering; dental engineering; biomedical materials; Industry 4.0; Dentistry 4.0; Bioengineering 4.0; engineers' ethics

check for
updates

Citation: Dobrzański, L.A.; Dobrzańska-Danikiewicz, A.D.; Dobrzański, L.B. Effect of Biomedical Materials in the Implementation of a Long and Healthy Life Policy. *Processes* **2021**, *9*, 865. https://doi.org/10.3390/pr9050865

Academic Editor: Selestina Gorgieva

Received: 23 March 2021
Accepted: 12 May 2021
Published: 14 May 2021

Publisher's Note: MDPI stays neutral with regard to jurisdictional claims in published maps and institutional affiliations.

1. Introduction

Since the dawn of time, health has been an inalienable and unquestionable human value. Doctors and philosophers have been dealing with the definition of health, diseases, and various disabilities and differences from normality in this area also for many millennia. It turns out, however, that defining these concepts is still difficult to define, although it may seem that they are intuitively close to every human being. Caring for health, from everyday elementary hygiene, through treatments aimed at maintaining physical condition, wide-ranging preventive actions to therapeutic treatments using natural or synthetic pharmacological agents, and decisive interventions in surgical and prosthetic procedures is a normal practice and involves billions of people in the world. Moreover, it is commonly known that the scope of officially organized medical care is strongly dependent on the organization's level of society and resources of the state budgets. It is the main reason for the differences in the standard of health care in different countries. Affluent countries are trying to reach many citizens with this care, and health care can be considered common there. Usually, this medical care is, to a large extent, supported, if necessary, from the citizens' own funds. The situation is completely different in economically underdeveloped countries, and political systems are far from liberal democracy. In such countries, medical care does not reach the majority of citizens, or hardly at all. Although the right to health is one of the basic human rights, the level of access to exercising this right is highly diversified depending on the geographic location. The level of health care may vary by several dozen or even several hundred times. It, of course, is closely related to well-being, which is also very diverse. This situation is of particular importance in the era of a severe global pandemic on the transmission of the SARS-CoV-2 coronavirus. The safety of patients and, above all, the safety of medical staff poses new challenges.

The modern level of medicine and dentistry, as its special area, is currently closely related to the development of numerous technical means and medical devices, which became possible thanks to the dynamic development of bioengineering along with medical, dental, and tissue engineering regenerative medicine and cell therapies. Both diagnostics and complex interventional procedures require very advanced medical devices. The development of these areas of knowledge, usually engineering and technical, is now a strong determinant of the development of medicine and medical science in general. Devices of this type are subject to all the industrial development rules; therefore, it is not surprising that these areas reach the Industry 4.0 stage. It is a necessity on the one hand and a challenge on the other. The development of the industry is based on the development of production, which, in turn, cannot exist without the use of engineering materials. In medical and dental devices, a very specific set of properties of the materials used is required, especially biocompatibility and ensuring the absence of adverse interactions with living tissues of the human body, and this group of materials is called biomedical materials or simply biomaterials. It is easy to demonstrate the dependence of the development of this group of materials and related global markets with the level of care for the level of healthcare.

The aim of this very extensive review of the literature is to analyze the importance of bioengineering in the context of the Industry 4.0 stage of the industrial revolution and the development of global production of biomaterials, their production, and processing for the implementation of a policy of long and healthy life and the care the level of health care in the world.

The paper is extensively illustrated with appropriate diagrams and figures drawn by the authors that complement the verbal content presented in the paper in a convincing and perceptive-friendly way. These figures are the quintessence of knowledge on specific issues, not an indirect transfer or quotation from any source. A detailed analysis of the content of these figures undoubtedly makes it easier to understand the complex content of the paper and is explaining and supplementing it. Carefully selected own photos illustrating, among others, materiallographic research results are presented, in the parts containing examples of practical and clinical activities of the authors.

2. General Description of Health as a Universal Human Value and Assumptions of a Long and Healthy Life Policy

The basic and currently common economic doctrine based on the Club of Rome's achievements in 1972 [1] and taking into account the theory of public good [2], assumes that sustainable development occurs. This concept's essence consists of meeting the current needs of the present generation to a degree that does not diminish future generations' chances to meet their needs appropriately [3]. Prosperity at the expected high level can be guaranteed if a balance is ensured between the sufficiently high economic status, ensuring the proper state of the environment, both natural and human-made, and properly shaped quality of life, health protection potential, the scope of education, the level of information flow and by implementing out of 17 Sustainable Development Goals designated by the United Nations UN (Figure 1) [4].

Figure 1. Diagram linking the UN Sustainable Development Goals with selected detailed aspects of good health and well-being of people, including the organization of medical care and the design and production of medical devices necessary in these activities.

The basis of this approach, which is of common interest to all societies, is the assumption that the standard and quality of life is provided to people through access to consumer goods and products. As any product requires the use of technical materials, most often engineering materials, although sometimes also natural ones, an important determinant of prosperity associated with high quality of life is development and research in materials engineering, including nanotechnology and surface engineering.

Among the UN's Sustainable Development Goals, good health and well-being occupy an important place, inextricably linked with the concern to extend human life and promote an aging society's health (Figure 1). It requires action in various medicine and dentistry areas and others that the general public is usually unaware of. The current crisis situation caused by the COVID-19 pandemic also requires an appropriate response in health security and the organization and modernization of health systems [5]. *"The enjoyment of the highest attainable standard of health is one of the fundamental rights of every human being without distinction of race, religion, political belief, economic or social condition. The health of all peoples is fundamental to the attainment of peace and security and depends on the fullest cooperation of individuals and States"* as established in the World Health Organization WHO Constitution of 1948 [6]. It is, therefore, obvious that access to healthcare should be universal. Taking responsibility for ensuring health and well-being requires the so-called, more and more often, the "whole of government", which boils down to working towards this goal by the general public [7]. Such actions are targeted at individual citizens by recognizing each of them's right to their own health and healthcare and protection against threats to their health and life. For example, these tasks are set out in detail in the European Health Strategy [8]. The EU4Health Program [5], to protect against serious health threats, especially such as diseases of affluence, pandemics, and bioterrorism by improving the health of society, medical care, and health safety, manifested in the prevention of health risks, early diagnosis of the disease, efficient and effective implementation of medical procedures, improvement of the effectiveness of health systems, modernization of health care infrastructure and comprehensive and developmental medical therapies [9].

Both good health and well-being and general social cohesion and equality are the determinants of economic development in individual countries and their regions. More equitable societies tend to be characterized by greater progress in various aspects of their development. The measure is the Human Development Index HDI [10,11], developed by Pakistani economist Mahbub ul Haq, calculated for each country and first time reported by the United Nations Development Program (UNDP)'s Human Development Report Office [3,12,13] in 1990. In 2010 it was recognized that while HDI stands for an index of "potential" human development (corresponding to the maximum level of HDI), it could be achieved if there were no social inequalities between individual countries and people in these countries. Since such an assumption does not correspond to the truth, an Inequality-adjusted Human Development Index IHDI was introduced [14], which, taking into account inequality, gives the actual level of human development (Figure 2) [1].

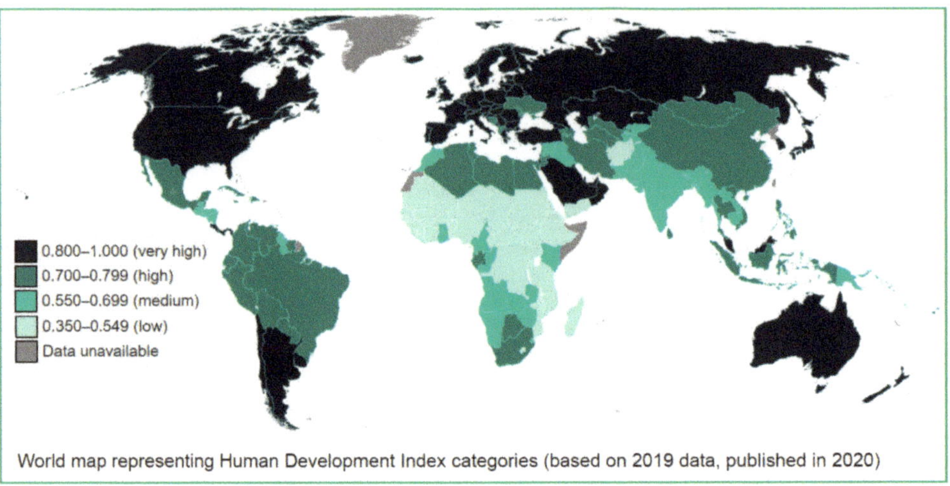

Figure 2. Geographical distribution of an Inequality-adjusted Human Development Index IHDI for individual countries.

These data were published firstly on 4 November 2010, but they were updated on 10 June 2011. Since then, the Human Development Report provides the calculated Inequality-adjusted Human Development Index IHDI values [15], taking into account three parameters to assess the quality of life which illustrate human development and well-being (Figure 3):

1. a long and healthy life index: life expectancy at birth,
2. an education index: expected years of schooling,
3. a decent standard of living index: gross national income (GNI) per capita at purchasing power parity (PP) (in international dollars).

Figures 2 and 4–6 show the indexes 'geographical distribution, as mentioned earlier', values according to the most available data. It is clear that there is an annual fluctuation in the details, but the overall view cannot be radically changed if the relevant data is from the past decade.

Thus, for many years, the health component has been included by the United Nations in assessing progress and the level of education and income assessment. There are many examples of it in many countries, including in the European Union, wherein the Treaty of Lisbon, Art. 3 states that *"the Union's aim is to promote peace, its values and the well-being of its citizens"*. The Sustainable Development Strategy's goal is to *"continuously improve the quality of life and prosperity on our planet for the present and future generations"*, while Art. 9 of the Treaty of Lisbon states that policies *"should take into account the requirements of social protection, the fight against exclusion, the promotion of education and training and the protection of human health"*. Undoubtedly, health is both a measure and a goal of progress and development.

Inclusive development reduces inequalities through universal access to healthcare [16]. Improving health reduces development burdens and costs to society, health systems, and the economy in general. The proportion of the population experiencing adequate health care in low and middle-income countries (LMIC) is generally lower than in high-income countries (HIC). In each country, the poorest quintiles have the most inadequate coverage of medical services. Diseases typically reduce life chances, diminish dignity, and reduce individuals' productive contribution to society as a whole and their local communities.

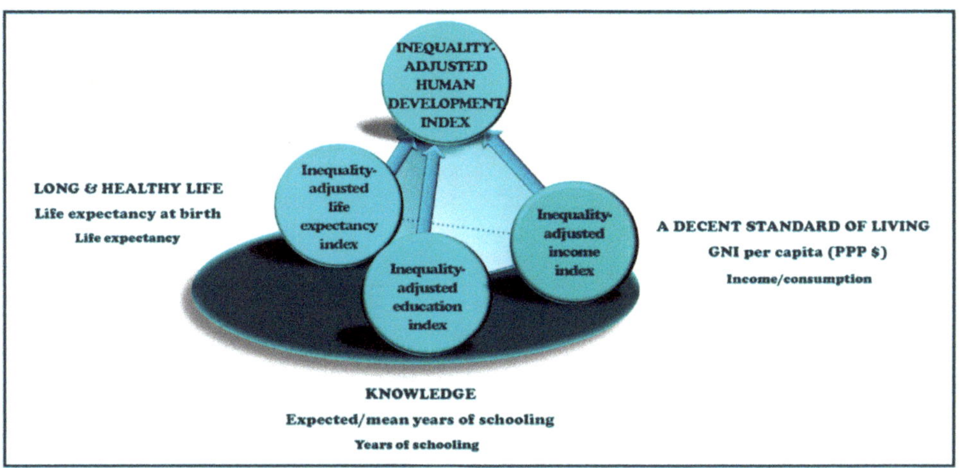

Figure 3. Diagram of the influence of various factors on Inequality-adjusted Human Development Index IHDI values.

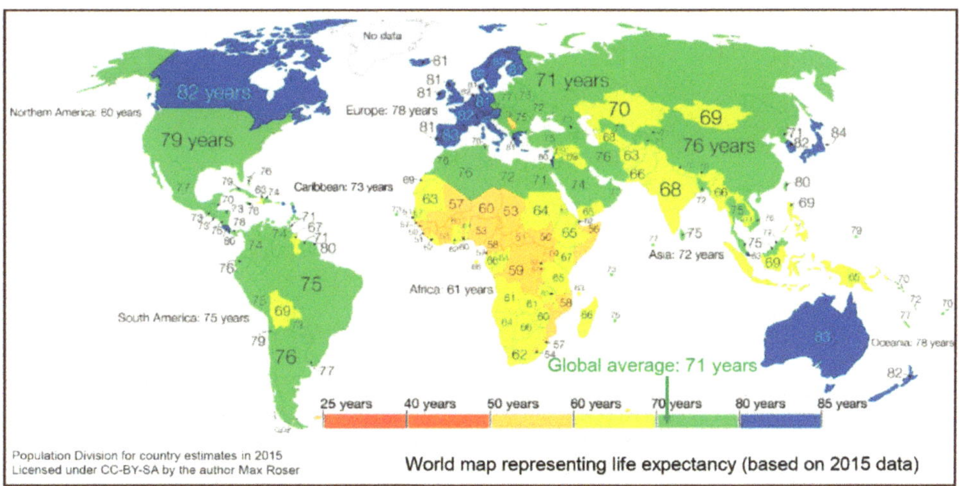

Figure 4. Geographical distribution of a long and healthy life index: life expectancy at birth.

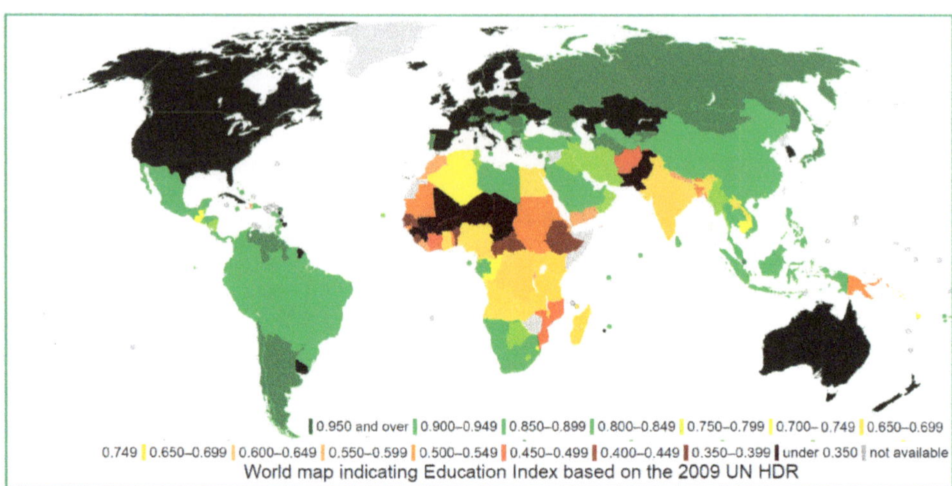

Figure 5. Geographical distribution of education index: mean years of schooling and expected years of schooling.

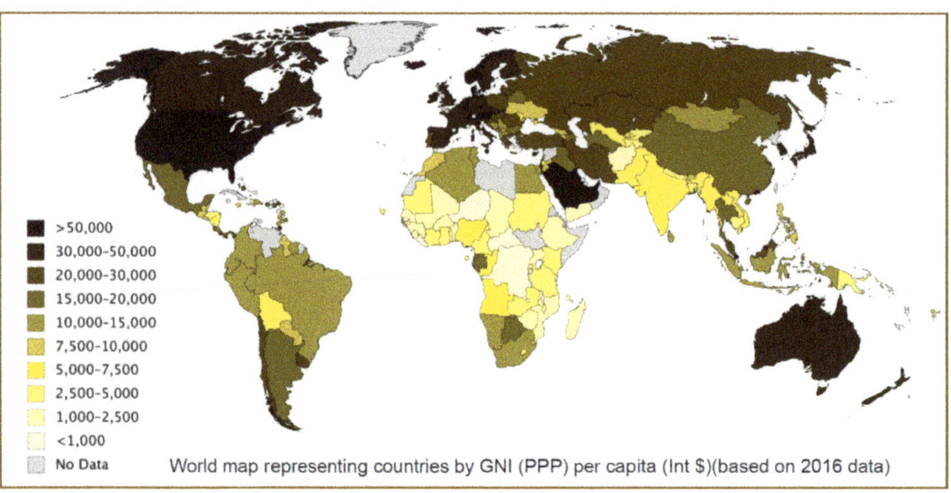

Figure 6. Geographical distribution of a decent standard of living index: gross national income (GNI) per capita at purchasing power parity (PPP) (in international dollars).

Figure 7 shows, for example, a comparison of mortality rates per 100,000 inhabitants by myocardial infarction and other ischemic heart diseases [17,18] in the member countries of the Organization for Economic Co-operation and Development (OECD), which includes 36 highly developed and democratic countries [16,17]. Although all these countries are undoubtedly among the relatively best-off, the mortality difference between first, Japan, and last, Lithuania, is almost twelve times. Figure 7b,c compares these countries in terms of spending per person on health care [19–21]. Here the differences are even greater, as the difference exceeds 50 times between the USA and India. Figure 7d compares the ranking of employment in health and social work as a share of total employment in these countries. There is an almost sevenfold difference between Norway and Mexico. It entails a variation in the number of doctors, which is already 20 times between Greece and Indonesia. One could conclude the creation of two-tier healthcare [17] was not because, apart from OECD countries, there are still many poorer countries. Therefore these differences are certainly much larger, although adequate statistical data for the whole world is not available. Undoubtedly, however, access to medical care varies greatly in different countries. Certainly, the situation is much worse in the LIC and LMIC countries. The lower and lower level of financing health care expenditure limits access to it and is the cause of deep social stratification in each country and differences between individual countries.

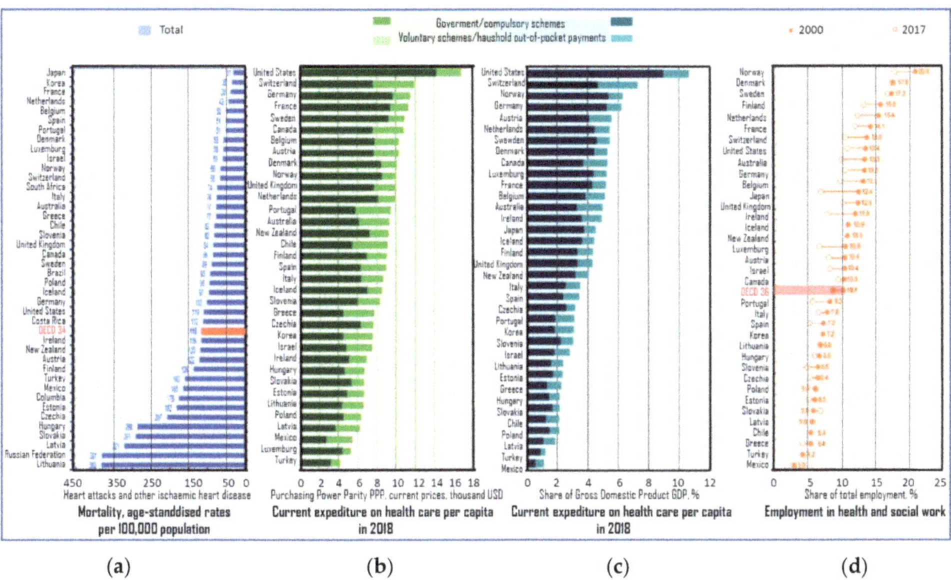

Figure 7. OECD countries compared in terms of (**a**) mortality rates per 100,000 inhabitants by a heart attack and other ischemic heart diseases in 2018; (**b**,**c**) health expenditure in 2018 per capita; (**d**) changes in employment in the health and social work as a share in total employment in 2000–2017.

However, when reaching the philosophical foundations of the concept of health, it should be noted that it is ambiguous and still raises numerous controversies. The ideas of health and disease are neither purely scientific nor solely part of common sense. They appear both in the conceptual apparatus of scientific theories and everyday thought. It cannot be concluded that the scientific and native application of the two concepts is completely independent.

The concept of disease evolves over time. It was believed that the disease was only an observable set of symptoms with a predictable course [22–24]. Its approach has become obsolete, and as medicine has developed, it has been established that disease occurs in the body as a set of destructive processes, "diverting part of the essence of an individual from actions that are natural to the species to another type of action" [25–28]. In the modernized, actuarial model [29], the disease is an increased risk, even without visible symptoms, in the event of a destructive pathological process, an example of which may be, for example, the hypertensive disease manifested by increased blood pressure [30]. The concept of disease in modern philosophy is more and more often seen as a set of empirical judgments on the one hand about human physiology and, on the other hand, normative judgments about human well-being or behavior [31–35]. The disease can be perceived as a process that repeats in different people in a slightly different form, so it is an abstract concept implemented in various ways [25,27]. In other fields of the humanities and social sciences, apart from philosophy, there is widespread support for the view that it is only normative judgments that determine the notions of health and disease [36,37]. Two basic ideas on the disease are emerged [38]. Objectivists, now more often referred to as naturalists [32,39–42], believe that the concept of disease is based on objective facts about the human body and that their correct interpretation and understanding in even the most difficult cases make it possible to define the boundary between health and disease. Assuming that the damage to health is a disease, an abnormality, or any biological defect, it can be concluded that health consists in the absence of any disease or such abnormality, and a healthy person is a person who has all biological systems fully functional. In the social consciousness functioning on an intercultural and over individual level, it is possible to mediate an individual's health objectively, e.g., with objective tests and research tools, as is usually the case in medicine. The beginning of medical objectivism was René Descartes's reflections on the mechanistic conception of the world and man [43]. The response to human diseases was reduced to repairing the damage to the biological machine man was considered to be, without considering it as a whole. In positivism, knowledge about man as a whole was considered the sum of knowledge about each of his components, which was separated for this purpose [44]. This approach's result was the perception of health as the absence of disease because the essence of health turned out to be too difficult to define. The same reasoning was adopted by WHO at its founding Assembly in 1948, accepting the International Lists of Diseases and Causes of Death [45]. Thus, health was defined as a condition in which none of the diseases thus defined occur. Quantitative indicators of disease occurrence or not, set by the physician, make it possible to judge the disease's state and by contrasting to determine the patient's health also, who is thus objectified. Therefore, one can adopt an external-conscious interpretation of both health and disease. Therefore, health objectivity, considering adopting various indicators or norms, enables the binary classification of a given person in one of the separate sets of healthy or sick people. Nevertheless, it is possible to formulate a broad spectrum of health for each person, from full health to terminal illness [46].

Constructivists take the opposite view, arguing that disease is an illusion and that disputed cases do not constitute ignorance of the facts but demonstrate a conflict of values between different social groups. Diseases damage them, for which a biological process is to blame, not their dysfunction [47]. Sometimes, however, there is a unification of views between these two views regarding the general acceptance of the system of values [38].

Philosophically or scientifically, it is very difficult to separate diseases and other ailments, including injury and/or disability [47]. Since the disease is a biological ailment, however, damages and/or disabilities can nevertheless be treated as events as they are not processes in essence. Similarly, it is impossible to separate disability from the issues of health and well-being. Usually, the problems of both disabilities, considered in the medical model as a product of physiological failure and/or human functional impairment, along with injuries and other diseases that do not require separation and the problem of diseases, are treated analogously. However, they are rarely analyzed jointly [48]. This approach to assessing organism disability is close to the naturalistic model of the disease. The condition of disabled people, similar to those affected by the disease, deteriorates due to functional impairments due to physiological or psychological inability to perform the organism's natural functions. In recent decades, a view has developed that disability cannot be treated as a dysfunction [11,22]. In the social model, disability is a simple difference between normal and healthy human functioning [49], and the diversity of these functions is a biological fact [50]. Therefore, the related differentiation does not result from physical disabilities but corresponds to social norms.

Consequently, it is necessary to analyze the health-disease-normality triad each time. According to [51], humans and their microbiomes are part of an ecosystem that can be assessed as healthy or not, so the analysis of health, disease, and normality should not be specific to individuals. Since it can be assumed that the human body belongs to the physical world, absolute health criteria can be determined by analogy to nature's laws due to the material from which it is made.

However, there is also a different approach to health, where it is determined not only by the absence of disease but also by its overall positive state. Therefore, total health can be understood as a multifaceted reflection of the synergistic impact of human psychology, physicality, and social relations that make up a specific person's well-being. In a relational approach, as stated in the first sentence of the WHO Constitution from 1948, *"health is a state of complete physical, mental and social well-being and not merely the absence of disease or infirmity"* [6]. Therefore, full health is guaranteed by the balance and integration of all these factors, complementary and autonomous at the same time. A person may feel sick, in the face of emotional problems or the case of social isolation, despite being fully physically fit. Conversely, he/she may consider himself healthy, despite his ailments or physical illness, in the event of inner mental strength and an atmosphere of social benevolence [52]. This ambiguity is the basis for criticizing the definition mentioned above formulated by WHO [53]. Based on the analyzes performed, three general categories of health can be formulated [54]:

- All healthy people must have the same health characteristics;
- Everyone is healthy in their own way;
- Health assessment criteria concern a set of universal and individual aspects.

"There is some intercultural agreement as to the basic characteristics of health, manifested primarily in the universal treatment of it as good, potency and activity" [55,56].

The unquestionable difficulty in establishing the essence of health is the dynamics of its changes over time. Moreover, it seems necessary to adopt the concept of health as a state of balance or a resultant balance between spiritual and bodily health [46]. There are doubts about whether the essence of health is related to man as such when it could only concern the civilization perspective. It is also worth noting that civilization's progress may introduce further changes and reevaluate the concept of health [46].

Regardless of the adopted philosophical view, the level of medicine, especially health care, significantly impacts a country's 'economic conditions and societies' health welfare. It raises certain serious professional and ethical obligations on the part of physicians and other healthcare professionals. Even in ancient times, the level of consciousness was so high that medics, and with time, doctors, were required to swear an oath regarding the morally appropriate level of performed tasks [57,58]. The Hippocratic Oath [59] has traditionally become a moral duty for all physicians to this day. Although its words changed, the meaning has remained the same for millennia. The doctor must help the sick, respecting the patient's autonomy [60,61], show charity [62], take actions that will not harm the patient under any circumstances [63], and be fair to the patient [64]. In this way, the generally recognized four basic ethical principles applicable to physicians were defined [65–68]. The doctor must also be truthful concerning the patient's health condition, the development and advancement of his disease, and the undertaken therapeutic activities. This principle has even been included as the fifth fundamental principle for dentists [69–72]. Doctors' ethical duties also include the systematic improvement of professional knowledge and the ongoing learning of scientific and clinical achievements, at least in their medical specialization. It is extended by the knowledge of appropriate technical solutions that can be used to improve the effectiveness of therapeutic activities undertaken by a physician. It applies in particular to modern diagnostic techniques, the possibility of using modern treatment techniques, especially surgical ones, the widespread use of the achievements of medical informatics and imaging methods, as well as modern surgical tools, as well as materials and technologies for the manufacturing of implants, but also, e.g., fillings used in dentistry and new materials and technologies for the production of prosthetic restorations.

Undoubtedly, the subject of this paper largely concerns the philosophical assessment of the current stage of development, goals, and threats of modern medicine. As mentioned earlier, the set as mentioned earlier of four basic ethical principles supplemented with the principle of truthfulness in relation to health and therapeutic activities constitutes the moral foundation of all medical activity undertaken over the millennia. Suppose you want to understand and at the same time explain any aspect of good health and well-being today. In that case, it is impossible not to refer to the moral imperative that every doctor and other person working in this important area of life must follow. It is obvious that every engineer who works with physicians to improve human health is also strongly morally bound to respect these principles. On the other hand, disregarding these issues would render such an analysis both incomplete and in many respects even non-purposeful.

It could also turn out that a medical device manufactured with or without the awareness of these aspects may lead to the fact that the undertaken therapy harms the patient directly in the short term, or systematically over many years, and in some extreme cases even lead to the death of the patient.

Therefore, these are only appearances that biomaterials are not related to ethics. Any biomaterial that in any aspect conflicts with any of the listed basic ethical principles can never be applied by a doctor, e.g., as an implant or other medical device, making engineering activities in this area completely pointless. Therefore, engineers working in this area are strictly subject to the moral imperatives in force and expected in this regard. One of the goals of this paper is to clearly emphasize the importance of these aspects in the context of the design, manufacturing, and application of biomaterials.

3. The Impact of the COVID-19 Coronavirus Pandemic on the Health Situation in the World

When analyzing the global health level in the second year of the severe acute respiratory syndrome pandemic caused by the SARS-CoV-2 COVID-19 coronavirus, this aspect's importance should be considered. The appearance of this creation of nature with a size of 100 trillion times shorter than the globe's size caused unimaginable changes in the world and resulted in virtually nothing returning to its original state from before the pandemic. As of the first identified cases in Wuhan, China, in December 2019, the disease has spread worldwide (Figure 8) [73–88], and lives have lost more than 2.5 million people (Figure 8).

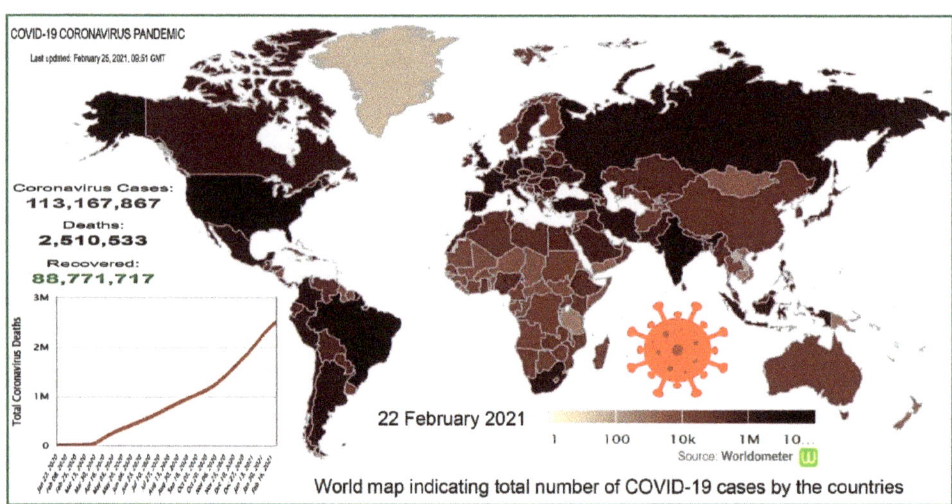

Figure 8. Spread of the COVID-19 disease in various countries of the world as of 22 February 2021.

For example, during a ceremony at the White House on 23 February 2021, US President, Joe Biden, in commemorating 500,000 people who died after being infected with COVID-19 in the US, stated that the disease in the US has so far claimed more lives than in World War I and II combined with Vietnam War. It illustrates the scale of this destruction.

Doctors, paramedics, nurses, and other health professionals, including dentists working in the respiratory tract of patients and ENT specialists, have become numerous victims of the SARS-CoV-2 coronavirus pandemic, many of whom, unfortunately, died in various countries. Figure 9 shows an example [89] of the spread of the coronavirus in a dental clinic.

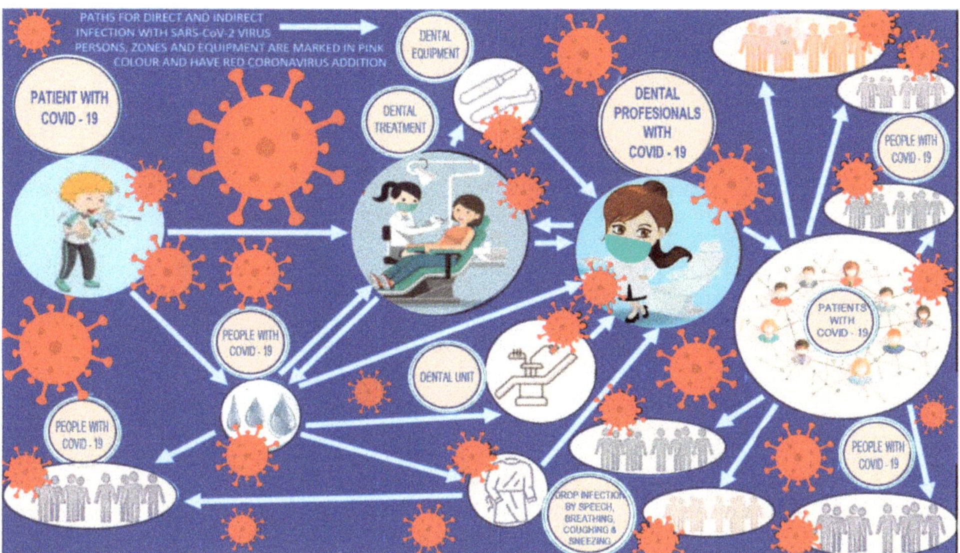

Figure 9. Scheme of SARS-CoV-2 coronavirus transmission as a result of primary infection of the dentist by one sick patient.

Figure 10 shows the life cycle of the SARS-CoV-2 coronavirus along with the potential induced immune responses [90]. Airborne droplets carry out the transmission of the SARS-CoV-2 coronavirus to infect the mucosa cells after reaching the oral cavity and respiratory tract. The angiotensin-converting-enzyme ACE-2, which is identified as a receptor by the coronavirus spike receptor protein, exists in the human lower respiratory tract. The interaction of the spike coronavirus receptor protein with human cells is the initial stage of infection. Then the ribonucleic acid RNA of the viral genome is released into the cytoplasm [91]. RNA occurs in the cell nucleus and the cytoplasm and is the carrier of information between deoxyribonucleic acid DNA and proteins. RNA components are nucleotides whose sequence depends on the order of these particles in DNA. RNA carries the genes of many viruses. RNA is typically single-stranded. In the case of single-stranded molecules, especially those having enzymatic functions or cooperating in these functions, the formation of double-stranded fragments by pairing different sections of the same strand determines the structure of the entire molecule. Their polarity characterizes RNA viruses containing a single strand of nucleic acid. The SARS-CoV-2 virus has a single-stranded RNA genome wrapped in a nucleocapsid (N) protein and three major surface proteins: the membrane (M), the envelope (E), and the spike (S), replicated to pass into the lower respiratory tract. It can result in severe pneumonia. The spike fusion protein of the virus plays an important role in penetrating the SARS-CoV-2 coronavirus into human cells [92]. In the first stage, the S1 glycoprotein determines the viral entry into the cell. In contrast, in the next stage, the human cell fuses with the viral envelope with the participation of S2 glycoprotein [93]. The human angiotensin-converting enzyme 2 hACE2 receptor protein acts as a receptor for this coronavirus, thanks to the SARS-CoV-2 Receptor-Binding Domain RBD [94–101]. It is the cause of the multiplication of the SARS-CoV-2 coronavirus, including in lung tissues, intestinal epithelial cells (enterocytes), kidneys, and blood vessels where hACE2 is present [102,103]. However, not only the hACE2 protein is a receptor in cell fusion because the SARS-CoV-2 coronavirus, to bind to the receptor of human cells, also activates them through their proteases [104]. Unlike the active standing position of the dynamic property of the RBD domain of other viruses, in the case of SARS-CoV-2, the RBD is usually in a prone position, which makes binding to a human cell receptor through it impossible. Endocytosis is the main way this coronavirus enters human cells. Endocytosis refers to the penetration of larger molecules inside the cell that are too large to be transported by protein transporters. Therefore, vacuoles are formed, and coronaviruses enter the cell along with fragments of the cell membrane. In addition to the hACE2 receptor, cell fusion activation can also be caused by the body's furin enzyme [100,105–109]. Inhibition of this enzyme may imply a potential antiviral strategy [105–108,110]. Necessary in this process is the phosphoinositide kinase enzyme PIKfyve containing five fingers, the TPCN2 dual-pore calcium channel protein, and the cathepsin CTS toll-like receptor 7 TLR7. TLR7 is one of the endolysosomal proteases whose role is to regulate innate and acquired immunity, including cell adhesion and migration, processing, antigen presentation, and resistance to various viral infections. It is necessary for the cleavage of TLRs and the formation of functional receptors that enable the detection of viral nucleic acids such as RNA or DNA [109]. The transmembrane protease enzyme serine 2 TMPRSS2 is also important in the pathogenesis of SARS-CoV-2, the protease responsible for cell fusion with the S protein of this coronavirus and induces the formation of syncytium [111,112]. The syncytium is a multinuclear cell (multinuclear mass of protoplasm) formed by the fusion of loose single mononuclear cells, with the simultaneous disappearance of the surrounding cell membranes. The SARS-CoV-2 coronavirus may most likely form a syncytium depending on the activity of the receptor but independently of the protease and also without the participation of trypsin [109]. Trypsin belongs to endopeptidases and is a digestive enzyme produced by the exocrine part of the pancreas in the form of a proenzyme—trypsinogen, an amount of pancreatic juice. Trypsin is also able to generate syncytium through activation of S-glycoprotein. The cumulative effect of TMPRSS2, lysosomal cathepsins, and furin on the cellular penetration of SARS-CoV-2 has again been proven. Thus, blocking TMPRSS2 may be one of the treatments for COVID-19 [112–114]. Blocking the cell surface receptors helps the virus replicate. The use of an angiotensin-converting enzyme may be one way of preventing viral infection [115,116].

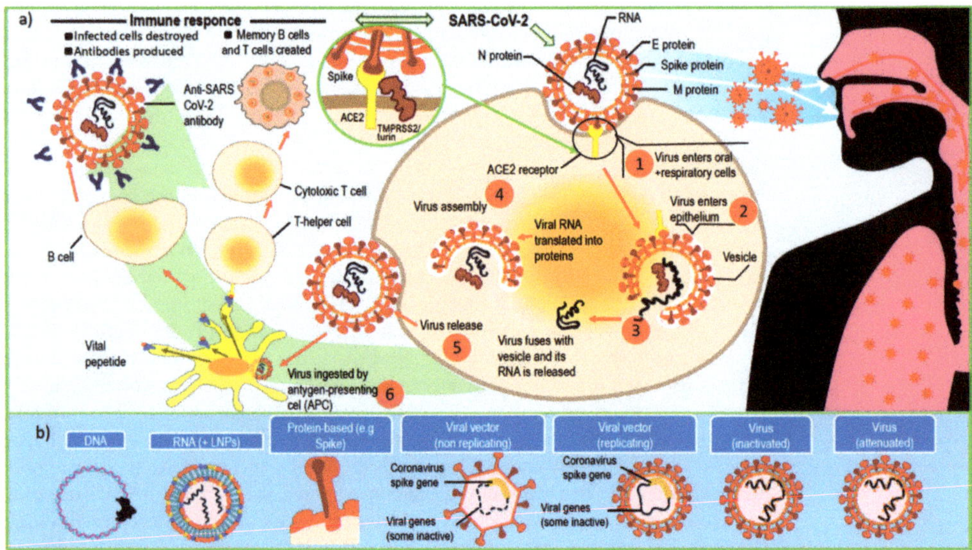

Figure 10. Diagram of (**a**) transmission and life cycle of the SARS-CoV-2 coronavirus; (**b**) preventing modes of SARS-CoV-2 coronavirus infection.

From the epidemiological point of view, patients who do not show clear or no disease symptoms are dangerous, especially as they constitute over 30% of the population, while effectively transmitting the disease [78,79]. Severe symptoms of the disease, manifested by dyspnea, hypoxia and resulting from the involvement of more than 50% of the lungs, occur statistically in about 14% of cases, and in another 5% even acute symptoms of the disease, apart from respiratory failure, manifested by shock or multi-organ dysfunction [78], which often leads to the death of the patient due to irreversible lung damage due to fibrotic pathology and the inability to withdraw from mechanical ventilation or extracorporeal life support. In some cases, it can be counteracted by lung transplantation [117]. Statistics show a very high recovery rate, although some convalescents experience so-called long COVID, causing damage to various organs over time [81], which is still under constant investigation.

Research activities have been undertaken to develop anti-virus drugs, and work is still ongoing, although there are no measurable effects to date. Therefore, almost exclusively symptomatic treatment, supportive care, isolation, and severe cases, oxygen therapy, special treatment methods, and even experimental methods are used [118].

In many countries, numerous preventive measures are taken to minimize virus transmission risk, especially in public places. These activities mainly include social distancing, disinfecting, washing hands, keeping unwashed hands away from the face, covering up coughing and sneezing, using masks or face shields, ventilating rooms, physical or social isolation, quarantine, shutting down various sectors of the economy, shutting down schools and universities and a total social and economic lockdown.

As the number of COVID-19 cases continues to increase, the professional healthcare system is facing growing challenges. One of the important elements of this approach is the protection of doctors and other medical staff because only their efficiency gives any chance of survival to millions of people in the world in the face of the threat of a fatal disease. Various practically used methods were compared by performing a benchmarking analysis and using the dendrological contextual matrix (Figure 11) [89,119–121], analogous to the matrices of Boston Consulting Group introduced in the 1960s [122]. Figure 11 also shows all types of applied solutions and their total assessment made based on the criteria of potential and attractiveness given in this figure.

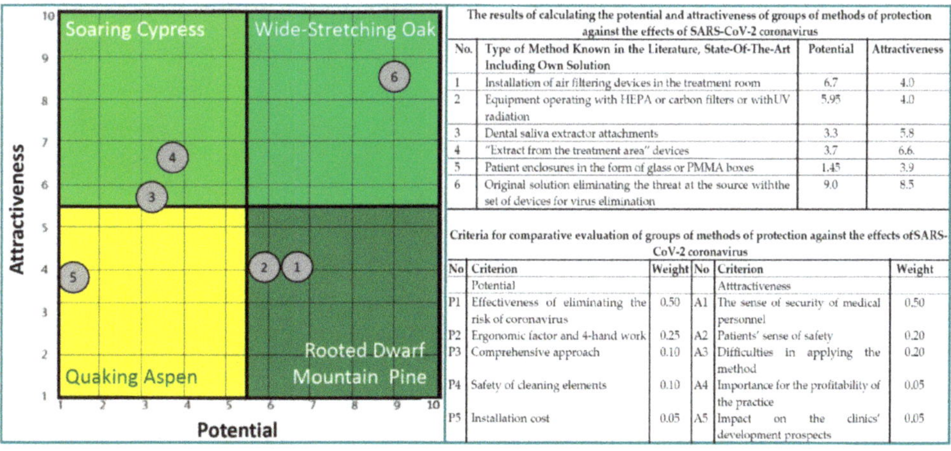

The results of calculating the potential and attractiveness of groups of methods of protection against the effects of SARS-CoV-2 coronavirus

No.	Type of Method Known in the Literature, State-Of-The-Art Including Own Solution	Potential	Attractiveness
1	Installation of air filtering devices in the treatment room	6.7	4.0
2	Equipment operating with HEPA or carbon filters or with UV radiation	5.95	4.0
3	Dental saliva extractor attachments	3.3	5.8
4	"Extract from the treatment area" devices	3.7	6.6.
5	Patient enclosures in the form of glass or PMMA boxes	1.45	3.9
6	Original solution eliminating the threat at the source with the set of devices for virus elimination	9.0	8.5

Criteria for comparative evaluation of groups of methods of protection against the effects of SARS-CoV-2 coronavirus

No	Criterion	Weight	No	Criterion	Weight
	Potential			Atttractiveness	
P1	Effectiveness of eliminating the risk of coronavirus	0.50	A1	The sense of security of medical personnel	0.50
P2	Ergonomic factor and 4-hand work	0.25	A2	Patients' sense of safety	0.20
P3	Comprehensive approach	0.10	A3	Difficulties in applying the method	0.20
P4	Safety of cleaning elements	0.10	A4	Importance for the profitability of the practice	0.05
P5	Installation cost	0.05	A5	Impact on the clinics' development prospects	0.05

Figure 11. Results of a comparative analysis of various methods of protecting medical staff against the harmful effects of the SARS-CoV-2 coronavirus.

Assessment results are plotted on a dendrological matrix. The quarter "Rooted dwarf mountain pine" includes proven, mature methods (1,2), which have great potential but may currently have complementary importance. "Soaring Cypress" is a quarter containing very attractive methods (3,4), with possibilities for the future but requiring consolidation to be able to apply them more widely. Weak in the "Quaking Aspen" quarter, there are low potential and attractive methods (5), replaced with better solutions. The authors' solution (6), which eliminates the threat at the source with a set of virus elimination devices, is located in the "Wide-stretching oak" quarter and is characterized by the greatest number of new products, objectively presented and groundbreaking meaning. However, the other methods are relatively but only seemingly much more attractive (subjective impression of the recipient) than their real potential (objective technical analysis). Therefore, their importance is overestimated, despite their objectively weak usefulness.

A common approach to solving the problem is the STOP strategy. With this approach, the protection of medical personnel against epidemiological threats consists in the successive implementation of more and more radical measures, from systemic solutions to technical and organizational solutions in the use of personal protective equipment. In the most radical approach, doctors and nurses equipped with appropriate hermetic coveralls, gloves, and helmets are completely isolated from the contaminated environment with infected patients. This method almost completely limits the freedom of movement, exposes to sweating and extreme thermal discomfort, shortens the effective working time, and greatly complicates the liberty to perform precise medical procedures, such as surgery or dentistry [89]. An almost opposite approach is represented by the proprietary SPEC strategy, which can be used in dentistry, ENT, and anesthesiology. This breakthrough strategy isolates the patient and the bioaerosol he exhales, as well as the clinical aerosol associated with dental procedures, from the physician and the medical treatment room (Figure 12). Bioaerosol and clinical aerosol are eliminated at the source. The solution hits the cause to eliminate the causative factor. A comprehensive set of devices for collecting the aerosol is right next to the patient's mouth, and complete deactivation of this aerosol after collecting in appropriate containers. The modular structure of the system protects medical staff against coronavirus and other pathogens. The face shield is an important element of this solution, and therefore one variant is shown in Figure 12. The study [89] also compared this solution with the concepts of all other available personal protective equipment for medical personnel against pathogens.

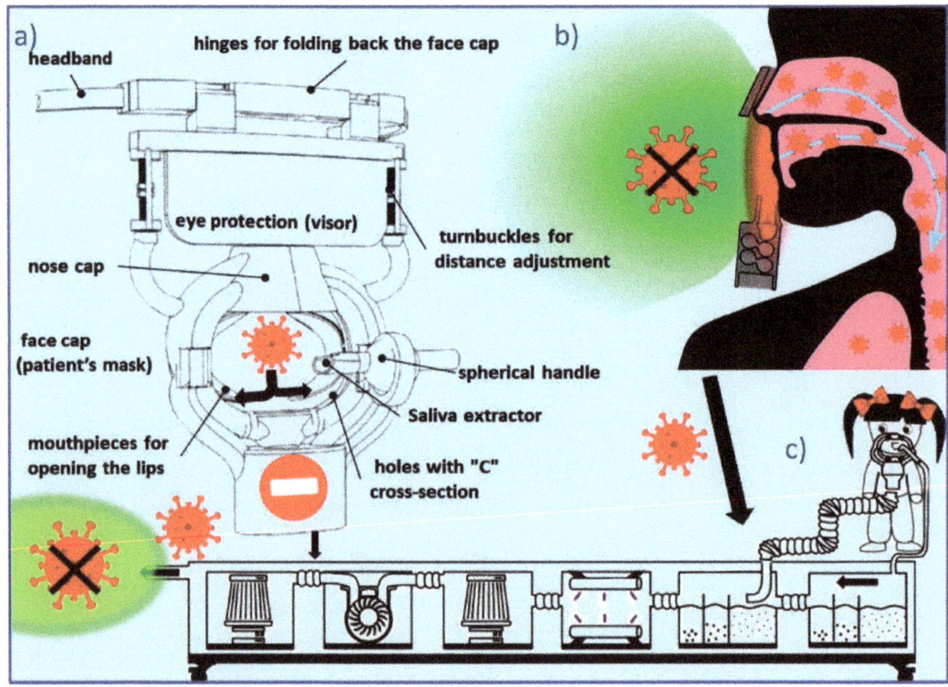

Figure 12. Diagram of the medical staff protection system against the transmission of the SARS-CoV-2 coronavirus: (**a**) diagram of the face shield with a full set of utility functions; (**b**) a source risk elimination scheme; (**c**) a set of devices for inactivation and decontamination of the coronavirus.

The importance of preventive vaccinations against the SARS-CoV-2 coronavirus cannot be overestimated. The different ways to prevent viral infection are given in Figure 10b, which breaks down all SARS-CoV-2 vaccine candidates into seven major platforms (DNA, RNA, protein-based, viral vector (non-replication), viral vector (with replication), virus (inactivated), and virus (activated)), represented by different regimens. As of February 2021, 66 vaccine candidates are undergoing clinical trials in various stages. As of February 2021, 11 vaccines have been approved for public use against COVID-19, which have gained acceptance in at least one country, and fall into four groups of vaccines:

- Two RNA vaccines (Pfizer—BioNTech and Moderna),
- Four conventional inactivated vaccines (BBIBP-CorV, Covaxin, CoronaVac, and CoviVac),
- Four vaccines with viral vectors (Oxford—AstraZeneca, Convidicea, Johnson & Johnson, and Sputnik V),
- One peptide vaccine (EpiVacCorona).

Many countries have implemented distribution plans that prioritize those at the highest risk of complications, including healthcare system staff due to the high probability of infection and transmission of the coronavirus and the elderly with the lowest individual immunity [123]. According to official data from national health agency reports, 216.17 million doses of the COVID-19 vaccine have been administered worldwide so far until 23 February 2021 [124]. The vaccination campaign is intense all over the world. However, it is worth noting that it is unfortunately impossible to guarantee full immunity to the COVID-19 disease. Simultaneously, the vaccination of about 60–70% of the world's population and the population of all countries in the world gives a great chance to weaken both the virus's aggressiveness and mortality. Unfortunately, the virus mutates, and it is not certain that any of the vaccines invented and implemented so far gives resistance to mutating variants of the virus.

It should be expected that a similar situation may occur as in the case of acquired immune deficiency syndrome AIDS, i.e., a disease caused by human immunodeficiency virus HIV, which is transmitted not only through sexual contact and perinatal, but also through exposure to secretions or tissues containing the virus, including, e.g., through the blood. Despite the fact that the epidemic conditions have subsided, all health care staff to this day use restrictive procedures to prevent accidental contact with blood, saliva, and other secretions or tissues of potentially infected patients with this virus, that is in practice against all patients for prevention [125–129].

It should be anticipated that, analogically, even in the event of an end to the COVID-19 pandemic and in the event of high effectiveness of vaccines, in the absence of certainty that the protection in this respect is fully effective, it will be necessary for doctors and other medical staff to apply preventive measures, according to appropriately adapted strategies STOP and SPEC, protecting against the transmission of SARS-CoV-2 coronavirus and its subsequent variants, including British, South African, Indian and Brazilian ones [130] and all others that may emerge in the future.

In the book "Science of Logic" from 1812, the German philosopher Georg Wilhelm Friedrich Hegel (1770–1831) formulated the thought that "Purely quantitative changes at some point turn into qualitative changes." The scale of the SARS-CoV-2 pandemic in a little more than a year around the world makes it a serious qualitative change. Millions of people worldwide get sick, millions of people have lost their lives because of it. Among them were numerous doctors, nurses, paramedics, therapists, and diagnosticians. They died because they saved the lives of others, their patients.

Since the first five cases of human immunodeficiency virus, HIV, infection causing acquired immune deficiency syndrome, AIDS, were recorded in 1981, according to estimates by the World Health Organization (WHO), the disease has affected approximately 80 million people since the beginning of the epidemic, of which over 35 million people died. Since then, the approach of all physicians to protect themselves from infection has changed. Infection with the hepatitis C virus HCV is similarly dangerous, for the discovery of which Harvey Alter, Michael Houghton, and Charles Rice received the Nobel Prize in medicine in 2020. It is estimated that about 170 million people are infected with HCV in the world. Blood may also be among the epidemiological routes of infection. Today, no one doctor or dentist will start the medical procedure unless they put on gloves, although not each patient can infect them through the blood. However, it will be enough if it is one in a thousand or even in several thousand. The infection will be fact and death could happen.

Routine safety and precautionary measures should be taken as each physician knows and follows them.

In essence, completely different measures must be taken than in HIV or HCV due to the other way of transmitting the SARS-CoV-2 coronavirus. The approach to continuous and systemic protection against this pathogenic and dangerous virus will remain unchanged, but the mode should be absolutely unique and different.

Airborne droplets spread the SARS-CoV-2 virus. It is virulent, lethal, and is very difficult to protect against. Doctors cannot risk their lives saving someone else's life and losing their own life. This fact completely changes the approach to medicine. As authors, we realized that ensuring the safety of doctors and other members of the medical staff is an important part of modern medicine and even becomes hierarchically superior to the normal routine preventive, diagnostic, therapeutic, and interventional activities, including implantology, and organ transplantation. The people should know it because everyone will one day be a patient. That is why the safety aspects of medicine need to be addressed in the paper on biomaterials. There is a need to change the approach. Appropriate protective equipment against SARS-CoV-2 according to authors' SPEC strategy should be produced, which will allow doctors to work normally, without having to wear personal protection equipment resembling a spaceman's outfit, and at the same time so restraining to movements and not providing any thermal comfort, and to perform many precise medical procedures. A new generation of preventive measures requires the use, and perhaps even the development and manufacturing, of a new generation of biomaterials for this type of medical device. Looking at the problem of biomaterials holistically, it is impossible to ignore this problem. It is the justification that one part of this paper is dedicated to this issue.

4. The Necessity for Physicians to Use Technical Solutions Supporting Medicine and Dentistry

The basic philosophical approach of medicine is disease prevention. Considering the relationship between health and economic well-being requires the support of scientific research and the development and application of advanced technologies related to medicine's technical support to ensure the fullest possible prevention of diseases. It requires a very wide-ranging and planned action by entire national health systems. However, practice shows that despite the great efforts, activities of many people, and huge financial outlays, such activities cannot protect all patients from the disease. An example is dentistry. It is estimated that 3.5–5 billion people worldwide, i.e., up to 60% of the population, suffer from caries, leading to toothlessness and many systemic complications. Many people also suffer from periodontal diseases, which leads to similar complications [16]. In such cases and many other cases, e.g., in orthopedics, heart diseases, and cancer, it is necessary to treat patients using interventional medicine methods safely. In traumatology, orthopedics, maxillofacial surgery, and dentistry, it is often required to replace or supplement organs or tissues to restore patients' vital functions and prevent their biological and social degradation today is both an avant-garde and an expensive branch of modern medicine [5]. The number of cases requiring this type of intervention is systematically increasing. The dynamics of this growth is closely related, inter alia, with surgical treatment of neoplastic tumors, often saving human life and removing inflammation, numerous road accidents and the resulting serious injuries, sports accidents related to, among others, with extending human life and practicing various sports by elderly people, as well as with the spread of caries, periodontal diseases and increasing the scale of edentulousness [16]. For example, in the European Union, around 1.7 million people are injured in road accidents each year, according to the Association for the Improvement of Road Safety. In comparison, an average of about 6.1 million people is hospitalized in hospitals each year. It is predicted that the number of patients requiring replacement or replacement of organs or tissues, including dental implantology, will systematically increase along with the extension of human life and increase people's participation over 65 in the total population, which will increase by approximately 70% by 2050 [9].

Achievements in the field of implantology, prosthetics, tissue engineering, and natural organ transplants, serving specific patients directly, setting the latest trends in medicine and interventional and regenerative dentistry, have their sources in the deep knowledge and experience of doctors and biologists, what is obvious. However, the enormous achievements in the field of engineering, generally referred to as bioengineering, with the participation of medical and dental engineering, including, among others, advanced modeling, computational science, and computer-aided engineering design, materials engineering, materials process technologies, and manufacturing engineering together with additive technologies, nanotechnology, and surface engineering (Figure 13). Bioengineering is richly collaborative and interdisciplinary and focuses on integrative applications and solves problems left unanswered by engineering and physical/life science disciplines [131,132]. The diagram shows the multidisciplinary nature of this engineering discipline in the form of a multicoloured flower. Each of the petals corresponds to a different area of knowledge contributing to it. In the central part, various areas of interest in science and engineering are specified as examples.

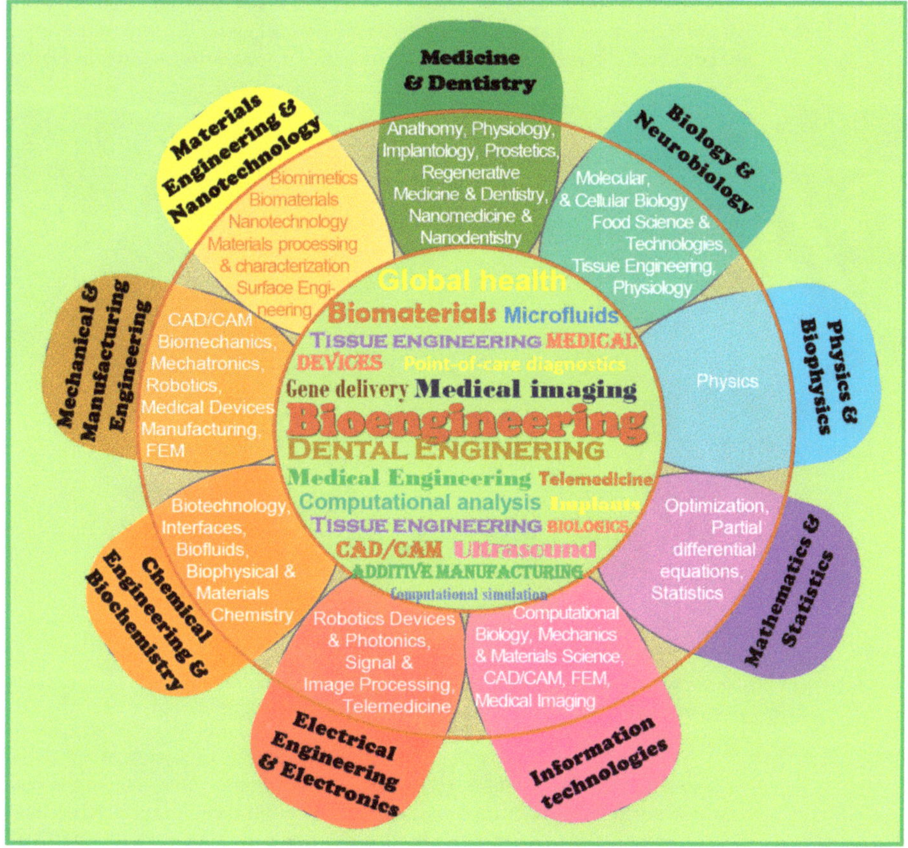

Figure 13. The flower of bioengineering showing a scheme of the scope of its interests, where each petal illustrates the interdisciplinary contribution of each discipline; the central circle corresponds to selected bioengineering problems; the middle circle determines the contribution of the detailed issues of each component disciplines.

Bioengineering tasks pose special challenges for the engineering staff of many specialties indicated in the diagram, which should meet, on the one hand, the expectations of doctors serving the avant-garde areas of medicine, regenerative dentistry, tissue engineering and corresponding to the challenges related to patients health, and on the other hand, fulfill the principles of the current development stage of Industry 4.0 [133–139] is very responsible research. According to complex and advanced original technologies, the most avant-garde trends concern the offering of personalized medical devices manufactured according to the patient's individual anatomical features and newer and newer biomaterials used in implantable devices. These issues are among the most avant-garde and constantly developing engineering problems of great importance for the development of medicine and dentistry and have been the authors' scientific interest for many years [9,140–159]. The collaboration of biomedical and dental engineers with doctors in patients' service and assisting them in removing or at least alleviating the disease's effects obligates doctors and engineers with a specific ethical approach. As part of the obligation to systematically expand their knowledge, this cooperation imposes on doctors to follow the latest scientific achievements to use them for their own patients. The far-reaching dependence of modern medicine and dentistry on technical progress and engineering achievements requires doctors to trust engineers and rely on their experience and knowledge and their entrepreneurship and initiative. From this, these particular ethical requirements arise, which must be met by every engineer working with a doctor, even though he/she is not officially obliged to take any oaths. However, this is an undisputed and unconditional moral imperative of an engineer. It is a schematic view in the form of the oppression of two hands symbolizing two different communities' cooperation, shown in Figure 14 [57,58].

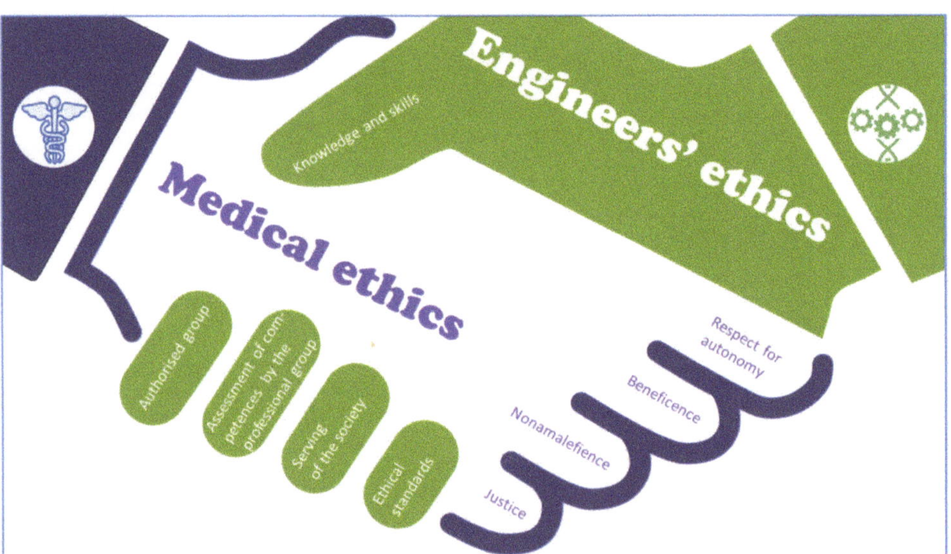

Figure 14. Diagram of the coincidence of ethical principles of the medical and engineering communities necessary for implementing therapeutic tasks requiring cooperation with bioengineers.

The requirements for this cooperation include the provision of advanced medical devices using the latest technology achievements, but also often, e.g., in the case of dentistry, trauma surgery, regenerative medicine, individual design and production of prosthetic restorations, implants, or rehabilitation devices personalized to the patient's anatomical features. It requires the use of extensive engineering knowledge in the field of modern design and manufacturing methods using computer-aided engineering works, including computer-aided design CAD and manufacturing CAM, computerization, robotization, and automation. It requires using the most advanced biomedical and other engineering materials and the most modern engineering technologies, including additive manufacturing and surface treatment. Although engineers generally do not participate in medical procedures involving patients, they are also guided by ethical principles [160–163] (Figure 14), which include a pentad, including improving specialist knowledge, achieving high competence, ensuring high standards, serving society in line with the ethos of the engineering profession, and designing and manufacturing products of high technical and utility values. The considerations are of professional ethics, which results directly from every person's internal imperative, which is a derivative of upbringing and worldview beliefs. Admittedly, many so-called ethical codes of various professional groups, but in fact, play a similar role to legal regulations. Sometimes the effect of such regulations coincides with the ethical imperative, but the motivations are completely different. Three orders are not mutually exclusive and can even complement each other. The motivation in each of them answers a different question. How to proceed is the ethical question. What people will say is a social category. What kind of punishment can be imposed is the legal order. Therefore, each code can be included in the second or even third of the groups mentioned above of regulations. On the other hand, the highest possible value is the inner conviction of the need to perform the duties properly. In the case of a biomedical or dental engineer, it is also a moral responsibility towards patients, despite the most frequent anonymity of the function performed by him/her in the process of treating patients.

However, the implementation of all avant-garde technical solutions is acceptable only when their use means an absolute risk that is less than the total benefits. In the latest technological solutions, sometimes, this rule should be exceptionally abandoned [164]. It will be an ethical tort, or even an offense, to avoid the latest technical solutions in clinical practice consciously.

A very wide range of medical products and devices is of interest to biomedical engineering (Figures 15 and 16). Due to the significant degree of possible health complications, the principles of manufacturing both medical products and devices are subject to strict legal regulations. For example, in the European Union, these issues are covered by the medical device regulation MDR and the in vitro diagnostic medical device regulation IVDR [165]. *"'Medical device' means any instrument, apparatus, appliance, software, implant, reagent, material or other article intended by the manufacturer to be used, alone or in combination, for human beings"*, instruments, bandages and splints, and treatment chairs and hospital beds. In turn, *"in vitro diagnostic medical device means any medical device which is a reagent, reagent product, calibrator, control material, kit, instrument, apparatus, piece of equipment, software or system, whether used alone or in combination, intended by the manufacturer to be used* in vitro *for the examination of specimens, including blood and tissue donations, derived from the human body."*

Examples include pregnancy tests and blood glucose monitors. Some products called borderline ones are difficult to put into any of these groups. For example, it is possible to specify medicated surgical dressings or head lice products. Specific legal regulations similar to those relating to medical devices also apply to aesthetic products, including non-corrective contact lenses, liposuction equipment, or equipment intended for brain stimulation. The classification of both basic groups of medical devices is risk-based, which dictates the appropriate conformity assessment procedures. It also applies to research works, when in a significant number of cases, the Bioethics Committee's consent is required. Class I, and as appropriate devices do not require release procedures. However, some require approval by a notified body for parts of the manufacturing process related to sterility or metrology if the medical device contains sterile products or has a measuring function. In all other cases, it is necessary to perform appropriate and prescribed by law con assessment activities, referred to as General Safety and Performance Requirements. These procedures are handled by the proper Competent Authority Institutions designated in each country.

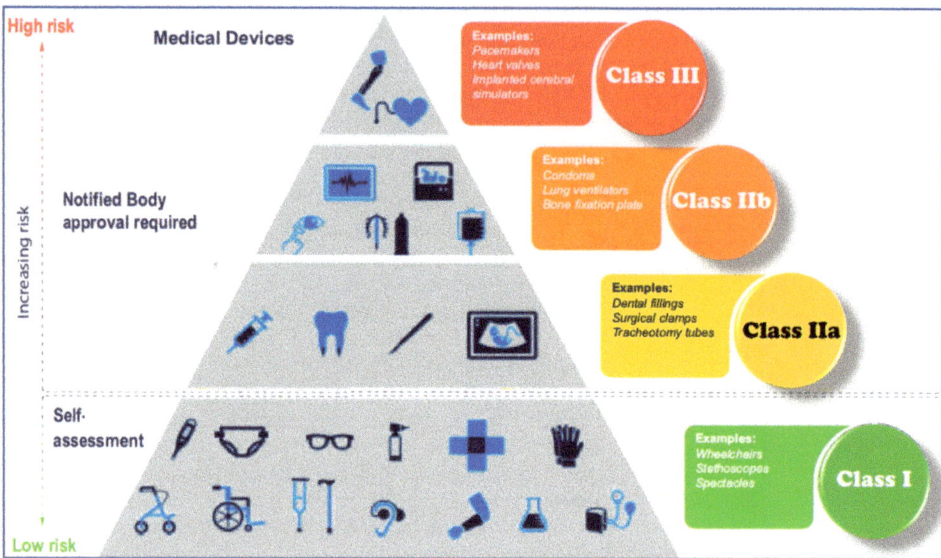

Figure 15. Scheme of the general classification of medical devices depending on the application risk.

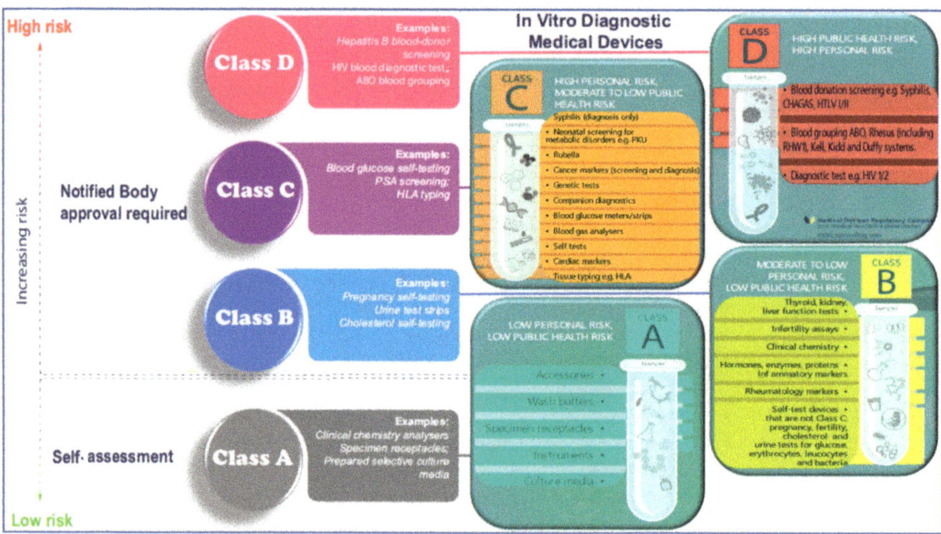

Figure 16. Scheme of the general classification of in vitro diagnostic medical devices depending on the application risk.

Figure 17 shows examples of artificial organs, implants, and other medical devices.

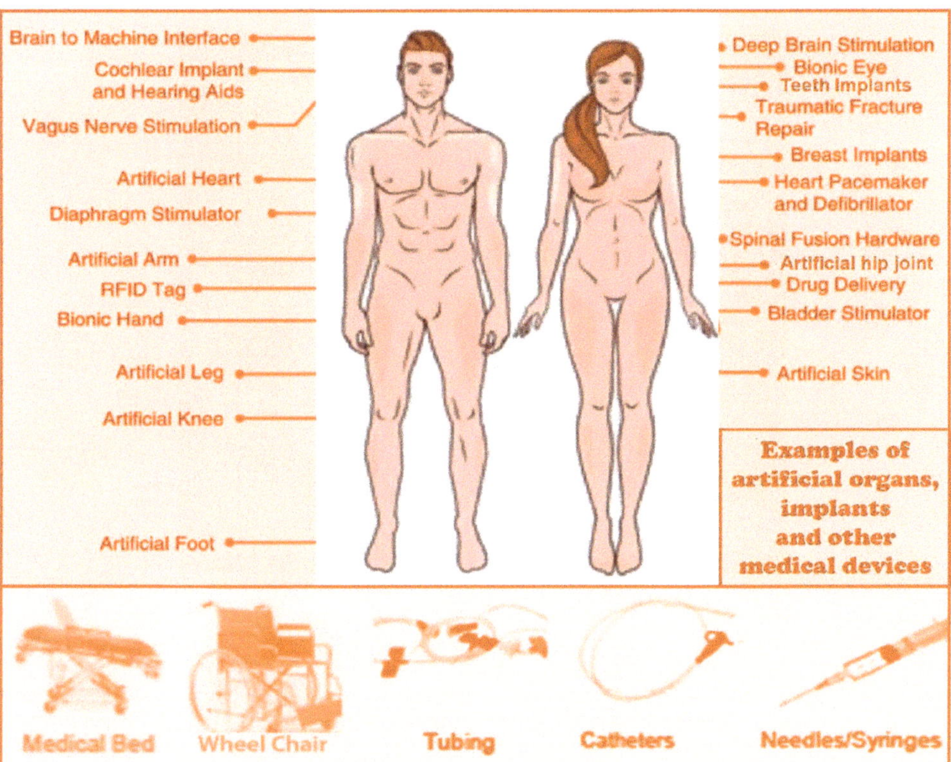

Figure 17. Examples of artificial organs, implants, and other medical devices to replace various tissues and organs of the human body and medical care.

The historical development of manufacturing processes is part of the development of human material culture. The most popular industrial development concept developed in Germany, popularized globally, and described in numerous authors' own works [133,134, 136,145,166–232]. The first step in this development of Industry 1.0 and the beginning of the industrial revolution was the steam engines' introduction at the end of the 18th century. In the classic model of the current stage of Industry 4.0 [133,134,136,179,185–187,190–193,196– 200,205–209,211,213,220,233,234] nine basic technologies described in the literature are indicated [133,134,166,167,174,178–233,235–238] creating cyber-physical systems (CPS), making smart decisions based on real-time communication with people, machines and sensors. It turned out that this model is one-sided and requires a significant extension, not just complex cyber-physical systems and very advanced tools and information systems. This model, however, gives the erroneous impression that the progress concerns only CPS systems, which is not true because it is necessary to include engineering materials, technological manufacturing processes, and technological machines in this model [133,134, 136,168–172,183,239] (Figure 18). Far-reaching simplification also boils down to limiting technological issues only to additive manufacturing. As the only technology included in the classic Industry 4.0 model cannot be considered competitive to other technologies necessary to use in many product manufacturing processes.

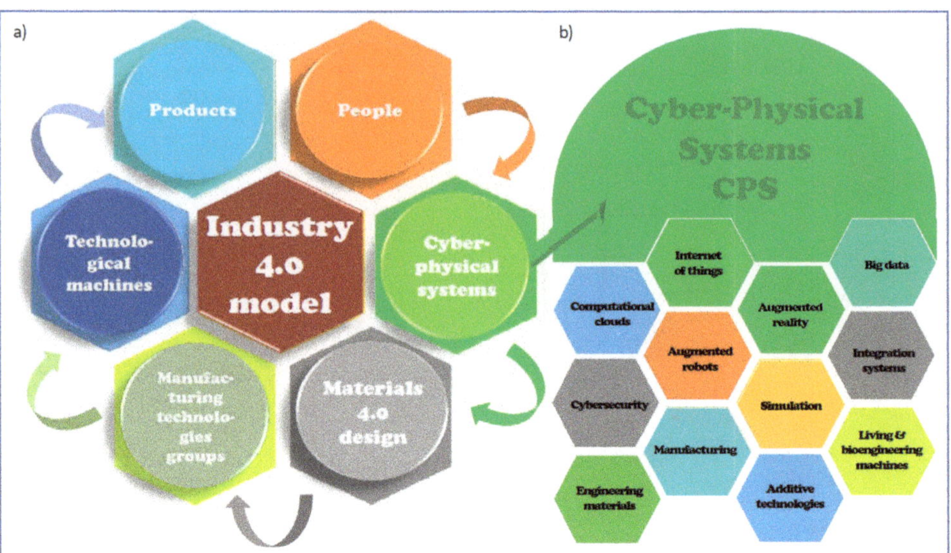

Figure 18. Diagram of (**a**) augmented elements of the holistic Industry 4.0 model; (**b**) extended scope of cyber-physical systems.

Manufacturing products is a technological process organized according to a well-designed plan, requires the use of engineering materials, as well as energy, capital, people, various machines, and is a comprehensive activity bringing together people who carry out different professions and activities, using different machines, equipment, and tools, to a varying degree automated, including computers and robots, most modernly included in the rules of the Industry 4.0 stage. The purpose of production is each time to satisfy the market needs of customers, following the developed strategy of the company or organization involved in the production, using the available possibilities and devices. The technical aspects of introducing a given product to the market, including medical devices, by the manufacturing organization relate to industrial design, engineering design, production preparation, manufacturing, and servicing (Figure 19).

Thus, in introducing products to the market, three main spheres can be selected, i.e., marketing and sales, product development, including the design of the product, and manufacturing [240]. The product design phase is concerned with industrial design, engineering design, and production preparation. Engineering design, in which the design of the production system and product design can be mentioned, is not an isolated activity because it affects all other phases of introducing a given product to the market. It is also dependent. Modern engineering design of products is a complex process, covering three equally important and inseparable and closely interrelated and complementary areas of activities related to material design, structural design of product forms, and technological design of product manufacturing, in a wide range including, inter alia, materials processing technologies. The basic premise is the $6 \times E$ (expectations) rule [133,136,241]. The functional features of the products expected by the customer will be ensured if the expected engineering material is used for their production, processed with the use of technology with the expected quality, to obtain the expected shape of the final product with the expected structure and expected mechanical and functional properties. New engineering materials and manufacturing processes are subordinated to the customer's needs and the products' needed functions, including medical devices. The material design methodology shaped in this way is associated with numerous activities related to the modeling and simulation of manufacturing processes and the prediction of the operational properties of materials, the development of safe material technologies and products composed of nanostructured elements, the standardization of research on the properties of materials, especially nanostructured ones, and the development of a methodology for predicting the behavior of new materials during operation. The basis of material design in any practical application is multi-criteria optimization related to belonging to one of the main groups of engineering materials or the group of natural materials, with the chemical composition, manufacturing conditions, operating conditions, and the method of removing material waste in the post-use phase, as well as price conditions related to the acquisition of materials, their processing into products, the products themselves, as well as the costs of post-production and post-exploitation waste disposal, as well as modeling all processes and properties related to materials. The most advanced stage of selecting materials understood in this way is defined as Material 4.0 [133–136,173,241], together with the possibility of doing virtual experiments under the idea of "digital twins".

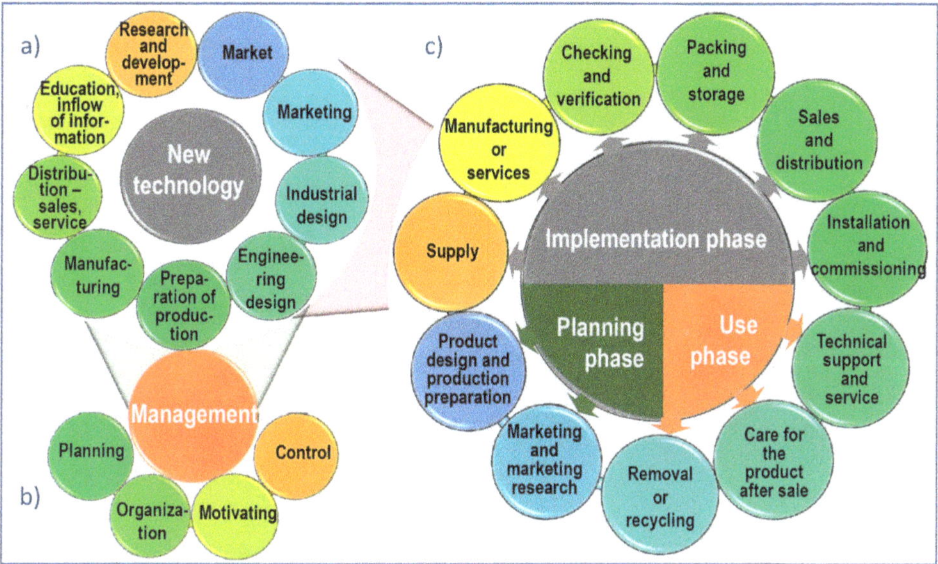

Figure 19. Diagram of the (**a**) interdependence of factors related to the introduction of a new product and a new technology to the market, (**b**) with production management, (**c**) and the product life cycle included in the quality loop.

Designing and manufacturing products with all expected utility values and properties required for technical reasons is the basic goal of all engineering activities. Engineering design is related to determining the shape of a product and its components to ensure the appropriate operation mode, taking into account the costs. The designed product must meet the parameters fully corresponding to its intended functional features, shape requirements, and dimensional tolerances. Apart from the list of materials and production methods, the project must consider the consequences and risk of damage to the product due to foreseeable but probable misuse or imperfection in the manufacturing process. Possible consequences of a product failure affect the assessment of the significance of its assumed reliability. The economic considerations do not impose overly high requirements on reliability if there is no injury to people and no significant loss to occur due to damage to this product during use. Each version of the shape of the product imposes material requirements, which may include the relationship between the stresses associated with the shape of the product and its load and the strength of the material. Process change can change material properties, and some product shapes and material combinations may not be feasible with some technological processes. Deterministic or probabilistic methods can determine the relationship between the requirements for the shape of the product and the material's characteristics. In deterministic methods, calculations are based on nominal or average values of stresses, dimensions, and strength. Also, appropriate safety factors are used, whose task is to consider the expected variability of these design parameters. In the probabilistic approach, individual parameters are assigned a proper variability distribution. Using these distributions and an acceptable margin of safety is possible to determine either the minimum adequate critical sections or the minimum strength of the product's key elements to be designed. It is then necessary to use more sophisticated calculation methods, resulting in a more compact structure of elements, requiring a smaller material mass. Regardless of the approach to the engineering design of product elements, the influence of notch and stress concentration should increase the damage's sensitivity. The cyclic loads, high or low-temperature operation and the presence of general corrosion or stress corrosion cracking are special hazards that must be taken into account in the material selection process.

In engineering design, quality management issues should be taken into account. Due to the disastrous consequences of placing defective products or services on the market, in each production process very important should be quality assurance. The constantly increasing quality requirements set by customers force manufacturers to focus their efforts in a pro-quality manner. Nowadays, the problem does not come down to detecting product defects but avoiding them or eliminating them during the manufacturing process. Therefore, it requires manufacturing and offering products for sale and services meeting the requirements of high and repeatable quality, modern, delivered on time, and provided at an affordable and competitive price. The conditions mentioned above pose certain expectations at the engineering design stage concerning the required quality of production and products. Failure Mode, Effect, and Causes Analysis (FMEA) is used more and more as an analytical method for both product design and manufacturing to avoid occurring or potential product defects. It is especially recommended in the development and production of a new product because it allows for identifying possible defects in advance, which enables their elimination as a result of applying preventive measures, even before the production of, for example, a new product [121,242].

In addition to the manufacturing processes, the engineering design process itself requires quality assurance, which is why it is planned, controlled, and supervised by clearly and comprehensively established goals, proper documentation of each stage of design, as well as by checking the results of activities by highly qualified staff. Such actions aim to change the previously common situation, where the design product was verified during the production or even operation of the product. Most errors and defects were revealed. Therefore, the applicable organizations and engineering design methods should reduce the number of design errors, detect them, and eliminate them already at the design time. Planning activities related to engineering design consist of dividing the process into stages, determining the design process points where project reviews and verifications are planned, linking and coordinating component activities, and determining the group of employees implementing the project and the means of enabling this activity. is essential to consider the complete product lifecycle when designing a product from the concept development stage. Due to the identification of threats related to the prospect of disturbing the ecological balance as a result of human activity, following the concept of sustainable development, ensuring the necessary balance between the interests of modern and future generations, the goal of which is a man and not material goods. Sustainable development on a global scale is the sum of local events, which requires searching for detailed design and technological solutions. Hence the emerging terms of sustainable technology or sustainable product or sustainable management. The fundamental role of material and technology design in the product life cycle's ecological course directly follows. It deals with two aspects: the market duration cycle and the technical duration cycle. The quality loop (Figure 19c) covers all phases of the project that enable shaping the product quality, from the initial determination of its parameters and description of needs to the final fulfillment of the recipient's requirements. Quality issues in the production, distribution, and consumption processes of a product must be considered comprehensively.

The product should be designed considering the possibility of reusing its elements that are not subject to wear or recovering the materials from which it is made. The re-use of these elements and materials reduces the consumption of raw materials and allows you to save energy needed to obtain engineering materials, e.g., metal from ore. The energy consumption is significantly reduced when using materials that do not require heat treatment with good strength properties. During operation, the energy consumed also has a significant share in the cost of a product's full life cycle [121,242].

The production of many medical devices, including orthopedic implants and standard mass-produced dental implants, in terms of organization, does not differ in any way from the production of other products, e.g., household appliances or cars. A smart factory is the basis of the modern manufacturing process [204,206], with a high degree of automation, robotization, and computerization based on the principles of Industry 4.0 for the development of industrial production (Figure 20).

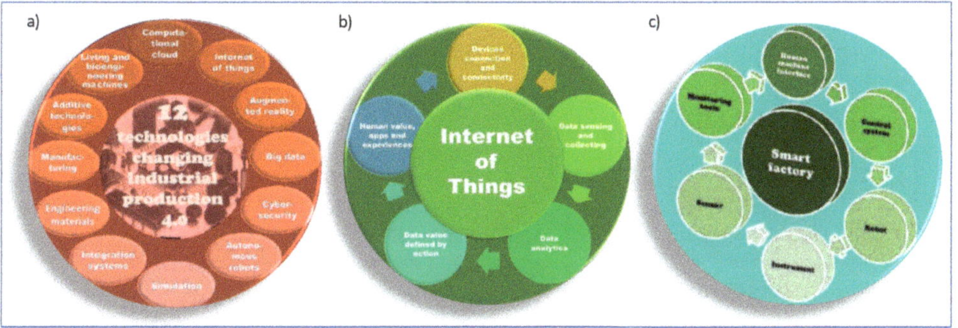

Figure 20. Diagram: (**a**) twelve technologies underlying the assumptions of Industry 4.0; (**b**) the Internet of Things; (**c**) interrelationships between smart elements of the Industry 4.0 system inside the Smart Factory.

Smart factories producing smart products use the embedded cyber-physical system to create value and exchange smart data via smart grids. Intelligent decisions regarding both the conditions of production and the value chain are made due to smart cooperation and information exchange between people and machines using a set of smart sensors. The exchange of information is ensured by the collaboration of the Internet of Things, people, and services. Relationships between stakeholders, i.e., employees, suppliers, and customers, production devices, and products in the product life cycle, are realized due to the exchange of information located in a virtual network using the computing cloud [190,191,211]. The smart data collected in this way influences the actuators in real-time due to which they change processes, products, and people's reactions. According to the augmented holistic model of Industry 4.0, twelve key technologies are used, including cyber-physical systems and engineering materials, and a full set of manufacturing technologies and machines, including additive technologies (Figure 20). Undoubtedly, this approach concerning the discussed problem should be considered Bioengineering 4.0, as a detailed case of Industry 4.0. On the other hand, this requires strict adherence to the achievements in this area in all phases of the manufacturing process, starting from market analysis, through industrial design, engineering design, production preparation, manufacturing, distribution, and maintenance, as well as removal or processing in the post-consumer phase of the product life cycle (Figure 21).

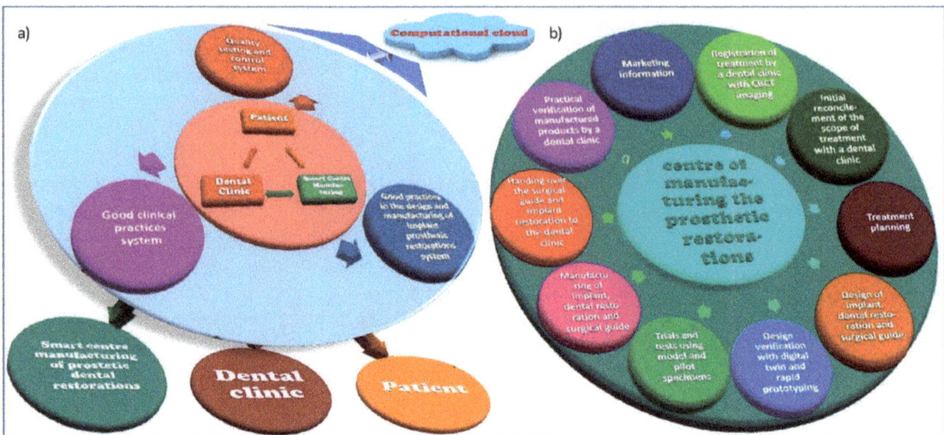

Figure 21. Diagram (**a**) relationship between the patient's expectations, design, and manufacturing capabilities of the center for the manufacture of dental prosthetic restorations and (**b**) the medical tasks of a dental clinic according to the concept of Dentistry 4.0.

A quite different situation is in the production of fully personalized implants and prosthetic restorations, which usually occur during implantological and prosthetic treatment in dentistry. In this case, the implementation of the achievements of Dentistry 4.0 corresponding to the current stage of Industry 4.0 development takes place in the center of manufacturing of prosthetic dental restoration, cooperating directly with the dental clinic. Figure 21 shows the role and the most important factors and tasks of the three basic subjects participating in modern implant prosthetic treatment. In the previous historical stage of Dentistry 3.0, the dominant achievement was improving conservative dentistry's effectiveness and disseminating X-ray images in diagnosing the dental condition [243]. The noticeable effects at the stage of Dentistry 4.0 corresponding to the smart factory standards include 3D imaging with the use of cone-beam computer tomography CBCT, data manipulation, and cloud computing, and above all, the use of computer-aided design CAD and manufacturing CAM methods for the production of personalized prosthetic restorations and it is not only by milling with the use of center numerically controlled CNC but above all by additive manufacturing AM, the so-called 3D printing technology. The numerous benefits include improved oral state and a significant reduction in the total time of medical procedures, in some cases from several months to one day, after prior diagnosis with CBCT tomography [243]. The benefits also include reducing manufacturing costs and the overall costs of implant-prosthetic treatment and the costs of the dentist, medical and engineering staff, and the patient's time [133,134]. The benefits of Dentistry 4.0 also include the integration of supplier and patient networks (Figure 22).

As the manufacturing centers for prosthetic restorations belong to micro or small enterprises, and the production is unique and personalized, automated guided AGVs are usually not used. Still, smart entry and exit logistics are applied. These centers collaborate with numerous dental clinics, from where they receive electronic data on the diagnosis of individual patients, including CBCT photos, via the network or cloud computing. On this basis, prosthetic restorations, models, and templates are designed and manufactured through computer-aided design CAD and production planning CAPP with the use of centers numerically controlled CNC, machines for additive manufacturing, e.g., the selective laser sintering SLS method, as well as surface coating, including internal surface coating, by atomic layer deposition ALD, often involving robots. A cyber-physical system enables creating a value chain. Sensor systems of manufacturing machines identify the factors of creating value and monitor manufacturing processes, and smart data activates the appropriate actuators. Prosthetic restorations from essence are personalized according to the principle of demand in every case so that a prototype is always produced. Still, it can be verified in virtual reality using the method of "digital twins". It is also possible to apply artificial intelligence and machine learning methods, which enable machines to be smart to react to various production situations.

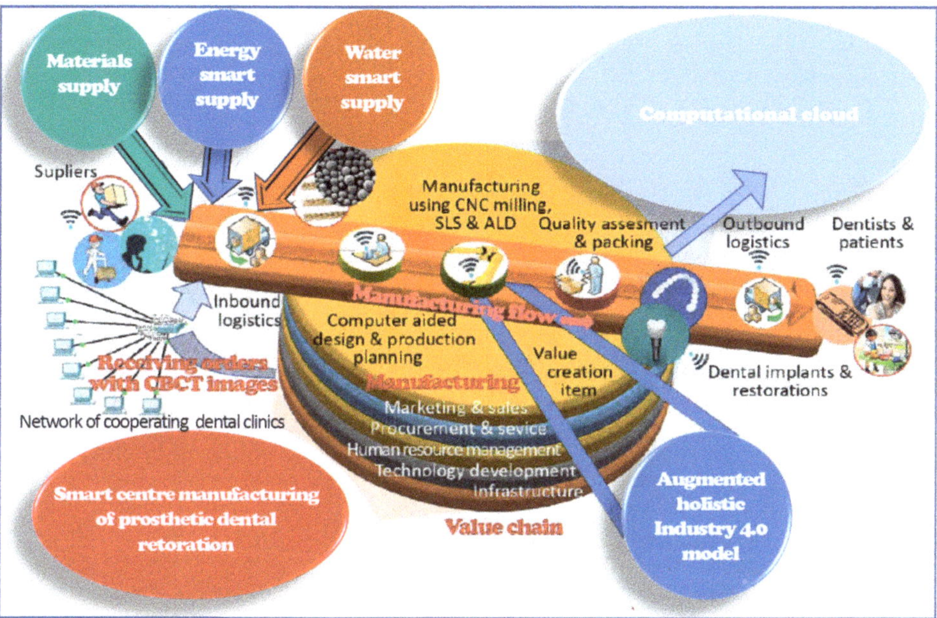

Figure 22. Diagram of an example technological line in intelligent centers produces dental prosthetic restorations, including horizontal and vertical integration.

Since ancient times, numerous medical procedures have been performed, including surgical ones requiring the use of various tools, documented, among others, by still in ancient Egypt recorded on artifacts and inscriptions discovered in Thebes and Rome and Greece. Even then, the relationship between technology and medicine became stronger with time. Many medical achievements have become possible only thanks to the possibilities offered by modern computer science, image analysis, electronics, and numerous medical devices, manufactured using the latest achievements of modern industry. From the point of view of the essence of the design and manufacturing processes, there is no difference between the production of an implant or dental prosthetic restoration and the production of a modern car or large passenger aircraft. The old, masterful approach has long gone down in history. The latest achievements of the current stage of the Industry 4.0 industrial revolution also apply to medical devices. The awareness of this fact must be disseminated among engineers dealing with bioengineering issues, and, importantly, among physicians who must have such advanced knowledge to be able to make appropriate demands with engineers cooperating with them. The extremely high and advanced level of engineering knowledge enables the achievement of such goals, but with full respect for the ethical principles that apply to all possible aspects of medicine. Biomaterials, without which any medical device cannot be made, play an extremely important role in this process.

Hence, this part of this paper focuses on the processes of designing and manufacturing biomaterials that ensure the advancement of medicine for a stage that is generally suggested name as Bioengineering 4.0. It should be noted that it is possible and necessary to use the idea of "digital twins fully". With extensive experiments in virtual reality, the costs and the number of practically performed laboratory and/or industrial trials can be reduced to a minimum.

5. Examples of Application of Various Biomaterials in Medicine and Dentistry

All products, including implants, implant-scaffolds, prosthetic restorations, and elements of medical devices, appliances, and apparatus, including electronic components in diagnostic devices and computer equipment used in medicine and dentistry, are produced of engineering materials. Without engineering materials, there are no products. Hence is the importance of engineering materials in any type of human production. In the context of health care, the group of biomedical materials [146,152,244–255], also often referred to as biomaterials, is of special importance. A biomaterial (biomedical material) is any substance that can be used at any time to supplement or replace the tissues of an organ or part of it to fulfill its function [150,242,256–258], provided that it is neither a drug nor a combination of synthetic or natural substances [258]. Biomaterials are characterized by the required biotolerance (biocompatibility), i.e., biological compatibility and harmony of interaction with living matter. Biotolerance (biocompatibility) is defined as biocompatibility. It means the harmony of interactions within living matter. The biomaterial with optimal biotolerance does not cause acute or chronic reactions or inflammation of the surrounding tissues. It does not interfere with the proper differentiation of the amputated tissue surroundings. The concept of biotolerance is often associated with preventing the initiation of toxicological and immunological reactions and the effects of tissue irritation [256]. Figure 23 shows the general classification of implants. A few criteria for the division of this group of materials can be adopted [152,250,253,259–280]. Therefore in this figure, a different graphic form was used for each criterion.

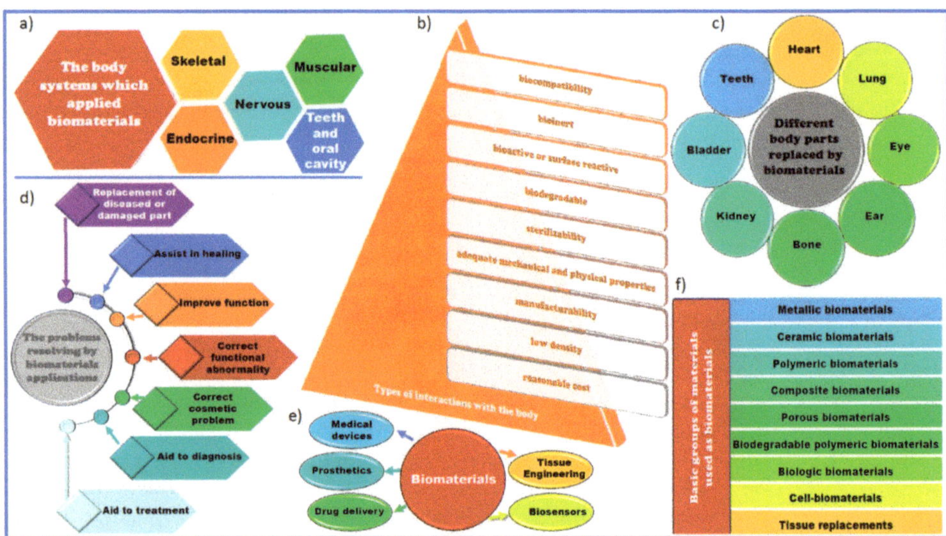

Figure 23. The general classification of biomedical materials according to various criteria: (**a**) the body systems which applied biomaterials; (**b**) types the interactions with the body; (**c**) different body parts replaced by biomaterials; (**d**) the problems resolving by biomaterials application; (**e**) applications areas of biomaterials; (**f**) basic groups of materials used as biomaterials.

This group of materials includes both metal materials, including pure metals and their alloys, and ceramic, polymer, and composite materials, and porous materials. Table 1 lists the most important criteria for the quality of biomaterials, including a set of requirements for implants [146].

Table 1. Examples of requirements for materials used for surgical implants.

Mechanical Properties	Technological Properties	Biotolerance
• tensile strength, • yield point, • fatigue strength, • hardness, • abrasion resistance, • stiffness, • plasticity (elongation, contraction), • ductility (resistance to fracture).	• ensuring the assumed quality of the biomaterial, • ensuring the required quality of the surface and the implant, • suitability of the material and product for effective sterilization, • minimal manufacturing costs.	• reactions with tissues and body fluids, • stability of ownership: • mechanical, • physical, • chemical, • degradation related to: • local damage to the implant (harmful changes), • systematic corrosion effects (harmful damage).

Biomedical materials usually come into direct contact with living cells, tissues, proteins, and even organs and organ systems. The criteria for classifying medical devices include [146]:

- contact or interaction with the body,
- contact with injured skin,
- contact with internal organs (e.g., heart, circulatory system),
- invasive nature with the holes in the body,
- implantation into the body,
- donating energy or substances to the body,
- period of application.

Table 2 presents the general classification of medical devices [146].

Biomedical materials are usually used to make the various medical devices listed in Table 2, which corresponds to the definition of engineering material. Sometimes, however, some chemicals, such as collagen injected to remove soft tissue defects, are used alone. Although they do not meet the definition of engineering materials, they are broadly referred to together with the relevant biomedical materials.

Technological progress has influenced the spread of biomaterials in various healthcare areas, including bioengineering, regenerative medicine, and tissue engineering. The authors previously presented their views and gathered extensive knowledge in this regard in previously published scientific books and academic textbooks [146,150,152,154,242,281]. Simultaneously, the biomaterials' usability and application versatility increased, regardless of the wide range of unique applications. There is a continuous development of novel methods for the controlled delivery and release of drugs. Smart biomaterials have been developed, including in some applications; they allow interaction with biological systems, such as bioactive molecules' transport. Smart biomaterials generate and transmit bioelectric signals analogous to those generated in original tissues, ensuring proper physiological functions. Piezo scaffolds are smart materials that play a significant role in tissue engineering. They stimulate signaling pathways and consequently improve tissue regeneration in the damaged area. The importance of partially or fully porous orthopedic and dental and craniofacial implants increases, including those with layers inside the pores, both biocompatible and preventing toxic spinal diffusion of certain metals, preventing metallosis. Porous structures stimulate bone growth around the implant and reduce the modulus of elasticity. Scaffolds enable the production of functional tissues. Powder metallurgy, 3D printing, and additive manufacturing are just some of the potential manufacturing techniques. Porous implants and implant-scaffolds, both metal, and ceramic, can be designed and manufactured using additive methods, the so-called 3D printing [139,282–291]. However, powder engineering is still offered wide possibilities, traditionally called powder metallurgy [137,138,152,292,293]. Additive methods are gaining importance among powder engineering methods applied to medical and dental devices made of biomaterials. These methods are comprised of others, in which metal and ceramic powders are used, in Figure 24. The method of procedural benchmarking was used [119,121,138]. The figure also lists the criteria for the potential and attractiveness of each technology. Marked as maroon-violet additive methods, including selective laser sintering and related methods, are located in the most favorable quadrant of the dendrological matrix–wide-stretching oak. Additive technologies concerning polymer materials used, among others, are also very possible on surgical templates and epitheses.

For these reasons, additive manufacturing technology is becoming more common concerning biomaterials and various medical devices, especially implants, both orthopedic, surgical, and dental [142,294–314].

Table 2. The general classification of medical devices.

The Main Criterion for Classification	Medical Device Groups	Medical Devices Subgroups	Comment
period of use of medical devices	transient (<60 min)		
	short-term (<30 days)		
	long-term (>30 days)		
degree of invasiveness	invasive devices (penetrating deep into the body through an opening in the body or its surface)	surgical	as a result of a surgical procedure, they are introduced inside the body or under its surface
		implanted	intended to be completely introduced into the body or to replace the epithelial surface or the surface of the eye as a result of a surgical intervention
	surgical instruments		
	active devices	medical	their operation depends on the conversion of feed energy other than directly generated by the body or gravity
		therapeutic	
		diagnostic	
	Implants	surgical	placed in the intended place in the body by surgical methods
		other	for example, needles, drains, filters
		implanted prostheses	internal prostheses or endoprostheses that physically replace an organ or tissue
		artificial organs	replacing wholly or partially the function of one of the main organs, often in a non-anatomical way
a field of medical use or a specific location in the body	Implants	orthopedic	used to support, replace or supplement temporarily or permanently bone, cartilage, ligaments, tendons or associated tissues
		oral	used to improve, enlarge, or replace any hard or soft tissue in the mouth involving the maxilla, mandible, or temporomandibular joint
		craniofacial	used to correct or replace hard or soft tissues in the craniofacial area except for the brain, eyes and inner ear
		dental	used to replace missing teeth

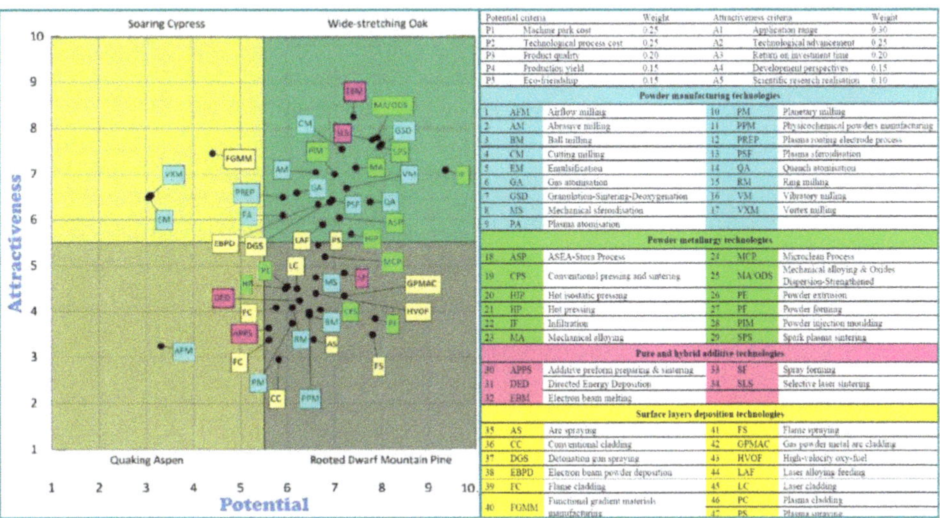

Potential criteria		Weight	Attractiveness criteria		Weight
P1	Machine park cost	0.25	A1	Application range	0.30
P2	Technological process cost	0.25	A2	Technological advancement	0.25
P3	Product quality	0.20	A3	Return on investment time	0.20
P4	Production yield	0.15	A4	Development perspectives	0.15
P5	Eco-friendship	0.15	A5	Scientific research realisation	0.10
Powder manufacturing technologies					
1	AFM	Airflow milling	10	PM	Planetary milling
2	AM	Abrasive milling	11	PPM	Physicochemical powders manufacturing
3	BM	Ball milling	12	PREP	Plasma rotating electrode process
4	CM	Cutting milling	13	PSF	Plasma sferoidisation
5	EM	Emulsification	14	QA	Quench atomisation
6	GA	Gas atomisation	15	RM	Ring milling
7	GSD	Granulation-Sintering-Deoxygenation	16	VM	Vibratory milling
8	MS	Mechanical sferoidisation	17	VXM	Vortex milling
9	PA	Plasma atomisation			
Powder metallurgy technologies					
18	ASP	ASEA-Stora Process	24	MCP	Microclean Process
19	CPS	Conventional pressing and sintering	25	MA/ODS	Mechanical alloying & Oxides Dispersion-Strengthened
20	HIP	Hot isostatic pressing	26	PE	Powder extrusion
21	HP	Hot pressing	27	PF	Powder forming
22	IF	Infiltration	28	PIM	Powder injection moulding
23	MA	Mechanical alloying	29	SPS	Spark plasma sintering
Pure and hybrid additive technologies					
30	APPS	Additive preform preparing & sintering	33	SF	Spray forming
31	DED	Directed Energy Deposition	34	SLS	Selective laser sintering
32	EBM	Electron beam melting			
Surface layers deposition technologies					
35	AS	Arc spraying	41	FS	Flame spraying
36	CC	Conventional cladding	42	GPMAC	Gas powder metal arc cladding
37	DGS	Detonation gun spraying	43	HVOF	High-velocity oxy-fuel
38	EBPD	Electron beam powder deposition	44	LAF	Laser alloying feeding
39	FC	Flame cladding	45	LC	Laser cladding
40	FGMM	Functional gradient materials manufacturing	46	PC	Plasma cladding
			47	PS	Plasma spraying

Figure 24. Results of the comparatist analysis of various powder engineering technologies for the application for medical device production.

Figure 25 shows an example of a full-arch dental bridge manufactured by selective laser sintering SLS from Co25Cr5W5MoSi alloy after individual design using CAD methods.

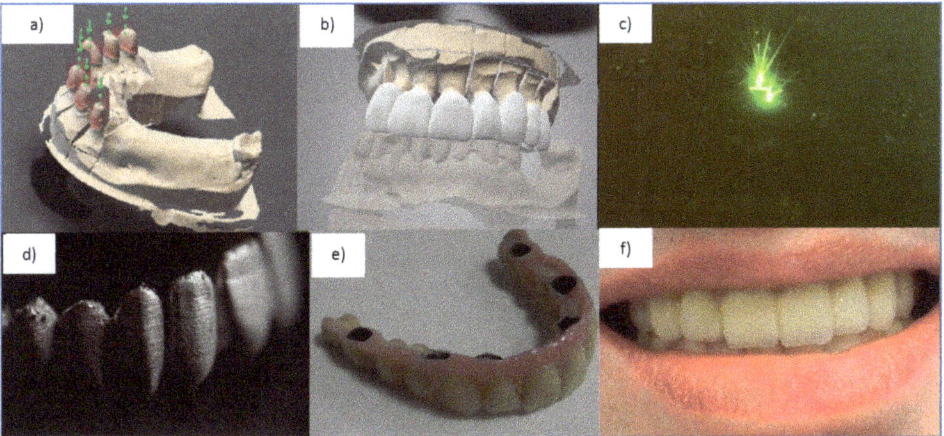

Figure 25. The examples of different stages of design and manufacturing by the selective laser sintering SLS of dental bridges from Co-Cr powders: (**a**) virtual tooth model and the verification of the parallelism of the pillars; (**b**) the finished foundation design for transferring from CAD to CAM; (**c**) selective laser sintering process photo; (**d**) the finished element after SLS; (**e**) the prosthetic restoration veneered with porcelain; (**f**) the bridge installed in the patient mouth.

Figure 26 shows the research results on selecting the conditions for selective laser sintering of Co25Cr5W5MoSi [291] and Ti6Al4V [139] alloys and their influence on the structure and properties of these alloys. These metal materials can be used in surgery and implantology, both in dentistry and orthopedics, and for other types of implants and prosthetic restorations. It turns out that the technological porosity depends, among others, on on the laser power density resulting from the interaction of the laser power and the diameter of the laser beam, as well as on the flow of protective gas and a dozen or so variables, it can vary from 0.06 to over 10%, which causes almost 2.5 times differentiation in the strength of additive manufactured materials. Structural studies show that too low laser power density results in many powder particles did not melt during sintering. Sintering with the liquid phase's participation cannot occur, while diffusion processes do not occur [315–318]. Therefore, the technological process carried out in these conditions is far from correct.

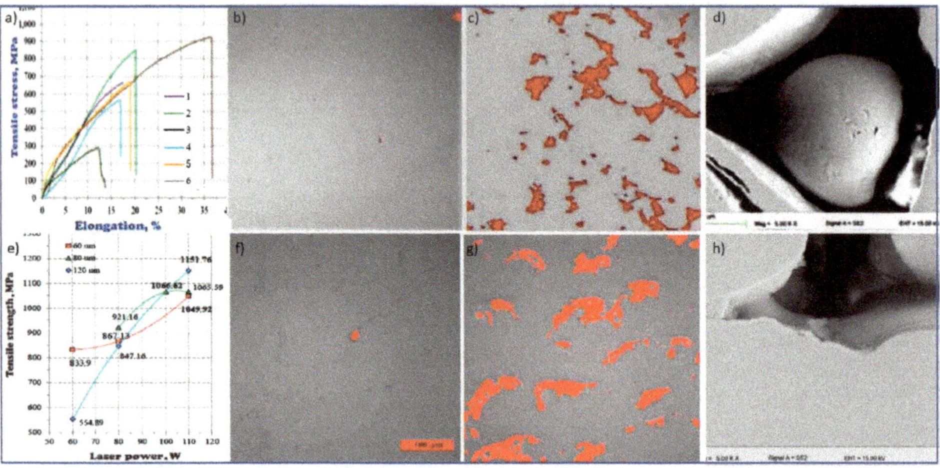

Figure 26. Influence of manufacturing conditions during selective laser sintering of powder on the tensile strength and structure of: (**a–d**) Co25Cr5Mo5WSi alloy; (**e–h**) Ti6Al4V alloy: (**a**) tensile strength (powder thickness 25 μm; laser power 105 W; laser spot diameter 10 μm (variant 4—80 μm), scan speed 630 mm/s; laser beam width variant 1,4,5—120 μm, variant 2,3—80 μm, variant 6—100 μm, allowance, variant 1,2,4.5—0, variant 3.6—10%; variant 6—incorrect flow of protective gas); (**b**) porosity 0.08%—option 3; (**c,d**) porosity 9.87%—variant 1; (**e**) tensile strength; (**f–h**) porosity (**f**) correctly sintered alloy 0.06%; (**g,h**) under extremely inadequate conditions 10.53%; (**d,h**) pores resulting from incompleteness of liquid phase sintering; (**b,c,f,g**) LM; (**d,h**) SEM.

It turns out that the technological porosity depends, among others, on the laser power density resulting from the interaction of the laser power and the diameter of the laser beam, as well as on the flow of protective gas and a dozen or so variables, it can vary from 0.06 to over 10%, which causes almost 2.5 times differentiation in the strength of additive manufacturing materials. Structural studies show that too low laser power density results in the fact that many powder particles do not melt during sintering. Sintering with the liquid phase's participation cannot occur, while diffusion processes do not occur [315–318]. Therefore, the technological process carried out under these conditions is far from correct.

Figure 27 shows a diagram of the SLS selective laser sintering process [137,293,319,320]. It is worth noting that this technology belongs to the group of Powder Bed Fusion (PBF) processes. The literature has different names for additive manufacturing AM's technological processes, which are identical or very similar in principle. Such processes include, among others [293,319]: Selective Laser Melting (SLM), direct metal laser sintering (DMLS), or high-temperature laser sintering (HTLS). As these processes are similar in terms of basic operation principles, it is advisable to use SLS. The study of the differences between these processes practiced, e.g., in dentistry, is unjustified and pointless. The essence of the process is sintering with the participation of the liquid phase sintering LPS. It means that each powder particle is melted or, when it is very fine, even completely melted. Generally, it does not completely melt, remelting the joined material and even any of its successive layers of powder (Figure 27) [137,315]. Liquid phase sintering (LPS) consists of sintering with the coexistence of a liquid and a solid during all the sintering process or its part [321], which then arises, ensures that the undissolved powder particles or their cores contained in the mixture are wetted [322,323]. The liquid is subject to capillary forces [324,325], which results in densification, a further fusion of the solid with increased surface energy, and subsequent solidification of the liquid phase [321]. Mass transport speed in the liquid phase is hundreds of times faster than in solid sintering [315]. The improvement of the liquid phase's wettability and the change of the surface energy is favored by increasing the system temperature, thanks to which the sintering intensity increases [326]. This mechanism practically does not occur in pure elements and is appropriate for multi-phase steppes, especially for multi-component powders. The process consists of melting at the boundaries of each powder particle, and the share of solid's diffusion processes is then of marginal importance. Such a liquid phase sintering mechanism is dominant in 90% of all sintered materials' commercial value [137,138,292,315]. The phenomena occurring are similar to those described during the heating of the supersolidus and shown in [137].

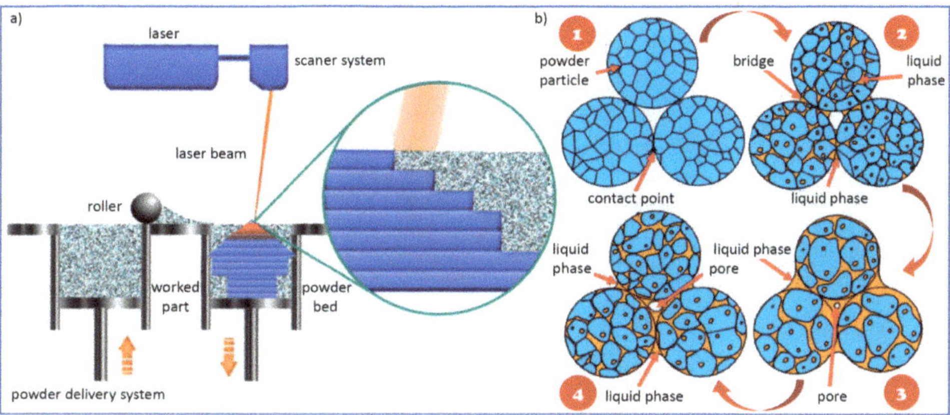

Figure 27. Diagram (**a**) Selective laser sintering methods; (**b**) the successive stages of liquid phase sintering.

Unfortunately, this obvious information is unknown in many centers trying to apply additive manufacturing technology in medicine and dentistry. Hence, there will be publications [327,328] in which it is generalized that the properties of additive manufacturing or conventionally cast materials do not differ from each other. It seems to indicate that about 60% of the possibilities provided by the material used and the additive manufacturing process are used in such a technological process. It is just such a lowering of the manufactured medical and dental devices' properties due to the uncritical application of factory recommendations for additive manufacturing. This type of practice used in medicine and dentistry bears the features of an ethical tort committed both by doctors who apply such manufactured implants or prosthetic restorations and engineers who design and manufacture medical devices with unacceptably low properties exposing patients to significant health damage. It is hard to imagine, but the reason may be even more prosaic, that doctors stock up on an expensive machine and, without an engineering background, do the work on their own. It is an even greater ethical delict that violates the Hippocratic oath principles [57,58]. In the name of well-understood patients' expectations, substantive attention should be paid to the irregularities in such papers to avoid false information that cannot be used for patients.

Figure 28 shows a comparison of the properties of metal materials, including titanium, Ti6Al4V alloy, and Co-Cr alloy produced by selective laser sintering and by milling a conventionally cast disc in a CNC milling center, as well as ZrO_2 material milled in the pressed state in a CNC milling center from the disc and then sintered. These studies indicate very low strength properties of this sintered material, indicating the maintenance of far-reaching caution in the use of this material in prosthetics and implantology, including dentistry [329].

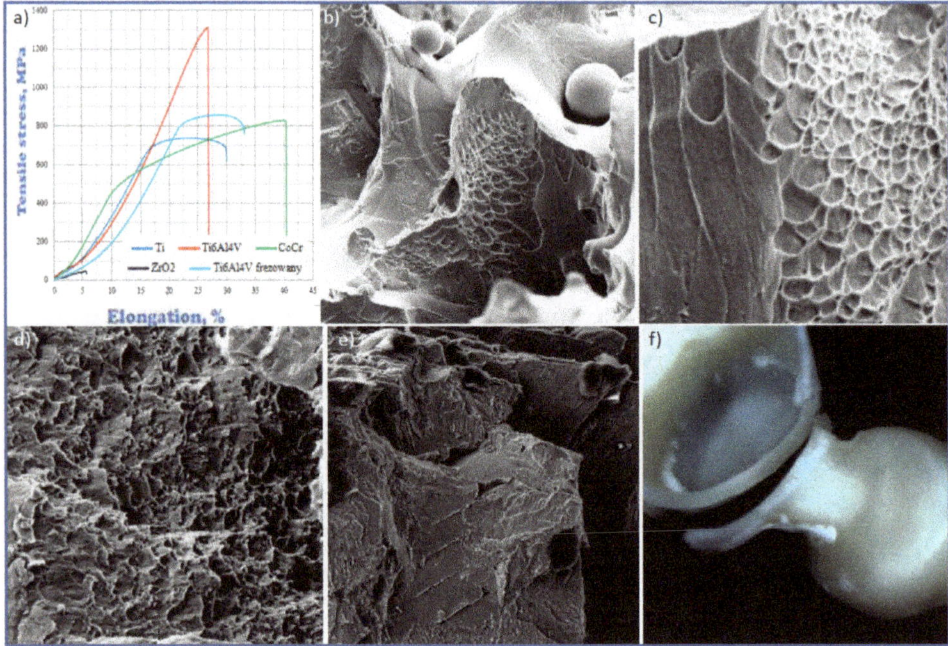

Figure 28. Comparison of the structure and properties of sintered Ti, sintered Ti6Al4V alloy, milled Co-Cr alloy, milled ZrO$_2$ sinter, and milled Ti6Al4V alloy: (**a**) comparison of the graphs of tensile stress versus elongation; (**b–e**) fracture structure after static bending test (**b,c**) Ti6Al4V solid alloy samples produced by selective laser sintering using a 50 μm laser spot after sintering at laser power (**b**) 70 W; (**c**) 110 W; (**d,e**) samples produced by the CAD/CAM method by milling disks from solid alloys: (**d**) Ti6Al4V, (**e**) Co-Cr; (**f**) view of a dental bridge made of ZrO$_2$ sinter produced by the CAD/CAM method by milling the disc and then sintered after a static bending test; (**b–e**) scanning electron microscope (SEM); (**f**) stereoscopic microscope.

For this reason, Co-Cr alloy is still of great importance in dental prosthetics. Figure 29 shows an example of manufacturing a full-arch prosthetic bridge based on tooth abutments through milling in a CNC center from conventionally cast discs.

Porous materials made from powders by additive methods are very useful in some applications [137,138,282–288,292,293]. Figure 30 shows the results of studies on selectively laser sintered porous titanium [144], including the structure observed in the high-resolution transmission electron microscope HRTEM.

The original concept of porous materials or porous zones on a solid substrate was used, among others, in metal implant-scaffolds produced by selective laser sintering [330–334]. Implementing this concept, both in dentistry and orthopedics, is possible to design and manufacture a solid core and a porous surface layer. Living cells, most often osteoblasts, can grow. Surface layers also play an important role, making engineering materials useful in these applications, which meet the stringent requirements to a lesser extent. Figure 31 shows the research results on the application of 1000–1500 layers of Al$_2$O$_3$ and TiO$_2$ using the atomic layers deposition method with a total thickness of 1–3 μm. It is the only method that allows the coatings to be applied inside the porous structure without producing a shadow effect. HRTEM studies show that the nanostructured layer is situated on the crystalline substrate without any discontinuities and gaps.

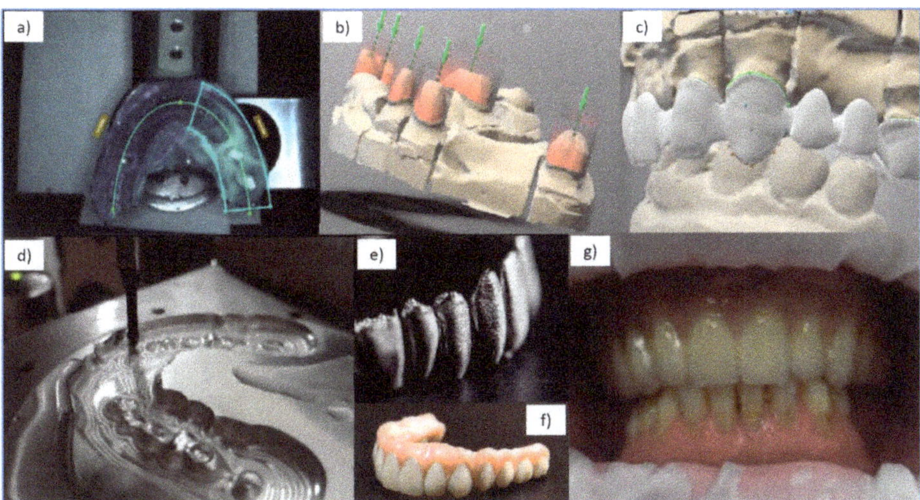

Figure 29. The procedure for manufacturing dental bridges using solid metal block milling technology; (**a**) a dental impression which is then scanned directly from the impression spoon; (**b**) the verification of the parallelism of the pillars; (**c**) the finished foundation design for transferring from CAD to CAM; (**d**) using the CNC milling center, the foundation is produced; (**e**) the finished element after cutting; (**f**) the prosthetic restoration veneered with porcelain; (**g**) the bridge installed in the patient mouth.

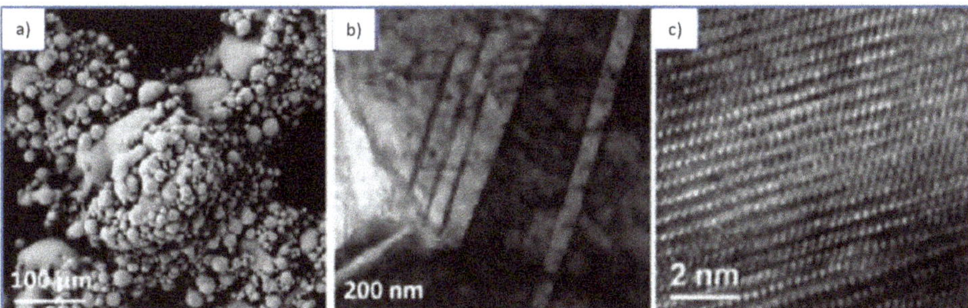

Figure 30. Structure of selectively laser-sintered titanium; (**a**) skeleton topography with a pore diameter of ~450 μm (SEM); (**b**) Ti-α martensite crystal structure in the thin foil (TEM); (**c**) distribution of Ti atoms in Ti-α martensite (HRTEM).

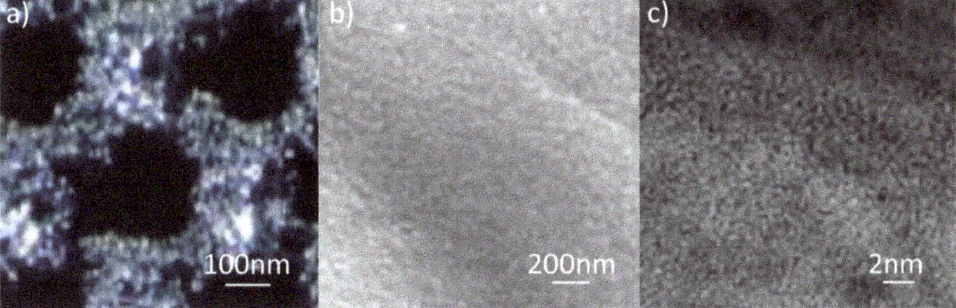

Figure 31. The surface structure of SLS titanium coated with TiO_2 coating during 1500 cycles; (**a**) Surface topography (stereoscopic microscope); (**b**) surface image (SEM); (**c**) amorphous TiO_2 coating layer (bottom) and titanium crystal structure (top) (HRTEM).

In this context, biological research plays an important role. Many original studies have been carried out [289,290], which indicate alloys' favorable behavior with a titanium and cobalt matrix. It concerns biocompatibility, toxicity, corrosion resistance, and tribology. It is very advantageous to use surface layers that favor the growth of living cells, as shown in Figure 32.

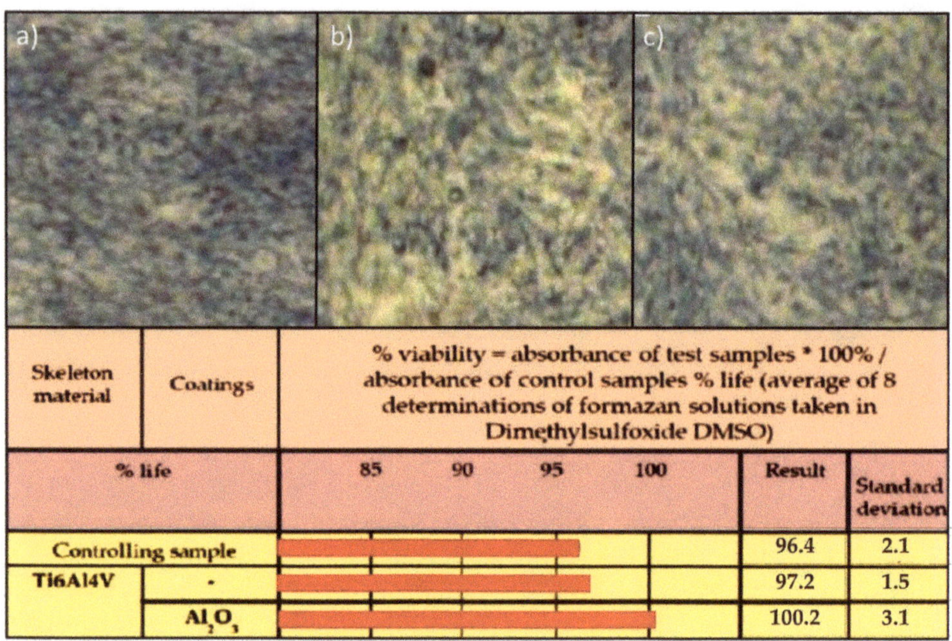

Skeleton material	Coatings	% viability = absorbance of test samples * 100% / absorbance of control samples % life (average of 8 determinations of formazan solutions taken in Dimethylsulfoxide DMSO)				Result	Standard deviation
% life		85	90	95	100		
Controlling sample						96.4	2.1
Ti6Al4V	-					97.2	1.5
	Al$_2$O$_3$					100.2	3.1

Figure 32. Images of specimen surfaces (light microscope, fluorescent contrast); (**a**) control glassware; (**b**) Ti$_6$Al$_4$V alloy without surface layer; (**c**) Ti$_6$Al$_4$V alloy with Al$_2$O$_3$ coating applied by the ALD method.

Nanostructured materials are increasingly used, or this group of materials is used in biomedical materials and devices, e.g., as reinforcement of biomedical composite materials or as elements of layers used in medical devices, shown in Figure 31. Sometimes this group also includes inorganic substances and natural macromolecules (biopolymers), useful for various medical devices, including prostheses, even if they do not meet the general definition of engineering materials. It should be noted that the materials of this group can often be used in veterinary medicine and animal husbandry. The requirements in such cases are similar to those for medicine and dentistry, although the legal provisions regulating these different application aspects are usually completely other. The most common types of biomedical materials were engineering materials used in varieties requiring exceptional purity due to their medicine or dentistry applications. Currently, many materials are specially developed to a much greater extent, e.g., polymers, biopolymers, and various inorganic substances dedicated exclusively to medical applications and not applicable in other areas of materials science. Examples of the use of gutta-percha in dental endodontic treatment are presented (Figure 33) [16,335,336].

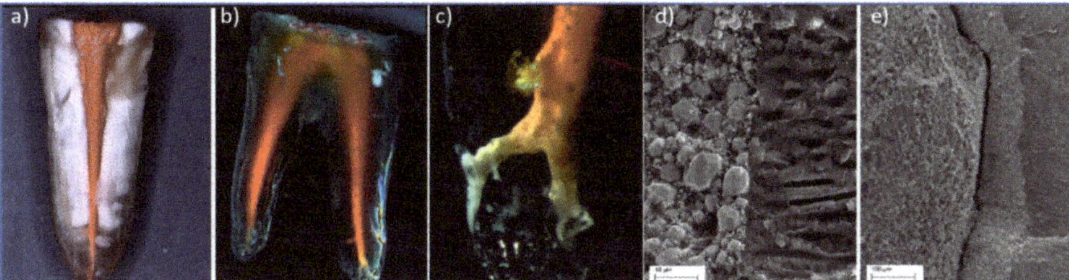

Figure 33. Examples of endodontically treated tooth with root canal filled with gutta-percha material with AH Plus sealant by the thermoplastic method: (**a**) longitudinal fracture of the tooth; (**b,c**) decalcified teeth; (**c**) root delta of the transverse fracture of the canal dentin; (**d**) in all sections of the root canal the material based on gutta-percha and sealant is tightly connected; (**e**) the border of three layers: a sealant layer covering the material on the gutta-percha matrix in a leak between a thick intermediate layer of sealant and the sealant with tightly connected dentine of the root canal after an incorrectly performed procedure; (**c**) atomic force microscope; (**d,e**)—scanning electron microscope.

Another example of polymeric materials is the design and manufacture of surgical guides used in dentistry, shown in Figure 34 [337,338]. The template is made of light-cured polymeric materials using the SLA stereolithography technology. According to the own invention, a special type is a template containing the accessory enabling implants' insertion [339].

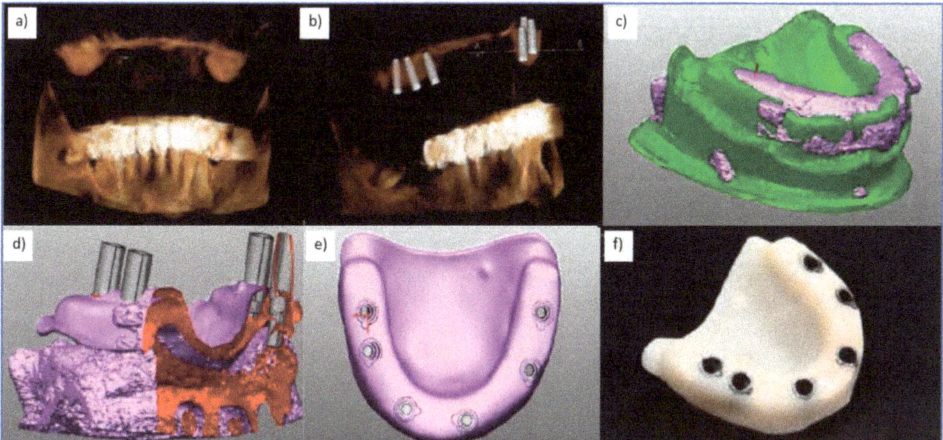

Figure 34. The procedure for making the template for implantation of six implants; (**a**) a 3-D model created based on CBCT; (**b**) planed the placement of implants; (**c**) the ready bone base model folded together with the intraoral model using an individual positioning plate; (**d,e**) two views of a model of the drill guide tubes; (**f**) the ready template plate using the 3-DP SLA technology.

Another example of polymeric materials is nanofibers obtained by electrospinning, especially long-absorbable composite nanofibers with a bioactive core and a bactericidal coating. Polymer nanofibers can find application in medicine provided that their potential toxicity is eliminated, starting with the materials used for obtaining solutions and taking into account a set of other factors [154]. In coaxial electrospinning (Figure 35a), moderately volatile solvents, such as a mixture of formic acid and hydrochloric acid of the rotary collector, are used to prevent the tendency of nanofibers to stick together. Composite nanofibers of the core-shell type (Figure 35b), combining the antibacterial properties of the coating with the core's bioactive properties, are an attractive material for a three-dimensional tissue scaffold, porous implant-scaffolds, for an innovative generation of flexible composite materials for regenerative medicine. When silver nanoparticles are deposited on nanofibers' surface (Figure 35c), high bactericidal properties can be ensured. Silver nitrate and AlphaSan are highly effective in combating Gram +, Gram- bacteria, and fungi, and the high-molecular chitosan has no antibacterial and antifungal properties. Such materials can also serve as a carrier for drugs and, in the form of soft mats, as dressings, e.g., in the event of burns, or as cells' carriers attached to the burned skin.

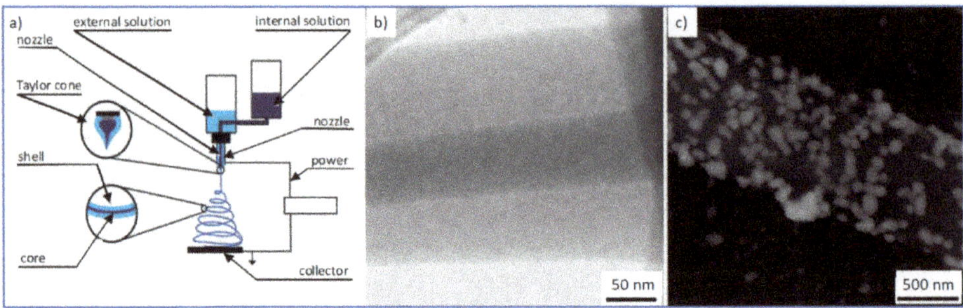

Figure 35. Manufacturing of polymer nanofibers by electrospinning for application in regenerative medicine: (**a**) scheme of the manufacturing method; (**b**,**c**) geometrical characteristics of the core-shell composite obtained from (**b**) a coating solution of 10% polycaprolactone without additives and inner core solution of 4% of polycaprolactone without additives (TEM), (**c**) double-component fibers obtained by dissolving PCL granulate with the molecular mass of Mw = 70,000–90,000 g/mol using 10% mixture of hydrochloric acid and formic acid with a mass ratio of 70:30 with the addition of 25% of AgNO₃; the photograph was taken after precipitation of silver in 2% ascorbic acid solution (SEM).

In 1992, the development of regenerative medicine began, closely related to the subject of biomaterials and implantology, especially with the use of bionic medical implants [340]. Bionics is concerned with designing, manufacturing, and studying artificial engineering systems that can restore lost functions of biological systems [341,342]. The replacement of organs and tissues occurs by their replacement by autographs, allografts, or devices made of various biomaterials [270]. Methods of treating patients or regenerating the body through cell-based therapies involving the replacement of diseased and old cells and tissues with young cells and gene therapy and tissue engineering methods are regenerative medicine goals. It makes it possible to counteract the causes, symptoms, and sequelae of various diseases [343–346]. Development of biological materials for the reconstruction and improvement of the functions of tissues and/or organs and their substitution with the interdisciplinary use of the principles of natural sciences and engineering, mainly materials, has been the essence of tissue engineering since 1985, constituting a field of technical sciences supporting regenerative medicine [347–352]. The history of the use of live cell therapy in medicine dates back to the first successful allogeneic human hematopoietic stem cell (HSC) transplantation in 1968 [353], and subsequent studies intensified in the last quarter of a century [354–357] relating to, among other things, for cartilage and skin diseases. Adult and embryonic multipotent and self-renewing stem cells (MSCs) [355,358,359] found in umbilical cord blood [360,361], placenta [258], amniotic fluid [362], or the pulp of infant milk teeth are used. Most preferably, to a limited extent, to use autologous cells [355,363] not requiring immunosuppressive treatment [362,364–366].

Somatic, and especially hematopoietic stem cells (HSCs) and bone marrow stromal cells (BMSCs) [355,367–369], and adult stem cells from synovial fluid, tendons, skeletal muscles [369,370] and adult muscles, adipose tissue ASC can be used more commonly (fat stem cells) [369,370], corneal stroma [361], peripheral blood, nerve tissue, and dermis can transform into different tissue types. A prerequisite for cell implantation is preventing ischemia and necrosis [371–376] from maintaining viability [377,378].

In addition to the so-called pure cell therapies consisting of direct injection of stem cells into the peripheral circulation or specific tissues, stem cell carriers are often used for their transport, and scaffolds are most often used to group them in selected places in the body [152,154,156,379–388]. A microscopic, porous structure of the scaffolds must ensure diffusion of nutrients and metabolic products, enabling the formation of a three-dimensional tissue structure simulating natural solutions [387,389] and ensuring proper vascularization [390–393]. It is needed to develop the technology of producing materials with the required pore structure and select engineering materials that provide the required properties and biocompatibility.

This part of the paper indicates, through numerous examples from the authors' own clinical and manufacturing practice, how important the success of countless clinical procedures is the correct design of the concept of the medical devices, the biomaterials used for their production manufacturing processes. Increasingly better quality of biomaterials and engineering materials used in the production of numerous medical devices, ranging from implants, through surgical tools, to hospital beds and rehabilitation equipment, is an extremely important factor. Any research to improve the properties of biomaterials and products derived from them serving improve patients' health, relieve pain during illness, and often allowing them to return to everyday life and ensure normal living and working conditions are highly desirable. The engineering activities in bioengineering have a fundamental impact on the level and quality of manufactured medical devices and the fulfillment of the described expectations and requirements set by the Industry 4.0 stage. Technological processes of manufacturing and processing biomaterials are extremely important, including additive processes, electrospinning, and sophisticated surface engineering methods, e.g., atomic layers deposition.

These aspects have an immediate effect on the expected high standard of living, good health, and well-being when they require improvement through medical intervention. For this reason, a lot of attention has been paid to these aspects in this part of the paper.

6. Economic Conditions for the Implementation of Various Groups of Biomedical Materials

The above-mentioned technical achievements and the systematic increase of the application areas are accompanied by the dynamic growth of the biomaterials market and its value. The development trend of the biomaterials market has continued over the last two decades and is expected to continue in the 2020s. The market is divided into five main categories, depending on the type of products, technology of manufacture, and type of medical condition, with regard to the eyes, ears, orthopedics, heart, and nervous system, and brain. The market for dental implants and dentures is separate. Safe, reliable, and inexpensive biomaterials include metals, ceramics, polymers, and naturally derived biomaterials. So far, metal materials have had the largest share in the global biomaterials market, with the rapidly growing share of polymeric biomaterials. It should be noted that among the metal materials used as biomaterials, the use of austenitic steels resistant to corrosion, popular in the previous decades of the 20th century, has virtually been eliminated due to the identified allergic effect of nickel present in these steels and the restrictive directives introduced by the European Union in this regard. Instead, their use of titanium and magnesium and their alloys have increased significantly.

The value of the global biomaterials market in 2019 was estimated at USD 106.5 billion in 2019, and it was estimated that in 2020–2027 compound annual growth rate, CAGR, will amount to 15.97% (Table 3). The undoubted cause of such dynamic growth is, among others, the increasing frequency of diseases of the musculoskeletal system and chronic diseases of the skeletal system, as well as the very widespread prevalence of diseases of the oral cavity. The structure of biomaterial applications is illustrated in Figure 36a. Figure 37 shows the dynamics of growth in this market's value in the United Kingdom. It should be noted that the markets related to biomaterials and other avant-garde material technologies discussed here only partially overlap. Therefore, apart from analyzing the overall market, it is worth paying attention to a few detailed ones.

Table 3. The current value of various global biomaterial markets and the related forecast compound annual growth rates CAGR.

Attribute	Details					
Type of Global Market	**Biomaterials**	**Dental Biomaterials**	**Orthopedic Biomaterials**	**Tissue Engineering**	**Regenerative Medicine**	**Cell Therapy**
Unit	USD billion	USD million	USD billion	USD billion	USD million	USD billion
Base year for estimation	2020	2020	2018	2019	2016	2020
Market size value in based year	121.1	7.983	11.96	9.9	5444	6.09
Forecast period	2020–2027	2019–2025	2019–2025	2020–2027	2017–2023	2020–2027
Growth Rate CAGR, %	15.97	7.41	10.3	14.2	32.2	5.4
Literature	[394]	[395]	[396]	[397]	[398]	[399]

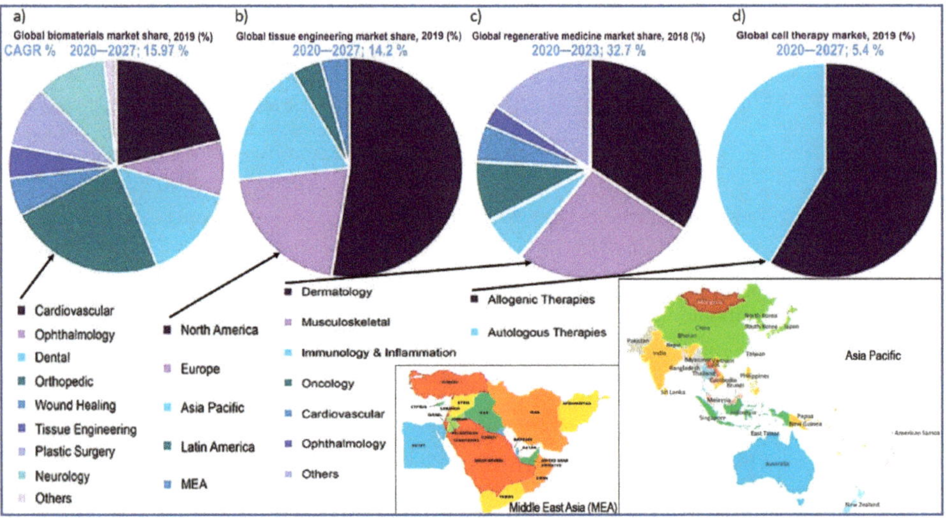

Figure 36. The structure of global markets due to (**a**) the use of biomaterials for the treatment of various diseases; (**b**) involvement of individual geographic regions in tissue engineering technologies; (**c**) the suitability of regenerative medicine methods for treating multiple diseases; (**d**) the contribution of allogeneic and autologous cell-mediated therapies.

The increasing scale of oral diseases leads to toothlessness and, consequently, to advanced implantology and dental prosthetics, which results in the forecast of an increase in the value of the dental biomaterials market from USD 7983 billion in 2020 in the period 2019–2025 with a CAGR of 7.41%. The popularity of dental biomaterials is due to the increasingly common problems with oral health and the population's progressive aging. It is not without significance that dental care has been privatized quite commonly and in many countries. Therefore the range of expectations of many people in dental care is increasing due to the growing wealth of many people. The dental biomaterials market is divided into metal, ceramic, polymers, and natural materials. The latter segment is to record the value of the incremental value in the string of means of payment. They will impact nanotechnology's latest achievements, leading, among other things, to the work of dental autogenous bone grafts.

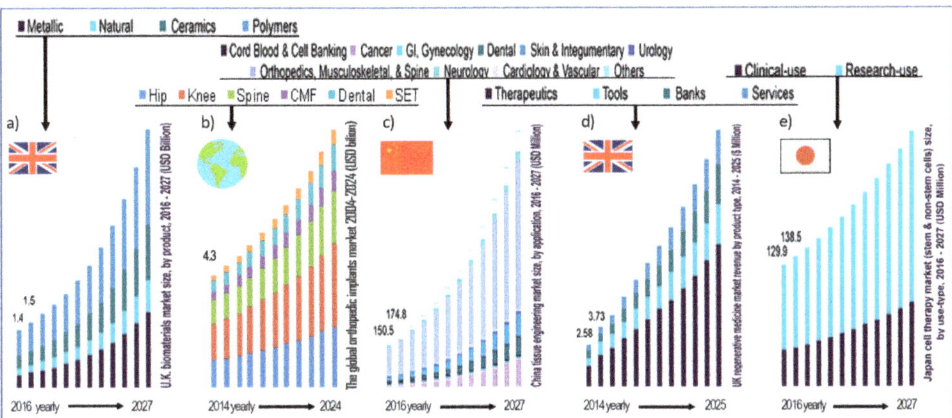

Figure 37. Forecasts of market growth until 2024–2027, respectively: (**a**) types of biomaterials in the UK; (**b**) the purpose of the main groups of orthopedic implants in the world; (**c**) the allocation of various tissue engineering methods in China; (**d**) different application groups of biomaterials in regenerative medicine in the UK; (**e**) the scope of the therapy goal in Japan.

The increased risk of osteoarthritis, osteoporosis, and other musculoskeletal disorders that come with adulthood, the social scale of which is increasingly widespread due to the growing geriatric population globally, is why the demand for orthopedic implants. It is directly related to the increase in the orthopedic biomaterials market's value from USD 11.96 billion with a CAGR of 10.3% (Table 3). The dynamics of the increase in the value of this market are shown in Figure 37b.

The main market for orthopedic joint replacement implants is those used in knee and hip replacement surgery (Figure 38), mainly aimed at alleviating patients' pain and restoring limb function. This global market reached over USD 16.5 billion in 2017, and its CAGR has increased by more than 3% since 2013 [400] and is broken down by common type for knee or hip replacement implantation. In the case of the knee replacement implant, the CAGR was relatively higher, and in the historical period, it reached a CAGR of nearly 4%. The largest market for major orthopedic joint replacement implants in North America with 54% of the world market, followed by Western Europe and the Asia-Pacific region, with this region and South America expected to develop the fastest. It increase was justified by the increase in the number of older people, the increase in the prevalence of obesity and arthritis, and the increasing level of knowledge about the positive importance of joint replacement surgery, taking into account the high cost of surgery and the relatively high share of failure, as well as the significant development of novel conservative therapies for arthritis.

However, significant advances in surgical techniques have been noted, e.g., the implementation in 2015 of a robotic total hip arthroplasty system [401–419], which improves the procedure's accuracy and precision to four times greater than in the case of manual implant placement. Increasingly, to increase the speed and efficiency of joint operations, additive methods, the so-called 3D printing, manufacturing of implants [420–423] after initial replication of components used in the process using patient's X-rays or computed tomography. Implants can be manufactured from solid and porous pieces fabricated in the same element or product to provide patients with individual knee or hip replacement. Metals and ceramics are the largest segment of orthopedic implants in the hips and knees, respectively, although polymeric materials also play a role. The most common, however, is a combination of cooperating elements in both types of these implants. By type of material combination—the hip replacement implant market is broken down by material combination type into [400]: Metal on Polymer (MOP), Ceramic on Polymer (COP), Ceramic on Ceramic (COC), Metal on Metal (MOM) and Ceramic on Metal (COM), although for the knee replacement implants these combinations are limited to MOP and COP only. The MOP market is the largest for both ponds. There are three possibilities of fixing both implants' types without cement, either cemented or hybrid with partial cementation. Cementless fixation is more commonly used for hip implants, while for the knee joint, it is more widely used for cement fixation.

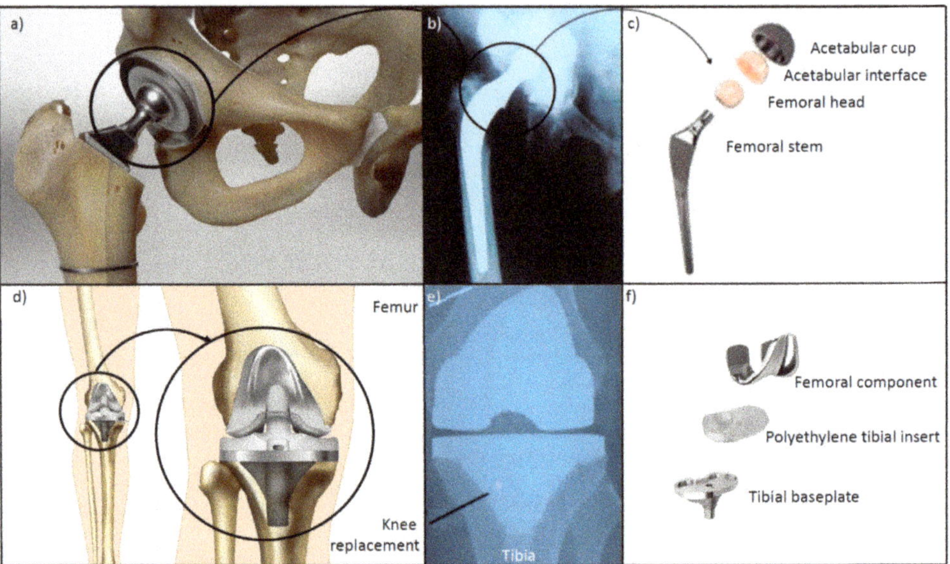

Figure 38. Scheme of application of orthopedic implants (**a–c**) of the hip joint; (**d–f**) knee joint; (**a,d**) general view of the application method; (**b,e**) X-ray pictures after installing the implant in the human body; (**c,f**) assembly diagram for implants' construction.

For hip replacement, the global market size was estimated at USD 7.13 billion in 2018, forecasting to grow to USD 10.51 billion in 2026 with a CAGR of 5.0% during the forecast period [424]. It must be stated that the information provided in other reports is less optimistic [424–427]. This market is divided into total hip implants, which had the largest market share since 2018, and the revision hip. The growing demand for minimally invasive surgeries and continuous technological advances favor the ever-increasing importance of other types of partial femoral head replacement and hip resurfacing. Hip arthroplasty surgery is an alternative to total hip replacement, allowing for alignment rather than a complete replacement, increasing the market share of hip replacement implants. New technologies and navigation systems are constantly being implemented to facilitate and simplify surgical procedures. For example, in 2018, 3D total hip arthroplasty was performed for the first time [428], and in 2019, an intelligent navigation application for hip surgery was implemented [429,430]. In the hip joint replacement market, the share of MOP implants is the largest because it is the most cost-effective, causes fewer complications than other types of materials, and shows better tribological properties. MOM, on the other hand, is widely used for total hip replacement and revision hip replacement surgery. However, too frequent postoperative complications may cause a decline in interest in this type of implant in the near term, unlike the COP market, which is expected to grow dynamically due to the low complication rate postoperative and the lowest wear rate.

The global knee replacement implant market is projected to grow at a rate of 5.3%, from USD 10.01 billion in 2019 to USD 15.33 billion in 2027 [431]. However, information on this in other reports is ambiguous [431–434] and generally less optimistic. Knee arthroplasty, also known as knee arthroplasty, alleviates pain and disability in patients with knee deformities in the face of an increase in the geriatric population, combined with an increase in obesity and osteoporosis. Total knee prostheses had the highest share of around 75% in 2019, but the percentage of partial knee replacement increases with a CAGR of 6.7% as only the damaged part of the knee is replaced. A knee revision means the amount needs to be replaced. The artificial material used in arthroplasty as a knee joint implant restores the function of the knee. The United States is the largest market for these products due to the use of technologies such as 3D printing, and the CAGR of this market is 5.0% in the US. Total knee arthroplasty is performed in 75% of cases. In Germany, the CAGR is 4.1%. Among the causes of these operations, the main reasons are osteoarthritis and ligament rupture, road and sports accidents, and related injuries. Usually, the recovery period after surgery takes six weeks. Still, efforts are being made to shorten the healing period by introducing innovative devices and technologies, including the so-called 3D printing increase in osteoarthritis incidence among adults. The high cost of implants reduces the pace of market growth, which are favored by the popularization of minimally invasive operations, the implementation of advanced technologies, and the improvement of the materials' strength properties from which the implants are made. Due to nickel toxicity, stainless steel with an austenitic structure has completely lost its importance, while the most commonly used metals for knee prostheses are Co-Cr alloys. Pure titanium is biocompatible, as is its Ti6AI4V alloy, and is, therefore, the most desirable material in this application.

Tissue engineering is an alternative to mechanical devices used to treat damaged tissues, surgical reconstruction, and transplants. Cell therapies, mainly with stem cells, are gaining more and more therapeutic importance in various clinical areas and the management of numerous diseases, including cancer, diabetes, obesity, and other chronic diseases caused by an aging population, lifestyle changes, and the growing number of, among other things, as a result of traffic and sports accidents. With the use of stem cells, it is possible to functionally revitalize an organ or its substitution by transplanting an artificial organ with the help of cultured stem cells, such as in a pacemaker produced in this way. Technological progress in this area is also associated with the automation of archiving and processing stem cells, which is undoubtedly one of the key factors of tissue engineering progress. The value of the global tissue engineering market in 2019 was USD 9.9 billion. It was also estimated that in the years 2020–2027, the CAGR will amount to 14.2% (Table 3). Figure 36a shows the geographical distribution of tissue engineering in important economic areas in the world. Figure 37c gives an example of a very high estimated increase in tissue engineering in China.

It has been estimated that the global regenerative medicine market is developing the most intensively in terms of value in 2017–2023, which in 2016 reached USD 5.44 billion, and the CAGR in 2017–2023 was estimated at an extremely high level of 32.2% (Table 3). Figure 36c shows the role of regenerative medicine methods in the treatment of various diseases. The increase in regenerative medicine in the United Kingdom is given as an example (Figure 37d). In the regenerative medicine market, key players focused on new products to treat cardiovascular diseases, diabetes, cancer, neurological disorders, and regenerative drugs. Figure 39 shows the share of various materials, including synthetic, genetically engineered, and pharmaceuticals and dynamic forecasted growth, and the increased importance of biologically derived materials in 2016–2023, a strong development determinant of regenerative medicine in general.

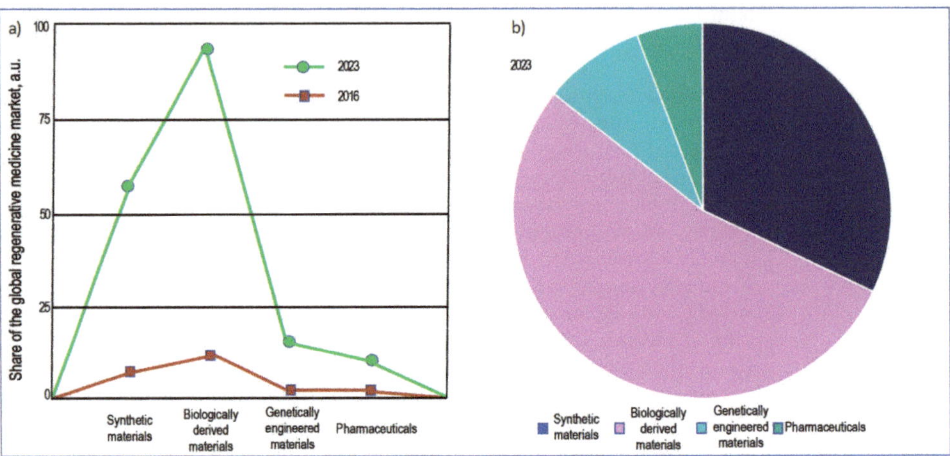

Figure 39. Scheme of forecasting the increase in the share of various groups of biomaterials used in regenerative medicine in the world in 2016–2023: (**a**) bar chart; (**b**) pie chart.

Advances in stem cell therapies such as the treatment of leukemia and blood diseases with hematogenic stem cells support regenerative medicine. The value of the cell therapy market in 2020 was USD 6.09 billion, and a CAGR of 5.4% is forecasted for 2020–2027 (Table 3). Figure 37e shows an example of the cell therapy market's growth in Japan, while Figure 36d shows cell therapies' division into predominantly, slightly allogeneic, and autologous.

In light of changes in the world caused by the SARS-CoV-2 coronavirus pandemic, adjustments to the forecasts described above can be expected, which will not change the global markets' general development trend. However, the high CAGR forecasts will decrease. Relevant reports in this regard will most likely be published, although the scale of the correction will be indicated by life and the passage of time.

This chapter analyzes in detail the expected growth of the biomaterials, both orthopedic and dental markets in the coming years, and the expected growth of tissue engineering, regenerative medicine, and related cell therapies markets. The compound annual growth rate CAGR of each of these markets is diversified, as the lowest CAGR is approx. 5.5%, and the highest CAGR is approx. 32%. Most of the available forecasts were made before the SARS-CoV-2 pandemic broke out, so it can be expected that there will be fluctuations in this range. However, they probably cannot be long-term, as people develop various diseases regardless of the pandemic. However, extraordinary demand for materials should be expected, including in part biomaterials needed to produce security measures to prevent droplet infection of coronavirus. As public awareness grows, mainly among doctors, the market for these products will systematically grow. As expected, the demand for these protection measures will linger on the market for the long term. It is obviously related to the systematic increase in the demand for appropriate biomaterials, which will be necessary to produce such means of protecting the epidemiological safety of doctors, which will have to be used long after the official end of the pandemic. This part of this paper is dedicated to the issues of economic forecasts.

7. Recapitulation and Final Remarks

Health has been a concern of humankind since the beginning. Figure 40 symbolically illustrates medicine and dentistry's progress as its specific part dealing with the oral cavity. The dawn of medicine is depicted in an ancient votive relief ca 470–450 BC found in Thyrea, Greece, now in the National Archeological Museum in Athens, Greece, showing Asclepius with his sons Podalairius and Machaon and three daughters (Figure 40) [435,436]. In

Greek mythology, Asclepius was the god of medical art, one of the heroes, as the son of the god Apollo and the nymph Koronis, husband of Epione (Soothing Pain) and father of Hygiea, Panacea, Iaso, Ajgle, Akeso, and the doctors Machaon, the internist and the surgeon Podalejrios. He is credited with raising the dead with a herb torn from a snake's mouth. Hence the staff of Asclepius entwined with a serpent is a symbol of medicine to this day. His Roman counterpart was Aesculapius. One of the first material pieces of evidence of deliberate medical intervention is the tooth of the lower right third molar (RM3) of the man from Villabruna, today in Italy, from the Late Upper Palaeolithic 14,000 years ago [437]. Hippocrates born c. 460 BC on the island of Kos, and died around 375 BC, was a Greek physician, nicknamed "the father of medicine". He is credited with the authorship, if not all. At least part of the extant work of the "Corpus Hippocraticum", a collection of about seventy medical writings from before the middle of the 4th century BC. collected 100 years after the death of Hippocrates. Originally was written in Ionic Greek. The works of Hippocrates were first translated and printed in 1525 in Venice, Italy. Among the numerous papyri with medical content found at Oxyrnchos, Egypt, papyrus 4970 was found that testified to the use of the Hippocratic Oath in antiquity. A fragment of the oath on the 3rd-century Papyrus Oxyrhynchus 2547 [438–440] in English translation begins with the words: "I swear by Apollo Healer, by Asclepius, by Hygieia, by Panacea, and by all the gods and goddesses, making them my witnesses, that I will carry out, according to my ability and judgment, this oath and this indenture". A fragment of the original papyrus is shown in the picture. During the Renaissance in Italy, there was also an intense interest in the human body structure and medical knowledge progress. It was then that the first public postmortem examination took place in a purpose-built anatomical theatre at the University of Bologna, Italy by Mondino de Luzzi in 1315. His book "Anathomia corporis humani" written around 1316 [441], was the first in Europe since ancient times, starting with Galen, Aristotle, and Hippocrates. The knowledge accumulated in it was only verified over 200 years later in 1543 by a Flemish anatomist Andreas Vesalius in "De Humani Corporis Fabrica" [442,443]. At the turn of the 15th and 16th centuries, Leonardo Da Vinci became interested in anatomy through painting. The Renaissance celebrated the beauty of the human body, and Da Vinci wanted to understand its structure, function, and proportions and present it in his paintings as realistically as possible. From 1489, he performed the first anatomical examinations. In 1506 he performed the first autopsy, then examined over 30 corpses, and between 1510 and 1511, he prepared a series of about 240 drawings known as the "Anatomical Manuscript" [444]. Da Vinci's discoveries may have revolutionized medicine, but unfortunately, he did not publish his notes, and after his death, many drawings were lost and were not discovered until the 20th century. This cartridge is symbolically represented by the famous drawing "Le proporzioni del corpo umano secondo Vitruvio", disseminated around 1490, created in pencil and ink on paper [445,446]. The drawing shows a figure of a naked man in two superimposed positions, inscribed in a circle and a square, and a text drawn up in the so-called mirror writing. The last item in this figure is a copy of the Nobel Prize in Physiology or Medicine 2020 Diploma recently awarded to Prof. Michael Houghton of the University of Alberta in Edmonton, Canada "for the discovery of Hepatitis C virus." [447]. It symbolically closes the multi-thousand-year achievements of medicine that invariably serves human health.

Health and well-being have now become one of the 17 sustainable development goals set by the United Nations. It was included in the sphere of interests of many countries and international organizations, including the European Union. Progress is determined by the Human Development Index, significantly diversified in different countries, including the life expectancy index, which varies depending on the country from slightly over 50 to 84 years from birth. Even in the Organization for Economic Co-operation and Development OECD countries, the variation in different detailed indicators assessing health and health care level may differ fifty-fold. The differentiation in access to health care and the possibilities of prevention and treatment are much more varied. Due to many factors, including the index of the number of doctors related to the population and the degree of saturation of health care facilities with medical devices and products, and the level of technology supporting medicine. Despite such a long tradition and universal concern for societies' health and the great involvement of the intellect of many generations of philosophers and doctors, an unambiguous definition of health and disease has not been developed. It seems that the provision from the World Health Organization WHO constitution intuitively corresponds to the expectations, saying that health is "a state of complete physical, mental, and social well-being and not merely the absence of disease or infirmity". The COVID-19 pandemic, which significantly determines the organization of health care and the access of individual people to medical services, has a huge impact on societies' general health conditions worldwide. It is important to ensure proper safety for all people, especially doctors and medical staff, actively fighting the pandemic. It is achieved by two mutually complementary strategies—passive STOP and active SPEC created by authors, eliminating the threat at source, and the extensive worldwide vaccination campaign against the pandemic effects of the SARS-CoV-2 coronavirus.

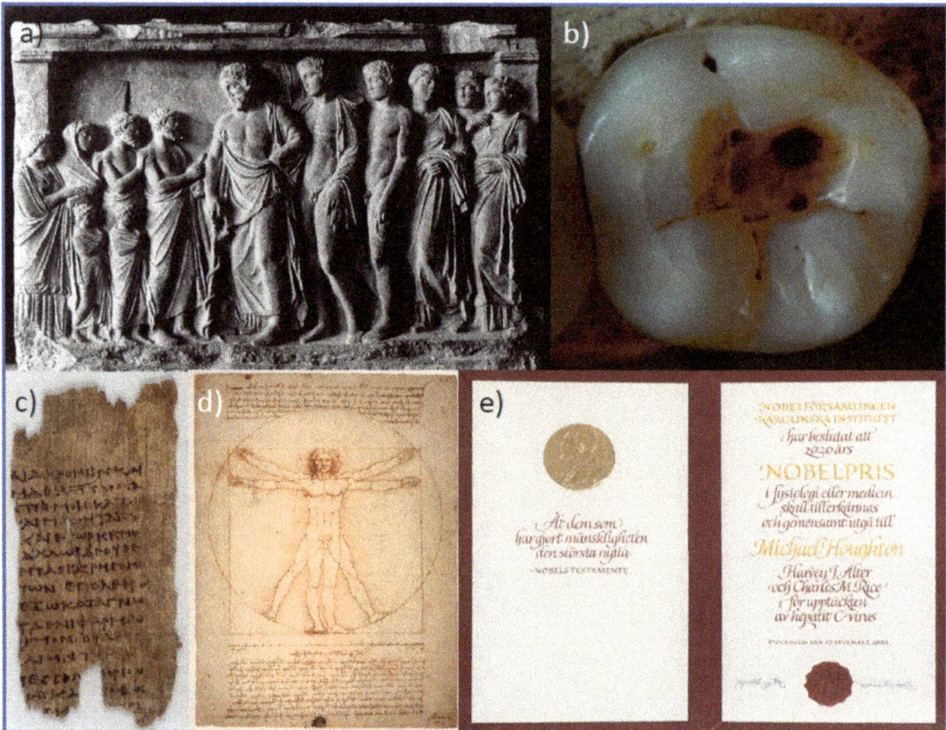

Figure 40. Diagram of the progress of medicine throughout history: (**a**) Greek votive relief ca 470–450 BC, found at Thyrea, Greece, now in the National Archeological Museum in Athens, Greece, depicting Asclepius as the god of medical art in Greek mythology with his sons Podalirius and Machaon and his three daughters, with supplicants; (**b**) tooth the lower right third molar (RM3) of the man from Villabruna, today in Italy, from the Late Upper Palaeolithic about 14,000 years ago with traces of intervention to repair the cavity and relieve pain; (**c**) a fragment of Papyrus Oxyrhynchus 2547, found in Egypt from the oath of Hippocrates dating from the 3rd-century BC; (**d**) the drawing "Le proporzioni del corpo umano secondo Vitruvio", created in pencil and ink on paper by Leonardo da Vinci and disseminated around 1490; (**e**) a copy of the Nobel Prize in Physiology or Medicine 2020 diploma recently awarded to Prof. Michael Houghton of the University of Alberta in Edmonton, Canada "for the discovery of Hepatitis C virus.".

The current level of medical procedures, including diagnostics, prevention, and supportive convalescence procedures, requires an extremely significant share of medical devices, including many that are temporarily or permanently installed or implanted into the human body. Bioengineering, including medical, dental, and tissue engineering, combines collaborative and interdisciplinary issues focused on integrative applications solutions of the problems unanswered independently by engineering and physical/life science disciplines. Since this is an important area of engineering sciences, it is necessary to meet the current stage of the technological revolution of Industry 4.0. By analogy to the idea of Dentistry 4.0, the problem can be generalized to Bioengineering 4.0. It requires widespread use in designing and manufacturing medical devices, modern achievements of computerization, automation, robotization, and the use of all possibilities of computer-aided engineering, transmission, collection, and security of cyber-physical systems, resulting from the augmented holistic model Industry 4.0.

It should be noted that achieving such a high level in patients' health care requires constant cooperation between doctors and biomedical and dental engineers. A strong imperative resulting from this cooperation is mutual respect for the moral principles applicable to both these circles. The four basic ethical principles for doctors are inextricably intertwined with the five moral principles for engineers. The obligation to assist patients in treating their diseases requires using the latest technological achievements that can serve this purpose. On the contrary, not keeping up with technological progress in this area must be treated as an ethical tort. The use of various groups of medical devices requires careful compliance with the applicable legal provisions, referred to as notified body approval requirements. Only in a limited number of these devices is enough self-assessment sufficient.

Any product's manufacturing requires appropriate engineering design, currently most often computer-aided, with proper engineering material and properly designed technological processes. The manufacturing of medical devices most often involves the use of biomedical materials, simply called biomaterials. Biomaterials are a wide group of engineering materials, and less often also some natural ones, including metals, ceramics, polymers, composites, and porous materials with the required biocompatibility and harmony of interaction with living matter and the related prevention of the initiation of toxicological and immunological reactions and effects of irritation of the living tissues of the body. Depending on the expected functions and the medical device's location in the human body, including the implants or prosthetic restorations and the drug distribution systems, biomaterials can be classified according to various criteria. Additive manufacturing plays a special role in technological processes, as is illustrated in this paper by numerous examples from clinical practice.

It should be noted that the use of these advanced technologies requires detailed knowledge. It turns out that there are publications that prove that the technological process is being carried out incorrectly. It mainly concerns the use of too low a laser or electron beam density adequately and improper flow of protective gases. Such reasons can cause even a 2.5-fold reduction in the strength properties of medical devices, e.g., implants. As a result, hardly any materials are obtained with better properties than those manufactured conventionally, and on the contrary, these properties are even lowered below an acceptable level. Such activities, causing the patient to be exposed to unforeseen and/or premature wear of elements, e.g., implantable components manufactured inappropriately, must be treated as an ethical tort, clearly incriminating the physician and the engineer cooperating with him/her, unless a physician without proper engineering knowledge, tried to manufacture such elements himself/herself, taking full responsibility. On the other hand, appropriate technology brings great benefits to patients, consisting of simplifying and shortening burdensome medical procedures, reducing pain sensations, and proper extension of failure-free use of implanted medical and dental devices.

The biomaterials market, both in orthopedic, dental, tissue engineering, regenerative medicine, and related cell therapies, is systematically growing. Depending on the sector, the compound annual growth rate CAGR in the next at least five years ranges from approx. 5.5 to approx. 32%, and for the total of biomaterials, is expected at the level of almost 16%. Certainly, the COVID-19 pandemic and the related decline in the economy in many countries and local lockdowns will weaken the growth rate. Still, in turn, the need for new safety measures resulting from the development of the pandemic and necessary to be used long after its end will increase.

Each part of this paper ends with a set of thoughts that indicate the current state of the considered aspects and trends and development prospects. Elements of the policy of a long and healthy life were indicated in the context of the ethical aspects of doctors' mission and cooperating with them engineers. An important element is the dynamically growing need to protect doctors and medical staff against the risk of SARS-CoV-2 coronavirus infection, which tends to persist for many years to come. It is related to the growing demand for biomaterials. Subsequently, the focus was on the relationship between the development of biomaterials and the current stage of progress of Industry 4.0 of the industrial revolution with the proposed general approach of Bioengineering 4.0. Numerous examples of clinical applications of biomaterials and advanced technologies, such as additive manufacturing, electrospinning, and atomic layer deposition, have been shown, richly illustrated by the results of materiallographic research. Forecasts for the development of global markets for biomaterials, regenerative medicine and tissue engineering were analyzed. The technological processes of producing and processing biomedical materials and the systematic increase in their global production are the determinants of implementing a long and healthy life policy.

Summing up the considerations presented in this paper, it should be stated that the implementation of the long and healthy life policy, declared by the United Nations and by many countries, requires very intensive development of bioengineering under the rules of the Industry 4.0 stage, which in turn is very dependent closely on the development of and increasing the global production of biomaterials and technological processes development of their production and processing.

Author Contributions: Conceptualization, literature review, presentation design, resources, data curation, software, formal analysis, practical verification, writing—original draft preparation, visualization, writing—review and editing, supervision, project administration—L.A.D., A.D.D.-D., L.B.D., funding acquisition—L.A.D. and L.B.D. All authors have read and agreed to the published version of the manuscript.

Funding: This research was not directly financed by external funding.

Note: The paper is prepared due to the sending of the projects applications no.: POIR.01.01.01-00-2393/20 cleanDENTworkspace—Antiviral (SARS-CoV-2) and antimicrobial repeatedly awarded at the international innovation fair, a breakthrough system for ensuring the asepsis of the dental, ENT and general medical practice, as well as full protection of medical personnel and patients undergoing oral procedures, 14.12.2020, and no: POIR.01.01.01-00-0694/20 ASKLEPIOS-IFIS/Implantological final immediate reconstruction—a system based on implants, in particular individual ones, manufactured based on own innovative design and production technology using 3D printing, 17.06.2020, and due to actually implement Project POIR.01.01-00-0485/16-00 on "IMSKA-MAT Innovative dental and maxillofacial implants manufactured using the innovative additive technology supported by computer-aided materials design ADD-MAT" realized by the Medical and Dental Engineering Center for Research, Design, and Production ASKLEPIOS in Gliwice, Poland. The project was implemented in 2017–2021 and is co-financed by the Operational Program Intelligent Development of the European Union.

Conflicts of Interest: The authors declare no conflict of interest.

References

1. Meadows, D.H.; Meadows, D.L.; Randers, J.; Behrens, W.W., III. *The Limits to Growth: A Report for the Club of Rome's Project on the Predicament of Mankind*; A Potomac Associates Book; Universe Books: New York, USA, 1972. Available online: https://collections.dartmouth.edu/teitexts/meadows/diplomatic/meadows_ltg-diplomatic.html (accessed on 1 March 2021).
2. Holcombe, R.G. A theory of the theory of public goods. *Rev. Austrian Econ.* **1997**, *10*, 1–22. [CrossRef]
3. Brundtland, G.H. (Ed.) Report of the World Commission on Environment and Development (WCED): Our Common Future (Also Known As the Brundtland Report). United Nations. 1987. Available online: https://sustainabledevelopment.un.org/content/documents/5987our-common-future.pdf (accessed on 1 March 2021).
4. The 17 Goals. Available online: https://sdgs.un.org/goals (accessed on 1 March 2021).
5. Proposal for a Regulation of the European Parliament and of the Council on the Establishment of a Programme for the Union's Action in the Field of Health—for the Period 2021–2027 and Repealing Regulation (EU) No 282/2014 ("EU4Health Programme"). Available online: https://eur-lex.europa.eu/legal-content/EN/TXT/?uri=CELEX:52020PC0405 (accessed on 1 March 2021).
6. Constitution of the World Health Organization. Available online: https://www.who.int/governance/eb/who_constitution_en.pdf (accessed on 1 March 2021).
7. Jakab, Z. Designing the road to better health and well-being in Europe. In Proceedings of the 14th European Health Forum Gastein, Bad Hofgastein, Austria, 7 October 2011. Available online: https://www.euro.who.int/__data/assets/pdf_file/0003/152184/RD_Dastein_speech_wellbeing_07Oct.pdf (accessed on 1 March 2021).
8. EU Health Policy. Available online: https://ec.europa.eu/health/policies/overview_pl (accessed on 1 March 2021).
9. Dobrzański, L.A. The concept of biologically active microporous engineering materials and composite biological-engineering materials for regenerative medicine and dentistry. *Arch. Mater. Sci. Eng.* **2016**, *80*, 64–85. [CrossRef]
10. United Nations Development Programme, Human Development Reports, Human Development Index (HDI). Available online: http://hdr.undp.org/en/indicators/138806# (accessed on 1 March 2021).
11. Human Development Report 2020. The Next Frontier: Human Development and the Anthropocene. United Nations Development Programme. Available online: http://hdr.undp.org/sites/default/files/hdr2020.pdf (accessed on 1 March 2021).
12. Malik, K. Director, Human Development Report Office, UNDP. Inaugural Mahbub ul Haq-Amartya Sen Lecture, University De Geneve. Advancing, Sustaining Human Progress: From Concepts to Policies. Available online: http://hdr.undp.org/sites/default/files/malik_mahbubulhaqlecture_2014.pdf (accessed on 1 March 2021).
13. UNDP. Human Development Report 1990: Concept and Measurement of Human Development. 1990. Available online: http://www.hdr.undp.org/en/reports/global/hdr1990 (accessed on 1 March 2021).
14. UNDP. Human Development Report 2010: The Real Wealth of Nations—Pathways to Human Development. 2010. Available online: http://hdr.undp.org/en/content/human-development-report-2010 (accessed on 1 March 2021).
15. Inequality-Adjusted Human Development Index (IHDI). Available online: http://hdr.undp.org/en/content/inequality-adjusted-human-development-index-ihdi (accessed on 1 March 2021).
16. Dobrzański, L.A.; Dobrzański, L.B.; Dobrzańska-Danikiewicz, A.D.; Dobrzańska, J. The Concept of Sustainable Development of Modern Dentistry. *Processes* **2020**, *8*, 1605. [CrossRef]
17. OECD. *Health at a Glance 2019: OECD Indicators*; OECD Publishing: Paris, France, 2019. [CrossRef]
18. Bernabé, E.; Masood, M.; Vujicic, M. The impact of out-of-pocket payments for dental care on household finances in low and middle income countries. *BMC Public Health* **2017**, *17*, 109. [CrossRef] [PubMed]
19. FDI World Dental Federation. Available online: https://www.fdiworlddental.org/about-fdi (accessed on 1 March 2021).
20. Holst, D.; Sheiham, A.; Petersen, P. Regulating entrepreneurial behaviour in oral care health services. In *European Observatory on Health Care Systems Series: Regulating Entrepreneurial Behaviour in European Care Health Systems*; Saltman, R., Busse, R., Mossialos, E., Eds.; Open University Press: Philadelphia, PA, USA, 2002; pp. 215–231.
21. Statistics Poland. Health and Health Care in 2019. Available online: https://stat.gov.pl/obszary-tematyczne/zdrowie/zdrowie/zdrowie-i-ochrona-zdrowia-w-2019-roku,1,10.html (accessed on 1 March 2021).
22. Concepts of Disease and Health. Available online: https://plato.stanford.edu/entries/health-disease/ (accessed on 1 March 2021).
23. Cain, A.J. Thomas Sydenham, John Ray, and some contemporaries on species. *Arch. Nat. Hist.* **1999**, *26*, 55–83. [CrossRef] [PubMed]
24. Hippius, H.; Müller, N. The work of Emil Kraepelin and his research group in München. *Eur. Arch. Psychiatry Clin. Neurosc.* **2008**, *258*, 3–11. [CrossRef] [PubMed]
25. Whitbeck, C. Causation in Medicine: The Disease Entity Model. *Philos. Sci.* **1977**, *44*, 619–637. [CrossRef]
26. Snow, J. On Continuous Molecular Changes, More Particularly in their Relation to Epidemic Diseases. In *Snow on Cholera*; Frost, W.H., Ed.; Hafner: New York, NY, USA, 1965; pp. 147–175.
27. Carter, K.C. *The Rise of Causal Theories of Disease*; Ashgate: Burlington, VT, Canada, 2003.
28. Broome, M. Taxonomy and Ontology in Psychiatry: A survey of recent literature. *Philos. Psychiatry Psychol.* **2006**, *13*, 303–319. [CrossRef]
29. Green, J.A. *Prescribing by Numbers*; Johns Hopkins: Baltimore, MD, USA, 2007; Chapter 2.
30. Plutynski, A. *Explaining Cancer*; Oxford University Press: New York, NY, USA, 2018; Chapter 2.
31. Bloomfield, P. *Moral Reality*; Oxford University Press: New York, NY, USA, 2001.

32. Boorse, C. On The Distinction Between Disease and Illness. *Philos. Pub. Aff.* **1975**, *5*, 49–68.
33. Ereshefsky, M. Defining 'Health' and 'Disease'. *Stud. Hist. Philos. Biol. Biomed. Sci.* **2009**, *40*, 221–227. [CrossRef] [PubMed]
34. Culver, C.M.; Gert, B. *Philosophy in Medicine*; Oxford University Press: New York, NY, USA, 1982.
35. Thagard, P. *How Scientists Explain Disease*; Princeton University Press: Princeton, NJ, USA, 1999.
36. Kennedy, I. *The Unmasking of Medicine*; Allen and Unwin: London, UK, 1983.
37. Brown, P. The Name Game: Toward A Sociology of Diagnosis. *J. Mind Behav.* **1990**, *11*, 385–406.
38. Kitcher, P. *The Lives To Come: The Genetic Revolution and Human Possibilities, Revised Ed.*; Simon & Schuster: New York, NY, USA, 1997; pp. 208–209.
39. Boorse, C. What A Theory of Mental Health Should Be. *J. Theory Soc. Behav.* **1976**, *6*, 61–84. [CrossRef]
40. Boorse, C. Health as a Theoretical Concept. *Philos. Sci.* **1977**, *44*, 542–573. [CrossRef]
41. Boorse, C. A rebuttal on health. In *What Is Disease?* Humber, J.M., Almeder, R.F., Eds.; Humana Press: Totowa, NJ, USA, 1997; pp. 3–143.
42. Murphy, D. *Psychiatry in the Scientific Image*; MIT Press: Cambridge, MA, USA, 2006.
43. Hatfield, G. René Descartes. In *Stanford Encyclopedia of Philosophy*; Winter 2017, Edition; Zalta, E.N., Ed.; Metaphysics Research Lab, Stanford University: Stanford, CA, USA, 2017. Available online: https://plato.stanford.edu/archives/win2017/entries/descartes/ (accessed on 1 March 2021).
44. Kulik, T.B. Zdrowie—Kategoria uniwersalna. *Wych Fiz Zdrow* **1996**, *3*, 103–105.
45. Manual of the International Statistical Classification of Diseases, Injuries, and Causes of Death: Based on the Recommendations of the Ninth Revision Conference, 1975, and Adopted by the Twenty-Ninth World Health Assembly, 1975 Revision. Available online: https://apps.who.int/iris/handle/10665/40492 (accessed on 1 March 2021).
46. Firek, W. W stronę filozofii zdrowia. In *Zdrowie I Jego Uwarunkowania*; Mucha, D., Zięba, H., Eds.; PPWSZ: Nowy Targ, Poland, 2011; pp. 22–33.
47. Reznek, L. *The Nature of Disease*; Routledge: New York, NY, USA, 1987.
48. Glackin, S. Tolerance and illness: The politics of medical and psychiatric classification. *J. Med. Philos.* **2010**, *35*, 449–465. [CrossRef]
49. Barnes, E. *The Minority Body*; Oxford University Press: New York, NY, USA, 2016.
50. Amundson, R. Against Normal Function. *Stud. Hist. Phil. Biol. Biomed. Sci.* **2000**, *31*, 33–53. [CrossRef]
51. Inkpen, S.A. Health, Ecology and the Microbiome. *eLife* **2019**, *8*, e47626. [CrossRef] [PubMed]
52. Maszczak, T. Zdrowie jako wartość. *Kult Fiz* **2006**, *9–12*, 23.
53. Puchalski, K. Pojecie Świadomości Zdrowotnej. In *Zdrowie W Świadomości Społecznej*; KCPZMP: Łódź, Poland, 1997; p. 18.
54. Tatarkiewicz, W. *O szczęściu*; PWN: Warszawa, CA, USA, 1979; pp. 216–217.
55. Puchalski, K. Zdrowie Jako Wartość. In *Zdrowie W Świadomości Społecznej*; KCPZMP: Łódź, Poland, 1997; p. 71.
56. Díaz-Guerrero, R. *Psychology of the Mexican: Culture and Personality*; University of Texas Press: Austin, TX, USA, 1975.
57. Dobrzański, L.A.; Dobrzański, L.B.; Dobrzańska-Danikiewicz, A.D.; Dobrzańska, J.; Kraszewska, M. The synergistic ethics interaction with nanoengineering, dentistry and dental engineering. In *Ethics in Nanotechnology*; Jeswani, G., Van De Voorde, M., Eds.; De Gruyter: Berlin, Germany, 2020; (prepared for printing).
58. Dobrzański, L.A.; Dobrzański, L.B.; Dobrzańska-Danikiewicz, A.D.; Dobrzańska, J.; Kraszewska, M. Development prospects of the dentistry 4.0, dental engineering and nanotechnology triad towards ethical imperatives. In *Ethics in Nanotechnology*; Jeswani, G., Van De Voorde, M., Eds.; De Gruyter: Berlin, Germany, 2020; (prepared for printing).
59. Steinke, H. Der Hippokratische eid: Ein schwieriges erbe. Horizonte medizingeschichte, schweizerische ärztezeitung. *Bull. Médecins Suisses Boll. Med. Svizz* **2016**, *97*, 1699–1701.
60. Murgic, L.; Hébert, P.C.; Sovic, S.; Pavlekovic, G. Paternalism and autonomy: Views of patients and providers in a transitional (post-communist) country. *BMC Med. Ethics.* **2015**, *29*, 65. [CrossRef] [PubMed]
61. Will, J.F. A brief historical and theoretical perspective on patient autonomy and medical decision making: Part II: The autonomy model. *Chest* **2011**, *139*, 1491–1497. [CrossRef] [PubMed]
62. Baumann, A.; Audibert, G.; Guibet Lafaye, C.; Puybasset, L.; Mertes, P.M.; Claudot, F. Elective non-therapeutic intensive care and the four principles of medical ethics. *J. Med. Ethics.* **2013**, *39*, 139–142. [CrossRef] [PubMed]
63. Jotterand, F. The hippocratic oath and contemporary medicine: Dialectic between past ideals and present reality? *J. Med. Philos.* **2005**, *30*, 107–128. [CrossRef]
64. Gillon, R. Medical ethics: Four principles plus attention to scope. *BMJ* **1994**, *309*, 184. [CrossRef] [PubMed]
65. Beauchamp, T.L.; Childress, J.F. *Principles of Biomedical Ethics*, 8th ed.; Oxford University Press: New York, NY, USA, 2019.
66. Holm, S. Not just autonomy—The principles of American biomedical ethics. *J. Med. Ethics.* **1995**, *21*, 6. [CrossRef] [PubMed]
67. Beauchamp, T.L.; Childress, J.F. *Principles of Biomedical Ethics*, 1st ed.; Oxford University Press: New York, NY, USA, 1979.
68. Veatch, R.M. The impossibility of a morality internal to medicine. *J. Med. Philos. A Forum. Bioeth. Philos. Med.* **2001**, *26*, 621–642. [CrossRef]
69. Nash, D.A. Ethics in dentistry: Review and critique of principles of ethics and code of professional conduct. *J. Am. Dent. Assoc.* **1984**, *109*, 597–603. [CrossRef] [PubMed]
70. Seear, J.E.; Walters, L. *Law and Ethics in Dentistry Paperback*, 3rd ed.; Butterworth-Heinemann Ltd.: Oxford, UK, 1991.
71. Jessri, M.; Fatemitabar, S.A. Implication of ethical principles in chair-side dentistry. *Iran. J. Allergy Asthma Immunol.* **2007**, *6* (Suppl. S5), 53–59.

72. Nash, D.A. On ethics in the profession of dentistry and dental education. *Eur. J. Dent. Educ.* **2007**, *11*, 64–74. [CrossRef] [PubMed]
73. Covid-19. Available online: https://www.oed.com/view/Entry/88575495 (accessed on 1 March 2021).
74. Symptoms of Coronavirus. Available online: https://www.cdc.gov/coronavirus/2019-ncov/symptoms-testing/symptoms.html (accessed on 1 March 2021).
75. Coronavirus Disease (COVID-19). Available online: https://www.who.int/emergencies/diseases/novel-coronavirus-2019/question-and-answers-hub/q-a-detail/coronavirus-disease-covid-19 (accessed on 1 March 2021).
76. Nussbaumer-Streit, B.; Mayr, V.; Dobrescu, A.I.; Chapman, A.; Persad, E.; Klerings, I.; Wagner, G.; Siebert, U.; Christof, C.; Zachariah, C.; et al. Quarantine alone or in combination with other public health measures to control COVID-19: A rapid review. *Cochrane Database Syst. Rev.* **2020**, *4*, 1–45. [CrossRef]
77. COVID-19 Dashboard by the Center for Systems Science and Engineering (CSSE) at Johns Hopkins University (JHU). Available online: https://gisanddata.maps.arcgis.com/apps/opsdashboard/index.html#/bda7594740fd40299423467b48e9ecf6 (accessed on 1 March 2021).
78. Interim Clinical Guidance for Management of Patients with Confirmed Coronavirus Disease (COVID-19). Available online: https://www.cdc.gov/coronavirus/2019-ncov/hcp/clinical-guidance-management-patients.html (accessed on 1 March 2021).
79. Oran, D.P.; Topol, E.J. The Proportion of SARS-CoV-2 Infections That Are Asymptomatic. *A Systematic Review. Ann. Intern. Med.* **2021**, *M20*, 6976. [CrossRef]
80. Transmission of COVID-19. Available online: https://www.ecdc.europa.eu/en/covid-19/latest-evidence/transmission (accessed on 1 March 2021).
81. Long-Term Effects of COVID-19. Available online: https://www.cdc.gov/coronavirus/2019-ncov/long-term-effects.html (accessed on 1 March 2021).
82. Coronavirus Disease: COVID-19—Appendices. Available online: https://www.cdc.gov/coronavirus/2019-ncov/php/contact-tracing/contact-tracing-plan/appendix.html#contact (accessed on 1 March 2021).
83. Quarantine for Coronavirus (COVID-19). Available online: https://www.health.gov.au/news/health-alerts/novel-coronavirus-2019-ncov-health-alert/how-to-protect-yourself-and-others-from-coronavirus-covid-19/quarantine-for-coronavirus-covid-19#what-is-a-close-contact (accessed on 1 March 2021).
84. How COVID-19 Spreads. Available online: https://www.cdc.gov/coronavirus/2019-ncov/prevent-getting-sick/how-covid-spreads.html (accessed on 1 March 2021).
85. Coronavirus Disease (COVID-19): How Is It Transmitted? Available online: https://www.who.int/news-room/q-a-detail/coronavirus-disease-covid-19-how-is-it-transmitted (accessed on 1 March 2021).
86. Coronavirus Disease (COVID-19): Frequently Asked Questions. Available online: https://www.cdc.gov/coronavirus/2019-ncov/faq.html (accessed on 1 March 2021).
87. Transmission of SARS-CoV-2: Implications for Infection Prevention Precautions. Available online: https://www.who.int/news-room/commentaries/detail/transmission-of-sars-cov-2-implications-for-infection-prevention-precautions (accessed on 1 March 2021).
88. Clinical Questions about COVID-19: Questions and Answers. Available online: https://www.cdc.gov/coronavirus/2019-ncov/hcp/faq.html (accessed on 1 March 2021).
89. Dobrzański, L.A.; Dobrzański, L.B.; Dobrzańska-Danikiewicz, A.D.; Dobrzańska, J.; Rudziarczyk, K.; Achtelik-Franczak, A. Non-Antagonistic Contradictoriness of the Progress of Advanced Digitized Production with SARS-CoV-2 Virus Transmission in the Area of Dental Engineering. *Processes* **2020**, *8*, 1097. [CrossRef]
90. Funk, C.D.; Laferrière, C.; Ardakani, A. A Snapshot of the Global Race for Vaccines Targeting SARS-CoV-2 and the COVID-19 Pandemic. *Front. Pharm.* **2020**, *11*, 937. [CrossRef] [PubMed]
91. Jia, Z.; Barbier, L.; Stuart, H.; Amraei, M.; Pelech, S.; Dennis, J.W.; Metalnikov, P.; O'Donnell, P.; Nabi, I.R. Tumor cell pseudopodial protrusions. Localized signaling domains coordinating cytoskeleton remodeling, cell adhesion, glycolysis, RNA translocation, and protein translation. *J. Biol. Chem.* **2005**, *280*, 30564–30573. [CrossRef] [PubMed]
92. Li, X.; Geng, M.; Peng, Y.; Meng, L.; Lu, S. Molecular immune pathogenesis and diagnosis of COVID-19. *J. Pharm. Anal.* **2020**, *10*, 102–108. [CrossRef] [PubMed]
93. How Furin and ACE2 Interact with the Spike on SARS-CoV-2. Available online: https://www.assaygenie.com/how-furin-and-ace2-interact-with-the-spike-on-sarscov2 (accessed on 1 March 2021).
94. Zhou, P.; Yang, X.L.; Wang, X.G.; Hu, B.; Zhang, L.; Zhang, W.; Si, H.-R.; Zhu, Y.; Li, B.; Huang, C.-L.; et al. A pneumonia outbreak associated with a new coronavirus of probable bat origin. *Nature* **2020**, *579*, 270–273. [CrossRef] [PubMed]
95. Letko, M.; Munster, V. Functional assessment of cell entry and receptor usage for lineage B β-coronaviruses, including 2019-nCoV. *bioRxiv* **2020**, 1–26. [CrossRef]
96. Letko, M.; Marzi, A.; Munster, V. Functional assessment of cell entry and receptor usage for SARS-CoV-2 and other lineage B betacoronaviruses. *Nat. Microbiol.* **2020**, *5*, 562–569. [CrossRef] [PubMed]
97. Genomic Characterization of the 2019 Novel Coronavirus. Available online: https://www.jwatch.org/na50823/2020/02/06/genomic-characterization-2019-novel-coronavirus (accessed on 1 March 2021).
98. Gralinski, L.E.; Menachery, V.D. Return of the Coronavirus: 2019-nCoV. *Viruses* **2020**, *12*, 135. [CrossRef] [PubMed]
99. Lu, R.; Zhao, X.; Li, J.; Niu, P.; Yang, B.; Wu, H.; Wang, W.; Song, H.; Huang, B.; Zhu, N.; et al. Genomic characterisation and epidemiology of 2019 novel coronavirus: Implications for virus origins and receptor binding. *Lancet* **2020**, *395*, 565–574. [CrossRef]

100. Walls, A.C.; Park, Y.J.; Tortorici, M.A.; Wall, A.; McGuire, A.T.; Veesler, D. Structure, Function, and Antigenicity of the SARS-CoV-2 Spike Glycoprotein. *Cell* **2020**, *181*, 281–292.e6. [CrossRef] [PubMed]

101. Zheng, Y.Y.; Ma, Y.T.; Zhang, J.Y.; Xie, X. COVID-19 and the cardiovascular system. *Nat. Rev. Cardiol.* **2020**, *17*, 259–260. [CrossRef] [PubMed]

102. Lamers, M.M.; Beumer, J.; van der Vaart, J.; Knoops, K.; Puschhof, J.; Breugem, T.I.; Ravelli, R.B.G.; Paul van Schayck, J.; Mykytyn, A.Z.; Duimel, H.Q.; et al. SARS-CoV-2 productively infects human gut enterocytes. *Science* **2020**, *369*, 50–54. [CrossRef] [PubMed]

103. Monteil, V.; Kwon, H.; Prado, P.; Hagelkrüys, A.; Wimmer, R.A.; Stahl, M.; Leopoldi, A.; Garreta, E.; Hurtado Del Pozo, C.; Prosper, F.; et al. Inhibition of SARS-CoV-2 Infections in Engineered Human Tissues Using Clinical-Grade Soluble Human ACE2. *Cell* **2020**, *181*, 905–913.e7. [CrossRef]

104. Shang, J.; Wan, Y.; Luo, C.; Ye, G.; Geng, Q.; Auerbach, A.; Li, F. Cell entry mechanisms of SARS-CoV-2. *Proc. Natl. Acad. Sci. USA* **2020**, *117*, 11727–11734. [CrossRef] [PubMed]

105. Coutard, B.; Valle, C.; de Lamballerie, X.; Canard, B.; Seidah, N.G.; Decroly, E. The spike glycoprotein of the new coronavirus 2019-nCoV contains a furin-like cleavage site absent in CoV of the same clade. *Antivir. Res.* **2020**, *176*, 104742. [CrossRef] [PubMed]

106. Li, H.; Wu, C.; Yang, Y.; Liu, Y.; Zhang, P.; Wang, Y.; Wang, Q.; Xu, Y.; Li, M.; Zheng, M.; et al. Furin, a potential therapeutic target for COVID-19. *ChinaXiv* **2020**, 1–34. [CrossRef]

107. Thailand Medical News. Breaking! Latest Coronavirus Research Reveals That The Virus Has Mutated Gene Similar To HIV and Is 1,000 Times More Potent. Available online: https://www.thailandmedical.news/news/breaking!-latestcoronavirus-research-reveals-that-the-virus-has-mutated-gene-similar-to-hiv-and-is-1,000-times-more-potent- (accessed on 1 March 2021).

108. Harcourt, J.; Tamin, A.; Lu, X.; Kamili, S.; Sakthivel, S.K.; Murray, J.; Queen, K.; Tao, Y.; Paden, C.R.; Zhang, J.; et al. Severe Acute Respiratory Syndrome Coronavirus 2 from Patient with Coronavirus Disease, United States. *Emerg. Infect. Dis.* **2020**, *26*, 1266–1273. [CrossRef] [PubMed]

109. Ou, X.; Liu, Y.; Lei, X.; Li, P.; Mi, D.; Ren, L.; Guo, L.; Guo, R.; Chen, T.; Hu, J.; et al. Characterization of spike glycoprotein of SARS-CoV-2 on virus entry and its immune cross-reactivity with SARS-CoV. *Nat. Commun.* **2020**, *11*, 1620. [CrossRef] [PubMed]

110. Jasiński, G. Koronawirus Zakaża skuteczniej niż SARS? Nowe fakty ws. wirusa z Chin. Available online: https://www.rmf24.pl/raporty/raport-koronawirus-z-chin/najnowsze-fakty/news-koronawirus-zakaza-skuteczniej-niz-sars-nowe-fakty-ws-wirusa,nId,4352687 (accessed on 1 March 2021).

111. Huggins, D.J. Structural analysis of experimental drugs binding to the SARS-CoV-2 target TMPRSS2. *J. Mol. Graph. Model.* **2020**, *100*, 107710. [CrossRef] [PubMed]

112. Hoffmann, M.; Kleine-Weber, H.; Schroeder, S.; Krüger, N.; Herrler, T.; Erichsen, S.; Schiergens, T.S.; Herrler, G.; Wu, N.H.; Nitsche, A.; et al. SARS-CoV-2 Cell Entry Depends on ACE2 and TMPRSS2 and Is Blocked by a Clinically Proven Protease Inhibitor. *Cell* **2020**, *181*, 271–280.e8. [CrossRef] [PubMed]

113. Harrison, C. Coronavirus puts drug repurposing on the fast track. *Nat. Biotechnol.* **2020**, *38*, 379–381. [CrossRef] [PubMed]

114. Thailand Medical News. Coronavirus Drug Research: German Researchers Identify Japanese Drug, Camostat Mesylate That Could Be Repurposed To Treat Covid-19. Available online: https://www.thailandmedical.news/news/coronavirus-drug-research-german-researchers-identify-japanese-drug,-camostat-mesylate-that-could-be-repurposed-to-treat-covid-19 (accessed on 1 March 2021).

115. Xu, H.; Zhong, L.; Deng, J.; Peng, J.; Dan, H.; Zeng, X.; Li, T.; Chen, Q. High expression of ACE2 receptor of 2019-nCoV on the epithelial cells of oral mucosa. *Int. J. Oral. Sci.* **2020**, *12*, 8. [CrossRef] [PubMed]

116. Yan, T.; Xiao, R.; Lin, G. Angiotensin-converting enzyme 2 in severe acute respiratory syndrome coronavirus and SARS-CoV-2: A double-edged sword? *FASEB J.* **2020**, *34*, 6017–6026. [CrossRef] [PubMed]

117. Bharat, A.; Querrey, M.; Markov, N.S.; Kim, S.; Kurihara, C.; Garza-Castillon, R.; Manerikar, A.; Shilatifard, A.; Tomic, R.; Politanska, Y.; et al. Lung transplantation for patients with severe COVID-19. *Sci. Transl. Med.* **2020**, *12*, eabe4282. [CrossRef] [PubMed]

118. WHO Europe. Postępowanie Kliniczne W Ostrym Zakażeniu Dróg Oddechowych O Ciężkim Przebiegu (Sari) W Przypadku Podejrzenia Choroby COVID-9. Available online: https://apps.who.int/iris/bitstream/handle/10665/331809/WHO-2019-nCoV-clinical-2020.4-pol.pdf (accessed on 1 March 2021).

119. Dobrzańska-Danikiewicz, A.D. Metodologia komputerowo zintegrowanego prognozowania rozwoju inżynierii powierzchni materiałów. In *Open Access Library 1(7)*; Dobrzański, L.A., Ed.; International OCSCO World Press: Gliwice, Poland, 2012; pp. 1–289.

120. Dobrzańska-Danikiewicz, A.D. Księga technologii krytycznych kształtowania struktury i własności powierzchni materiałów inżynierskich. In *Open Access Library 8(26)*; Dobrzański, L.A., Ed.; International OCSCO World Press: Gliwice, Poland, 2013; pp. 1–823.

121. Dobrzański, L.A.; Dobrzańska-Danikiewicz, A.D. Inżynieria powierzchni materiałów: Kompendium wiedzy i podręcznik akademicki. In *Open Access Library VIII(1)*; Dobrzański, L.A., Ed.; International OCSCO World Press: Gliwice, Poland, 2018; pp. 1–1138.

122. Morrison, A.; Wensley, R. Boxing up or Boxed in?: A Short History of the Boston Consulting Group Share/Growth Matrix. *J. Mark. Manag.* **1991**, *7*, 105–129. [CrossRef]

123. Beaumont, P. Covid-19 Vaccine: Who Are Countries Prioritising for First Doses? Available online: https://www.theguardian.com/world/2020/nov/18/covid-19-vaccine-who-are-countries-prioritising-for-first-doses (accessed on 1 March 2021).

124. Coronavirus (COVID-19) Vaccinations—Ststistics and Research. Available online: https://ourworldindata.org/covid-vaccinations (accessed on 23 February 2021).

125. Geter, A.; Herron, A.R.; Sutton, M.Y. HIV-Related Stigma by Healthcare Providers in the United States: A Systematic Review. *Aids. Patient Care Stds.* **2008**, *32*, 418–424. [CrossRef] [PubMed]

126. Krishnaratne, S.; Hensen, B.; Cordes, J.; Enstone, J.; Hargreaves, J.R. Interventions to strengthen the HIV prevention cascade: A systematic review of reviews. *Lancet Hiv.* **2016**, *3*, e307–e317. [CrossRef]

127. Coates, T.J.; Richter, L.; Caceres, C. Behavioural strategies to reduce HIV transmission: How to make them work better. *Lancet* **2008**, *372*, 669–684. [CrossRef]

128. Hurley, S.F.; Jolley, D.J.; Kaldor, J.M. Effectiveness of needle-exchange programmes for prevention of HIV infection. *Lancet* **1997**, *349*, 1797–1800. [CrossRef]

129. Vlahov, D.; Junge, B. The role of needle exchange programs in HIV prevention. *Public Health Rep.* **1998**, *113* (Suppl. S1), 75–80. [PubMed]

130. Wszystko O Wariantach Brytyjskim, Brazylijskim I Południowoafrykańskim. Available online: https://www.medonet.pl/koronawirus/to-musisz-wiedziec,wariant-covid-19-brytyjski--brazylijski-i-poludniowoafrykanski--co-warto-wiedziec-,artykul,88341446.html (accessed on 1 March 2021).

131. Bioengineering. Available online: https://bioe.uw.edu/wp-content/uploads/2013/05/bioengineering-vs-other-disciplines-alt1.png (accessed on 1 March 2021).

132. Dobrzański, L.A. Introductory Chapter: Multi-Aspect Bibliographic Analysis of the Synergy of Technical, Biological and Medical Sciences Concerning Materials and Technologies Used for Medical and Dental Implantable Devices. In *Biomaterials in Regenerative Medicine*; Dobrzański, L.A., Ed.; IntechOpeen: Rijeka, Croatia, 2018; pp. 1–43. [CrossRef]

133. Dobrzański, L.A.; Dobrzański, L.B. Approach to the Design and Manufacturing of Prosthetic Dental Restorations According to the Rules of Industry 4.0. *Mater. Perform. Charact* **2020**, *9*, 394–476. [CrossRef]

134. Dobrzański, L.A.; Dobrzański, L.B. Dentistry 4.0 Concept in the Design and Manufacturing of Prosthetic Dental Restorations. *Processes* **2020**, *8*, 525. [CrossRef]

135. Dobrzański, L.A. Role of materials design in maintenance engineering in the context of industry 4.0 idea. *J. Achiev. Mater. Manuf. Eng.* **2019**, *96*, 12–49. [CrossRef]

136. Dobrzański, L.A.; Dobrzańska-Danikiewicz, A.D. Why Are Carbon-Based Materials Important in Civilization Progress and Especially in the Industry 4.0 Stage of the Industrial Revolution. *Mater. Perform. Charact* **2019**, *8*, 337–370. [CrossRef]

137. Dobrzański, L.A.; Dobrzański, L.B.; Dobrzańska-Danikiewicz, A.D. Overview of conventional technologies using the powders of metals, their alloys and ceramics in Industry 4.0 stage. *J. Achiev. Mater. Manuf. Eng.* **2020**, *98*, 56–85. [CrossRef]

138. Dobrzański, L.A.; Dobrzański, L.B.; Dobrzańska-Danikiewicz, A.D.; Kraszewska, M. Manufacturing powders of metals, their alloys and ceramics and the importance of conventional and additive technologies for products manufacturing in Industry 4.0 stage. *Arch. Mater. Sci. Eng.* **2020**, *102*, 13–41. [CrossRef]

139. Dobrzański, L.A.; Dobrzański, L.B.; Achtelik-Franczak, A.; Dobrzańska, J. Application Solid Laser-Sintered or Machined Ti6Al4V Alloy in Manufacturing of Dental Implants and Dental Prosthetic Restorations According to Dentistry 4.0 Concept. *Processes* **2020**, *8*, 664. [CrossRef]

140. Dobrzański, L.A. Autorska koncepcja rozwoju implanto-skafoldów oraz materiałów biologiczno-inżynierskich do aplikacji w medycynie i stomatologii. In *Metalowe Materiały Mikroporowate I Lite Do Zastosowań Medycznych I Stomatologicznych*; Open Access Library VII(1); Dobrzański, L.A., Dobrzańska-Danikiewicz, A.D., Eds.; International OCSCO World Press: Gliwice, Poland, 2017; pp. 535–580.

141. Dobrzański, L.A.; Dobrzańska-Danikiewicz, A.D.; Achtelik-Franczak, A.; Szindler, M. Structure and properties of the skeleton microporous materials with coatings inside the pores for medical and dental applications. In *Frontiers in Materials Processing, Applications, Research and Technology*; Muruganant, M., Chirazi, A., Raj, B., Eds.; Springer Nature: Singapore, 2018; pp. 297–320. [CrossRef]

142. Dobrzański, L.A.; Dobrzańska-Danikiewicz, A.D.; Gaweł, T.G.; Achtelik-Franczak, A. Selective laser sintering and melting of pristine titanium and titanium Ti6Al4V alloy powders and selection of chemical environment for etching of such materials. *Arch. Metall. Mater.* **2015**, *60*, 2039–2045. [CrossRef]

143. Dobrzański, L.A. Overview and general ideas of the development of constructions, materials, technologies and clinical applications of scaffolds engineering for regenerative medicine. *Arch. Mater. Sci. Eng.* **2014**, *69*, 53–80.

144. Dobrzański, L.A.; Achtelik-Franczak, A. Struktura i własności tytanowych szkieletowych materiałów mikroporowatych wytworzonych metodą selektywnego spiekania laserowego do zastosowań w implantologii oraz medycynie regeneracyjnej. In *Metalowe Materiały Mikroporowate I Lite Do Zastosowań Medycznych I Stomatologicznych*; Open Access Library VII(1); Dobrzański, L.A., Dobrzańska-Danikiewicz, A.D., Eds.; International OCSCO World Press: Gliwice, Poland, 2017; pp. 186–244.

145. Dobrzański, L.A.; Dobrzańska-Danikiewicz, A.D.; Malara, P.; Gaweł, T.G.; Dobrzański, L.B.; Achtelik-Franczak, A. Fabrication of scaffolds from Ti6Al4V powders using the computer aided laser method. *AMM* **2015**, *60*, 1065–1070. [CrossRef]

146. Dobrzański, L.A. Metale i ich stopy. In *Open Access Library VII(2)*; Dobrzański, L.A., Ed.; International OCSCO World Press: Gliwice, Poland, 2017; pp. 1–982.

147. Dobrzański, L.A.; Achtelik-Franczak, A. Struktura i własności materiałów kompozytowych do zastosowań medycznych o osnowie odlewniczych stopów aluminium wzmacnianych tytanowymi szkieletami wytworzonymi metoda selektywnego spiekania laserowego. In *Metalowe Materiały Mikroporowate I Lite Do Zastosowań Medycznych I Stomatologicznych*; Open Access Library VII(1); Dobrzański, L.A., Dobrzańska-Danikiewicz, A.D., Eds.; International OCSCO World Press: Gliwice, Poland, 2017; pp. 376–433.
148. Dobrzański, L.A. Stan wiedzy o materiałach stosowanych w implantologii oraz medycynie regeneracyjnej. In *Metalowe Materiały Mikroporowate I Lite Do Zastosowań Medycznych I Stomatologicznych*; Open Access Library, VII(1); Dobrzański, L.A., Dobrzańska-Danikiewicz, A.D., Eds.; International OCSCO World Press: Gliwice, Poland, 2017; pp. 117–120.
149. Dobrzański, L.A. Applications of newly developed nanostructural and microporous materials in biomedical, tissue and mechanical engineering. *Arch. Mater. Sci. Eng.* **2015**, *76*, 53–114.
150. Nowacki, J.; Dobrzański, L.A.; Gustavo, F. Implanty śródszpikowe w osteosyntezie kości długich. In *Open Access Library 11(17)*; Dobrzański, L.A., Ed.; International OCSCO World Press: Gliwice, Poland, 2012; pp. 1–150.
151. Dobrzańska-Danikiewicz, A.D.; Gaweł, T.G.; Kroll, L.; Dobrzański, L.A. Nowe porowate materiały kompozytowe metalowo-polimerowe wytwarzane z udziałem selektywnego stapiania laserowego. In *Metalowe Materiały Mikroporowate I Lite Do Zastosowań Medycznych I Stomatologicznych*; Open Access Library, VII(1); Dobrzański, L.A., Dobrzańska-Danikiewicz, A.D., Eds.; International OCSCO World Press: Gliwice, Poland, 2017; pp. 245–288.
152. Dobrzański, L.A.; Dobrzańska-Danikiewicz, A.D. (Eds.) *Metalowe Materiały Mikroporowate I Lite Do Zastosowań Medycznych I Stomatologicznych*; Open Access Library VII(1); International OCSCO World Press: Gliwice, Poland, 2017; pp. 1–580.
153. Dobrzański, L.A.; Matula, G.; Dobrzańska-Danikiewicz, A.D.; Malara, P.; Kremzer, M.; Tomiczek, B.; Kujawa, M.; Hajduczek, E.; Achtelik-Franczak, A.; Dobrzański, L.B.; et al. Composite materials infiltrated by aluminium alloys based on porous skeletons from alumina, Mullite and titanium produced by powder metallurgy techniques. In *Powder Metallurgy—Fundamentals and Case Studies*; Dobrzański, L.A., Ed.; IntechOpen: Rijeka, Croatia, 2017; pp. 95–137. [CrossRef]
154. Dobrzański, L.A. (Ed.) *Polymer Nanofibers Produced by Electrospinning Applied in Regenerative Medicine*; Open Access Library V(3); International OCSCO World Press: Gliwice, Poland, 2015; pp. 1–168.
155. Dobrzański, L.A.; Dobrzańska-Danikiewicz, A.D.; Achtelik-Franczak, A.; Dobrzański, L.B.; Hajduczek, E.; Matula, G. Fabrication Technologies of the Sintered Materials Including Materials for medical and dental application. In *Powder Metallurgy—Fundamentals and Case Studies*; Dobrzański, L.A., Ed.; IntechOpen: Rijeka, Croatia, 2017; pp. 17–52. [CrossRef]
156. Dobrzański, L.A. (Ed.) *Powder Metallurgy—Fundamentals and Case Studies*; IntechOpen: Rijeka, Croatia, 2017; pp. 1–392.
157. Dobrzańska-Danikiewicz, A.D.; Dobrzański, L.A.; Szindler, M.; Achtelik-Franczak, A.; Dobrzański, L.B. Obróbka powierzchni materiałów mikroporowatych wytworzonych metodą selektywnego spiekania laserowego w celu uefektywnienia proliferacji żywych komórek. In *Metalowe Materiały Mikroporowate I Lite Do Zastosowań Medycznych I Stomatologicznych*; Open Access Library, VII(1); Dobrzański, L.A., Dobrzańska-Danikiewicz, A.D., Eds.; International OCSCO World Press: Gliwice, Poland, 2017; pp. 289–375.
158. Dobrzański, L.A. Korszerű mérnöki anyagok kutatásának nanotechnológiai aspektusai. *Miskolci Egy. Multidiszcip. Tudományok* **2016**, *6*, 21–44.
159. Dobrzański, L.A.; Dobrzańska-Danikiewicz, A.D.; Achtelik-Franczak, A.; Dobrzański, L.B.; Szindler, M.; Gaweł, T.G. Porous selective laser melted Ti and Ti6Al4V materials for medical applications. In *Powder Metallurgy—Fundamentals and Case Studies*; Dobrzański, L.A., Ed.; IntechOpen: Rijeka, Croatia, 2017; pp. 161–181. [CrossRef]
160. Davis, M. *Thinking Like an Engineer: Studies in the Ethics of a Profession*; Oxford University Press: New York, NY, USA, 1998.
161. Davis, M. *Engineering Ethics*; Routledge: London, UK, 2005. [CrossRef]
162. Martin, M.W.; Schinzinger, R. *Ethics in Engineering*, 4th ed.; McGraw-Hill: Boston, MA, USA, 2005.
163. Harris, C.E.; Pritchard, M.S.; Rabins, M.J. *Engineering Ethics: Concepts and Cases*, 4th ed.; Wadsworth: Belmont, TN, USA, 2008.
164. Hansson, S.O. Ethical Criteria of Risk Acceptance. *Erkenntnis* **2003**, *59*, 291–309. [CrossRef]
165. Regulation (EU) 2017/745 of the European Parliament and of the Council on Medical Devices, Amending Directive 2001/83/EC, Regulation (EC) No 178/2002 and Regulation (EC) No 1223/2009 and Repealing Council Directives 90/385/EEC and 93/42/EEC. Available online: https://eur-lex.europa.eu/legal-content/EN/TXT/PDF/?uri=CELEX:32017R0745 (accessed on 1 March 2021).
166. Dobrzański, L.A.; Dobrzańska-Danikiewicz, A.D. Applications of Laser Processing of Materials in Surface Engineering in the Industry 4.0 Stage of the Industrial Revolution. *Mater. Perform. Charact* **2019**, *8*, 1091–1129. [CrossRef]
167. Dobrzański, L.A. Effect of heat and surface treatment on the structure and properties of the Mg-Al-Zn-Mn casting alloys. In *Magnesium and Its Alloys*; Dobrzański, L.A., Totten, G.E., Bamberger, M., Eds.; CRC Press: Boca Raton, FL, USA, 2019; pp. 91–202.
168. Dobrzański, L.A. Structural Phenomena Accompanying the Production of Composite and Nanocomposite Materials Using Selected Technologies. In Proceedings of the 22nd Physical Metallurgy and Materials Science Conference: Advanced Materials and Technologies, Bukowina Tatrzańska, Poland, 9–12 June 2019.
169. Dobrzański, L.A. *Stage 4.0 of the Technological Revolution in the Context of the Development of Engineering & Dental Materials*; Lviv Polytechnic National University: Lviv, Ukraine, 2019.
170. Dobrzański, L.A. Rola inżynierii materiałowej w etapie 4.0 rewolucji technologicznej. In Proceedings of the the the 24th Seminar of the Polish Society of Materials Science, Jachranka, Poland, 12–15 May 2019.
171. Dobrzański, L.A. *The Importance of Materials Engineering in Stage 4.0 of the Technological Revolution*; Institute of Fundamental Technological Research: Warsaw, Poland, 27 May 2019.

172. Dobrzański, L.A. Role of Materials Design in Maintenance Engineering in the Context of Industry 4.0 Idea. In Proceedings of the the International Maintenance Technologies Congress and Exhibition, Denizli, Turkey, 26–28 September 2019.
173. Jose, R.; Ramakrishna, S. Materials 4.0: Materials Big Data Enabled Materials Discovery. *Appl. Mat. Today* **2018**, *10*, 127–132. [CrossRef]
174. Dobrzański, L.A. Komputerowa nauka o materiałach jako metoda projektowania współczesnych materiałów i produktów inżynierskich. In *Seminary of Research and Education Programmes in Materials Engineering*; Polish Society of Materials Science: Myczkowce, Poland, 2004; pp. 55–88.
175. Ruehle, M.; Dosch, H.; Mittemeijer, E.; van de Voorde, M.H. (Eds.) *European White Book on Fundamental Research in Materials Science*; Max-Planck-Institute für Metallforschung: Stuttgart, Germany, 2001.
176. Kagermann, H.; Wahlster, W.; Helbig, J. *Recommendations for Implementing the Strategic Initiative INDUSTRIE 4.0: Final Report of the Industrie 4.0 Working Group*; Federal Ministry of Education and Research: Bonn, Germany, 2013.
177. Kagermann, H. Chancen Von Industrie 4.0 Nutzen. In *Industrie 4.0 in Produktion, Automatisierung und Logistik*; Springer Fachmedien Wiesbaden: Wiesbaden, Germany, 2014; pp. 603–614.
178. Hermann, M.; Pentek, T.; Otto, B. *Design Principles for Industrie 4.0 Scenarios: A Literature Review*; Technische Universität Dortmund: Dortmund, Germany, 2015.
179. Rüßmann, M.; Lorenz, M.; Gerbert, P.; Waldner, M.; Justus, J.; Engel, P.; Harnisch, M. *Industry 4.0: The Future of Productivity and Growth in Manufacturing Industries*; Boston Consulting Group: Boston, MA, USA, 2015.
180. European Commission. Commission Sets Out Path to Digitise European Industry. Available online: https://ec.europa.eu/growth/content/commission-sets-out-path-digitise-european-industry-0_en (accessed on 1 March 2021).
181. Danieluk, K. Digital Innovation Hubs on Smart Factories in New EU Member States. Available online: https://ec.europa.eu/futurium/en/implementing-digitising-european-industry-actions/digital-innovation-hubs-smart-factories-new-eu (accessed on 1 March 2021).
182. Lu, B.H.; Bateman, R.J.; Cheng, K. RFID Enabled Manufacturing: Fundamentals, Methodology and Applications. *Int. J. Agil. Syst. Manag.* **2006**, *1*, 73–92. [CrossRef]
183. Giusto, D.; Iera, A.; Morabito, G.; Atzori, L. (Eds.) *The Internet of Things*; Springer: New York, NY, USA, 2010.
184. Zhu, Q.; Wang, R.; Chen, Q.; Liu, Y.; Qin, W. IOT Gateway: Bridging Wireless Sensor Networks into Internet of Things. In *Proceedings of the 2010 IEEE/IFIP International Conference on Embedded and Ubiquitous Computing, Hong Kong, China, 11–13 December 2010*; The Institute of Electrical and Electronics Engineers: Piscataway, NJ, USA, 2010; pp. 347–352.
185. Wan, J.; Yan, H.; Liu, Q.; Zhou, K.; Lu, R.; Li, D. Enabling Cyber-Physical Systems with Machine-to-Machine Technologies. *Int. J. Ad. Hoc. Ubiquitous Comput.* **2013**, *13*, 187–196. [CrossRef]
186. Gubbi, J.; Buyya, R.; Marusic, S.; Palaniswami, M. Internet of Things (IoT): A Vision, Architectural Elements, and Future Directions. *Future Gener. Comput. Syst.* **2013**, *29*, 1645–1660. [CrossRef]
187. Zhong, R.Y.; Li, Z.; Pang, L.Y.; Pan, Y.; Qu, T.; Huang, G.Q. RFID-Enabled Real-Time Advanced Planning and Scheduling Shell for Production Decision Making. *Int. J. Comput. Integr. Manuf.* **2013**, *26*, 649–662. [CrossRef]
188. Wu, D.-Z.; Greer, M.J.; Rosen, D.W.; Schaefer, D. Cloud Manufacturing: Strategic Vision and State-of-the-Art. *J. Manuf. Syst.* **2013**, *32*, 564–579. [CrossRef]
189. Moreno-Vozmediano, R.; Montero, R.S.; Llorente, I.M. Key Challenges in Cloud Computing: Enabling the Future Internet of Services. *IEEE Internet Comput.* **2013**, *17*, 18–25. [CrossRef]
190. Lee, J.; Kao, H.-A.; Yang, S. Service Innovation and Smart Analytics for Industry 4.0 and Big Data Environment. *Proc. Cirp* **2014**, *16*, 3–8. [CrossRef]
191. Bi, Z.; Xu, L.D.; Wang, C. Internet of Things for Enterprise Systems of Modern Manufacturing. *IEEE Trans. Indust. Inf.* **2014**, *10*, 1537–1546. [CrossRef]
192. Brettel, M.; Friederichsen, N.; Keller, M.; Rosenberg, M. How Virtualization, Decentralization, and Network-Building Change the Manufacturing Landscape: An Industry 4.0 Perspective. *Int. J. Mech. Aerospac. Indust. Mechatron. Manuf. Eng.* **2014**, *8*, 37–44.
193. Buer, S.-V.; Strandhagen, J.O.; Chan, F.T.S. The Link between Industry 4.0 and Lean Manufacturing: Mapping Current Research and Establishing a Research Agenda. *Int. J. Prod. Res.* **2018**, *56*, 2924–2940. [CrossRef]
194. Patel, P.; Cassou, D. Enabling High-Level Application Development for the Internet of Things. *J. Syst. Softw.* **2015**, *103*, 62–84. [CrossRef]
195. Zhang, Y.; Zhang, G.; Wang, J.; Sun, S.; Si, S.; Yang, T. Real-Time Information Capturing and Integration Framework of the Internet of Manufacturing Things. *Int. J. Comput. Integr. Manuf.* **2015**, *28*, 811–822. [CrossRef]
196. Qiu, X.; Luo, H.; Xu, G.; Zhong, R.-Y.; Huang, G.Q. Physical Assets and Service Sharing for IoT-Enabled Supply Hub in Industrial Park (SHIP). *Int. J. Prod. Econ.* **2015**, *159*, 4–15. [CrossRef]
197. Posada, J.; Toro, C.; Barandiaran, I.; Oyarzun, D.; Stricker, D.; de Amicis, R.; Pinto, E.B.; Eisert, P.; Döllner, J.; Vallarino, I. Visual Computing as a Key Enabling Technology for Industrie 4.0 and Industrial Internet. *IEEE Comp. Graph. Appl.* **2015**, *35*, 26–40. [CrossRef] [PubMed]
198. Farooq, M.U.; Waseem, M.; Mazhar, S.; Khairi, A.; Kamal, T. A Review on Internet of Things (IoT). *Int. J. Comp. Appl.* **2015**, *113*, 1–7. [CrossRef]
199. Lee, J.; Bagheri, B.; Kao, H.-A. A Cyber-Physical Systems Architecture for Industry 4.0-Based Manufacturing Systems. *Manuf. Lett.* **2015**, *3*, 18–23. [CrossRef]

200. Almada-Lobo, F. The Industry 4.0 Revolution and the Future of Manufacturing Execution Systems (MES). *J. Innov. Manag.* **2015**, *3*, 16–21. [CrossRef]

201. Wamba, S.F.; Akter, S.; Edwards, A.; Chopin, G.; Gnanzou, D. How 'Big Data' Can Make Big Impact: Findings from a Systematic Review and a Longitudinal Case Study. *Int. J. Prod. Econ.* **2015**, *165*, 234–246. [CrossRef]

202. Yin, Y.H.; Nee, A.Y.C.; Ong, S.K.; Zhu, J.Y.; Gu, P.H.; Chen, L.J. Automating Design with Intelligent Human-Machine Integration. *Cirp Ann.* **2015**, *64*, 655–677. [CrossRef]

203. Colin, M.; Galindo, R.; Hernández, O. Information and Communication Technology as a Key Strategy for Efficient Supply Chain Management in Manufacturing SMEs. *Proc. Comp. Sci.* **2015**, *55*, 833–842. [CrossRef]

204. Hozdic, E. Smart Factory for Industry 4.0: A Review. *Int. J. Mod. Manuf. Tech.* **2015**, *7*, 28–35.

205. Zhong, R.Y.; Huang, G.Q.; Lan, S.; Dai, Q.Y.; Zhang, T.; Xu, C. A Two-Level Advanced Production Planning and Scheduling Model for RFID-Enabled Ubiquitous Manufacturing. *Adv. Eng. Inf.* **2015**, *29*, 799–812. [CrossRef]

206. Wang, S.; Wan, J.; Zhang, D.; Li, D.; Zhang, C. Towards Smart Factory for Industry 4.0: A Self-Organized Multi-Agent System with Big Data Based Feedback and Coordination. *Comp. Netw.* **2016**, *101*, 158–168. [CrossRef]

207. Monostori, L.; Kádár, B.; Bauernhansl, T.; Kondoh, S.; Kumara, S.; Reinhart, G.; Sauer, O.; Schuh, G.; Sihn, W.; Ueda, K. Cyber-Physical Systems in Manufacturing. *Cirp. Ann.* **2016**, *65*, 621–641. [CrossRef]

208. Zhong, R.Y.; Lan, S.; Xu, C.; Dai, Q.; Huang, G.Q. Visualization of RFID-Enabled Shopfloor Logistics Big Data in Cloud Manufacturing. *Int. J. Adv. Manuf. Technol.* **2016**, *84*, 5–16. [CrossRef]

209. Georgakopoulos, D.; Jayaraman, P.P.; Fazia, M.; Villari, M.; Ranjan, R. Internet of Things and Edge Cloud Computing Roadmap for Manufacturing. *IEEE Cloud Comp.* **2016**, *3*, 66–73. [CrossRef]

210. Zhong, R.Y.; Newman, S.T.; Huang, G.Q.; Lan, S. Big Data for Supply Chain Management in the Service and Manufacturing Sectors: Challenges, Opportunities, and Future Perspectives. *Comp. Indust. Eng.* **2016**, *101*, 572–591. [CrossRef]

211. Misra, G.; Kumar, V.; Agarwal, A.; Agarwal, K. Internet of Things (IoT)—A Technological Analysis and Survey on Vision, Concepts, Challenges, Innovation Directions, Technologies, and Applications (An Upcoming or Future Generation Computer Communication System Technology). *Am. J. Electr. Electro. Eng.* **2016**, *4*, 23–32. [CrossRef]

212. Schumacher, A.; Erol, S.; Sihn, W. A Maturity Model for Assessing Industry 4.0 Readiness and Maturity of Manufacturing Enterprises. *Proc. Cirp* **2016**, *52*, 161–166. [CrossRef]

213. Sipsas, K.; Alexopoulos, K.; Xanthakis, V.; Chryssolouris, G. Collaborative Maintenance in Flow-Line Manufacturing Environments: An Industry 4.0 Approach. *Proc. Cirp* **2016**, *55*, 236–241. [CrossRef]

214. Qin, J.; Liu, Y.; Grosvenor, R. A Categorical Framework of Manufacturing for Industry 4.0 and Beyond. *Proc. Cirp* **2016**, *52*, 173–178. [CrossRef]

215. *Referenzarchitekturmodell Industrie 4.0 (RAMI4.0)*; DIN SPEC 91345:2016-04; Beuth Verlag: Berlin, Germany, 2016.

216. Dobrzański, L.A. Comparative Analysis of Mechanical Properties of Scaffolds Sintered from Ti and Ti6Al4V Powders. In Proceedings of the Winter International Scientific Conference on Achievements in Mechanical and Materials Engineering, Zakopane, Poland, 6–9 December 2015.

217. Adolphs, P.; Auer, S.; Bedenbender, H.; Billmann, M.; Hankel, M.; Heidel, R.; Hoffmeister, M.; Huhle, H.; Jochem, M.; Kiele-Dunsche, M.; et al. *Structure of the Administration Shell: Continuation of the Development of the Reference Model. for the Industrie 4.0 Component*; Federal Ministry for Economic Affairs and Energy: Berlin, Germany, 2016.

218. Bahrin, M.A.K.; Othman, M.F.; Azli, N.H.N.; Talib, M.F. Industry 4.0: A Review on Industrial Automation and Robotic. *J. Tekno.* **2016**, *78*, 137–143.

219. Mosterman, P.J.; Zander, J. Industry 4.0 as a Cyber-Physical System Study. *Softw. Syst. Model.* **2016**, *15*, 17–29. [CrossRef]

220. Harrison, R.; Vera, D.; Ahmad, B. Engineering Methods and Tools for Cyber-Physical Automation Systems. *Proc. IEEE* **2016**, *104*, 973–985. [CrossRef]

221. Stock, T.; Seliger, G. Opportunities of Sustainable Manufacturing in Industry 4.0. *Proc. Cirp* **2016**, *40*, 536–541. [CrossRef]

222. Pfeiffer, S. Robots, Industry 4.0 and Humans, or Why Assembly Work Is More than Routine Work. *Societies* **2016**, *6*, 16. [CrossRef]

223. Gorkhali, A.; Xu, L.D. Enterprise Application Integration in Industrial Integration: A Literature Review. *J. Indust. Integr. Manag.* **2016**, *1*, 1650014. [CrossRef]

224. Zhong, R.Y.; Xu, X.; Klotz, E.; Newman, S.T. Intelligent Manufacturing in the Context of Industry 4.0: A Review. *Engineering* **2017**, *3*, 616–630. [CrossRef]

225. Thoben, K.-D.; Wiesner, S.; Wuest, T. 'Industrie 4.0' and Smart Manufacturing—A Review of Research Issues and Application Examples. *Int. J. Autom. Techn.* **2017**, *11*, 4–16. [CrossRef]

226. Xu, X. Machine Tool 4.0 for the New Era of Manufacturing. *Int. J. Adv. Manuf. Techn.* **2017**, *92*, 1893–1900. [CrossRef]

227. Li, B.-H.; Hou, B.-C.; Yu, W.-T.; Lu, X.-B.; Yang, C.-W. Applications of Artificial Intelligence in Intelligent Manufacturing: A Review. *Front. Infor. Techn. Electro. Eng.* **2017**, *18*, 86–96. [CrossRef]

228. Lu, Y. Industry 4.0: A Survey on Technologies, Applications and Open Research Issues. *J. Indust. Infor. Integr.* **2017**, *6*, 1–10. [CrossRef]

229. Vaidya, S.; Ambad, P.; Bhosle, S. Industry 4.0—A Glimpse. *Proc. Manuf.* **2018**, *20*, 233–238. [CrossRef]

230. Kumar, K.; Zindani, D.; Davim, J.P. *Industry 4.0: Developments towards the Fourth Industrial Revolution*; Springer Nature: Singapore, 2019.

231. Łobaziewicz, M. *Zarządzanie Inteligentnym Przedsiębiorstwem W Dobie Przemysłu 4.0*; Towarzystwo Naukowe Organizacji i Kierownictwa: Toruń, Poland, 2019.
232. Ardito, L.; Petruzzelli, A.M.; Panniello, U.; Garavelli, A.C. Towards Industry 4.0: Mapping Digital Technologies for Supply Chain Management-Marketing Integration. *Bus. Proc. Manag. J.* **2019**, *25*, 323–346. [CrossRef]
233. Boston Consulting Group. Putting Industry 4.0 to Work. Available online: https://www.bcg.com/capabilities/manufacturing/industry-4.0 (accessed on 1 March 2021).
234. Tietenberg, T.; Lewis, L. *Environmental and Natural Resource Economics*; Pearson Education: London, UK, 2003.
235. Dobrzański, L.A. Significance of Materials Science and Engineering for Advances in Design and Manufacturing Processes, Computer Integrated Manufacturing. In *Advanced Design and Management*; Skołud, B., Krenczyk, D., Eds.; WNT: Warsaw, Poland, 2003; pp. 128–140.
236. Dobrzański, L.A. The importance of the development of materials science and materials engineering to improve the quality of life of modern societies. *Wisnyk Technol. Univ. PodillaKhmelnytskyUkr.* **2003**, *I*, 34–47.
237. Dobrzański, L.A. Heat Treatment as Fundamental Technological Process of Formation of Structure and Properties of Metals. In *Proceedings of the 8th Seminar of the International Federation for Heat Treatment and Surface Engineering, Cavtat, Croatia, 12–14 Semptember 2001*; International Federation for Heat Treatment and Surface Engineering: Winterthur, Switzerland, 2001; pp. 1–12.
238. Dobrzański, L.A. *Significance of Materials Science for Advances in Products Design and Manufacturing*; General Assembly of the International Academy for Production Engineering: Cracow, Poland, 2004.
239. Tay, S.I.; Lee, T.C.; Hamid, N.A.A.; Ahmad, A.N.A. An Overview of Industry 4.0: Definition, Components, and Government Initiatives. *J. Adv. Res. Dyn. Control. Syst.* **2018**, *10*, 1379–1387.
240. Dobrzański, L.A. *Podstawy Metodologii Projektowania Materiałowego*; Wydawnictwo Politechniki Śląskiej: Gliwice, Poland, 2009.
241. Dobrzański, L.A. Significance of materials science for the future development of societies. *J. Mater. Process. Technol.* **2006**, *175*, 133–148. [CrossRef]
242. Dobrzański, L.A. *Materiały Inżynierskie I Projektowanie Materiałowe. Podstawy Nauki O Materiałach I Metaloznawstwo*, 2nd ed.; suppl. & changed; WNT: Warsaw, Poland, 2006.
243. Al-Fadda, S.A.; Zarb, G.A.; Finer, Y.A. A comparison of the accuracy of fit of 2 methods for fabricating implant-prosthodontic frameworks. *Int. J. Prosthodont.* **2007**, *20*, 125–131. [PubMed]
244. Helmus, M.N. Overview of Biomedical Materials. *Bullet* **1991**, *16*, 33–38. [CrossRef]
245. Singh, P.; Kumar, P. An Overview of Biomedical Materials and Techniques for Better Functional Performance, Life, Sustainability and Biocompatibility of Orthopedic Implants. *Indian J. Sci. Tech.* **2018**, *11*, 1–7. [CrossRef]
246. Hin, T.S. (Ed.) *Engineering Materials for Biomedical Applications*; World Scientific: Singapore, 2004. [CrossRef]
247. Wang, X. Overview on Biocompatibilities of Implantable Biomaterials. In *Advances in Biomaterials Science and Biomedical Applications*; Pignatello, R., Ed.; IntechOpen: Rijeka, Croatia, 2013. [CrossRef]
248. Pignatello, R. (Ed.) *Advances in Biomaterials Science and Biomedical Applications*; IntechOpen: Rijeka, Croatia, 2013.
249. Bronzino, J.D.; Peterson, D.R. *The Biomedical Engineering Handbook, Volume 1: Biomedical Engineering Fundamentals. Physiologic Systems. Biomechanics. Biomaterials. Bioelectric Phenomena. Neuroengineering*, 4th ed.; CRC Press: Boca Raton, FL, USA, 2019.
250. Bronzino, J.D.; Peterson, D.R. *The Biomedical Engineering Handbook, Volume 2: Medical Devices and Systems. Biomedical Sensors. Medical Instrumentation and Devices. Human Performance Engineering. Rehabilitation Engineering. Clinical Engineering*, 4th ed.; CRC Press: Boca Raton, FL, USA, 2019.
251. Bronzino, J.D.; Peterson, D.R. *The Biomedical Engineering Handbook, Volume 3: Biomedical Signals, Imaging and Informatics. Biosignal Processing. Medical Imaging. Infrared Imaging. Medical Informatics*, 4th ed.; CRC Press: Boca Raton, FL, USA, 2019.
252. Bronzino, J.D.; Peterson, D.R. *The Biomedical Engineering Handbook, Volume 4: Molecular, Cellular and Tissue Engineering. Molecular Biology. Transport. Phenomena and Biomimetic Systems. Physiological Modeling, Simulation and Control. Stem Cell Engineering; An. Introduction. Tissue Engineering. Artificial Organs. Drug Design, Delivery Systems and Devices. Personalized Medicine. Ethics*, 4th ed.; CRC Press: Boca Raton, FL, USA, 2019.
253. Kokubo, T. (Ed.) *Biocermics and Their Clinical Applications*; Woodhead Publishing: Cambridge, England, 2008; pp. 1–784.
254. Curtis, R.V.; Watson, T.F. (Eds.) *Dental Biomaterials*; Woodhead Publishing: Cambridge, UK, 2008; pp. 1–528.
255. Jenkins, M. (Ed.) *Biomedical Polymers*; Woodhead Publishing: Cambridge, UK, 2007; pp. 1–236.
256. Wiliams, D.F. (Ed.) *Definitions in Biomaterials*; Elsevier: Amsterdam, The Netherlands, 1987.
257. Dobrzański, L.A. *Metaloznawstwo Opisowe*; Wydawnictwo Politechniki Śląskiej: Gliwice, Poland, 2013.
258. Wang, S.; Qu, X.; Zhao, R.C. Clinical applications of mesenchymal stem cells. *J. Hematol. Oncol.* **2012**, *5*, 1–9. [CrossRef] [PubMed]
259. Heness, G.; Ben-Nissan, B. Innovative bioceramics. *Mater. Forum.* **2004**, *27*, 104–114.
260. Parida, P.; Behera, A.; Mishra, S.C. Classification of Biomaterials used in Medicine. *Int. J. Adv. Appl. Sci.* **2012**, *1*, 31–35. [CrossRef]
261. Bronzino, J.D.; Peterson, D.R. *The Biomedical Engineering Handbook*, 4th ed.; CRC Press: Boca Raton, FL, USA, 2019.
262. Ratner, B.D.; Hoffman, A.S.; Schoen, F.J.; Lemons, J.E. (Eds.) *Biomaterials Science: An. Introduction to Materials in Medicine*, 3rd ed.; Academic Press: Oxford, UK, 2013.
263. Helmus, M.N.; Gibbons, D.F.; Cebon, D. Biocompatibility: Meeting a key functional requirement of next-generation medical devices. *Toxicol. Pathol.* **2008**, *36*, 70–80. [CrossRef] [PubMed]
264. Parida, P.; Mishra, S.C. *Biomaterials in Medicine. Proceedings of the UGC Sponsored National Workshop on Innovative Experiments in Physics*; Neelashaila Mahabidyalaya: Rourkela, India, 2012.

265. Pramanik, S.; Agarwal, A.K.; Rai, K.N. Chronology of Total Hip Joint Replacement and Materials Development. *Trends Biomater. Artif. Organs.* **2005**, *19*, 15–26.

266. Chakrabarty, G.V. Biomaterials: Metallic Implant. Materials, Technology, Review Essays. *IJAAS* **2011**, *1*, 125–129.

267. Srivastav, A. An Overview of Metallic Biomaterials for Bone Support and Replacement. In *Biomedical Engineering, Trends in Materials Science*; Laskovski, A., Ed.; IntechOpen: Rijeka, Croatia, 2011; pp. 153–168. [CrossRef]

268. Yoo, Y.R.; Cho, H.H.; Jang, S.G.; Lee, K.Y.; Son, H.Y.; Kim, J.G.; Kim, Y.S. Effect of Co Content on The Corrosion of High Performance Stainless Steels In Simulated Bio-solutions. *Key Eng. Mater.* **2007**, *342–343*, 585–588. [CrossRef]

269. Billotte, W.G. Ceramic biomaterials. In *The Biomedical Engineering Handbook*, 2nd ed.; Bronzino, J.D., Ed.; CRC Press: Boca Raton, FL, USA, 2000; Volume 1, pp. 1–33.

270. Atala, A.; Lanza, R.; Nerem, R.; Thomson, J. *Principles of Regenerative Medicine*; Academic Press: Cambridge, MA, USA, 2007; pp. 1–1472.

271. Lee, H.B.; Khang, G.; Lee, J.H. Polymeric Biomaterials. In *Biomaterials*; CRC Press: Boca Raton, FL, USA, 2007; pp. 55–78. [CrossRef]

272. Ehrlich, H. *Biological Materials of Marine Origin*; Springer: Dordrecht, The Netherlands, 2014. [CrossRef]

273. Jockisch, K.A.; Brown, S.A.; Bauer, T.W.; Merritt, K. Biological response to chopped-carbon-fiber-reinforced peek. *J. Biomed. Mater. Res.* **1992**, *26*, 133–146. [CrossRef]

274. Bos, R.R.; Rozema, F.R.; Boering, G.; Nijenhuis, A.J.; Pennings, A.J.; Jansen, H.W. Bone-plates and screws of bioabsorbable poly (L-lactide)—An animal pilot study. *Br. J. Oral Maxillofac. Surg.* **1989**, *27*, 467–476. [CrossRef]

275. Albertsson, A.C.; Varma, I.K. Aliphatic Polyesters: Synthesis, Properties and Applications. In *Degradable Aliphatic Polyesters. Advances in Polymer Science*; Albertsson, A.-C., Ed.; Springer: Berlin/Heidelberg, Germany, 2002; Volume 157, pp. 1–40. [CrossRef]

276. Zhu, S.L.; Wang, X.M.; Qin, F.X.; Inoue, A. A new Ti-based bulk glassy alloy with potential for biomedical application. *Mater. Sci. Eng. A* **2007**, *459*, 233–237. [CrossRef]

277. Zhang, E.; Yin, D.; Xu, L.; Yang, L.; Yang, K. Microstructure, mechanical and corrosion properties and biocompatibility of Mg-Zn-Mn alloys for biomedical application. *Mater. Sci. Eng. C* **2009**, *29*, 987–993. [CrossRef]

278. Li, S.J.; Niinomi, M.; Akahori, T.; Kasuga, T.; Yang, R.; Hao, Y.L. Fatigue characteristics of bioactive glass-ceramic-coated Ti-29Nb-13Ta-4.6Zr for biomedical application. *Biomaterials* **2004**, *25*, 3369–3378. [CrossRef]

279. Rajendran, V.; Bhandari, S.K. Bioactive Glasses for Implant Applications. Available online: https://www.drdo.gov.in/monograph/bioactive-glasses-implant-applications (accessed on 1 March 2021).

280. Black, J. The education of the biomaterialist: Report of a survey, 1980–1981. *J. Biomed. Mater. Res.* **1982**, *16*, 159–167. [CrossRef] [PubMed]

281. Dobrzańska-Danikiewicz, A.D.; Łukowiec, D.; Cichocki, D.; Wolany, W. Nanokompozyty złożone z nanorurek węglowych pokrytych Nanokryształami Metali Szlachetnych. In *Open Access Library V(2)*; Dobrzański, L.A., Ed.; International OCSCO World Press: Gliwice, Poland, 2015; pp. 1–131.

282. Qiu, G.; Ding, W. Editorial for the Special Issue on Medical Additive Manufacturing. *Engineering* **2020**, *6*, 1205–1206. [CrossRef]

283. Qiu, G.; Ding, W.; Tian, W.; Qin, L.; Zhao, Y.; Zhang, L.; Lu, J.; Chen, D.; Yuan, G.; Wu, C.; et al. Medical Additive Manufacturing: From a Frontier Technology to the Research and Development of Products. *Engineering* **2020**, *6*, 1217–1221. [CrossRef]

284. Wang, X.; Zhang, M.; Ma, J.; Xu, M.; Chang, J.; Gelinsky, M.; Wu, C. 3D Printing of Cell-Container-Like Scaffolds for Multicell Tissue Engineering. *Engineering* **2020**, *6*, 1276–1284. [CrossRef]

285. Wang, Y.; Tan, Q.; Pu, F.; Boone, D.; Zhang, M. A Review of the Application of Additive Manufacturing in Prosthetic and Orthotic Clinics from a Biomechanical Perspective. *Engineering* **2020**, *6*, 1258–1266. [CrossRef]

286. Liu, G.; He, Y.; Liu, P.; Chen, Z.; Chen, X.; Wan, L.; Li, Y.; Lu, J. Development of Bioimplants with 2D, 3D, and 4D Additive Manufacturing Materials. *Engineering* **2020**, *6*, 1232–1243. [CrossRef]

287. Li, C.; Pisignano, D.; Zhao, Y.; Xue, J. Advances in Medical Applications of Additive Manufacturing. *Engineering* **2020**, *6*, 1222–1231. [CrossRef]

288. Wang, Y.; Fu, P.; Wang, N.; Peng, L.; Kang, B.; Zeng, H.; Yuan, G.; Ding, W. Challenges and Solutions for the Additive Manufacturing of Biodegradable Magnesium Implants. *Engineering* **2020**, *6*, 1267–1275. [CrossRef]

289. Dobrzański, L.A.; Dobrzańska-Danikiewicz, A.D.; Czuba, Z.P.; Dobrzański, L.B.; Achtelik-Franczak, A.; Malara, P.; Szindler, M.; Kroll, L. The new generation of the biological-engineering materials for applications in medical and dental implant-scaffolds. *Arch. Mater. Sci. Eng.* **2018**, *91*, 56–85. [CrossRef]

290. Dobrzański, L.A.; Dobrzańska-Danikiewicz, A.D.; Czuba, Z.P.; Dobrzański, L.B.; Achtelik-Franczak, A.; Malara, P.; Szindler, M.; Kroll, L. Metallic skeletons as reinforcement of new composite materials applied in orthopaedics and dentistry. *Arch. Mater. Sci. Eng.* **2018**, *92*, 53–85. [CrossRef]

291. Dobrzański, L.B.; Achtelik-Franczak, A.; Dobrzańska, J.; Dobrzański, L.A. Comparison of the Structure and Properties of the Solid Co-Cr-W-Mo-Si Alloys Used for Dental Restorations CNC Machined or Selective Laser-Sintered. *Mater. Perform. Charact* **2020**, *9*, 556–578. [CrossRef]

292. Dobrzański, L.A.; Dobrzański, L.B.; Dobrzańska-Danikiewicz, A.D. Manufacturing technologies thick-layer coatings on various substrates and manufacturing gradient materials using powders of metals, their alloys and ceramics. *J. Achiev. Mater. Manuf. Eng.* **2020**, *99*, 14–41. [CrossRef]

293. Dobrzański, L.A.; Dobrzański, L.B.; Dobrzańska-Danikiewicz, A.D. Additive and hybrid technologies for products manufacturing using powders of metals, their alloys and ceramics. *Arch. Mater. Sci. Eng.* **2020**, *102*, 59–85. [CrossRef]
294. Liao, B.; Xia, R.F.; Li, W.; Lu, D.; Jin, Z.M. 3D-Printed Ti6Al4V Scaffolds with Graded Triply Periodic Minimal Surface Structure for Bone Tissue Engineering. *J. Mater. Eng. Perform.* 2021. [CrossRef]
295. Xiong, Y.-Z.; Gao, R.-N.; Zhang, H.; Dong, L.-L.; Li, J.-T.; Li, X. Rationally designed functionally graded porous Ti6Al4V scaffolds with high strength and toughness built via selective laser melting for load-bearing orthopedic applications. *J. Mech. Behav. Biomed. Mater.* **2020**, *104*, 103673. [CrossRef] [PubMed]
296. Tripathi, Y.; Shukla, M.; Bhatt, A.D. Implicit-Function-Based Design and Additive Manufacturing of Triply Periodic Minimal Surfaces Scaffolds for Bone Tissue Engineering. *J. Mater. Eng. Perform.* **2019**, *28*, 7445–7451. [CrossRef]
297. Kelly, C.N.; Francovich, J.; Julmi, S.; Safranski, D.; Guldberg, R.E.; Maier, H.J.; Gall, K. Fatigue behavior of As-built selective laser melted titanium scaffolds with sheet-based gyroid microarchitecture for bone tissue engineering. *Acta Biomater.* **2019**, *94*, 610–626. [CrossRef]
298. Ma, S.; Tang, Q.; Feng, Q.; Song, J.; Han, X.; Guo, F. Mechanical behaviours and mass transport properties of bone-mimicking scaffolds consisted of gyroid structures manufactured using selective laser melting. *J. Mech. Behav. Biomed. Mater.* **2019**, *93*, 158–169. [CrossRef] [PubMed]
299. Yadroitsava, I.; du Plessis, A.; Yadroitsev, I. Bone regeneration on implants of titanium alloys produced by laser powder bed fusion: A review. In *Titanium for Consumer Applications*; Froes, F., Qian, M., Niinomi, M., Eds.; Elsevier: Amsterdam, The Netherlands, 2019; pp. 197–233. [CrossRef]
300. Zhang, L.; Yang, G.; Johnson, B.N.; Jia, X. Three-dimensional (3D) printed scaffold and material selection for bone repair. *Acta Biomater.* **2019**, *84*, 16–33. [CrossRef]
301. Ataee, A.; Li, Y.; Fraser, D.; Song, G.; Wen, C. Anisotropic Ti-6Al-4V gyroid scaffolds manufactured by electron beam melting (EBM) for bone implant applications. *Mater. Des.* **2018**, *137*, 345–354. [CrossRef]
302. Surmeneva, M.A.; Surmenev, R.A.; Chudinova, E.A.; Koptioug, A.; Tkachev, M.S.; Gorodzha, S.N.; Rännar, L.-E. Fabrication of multiple-layered gradient cellular metal scaffold via electron beam melting for segmental bone reconstruction. *Mater. Des.* **2017**, *133*, 195–204. [CrossRef]
303. Fousová, M.; Vojtěch, D.; Kubásek, J.; Jablonská, E.; Fojt, J. Promising characteristics of gradient porosity Ti-6Al-4V alloy prepared by SLM process. *J. Mech. Behav. Biomed. Mater.* **2017**, *69*, 368–376. [CrossRef]
304. Zhao, S.; Li, S.J.; Hou, W.T.; Hao, Y.L.; Yang, R.; Misra, R.D.K. The influence of cell morphology on the compressive fatigue behavior of Ti-6Al-4V meshes fabricated by electron beam melting. *J. Mech. Behav. Biomed. Mater.* **2016**, *59*, 251–264. [CrossRef] [PubMed]
305. Arahira, T.; Maruta, M.; Matsuya, S.; Todo, M. Development and characterization of a novel porous β-TCP scaffold with a three-dimensional PLLA network structure for use in bone tissue engineering. *Mater. Lett.* **2015**, *152*, 148–150. [CrossRef]
306. Yan, C.; Hao, L.; Hussein, A.; Young, P. Ti–6Al–4V triply periodic minimal surface structures for bone implants fabricated via selective laser melting. *J. Mech. Behav. Biomed. Mater.* **2015**, *51*, 61–73. [CrossRef] [PubMed]
307. Ahmadi, S.M.; Yavari, S.A.; Wauthle, R.; Pouran, B.; Schrooten, J.; Weinans, H.; Zadpoor, A.A. Additively Manufactured Open-Cell Porous Biomaterials Made from Six Different Space-Filling Unit Cells: The Mechanical and Morphological Properties. *Materials* **2015**, *8*, 1871–1896. [CrossRef] [PubMed]
308. Cheng, A.; Humayun, A.; Cohen, D.J.; Boyan, B.D.; Schwartz, Z. Additively manufactured 3D porous Ti-6Al-4V constructs mimic trabecular bone structure and regulate osteoblast proliferation, differentiation and local factor production in a porosity and surface roughness dependent manner. *Biofabrication* **2014**, *6*, 045007. [CrossRef] [PubMed]
309. Scherer, M.R.J. *Double-Gyroid-Structured Functional Materials. Synthesis and Applications*; Springer: Cham, Switzerland, 2013. [CrossRef]
310. Yan, C.; Hao, L.; Hussein, A.; Young, P.; Raymont, D. Advanced lightweight 316L stainless steel cellular lattice structures fabricated via selective laser melting. *Mater. Des.* **2014**, *55*, 533–541. [CrossRef]
311. Vrancken, B.; Thijs, L.; Kruth, J.-P.; Van Humbeeck, J. Heat treatment of Ti6Al4V produced by Selective Laser Melting: Microstructure and mechanical properties. *J. Alloys Compd.* **2012**, *541*, 177–185. [CrossRef]
312. Zhang, L.C.; Klemm, D.; Eckert, J.; Hao, Y.L.; Sercombe, T.B. Manufacture by selective laser melting and mechanical behavior of a biomedical Ti–24Nb–4Zr–8Sn alloy. *Scr. Mater.* **2011**, *65*, 21–24. [CrossRef]
313. Thijs, L.; Verhaeghe, F.; Craeghs, T.; Van Humbeeck, J.; Kruth, J.-P. A study of the microstructural evolution during selective laser melting of Ti–6Al–4V. *Acta Mater.* **2010**, *58*, 3303–3312. [CrossRef]
314. Parthasarathy, J.; Starly, B.; Raman, S.; Christensen, A. Mechanical evaluation of porous titanium (Ti6Al4V) structures with electron beam melting (EBM). *J. Mech. Behav. Biomed. Mater.* **2010**, *3*, 249–259. [CrossRef] [PubMed]
315. German, R.M. *Sintering Theory and Practice*; John Wiley & Sons: New York, NY, USA, 1996.
316. German, R.M. *Powder Metallurgy Science*; Metal Powder Industries Federation: Princeton, NJ, USA, 1984.
317. German, R.M. *Injection Molding of Metals and Ceramics*; Metal Powder Industries Federation: Princeton, NJ, USA, 1997.
318. German, R.M. *Powder Injection Molding*; Metal Powder Industries Federation: Princeton, NJ, USA, 1990.
319. Sun, S.; Brandt, M.; Easton, M. Powder bed fusion processes: An overview. In *Laser Additive Manufacturing; Materials, Design, Technologies, and Applications*; Brandt, M., Ed.; Woodhead Publishing: Cambridge, UK, 2017; pp. 55–77. [CrossRef]

320. Sahasrabudhe, H.; Bose, S.; Bandyopadhyay, A. Laser-based additive manufacturing processes. In *Advances in Laser Materials Processing*; Lawrence, J., Ed.; Woodhead Publishing: Cambridge, UK, 2018; pp. 507–539. [CrossRef]

321. Huo, S.H.; Qian, M.; Schaffer, G.B.; Crossin, E. Aluminium powder metallurgy. In *Fundamentals of Aluminium Metallurgy*; Production, Processing and Applications; Lumley, R., Ed.; Woodhead Publishing Limited: Cambridge, UK, 2011; pp. 655–701. [CrossRef]

322. German, R.M. Phase diagrams in liquid phase sintering treatments. *JOM* **1986**, *38*, 26–29. [CrossRef]

323. Schaffer, G.B.; Sercombe, T.B.; Lumley, R.N. Liquid phase sintering of aluminium alloys. *Mater. Chem. Phys.* **2001**, *67*, 85–91. [CrossRef]

324. Kang, S.-J.L. Basis of Liquid Phase Sintering. In *Sintering, Densification, Grain Growth and Microstructure*; Lumley, R., Ed.; Butterworth-Heinemann: Amsterdam, The Netherlands, 2005; pp. 199–203. [CrossRef]

325. Kang, S.-J.L. Liquid phase sintering. In *Sintering of Advanced Materials*; Fang, Z.Z., Ed.; Woodhead Publishing Limited: Cambridge, UK, 2010; pp. 110–129. [CrossRef]

326. Lumley, R.N.; Schaffer, G.B. The effect of solubility and particle size on liquid phase sintering. *Scr. Mater.* **1996**, *35*, 589–595. [CrossRef]

327. Kim, K.-B.; Kim, W.-C.; Kim, H.-Y.; Kim, J.-H. An evaluation of marginal fit of three-unit fixed dental prostheses fabricated by direct metal laser sintering system. *Dent. Mater.* **2013**, *29*, e91–e96. [CrossRef] [PubMed]

328. Barro, Ó.; Arias-González, F.; Lusquiños, F.; Comesaña, R.; del Val, J.; Riveiro, A.; Badaoui, A.; Gómez-Baño, F.; Pou, J. Effect of four manufacturing techniques (casting, laser directed energy deposition, milling and selective laser melting) on microstructural, mechanical and electrochemical properties of co-CR dental alloys, before and after PFM firing process. *Metals* **2020**, *10*, 1291. [CrossRef]

329. Dobrzański, L.B. Porównanie metod przyrostowych i ubytkowych wytwarzania uzupełnień protetycznych układu stomatognatycznego. In *Metalowe Materiały Mikroporowate I Lite Do Zastosowań Medycznych I Stomatologicznych*; Open Access Library, VII(1); Dobrzański, L.A., Dobrzańska-Danikiewicz, A.D., Eds.; International OCSCO World Press: Gliwice, Poland, 2017; pp. 434–499.

330. Dobrzański, L.A.; Dobrzańska-Danikiewicz, A.D.; Malara, P.; Achtelik-Franczak, A.; Dobrzański, L.B.; Gaweł, T.G. Implanto-Skafold Lub Proteza Elementów Anatomicznych Układu Stomatognatycznego Oraz Twarzoczaszki. Patent nr PL229148, 9 January 2018.

331. Dobrzański, L.A.; Dobrzańska-Danikiewicz, A.D.; Malara, P.; Achtelik-Franczak, A.; Dobrzański, L.B.; Gaweł, T.G. Implanto-Skafold Kostny. Patent nr PL229149, 9 January 2018.

332. Dobrzański, L.A.; Dobrzańska-Danikiewicz, A.D.; Malara, P.; Dobrzański, L.B.; Achtelik-Franczak, A.; Kremzer, M. Sposób wytwarzania materiałów kompozytowych o mikroporowatej szkieletowej strukturze wzmocnienia. Patent nr PL236090, 16 December 2019.

333. Dobrzański, L.B.; Dobrzański, L.A.; Dobrzańska, J.; Achtelik-Franczak, A. Wysokorozwinięty Powierzchniowo Implant Stomatologiczny. Zgłoszenie patentowe P.434312, 15 June 2020.

334. Dobrzański, L.B.; Dobrzański, L.A.; Dobrzańska, J.; Achtelik-Franczak, A. Wysokorozwinięty Powierzchniowo Implant Kostny W Tym Twarzoczaszki. Zgłoszenie patentowe P.434314, 15 June 2020.

335. Dobrzańska, J. Analiza Szczelności Wypełnień Kanałów Korzeniowych. Ph.D. Thesis, Śląski Uniwersytet Medyczny, Wydział Lekarski z Oddziałem Lekarsko-Dentystycznym, Zabrze, Poland, 2012.

336. Dobrzańska, J.; Gołombek, K.; Dobrzański, L.B. Polymer materials used in endodontic treatment—In vitro testing. *Arch. Mater. Sci. Eng.* **2012**, *58*, 110–115.

337. Dobrzański, L.B.; Achtelik-Franczak, A.; Dobrzańska, J.; Pietrucha, P. Application of polymer impression masses for the obtaining of dental working models for the stereolithographic 3D printing. *Arch. Mater. Sci. Eng.* **2019**, *95*, 31–40. [CrossRef]

338. Dobrzański, L.B.; Malara, P. Metodologia komputerowo wspomaganego projektowania i wytwarzania stomatologicznych uzupełnień protetycznych z litych materiałów inżynierskich. In *Metalowe Materiały Mikroporowate I Lite Do Zastosowań Medycznych I Stomatologicznych*; Open Access Library, VII(1); Dobrzański, L.A., Dobrzańska-Danikiewicz, A.D., Eds.; International OCSCO World Press: Gliwice, Poland, 2017; pp. 500–534.

339. Dobrzański, L.B.; Dobrzański, L.A.; Dobrzańska, J.; Achtelik-Franczak, A.; Rudziarczyk, K. Akcesorium Montażowe Implantu Stomatologicznego. Zgłoszenie patentowe P.434313, 15 June 2020.

340. Kaiser, L.R. The future of multihospital systems. *Top. Health Care Financ.* **1992**, *18*, 32–45. [PubMed]

341. Arsiwala, A.; Desai, P.; Patravale, V. Recent advances in micro/nanoscale biomedical implants. *J. Control. Release* **2014**, *189*, 25–45. [CrossRef]

342. Yang, F.; Williams, C.G.; Wang, D.A.; Lee, H.; Manson, P.N.; Elisseeff, J. The effect of incorporating RGD adhesive peptide in polyethylene glycol diacrylate hydrogel on osteogenesis of bone marrow stromal cells. *Biomaterials* **2005**, *26*, 5991–5998. [CrossRef]

343. US National Institutes of Health. Regenerative Medicine 2006. Available online: https://stemcells.nih.gov/sites/all/themes/stemcells_theme/stemcell_includes/Regenerative_Medicine_2006.pdf (accessed on 1 March 2021).

344. Cogle, C.R.; Guthrie, S.M.; Sanders, R.C.; Allen, W.L.; Scott, E.W.; Petersen, B.E. An overview of stem cell research and regulatory issues. *Mayo Clin. Proc.* **2003**, *78*, 993–1003. [CrossRef]

345. Metallo, C.M.; Azarin, S.M.; Ji, L.; De Pablo, J.J.; Palecek, S.P. Engineering tissue from human embryonic stem cells. *J. Cell Mol. Med.* **2008**, *12*, 709–729. [CrossRef] [PubMed]

346. Placzek, M.R.; Chung, I.M.; Macedo, H.M.; Ismail, S.; Mortera Blanco, T.; Lim, M.; Cha, J.M.; Fauzi, I.; Kang, Y.; Yeo, D.C.; et al. Stem cell bioprocessing: Fundamentals and principles. *J. R Soc. Interface* **2009**, *6*, 209–232. [CrossRef] [PubMed]

347. Langer, R.; Vacanti, J.P. Tissue engineering. *Science* **1993**, *260*, 920–926. [CrossRef]

348. Viola, J.; Lal, B.; Grad, O. The Emergence of Tissue Engineering as a Research Field. Available online: https://www.nsf.gov/pubs/2004/nsf0450/start.htm (accessed on 1 March 2021).

349. Fung, Y.C. *A Proposal to the National Science Foundation for An Engineering Research Center at UCSD: Center for the Engineering of Living Tissues. UCSD #865023*; University of California: San Diego, CA, USA, 2001.

350. Lanza, R.P.; Langer, R.; Vacanti, J. (Eds.) *Principles of Tissue Engineering*; Academic Press: San Diego, CA, USA, 2000.

351. Atala, A.; Lanza, R.P. (Eds.) *Methods of Tissue Engineering*; Academic Press: San Diego, CA, USA, 2002.

352. MacArthur, B.D.; Oreffo, R.O. Bridging the gap. *Nature* **2005**, *433*, 19. [CrossRef] [PubMed]

353. Tavassoli, M.; Crosby, W.H. Transplantation of marrow to extramedullary sites. *Science* **1968**, *161*, 54–56. [CrossRef] [PubMed]

354. Caplan, A.I. Mesenchymal stem cells. *J. Orthop. Res.* **1991**, *9*, 641–650. [CrossRef]

355. Pittenger, M.F.; Mackay, A.M.; Beck, S.C.; Jaiswal, R.K.; Douglas, R.; Mosca, J.D.; Moorman, M.A.; Simonetti, D.W.; Craig, S.; Marshak, D.R. Multilineage potential of adult human mesenchymal stem cells. *Science* **1999**, *284*, 143–147. [CrossRef] [PubMed]

356. Culme-Seymour, E.J.; Davie, N.L.; Brindley, D.A.; Edwards-Parton, S.; Mason, C. A decade of cell therapy clinical trials (2000–2010). *Regen Med.* **2012**, *7*, 455–462. [CrossRef]

357. Trounson, A.; Thakar, R.G.; Lomax, G.; Gibbons, D. Clinical trials for stem cell therapies. *BMC Med.* **2011**, *9*, 52. [CrossRef] [PubMed]

358. Thomson, J.A.; Itskovitz-Eldor, J.; Shapiro, S.S.; Waknitz, M.A.; Swiergiel, J.J.; Marshall, V.S.; Jones, J.M. Embryonic stem cell lines derived from human blastocysts. *Science* **1998**, *282*, 1145–1147, Erratum in *Science* **1998**, *282*, 1827. [CrossRef] [PubMed]

359. Shamblott, M.J.; Axelman, J.; Wang, S.; Bugg, E.M.; Littlefield, J.W.; Donovan, P.J.; Blumenthal, P.D.; Huggins, G.R.; Gearhart, J.D. Derivation of pluripotent stem cells from cultured human primordial germ cells. *Proc. Natl. Acad. Sci. USA* **1998**, *95*, 13726–13731. [CrossRef] [PubMed]

360. Cetrulo, C.L., Jr. Cord-blood mesenchymal stem cells and tissue engineering. *Stem. Cell Rev.* **2006**, *2*, 163–168. [CrossRef] [PubMed]

361. Branch, M.J.; Hashmani, K.; Dhillon, P.; Jones, D.R.; Dua, H.S.; Hopkinson, A. Mesenchymal stem cells in the human corneal limbal stroma. *Invest. Ophthalmol. Vis. Sci.* **2012**, *53*, 5109–5116. [CrossRef] [PubMed]

362. Americord. What Differentiates This Fast-Growth Cord Blood Bank? Available online: https://bioinformant.com/americord-what-differentiates-this-fast-growth-company-within-the-cord-blood-marketplace/ (accessed on 1 March 2021).

363. Lee, E.H.; Hui, J.H. The potential of stem cells in orthopaedic surgery. *J. Bone Jt. Surg. Br.* **2006**, *88*, 841–851. [CrossRef] [PubMed]

364. Le Blanc, K.; Tammik, L.; Sundberg, B.; Haynesworth, S.E.; Ringdén, O. Mesenchymal stem cells inhibit and stimulate mixed lymphocyte cultures and mitogenic responses independently of the major histocompatibility complex. *Scand. J. Immunol.* **2003**, *57*, 11–20. [CrossRef] [PubMed]

365. Maitra, B.; Szekely, E.; Gjini, K.; Laughlin, M.J.; Dennis, J.; Haynesworth, S.E.; Koç, O.N. Human mesenchymal stem cells support unrelated donor hematopoietic stem cells and suppress T-cell activation. *Bone Marrow Transpl.* **2004**, *33*, 597–604. [CrossRef] [PubMed]

366. Aggarwal, S.; Pittenger, M.F. Human mesenchymal stem cells modulate allogeneic immune cell responses. *Blood* **2005**, *105*, 1815–1822. [CrossRef]

367. Raff, M. Adult stem cell plasticity: Fact or artifact? *Annu. Rev. Cell Dev. Biol.* **2003**, *19*, 1–22. [CrossRef]

368. Morrison, S.J.; Weissman, I.L. The long-term repopulating subset of hematopoietic stem cells is deterministic and isolatable by phenotype. *Immunity* **1994**, *1*, 661–673. [CrossRef]

369. Bajada, S.; Mazakova, I.; Richardson, J.B.; Ashammakhi, N. Updates on stem cells and their applications in regenerative medicine. *J. Tissue Eng. Regen Med.* **2008**, *2*, 169–183. [CrossRef] [PubMed]

370. Johal, K.S.; Lees, V.C.; Reid, A.J. Adipose-derived stem cells: Selecting for translational success. *Regen Med.* **2015**, *10*, 79–96. [CrossRef]

371. Helmlinger, G.; Yuan, F.; Dellian, M.; Jain, R.K. Interstitial pH and pO2 gradients in solid tumors in vivo: High-resolution measurements reveal a lack of correlation. *Nat. Med.* **1997**, *3*, 177–182. [CrossRef]

372. Folkman, J. Tumor Angiogenesis: Therapeutic Implications. *N. Engl. J. Med.* **1971**, *285*, 1182–1186. [CrossRef]

373. Mooney, D.J.; Organ, G.; Vacanti, J.P.; Langer, R. Design and fabrication of biodegradable polymer devices to engineer tubular tissues. *Cell Transpl.* **1994**, *3*, 203–210. [CrossRef] [PubMed]

374. Li, G.; Virdi, A.S.; Ashhurst, D.E.; Simpson, A.H.; Triffitt, J.T. Tissues formed during distraction osteogenesis in the rabbit are determined by the distraction rate: Localization of the cells that express the mRNAs and the distribution of types I and II collagens. *Cell Biol. Int.* **2000**, *24*, 25–33. [CrossRef]

375. Sanders, J.E.; Malcolm, S.G.; Bale, S.D.; Wang, Y.N.; Lamont, S. Prevascularization of a biomaterial using a chorioallontoic membrane. *Microvasc. Res.* **2002**, *64*, 174–178. [CrossRef]

376. Muschler, G.F.; Nakamoto, C.; Griffith, L.G. Engineering principles of clinical cell-based tissue engineering. *J. Bone Jt. Surg. Am.* **2004**, *86*, 1541–1558. [CrossRef]

377. Wilson, C.E.; Dhert, W.J.A.; van Blitterswijk, C.A.; Verbout, A.J.; de Bruijn, J.D. Evaluating 3D bone tissue engineered constructs with different seeding densities using the alamarBlue™ assay and the effect on in vivo bone formation. *J. Mater. Sci. Mater. Med.* **2002**, *13*, 1265–1269. [CrossRef] [PubMed]

378. Kruyt, M.C.; de Bruijn, J.D.; Wilson, C.E.; Oner, F.C.; van Blitterswijk, C.A.; Verbout, A.J.; Dhert, W.J. Viable osteogenic cells are obligatory for tissue-engineered ectopic bone formation in goats. *Tissue Eng.* **2003**, *9*, 327–336. [CrossRef]

379. Peppas, N.A.; Langer, R. New challenges in biomaterials. *Science* **1994**, *263*, 1715–1720. [CrossRef] [PubMed]

380. Hubbell, J.A. Biomaterials in tissue engineering. *Biotechnnology* **1995**, *13*, 565–576. [CrossRef] [PubMed]

381. Hubbell, J.A. Bioactive biomaterials. *Curr. Opin. Biotechnol.* **1999**, *10*, 123–129. [CrossRef]

382. Healy, K.E.; Rezania, A.; Stile, R.A. Designing biomaterials to direct biological responses. *Ann. N. Y Acad. Sci.* **1999**, *875*, 24–35. [CrossRef] [PubMed]

383. Langer, R.; Tirrell, D.A. Designing materials for biology and medicine. *Nature* **2004**, *428*, 487–492. [CrossRef] [PubMed]

384. Lutolf, M.P.; Hubbell, J.A. Synthetic biomaterials as instructive extracellular microenvironments for morphogenesis in tissue engineering. *Nat. Biotechnol.* **2005**, *23*, 47–55. [CrossRef]

385. Elisseeff, J.; Ferran, A.; Hwang, S.; Varghese, S.; Zhang, Z. The role of biomaterials in stem cell differentiation: Applications in the musculoskeletal system. *Stem. Cells Dev.* **2006**, *15*, 295–303. [CrossRef] [PubMed]

386. Hwang, N.S.; Kim, M.S.; Sampattavanich, S.; Baek, J.H.; Zhang, Z.; Elisseeff, J. Effects of three-dimensional culture and growth factors on the chondrogenic differentiation of murine embryonic stem cells. *Stem. Cells* **2006**, *24*, 284–291. [CrossRef] [PubMed]

387. Yang, F.; Neeley, W.L.; Moore, M.J.; Karp, J.M.; Shukla, A.; Langer, R. Tissue Engineering: The Therapeutic Strategy of the Twenty-First Century. In *Nanotechnology and Tissue Engineering: The Scaffold*; Laurencin, C.T., Nair, L.S., Eds.; CRC Press Taylor & Francis Group: Boca Raton, FL, USA, 2008; pp. 3–32.

388. Bettinger, C.J.; Borenstein, J.T.; Langer, R. Microfabrication Techniques in Scaffold Development. In *Nanotechnology and Tissue Engineering: The Scaffold*; Laurencin, C.T., Nair, L.S., Eds.; CRC Press Taylor & Francis Group: Boca Raton, FL, USA, 2008; pp. 87–122.

389. Noga, M.; Pawlak, A.; Dybala, B.; Dabrowski, B.; Swieszkowski, W.; Lewandowska-Szumiel, M. Biological Evaluation of Porous Titanium Scaffolds (Ti-6Al-7Nb) with HAp/Ca-P Surface Seeded with Human Adipose Derived Stem Cells. In Proceedings of the E-MRS Fall Meeting, Warszawa, Poland, 16–20 September 2013.

390. Rouwkema, J.; Rivron, N.C.; van Blitterswijk, C.A. Vascularization in tissue engineering. *Trends Biotechnol.* **2008**, *26*, 434–441. [CrossRef]

391. Bramfeldt, H.; Sabra, G.; Centis, V.; Vermette, P. Scaffold vascularization: A challenge for threedimensional tissue engineering. *Curr. Med. Chem.* **2010**, *17*, 3944–3967. [CrossRef] [PubMed]

392. Jain, R.K.; Au, P.; Tam, J.; Duda, D.G.; Fukumura, D. Engineering vascularized tissue. *Nat. Biotechnol.* **2005**, *23*, 821–823. [CrossRef] [PubMed]

393. Bose, S.; Roy, M.; Bandyopadhyay, A. Recent advances in bone tissue engineering scaffolds. *Trends Biotechnol.* **2012**, *30*, 546–554. [CrossRef] [PubMed]

394. Biomaterials Market Size, Share & Trends Analysis Report by Product (Natural, Metallic, Polymer), by Application (Cardiovascular, Orthopedics, Plastic Surgery), by Region, and Segment Forecasts, 2020–2027. Available online: https://www.grandviewresearch.com/industry-analysis/biomaterials-industry (accessed on 12 March 2021).

395. Dental Biomaterial Market—Forecasts from 2020 to 2025. Available online: https://www.researchandmarkets.com/reports/5125071/dental-biomaterial-market-forecasts-from-2020#rela0-4989738 (accessed on 12 March 2021).

396. Orthopedic Biomaterials Market Size, Share & Trends Analysis Report by Material Type (Ceramics & Bioactive Glasses, Polymers), by Application (Orthobiologics, Orthopedic Implants), by Region, and Segment Forecasts, 2019–2025. Available online: https://www.grandviewresearch.com/industry-analysis/orthopedic-biomaterials-market (accessed on 12 March 2021).

397. Tissue Engineering Market Size, Share & Trends Analysis Report by Application (Cord Blood & Cell Banking, Cancer, GI & Gynecology, Dental, Orthopedics, Musculoskeletal, & Spine), by Region, and Segment Forecasts, 2020–2027. Available online: https://www.grandviewresearch.com/industry-analysis/tissue-engineering-and-regeneration-industry (accessed on 12 March 2021).

398. Regenerative Medicine Market Size, Share & Trends Analysis by Product (Primary Cell-based, Stem & Progenitor Cell-based), by Therapeutic Category (Dermatology, Oncology) and Segment Forecasts, 2019–2025. Available online: https://www.grandviewresearch.com/industry-analysis/regenerative-medicine-market (accessed on 12 March 2021).

399. Cell Therapy Market Size, Share & Trends Analysis Report by Use-type (Research, Commercialized, Musculoskeletal Disorders), by Therapy Type (Autologous, Allogeneic), by Region, and Segment Forecasts, 2020–2027. Available online: https://www.grandviewresearch.com/industry-analysis/cell-therapy-market (accessed on 12 March 2021).

400. Major Orthopedic Joint Replacement Implants Market—By Type (Knee Replacement Implants and Hip Replacement Implants), by Geography, and By Region, Opportunities And Strategies—Global Forecast To 2022. Available online: https://www.thebusinessresearchcompany.com/report/major-orthopedic-joint-replacement-implants-market (accessed on 12 March 2021).

401. Illgen, R.; Bukowski, B.; Abiola, R.; Anderson, P.; Chughtai, M.; Khlopas, A.; Mont, M. Robotic-assisted total hip arthroplasty: Outcomes at minimum two year follow up. *Surg. Technol. Int.* **2017**, *30*, 365–372.

402. Kayani, B.; Konan, S.; Tahmassebi, J.; Pietrzak, J.R.T.; Haddad, F.S. Robotic-arm assisted total knee arthroplasty is associated with improved early functional recovery and reduced time to hospital discharge compared with conventional jig-based total knee arthroplasty: A prospective cohort study. *Bone Jt. J.* **2018**, *100*, 930–937. [CrossRef] [PubMed]

403. Kleeblad, L.J.; Borus, T.; Coon, T.; Dounchis, J.; Nguyen, J.; Pearle, A. Midterm survivorship and patient satisfaction of robotic-arm assisted medial unicompartmental knee arthroplasty: A multicenter study. *J. Arthroplast.* **2018**, *33*, 1719–1726. [CrossRef]

404. Dretakis, K.; Igoumenou, V.G. Outcomes of robotic-arm-assisted medial unicompartmental knee arthroplasty: Minimum 3-year follow-up. *Eur. J. Orthop. Surg. Traumatol.* **2019**, *29*, 1305–1311. [CrossRef]

405. Nawabi, D.; Conditt, M.; Ranawat, A.; Dunbar, N.; Jones, J.; Banks, S.; Padgett, D. Haptically guided robotic technology in total hip arthroplasty: A cadaveric investigation. *Proc. Inst. Mech. Eng. H* **2013**, *227*, 302–309. [CrossRef]

406. Suarez-Ahedo, C.; Gui, C.; Martin, T.; Chandrasekaran, S.; Domb, B. Robotic arm assisted total hip arthoplasty results in smaller acetabular cup size in relation to the femoral head size: A Matched-Pair Controlled Study. *Hip. Int.* **2017**, *27*, 147–152. [CrossRef]

407. Kayani, B.; Konan, S.; Pietrzak, J.R.T.; Haddad, F.S. Iatrogenic Bone and Soft Tissue Trauma in Robotic-Arm Assisted Total Knee Arthroplasty Compared with Conventional Jig-Based Total Knee Arthroplasty: A Prospective Cohort Study and Validation of a New Classification System. *J. Arthroplast.* **2018**, *33*, 2496–2501. [CrossRef]

408. Hampp, E.; Chang, T.C.; Pearle, A. Robotic partial knee arthroplasty demonstrated greater bone preservation compared to robotic total knee arthroplasty. In Proceedings of the Annual Orthopaedic Research Society Meeting, Austin, TX, USA, 2–5 February 2019. [CrossRef]

409. Kayani, B.; Konan, S.; Tahmassebi, J.; Rowan, F.E.; Haddad, F.S. An assessment of early functional rehabilitation and hospital discharge in conventional versus robotic-arm assisted unicompartmental knee arthroplasty. *Bone Jt. J.* **2019**, *101-B*, 24–33. [CrossRef]

410. Bell, S.W.; Anthony, I.; Jones, B.; MacLean, A.; Rowe, P.; Blyth, M. Improved Accuracy of Component Positioning with Robotic-Assisted Unicompartmental Knee Arthroplasty: Data from a Prospective, Randomized Controlled Study. *J. Bone Jt. Surg. Am.* **2016**, *98*, 627–635. [CrossRef] [PubMed]

411. Herry, Y.; Batailler, C.; Lording, T.; Servien, E.; Neyret, P.; Lustig, S. Improved joint-line restitution in unicompartmental knee arthroplasty using a robotic-assisted surgical technique. *Int. Orthop.* **2017**, *41*, 2265–2271. [CrossRef] [PubMed]

412. Batailler, C.; White, N.; Ranaldi, F.M.; Neyret, P.; Servien, E.; Lustig, S. Improved implant position and lower revision rate with robotic-assisted unicompartmental knee arthroplasty. *Knee Surg. Sports Traumatol. Arthrosc.* **2019**, *27*, 1232–1240. [CrossRef] [PubMed]

413. Jaramaz, B.; Nikou, C.; Casper, M.; Grosse, S.; Mitra, R. Accuracy validation of semi-active robotic application for patellofemoral arthroplasty. In Proceedings of the International Society for Computer Assisted Orthopaedic Surgery, Vancouver, BC, Canada, 17–20 June 2015.

414. Jaramaz, B.; Mitra, R.; Nikou, C.; Kung, C. Technique and Accuracy Assessment of a Novel Image-Free Handheld Robot for knee Arthroplasty in Bi-Cruiciate Retaining Total Knee Replacement. In *Proceedings of the CAOS 2018. The 18th Annual Meeting of the International Society for Computer Assisted Orthopaedic Surgery, Beijing, China, 6–9 June 2018*; Tian, W., Baena, F.R.Y., Eds.; EasyChair: Oxford, UK, 2018; Volume 2, pp. 98–101.

415. *Data on File Smith & Nephew. Sg2 Healthcare Intelligence: Technology Guide*; Sg2: Skokie, IL, USA, 2014.

416. Gregori, A.; Picard, F.; Bellemans, J.; Smith, J.; Simone, A. Handheld precision sculpting tool for unicondylar knee arthroplasty. A clinical review. In Proceedings of the 15th EFORT Congress, London, UK, 4–6 June 2014.

417. Smith, J.R.; Picard, F.; Lonner, J.; Hamlin, B.; Rowe, P.; Riches, P.; Deakin, A. The accuracy of a robotically-controlled freehand sculpting tool for unicondylar knee arthroplasty. In Proceedings of the Congress of the International Society of Biomechanics, Natal, Brazil, 4–9 August 2013.

418. Gustke, K.; Golladay, G.; Roche, M.W.; Jerry, G.; Elson, L.C.; Anderson, C.R. Increased Patient Satisfaction After Total Knee replacement using sensor-guided technology. *Bone Jt. J.* **2014**, *96*, 1333–1338. [CrossRef] [PubMed]

419. Gregori, A.; Picard, F.; Lonner, J.; Smith, J.; Jaramaz, B. Accuracy of imageless robotically assisted unicondylar knee arthroplasty. In Proceedings of the International Society for Computer Assisted Orthopaedic Surgery, Vancouver, BC, Canada, 17–20 June 2015.

420. 3D Printing in the Medical Industry: The 3D Printed Knee Replacement. Available online: https://www.sculpteo.com/blog/2018/05/04/3d-printing-in-the-medical-industry-the-3d-printed-knee-replacement/ (accessed on 12 March 2021).

421. How 3D Printing and Modeling are Changing Joint Replacement Surgery. Available online: https://www.yalemedicine.org/news/3d-joint-replacement (accessed on 12 March 2021).

422. Replacement Knee Surgery: 3D Printing … Science Fiction or Remarkable Reality? Available online: https://www.bennettorthosportsmed.com/replacement-knee-surgery-3d-printingscience-fiction-remarkable-reality/ (accessed on 12 March 2021).

423. Knee Replacement Using State-Of-The-Art 3D Printer Technology. Available online: https://www.fairfield.org.uk/knee-replacement-using-state-of-the-art-3d-printer-technology/ (accessed on 12 March 2021).

424. Hip Replacement Market Size, Share & Industry Analysis, by Procedure (Total Hip Replacement, Partial Hip Replacement, and Revision & Hip Resurfacing), by End User (Hospitals & Ambulatory Surgery Centers, Orthopedic Clinics, and Others) and Regional Forecast, 2019–2026. Available online: https://www.fortunebusinessinsights.com/industry-reports/hip-replacement-implants-market-100247 (accessed on 12 March 2021).

425. Hip Replacement Implants Market Size, Share, & Trends Analysis Report by Product (Total Hip, Partial Femoral Head), By Application, (MOM, MOP, COP), By End Use (Orthopedic Clinics), By Region, And Segment Forecasts, 2019–2026. Available online: https://www.grandviewresearch.com/industry-analysis/hip-replacement-implants-market (accessed on 12 March 2021).

426. Global Hip Replacement Implants Market Size, Market Share, Application Analysis, Regional Outlook, Growth Trends, Key Players, Competitive Strategies and Forecasts, 2018 to 2026. Available online: https://www.researchandmarkets.com/reports/4760571/global-hip-replacement-implants-market-size (accessed on 12 March 2021).

427. Hip Replacement Implants Market Growth Projection, Sales Statistics, Size Value, Latest Trends, Future Insights and Share Estimation by 2025. Available online: https://www.medgadget.com/2021/02/hip-replacement-implants-market-growth-projection-sales-statistics-size-value-latest-trends-future-insights-and-share-estimation-by-2025.html (accessed on 12 March 2021).

428. Conformis Announces the first 3D Total Hip Replacement Surgeries performed at JFK Medical Center in Florida. Available online: https://www.globenewswire.com/news-release/2018/08/01/1545816/0/en/Conformis-Announces-the-first-3D-Total-Hip-Replacement-Surgeries-performed-at-JFK-Medical-Center-in-Florida.html (accessed on 12 March 2021).

429. Sasazawa, F. Case Study: The Simple and Accurate Cup Setting with Portable Navigation System in Total Hip Arthroplasty. Available online: http://www.orthalign.com/wp-content/uploads/2018/08/Case-STudy-2017_HipAlign_Sasazawa.pdf (accessed on 12 March 2021).

430. Durbhakula, S. Case Study: OrthAlign®Technology Was Imperative for This Patient with a Prior Spinal Fusion Undergoing Right THA. Available online: http://www.orthalign.com/wp-content/uploads/2018/04/HipAlign_case-study-durbhakula3.pdf (accessed on 12 March 2021).

431. Knee Replacement Implants Market Size, Share & Analysis, by Procedure Type (Total Knee Replacement, Partial Knee Replacement, Revision Knee), by Material, by Component, by End Users (Clinic, Hospital, Ambulatory Service Centers), and Segment Forecasts to 2027. Available online: https://www.reportsanddata.com/report-detail/knee-replacement-implants-market (accessed on 12 March 2021).

432. Knee Replacement Market Size, Share & COVID-19 Impact Analysis, by Procedure (Total Knee Arthroplasty, Partial Knee Arthroplasty, and Revision Arthroplasty) by Implant Type (Fixed Bearing, Mobile Bearing, and Others) End-user (Hospitals, Orthopedic Clinics, Ambulatory Surgical Centers, and Others) and Regional Forecast, 2020–2027. Available online: https://www.fortunebusinessinsights.com/industry-reports/knee-replacement-implants-market-101244 (accessed on 12 March 2021).

433. Knee Implants Market Forecast to 2027 - Covid-19 Impact and Global Analysis - by Procedure (Total Knee Replacement, Partial Knee Replacement and Revision Knee Replacement), Implant Type (Fixed Bearing Implant and Mobile Bearing Implant), Material (Stainless Steel, Cobalt & Chromium Alloys, Titanium & Titanium Alloys, Tantalum, Zirconium and Others) and By End User (Hospitals, Ambulatory Surgical Centers and Others). Available online: https://www.theinsightpartners.com/reports/knee-implants-market (accessed on 12 March 2021).

434. Global Knee Replacement Market Research Report: Information by Product Type (Total Knee Reconstructive Implants, Partial Knee Reconstructive Implants, Revision Knee Reconstructive Implants), Material (Metal Alloy, Ceramic Material, Strong Plastic Parts and Others), End User (Hospitals and Clinics, Specialty Centers, and Rehabilitation Centers), and Region (Americas, Europe, Asia-Pacific, and the Middle East & Africa)—Forecast till 2025. Available online: https://www.marketresearchfuture.com/reports/knee-replacement-market-1578 (accessed on 12 March 2021).

435. Aesculapius with His Sons Podalairius and Machaon and Three Daughters—With Supplicants. Greek Relief Found at Thyrea. at National Museum Athens. Available online: https://wellcomecollection.org/works/ekhkh79b (accessed on 12 March 2021).

436. Filippou, D.; Tsoucalas, G.; Panagouli, E.; Thomaidis, V.; Fiska, A. Machaon, Son of Asclepius, the Father of Surgery. *Cureus* **2020**, *12*, e7038. [CrossRef]

437. Oxilia, G.; Peresani, M.; Romandini, M.; Matteucci, C.; Debono Spiteri, C.; Henry, A.G.; Schulz, D.; Archer, W.; Crezzini, J.; Boschin, F.; et al. Earliest evidence of dental caries manipulation in the Late Upper Palaeolithic. *Sci. Rep.* **2015**, *5*, 12150. [CrossRef]

438. Oxyrhynchus Papyrus 2547. Available online: http://archives.wellcomelibrary.org/DServe/dserve.exe?dsqIni=Dserve.ini&dsqApp=Archive&dsqCmd=Show.tcl&dsqDb=Catalog&dsqPos=47&dsqSearch=%28Sources_guides_used%3D%27Near%20and%20Middle%20East%27%29 (accessed on 12 March 2021).

439. Barns, J.W.B. *The Oxyrhynchus Papyri*; Part 31; Cambridge University Press: London, UK, 1966; pp. 62–65.

440. Nutton, V. Marie-Helene Marganne, Inventaire analytique des papyrus grecs de médecine, Geneva, Librairie Droz, 1981, 8vo, pp. x, 409. *Med. Hist.* **1983**, *27*, 97. [CrossRef]

441. Miranda, E.A. Mondino de Luzzi. Medical Terminology Dictionary; Clinical Anatomy Associates: Ohio. Available online: https://clinicalanatomy.com/mtd/548 (accessed on 12 March 2021).

442. Nierzwicki, K. Warszawski egzemplarz De humani corporis fabrica Andreasa Vesaliusa (Bazylea 1555) ze zbiorów Biblioteki Narodowej. Przyczynek do dziejów recepcji anatomii wesaliańskiej w Polsce. Available online: https://repozytorium.umk.pl/bitstream/handle/item/3030/Warszawski%20egzemplarz%20A.%20Vesaliusa%20%E2%80%94%20ksi%C4%99ga%20jubileuszowa%20%E2%80%94%20z%20ilustracjami.pdf?sequence=1 (accessed on 12 March 2021).

443. Kleczek, K. De humani corporis fabrica libri septem. Najsłynniejsze XVI-wieczne dzieło o anatomii człowieka. Available online: https://kc-cieszyn.pl/de-humani-corporis-fabrica-libri-septem-najslynniejsze-xvi-wieczne-dzielo-o-anatomii-czlowieka/ (accessed on 12 March 2021).

444. Szkice Anatomiczne. Leonardo da Vinci. SERIA I. Available online: https://manuscriptum.pl/store/product/szkice-anatomiczne-leonardo-da-vinci-seria-i (accessed on 12 March 2021).
445. Landrus, M. *The Treasures of Leonardo da Vinci: The Story of His Life & Work*; Andre Deutsch: London, UK, 2014.
446. Lugli, E. In cerca della perfezione: Nuovi elementi per l'Uomo vitruviano di Leonardo Da Vinci. In *Leonardo e Vitruvio: Oltre Il Cerchio E Il Quadrato*; Borgostr, F., Ed.; Marsilio Books: Venice, Italy, 2019; pp. 69–91.
447. Press Release: The Nobel Prize in Physiology or Medicine 2020. Available online: https://www.nobelprize.org/prizes/medicine/2020/press-release/ (accessed on 12 March 2021).

Review

The Concept of Sustainable Development of Modern Dentistry

Leszek A. Dobrzański [1,*], **Lech B. Dobrzański** [1], **Anna D. Dobrzańska-Danikiewicz** [2] and **Joanna Dobrzańska** [1]

[1] Medical and Dental Engineering Centre for Research, Design and Production ASKLEPIOS, 13 D Królowej Bony St., 44-100 Gliwice, Poland; dobrzanski@centrumasklepios.pl (L.B.D.); joanna.dobrzanska@centrumasklepios.pl (J.D.)

[2] Department of Mechanical Engineering, University of Zielona Góra, 4 Prof. Z. Szafrana St., 65-516 Zielona Góra, Poland; anna.dobrzanska.danikiewicz@gmail.com

* Correspondence: leszek.dobrzanski@centrumasklepios.pl

Received: 8 September 2020; Accepted: 2 December 2020; Published: 6 December 2020

Abstract: This paper concerns the assessment of the current state of dentistry in the world and the prospects of its sustainable development. A traditional Chinese censer was adopted as the pattern, with a strong and stable support on three legs. The dominant diseases of the oral cavity are caries and periodontal diseases, with the inevitable consequence of toothlessness. From the caries 3.5–5 billion people suffer. Moreover, each of these diseases has a wide influence on the development of systemic complications. The territorial range of these diseases and their significant differentiation in severity in different countries and their impact on disability-adjusted life years index are presented (DALY). Edentulousness has a significant impact on the oral health-related quality of life (OHRQoL). The etiology of these diseases is presented, as well as the preventive and therapeutic strategies undertaken as a result of modifying the Deming circle through the fives' rules idea. The state of development of Dentistry 4.0 is an element of the current stage of the industrial revolution Industry 4.0 and the great achievements of modern dental engineering. Dental treatment examples from the authors' own clinical practice are given. The systemic safety of a huge number of dentists in the world is discussed, in place of the passive strategy of using more and more advanced personal protective equipment (PPE), introducing our own strategy for the active prevention of the spread of pathogenic microorganisms, including SARS-CoV-2. The ethical aspects of dentists' activity towards their own patients and the ethical obligations of the dentist community towards society are discussed in detail. This paper is a polemic arguing against the view presented by a group of eminent specialists in the middle of last year in *The Lancet*. It is impossible to disagree with these views when it comes to waiting for egalitarianism in dental care, increasing the scope of prevention and eliminating discrimination in this area on the basis of scarcity and poverty. The views on the discrimination of dentistry in relation to other branches of medicine are far more debatable. Therefore, relevant world statistics for other branches of medicine are presented. The authors of this paper do not agree with the thesis that interventional dental treatment can be replaced with properly implemented prophylaxis. The final remarks, therefore, present a discussion of the prospects for the development of dentistry based on three pillars, analogous to the traditional Chinese censer obtaining a stable balance thanks to its three legs. The Dentistry Sustainable Development (DSD) > 2020 model, consisting of Global Dental Prevention (GDP), Advanced Interventionist Dentistry 4.0 (AID 4.0), and Dentistry Safety System (DSS), is presented.

Keywords: dentistry sustainable development; dental prophylaxis; dental interventionistic treatment; caries; periodontology; toothlessness; endodontics; Dentistry 4.0; dental implantology; dental prosthetics; dentist safety; dentist ethics

1. Introduction

In order to introduce the subject of this paper, several concepts are presented that seemingly do not combine into a whole, but are necessary in order to understand the view and content presented in this paper. It is obvious that the main content of this paper will discuss all issues that relate exclusively to dentistry, but the broad context presented in this introduction is necessary.

The first of these notions is balance. In the Dictionary of the Polish Language, we find a definition of balance as a stable system of opposing forces and values.

> *"Balance appears as the basic nature of all things, no matter what level we pay attention to. It applies to the same degree to elementary particles, the functioning of the organism, social phenomena, or the movement of galaxies. Opposites strive towards each other, they try to complement and complete each other because in themselves they do not actually exist. This process is constantly accompanied by movement, which is a tool of balance. It is because of him that everything that exists is constantly changing and tends to the perfect point that can never be achieved. Because balance is not so much a point, an ideal, but a constant balancing around that point. Balancing unity is the ever-pulsating dance of everything, it is the beating heart of the Cosmos, it is the very breath of God." [1].*

It has been known in statics for centuries that a stable, balanced position is provided by three points of support. Examples of this principle have been in existence for centuries. Figure 1 shows a Chinese censer dating back to a centuries-old tradition made at the turn of the 19th and 20th centuries as a bronze cast in the form of a spherical vase on three animal legs, with mascarons at the base, a belly decorated in relief with two pairs of dragons holding a fireball, ears with dragon heads on the sides, and a meander pattern strip around the top edge. The cover is strongly arched, openwork, with four handles in the form of elephant heads; at the top, there is a figure of a lying Chinese lion. This piece of Chinese art can be considered a symbol of balance.

Figure 1. Chinese censer from the turn of the 19th and 20th centuries with three legs to ensure stability.

Another defined term is sustainable development. It is necessary to distinguish between instantaneous equilibrium and a process based on equilibrium. One such notion is sustainable development.

The sources of this concept are ecological, because Hans Carl von Carlowitz in his book [2], published firstly in 1713, which was a comprehensive treatise on forestry, introduced the concept of sustainable yield. The view outlined by him assumed forest management in such a way that the forest could always rebuild itself, which required planting as many trees as had just been cut down. The concept spread in many European countries and gained the English term Sustained Yield Forestry, and in the 1980s it was spread by environmental movements to finally evolve into sustainable development in a very broad context.

Currently, the content of this notion is determined by the doctrine of economics, assuming the quality of life at the level allowed by the current civilization development provided in the preamble to the World Commission on Environment and Development (WCED) report: "At the current civilization level, sustainable development is possible (...) in which the needs of the present generation can be satisfied without reducing future opportunities generations to satisfy them" [3]. The achieved level of prosperity is possible to maintain, provided that the relationship between economic growth, care for the environment, including the man-made environment, and quality of life, including human health, is appropriately and consciously formed. This document draws on a study by the Club of Rome dating back to 1972 [4]. The basis of all activities in this area is included in the theory of public good [5].

One of the indisputable goods that sustainable development guarantees to people in the world is health, as a necessary condition of well-being, and the concern of all societies is to extend human life. Health, including oral health, is therefore a human right, and inclusive development reduces inequalities, provided by universal access to healthcare, including primary dental care. Improving oral health reduces development burdens and costs to society, health systems, and the economy in general. Oral disease affects approximately 3 to 5 billion people worldwide [6,7]. The proportion of the population receiving oral health care in Low and Middle-Income Countries (LMIC) is generally lower than in High-Income Countries (HIC), with median estimates ranging from 35% in Low-Income Countries (LIC) to 60% in lower-income low-middle-income countries, 75% in upper-income low-middle-income countries, and 82% in high-income countries [7]. Within each country, the poorest quintiles have the lowest service coverage. Oral diseases reduce life chances, diminish the sense of dignity, and reduce the productive contribution of individuals to their local communities and to society as a whole.

The aim of this paper is to analyze the state of modern dentistry and indicate the conditions for ensuring its current balance and sustainable development, which on the one hand must always involve an analysis of the needs, but on the other hand must take into account the possibility of indicating benefits but without exposing anyone to any loss or harm. The dominant diseases of the oral cavity are caries and periodontal diseases. The strategies for treating caries and periodontal diseases require detailed analysis, which is presented in this paper. The territorial range of these diseases and their significant differentiation in severity in different countries as well as their impact on the disability-adjusted life years (DALY) index are presented. Across the world, 3–5 billion people suffer from tooth decay. Caries very often in adults is accompanied by periodontal diseases, which are also very widespread across the world. Toothlessness, an inevitable consequence of both of these diseases, especially if not treated at all and in the case of bad habits, has a significant impact on the oral health-related quality of life (OHRQoL). Oral diseases have a wide impact on other organs, even distantly, and are a direct cause of systemic diseases, which not all people are aware of. This paper was inspired because it is a polemic arguing against the views presented by a group of eminent specialists in the middle of last year in a series of two articles [6,7]. It is impossible to disagree with these views when it comes to waiting for egalitarianism in dental care, increasing the scope of prevention, and eliminating social exclusion in this area due to scarcity and poverty. The views on the discrimination of dentistry as compared to other fields of medicine are much more debatable. Relevant statistical data concerning, inter alia, the mortality from heart disease, stroke, and cancer and human health expenditure, for example, can vary between OECD countries by more than 50 times. This situation is largely shaped by politicians in individual countries and the extent to which individual societies enforce their democratic rights, including the share of health care and social security costs in the gross domestic product (GDP). The most debatable view is that prevention in the field of dentistry is treated as an opposite approach to interventional dentistry and is mutually exclusive with the most advanced stage of Dentistry 4.0. There are many examples of the development of Dentistry 4.0 as great achievements, based on the development of computer imaging methods, computer-aided design, and manufacturing of implants and prosthetic restorations, including the use of cloud computing, automation, and robotization, as well as additive manufacturing technologies determining the advanced state of modern dental engineering. The cooperation between the dentist and the dental engineer plays an important role here, which is a prerequisite for success in this field. The level of modern dentistry is also determined by modern endodontic treatment methods and the possibility of using avant-garde implant scaffolds and implant scaffolds. In the field of periodontitis, an important role is played by the TIPPS strategy (Talk-Instruct-Practise-Plan-Support) [8], which draws on the traditional quality management solutions known as the Deming circle [9]. Therefore, this case requires a very broad and multifaceted analysis, which is also the purpose of this paper. Caries treatment, including endodontic and periodontal treatment, eliminates many diseases of distant organs of the body, as well as the implantological and prosthetic treatment of toothlessness. An analysis of the contexts of these systemic interactions is also one of the purposes of this paper. This significantly relieves hospital treatment and the social insurance system from eliminating long-term illnesses. The authors of this paper do not agree with the thesis that interventional dental treatment can be replaced with properly implemented prevention and, in broadly discussing this problem, indicate that sustainable development is essential in both of these areas. Both are important and require equal effort for the sustainability of dentistry. The third important development pillar of dentistry, and therefore another goal of the analyses presented in this paper, is the systemic epidemiological safety of a huge number of dentists in the world, an issue the current COVID-19 pandemic has highlighted. In place of the passive STOP strategy (system, technical, organizational, personal) used so far, which determines the introduction of more and more advanced personal protective equipment (PPE), assuming that the entire environment around the dentist and the patient are infected, the proprietary system, prevention, efficiency, cause (SPEC) strategy is introduced, consisting of the active prevention of the spread of pathogenic microorganisms, including SARS-CoV-2, in a dental practice environment. The point is to eliminate the threat directly at the source. An example of a proprietary accessory that meets these requirements is presented. In the course of this paper, an attempt is made to prove this thesis, supported by extensive literature studies, showing that the developmental balance of modern dentistry requires three equal elements. The association and the similarity to the case of the Chinese censer, which stands steadily on three legs, is clear. Hence, the aim of this paper is to support the thesis that the stability of dentistry development depends on three thematic areas that determine its even development, with numerous pieces of evidence from the literature. The thesis includes prevention and interventionist treatment in the Dentistry 4.0 system, including, among other things, endodontic and periodontal treatment, as well as an efficient system of epidemiological safety related to the prevention of the spread of pathogenic pathogens at the source. These analyses were preceded by a brief analysis of the condition of oral cavity diseases. The final remarks present a discussion of the prospects for the development of dentistry based on the three above-mentioned pillars, the prototype of which is the presented Chinese censer.

2. Analysis of the Contemporary State of Oral Cavity Diseases in the World

In order to draw any conclusions about the development of dentistry in the coming decades, it is necessary to analyze its current state.

Mature men and women have 32 teeth. Their condition in each case depends on many factors, including genetic conditions; developmental conditions; age; oral hygiene in relation to life history, diet, and nutritional preferences; interactions with other diseases; possible sports; transport's and professional accidents; treatment history and possible errors in this regard; and many other factors (Figure 2) [6].

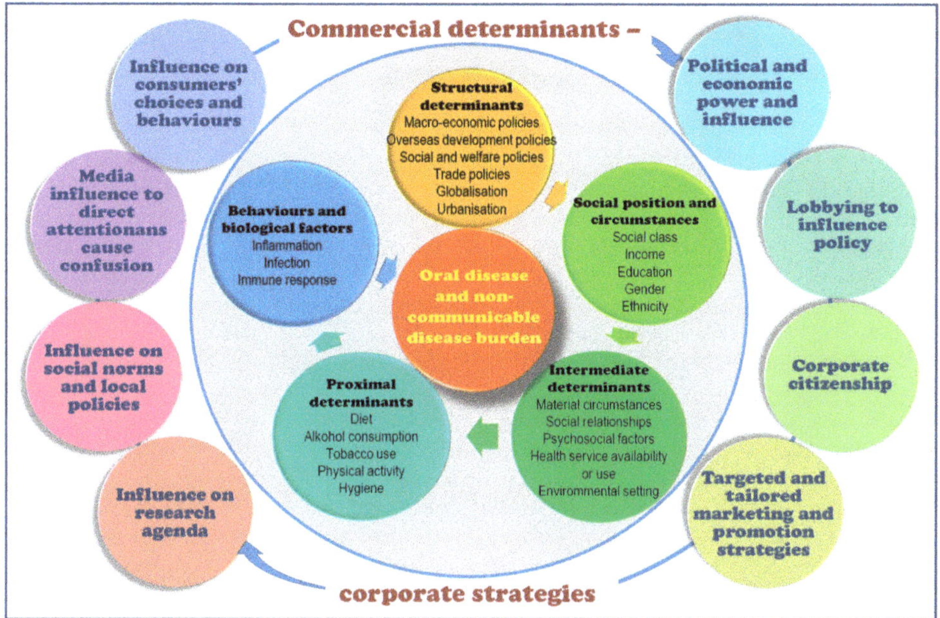

Figure 2. Social and commercial determinants of oral diseases.

The stomatognathic system, especially the dentition, is also subject to numerous disease processes, under the influence both of internal and external factors. Oral diseases affect approximately 3 to 5 billion people worldwide [6,7]. Pulp diseases are common, mainly causing inflammation and pulp necrosis due to bacterial, thermal, mechanical, chemical, and electrical factors, as well as due to infection by blood vessels that penetrate the pulp through the apical foramen as well as due to cavities originated by caries and by non-carious factors and also by iatrogenic agents created during therapeutic procedures. The causes of oral disease, often leading to tooth loss, vary, although microbes are the most common cause of pulpitis. Hence, according to epidemiological data, the most common cause of tooth extraction is caries [10–25]. Carefully performed conservative treatment, especially endodontic treatment, allows for a relatively long time to keep the healed teeth in the mouth. Periodontal disease has also reached epidemic proportions [26–34], although this has only recently been recognized. This is the most common inflammation, affecting almost 50% of adults in the world, and can lead to complete toothlessness if left untreated. Up to 83% of the population develop some degree of bleeding from the gums during their lifetime. Careful hygiene and timely and carefully performed periodontal treatment also allow you to preserve your own teeth for a long time. Both caries and periodontal disease can lead to the extraction of single teeth and even complete toothlessness. Often, it is impossible to avoid tooth extraction, which, although allowing for immediate problem solving, creates further problems. The consequence of extracting even one tooth is an imbalance in the stomatognathic system, which ceases to function efficiently in the absence of any component. Oral neoplasms were excluded from consideration in this paper; although they occur quite frequently, they require separate analysis and usually fall within the scope of oncology or/and craniofacial surgery.

Oral diseases, both caries and periodontitis, as well as toothlessness, which after some time is an inevitable consequence of both of these diseases, are serious problems in modern medicine, not only because of local diseases of the oral cavity but above all because of the high risk of systemic complications.

Caries, as the primary pulp disease in the oral cavity, in the case of delayed, ineffective, or not-performed endodontic treatment is the direct cause of bacterial pneumonia, nephritis, rheumatoid arthritis, osteoporosis, abscess, stroke, endocarditis, ischemic heart disease, the decreased birth mass of infants, premature birth, and even the patient's death [35–40]. Carefully performed endodontic treatment eliminates these risks and allows for a relatively long time to preserve healed teeth in the mouth.

Periodontal disease has also reached epidemic proportions [26–30], although this has only recently been recognized. It are the most common inflammation, affecting almost 50% of adults in the world, and can lead to complete toothlessness if left untreated. Up to 83% of the population develop some degree of bleeding from the gums during their lifetime. Periodontal disease, like tooth decay, is a direct cause of many long-term systemic diseases and complications, including diseases of the kidneys, lungs, cardiovascular system, pancreatic and oral cancer, as well as strokes and dementia [26–30,41]. The group of diseases commonly associated with periodontal diseases include cardiovascular diseases, heart failure, ischemic disease, and hypertension [42–47]. Numerous literature reports indicate erectile dysfunction in men and other complications of the reproductive system [48–74], which may also affect women [75] who are also pregnant [76].

Nevertheless, tooth extraction is often unavoidable. Both caries and periodontitis can even lead to complete toothlessness.

Lack of teeth causes not only a loss of aesthetic value, as many think, but is associated with some discomfort associated with eating difficulties and is also the cause of many diseases throughout the body due to chewing dysfunction or a lack of it [27,77–90]. Toothlessness can often be a direct cause of shortening the human life and is even a predictor of mortality, mainly from cardiovascular causes. Missing teeth cause coronary plaque formation and aortic hardening, increased susceptibility to electrocardiographic abnormalities, heart failure, ischemic heart disease, hypertension, and stroke. Diseases associated with edentulousness include certain cancers, diabetes mellitus, insulin independence, kidney disease, and rheumatoid arthritis. Often, there are diseases of the stomach; gastritis; duodenal ulcer; diseases of the pancreas, including tumors; as well as neoplastic changes in the esophagus and upper gastrointestinal tract. Chewing dysfunction, including it being impossible to grind and chew food, also causes unfavorable results in pregnancy. Mood and neurocognitive disorders, obstructive sleep apnea, cervical spine pain, migraine, and headaches, the direct cause of which are malocclusions due to tooth loss, determining the metabolism of the cerebrospinal fluid, may appear. Malocclusions and masticatory dysfunctions cause functional and morphological changes in the hippocampus in the temporal lobe of the forebrain. They negatively affect the hippocampus in the elderly [91–97] and cause associated spatial and episodic memory disturbances, as well as exacerbating symptoms of multiple sclerosis and increasing the risk of general dementia. Improvement in chewing conditions as a result of implantological and prosthetic treatment prevents hippocampus dysfunction.

It is also worth noting that the treatment of these pathologies requires high financial outlays and entails high costs. Undoubtedly, this makes it difficult to access both preventive and curative dental care in a lot of countries, mainly in LICs or LMICs. As a result, the scale of damage in the oral cavity in many patients throughout their life is catastrophically increased [98].

A measure of disease prevalence is the disability-adjusted life-years (DALY) indicator, often referring 100,000 inhabitants, which is used to determine the health condition of a given society. It expresses the total number of years of life lost as a result of premature death or damage to health as a result of an injury or disease (Figure 3) [99].

Caries is widespread practically all over the globe. It is considered the most common and costly chronic social disease, depending on the conditions of civilization. Figure 4 [100] shows the spread of caries in 2017 in different countries of the world and its diversity. The countries with the most widespread caries in the 5–14 age group are Kyrgyzstan, Uzbekistan, Romania, Bulgaria, Poland, and Ecuador. In addition, in the entire age range, these countries include France, Spain, Iceland, Greece, and Georgia. Figure 4 illustrates also the spread of edentulousness and severe tooth loss in different countries and the spread of periodontal disease. There seems to be a clear correlation between edentulousness and the prevalence of caries, with Ukraine, Russia, and some Balkan countries in

Europe among the infamous leaders. In some cases, such as Australia and Brazil, edentulousness is relatively greater than the spread of caries. In almost all of the above-mentioned cases, as well as in others, where edentulousness is relatively greater than the prevalence of caries, endodontic treatment on an appropriate scale is most likely neglected or even not implemented. It seems that the situation is quite the opposite in countries such as France and Spain, where, despite a very high spread of caries, the share of edentulousness is relatively small. In countries such as Chile, Norway, Germany, and Denmark, periodontal diseases have a significant impact on the level of edentulousness. It should be noted that these observations are only correct to the extent that they are adequate to the actual states of the cited statistical studies. It seems, for example, in the case of Ukraine, Uzbekistan, and Kyrgyzstan, that these data do not correspond.

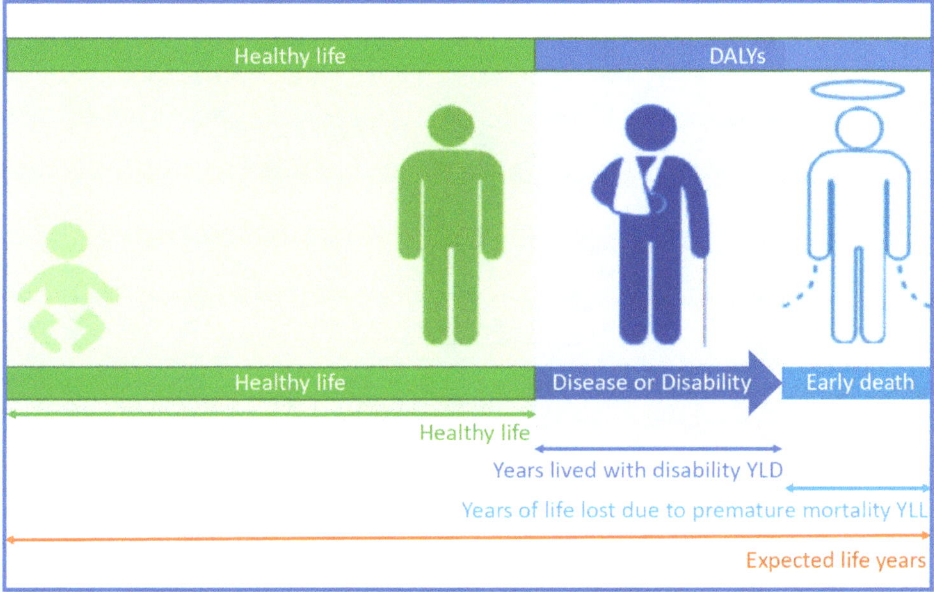

Figure 3. Diagram explaining the meaning of the disability-adjusted life-years (DALY) indicator.

The incidence of severe tooth loss in the world in 1990 to 2010 is not gender-dependent, but increases gradually with age, showing a sharp increase around the seventh decade, with a peak incidence after the age of 65. On the other hand, there are clear geographical differences in the prevalence, incidence, and rate of improvement in the indicated period [101]. It is obvious that missing teeth, especially large ones, have a significant impact on the oral health-related quality of life (OHRQoL). People with shortened dental arches (SDAs) do not show OHRQoL worse than people with removable dentures. The number of closing pairs and the location of the remaining teeth have a great influence on the OHRQoL [102].

The incidence of severe tooth loss in the world in the period from 1990 to 2010 is not gender-dependent, but increases gradually with age, showing a sharp increase around the seventh decade, with a peak incidence after the age of 65.

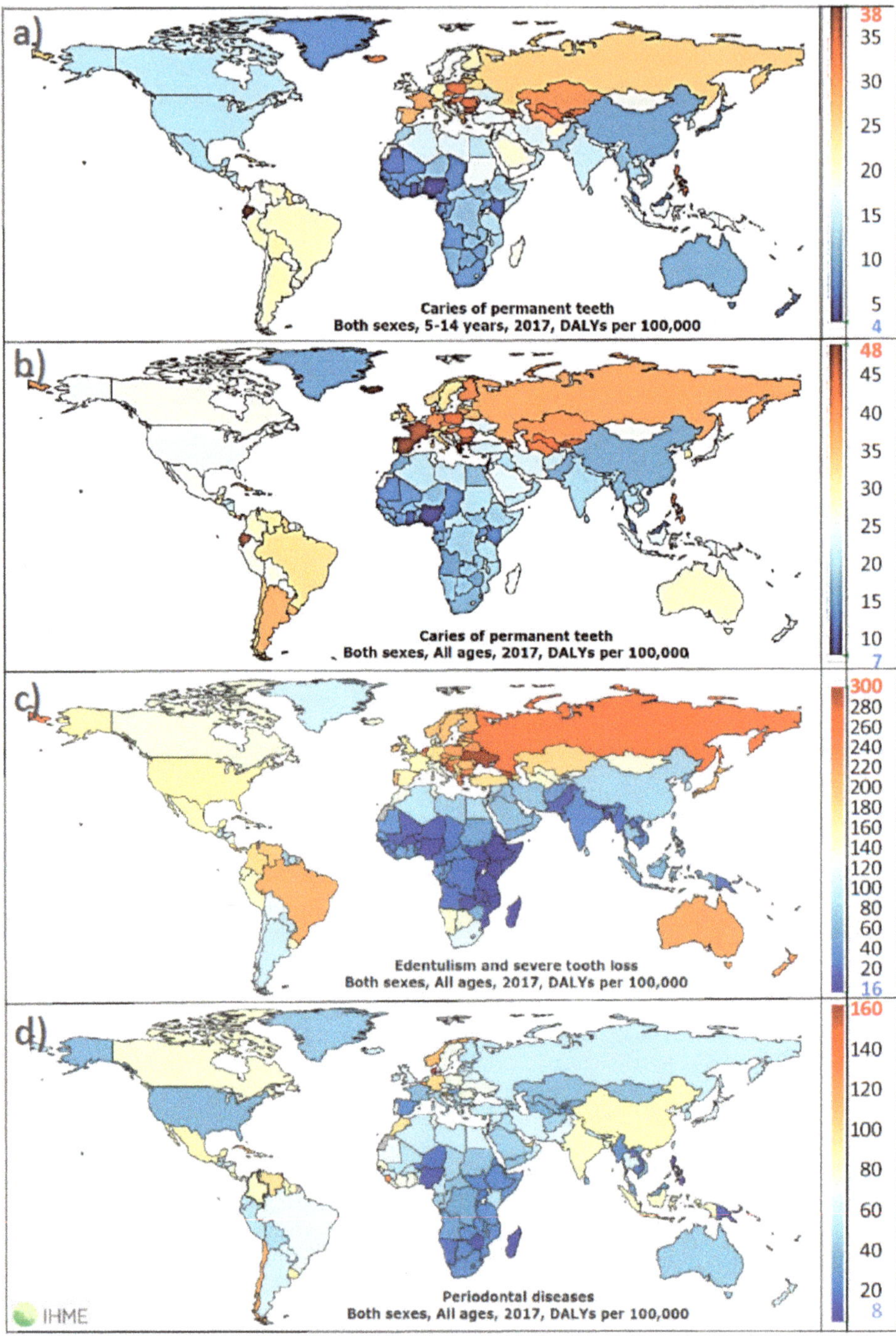

Figure 4. The spread of caries in 2017 in various countries of the world (no newer analyses); (**a**) in the 5–14 age group and (**b**) regardless of age; (**c**) The spread of edentualism and severe tooth loss; (**d**) The spread of periodontal diseases regardless of age.

On the other hand, there are clear geographical differences in the prevalence, incidence, and rate of improvement in the indicated period [101]. It is obvious that missing teeth, especially large ones, have a significant impact on the oral health-related quality of life (OHRQoL). People with shortened dental arches (SDAs) do not show a OHRQoL worse than people with removable dentures. The number of closing pairs and the location of the remaining teeth have a great influence on the OHRQoL [102]. Tooth loss causes malocclusion and, apart from aesthetic considerations, causes serious gastric diseases and has an adverse effect on the metabolism of the cerebrospinal fluid, which nourishes the brain. The consequence of this may be headaches, migraines, pains in the cervical spine, mood disorders, the intensification of the symptoms of multiple sclerosis, an unfavorable effect on the hippocampus in the temporal lobe of the cerebral cortex and other effects responsible among others on spatial memory. In a particularly unfavorable situation, caries and periodontal diseases may directly contribute to the patient's death. Counteracting the consequences of caries and periodontal diseases requires replacing the missing teeth with implants and prosthetic restorations, including bridges and crowns. Dental prosthetics are used to rebuild the aesthetic value of the appearance, improve the psychological comfort of patients, and restore the appropriate functions of the digestive system related to taking and chewing food. This is a branch of dentistry dealing with the restoration of the original occlusal conditions after the loss of natural teeth or after their massive damage. For reconstruction, a prosthetic restoration is used, requiring the use of various biomaterials. Dental surgery is often associated with prosthetics, which deals with the surgical treatment of the oral cavity and adjacent areas, especially in the field of implantology. Dental implant prosthetics cover the reconstruction of teeth on implants and allow them to achieve a permanent aesthetic effect by improving the mechanics of the masticatory apparatus, disturbed as a result of tooth loss. The reconstruction of the structures and functions of lost periodontal tissues or of the alveolar bone itself often requires controlled tissue regeneration with the use of non-resorbable or resorbable materials, stimulating fibroblasts to grow and separating bone tissue from rapidly growing epithelial tissue. Modern dentistry allows us to perform advanced prosthetic restorations that rebuild all kinds of missing teeth, both in the case of complete tooth lessness and partially missing ones.

3. Dentistry 4.0 as an Avant-Garde Stage of Dental Treatment

Dental diseases have accompanied man since the dawn of time. As the human species developed and with the discovery of fire, humans began to cook at least some of their food, including fish and shellfish, nuts, fruit, and animal meat, which became the staple diet. Over time, crops such as rice, wheat, and barley began to be cultivated, which caused the proportion of sugars in food to increase, favoring humus. Such ailments can be seen, among others, in as member of the subspecies of Homo sapiens called the Cro-Magnon man from the Upper Paleolithic era, who lived in Europe between 43 thousand and 10 thousand BC and was found in the rock shelter of Abri de Cro-Magnon in Les Eyzies near the Dordogne in France. Traces of dental treatment were found in the Gaione cemetery in Italy in pre-Neolithic finds, and the removal of food debris with wooden or bone toothpicks has been confirmed as common among Neanderthals and Paleolithic people [103,104]. Skulls with traces of caries were found in the Egyptian pyramids of Giza, other places in Africa, America, and Asia, including in the village of Mehrgarh in a neolithic urban settlement from the 7th millennium BC belonging to the Indus Valley Civilization, located on the Kacchi plateau in present-day Balochistan in Pakistan [105,106]. It is reported that a tooth root abscess was removed through the skull in around 2500 BC in prehistoric Malta. However, confirmations of dental interventions from the neolithic period are rare [107]. Studies of the remains of ancient Egyptians, Greeks, and Romans also confirm early practices in the field of dental prosthetics, although perhaps not performed in life but after death—e.g., as part of the mummification of a corpse. At Gebel Ramlah cemetery in Egypt, an artificial tooth has been identified that can be used as a dental prosthesis [108].

Figure 5 shows [109] an infected molar with a large cavity, partially cleaned with flint tools from the skeleton of a ca. 25-years-old man, from the rocky shelter Ripari Villabruna in the Italian

Dolomites of Veneto near Belluno [109]. This is the first-ever evidence of intentional dental intervention, although it is suggested [110] that research carried out in 2017 indicates that, as early as 130,000 years ago, Neanderthals used dental tools known to them [111]. The earliest beeswax tooth fillings dating back 6500 years were found in Slovenia [112]; however, a mixture of beeswax mixed with powdered minerals was previously used in ancient Egypt to repair loose teeth, as reported in the 16th century BC Ebers Papyrus [113]. There are many other written documents confirming various dental activities in the past, including the so-called Edwin Smith papyrus [110,114], and the Ebers, Kahun, Brugschu, and Hearst papyri from ancient Egypt [115–117]. Dental ailments were reported by the Chinese as long as 3000 years ago [110]. An ancient Sumerian text describes "dental worm" as the cause of caries [109,118], which Homer also writes about, and this is even repeated in late medieval documents.

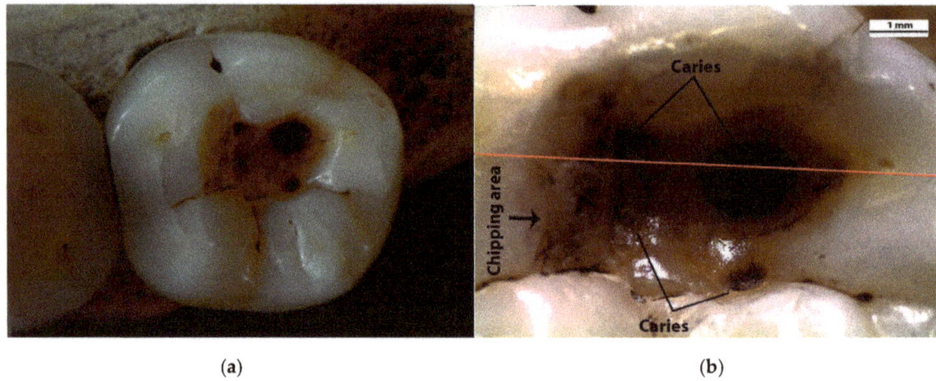

Figure 5. Top view the lower right third molar (RM3) of the man from Villabruna from the Late Upper Palaeolithic about 14,000 years ago; (**a**) Occlusal view; (**b**) detailed view of the occlusal cavity, with the four carious lesions and the chipping area on the mesial wall; red line is directed mesio-distally, passing through the larger carious lesion [109].

Oral diseases, mainly caries and periodontitis, are diseases that are spread all over the world in the 21st century. Dental treatment includes several specialties, depending on whether the disease is of the hard or soft tissues of the mouth.

Tooth caries is the most common infectious disease in the world and includes interactions between many factors, including the structure of the tooth; the bacterial biofilm formed on its surface; the interaction of saliva and genes; dietary carbohydrates, especially sugars, but also starches, though to a lesser extent; as well as behavioral, social, and psychological factors [16,19,119].

Figure 6 illustrates the dynamics of interactions—e.g., factors, some of which are protective and some pathological, which can shift the balance in the mouth or individual teeth towards health or disease, respectively. Tooth decay is the result of an unfavorable disturbance of the dynamic balance between two opposing processes of demineralization and remineralization, which are influenced by these factors [23,120,121]. A healthy state corresponds to the dynamic balance of the tooth surface with the local environment of the oral cavity and the conditions ensuring the balance of demineralization and remineralization processes or the advantage of remineralization [122–124]. The lack of this balance favors the demineralization of the tooth surface, because then organic acids, mainly lactic, are formed as a result of sugar metabolism [125]. Under these conditions, the pH decreases, and hydroxyapatite ($Ca_{10}(PO_4)_6(OH)_2$), which is the mineral phase of the tooth, dissolves, diffusing outwards from the tooth [23]. The speed of demineralization is the greater the greater the degree of unsaturation and the lower the pH [126], which inevitably favors the development of caries. The average pH = 5.5 is considered the critical point when the tooth enamel begins to demineralize, however it depends on the concentration of fluoride, calcium, and phosphate ions [127]. For dentin, the critical pH = 6 is higher. The diffusion of mineral ions outside the tooth towards the surface determines the rate of demineralization [128]. Demineralization is greater in subsurface enamel than on the surface. Single acts of demineralization and remineralization are dynamic, so they can take place simultaneously in different places depending on the local pH values. In the presence of fluoride ions [129–133], remineralization processes are activated, consisting of the re-deposition of calcium [134,135], sodium [136], and phosphate ions [129]. Enamel crystallites containing fluorapatite have a lower solubility than hydroxyapatite, and the average critical pH = 4.5 is lowered when tooth enamel begins to demineralize; therefore, susceptibility to caries development is much lower [10]. The result of subsurface demineralization is the progression of humus in this zone and the formation of the so-called white spot with no cavitation on the enamel surface. It is considered to be a very early stage of the subclinical course of caries, and the unstable equilibrium of demineralization and remineralization phenomena will persist at this level over the period from several weeks to many years. Fluoride activity may counteract this, stopping, slowing down, or even reversing the course of caries at this early stage, while if the predominance of demineralization processes is maintained, more advanced symptoms of the disease, including cavitation, will appear [10,137].

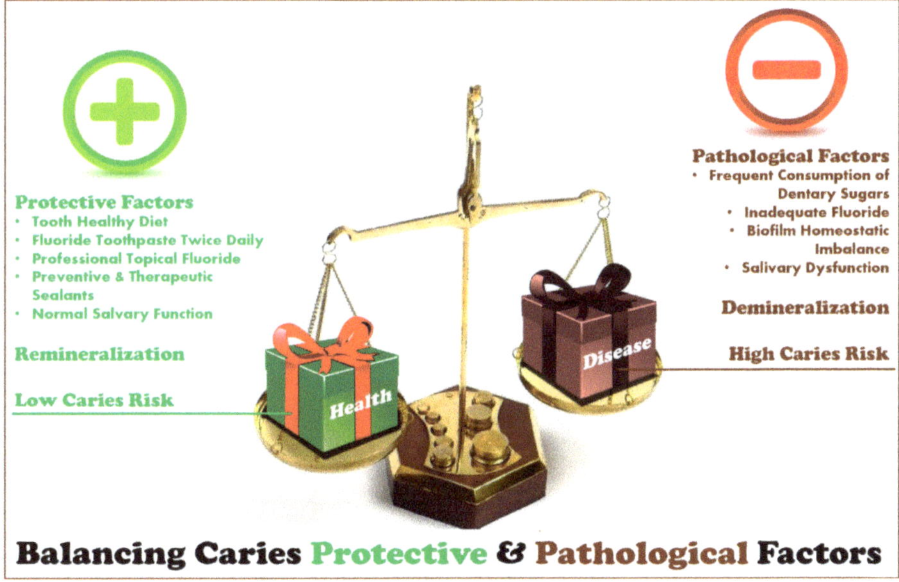

Figure 6. Diagram of the dynamics of interactions between protective and pathological factors that can shift the balance in the oral cavity towards health or disease, respectively.

The retention areas for biofilm formation and the deposition of food debris are occlusal cavities and molar grooves related to the morphology of the teeth, becoming the most vulnerable parts of the teeth, especially in children, as the solubility of the teeth decreases with post-eruptive age [23]. The clearance of cariogenic foods and drinks understood as the purification factor, which is the volume completely cleared of a given substance per unit time, depends on the flow rate and thickness of the salivary membrane, its bicarbonate buffering capacity, and the proximity of the teeth to the salivary glands [23] and the modification of the biofilm pH by saliva [138], which is responsible for maintaining a neutral pH of the biofilm, despite exposure to sugars in the diet. The course of the disease, as well as the increased susceptibility of some teeth and even places in them to it, depend on the environmental conditions of the postoperative age, ultrastructure, location, morphology, and composition of individual teeth [23,139]. The cause of successive damage to the hard tissues of the tooth; pain; abscesses; and, ultimately, the possibility of losing teeth is the influence of sugar-metabolizing biofilm bacteria, resulting in the formation of acids, which lower the pH of the biofilm. The resulting unsaturated conditions systematically demineralize the enamel and break down the dentin. The situation changes in the presence of fluoride, which delays cavitation, even if the sugar level in the diet is elevated [140], but this is often not a sufficient condition as, despite the extensive use of fluoride in many countries, the proportion of caries is still at an unacceptably high level [141]. This issue is described in more detail later in this paper.

In recent decades, the importance of bacterial biofilm on the tooth surface, which also includes saliva and protein glycoproteins, called the enamel epithelium, and its microorganisms, mainly Streptococcus mutans, and their important role in the etiology of caries have been explained. Biofilm is a set of endogenous microorganisms which, under unfavorable environmental conditions in the oral cavity resulting from a relatively high level of sugars in the diet with a simultaneous low flow of saliva, may shift the equilibrium from a healthy state to a state conducive to the development of caries [125,142,143]. Therefore, biofilm can be a binding site and a conditioning layer for primary bacterial colonizers, and then the formation of tooth biofilm is an important stage in the generation of caries. Dietary sugars are readily metabolized by biofilm microorganisms to produce organic acids (mainly lactic acid), which lower the biofilm pH, while the temporary reduction in biofilm pH cannot directly cause permanent demineralization. Biofilm plays a pathogenic role when there is a long-term decrease in the pH of the tooth biofilm—i.e., its acidification—and the endogenous microorganisms that are tolerant to the increasingly acidic environment determine the increased tendency of acid production, multiplying demineralization and causing the simultaneous development of caries [125,142,143]. Many of the microorganisms in the biofilm, such as *S. mutans* and *lactobacilli*, can therefore produce acids, unlike the others such as *Veillonella, Lactobacillus, Bifidobacterium* and *Propionibacterium*, streptococci nonmutans with a low pH, *Actinomyces* spp., and *Atopobium* spp. [144]. Biofilm can therefore play a twofold role depending on the composition and environmental conditions in the oral cavity [142,144]. Bacteria are able to form bases, especially in the presence of nitrogen (e.g., in peptides or proteins) [145], and for some bacteria, such as Veillonella, lactic acid, once it is produced, is a source of energy, so they consume it [146]. In this way, a decrease in pH is prevented [145]. Thus, the presence of a biofilm on the tooth surface does not mean disease in essence, as it may even constitute a physical barrier to acid diffusion.

The indicated information clearly shows the necessary and synergistic interaction of both factors—i.e., tooth biofilm and the presence of sugars in the diet—for the development of caries [19,141,147,148], which occurs when the pH of the tooth biofilm is lowered. Thus, the modification of the composition and metabolic activity of the tooth biofilm indirectly depends on dietary factors. A protein-fat diet ensures a neutral pH of the biofilm, as it increases the proportion of urea in saliva, converted into ammonia by ureolytic bacteria, blocking the development of caries [149]. On the other hand, the pH value decreases when the diet consists mainly of carbohydrates, especially sugars, which, when present in the diet, are one of the most important factors causing caries [140,141]. It should be taken into account that the limit of the negative effect of sugars promoting caries is a share of sugars in the diet exceeding 5% of the daily energy intake [147], or slightly more in the presence of fluoride [150–152], which will also be discussed in more detail later in this paper. Apart from fluoride, protective factors in food also include calcium and phosphates [153].

Extending the time in the mouth of foods containing fermentable carbohydrates promotes the extension of the actual time of acid formation, making it an important cariogenic factor [154]. Cariogenic monosaccharides, such as glucose, fructose, and sucrose—which is the most cariogenic among them due to the synthesis of soluble and insoluble extracellular glucan by glucosyltransferases (GTF) from *S. mutans*—are more harmful [140,155]. The appearance of insoluble glucan increases the extent of demineralization [23,156], causing the deeper penetration of carbohydrates from food [23]. Complex carbohydrates, such as starches [141,153], are more difficult to dissolve in fluids in the oral cavity, and their penetration rate into dental biofilms is slow. The biofilm bacteria require the starch to be broken down by saliva amylase to maltose first, which is usually not the case when starch is removed from the mouth by swallowing, which is not the case with the combination of starch and sugar found in most modern processed foods [157]. The intensity of the flow and the composition of saliva also depend on the composition of the diet, because with a liquid or even a soft diet salivary glands weaken or atrophy, while in the case of a coarse-grain diet the activity of the salivary glands is favorably stimulated [158].

The classic but now archaic approach to caries treatment was to remove demineralized tissue. Currently, such exclusivity may raise legitimate objections, as one would expect the knowledge of this regulatory mechanism to be knowingly used in dental practice. It can be concluded with a high degree of probability that the essence of these processes is well known to practicing dentists. They can, of course, have an ethical duty to spread this knowledge to people who come to them for advice or to treat oral diseases. However, there is a huge group of people in many countries of the world, especially those with a low income and in high poverty spheres, especially in LIC and LMIC countries, where a large proportion of people never go to a dentist. Therefore, they can only go there when forced to do so by pain associated with a very advanced stage of the disease. It is difficult to imagine that in such conditions it would be possible to implement any policy of making patients aware of prevention and thus provide prophylaxis. The general, relatively high level of knowledge in this area among dentists cannot, therefore, constitute any guarantee for a preventive policy. Regardless of these objective difficulties, it is imperative that caries be detected as early as possible in the disease, preferably when cavitation is not present, and a risk assessment is highly desirable in order to determine the appropriate preventive intervention and the desired recall frequency [10,159–161]. Undoubtedly, in an ideal understanding of the problem, it would be extremely beneficial to move from caries treatment to its early detection and the implementation of appropriate preventive measures. This issue is described in detail later in this paper.

In the field of epidemiology and scientific research, three main grades of caries severity are distinguished, distinguishing visually changes in enamel, dentin, and pulp [162,163]. The caries classification system is derived from the works of G.V. Black from 1917 [164]. The next stages of caries development are well illustrated by the so-called caries continuum, shown in Figure 7, which also shows the International Caries Detection and Classification System (ICDAS), which was initially used in epidemiology, practice, research, and education [21,120,165] and is now available on the website [166]. In [167], the criteria according to which carries should be classified (Table 1) are given.

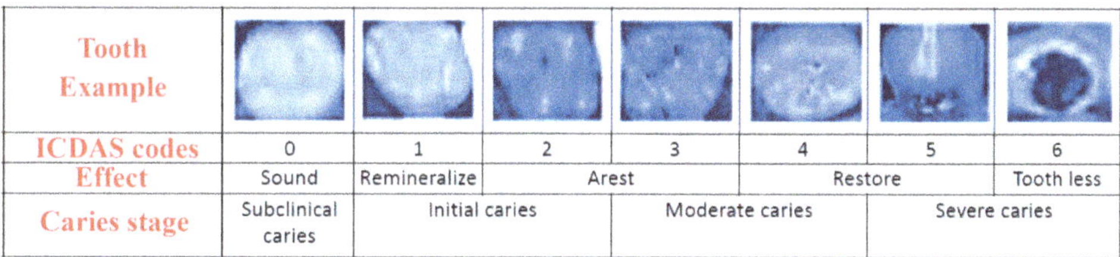

Tooth Example							
ICDAS codes	0	1	2	3	4	5	6
Effect	Sound	Remineralize	Arest		Restore		Tooth less
Caries stage	Subclinical caries	Initial caries		Moderate caries		Severe caries	

Figure 7. Scheme of the so-called caries continuum with codes from the ICDAS system.

In turn, Table 1, drawing on the work [168], presents the International Dentist Federation (FDI) World Dental Federation Caries Matrix as a framework (not a new system), integrating as much as possible at least some of the previously used classification systems, incl. those of clinicians, researchers, teachers, healthcare professionals, and policymakers.

Works in this area are still being carried out, which has resulted in, among other thing, the International Caries Classification and Management System (ICCMS) [169–172]. While the ICDAS uses an evidence-based and prophylaxis-oriented approach to detect and evaluate caries steps by their histology and activity, supporting decision-making at both the individual and public health levels, ICCMS can help to improve the long-term results of caries treatment. ICCMS enables dentists to integrate and synthesize dental and patient information, including caries risk indicators, to plan, manage, and evaluate caries in clinical and public health practice. The principles of the ICCMS system are presented in Figure 8.

Table 1. Principal carious lesion classification criteria.

No.	Classification Basis	Classification
1	Treatment of Caries	D = decayed or caries lesions, M = missing owing to extraction, F = filled or restored caries lesions
2	Morphology (Location of the Lesion)	Occlusal caries, smooth-surface caries and root caries
3	Prior Condition of the Tooth	Primary caries, secondary (recurrent) caries
4	Severity and Rate of Caries Progression	Acute caries, chronic caries, active caries and arrested caries
5	Extent of the Lesion	Incipient caries, advanced caries
6	Chronology or Age	Early childhood caries, adolescent caries, adult caries
7	Etiology (Causes or Origins of Caries)	Baby bottle tooth decay
8	Affected tissues	Enamel, dentin, cementum

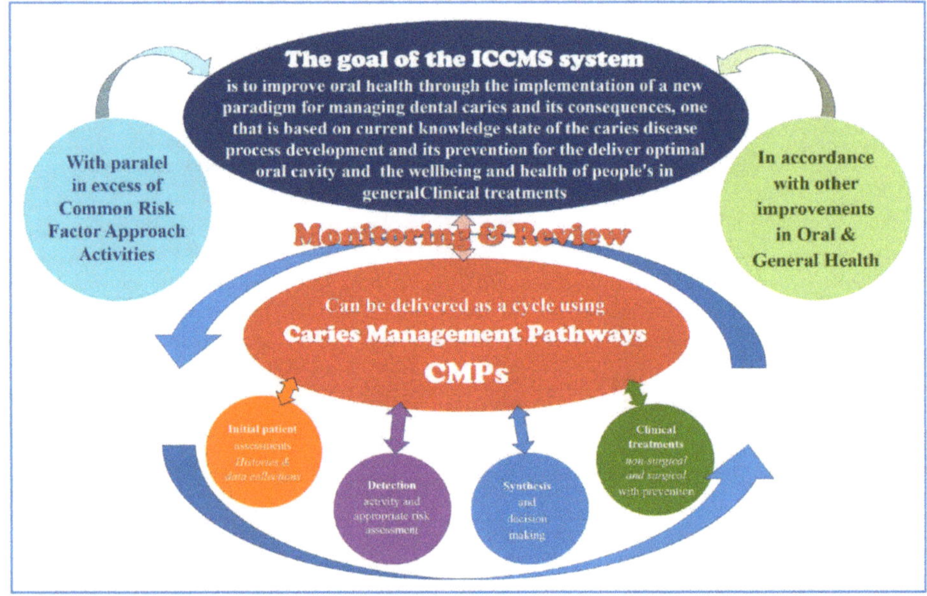

Figure 8. Scheme of the International Caries Classification and Management System (ICCMS).

Over time, the "iceberg" model was developed [173–175], which was supposed to be very suggestive, but, disregarding the substantive aspects, it should be stated that the analogy cannot be considered accurate, despite the fact that it has survived for over a quarter of a century.

Figure 9 shows a schematic image of the iceberg for comparison, the greater part of which is always submerged below the water level, while the entire caries classification scheme should, in principle, concern the visible part above the water level. The D3 carious threshold was defined as large lesions with open cavities extending to the pulp of the tooth, together with more limited open cavities in the dentin visible using organoleptic methods "above the waterline" [176]. In order to account for clinically detectable enamel defects with the same methods, dentin lesions not visualized were assessed as the D1 threshold, including lesions inherent to the D3 threshold with enamel lesions, but this threshold was already largely below the water level in this model. Of course, the use of microscopic techniques and other modern diagnostic methods, including radiological methods, detects the number of subclinical carious lesions, which will turn out to be greater [16], and the classification becomes ambiguous; as a result, it is of little applicability in clinical practice. It is not true that changes above the water level in this iceberg diagram correspond to the development of caries when those below the water level are decay-free stages. This does not correspond to the separation between "obvious distribution" and "no clear distribution", as discussed in [16]. In fact, all the changes shown in the diagram of the iceberg concern the part of the mountain visible above the water level, which clearly proves that this comparison is not accurate. It must certainly be recognized that this model has already fulfilled its historic role. Therefore, this paper presents the scheme of the caries development pyramid (CDP) from the most common (layer line) and the least advanced subclinical stages of lesions (height) to deep lesions extending into the pulp, relatively less frequent in relation to the developed hygienic activities of patients and the therapeutic activities undertaken by dentists in the previous, less advanced stages of the disease. This diagram also shows the caries development pyramid (CDP) individual methods of the methodological approach, ranging from background-level care (BLC) through to preventive treatment options (PTOs) to the operative treatment option (OTO), all distinguished in the works [174,175,177].

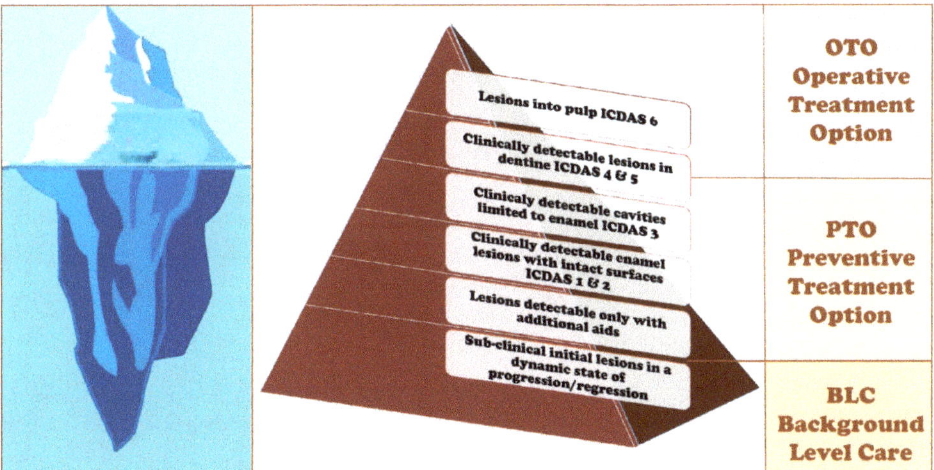

Figure 9. Scheme of the caries development pyramid (CDP) with indications ranging from background-level care (BLC) to preventive treatment options (PTOs) and the operative treatment option (OTO) in comparison with the general view of an iceberg submerged largely in water.

Assessing the risk of developing caries in a given patient is an essential element of good professional dental care. Of the many caries risk factors that can shift the balance towards health or disease and can be rated as low, moderate, or high, the most important are:

- The frequent and/or long-term consumption of sugars, when they are a large part of the diet;
- insufficient saliva flow rate;
- poor oral hygiene;
- suboptimal exposure to fluoride;
- tooth deformities caused by poor nutrition due to socio-economic degradation;
- a lack or low level of dental care.

This paper presents a few of the "Fives' rules" several times, wherein this detailed case 5D Caries Management Cycle Rules (CMCRs) is the first such example. It should be noted that each time such a mnemonic approach to selected specific aspects is inspired that is not necessarily a repeated approach resulting from literature studies, it is usually a modification of the assumptions of the behavioral strategy based on the idea of Deming's circle. Deming's cycle of Plan-Do-Check-Act (PDCA) illustrates the basic principle of continual improvement [9]. In the case of evidence-based clinical caries management, especially in the initial stages, this cycle includes the essential cyclical steps of the International Caries Classification and Management System (ICCMS) and shows how this system can be implemented as the 5D Caries Management Cycle Rules 5D (CMCRs) (Figure 10). The essential 5Ds are implemented cyclically and concern the following assessment and therapeutic activities:

1. Detect lesions;
2. Determine lesions' activity;
3. Dispense and assess the scale of lesions and their activity;
4. Decide on the patient's personalized care plan;
5. Conduct the right intervention at the right time.

Figure 10. Diagram of Deming's circle adopted for the fives' rule on the 5D Caries Management Cycle Rules 5D (CMCRs) for the continuous improvement of the level of the patient's oral health state.

In general, the preventive and therapeutic strategies for the treatment of dental caries cover the three most important stages—i.e., primary, secondary, and tertiary. When lesions in the ICDAS 1–4 range are identified, the most commonly used approach is the 5D Caries Management Cycle Rules 5D (CMCRs), which were presented previously (Figure 10). Secondary prevention strategies are aimed at stopping or reversing the progression of caries at the stage of identified changes that are not manifested by cavitation, mainly referred to as the moderate state (Figure 7). An important goal of this procedure is to eliminate the need for caries surgical intervention by placing fillings, which may result in the long-term preservation of the patient's own teeth. The accurate detection and evaluation of the early stages of non-cavitated caries require immediate intervention through fluoride therapy and the use of sealants to reverse or halt the progression of caries [178]. Due to the possibility of subclinical active caries lesions, it is impossible to establish a precise boundary between primary and secondary caries prevention, and many of the same interventions, such as biofilm control and topical fluoride and sealant application, apply to both prevention groups. Similarly, the transition between secondary and tertiary prophylaxis is blurred due to the possibility of the biological non-invasive treatment of cavitation changes in both cases, both in deciduous and permanent teeth [179]. Precision Caries Management (PCM), the assumptions of which are presented by analogy to the Deming circle in the form of the complex developed concentric circles, provides a wide range of possible therapeutic decisions (Figure 11). Each decision is based on evidence and the best decision from the point of view of the patient's oral wellbeing and wellbeing in general, taking into account the severity and activity of the lesions, the assessed risk of caries, and the prediction of patient compliance and preferences [22].

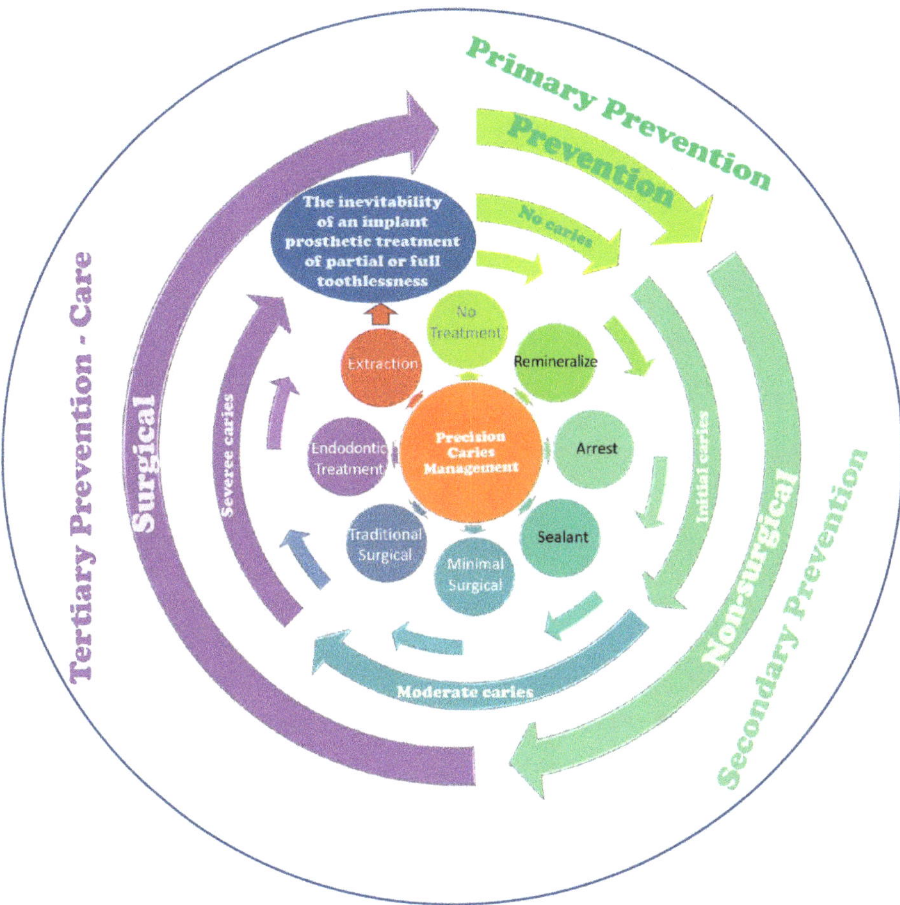

Figure 11. Precision Caries Management (PCM) schematic diagram developed in the form of the complex developed concentric circles by analogy to the Deming circle with a wide range of possible therapeutic decisions depending on the stage of caries development.

The most popular and most frequently performed procedure is conservative treatment, which deals with the preparation of carious lesions and then the direct reconstruction of hard tooth tissues with the use of filling material. Progress in this regard is obvious. Currently, composite materials, glass-ionomer types of cement, compomers (a mixture of composite and glass-ionomer materials), silicon types of cement, and silicon-phosphate types of cement are used for this purpose. Composite dental restorations are used in direct fillings and are tooth-colored. Dental composite materials usually consist of a resin-based matrix, which contains a modified methacrylate or acrylate—for example, bisphenol A-glycidyl methacrylate (BISMA) and urethane dimethacrylate (UDMA), together with tri-ethylene glycol dimethacrylate (TEGMA). A coupling agent such as a silane is used to strengthen the bond between the resin matrix and the filler particles. The initiator pack starts the polymerization reaction of the resins when external energy (light/heat, etc.) is applied. For example, camphorquinone can be excited with visible blue light with a critical wavelength of 460–480 nm to obtain the free radicals necessary to initiate the process. After the teeth are prepared, a thin primer or bonding agent is used. Photo-polymerized composites are used as thin layers determined by their opacity. The final surface is shaped and polished after some curing. Glass ionomer cement (GIC) and resin-modified glass ionomer cement (RMGIC), which is a composite technology with the properties of glass ionomer cement, may be used for forming dental compomers by modifying dental composites with poly-acid.

The most doubts have been raised by the use of amalgam as a dental filling, to such an extent that there are so-called "amalgam wars" between the supporters and opponents of amalgams. Amalgams are metal alloys in which mercury is the primary component, forming solutions of other metals in mercury under ambient conditions. Most metals form amalgams, with the exception of iron. Dental amalgams have been used almost since the dawn of conservative dentistry, because they show good properties, despite their lack of aesthetic values, and due to their relatively low costs they are widely used and there is concern that they are the only ones in Low and Middle-Income Countries (LMIC), as well as among people with lower status in other countries, which can be considered a cause for concern and a sign of failure to follow development trends in these cases. The composition of dental amalgam is given in the ISO 1559 standard [180] of 18 July 2001, which, however, was finally withdrawn on 26 February 2008. With mercury, amalgam includes silver, tin, and copper, and, in low concentrations, zinc, palladium, platinum, and indium. In conventional grades of dental amalgams, there was at least 65% wt. Ag, while after 1986 the concentration was increased to 30% wt. Cu, and the concentration were reduced to at least 40% wt. Ag. Amalgam is recognized by the FDA as safe in the US for adults and children over six [181], while, in the European Union, based on the opinion of Scientific Committee on Emerging and Newly Identified Health Risks (SCENIHR) on "The safety of dental amalgam and alternative dental restoration materials for patients and users" from 29 April 2015 [182], the Minamata Convention on the Reduction of Mercury was adopted in 2017 [183], according to which it is necessary, inter alia, to gradually phase out the use of amalgam fillings so that they are not completely used in dentistry by 2030. The most susceptible to the harmful effects of unbound mercury are dentists, as they have the closest and longest contact with the vapors of this element. In order to reduce the possibility of negative consequences for the patient and dentist, encapsulated amalgams are currently used during filling, reducing the exposure to mercury. Considering the above-mentioned documents, the use of this material should be considered high risk, and in the case of the exclusive use in the aforementioned countries and towards the social groups mentioned above, as a probable element of exclusion or even unequal treatment, contrary to the Convention on Human Rights [184]. This problem seems to be important on a global scale.

An element of noticeable progress in this area, as well as an alternative to the traditional direct filling in dental conservative treatment, is the indirect method with the use of a system of cosmetic ceramic inlays and crowns prepared in the dental engineering laboratory on the basis of the impressions taken and used in the event of large cavities. Inlays and crown onlays enable the exact reconstruction of the anatomical shape of the teeth, including the cusps and fissures of the contact points between the teeth. Most often, they are made of porcelain, but they can also be made of gold alloys, composite materials, or gold alloys and ceramics. Inlays and onlays, made in a dental engineering laboratory, are glued to a properly prepared cavity in a tooth with the use of a bonding type of cement. Inlays and onlays are durable, aesthetic, and resistant to abrasion, and the reconstructed tooth, thanks to the precise reconstruction of the anatomical shape, can fulfill its functions and look very natural.

The attack of the dental pulp by bacteria of the carious cavity results in pain, which requires dental intervention in the form of endodontic treatment [185–187]. This treatment consists of the removal of the dental pulp and the preparation and filling of the entire root canal system with replacement material, thanks to which the tooth, devoid of the living intra-root structure, can continue to function in the oral cavity. The development of root canals consists of removing the contents inside the root canal, disinfecting it, and giving it a shape that will ensure the best hermitization of the root canal with filling material. Root canal filling material, filling the root canal space during endodontic treatment, replaces the living tissue. The most popular material for filling root canals at present is gutta-percha [188–193], which is a trans polymer with 1.4 polyisoprene in three isomeric forms, α, β, and γ, of natural origin, obtained from the milk juice of Palaquium gutta and Palaquium oblongifolia plants. In dentistry, only two types are used—β, which has the form of a solid and turns into the α form of gutta-percha under the influence of 48.6–55.7 °C. The melting point of gutta-percha is 64 °C. Gutta-percha used in endodontics is most often in the form of studs or pellets, in which pure β gutta-percha is only 18–22% and the rest is zinc oxide at 59–75%, metal compounds such as barium and strontium sulfate at 1.1–31.2%, as well as wax and other polymers at 1–4.1%. Gutta-percha is always used in combination with a sealant. Both in cold and thermoplastic condensation methods, sealants based on synthetic resins are recommended. There is also a liquid gutta-percha available on the market for filling cold canals, which is in a powdered form with a particle size of less than 30 μm with a sealant and nanosilver particles to prevent infections; it is packaged in capsules. An alternative to gutta-percha is a material based on polymeric polyester materials, composed of an organic part constituting a resin polymer matrix and inorganic fillers (Figure 12). The quality of endodontic treatment depends mainly on the tightness of the root canal filling and the tight fusion of the root canal dentin with the filling material. Polymer materials with a matrix of gutta-percha and polyester polymeric materials enable the 3D filling of the entire inner space of the root canal as a result of thermal plasticization. In this regard, they are far from the other materials used for tooth fillings during endodontic treatment. The material with a gutta-percha matrix together with the applied sealant ensures close bonding with the root canal wall. In the case of the matrix of materials, leaks at the boundary of the polyester-polymer material with the walls of the root canal were found much more often, and are caused, among other things, by polymerization shrinkage [192,193].

Carefully endodontic treated teeth remain in the patient's mouth for a relatively long time, although in 3% of cases, 8 years after the end of treatment, periapical changes in the form of cysts or granulomas may occur [194]. However, in over 90% of cases, the causes of possible treatment failure, in this case, are errors and inaccuracies in the preparation and filing of root canals [195–199], as well as iatrogenic causes, mainly crown or root fractures [200]. In such cases, tooth extractions are most often unavoidable, which results in the need for prosthetic and/or implantology treatment.

Figure 13 shows an example of a circumferential bridge when the patient's teeth were removed as a result of his neglect of caries treatment and subsequent local extractions in the last dozen or so years. The example concerns a fully digitized model in the Dentistry 4.0 standard and the manufacturing of dental bridges on this basis, alternatively using the technology of milling blocks from solid metals or selective laser sintering (SLS) from metal powders. Co25Cr5W5MoSi-type alloy was used. In the dentist's clinic, the dentist takes an impression, which is then scanned together with the impression tray, (Figure 13a), which is the basis for the development of a model with digital dentition with an occlusal relationship (Figure 13b). After the tooth stumps are separated and the edge is marked by a dental engineer, the parallelism of the pillars is verified (Figure 13c) and a tooth design is generated using virtual ready-made libraries of shapes, and the required corrections of these shapes are made to adapt them to the situation in the oral cavity of the patient. The finished framework design (Figure 13d) is transferred from the computer-aided design (CAD) software to the computer-aided manufacturing (CAM) software, in which the dentist engineer virtually selects the position of the model on the disk in the case of the removal method by milling, or on the worktable in the case of making a bridge using the method of additive manufacturing. The bridge substructure can be produced by milling in a controlled numerically center (CNC) (Figure 13e) or, alternatively, in additive selective laser sintering technology, where, of course, it is also necessary to design brackets (Figure 13f). The framework of the prosthetic restoration, produced in one of the methods given above (Figure 13g), is veneered with ceramics (Figure 13h), after which the bridge can be installed in the patient's mouth (Figure 13i).

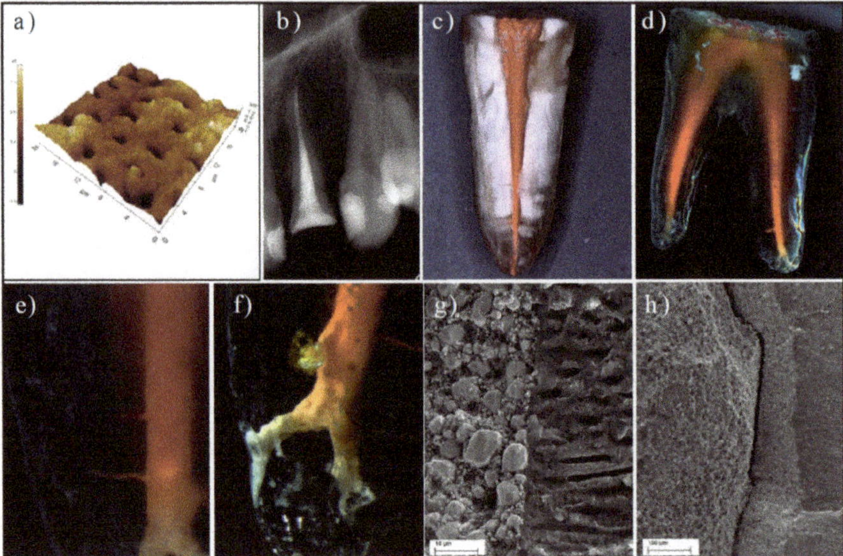

Figure 12. Examples of endodontically treated teeth: (**a**) morphology of the transverse fracture of the canal dentin (atomic force microscope); (**b–g**) root canal filled with gutta-percha material with AH Plus sealant by the thermoplastic method: (**b**) X-ray of the patient; (**c**) longitudinal fracture of the tooth; (**d–f**) decalcified teeth; (**e**) filling the side branches of the main canal; (**f**) root delta ((**c–f**) Stereo Discovery stereo light microscope with AxioCam HRC digital camera by Zeiss at 8–50× magnification); (**g**) tight connection of the root canal with the material based on gutta-percha and sealant in all sections of the root canal; (**h**) after an incorrectly performed procedure, border of three layers, tightly connected dentine of the root canal with a thick intermediate layer of sealant and a leak between the sealant and the material on the gutta-percha matrix covered with a sealant layer (**g,h**)—scanning electron microscope).

A common cause of dental extractions is periodontal diseases, sometimes requiring even all teeth to be removed, sometimes when the patient is not more than 40 years old. Periodontology concerns the suspension apparatus of the teeth—i.e., the periodontium. As a field of dentistry, it deals with the prevention and treatment of periodontal and oral mucosa diseases. Despite great progress in this area, in many cases, patients cannot be prevented from the tooth extraction. Rarely, orthodontic considerations also require the extraction of certain teeth.

Dental extractions require prosthetic and/or implantological treatment. The causes of tooth extraction include not only caries and iatrogenic causes during treatment, but also periodontal disease. Sometimes it is even necessary to remove all teeth, and sometimes at the age of 40. Periodontics—i.e., the suspension of the teeth—is dealt with by periodontics. As a field of dentistry, it deals with the prevention and treatment of periodontal and oral mucosa diseases.

Periodontal disease is one of the major clinical problems in the field of oral cavity diseases worldwide. This group of acute and chronic diseases that manifest as inflammation of the periodontal tissues in the presence of plaque and are interrelated collectively include periodontal diseases. Risk factors for periodontal disease include those that depend on lifestyle, including smoking and alcohol consumption, as well as diseases and conditions such as diabetes, obesity, metabolic syndrome, osteoporosis, low dietary calcium and vitamin D, osteoporosis in postmenopausal women [201], as well as bruxism [202]. The listed risk factors are modifiable, which gives modern dentistry a chance to treat periodontal diseases. Genetic factors (i.e., specific genes) also play a role in aggressive periodontitis and most likely in chronic adult periodontitis. However, there is no clear evidence for this in the general population so far. The etiological microorganisms *P. gingivalis*, *T. forsythia*, and *A. actinomycetemcomitans* are risk indicators [203] for periodontal disease. Other factors contributing to or supporting symptoms of periodontal disease include age, oral hygiene [204], diet [205], and its supplementation [206].

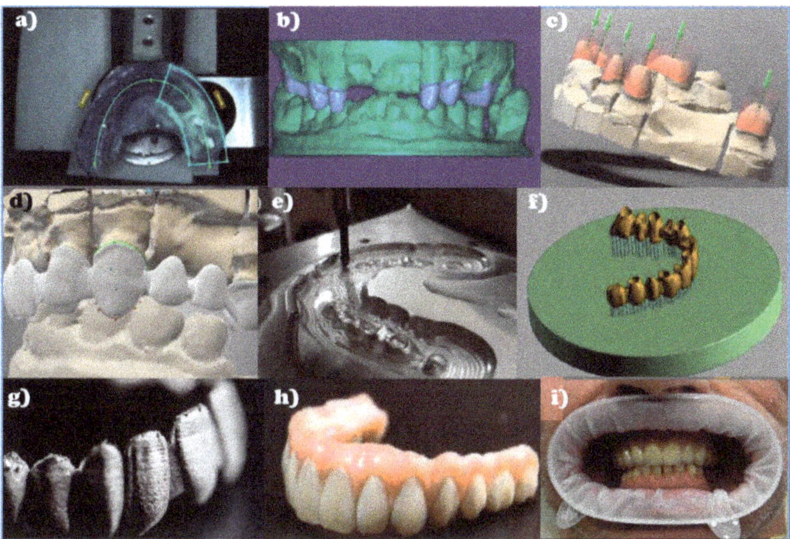

Figure 13. An example of a circumferential bridge after the extraction of the patient's teeth as a result of neglecting the treatment of caries, alternatively manufactured using the technology of milling solid-state metal blocks or selective laser sintering (SLS) from powders of the Co25Cr5W5MoSi alloy: (**a**) scan of the impression, (**b**) model with finger dentition occlusion, (**c**) verification of the parallelism of the pillars, (**d**) finished design of the skeleton, € milling of the bridge in a controlled numerically controlled center (CNC), (**f**) layout scheme of the bridge selectively sintered with a set of supports on the work table, (**g**) foundation of the bridge manufactured using one of the given technological methods, (**h**) finished bridge veneered with ceramics, (**i**) bridge installed in the patient's mouth.

The immediate cause of periodontal disease is the deposition of dental plaque and tartar as a result of the coalescence and biofilm accumulation of complex bacteria, combined with the immunoinflammatory mechanism, other risk factors, and the disruption of host–microbial interactions near the gingiva [207]. Tartar is a mineralized, hardened plaque that contains bacteria, calcium and phosphorus compounds, food debris, and substances contained in saliva. Advanced stone becomes darker in color, which is due to the consumption of large amounts of coffee, tea, red wine, and smoking. The supragingival calculus is deposited on the surface of the tooth crown, above the gum line, and is more visible and more comfortable to remove. Stone subgingival is deposited below the gum line, making it more difficult to notice, and, because it is harder than stone supragingival, its elimination is more time-consuming.

Untreated, periodontal disease results in the loss of alveolar attachment and bone destruction [208], and can also lead to tooth loss. Detailed studies have shown that periodontal disease is the cause of tooth loss in as many as 70.8% [209] of cases and that, after caries, it is the second most common cause of tooth loss [210]. The symptoms of periodontal disease include bleeding gums and gum recession, bad breath, as well as moving teeth and poorly fitting dentures. The subsequent stages and forms of the disease development can be defined as follows: parodonthopaties periodontitis, chronic periodontitis, aggressive periodontitis, periodontitis as a manifestation of systemic disease, necrotizing periodontal diseases, abscesses of the periodontium, periodontitis associated with endodontic lesions, development or acquired deformities and conditions, periodontitis, inflammation in deeper structures of the periodontium, periodontitis, gingivitis, pocketing, loss of alveolar bone, drifting and mobility, exposure to furcation, and recession. Furcation covers the anatomical inter-root space within multi-rooted teeth, located at the base of the root bifurcation. Pathological changes involving the periodontium reveal it, creating favorable conditions for pathology [211]. Furcations constitute a specific, separately classified group of intramedullary defects in the alveolar bone [212]. The exposure of the inter-root area and the formation of a defect in the furcation area, related to the loss of attachment, is an area where adequate therapeutic management is difficult and maintaining hygiene causes many problems for patients, worsening the long-term prognosis [213–215].

One of the common forms of periodontal disease is the inflammation of the gums or the mucosa tissue that surrounds the teeth. Although this disease does not cause permanent changes or damage to the periodontium, there is reddening of the mucosa, swelling, and generalized inflammation, and this is usually accompanied by pain that is burdensome for the patient and typically spontaneous bleeding during eating or brushing the teeth. Gingivitis requires the removal of the causative agents and, ultimately, treatment to prevent complications of other diseases, including periodontitis, of greater health risk [204,216]. Reversible gingivitis is easy to treat by the patients themselves. Without adherence to the dentist's recommendations and the continued removal of plaque and tartar, gingivitis cannot be completely cured and may develop into irreversible periodontitis [217].

Periodontal diseases, but also growths, inflammations, improper flossing, aggressive tooth brushing, abnormal occlusal conditions, and dominant roots can cause local or generalized gingival recession. Gingival recession is a peak migration of gingival margins into the cement-enamel junction (CEJ), with or without loss of attached tissue, and affects almost all middle-aged and older people to some extent. Gingival recession causes the increased sensitivity of exposed dentin [218], including to temperature and the environment. Figure 14 shows a simplified longitudinal section through the periodontal pocket. The normal position of the gingival margin is at the cement-enamel junction (CEJ), but in healthy young patients, or in cases where the gingival margin is swollen, it may be coronal to the CEJ. If the gingival margin is at the apex of the CEJ, there is a gingival recession—i.e., "receding gums"—and the exposure of the root surfaces, which is usually accompanied by excessive sensitivity to various external stimuli, mainly temperature changes. However, Table 2 lists also several different basic symptoms of the various stages of periodontal disease [219].

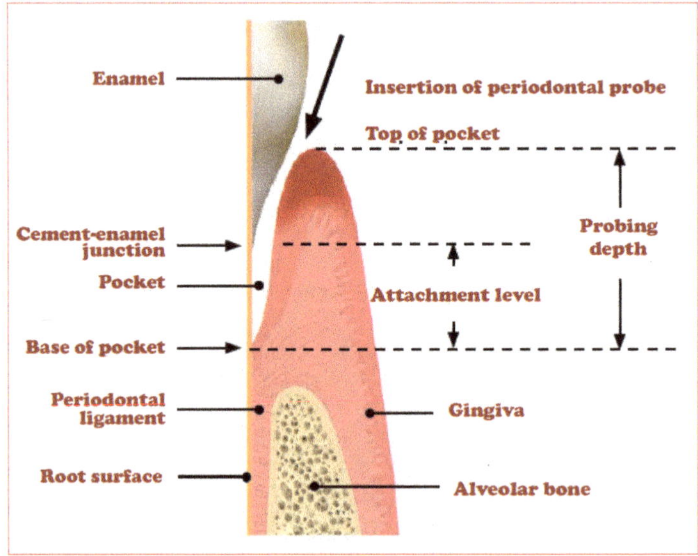

Figure 14. Diagram of a longitudinal section through the periodontal pocket illustrating the gingival recession and the exposure of the tooth root surface.

Table 2. Presentation of the basic symptoms accompanying different stages of the development of plaque diseases.

Disease Development Stages	Basic Symptoms in the Subsequent Stages of the Disease Development
Gingivitis	This is caused by plaque; there are red, swollen tissues that bleed when brushing and probing.
Chronic periodontitis	There is a slow destruction of the connective epithelium and attachment of the connective tissue of the tooth, bone destruction with a loss of bone mass, and the formation of periodontal pockets.
Aggressive periodontitis	The degree of destruction of the attachment of connective tissue and bone is severe at the rapid progression of the disease; severe condition in the group of younger patients is related to a family history of aggressive periodontitis.
Necrotizing ulcerative gingivitis (NUG)	Necrotizing ulcerative periodontitis (NUP) is present in the presence of the loss of connective tissue and bone destruction, manifested by the painful ulceration of the ends of the interdental papillae with visible gray necrotic tissue accompanied by halitosis.
Periodontal abscess	There is acute or chronic and asymptomatic freely draining infection of the periodontal pocket.
Perio-endo lesions	The source of the bacteria is from the periodontium or the root canal system, while lesions may heal or may be independent.
Gingival enlargement	Irritation from plaque or tartar, as well as repeated rubbing or trauma, as well as fluctuating hormone levels or the use of certain medications, cause the gingivae to thicken.

Periodontology is the branch of dentistry dealing with the prevention and treatment of periodontal disease and also includes the treatment of periimplantitis [220], which is similar to periodontal disease. A schematic classification of diseases, conditions, and periodontal issues surrounding implants was presented by the American Academy of Periodontology (AAP) in 1989 and revised ten years later. In detail, it was analyzed during a workshop with the participation of the European Federation of Periodontology [221,222] and it is necessary for clinicians for the correct diagnosis and treatment of patients, and for scientists studying the etiology, pathogenesis, and treatments of periodontal disease. The latest findings are the result of a joint work by the American Academy of Periodontology and the European Federation of Periodontology conducted at the end of 2017 [222] (Table 3).

The most effective method for preventing periodontal diseases is the self-discipline of the patient and the proper maintenance of oral hygiene by the proper cleaning and brushing of teeth and implants using appropriate devices for cleaning interdental spaces, including dental brushes and dental floss, and fluoridating the teeth with the use of a proper paste. Periodic checkup with a dentist is necessary. In addition to these approaches, which are well-known and confirmed by scientific research results, we also rely on the topical application of chemotherapeutic agents, reducing risk factors, including smoking, and numerous new methods are introduced and investigated, such as antioxidants, probiotics, vaccines, and alternative chemotherapeutic agents. The new strategies also provide for the appropriate shaping of a new approach and previously unknown preventive behaviors of patients [223].

Table 3. A new classification scheme for periodontal and peri-implant diseases and conditions.

Periodontal Diseases and Conditions		
Periodontal Health, Gingival Diseases and Conditions	**Forms of Periodontitis**	**Periodontal Manifestations of Systemic Diseases And Developmental and Acquired Conditions**

1. Periodontal health and gingival health
a. Clinical gingival health on an intact periodontium
b. Clinical gingival health on a reduced periodontium
   i. Stable periodontitis patient
   ii. Non-periodontitis patient
	1. Necrotizing Periodontal Diseases
	a. Necrotizing Gingivitis
	b. Necrotizing Periodontitis
	c. Necrotizing Stomatitis
		1. Systemie diseases or conditions affecting the periodontal supporting tissues
		
		2. Other Periodontal Conditions
		a. Periodontal Abscesses
		b. Endodontic-Periodontal Les ions

2. Gingivitis—dental biofilm-induced a. Associated with dental biofilm alone b. Mediated by systemie or locaI risk factors c. Drug-influenced gingival enlargement	**2. Periodontitis as Manifestation of Systemic Diseases** Classification of these conditions should be based on the primary systemic disease according to the International Statistical Classification of Diseases and Related Health Problems codes	**3. Mucogingival deformities and conditions around teeth** a. Gingival phenotype b. Gingivaljsoft tissue recession c. Lack of gingiva d. Decreased vestibular depth e. Aberrant frenum/muscle posit ion f. Gingiva l excess g. Abno rmal color h. Condition of the exposed root surface **4. Traumatic occlusal forces** a. Primary occlusal trauma b. Secondary occlusal trauma c. Orthodontic forces

Table 3. *Cont.*

Periodontal Diseases and Conditions		
Periodontal Health, Gingival Diseases and Conditions	**Forms of Periodontitis**	**Periodontal Manifestations of Systemic Diseases And Developmental and Acquired Conditions**
	3. **Periodontitis**	
	a. Stages: Based on Severity [1] and Complexity of Management [2]	
3. **Gingival diseases—non-dental biofilm induced** a. Genetic/developmental disorders b. Specific infections c. Inflammatory and immune conditions d. Reactive processes e. Neoplasms f. Endocrine, nutritional & metabolic diseases g. Traumatic lesions h. Gingival pigmentation	Stage I: Initial Periodontitis Stage II: Moderate Periodontitis Stage III: Severe Periodontitis with potential for additional tooth loss Stage IV: Severe Periodontitis with potential for loss of the dentition b. Extent and distribution [3]: localized; generalized; molar-incisor distribution c. Grades: Evidence or risk of rapid progression [4] anticipated treatment response [5] i. Grade A: Slow rate of progression ii. Grade B: Moderate rate of progression iii. Grade C: Rapid rate of progression	5. **Prostheses and tooth-related factors that modify or predispose to plaque- induced gingival diseases/periodontitis** a. Localized tooth -relat ed factors b. Localized dental prostheses-related factors

[1] Severity: Interdental clinical attachment level (CAL) at site with greatest loss; radiographic bone loss and tooth loss. [2] Complexity of management: probing depths, pattern of bone loss, furcation lesions, number of remaining teeth, tooth mobility, ridge defects, masticatory dysfunction. [3] Add to stage as descriptor: localized <30% teeth, generalized 30% teeth. [4] Risk of progression: direct evidence by paradiographs or CAL loss, or indirect (bone loss/age ratio). [5] Anticipated treatment response: case phenotype, smoking, hyperglycemia.

It turns out, however, that, for various reasons, including patient negligence and also objective difficulties in maintaining proper hygiene as the disease progresses, these activities become insufficient. Tartar that is not removed on an ongoing basis can be a severe threat to the health of the teeth, causing caries and cavities. It causes bleeding and gum disease over time and can also be one of the causes of bad breath. The subgingival calculus leads to the exposure of the necks of the teeth, which means that patients experience pain due to progressive hypersensitivity. The most severe consequence of unremoved tartar is periodontal disease, which promotes loosening in the sockets and the loss of teeth. A general dentist-assisted action should therefore be taken through routine scaling and cleaning using hand tools or an ultrasonic scaler, or a combination thereof. Individual techniques can be combined, depending on the type and position of the stone (supragingival or subgingival), and they are selected individually. Then, polishing is performed—i.e., the surface of the teeth is smoothed with a special paste, which additionally strengthens the teeth and prevents the development of caries. Many patients are concerned about pain during scaling, but local anesthesia may only be necessary when removing the subgingival calculus. The doctor may also use specialized methods of dental prophylaxis. Due to the possibility of releasing bacteria from deposits, the dentist may decide to use disinfectant rinses.

Diagnosis of periodontitis and associated conditions is made after a thorough examination of the patient's medical, dental, and social history, combined with the results of a detailed internal and oral investigation. The periodontal screening record (PSR) and the Community Periodontal Treatment Requirement Index (CPITN) are used to determine the stage of the disease [224]. In turn, a complete periodontal analysis is performed, with measurements of pocket depth, clinical attachment loss and recession, plaque, bleeding, furcation and motility involvement, and often also radiographs, assessing the bone level and alveolar destruction [225].

Comprehensive diagnostics always precedes the treatment of periodontal diseases to identify the causes and adequately plan the procedure precisely. Conical beam computed tomography (CBCT) is useful in diagnosis and periodontal treatment [226,227]. CBCT results have been found to have a high accuracy (80–84%) vs. interoperation statements attachment furcations. Comparing the CBCT of interoperation measurements, the vertical or horizontal accuracy bone loss was found to be between 58% and 93%. The CBCT is, therefore, an accurate diagnostic tool in periodontology [228]. Three-dimensional CBCT images are necessary for the diagnosis of inter bone defects and damaged furcation class cheekbone/language, therefore, the use of CBCT provides obvious advantages in periodontology [229]. However, despite this, the use of CBCT in periodontology is limited, especially for the assessment of furcations and congenital malformations [230].

The treatment is divided into four stages [219], starting with the hygienization phase and the initial treatment, through to the proper non-surgical treatment phase, the surgical treatment phase in a very advanced disease stage when non-surgical treatment has not brought good results, to the last maintenance phase for the long-term maintenance of the periodontal tissue condition achieved in the previous steps of treatment. The hygienization phase aims to remove the cause of infection; eliminate or at least reduce inflammation; remove plaque and tartar; clean the teeth; correct iatrogenic factors, carious cavities, and occlusion; and undertake minor orthodontic procedures. Very loose teeth are removed in this phase, while damaged or incorrect fillings and prosthetic restorations in the remaining teeth are corrected.

During this phase, the cooperation of the periodontist with the patient is necessary and efforts to convince the patient to take the correct approach to the strategy of solving his/her health problem. Figure 15 presents the assumptions of a behavioral strategy based on the idea of Deming's circle. Deming's cycle of (PDCA) Plan-Do-Check-Act illustrates the basic principle of continuous improvement [9]. The TIPPS concept comprises several successive steps: Talk-Instruct-Practise-Plan-Support [8]. It is another example of the use of the fives' rule concept in this paper.

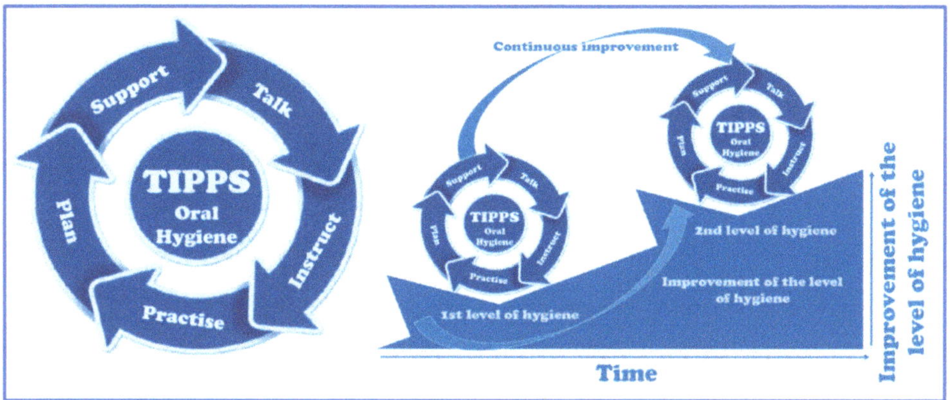

Figure 15. Diagram of Deming's circle adopted for the prophylaxis of periodontal diseases and a diagram of the continuous improvement of the level of oral hygiene with a TIPPS circle.

This cycle begins with the dentist talking to the patient about the causes of periodontal disease and the advisability of plaque removal, then instructing how to remove it effectively. The next step is to practice the application of the recommended methods. Then, together with the patient, a plan is developed by the dentist that the patient incorporates into their daily oral hygiene. The last stage of this cycle consists of supporting the patient in this action and eliminating the mistakes made, which results in the continuous improvement of the quality of activities undertaken in this phase of treatment from cycle to cycle.

The dentist's task is to continually control the indicators of oral hygiene and strive to obtain favorable values. It is necessary to determine several quantitative indicators based on a detailed analysis of each tooth and the space between them after appropriate measurements (Figure 16).

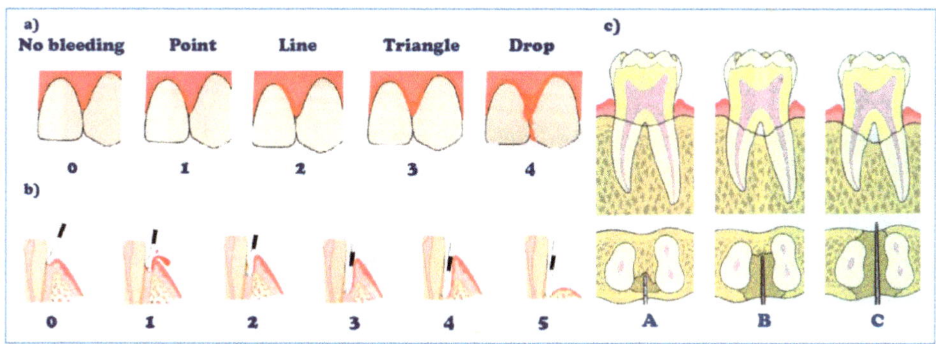

Figure 16. Characteristic damage caused by the development of the periodontal disease: (**a**) papilla bleeding; (**b**) depth of periodontal pockets; (**c**) classes of furcations.

The condition of the periodontium of patients in everyday clinical practice is assessed by Basic Periodontal Examination (BPE) indicators when the World Health Organization Probe is used. BPE was developed by the British Society of Periodontology in 1986, and the latest changes come from 2011 and mainly consist of * marking the furcations occurring in a given sextans immediately after the BPE code. BPE is a simple and quick screening tool that is used to indicate the periodontium condition, provide basic guidance on treatment needs, and help motivate patients to improve their oral hygiene. The accurate evaluation of periodontal tissues is therefore an important element of patient management. The procedure includes the operations shown in Table 4, and scoring codes are shown in Table 5 [231].

Table 4. Method of determining Basic Periodontal Examination (BPE) indicators.

No	Steps by the dentist to assess BPE
1	A World Health Organization WHO BPE probe is used. This has a "ball end" 0.5 mm in diameter, and a black band from 3.5 to 5.5 mm. Light probing force should be used (20–25 g).
2	The dentition is divided into 6 sextants: upper right (17 to 14), upper anterior (13 to 23), upper left (24 to 27). lower right (47 to 44), lower anterior (43 to 33), lower left (34 to 37).
3	All teeth in each sextant are examined (with the exception of 3rd molars).
4	For a sextant to qualify for recording, it must contain at least 2 teeth (if only 1 tooth is present in a sextant, the score for that tooth is included in the recording for the adjoining sextant).
5	The probe should be "walked around" the sulcus/pockets in each sextant, and the highest score recorded. As soon as a code 4 is identified in a sextant, the clinician may then move directly on to the next sextant, though it is better to continue to examine all sites in the sextant. This will help to gain a fuller understanding of the periodontal condition, and will make sure that furcation involvements are not missed. If a code 4 is not detected, then all sites should be examined to ensure that the highest score in the sextant is recorded before moving on to the next sextant.

Table 5. Method of determining the scoring codes value when evaluating Basic Periodontal Examination (BPE) indicators.

Code BPE	Description
0	No pockets > 3.5 mm, no calculus/overhangs, no bleeding after probing (*black band completely visible*).
1	No pockets > 3.5 mm, no calculus/overhangs, but bleeding after probing (*black band completely visible*).
2	No pockets > 3.5 mm, but supra- or subgingival calculus/overhangs (*black band completely visible*).
3	Probing depth 3.5–5.5 mm (*black band partially visible, indicating pocket of 4–5 mm*).
4	Probing depth > 5.5 mm (*black band entirely within the pocket, indicating pocket of 6 mm or more*).
*	Furcation involvement.

Comments: Both the number and the * should be recorded if a furcation is detected—e.g., the score for a sextant could be 3 * (e.g., indicating probing depth 3.5–5.5 mm plus furcation involvement in the sextant).

An example BPE score grid:			
	4	3	3 *
	-	2	4 *

In addition, the determinations or measurements of other indicators given in Table 6 are made respectively. Among other things, the Plaque Index (PI), Bleeding on Probing (BOP), Periodontal Probing Depths (PPDs), the Furcation Index (FUR), and the Gum Recession Index (REC) are determined to finally establish the Community Periodontal Index of Treatment Needs (CPITN).

The interpretation of the BPE score depends on many factors that are personalized for each patient. However, these results should be taken into account by the dentist among other factors when making decisions about the prophylaxis or therapy strategy for a given patient.

Table 6. Indicators determined in order to objectively assess the condition of the periodontium and its diseases.

Indicator Type	Description or Specification	
BOP Bleeding on Probing (Papilla Bleeding Index PBI)	Assesses the presence of bleeding gums; a patient with less than 10% bleeding is counted as having a healthy mouth cavity; this indicator is assessed using a WHO or North Carolina probe; the probe should "go around" the pocket on each side and if there is bleeding, the site is noted on the plot (at the mesial, distal, buccal, palatal and lingual surfaces); after using the appropriate computer software, the system calculates the proportion of bleeding areas.	
PI Plaque Index (Approximal Plaque Index API)	Assesses the presence of plaque; this indicator is assessed using a WHO or North Carolina probe or by plate staining; the probe should "go around" the tooth on each side and if plaque is present, the site is noted on the plot (at the mesial, distal, buccal, palatal and lingual surfaces); after using the appropriate computer software, the system calculates the proportion of biofilm area.	
PPDs Periodontal Probing Depths	Assesses the depth of the pockets; this index is assessed using the North Carolina probe; the probe should "go around" the pocket on each side and record the measurement results in millimeters on a plot with the location (at the mesial, distal, buccal, palatal and lingual surfaces).	
REC Gum Recession Index	Assesses gingival recessions; this index is assessed using the North Carolina probe; the results of measurements of the distance in millimeters from the cement-enamel junction (CEJ) [232] to the gingival margin should be made and recorded.	
MOB Mobility Index	Assesses the degree of tooth mobility: the other end of the mirror and a metal probe can be used to determine this index; While holding the lock with two tools, move it in all directions.	
	Code	Grade on a scale of 0–3.
	0	No movable property.
	1	Gentle mobility.
	2	The tooth moves horizontally >1 mm.
	3	The tooth moves horizontally and vertically >2 mm.
FUR Furcation Index	Assesses the degree of furcation using a furcation probe.	
	Code	Measure with an appropriate furcation probe.
	A	Measurement with a furcation probe 1–3 mm.
	B	Measurement with a furcation probe 4–6 mm.
	C	Measurement with furcation probe >7 mm.

The activities related to the prevention and treatment of periodontal diseases in primary care are schematically presented in Figure 17. These sequential activities begin with the assessment and diagnosis of the patient's oral cavity and soft and hard tissue condition, which is supported by a whole set of indicators, including BPE [219]. This enables the assessment and constitutes the basis for explaining to the patient about the state of his disease and the proposed, and indeed necessary, therapeutic measures. An important aspect of the entire therapy is the reliable and accurate keeping of medical records. As a result of the assessment, it is possible to use the TIPPS strategy and analyze all the risk factors for periodontal disease. In fact, every measure of assessing and diagnosing a patient's oral condition is a screening test. There are different methods of proceeding when the BPE indexes do not exceed 2, while the procedure is different when BPE is equal to 3 or 4 and when furcations occur. Then, a full periodontal examination should be performed. The next phase, when BPE indicators are relatively low, comes down to influencing the change in the patient's behavior, according to the TIPPS strategy. The next phase concerns treatment in the non-surgical periodontal therapy phase.

Long-term maintenance consists of dental prophylaxis for patients without periodontal diseases and supporting therapies for people with a history of periodontal diseases. In the case of patients receiving dental prophylaxis, plaque, tartar, supragingival discoloration, and subgingival deposits are removed when necessary. For patients undergoing periodontal supportive therapy, the instrumentation of the root surface takes place in places with a probing depth of ≥4 mm, where subgingival deposits are present or which bleed during probing. Dental implants must also be maintained. In some cases, it may be advisable to refer the patient to a specialist or primary care physician, especially when there is a reasonable suspicion of systemic complications or a need for more extensive surgery. This is particularly true for patients with a BPE 4 score in any sextant with an additional modifying factor, such as a disease state affecting periodontal tissues or when diagnosed with aggressive periodontitis.

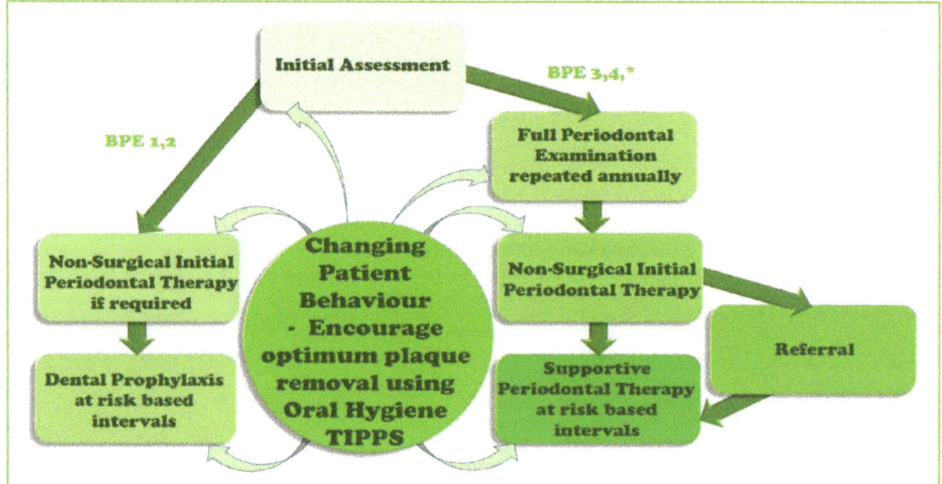

Figure 17. The scheme of the main elements of the prevention and treatment of periodontal diseases in primary care.

Table 7 shows the procedures that should be adopted depending on the results of activities carried out in the hygienization phase of the periodontal disease treatment.

The Approximal Plaque Index (API) is the ratio of the sum of positive findings/sum of investigated approximal spaces multiplied by ten, with one meaning that there is a change and 0 meaning that there is no change. If the API index is lower than 25%, the situation is correct; it is good if it does not exceed 39%; and it is insufficient if it exceeds 70%.

Very loose teeth are removed in this phase, while damaged or incorrect fillings and prosthetic restorations in the remaining teeth are corrected.

In the proper phase of treatment of periodontal diseases, defined as non-surgical, deep subgingival scaling, analysis and correction of the bite, and the permanent or temporary immobilization of teeth are usually performed. Mechanotherapy used in this phase is the most crucial method for the elimination of periodontal diseases or at least limiting their advancement. Mechanotherapy is both manual deep scaling and tooth-root polishing, related to the removal of subgingival calculus and contaminated cement from the root surface. It results in the elimination of pathogenic bacteria, preventing their re-colonization and the diminishing pockets of periodontal. Under these conditions, within six months to a year, new fibres of the gum attachment are formed, with time more and more resistant to the adverse effects of bacteria. Pharmacological treatment is also used, and targeted antibiotic therapy is preceded by non-invasive bacteriological tests in periodontal pockets.

Table 7. Preventive and/or therapeutic procedures depending on the results of activities carried out in the hygienization phase of periodontal disease treatment and the BPE score results.

Diagnosis		Preventive and/or Therapeutic Measures Necessary to Be Taken by a Dentist	Treatment Phase	
Code BPE	Description			
0	normal	No need for periodontal treatment.	0	No treatment required
1	bleeding after probing	OHI	I	Hygienic training
2	dental stone supra- or subgingival, iatrogenic injuries	OHI, the removal of plaque retentive factors, including all supra- and subgingival calculus.	II	I + scaling supra- and subgingival
3	depth of periodontal pockets 3.5–5.5 mm	OHI, RSD		
4	depth of periodontal pockets >6 mm	OHI, RSD. Assess the need for more complex treatment; referral to a specialist may be indicated.	III	I = II + surgical treatment
*	furcations	OHI, RSD. Assess the need for more complex treatment; referral to a specialist may be indicated.		

Comments: As a general rule, radiographs to assess alveolar bone levels should be obtained for teeth or sextants where BPE codes 3 or 4 are found.

Abbreviations: OHI—oral hygiene instruction; RSD—root surface debridement.

In the phase referred to as surgery, appropriate surgical resection procedures are performed related to the removal of diseased and/or regenerative tissues to rebuild damaged bone defects to restore the bone support of the teeth and soft tissue reconstruction. They are usually deep cleansing treatments for the flap pockets of the periodontal.

Despite periodontal treatment, various complications can occur, including reversible gingivitis [217], necrotizing ulcerative gingivitis, acute [233] gingivitis, and the irreversible destruction of the alveolar bone and surrounding tooth structures [234], aggressive periodontitis [235], and periodontitis as a manifestation of systemic disease [236]. In this case, it is necessary to extend the treatment to the proper phase and apply appropriate therapeutic methods.

The last maintenance phase after proper periodontal treatment is aimed at the long-term maintenance of the periodontal tissues improved as a result of the treatment performed in the previous steps. The treatment of periodontal disease continues as long as the patient has his/her own teeth.

Periodontological treatment is usually associated with implant-prosthetic procedures to supplement missing teeth and prevent tooth loss. Of course, if it is necessary to extract teeth, it is necessary to continue implant-prosthetic treatment.

An example of such treatment from the authors of this paper's own practice is given in Figure 18. An example is presented in which a 57-year-old female patient came to the dentist with a 3-degree unstable (MOB = 3) prosthetic restoration of the 12–24 segment. The pantomographic X-ray revealed significant bone loss in the area of all abutment teeth and the resorption of the root of tooth 21. The intraoral examination showed a 3-degree loosening of the teeth (MOB = 3) and periodontal disease. Due to irreversible inflammatory changes and the looseness of the teeth, the prosthetic restoration with the abutment teeth was qualified for removal and replacement with a restoration based on three implants located in the area of teeth 12, 22, and 24. Periodontal disease was also found in the area of the remaining teeth in the maxilla and all the mandibular teeth, requiring starting periodontal treatment and the constant monitoring of the progress of this disease according to the TIPPS strategy at a given stage of disease development. The remaining teeth were not loose. The procedure consisted of tooth extraction after the prior disassembly of the prosthetic restoration, the provision of a temporary restoration, and implantation performed after 10 weeks in a two-stage procedure—i.e., prosthetic loading after approx. 4 months after implantation with a cemented bridge based on individual abutments. In order to assemble the prosthetic bridge, a surgical template was previously made for the targeted drilling of holes for implants after a careful inventory of the remaining bone base. The remaining teeth in the mouth were subjected to periodontal procedures according to the TIPPS strategy.

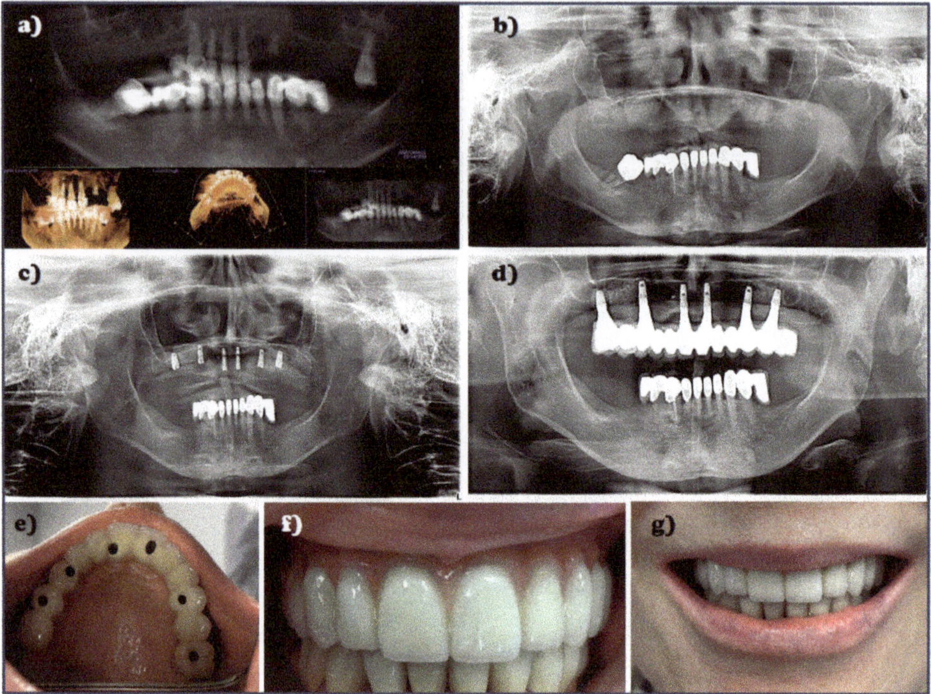

Figure 18. An example of an implant-prosthetics treatment of the female patient with 3-degree unstable (MOB = 3) prosthetic restoration of the 12–24 segment due to periodontal disease. (**a**–**d**) The pantographic X-ray images: (**a**) revealing significant bone loss; (**b**) after tooth extraction from periodontal diseases reasons; (**c**) after the insertion of the implants using a surgical template; (**d**) after installing the finished prosthetic restoration; (**e**–**g**) view of the prosthetic restoration: (**e**) in the oral cavity; (**f**) full-arch prosthetic bridge; (**g**) the patient's full smile after treatment.

The condition of the periodontium is a decisive factor in the prognosis in restorative treatment. A prerequisite is a stable gingival margin and no bleeding tissue [237]. Specific periodontal treatments are aimed at increasing the length of the teeth sufficient for retention. Failure to perform a restoration may lead to treatment failure—e.g., the inability to make an impression and prepare the restoration. Periodontal treatment should follow the restorative method; therefore, tooth repositioning as well as changes in the shape of soft tissues may occur.

Over the last several decades, huge progress has been made in the field of implantology and dental prosthetics. Undoubtedly, the breakthrough was the work of P.I. Brånemark [238–242] concerning cylindrical dental implants with a helical surface and secondary stabilization provided by osseointegration. Especially since then, modern dentistry has widely used engineering support. Most generally, the interdisciplinary branch of technology dealing with these issues is referred to as Dental Engineering. Engineering activities in this area fully correspond to contemporary trends in industrial development [108,243–248], and dental engineering has all the achievements of material engineering and material processing technologies, including additive technologies, manufacturing engineering with computer-aided design and computer-aided manufacturing (CAD/CAM), tissue engineering, as well as information technology and automation and robotics, taking into account the development of machines and technological devices used both by engineers producing elements of prosthetic restorations and implantable devices and by dentists directly during the implementation of medical procedures. An extremely important issue is the use of medical imaging methods, including intraoral and extraoral scanning, and mainly cone-beam computed tomography (CBCT), as the basis for treatment planning and designing implants and prosthetic restorations. A very large share of engineering works in modern dentistry determines that this area is also subject to general processes related to industrial development. In the previous stage, which can be referred to as Dentistry 3.0 and as noted previously, the most important attribute of real progress was the progress in conservative dentistry and the implementation of X-ray imaging of the patients' dentition. The current stage of Dentistry 4.0, described along with this newly introduced concept in the original works of L.A. Dobrzanski and L.B. Dobrzanski [108,248], is characterized by advances in cloud computing, 3D imaging with the use of CBCT, data manipulation, personalized additive technologies, so-called 3D printing. Information on digital dentistry can also be found in other publications [249–253], although the problem posed today in one of them [188] could be surprising—is digital dentistry disruptive or destructive? It is important to realize that this is not an academic problem, but a normal practice in numerous centers for the manufacture of prosthetic restorations and in numerous dental clinics. It is therefore about real activities, and not about hypothetical studies. Fortunately, the conclusion of this overview is positive because it proves that digital dentistry has not been destructive. The publication [254] provides a detailed division of tasks of the center for the manufacture of prosthetic restorations and the dental clinic, as well as patients' expectations regarding implantological treatment in accordance with the concept of Dentistry 4.0. The target diagram of horizontal and vertical integration in intelligent centers for the production of prosthetic restorations was also indicated, along with an example of a technological line. Dentistry 4.0 is accompanied by significant benefits and improvement in the care of the oral cavity, as well as minimizing the cost of manufacturing prosthetic restorations which are personalized and, extremely importantly, time saving for a dentist, medical staff, as well as the team of dental engineers designing and manufacturing prosthetic restorations. Finally, this is significantly and positively felt by the patient, who requires less time for visits and attempts to adjust prosthetic restorations. Thanks to digital technology, it is possible to reduce the production costs of some prosthetic restorations by over 70%. Systemic and computerized activities favor the integration of networks of suppliers and patients. The implementation of this concept, however, requires highly trained engineering staff and significant investments in technological machines, computer hardware, and specialized computer software, which is even several dozen times greater than in the case of previously used conventional methods and technologies.

The above-mentioned standards of the modern center for the manufacturing of prosthetic restorations fully correspond to the dental version of the smart factory, in accordance with the requirements of the modern stage of Industry 4.0 of the industrial revolution. The Industry 4.0 stage consists of systematically implemented cyber-physical systems, although the Industry 4.0 model included in the source reports [243–245] turned out to be incomplete, as it reduced the problem only to the important, albeit partial, issue of cyber-IT systems. The criticism of this model has led the authors to develop an augmented holistic Industry 4.0 model [108,246,247,255]. In the technological plane of this model, there is an appropriately developed current model, which is one of the four components of the technological plane (Figure 19).

This technological level also includes materials with the achieved material design stage Materials 4.0 [108,248,256] (Figure 20), technological machines and devices, as well as technological processes that cannot only take into account additive manufacturing.

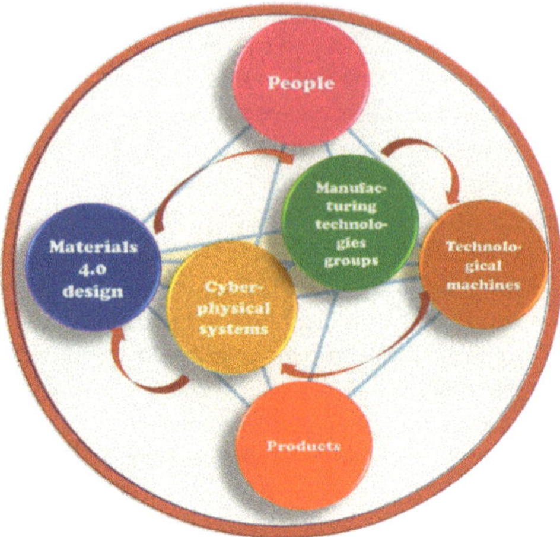

Figure 19. Schematic of the extended holistic Industry 4.0 model.

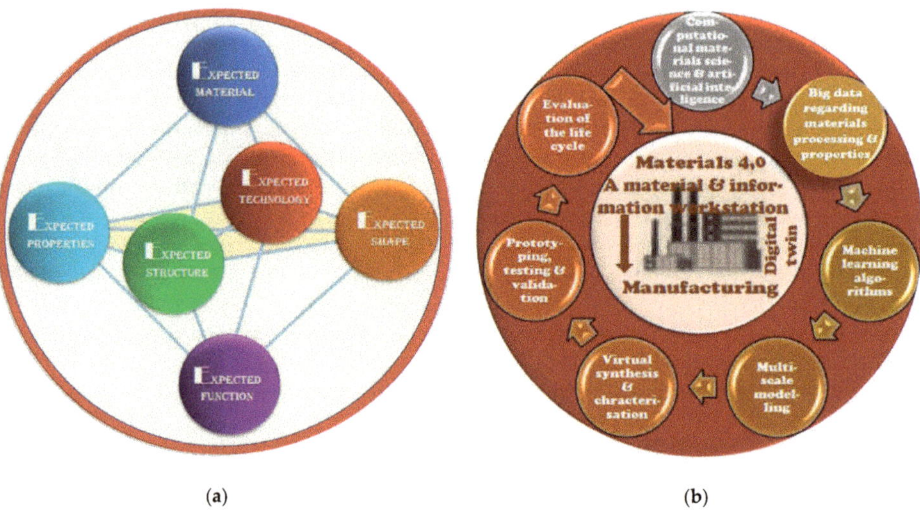

(a) (b)

Figure 20. Material engineering paradigm (**a**) and a diagram of the Materials 4.0 methodology of material design as part of engineering design (**b**).

Therefore, the material issues play an important role in the implementation of the humanistic mission and the tasks of the engineering community in general, including the implementation of the assumptions of Dentistry 4.0. The material engineering paradigm is defined by the 6xE principle [257] (Figure 20). The expected operational functions of the product, including the prosthetic restoration used in dentistry, are strictly dependent on the expected functional properties of the product, which can be achieved by designing the expected material, processed in the expected production process, which allows achieving both the expected shape and the expected geometric features of the product, as well as the expected structure of the material, which determines the set of the expected mechanical, physical, and/or chemical properties, on which the expected operational functions of the product are strictly dependent [257,258] (Figure 20). There are no starting preferences in the material design process, as a result of which all engineering materials, regardless of whether they are metals, ceramics, polymers, composites, or special materials—e.g., porous—are equal and are subject to multi-criteria optimization. Numerous criteria of varying weight are taken into account, starting with the determination of a set of requirements for functional properties in the procedure "on-demand" material: the possibilities and conditions of operation and maintenance, and the method of removing material waste in the post-consumer phase, through production conditions, price conditions, and cost related to the material acquisition and the entire material and product life cycle, through the structural and chemical composition and modeling of all processes and properties related to materials, inevitably computer-aided. With the passage of time, initially, the methodology of selecting and then more and more advanced methods of material design were developed (Figure 20) [108,246,248,257,258].

The initial trial and error method was replaced with a newer and newer approach, through the development and verification of the concept and prototyping and its evaluation in laboratory tests and testing in real working conditions, which is still often used today, Another method involves the use of advanced modeling methods physical and artificial intelligence and material data to the most advanced stage. Materials 4.0 widely uses cyber-physical systems, material data, artificial intelligence tools, smart machine learning algorithms, and cooperating human systems. It is no different when it comes to designing dental materials. It should be realized that, unfortunately, due to the poor substantive education of many people dealing with these issues in dentistry, in this professional group many stereotypes, simplifications bordering on untruth, and erroneous or anachronistic solutions in this area are spread. This results from reading numerous published papers and from reading the content of papers sent to various editorial offices of eminent scientific journals as part of the procedures for reviewing these proposals. These are the personal experiences of the authors of this paper. Material design is closely related to technological design, which is also related to the above-mentioned reservations about anachronisms, stereotypes, and unfortunately too-often-encountered untruths. There is no doubt that, in each such case, it is acting to the detriment and sometimes even to the harm of patients. Some of these glaring examples are given later in this paper, among the too many that have been identified through a literature study.

Numerous materials have been used in dental prosthetics, which results from many years of experience and multiple attempts over many years. Virtually all groups of biomaterials [259,260] are used in dentistry, including metal, ceramic, carbon, polymer, and composite materials. The metal biomaterials include [260] Cr-Ni-Mo steels with an austenitic structure; titanium and its alloys; cobalt-based alloys; tantalum, niobium, and their alloys; and noble metals. Among the noble metal alloys, mainly Au, Pd, and Ag, with other additives, including Pt, Cu, Zn, Sn, Ga, In, Re, Ir, and Ru, were used, although due to the high cost and high density of the alloys, they are used relatively less often or even rarely.

During the oxidation preceding the sintering of the applied dental covering porcelain, internal oxidation processes take place in the outer layer of the dental restoration made of these alloys, resulting in the formation of Pd, Cu, Ga, In, and Sn oxides, facilitating the diffusion connection of the substrate with the ceramic layer. Apart from noble metals and their alloys, alloys of other non-ferrous metals are of great practical importance, as they are used because of their satisfactory mechanical properties and appropriate biocompatibility, and above all, their price is much lower, especially than Au alloys. For many years, Ni-Cr alloys played a very important role in dental prosthetics, as well as steels and/or cast steels resistant to corrosion with austenitic structure type 18-8. Today these steels are completely withdrawn from use for prosthetic purposes, due to the relevant directive of the European Union, due to the harmful health effects of nickel. Although in some countries these materials are still used for prosthetic purposes, due to the very large proportion of people allergic to nickel in the general population, where the use of such materials has not been explicitly prohibited, they should be avoided all over the world. Co-Cr alloys, also containing various concentrations of W, Mo, and Si, play a fundamental role among the material used in dental prosthetics. These alloys are derivatives of the Vitalium group of alloys, known from American applications in aviation and energy since World War II. Grade 1 pure technical titanium has been used more and more often in these applications, especially for dental implants, although usually a higher metallurgical purity is required and hence most of the major manufacturers use Grade 4 technical titanium for this purpose. Titanium alloys with Al, V, Nb, and Ta, and, in particular, the Ti6Al4V alloy known as Grade 5, belong to the materials that meet the requirements regarding strength for use in dentistry as well as for the treatment of bone fractures very well. One of the essential advantages of titanium alloys is their relatively low density. It has been reported that V can cause aseptic abscesses and Al can cause scarring, while Ti, Zr, Nb, and Ta show excellent biocompatibility [261]. Some publications [262] contain even limited information on the toxic activity of V as an alloying element in the Ti6Al4V alloy and on the possible cytotoxicity of this alloy [261,263–265]. Therefore, some publications indicate that the unfavorable interaction of V in titanium alloys can be eliminated by replacing this element with Nb. The use of Ti24Nb4Zr8Sn, Ti7.5Mo, and Ti40Nb alloys with mechanical properties comparable to their traditionally produced counterparts [239–242] or other alloys above 7% Nb may be more advantageous because they can be used with selective laser sintering technologies, and their modulus elasticity is more similar to the bone than Ti6Al4V alloy. Comparative tests of the Ti6Al4V alloy and the Ti6Al7Nb alloy used as a replacement with allegedly better bioavailability and better corrosion resistance were carried out [266–269]. A direct comparison of these alloys under the same test conditions did not show significant differences [269], and even the Ti6Al4V alloy showed a higher antibacterial activity, resistance to Gram-positive bacteria, and thrombotic compatibility than the Ti6Al7Nb alloy [269–272], although the opposite is true for Gram-negative bacteria [269]. Titanium and Co-Cr alloys can be used as porous materials also [108,248,273,274].

Material design is closely related to technological design. It often happens that the decision regarding the choice of the material requires the correction of the decision regarding the correction or selection of the technological manufacturing process, and sometimes even the design assumptions. For example, after designing the shape and determining the type of material from which a given prosthetic restoration is to be made—e.g., Co-Cr or Ti6Al4V alloys—the selected material disc is machined on a CNC milling machine. In the case of using ceramic materials, coloring paints are applied to prosthetic restorations and then sintering in resistance or microwave ovens. The last stage of manufacturing prosthetic restorations is the superficial layering of veneering ceramics [108,248].

An alternative and increasingly used technology in dental prosthetics is additive manufacturing (AM) [275–305]. Among the many possible additive technologies, the most important in relation to metal prosthetic restorations are mainly selective laser sintering (SLS), also called selective laser melting (SLM) or DMLS (Direct Metal Laser Sintering), which it is not a separate method because, in both cases, there is the presence of sintering with liquid phase participation [306]. With regard to models and some movable and temporary restorations made of polymeric materials, stereolithography is most often used. In these applications, the use of additive technologies is unrivaled in relation to other technologies, as evidenced by the results obtained by the method of procedural benchmarking developed in the works [307–310] using the dendrological matrix of technology values [307–310]. The weighted scoring method was applied using a ten-point unipolar positive universal scale of relative states without zero, where 1 is the smallest rating and 10 is the largest possible rating [307–310]. The assumed criteria for assessing the attractiveness of subjective from the point of view of customers and an independent objective assessment of the technology potential were taken into account, taking into account the appropriate weights assigned to each criterion. The value of the additive technologies for the manufacturing of solid and microporous materials in medicine and dentistry was assessed compared to other technologies. The comparison of TAM additive manufacturing technologies is characterized by the coordinates given successively for potential and attractiveness (8.6; 6.6), the potential of which is much greater than that of other technologies, which indicates the purposefulness of their development. Other technologies do not show such favorable properties, including TPM powder metallurgy technologies (3.5; 5.0), TC casting technologies (4.9; 4.4), and TMF metal foams manufacturing technologies (6.5; 4.3).

Broadly following the general augmented holistic Industry 4.0 model and due to the very large share of digitization and computerized and robotic technology, the current stage of dentistry is rightly referred to as Dentistry 4.0 [256,311,312]. The role of computer-aided design/manufacturing

(CAD/CAM) in dental engineering has grown steadily over the past decade [108,255,313–318]. It is worth noting that information about the possibility of using additive manufacturing technologies, mainly selective laser sintering/melting SLS/SLM, despite the fact that they are in fact the same processes, sometimes called direct metal laser sintering DMLS, which is slightly differentiated from the previously mentioned, has appeared in numerous publications [108,248,295,318–328].

Probably in the majority of these publications, the authors did not pay attention to the appropriate selection of the entire set of technological conditions, which include, among others, laser power, the diameter of the laser spot, the partial overlapping of stitches or not with subsequent laser passes, scanning speed, powder diameter, and its spread, powder layer thickness, etc., which may decide about the porosity variation e.g., from 0.03–10% [329,330], and this, in turn, determines the approx. 2.5-fold differentiation of the mechanical properties. Figure 21 shows, for example, the dependence of the bending strength on micro samples made of Ti6Al4V alloy on the laser power and on the diameter of the laser spot according to [319]. The top line of photos shows the structure of these materials produced by the SLS method in appropriately selected conditions, guaranteeing a bending strength of approx. 2464 MPa. The bottom line of photos presents a structure of materials ensuring a strength of only approx. 1099 MPa. The first column of photos corresponds to the illustration in red of the pores at a magnification of 100×, in the first case with a pore area fraction of 0.03% and in the second case a fraction of 10.53%. In both cases, the width of the laser dot is the same and amounts to 120 μm, while the laser power is 110 and 60 W, respectively. The presence of these pores in the second case in the second column at the bottom is visualized in a scanning electron microscope. The third column shows the fractures of the tested specimens after bending. In the first case, there is a fracture with a homogeneous structure, without pores and visible boundaries of the sintered powder particles. In the lower photo, the structure of the fracture is heterogeneous, with visible fragments where there are open pores and spherical particles, completely or partially unsintered, are visible. This is obvious proof that a too low sintering power does not guarantee its full and proper course, and therefore leads to a reduction in strength of almost 2.5 times. It is important that the flow rate of the inert gas and possible significant differences in this rate and the unexpectedness or lack of control of this flow also significantly affect the meaningful changes in porosity and thus the differences in the strength of the materials thus produced.

In order to obtain the appropriate properties, the technological conditions must be precisely optimized. For example, if the density of the Co-Cr alloy varies between 8.52 and 8.66 g/cm^3 [331] or if it is stated that, regardless of whether the material is cast after making a die or laser sintered, the same properties are obtained [332], this most likely indicates the random selection of sintering conditions. It is possible to think, as the result of this, that firstly the conclusions are faulty, and secondly that the patients' service is with bad and technologically underdeveloped prosthetic restorations. It is enough that there is an error in one of the sintered layers for the product thus manufactured to be completely disqualified. Importantly, that factory settings suggested by manufacturers of technological devices for additive manufacturing differ far from those actually necessary for use. The Authors of this paper have this knowledge from their own technological practice [329,330]. It is a very high probability, and practically certainty, that the produced materials for dental application usually achieve about 60% of the properties that could be obtained if the process was performed correctly. It is a conjecture that, in most cases, this is the case in dental laboratories. this has led to the circulation of erroneous information among dentists on the possibility of using additive manufacturing technologies in dentistry, significantly harming the implementation of this modern technology, as well as, in fact, violating the interests of patients.

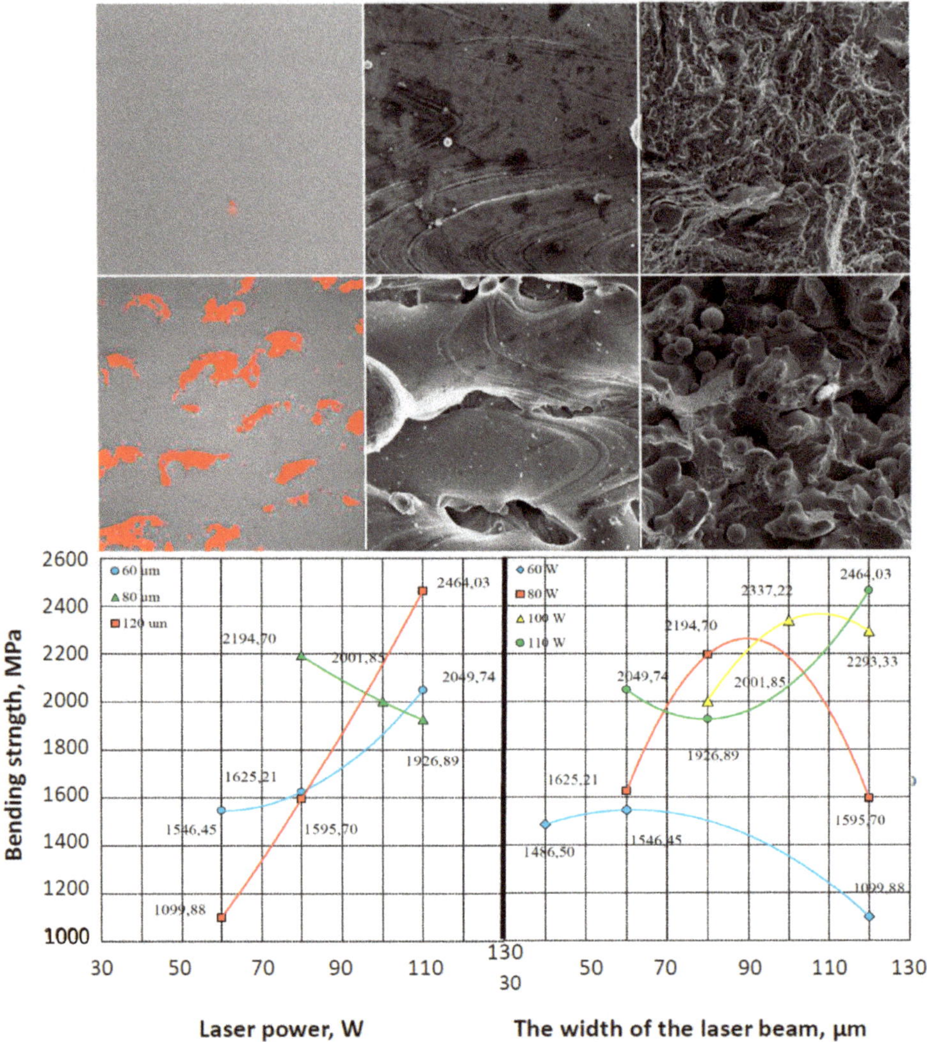

Figure 21. Influence of laser power and laser spot width on the bending strength and structure of a selectively laser-sintered Ti6Al4V alloy; the top-row of photos the structure of this material produced by the SLS method in appropriately selected conditions, guaranteeing a bending strength of approx. 2464 MPa; the middle-row of photos the structure of material ensuring the strength of only approx. 1099 MPa; the bottom-row with diagrams - the effect of the laser power (first) and the width of the laser spot (second) on the bending strength; first column of photos the illustration in red of the pores at a magnification of 100×, in the first-row with a pore area fraction of 0.03% and in the second-row a fraction of 10.53%; in both cases, the width of the laser dot is 120 μm, while the laser power is 110 and 60 W, respectively; the presence of these pores in the second case in the second column at the bottom is visualized in a scanning electron microscope; the third column—the fractures of the tested specimens after bending—in the first-row fracture with a homogeneous structure, without pores and visible boundaries of the sintered powder particles; in the second-row, the structure of the fracture heterogeneous, with visible fragments where there are open pores and spherical particles, completely or partially unsintered.

Usually, the appropriate prosthetic restoration is individually designed [101,187] based on the diagnosis of the state of damage to the patients' teeth using the cone-beam computed tomography (CBCT) method. The authors, together with the representatives of other centers, participate in the work and research on the dissemination of this diagnostic method and the development of the digitization of dental diagnostics [108,240,331–345]. Figure 22 schematically illustrates the synergistic interaction of the three pillars of the Dentistry 4.0 model, including dentistry, dental engineering, and materials engineering. The authors' view on this issue is presented in detail in [329].

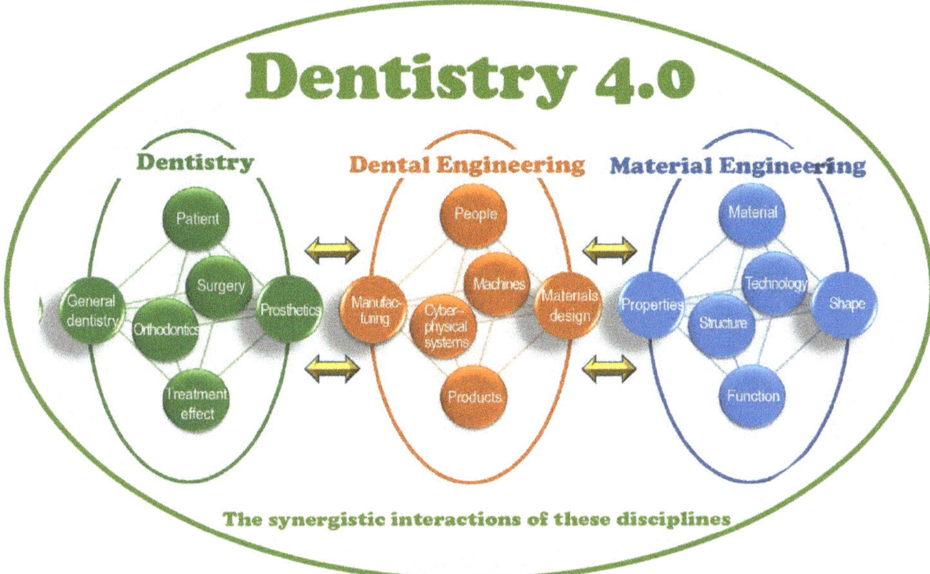

Figure 22. General diagram of the synergistic relations between dentistry, dental engineering, and materials engineering in the Dentistry 4.0 model.

Fulfilling the tasks of modern dentistry requires the use of various engineering devices to replace teeth removed due to disease or loss due to other causes, as well as those resulting from malformations. Such medical devices, including dental implants and other prosthetic restorations, are manufactured artificially in order to be placed wholly or partially under the epithelial surface in order to fulfill their intended functions for a long time.

A new class of implantable devices, the so-called implant-scaffolds (Figure 23), allow living cells to grow into pores or a surface composed of characteristic protrusions, and not only on their surface. Implant-scaffolds, which not only concern the teeth but can also apply to bones in general for applications such as orthopedics or maxillofacial surgery, are devices that, if only in part, are intended to weaken the artificial character of the parts implanted into the body. They consist of a solid part, typical for all implants, but also have porous parts [346,347] or protrusions on the surface [348,349] as scaffolds that create spaces of 400–800 nm in which living cells can proliferate either after implantation into the body, or prior to implantation, as autologous cultured under laboratory conditions. As a result, after implantation, engineering-biological materials are produced after implantation, coexisting with the body.

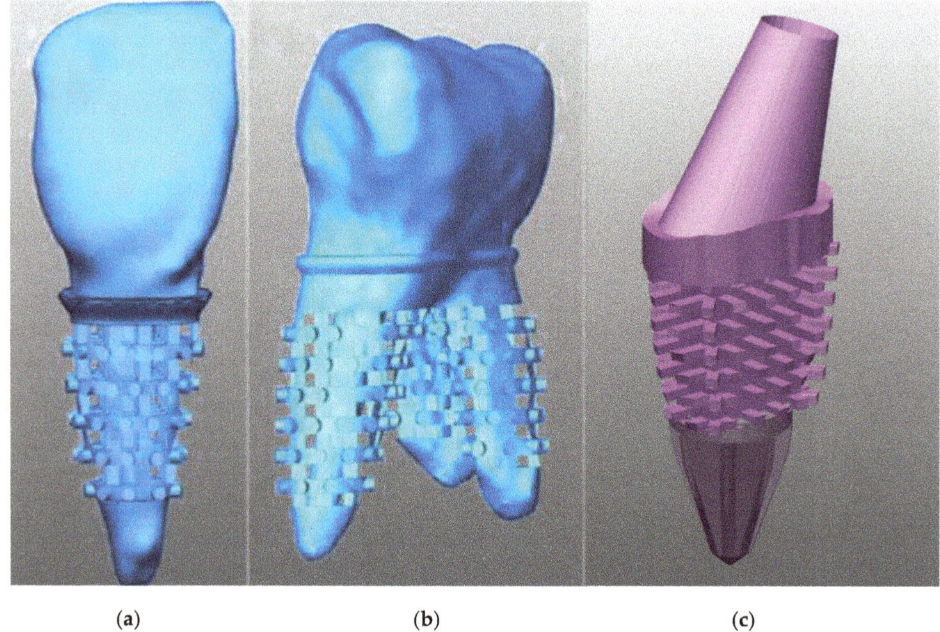

(a)   (b)   (c)

Figure 23. Examples of implant-scaffolds (**a**,**b**) with a porous structure and (**c**) with protrusions on the surface.

Both implants and implant-scaffolds are made of biomaterials, and can also have biocompatible surface coatings when applied to substrates of other materials. These layers are used in the manufacture of implants and other prosthetic restorations using some surface engineering technologies [334,350–352]—e.g., Atomic Layers Deposition (ALD), and less preferably Physical Vapor Deposition (PVD). These layers can prevent the re-diffusion of the base metal atoms into the ceramic surfaces of crowns and bridges, which detracts from the aesthetic effects of prosthetic restorations, especially those made of titanium and its alloys, and prevents cracking of the face layer of porcelain. In the case of the manufacturing of a selectively sintered laser skeleton porous structure—e.g., from titanium powders and its alloys or Co-Cr alloys, as well as other alloys used in dentistry—thin layers can be applied using the ALD method—e.g., TiO_2, ZrO_2, or Al_2O_3—to improve the proliferation and growth of living cells inside the pores [273,274].

Figure 24 shows as an example the models of fully porous implant-scaffolds with dimensions of $10 \times 10 \times 10$ mm, made of pure titanium, coated in the ALD process with TiO_2 nanolayers in 500, 1000 and 1500 cycles. The next column shows the surface morphology of these samples observed in the Atomic Force Microscope AFM corresponding to these thicknesses. The next column contains Fast Fourier Transform FFT of pure titanium sample film coated with a TiO_2 layer for 1500 cycles tested in a high-resolution transmission electron microscope confirming the crystalline Ti substrate structure and amorphous structure of TiO_2 layer.

Among the dental specializations in the Dentistry 4.0 model, dental and maxillofacial surgery with implantology and periodontics as well as prosthetics and implant-prosthetics are of particular importance. Other specialties in general dentistry and orthodontics are slightly less related to this concept, although they cannot be excluded.

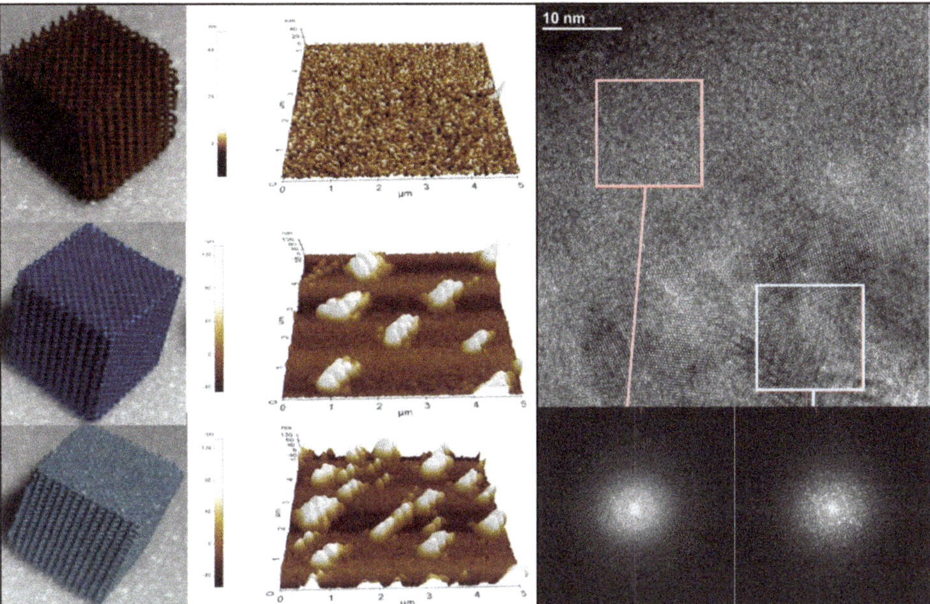

Figure 24. Examples of pure titanium porous samples manufactured by the selective laser sintering SLS method and coated with the atomic layers deposition ALD method inside the pores with a TiO$_2$ nano-layer; the first column—the models of fully porous implant-scaffolds with dimensions of $10 \times 10 \times 10$ mm, made of pure titanium, coated in the ALD process with TiO$_2$ nanolayers successively in 500, 1000 and 1500 cycles.; the middle column—the surface morphology of these samples observed in the Atomic Force Microscope AFM corresponding to these thicknesses; the last column—Fast Fourier Transform FFT of pure titanium sample film coated with a TiO$_2$ layer for 1500 cycles tested in a high-resolution transmission electron microscope HRTEM confirming the crystalline Ti substrate structure (pink) and amorphous structure of TiO$_2$ layer (violet).

It is necessary to note the differences between the classical approach and the Dentistry 4.0 concept. In the first situation, the work of a dental engineer began when the patient no longer had any teeth, which needed to be restored. Currently, it is necessary for a dental engineer to participate in treatment planning together with the dentist before any actions by the dentist in the patient's mouth, and for this purpose advanced computational methods are used for planning surgical and prosthetic procedures. In order to develop a treatment plan, the patient should first take impressions showing the condition of the dentition in the patient's mouth and perform a CBCT tomogram [337–339]. Based on the results of these activities performed jointly by the dentist realizing the treatment together with the dental engineer and after the diagnosis of the patient's dental condition, including fractures of the root or crown, irreversible pulp disease, and carious lesions or other important circumstances, the teeth are qualified for removal as a result of these diagnoses. The dental engineer and the dentist still make a plan for the type and method of installation of the intended prosthetic restoration, taking into account the existing teeth or dental implants that will need to be introduced. The dental engineer verifies the developed treatment plan, either with the digital twin technique in the virtual space or through experimental verification with the use of wax-up or mock-up restorations along with the production of PMMA models that faithfully imitate the final prosthetic restoration. It enables the dentist to make the necessary corrections before the complete manufacturing of the restoration, similar to assisting the dentist in assessing the parallelism of the ground pillars. The pillars teeth require graded preparation, and the verification of this occlusal position is therefore a guarantee of minimizing premature contacts, preventing porcelain deterioration after the final installation of the prosthetic restoration. In the case of work mounted on implants, very important support from the dental engineer is the design and production of surgical templates. After designing the placement of implants in the oral cavity, taking into account the reconstruction of the dental condition using medical imaging using CBCT methods, intraoral and extraoral scanning after making precise impressions made together with the dentist, the final design that is the basis for the creation of a surgical template is a specialized design made using CAD software. Oftenm these activities are also accompanied by the individual design and production of the necessary instruments. These activities are accompanied by manufacturing processes with the use of milling in a CNC machining center, and most often with the use of additive technologies, usually SLA for polymer materials or selective laser sintering SLS for metal alloys and composite materials. A successful design process requires close cooperation between the dentist and the dental engineer.

The widespread implementation of the Dentistry 4.0 assumptions still requires a continuous, extensive educational campaign among dentists to raise their level of awareness of the need, possibilities, scope, and benefits of joining these activities, as well as the education of new dental engineers with a hybrid knowledge of engineering and biomedicine, including the basics of dentistry and anatomy. Dental engineers are required to have proficiency in medical imaging techniques, including CBCT; very advanced IT tools, including programming skills; the use of cloud computing, Internet of Things, and augmented reality; as well as theoretical and practical knowledge of bioengineering materials and advanced manufacturing engineering, and using themed for technological machinery and equipment, including additive technologies. This knowledge must be supplemented by issues related to techniques and methods ensuring the aesthetic production of prosthetic restorations and methods of communicating with patients. The relationship between the dentist and the dental engineer undergoes qualitative changes. In fact, the entire program of designing and manufacturing a prosthetic restoration falls under the competence of a dental engineer who works with dentists to assemble the prosthetic restoration in the patient's mouth. The dentist invariably supervises the entire treatment from the medical point of view. The dentist diagnoses and documents the condition of the patient and his teeth with the use of CBCT. The entire prosthetic restoration and its design and manufacturing are planned by a dental engineer with the approval of the dentist. The dental engineer is fully responsible for the design and manufacturing of this restoration, which, after approval by the dentist, is installed in the patient's mouth. In addition to guaranteeing the high quality standards and tight dimensional tolerances of the manufactured prosthetic restorations, the dental engineer often designs and produces an implant template that allows the dentist to correctly make holes for mounting implants and/or implant-scaffolds, taking into account the condition of the maxillary and mandibular bone processes diagnosed by the dentist. Using the digital twin idea, virtual simulations are performed in order to optimize the design features of prosthetic restorations and dental templates, as well as experimental verification through rapid prototyping and the production of the model using lithographic methods and the assessment of the occlusal height with the use of an articulator.

Clinical activity in modern Dentistry 4.0 standards is synergistically dependent on highly advanced activities in the field of dental engineering. It applies equally to the design of prosthetic restorations and dental implants; material design, taking into account the advanced engineering materials used in these cases; the technological design of manufacturing processes using additive technologies; and structural design using computer-aided CAD/CAM engineering methods. There is no doubt that the Dentistry 4.0 stage is an important determinant of the current level of clinical possibilities offered by modern dentistry and in the logic of this paper; it is one of the legs (in the analogy with the Chinese censer) ensuring balance and sustainable development in the next few decades. Undoubtedly, it is also a reason for many dentists and dental engineers to be proud of the results achieved so far.

4. Development Strategy for the Prevention of Oral Cavity Diseases and Coordinated and Related Systemic and Political Actions

Reading two works by M.A. Peres et al. [6] and R. G. Watt et al. [7] that were developed by the same team of authors sheds a different light on the expected development of dentistry. Despite the fact that the content of these studies may seem controversial in many respects, it is impossible to disagree with the arguments discussed in them. In addition to the consistent implementation of the outlined trends resulting from the idea of Dentistry 4.0 following the modern stage of Industry 4.0 of the industrial revolution, in these works it is not enough that it is not considered a success, but it is even considered one of the symptoms of the crisis that has hit dentistry in the 21st century. The current dentistry is still unable to overcome the global challenge of oral diseases [353–355]. Despite significant advances in science in recent decades, among others in terms of explaining the pathogenesis and etiology of oral diseases, the global scope and scale of the problem are certainly not decreasing, and it can be said that it is even increasing. It is undeniable that tooth decay is the most serious social disease in the world. Oral diseases are therefore the most serious global public health problem due to their high prevalence and their detrimental effect on individuals but also on society as a whole.

Although in High-Income Countries (HIC) the incidence of caries in children has decreased, as age increases, the problem worsens, starting in childhood and progressing through adolescence, adulthood, and old age. The effects of the disease accumulate with age. In view of significant socio-economic inequalities, caries particularly affects poor and marginalized social groups in HIC countries and especially in Low and Middle-Income Countries (LMIC). Moreover, many countries have a significant predominance of dental services in major urban centers, where the private dental practice is more lucrative and services often partially or generally not reach patients living in more remote areas of the country [355,356]. The increase in the share of caries in these countries is favored by the growing consumption of free sugars, which also affects the morbidity of, among other things, obesity and diabetes. On the other hand, it should be noted that the distribution of caries (Figure 4) is much smaller in Africa and Southeast Asia. Therefore, it should be clarified that such generalizations apply to HIC countries, especially LMICs or LICs, where the severity of this disease is particularly high. This does not mean that in Africa or Southeast Asia the problem does not occur at all, but it is naturally less influential. In addition to the pain that affects many people in the world and a significant impact on reducing the quality of life, dealing with the effects of caries and the related costs of treatment have an impact on household budgets and the social security system, the costs of sickness absenteeism, and numerous systemic and related complications and hospital treatment costs.

In the cited studies, it was noted that the shortcomings of modern dentistry are caused by a defective general system approach, and often even its lack and the implemented largely imperfect models of dental care, and the study does not blame the individual dentists who care for individual patients. It was assessed that dentistry organized in the same way as before is not able to solve the problem of the spread of caries, let alone its elimination. This view is based on the statement that caries can be prevented and, if it occurs, even reversed with appropriate treatment [20,357]. An alternative is a minimal intervention (MI) and patient-centered care approach, based on an updated understanding of the histopathological process of caries, as well as the development of diagnostic and adhesive technologies and bioactive restorative materials and, finally, a revision consultations [357]. It was even written in [355] that, while tooth decay is the main reason for tooth extraction, it is dentists who actually contribute to tooth loss because they participate in the tooth repair cycle. It has been shown that up to 70% of cases there is an increase in the size of restored surfaces, and it has been suggested that dentists should try to avoid premature restoration, as this may speed up the repair cycle [358]. Therefore, it is necessary to abandon the cyclical and subsequent preparation of the cavity, starting with a very small one and placing more and more extensive fillings, which ultimately must result in failure and tooth extraction [359,360]. As an argument for the thesis put forward in the paper [355], it was shown that studies performed in populations with little or no access to dental care indicate that most people have most of their teeth throughout their lives, despite the lack of care for oral hygiene. It is important to record the early stages of disease symptoms—i.e., changes without cavitation. These changes result from the dynamic course of the processes of the net loss of minerals. Since the initiation and progression of lesions take place throughout life, the course of caries can be controlled in various ways, which, however, do not prevent the disease. This can only be prevented by controlling pathophysiological phenomena, preventing or at least limiting the net loss of minerals [19].

Therefore, it has been shown that topical fluoride application is different from the current repair approach to reduce caries [361–363]. The use of fluoride supplements prevents decay in deciduous teeth, although the clinical evidence in this regard is not entirely convincing. In contrast, there is strong evidence that fluoride supplements prevent caries in permanent teeth, although they may also cause mild to moderate dental fluorosis [362]. Therefore, low concentrations of fluoride have a beneficial effect on the counteraction of the removal and remineralization of enamel and dentin. When fluoride is applied topically, by rinsing or with dentifrice, the concentration of fluoride in saliva drops to very low levels within hours [363]. For treatments to be effective, fluoride must be deposited and slowly released. This is how tooth coatings with calcium fluoride or the like work. Undoubtedly, both the use of fluoride and the progress of conservative dentistry have a positive effect on reducing the extent of tooth decay at a given age and delaying the onset of the cavitation process. Such actions have been proven to be highly effective [358,364], and the topical application of fluoride is a proven clinical prophylactic method that should be promoted and improved [7]. This importance should not be underestimated, but the beneficial effects of fluoride as a fully effective means of caries prevention cannot be overstated. These actions, even systematically performed, cannot completely prevent tooth decay, which was confirmed in the WHO report [365]. Dental caries continues to be advanced in populations systematically using fluoride [364–367]. Tooth brushing and the use of fluoride in drinking water or toothpaste may only interfere with or delay the development of caries on a global scale [147].

The consumption of free and disaccharide sugars is essential and detrimental to the development of caries [368–378]. This group includes simple sugars added to food and beverages, regardless of whether they have been prepared by the producer, cook, or consumer personally, and natural sugars present in honey, syrups, fruit juices, and fruit juice concentrates [364]. Regardless of the topical application of fluoride, caries will continue to develop when sugar is consumed in excess of 10% of a person's daily energy requirement [147,364]. Meticulous Japanese data on caries incidence, assessed over several years in Japan on two types of teeth, show a log-linear dependence on sugar consumption in the range of 0–10% in excess of the daily energy requirement, with a 10-fold increase in the incidence of dental caries [148]. It has been found that nearly half of all tooth surfaces are affected by caries in adults aged 65 and over, despite living in areas where water is fluoridated and a large number of people use fluoridated toothpaste. Such a threat was not identified when sugar consumption was reduced to less than 3% of the daily energy requirement [148]. Moreover, even then, the level of fluoride use is appropriate, while the consumption of free sugars is greater than 2–3% of the total daily energy requirement, there is a significant risk of caries [364]. There is therefore a clear relationship between the over-consumption of free sugars and the development of caries. The WHO strongly recommends limiting the consumption of free sugars to a maximum of 10% of the daily energy requirement [141,148,364,379] and recommends reducing the consumption of these sugars throughout one's life [365]. It is noted that the reduction in free sugar consumption also results from coping with excessive body weight gain and obesity [379]. The limitation of excess body weight can occur by modifying consumption and is dependent on the change in energy intake, since the isoenergetic exchange of sugars for other carbohydrates does not change body weight [380]. Quantitative recommendations in this respect concern the consumption of sugars at below 5% of the daily energy requirement only due to the association with caries [381,382]. However, such limitations have been suggested [365], although there is no clear evidence of a correlation between the consumption of free sugars and the presence of caries [147,148], although such limitations are confirmed by the performed analyses [148,382]. It is indicated that immediate effects cannot be assessed because the negative effects of caries on the health of the teeth accumulate from childhood to adulthood [383,384]. For this reason, the consumption of free sugars should be minimized throughout one's life [364].

In the paper [7], it was pointed out that the issue of sugar consumption in the world was treated on a peripheral basis, and now this it is gaining strategic importance due to its strategic role in the field of public health. The voluntary arrangements made so far with the sugar-containing production and distribution industry have failed [385–387]. It is therefore necessary to develop a global strategy to reduce sugar consumption [384]. This also applies to ready-made food for infants [388,389], which largely contributes to the formation of eating habits that result in a predisposition to caries in adulthood. Among the various necessary legal, political, and organizational measures, it is suggested that the labeling of sugar-containing products, especially in excess, contains similar warnings, as is done in many countries for tobacco and alcohol. Similar actions are also proposed to those which led to a reduction in salt consumption [7].

The paper [390] emphasizes that, in almost every country, a different preventive approach should be applied and that there is no one universal preventive program. This is closely related to cultural and behavioral habits. Fluoridation can be accomplished by a variety of methods, including mouth rinsing, fluoridated toothpaste, water fluoridation, supervised tooth brushing, and other methods. Moreover, these methods should be developed and refined as the level of prevention progresses by each patient. Not only that, but it has also been shown that there are different oral health trajectories [391]. Starting on one of the possible trajectories, even in childhood or adolescence, practically excludes its change in later age. The idea is to start caries prevention according to a favorable trajectory before birth and follow it in the life cycle. Of course, health depends on many factors, including social, environmental, and economic ones. This requires constant supervision and care by primary health care staff also in the field of oral health. This applies to oral health screening, preventive education, and the use of fluoride, brushing teeth twice a day with fluoridated toothpaste and maintaining oral hygiene and patient self-care techniques. This method stimulates patients to self-control and plan appropriate actions with the support of the dentist. The dentist's activities and activity schedules are tailored to the goals of monitoring, motivating, and stimulating the patient [391].

Oral diseases are a neglected problem, mainly by governments, rarely seen in almost all countries (and perhaps even to a different degree in all countries without exception) as a health policy priority. Oral diseases are generally not mainstreamed in health policies and national health systems [392,393]. Similarly, in this sense, the profession and mission of the dentist have also been isolated, which should be changed [394,395]. As a result, dentistry does not currently meet the health needs of large social groups in many countries. It is assessed as unfavorable consumption motivations and the increasing concentration of dental treatment on the aesthetic context of the procedures performed, often combined with the motivation aimed at increasing the profits of people performing these works [355,396,397]. This thesis seems highly controversial due to the reliable performance of their tasks by many dentists, fully committed to their professional mission and in accordance with the Hippocratic Oath [398]. Consequently, this opinion appears to be unfair at least to this group of dentists, which may even include all medics.

Even assuming that, in fact, many millions of poor people in developing countries cannot afford basic dental treatment, to the point where they may never see a dentist, it does not seem justified that everyone else should give up proper care for this scope, consistent with the current state of knowledge and development at the stage of Dentistry 4.0. In the paper [397], it was written that wealthy members of society demand expensive, top-shelf treatment, mostly cosmetic, and not necessary to cure diseases. This opinion seems to be greatly exaggerated and not supported by sufficient evidence, as dental services mainly concern the treatment of diseases and the removal of their effects. It should be noted that such statements are also factually unreliable. In the case of objectively diagnosed caries, as presented in the second chapter of this article, there are numerous distant negative effects throughout the body caused directly by caries. In turn, many systemic diseases result directly from the fact that at least one tooth or rather a few teeth are missing in the stomatognathic system, which is also presented in chapter two. Removing the causes of all these diseases significantly reduces the social costs of health care and insurance for people who do not fall ill with these serious diseases. Therefore, it will not be necessary to bear the long-term costs of their hospital stays, as well as often the costs of sickness pensions and the elimination of these people from the labor market. These activities are obvious savings in countries' budgets and are by all means justified on each country scale, although it is true that many patients are motivated to undertake them in relation to themselves and often for aesthetic reasons. It cannot be denied that dentists, to some extent, perform cosmetic services also, which, however, are not dominant and are in no way financed by public funds. Regardless of the arguments presented, it is surprising in the scientific text that too many governments and dentists insist on an expensive and destructive regime of "drilling and filling and billing" [397]. Such slogans concerning social egalitarianism, bordering on populism, may be received positively by many people, but it could be no consolation that nobody, as a result of such an approach, will receive proper dental care. Unfortunately, it sounds like a policy that scientists should avoid, and communist ideas cannot meet with universal acceptance, as demonstrated by the social changes made over the last 40 years in many countries—e.g., in Europe in countries previously dominated by the former Soviet Union and in some countries, such as Belarus, where it is ongoing today. A reasonable compromise must be found, which inter alia it is the basis for the partial or complete privatization of dental services in many countries and their submission to market laws, such as in Europe [356,395]. In a discourse at this level, even serious factual arguments are lost in the climate of populism. On the other hand, it should not exempt the government of any country from its obligation to provide egalitarian oral health care for all citizens, under publicly available social security and a publicly accessible basket of basic and tax-financed health services. Nevertheless, it is shown below that this postulate is not very realistic. It is impossible to argue with the statement [397] that the aim of such activities should be to help people maintain healthy, natural teeth throughout their lives, as an important element in improving their general health. The issue, however, is far from medical considerations. In each country with a democratic system, the scope and development of policies in general, including health policy with oral health policy as part of it, are decided by electors. Elected politicians carry out all tasks in such a way as to stay in power, which may mean that this aspect is largely or completely ignored. The direct influence of doctors, including dentists, on these issues is negligible. This seems obvious. International conventions can help if they are signed and ratified in a given country. However, this is also a decision of politicians.

This is accompanied by the criticism contained in the cited studies [6,7] of an approach to dental care that is dominated by treatment, which, as described, is interventionist, specialized, and increasingly technology-driven. It should be noted that, in the previous subsection of this paper, these are the achievements deciding about reaching the stage of Dentistry 4.0, which the Authors of this paper classified as the greatest, leading achievements of dentistry in general. There is therefore an absolute double-voice in the formulated opinions. The question of whether this is a contradiction or an antagonistic contradiction, or whether the stated regularities should be interpreted as a form of coexistence or even synergy of both views. The authors dedicate the rest of their considerations in this paper to this problem.

It is estimated that dental treatment alone, regardless of the level achieved, will not solve the problem of caries removal in the world [6,7]. It is impossible to disagree with this opinion. A new and completely different approach is required to meet this global challenge. Therefore, effectively dealing with the global problem of oral cavity diseases requires a fundamentally different approach, which is the main thesis presented and demonstrated in these studies [6,7]. A systemic change is necessary and a decisive departure from the further application of the current interventionist approach. Current public health and healthcare responses are costly, largely inadequate, unfair, and billions of people in the world lack even basic oral health care. As the world intensifies its efforts to meet the Sustainable Development Goals over the next decade, oral health cannot lag behind and requires urgent, determined action. The authors of the cited studies [6,7] believe that the interventionist approach to the issues of oral cavity diseases, and mainly caries, could not effectively deal with global problems, and indeed it did not. Therefore, this approach should now be radically changed [6,7]. The main focus should be on prevention and focus all efforts on population-wide outcomes [394,399]. Even the concept of the so-called Western Dental Care, related to treatment, is more and more technologically advanced, interventionist, and specialized, which, according to the authors of this paper, corresponds to the current stage of Dentistry 4.0. This approach is undoubtedly dominant in the HCI countries, and in the cited studies it is accused of a lack of interest in the root causes of oral diseases and ineffectiveness in eliminating inequalities in access to dental care. The result is that in the LMIC countries, exclusions affect the majority of these populations, especially the poor from rural regions, and dental care is not only often financially unavailable, but very often also because of inadequate organization. The main goal of dental care systems should be maintaining oral health and promoting on achieving greater equity in oral health in socialites. The global view of dentistry covers three highly contrasting but interconnected realities. In high-income countries, the current dental care system—dominated by treatment and increasingly technology-centered—is a victim of an interventionist cycle that neither addresses the root causes of disease nor meets the needs of large sections of the population [359,394]. In many middle-income countries, the burden of oral disease is significant, and dental care systems are often underdeveloped and financially inaccessible to the majority of the population. The situation is worst in low-income countries. Although in these countries the overall burden of oral diseases is often still relatively low, the cited studies estimate that their incidence is increasing. As there are other needs with insufficient resources, investment in oral health is very limited, making dentistry a physically and financially inaccessible luxury reserved for the wealthy. As a result, most oral diseases remain untreated in the majority of the population, but especially among rural poor, who have very limited access to dental care. This, therefore, requires a change in the current individualistic clinical paradigm, as its application has not led to a lasting improvement in oral health in the global population, nor has it proven effective in removing the existing inequalities and exclusions consisting in the lack of access to the required level of dental care and dedicated in the just-in-time formula. The suggestions for changing the dental care system also concern changes in the academic and postgraduate education of dentists [354,397,400], which, as in the case of ENT and ophthalmologists, should be more related to the general education of doctors. In addition to changing the education profile in these papers are also suggestions for limiting the number of educated dentists and even limiting the number of universities or faculties educating in this area in some countries, as well as limiting the number of dental specialties.

It is impossible to resist the impression that the authors of these quoted studies [6,7] overestimate the roles of the medical staff, especially doctors, medical personnel, and even medical administration, because these otherwise legitimate objections do not so much apply to dentistry as a part of medicine, but to politicians responsible for the legal status, financial streams of the economy, and social security policies in each country. Therefore, in these studies, there is a postulate to abandon isolation and support the integration of dentistry with the mainstream of primary health care in individual countries. The global drive to provide universal health insurance provides an excellent opportunity to do so.

Undoubtedly, the problem requires a broader view of dentistry as part of the overall health care system. The thesis about particular discrimination of dentistry, due to the exceptional diversity of access and financing in different countries, is not confirmed by the facts. Figure 25 shows, for example, the prevalence of cancer, stroke, and heart disease [304,401]. These diseases, in many countries, and practically all over the world, affect millions of people. Data concern the member countries of the Organization for Economic Co-operation and Development (OECD) that works to build better policies for better lives. It is impossible to find analogies with the prevalence of oral diseases, especially caries.

On the other hand, there are interesting examples concerning expenditures per person on health care and the share of expenditures in the costs of gross domestic income GDP (Figure 26) [395,401]. Data concern the member countries of the OECD, which brings together 36 highly developed and democratic countries. Despite this fact, the health care expenditure per person differs in these countries up to more than 50 times.

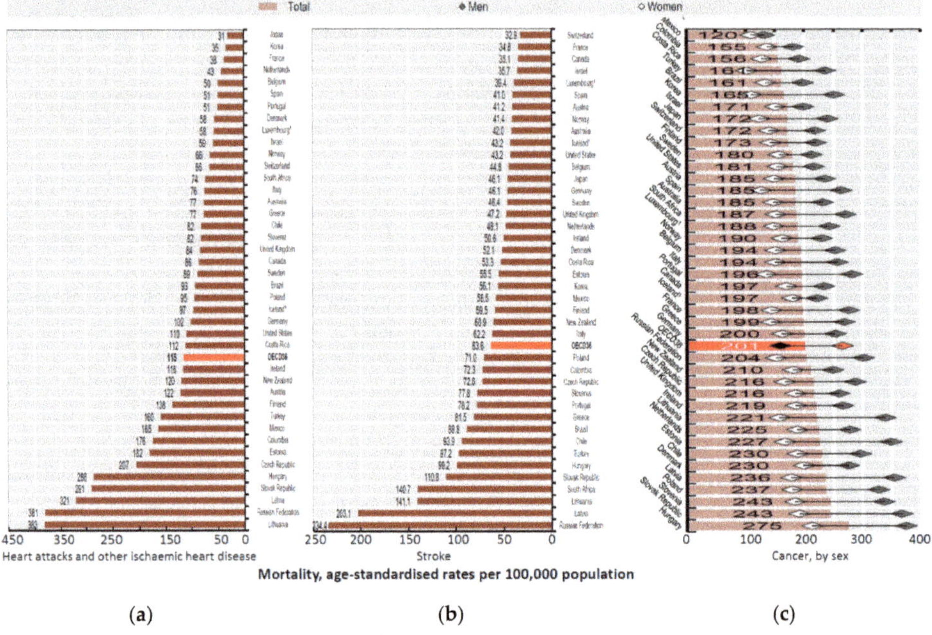

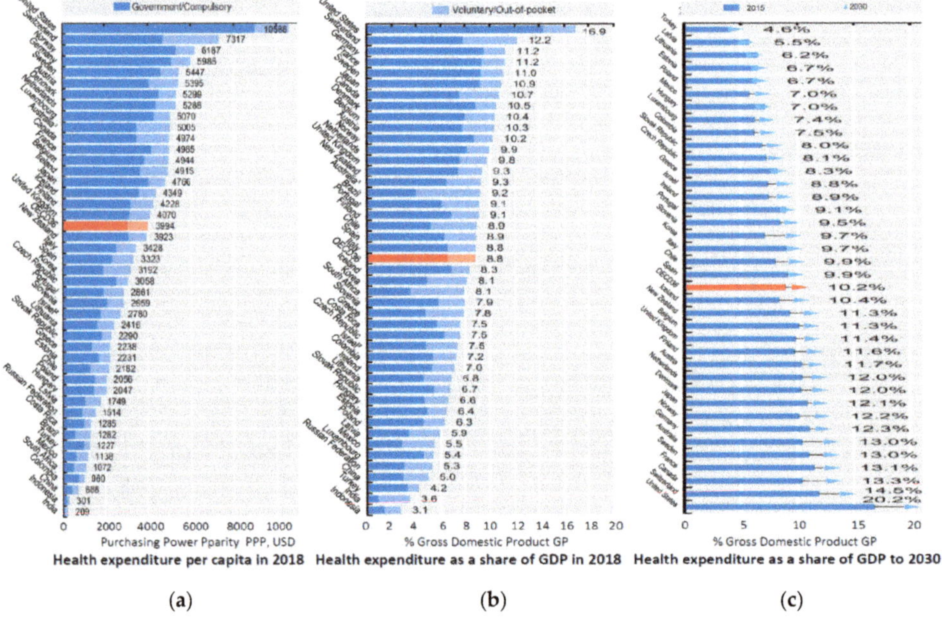

(a)   (b)   (c)

Figure 26. Comparison of the health expenditure in 2018 in the OECD countries (**a**) per capita, (**b**) as a share of GDP, (**c**) as a share of the GDP projected in 2030.

Figure 27, on the other hand, shows the variation in the number of doctors and auxiliary medical personnel referenced to 1000 inhabitants in different countries. The number of doctors varies from country to country by about 20 times. In fact, in [355], the current situation can be diagnosed as the creation of a two-tier health service: one for the rich, and the other, which is limited and often of worse quality, for the majority. This conclusion cannot be fully confirmed because, as stated above, the OECD represents a group of relatively highly developed countries but there are no general comparative statistics for all World. The indicated 50-fold and 20-fold differentiation undoubtedly confirm the thesis that access to medical care, including dental care, is very different in individual countries. It can be guessed that the lower the level of financing of this type of benefit within the budget of a given country, the worse the access to them, especially for the poor social classes. The situation is certainly much worse in the LIC and LMIC countries.

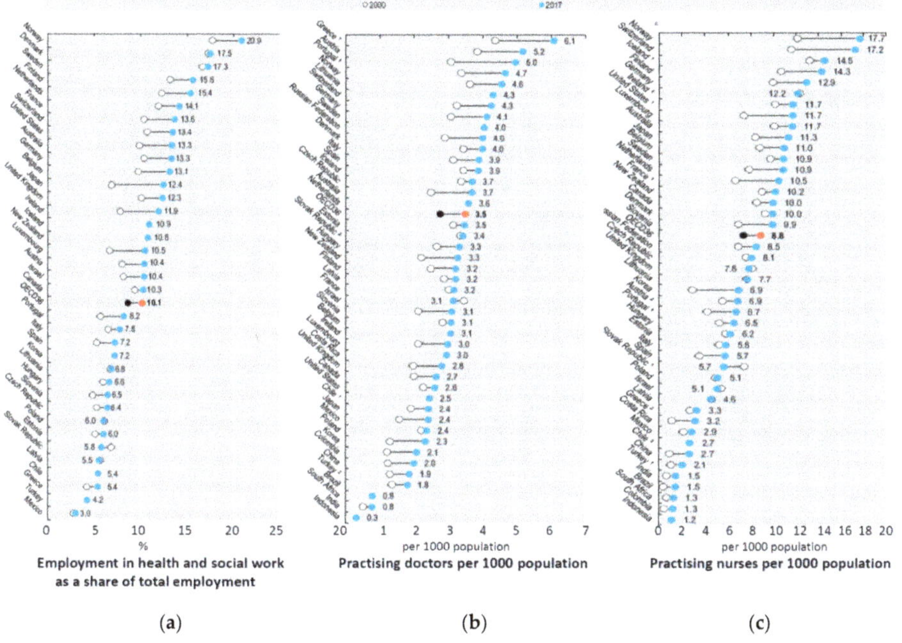

Figure 27. Changes in the years 2000–2017: (**a**) employment in health and social work as a share of total employment, (**b**) practising doctors per 1000 population, (**c**) practising nurses per 1000 population.

Figure 28 also shows the differences in the share of individual types of medical care, including dental care, covered by insurance paid by the Government. There is a significant variation in this regard depending of the country.

Figure 29 shows the statistics for visits to doctors and dentists. The information presented shows that the subsidies for health care and the possibility of using medical services, including dental services, are extremely diversified. In light of this information, the problem appears to be more general than that of dental care only. The authors of this article comment on the powers of politicians to make decisions in these matters. This applies in general to medical care and, therefore, to oral health issues. All postulates in this regard in each country should be addressed to the political circles that are shaped by democratic elections. Medical circles, including dentists, have the role of diagnosing the general condition of medical care and reporting postulates in this regard. To put it simply, not doctors, including dentists, are responsible for the organization of health care. However, they can only be a pressure or lobbying group.

	All services	Hospital care	Outpatient medical care	Dental care
OECD32	73%	88%	77%	29%
Australia	69%	68%	81%	23%
Austria	74%	87%	78%	45%
Belgium	77%	76%	76%	39%
Canada	70%	91%	87%	6%
Czech Republic	82%	95%	90%	48%
Denmark	84%	91%	92%	19%
Estonia	75%	98%	84%	25%
Finland	75%	91%	82%	30%
France	83%	96%	77%	N/A
Germany	84%	96%	89%	68%
Greece	61%	66%	62%	0%
Hungary	69%	91%	61%	36%
Iceland	82%	99%	78%	24%
Ireland	73%	70%	74%	N/A
Israel	63%	94%	62%	2%
Italy	74%	96%	58%	N/A
Japan	84%	93%	85%	78%
Korea	59%	65%	58%	33%
Latvia	57%	80%	61%	18%
Lithuania	67%	91%	77%	18%
Luxembourg	84%	92%	88%	43%
Mexico	52%	66%	85%	7%
Netherlands	82%	91%	84%	11%
Norway	85%	99%	86%	29%
Poland	69%	93%	67%	24%
Portugal	66%	85%	63%	N/A
Slovak Republic	80%	87%	98%	53%
Slovenia	72%	86%	76%	50%
Spain	71%	91%	76%	1%
Sweden	84%	99%	86%	40%
Switzerland	64%	84%	62%	6%
United Kingdom	79%	94%	85%	N/A
Costa Rica	75%	88%	59%	11%
Russian Federation	57%	82%	55%	N/A

Figure 28. The extent of coverage of financing medical services in 2017 by government and compulsory insurance including that dedicated to dental care.

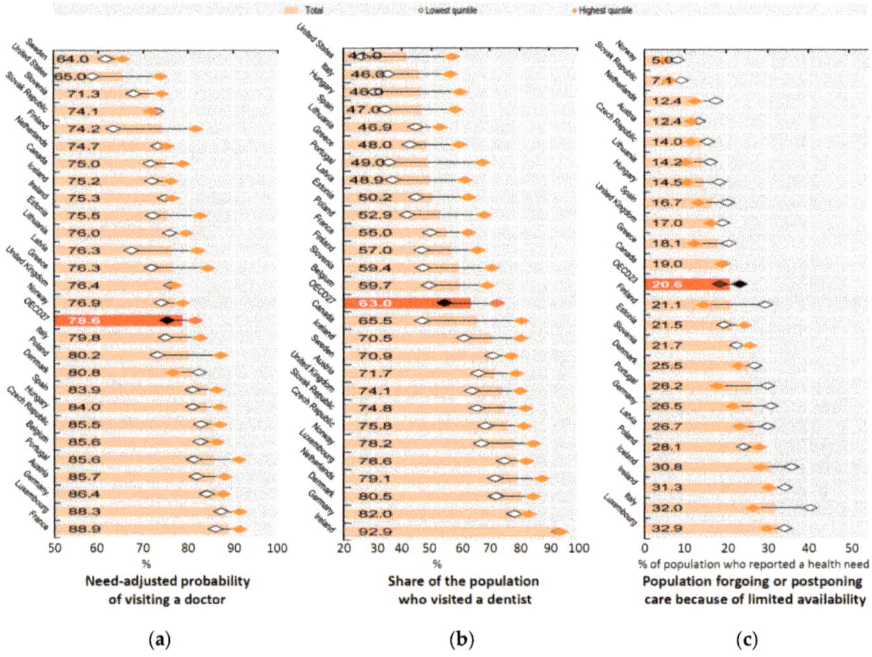

Figure 29. The statistics in 2014 for visits to (**a**) doctors and (**b**) dentists, and (**c**) the population forgoing or postponing care because of limited availability.

Figure 30 illustrates exemplary changes in the way health care financed in some European countries for all habitants, adults, or children. There are visible tendencies on the one hand in the privatization of medical services—e.g., in Central East Europe (CEE) countries but also in Sweden. In countries where medical services have already been privatized, shifts are also often made towards financing provisions towards greater private funds, such as in the United Kingdom, Denmark, Sweden, Germany, and the Netherlands. In some countries, such as Denmark, Norway, and France, all children's medical care is financed entirely from public funds. The figure is qualitative in nature, so no conclusions should be drawn from the position of the wheels represented on the figure for each country as to the level of funding, as it only concerns the principles [398].

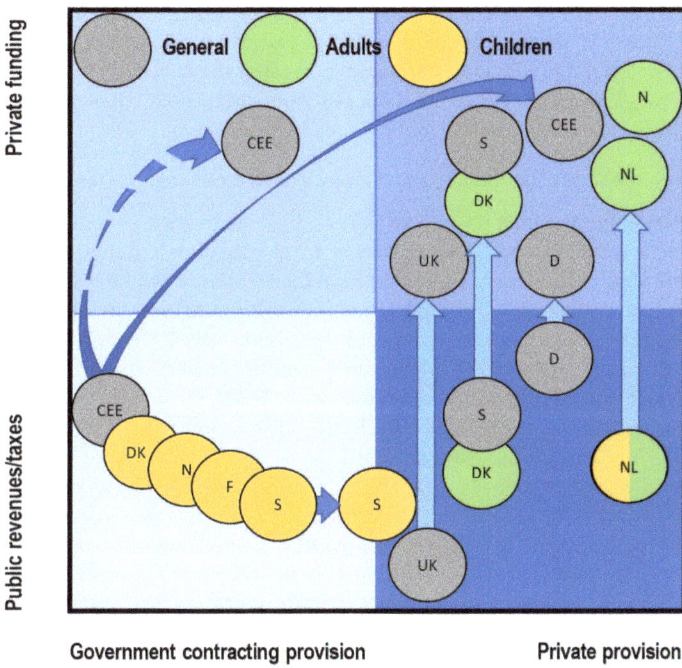

Figure 30. A scheme of exemplary changes in the method of financing health care in some European countries generally for all habitants, whether adults or children.

Politically coordinated programs with support from patients and communities in each country are needed to integrate health systems priorities, public health in general including oral health, and relevant activities related to research, education, and training in this field [6,7]. It is estimated that the currently chosen wrong paths to reach the so-called good dental form, consisting of reactive interventions when the disease manifests itself, instead of proactive action, consisting of the prevention of oral cavity diseases. Undoubtedly, it is a long-known principle in medicine in general that it is better and cheaper to prevent than to cure, only that it is an ideal, a kind of asymptote that everyone knows about, but its achievement is not and will never be possible. This is a kind of Holy Grail of medicine never found like the original in the Arthurian legend cycle. It is hard to resist the impression that this is an idealistic and somewhat romantic approach. One can quote the words of Carl Christian Schurz, an American statesman, who was not a romantic but wrote: "Ideals are like stars; you will not succeed in touching them with your hands, but like the sea fearing man on the desert of waters, you choose them as your guides, and following them, you reach your destiny". The criticism presented in these two cited studies [6,7] also concerns the biomedical and reductionist understanding of the causality of oral diseases and the belief that treatment and highly technical interventions will eventually restore oral health. These are the next arguments in this thoughts exchange, which accompanies the discussion of this problem in this subsection, but they do not remove the main controversies and analyses as to what extent such a view excludes the one considered to be the main one.

In the opinion of the authors of this paper, such a circumstance does not occur, as both views are not mutually exclusive. According to the authors of this paper, the views, opinions, and conclusions of both cited studies [6,7] should be considered as the second leg of the virtual Chinese censer, symbolizing the current balance of dentistry, and above all its sustainable development in the coming decades. However, it cannot be concluded that the view outlined in this way is an alternative or even antinomy to the current development reaching the stage of Dentistry 4.0. However, it must be remembered that the implementation of such an idea all over the world is a process that, once started, will take many years, or even decades. While the efforts in the field of systemic global prevention are fully justified, the interventionist and inherently individualistic removal of the effects of identified diseases is an indispensable synergistic action. Looking practically, it should be noted that even the most effective prevention is not able to effectively eliminate disease states in general.

5. Safety of Dentists and Dental Staff as a Development Determinant of Dentistry

The professional group of dentists is very large. However, there are no detailed statistical data. However, indirectly it can be established that there are at least over a million dentists in the world because this is the number of members affiliated with the FDI World Dental Federation as the largest membership-based dental organization in the world. The FDI's membership comprises approximately 200 national member dental associations and specialist groups in some 130 countries [402]. The situation is different, e.g., in Europe, where EUROSTAT keeps accurate statistics and the number of dentists in 2017 (the last year for which full data was provided) was 429,202, taking into account the amendment introduced by the Polish government [403]. The detailed distribution of dentists by country is given in Figures 31 and 32 show a map of Europe including the number of dentists per 100,000 inhabitants in 2017 and the rate of changes in this number in 2018 compared to the previous year, which is 1, that is, there are no significant changes in this respect [403]. Figure 33 shows the dynamics of changes in the number of dentists in the USA, which in 2017 reached 198,517 [404]. Such a large number of dentists in Europe and the USA seems to indicate that the number of dentists in the world certainly exceeds one million, which also indicates that many dentists in many countries in the world are not affiliated with any professional association.

The work of a dentist, just like an ENT and anesthesiologist, takes place in the patient's respiratory tract. Therefore, the patient breathing in always causes the dentist to inhale most of the exhaled bioaerosol. The use of a turbine or even a high-speed motor in the preparation of the teeth, and the associated intense cooling, mixes the resulting clinical aerosol with the respiratory bioaerosol, but also with the patient's saliva and blood, and the cooling stream is mixed with the infected saliva and the patient's blood intensely sprayed, disproportionately increasing the risk of the infection of the dentist and dental staff. Performing any dental treatment or preventive treatment requires bringing the dentist's face close to the patient's mouth at a distance of good vision (up to 20 cm), of which up to 5–10 cm is in the patient's mouth. This situation requires special care, because, in the event of any infection of the patient, there is a very high risk of infection of the dentist and the cooperating medical staff, because most procedures require assistance and so-called work for four hands. Therefore, the standard is to provide appropriate personal protective equipment, as well as protective gloves and masks with adequately effective filters, as well as highly demanding procedures for the disposal of biological waste, defined in each country by the relevant legal regulations.

The case gained special importance with the outbreak of the COVID-19 pandemic caused by the spread of the SARS-CoV-2 coronavirus. This is a virus with which the human population has not had any contact so far. For this reason, the virus has turned out to be particularly aggressive. The first information about its appearance comes from Wuhan, China, as of December 2019. On 20 March 2020, the World Health Organization declared COVID-19 a pandemic. By the time this paper was written, the disease has affected nearly 26 million people in 218 countries, the death toll was approaching 830,000 [405], and there were about 230,000 new infections per day. By the time the text of this article was revised, these numbers increased more than 2.3 times between 2 September 2020 and 25 November 2020. The disease has affected over 60.4 million people, and the death toll is approaching 1,306,500, and there are about 116,500 new infections per day and over 2750 deaths per day (Figure 34) [405].

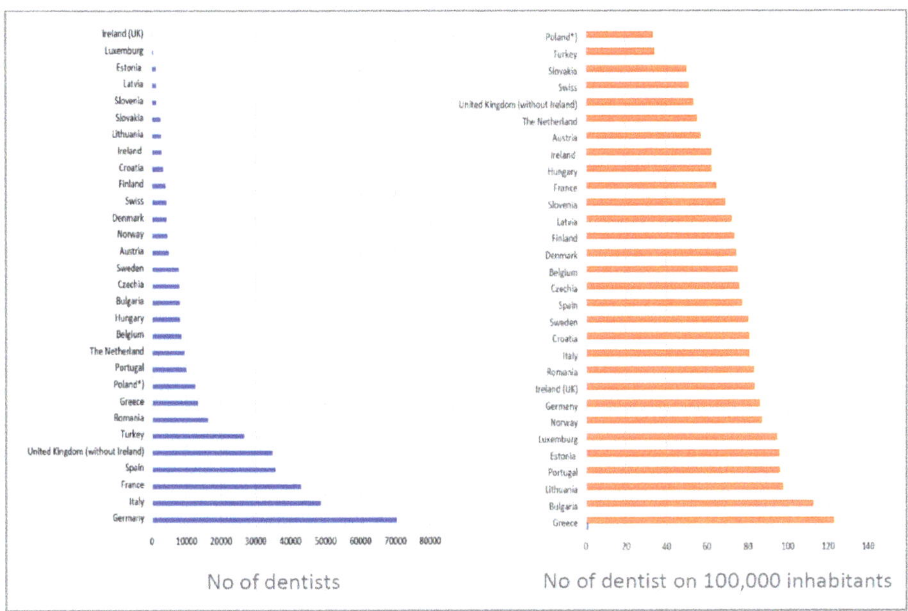

Figure 31. The detailed distribution of dentists in individual European countries (* Supreme Medical Chamber in Poland on behalf of the Government of Poland on 10 February 2020, sent a letter to Eurostat, correcting incorrect information because, in Poland, 42,425 people currently have the right to practice the profession of dentists in Poland, most of whom are dentists registered as practicing this profession. No. of dentist in 100,000 inhabitants should be 112.1).

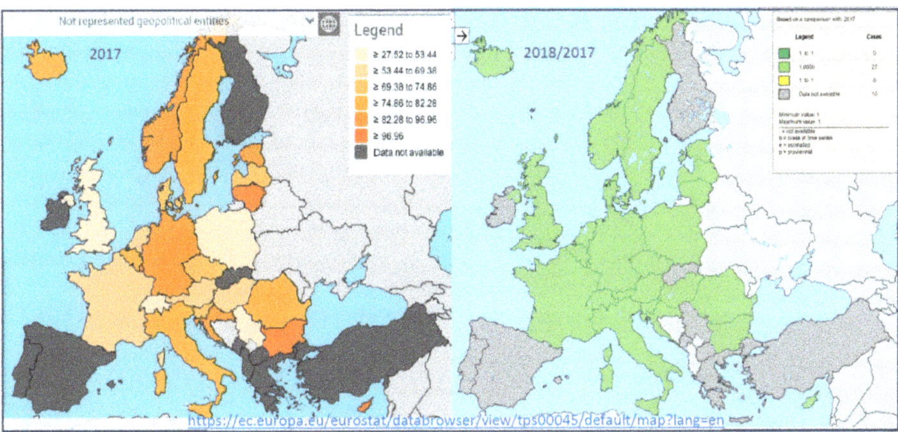

Figure 32. Number of practicing dentists in various European countries per 100,000 inhabitants in 2017 and the rate of changes in the number of dentists in 2018 compared to 2017.

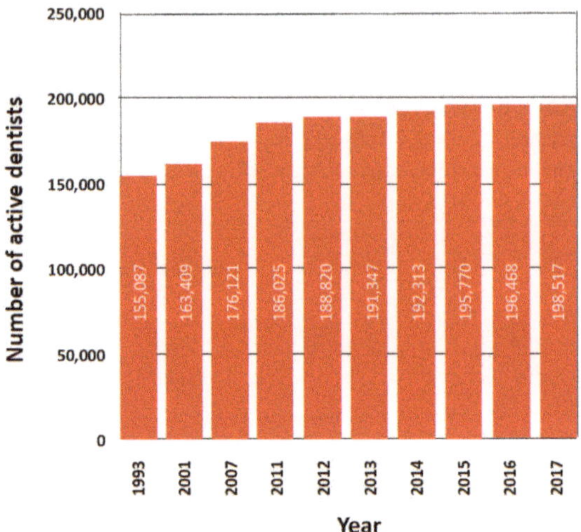

Figure 33. Changes in the number of practicing dentists in the USA in the period 1993–2017.

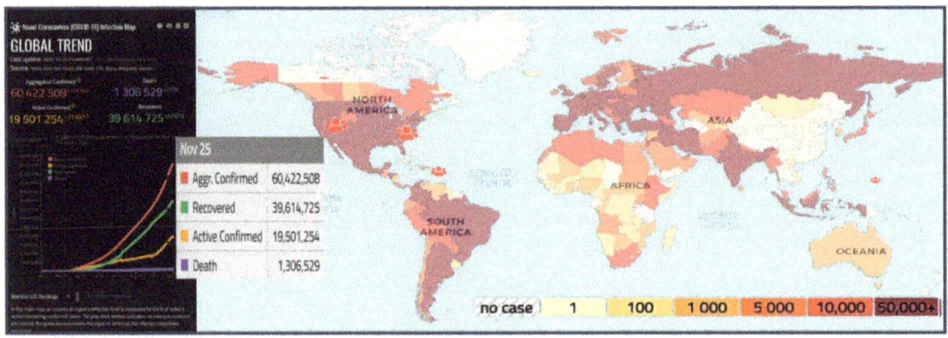

Figure 34. Epidemiological situation in various countries of the world on 25 November 2020 caused by the COVID-19 disease.

The disease spreads quickly by droplets. It is obvious that the infection is transmitted along with the virus contained in the colloidal particles of the secretion sprayed by the patient when coughing or sneezing, which are the colloidal particles of bioaerosol. It turns out, however, that transmission of the SARS-CoV-2 coronavirus in the form of bioaerosol takes place even with normal breathing [406]. However, the virus is viable and can survive on the ground and, for example, on clothing, on which it will settle for several or even several dozen hours. It is therefore easy to transfer with the hands after touching it and then touching the conjunctiva or mucosa of the nose or mouth, and most likely also an exposed wound or scratch. The virus incubation time has not been clearly established and maybe five or six days after infection, but may be up to fourteen days or even longer [407]. There are more and more reports about the possibility of reinfection by cured patients. This information is treated with considerable reserve, for example, because diagnostic errors may have occurred, especially in the case of asymptomatic disease.

The official statistics disseminated by Amnesty International indicate that at least 3000 medical workers in 79 countries around the world have died from the COVID-19 disease [408]. On the other hand, the World Health Organization indicates that dentists are included in the group of the highest risk of contracting SARS-CoV-2 coronavirus. Dental procedures are performed in the patient's open mouth, and the bioaerosol generated during the natural respiratory processes contains pathogenic viruses, bacteria, and fungi absorbed by them as a result of respiration, including the SARS-CoV-2 coronavirus. The dentist is therefore exposed to the direct transmission of these microorganisms multiplied in the upper and lower respiratory tract of the patient, especially since, as a result of the intensive cooling during tooth preparation, the pathogenic bioaerosol is mixed with the clinical aerosol, giving the intensive exposure. Therefore, it is necessary to strictly respect the rules of environmental and hand hygiene [409]. Any mistake made in this area results in exposure to the disease by dentists, as well as several hundred or even thousand patients as a result of the secondary infection of subsequent patients, who may be admitted by a dentist to several dozen or even over a hundred during the incubation of the disease or its asymptomatic course, while these patients, in turn, infect other people around them. A diagram of this process is shown in Figure 35.

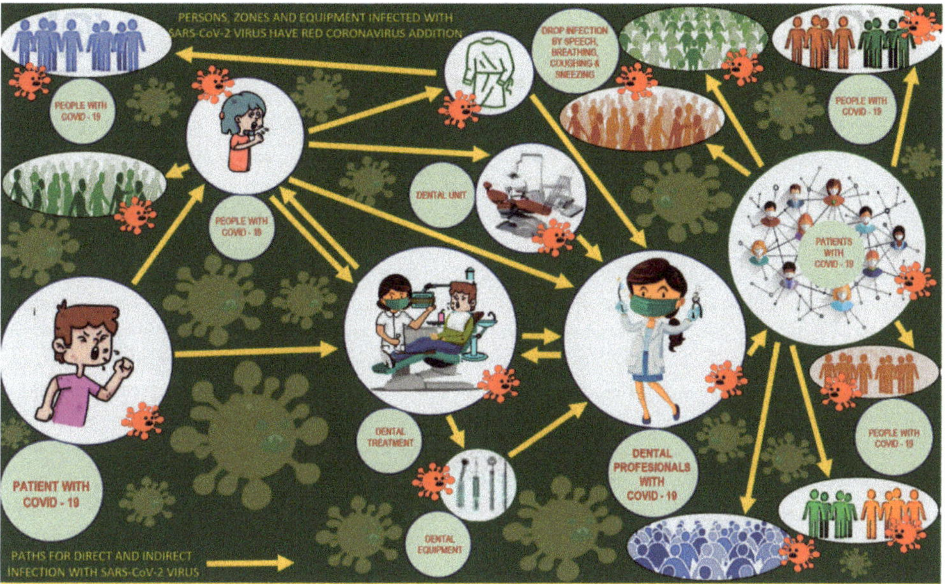

Figure 35. Diagram of SARS-CoV-2 transmission via a dentist in the event of one infected patient appearing in a dental clinic.

The spreading COVID-19 pandemic has caused an enormous disturbance in health care in many countries, mainly in infectious diseases hospitals, and also the abandonment of medical services in many other specializations and health care areas. In the case of dentistry in many countries, the governments of the USA, Europe, and Brazil have recommended closing offices in line with the "stay at home" strategy, adding to it "do nothing" to avoid becoming infected [410–414]. Several dental associations around the world have also ordered dentists to postpone scheduled procedures. For example, in Poland, on 15 July 2020, the decision to work in the emergency mode and the prohibition of normal dental practice to the full extent was renewed. Of course, such an attitude is insulting to the Hippocratic oath and dentists [398], dental staff, dental engineers, and the employees of prosthetic restoration manufacturing centers, depriving them of a chance to earn money.

The development of the COVID-19 pandemic has made dentists aware of the importance of the safety of dentists and dental staff, not only to dentists but also to the governments in many countries of the world. In addition to a development determinant of dentistry, this is on a par with the development strategies of Dentistry 4.0 and Global Dental Prevention, the third leg of the virtual Chinese censer guaranteeing the stable development of dentistry. Already at the beginning of the pandemic, the necessity of introducing a surgical mask and eye protection for the dentist with solid side shields or a face shield to protect the mucous membranes of the eyes, nose, and mouth during procedures, personal protective equipment for dentists and the requirement of strict hand hygiene, as well as in relation to rinsing the mouth of the patient before dental procedures, isolation by a cofferdam, but mainly in relation to the entire dental clinic, disinfection and ventilation of the entire clinic, and proper management of medical waste [415]. The problem is that the diameter of the SARS-CoV-2 coronavirus varies from 40–150 nm, and the most effective filters used in surgical masks have a mesh size of 300 nm, which indicates that surgical masks are not an effective protection in this situation. This approach is in line with the defensive STOP strategy (system, technical, organizational, personal), which accepts the fact that the SARS-CoV-2 virus is present throughout the clinic, requires passive protection of the patient by sterilizing all possible surfaces and air exchange at an appropriate frequency through ventilation and isolating the dentist and medical staff from this environment as much as possible, possibly testing each patient for the presence of coronavirus or antibodies. As a result, personal protective equipment PPE is intensified and practically impossible for the dentist to perform the work. Glasses and shields cause parallax and make it completely impossible to perform precise activities, and protective clothing causes sweating, fatigue, immobility, and a lack of precision in performing procedures that require it. Recommendations "do nothing" because you will get infected or if you have to dress as a result that you can do nothing, this approach is offered by medical administration almost all over the world. As a result, dentists and their mainly private clinics were practically deprived of a chance to return to normal functioning. This forces dentists to disregard the Hippocratic oath [398] because many patients cannot be treated because of the absolute organizational failure of the system. The context of the case is multifaceted, as there are violations of ethical principles, legal provisions in the field of protection of employee interests, and the protection of patients' interests, with violations in many countries of constitutional rights to protect the health, as well as a conflict with the European Union's priorities in terms of improving the comfort and extension of human health and longitude life and aspect of human rights violations. The matter is really dangerous from the point of view of every dentist, when the patient unknowingly infects the environment, including the dentist, in the case of asymptomatic COVID-19 disease or during the five or even fourteen days of incubation of the disease.

In the previous work of the authors of this paper [35], a comparative analysis was performed and the potential was analyzed as an objective measure resulting from technical analysis, and attractiveness as a subjective measure based on the recipient's impression, and the results of assessing the usefulness of all previous groups of methods of preventing the harmful effects of SARS-CoV-2 were presented. The results of these analyses were obtained using knowledge engineering tools [309,352]—i.e., procedural benchmarking with the use of ten detailed criteria and the corresponding weights—using the universal scale of relative states with positive values from 1 as the lowest value to 10 as the maximum possible value, and using the contextual matrix. The results are illustrated in the dendrological matrix presented in this paper. Table 8 summarizes the results of the analyses performed.

None of the currently used groups of methods obtained in any case a value greater than 7—that is, from quite low to at most moderate. Therefore, none of these groups of methods provides the expected effectiveness in removing the sources of the SARS-CoV-2 coronavirus threat. Moreover, the comparative analysis and its results indicate that the groups of methods known so far from reports on manufacturers' websites and from the literature usually show less potential than attractiveness, which is usually greater. This indicates that their importance is usually overestimated in practice, and their effectiveness in preventing the effects of the coronavirus is lower than it is believed. The methods marked as (1,2) have a high potential, but they cannot meet all the requirements on their own. On the other hand, methods (3,4) are highly attractive, but they can only prove successful if they are significantly strengthened in terms of design and technology, otherwise, they will not be successful in terms of application. On the other hand, method (5) has little potential and little attractiveness and is being replaced by other, better solutions.

Table 8. The potential and attractiveness of each group of methods of protection against the effects of the SARS-CoV-2 coronavirus.

No	Type of Method Known in the Literature, State-of-the-Art Including Own Solution	Potential	Attracti-Veness
1	Installation of air filtering devices in the treatment room	6.7	4.0
2	Equipment operating with HEPA or carbon filters or with UV radiation	5.95	4.0
3	Dental saliva extractor attachments	3.3	5.8
4	"Extract from the treatment area" devices	3.7	6.6
5	Patient enclosures in the form of glass or PMMA boxes	1.45	3.9
6	Original solution eliminating the threat at the source with the set of devices for virus elimination	9.0	8.5

In the cited author's work [35], a decidedly offensive SPEC (system, prevention, efficiency, cause) strategy was implemented and, as a result, a completely new breakthrough technical solution was proposed (Figure 36), consisting in the elimination of clinical aerosol mixed with the patient's pathogenic respiratory bioaerosol at the source, right next to the mouth of an infected patient, from the space of the dental clinic [416–419]. This is done by collecting this bioaerosol directly into an epidemiologically safe container and its complete deactivation and decontamination, rather than indirect isolation of medical personnel from the pathogenic aerosol and tolerating its presence in the entire space of the dental clinic. The design details and numerous design variants are given in [35,416–419], the solution is protected by a patent [416] and is started the commencement of serial market production of the device [417]. The own solution, according to the criteria relating to all the previously discussed groups of methods of preventing harmful effects of SARS-CoV-2 coronavirus, receives very high scores, respectively, 9 (potential) and 8.5 (attractiveness—there are difficulties at work) (Table 8). This demonstrates the breakthrough and full competitiveness of the developed solution. This solution is a set of many complex devices (Figure 36). As a result of negative pressure given by pomp, the bioaerosol exhaled by the patient and contaminated with the SARS-CoV-2 virus, mixed with clinical aerosol from the patient's respiratory tract is eliminated for to be completely deactivated and effectively decontaminated in a multi-stage procedure. An important part of the device is the face cap, which is available in several variants with an ever wider range of functions. The device also includes an extractor to aspirate the patient's saliva and blood, and a nasal cannula to allow oxygen to be delivered directly to the patient when needed.

The presented analysis shows that for the development of dentistry in the following decades, apart from other factors, it is necessary to pay special attention to the safety issues of dentists and medical staff during the performance of a full range of dental procedures. If a dentist is to do his job well, in order to effectively help the patient, he must have the inner conviction that he is safe if all applicable rules are complied with, in accordance with common sense. They cannot feel threatened, because in many cases they will not undertake the procedure, or in the event of excessive nervousness resulting from the fear of infection, they will do the job poorly, without due diligence, hurriedly, and under pressure. Everyone who works in direct contact with this deadly coronavirus is aware that after returning home from work, they can infect their family, children, elderly people who are at high risk, and other people around themselves. This is an enormous psychological pressure that many people, including dentists, cannot overcome. It cannot be assumed that every dentist who, due to their work in the patient's respiratory tract, is particularly exposed to droplet infection, will show heroism every day at their workplace, including the fact that they can literally die at this position. Currently, the world is struggling with a pandemic that has already claimed nearly 830,000 lives. Microbes, viruses, bacteria, and fungi have accompanied us for centuries, and many of them threaten with deadly droplet infection. The current SARS-CoV-2 coronavirus pandemic has made the entire medical community aware of the importance of ensuring safety and preventing a patient's droplet infection as an important development factor in dentistry. One solution is presented as an example, which is the result of an active strategy of removing threats at the source. It should be expected that over time, new ones that are equally good or better will emerge. It must be a stable direction of development, as the "stay at home" and "do nothing" strategies do not exist in the field of dentistry, and most likely in a much wider context.

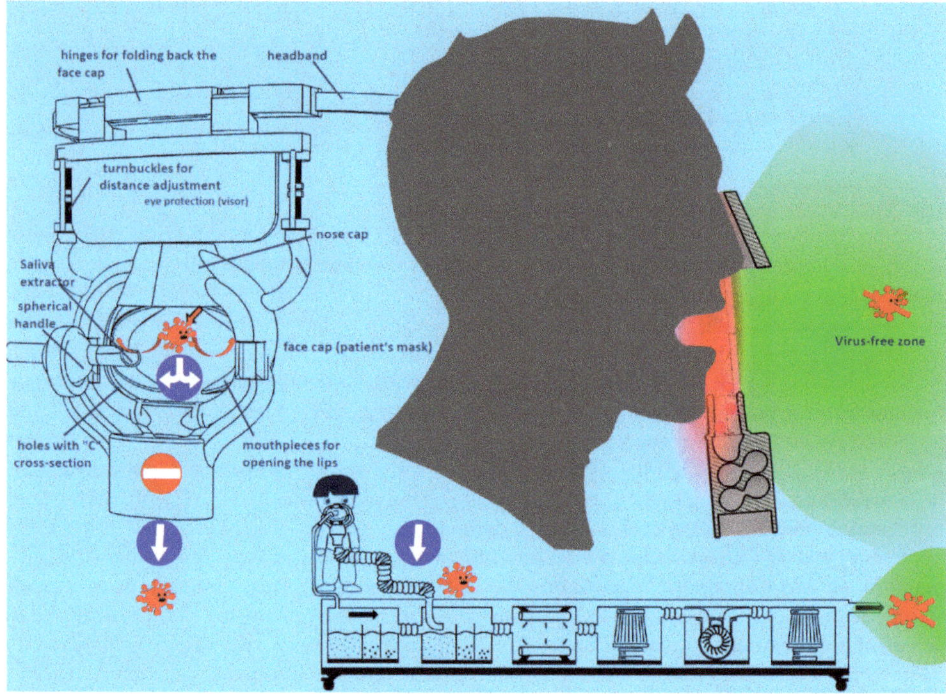

Figure 36. Diagram of one of the face cap solutions and an example of its use with a set of devices shown in the lower part of the drawing.

A separate, but equally important, is also the active strategy of protecting patients against infection during dental procedures. However, it should be noted that a healthy dentist and healthy uninfected medical staff in a dental clinic do not pose a threat to patients.

6. Ethical Context of the Treatment of Oral Diseases

The view that the so-called nice and above all healthy teeth is a private matter of each person is absolutely wrong. A nice appearance is of course important, but the matter has a much wider context and much, including social and economic issues. Numerous systemic diseases that are complications of caries, periodontitis and toothlessness significantly affect the burden on health care and social security systems in all countries. Therefore, it has a significant impact on the economic condition of states and health welfare of societies. Therefore, the dentists' community bears a special responsibility in a professional and ethical context. Ethical aspects constitute one of the important determinants of the sustainable development of dentistry in the future, and therefore this chapter of this paper is dedicated to these issues.

Since antiquity, every doctor of every specialty, and therefore also a dentist, is required to take and obey the Hippocratic Oath [420], the words of which has changed throughout history, but the ethical meaning has remained unchanged. Every doctor is supposed to help people with illness and in details he/she should respect the patient's autonomy [421,422] and show charity [423], harmlessness [424] and equity [425] towards him, defined as the four fundamental ethical principles [426–429] and the fifth general principle special for dentists, which additionally requires to be truthful in matters concerning the diagnosis and treatment of oral diseases [430–433] (Figure 37). These issues have been widely presented by the authors of this paper in two extensive literature studies [434,435]. Both parts of the figure show, in a form analogous to the Deming's circle [9], the basic and subsequent features that characterize the moral directives that characterize the professional community of dentists and engineers. These are other examples of the fives' rule idea contained in this paper.

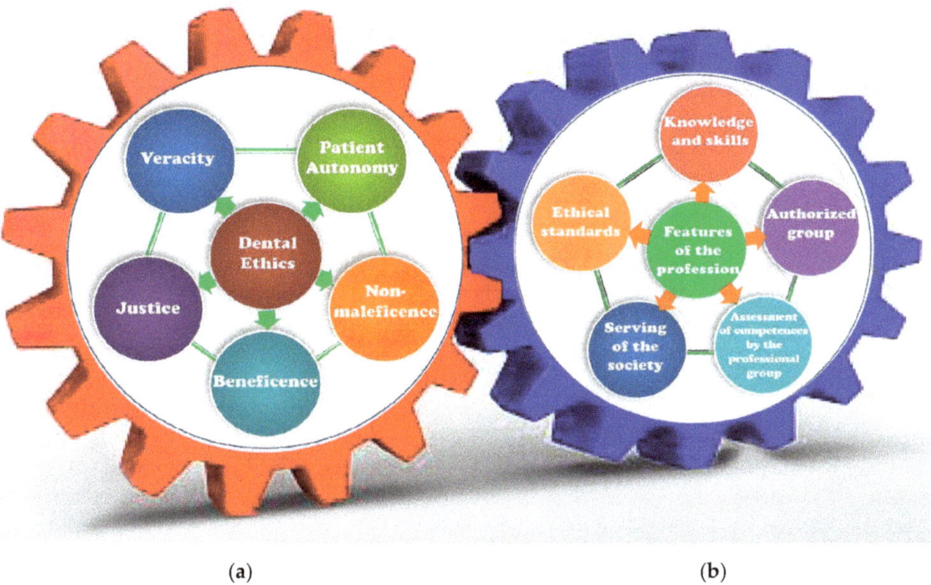

(**a**)   (**b**)

Figure 37. Diagram of the basic principles of dental (**a**) and engineering (**b**) ethics in the form of gears symbolizing the necessary cooperation between these areas.

An ethical requirement for doctors is the obligation to systematically expand their knowledge, and in particular to follow the latest scientific and clinical achievements, in order to quickly apply them to patients. The dentist's ethical obligation is also the cooperation with the center for the design and manufacturing of prosthetic restorations run by dental engineers. The requirements for this cooperation, illustrated by two gears' wheels, relate to advanced design and manufacturing methods using avant-garde methods of computer aided design and manufacturing, CAD/CAM, computerization, robotization and automation. The requirement, for obvious technical reasons, is to provide the most advanced materials and the most modern engineering technologies. This includes state-of-the-art additive manufacturing methods, as well as advanced surface treatment methods for implants and prosthetic restorations. Dental engineers working closely with dentists to meet the expectations of patients, similarly to engineers of all other specializations, including those who do not participate in medical procedures, also apply ethical principles [436–439] (Figure 37), which include improving specialist knowledge, achieving high competences and ensuring high standards, as well as serving society in accordance with the ideal of the profession of engineer and providing valuable products. It should be emphasized that the high standards of both dentists and dental engineers result from the internal moral imperatives of each of these people, but also from the entire their societies, and also they testify to the highest possible standards. It is also worth noting that among the three ethical orders, customary in a given community and legal in a given country, there are often analogous regulations, although they are undertaken for significantly different motivations. Therefore, various types of ethical codes usually belong to the second or even the third of the mentioned groups of regulations, which of course does not mean that they should not be respected, but on the contrary, the highest possible value is the inner conviction of both the dentist and the dental engineer about the need for proper fulfilling obligations. Therefore, one of the ethical requirements of dental engineers in such an approach to duties is moral responsibility towards patients. Both the dentist and, above all, the patients have the right to expect the implementation of advanced technologies in order to design and manufacture the necessary implants and prosthetic restorations. However, the implementation of such avant-garde solutions is morally acceptable only when the absolute risk of their application is lower than the total benefits, although new experimental technologies may be an exception [440]. A deliberate avoidance of the use of modern materials and available technological possibilities, and even the lack of sufficient knowledge in this field, should be treated as an ethical delict of a dental engineer and a dentist cooperating with him/her. It applies mainly to fillers, the achievements of nanodentistics, and above all, the conscious, but also unconscious underestimation of the achievements of the current stage of development of Dentistry 4.0, as well as the results of research obtained by tissue engineering, applicable in dentistry. It involves the need to implement new scaffolds designs for implants and biological-engineering materials [434,435].

Ethics has been considered paramount in all professional dental and engineering activities to provide each patient with the highest level of oral health care currently possible and in line with each country's political and economic trends and in line with all available strategies, essentially adapting the Deming Circle [9] to clinical issues and engineering activities in the field of dental engineering at the stage of Dentistry 4.0. In the works [434,435] the Authors of this paper schematically present the relationship between dentistry at the level of Dentistry 4.0 in connection with nanoengineering with nanodentistry, and dental engineering (Figure 38), showing development perspectives in relation to the current requirements of the present day and ethical imperatives from the point of view of the health well-being of each patient. In [441] seven aspects of sustainable health personal dimensions SHPD are defined. They include the personal values of each patient, which is certainly significantly influenced by the condition of the dentition and the well-being of the oral cavity, as well as the elimination of all systemic complications of oral cavity diseases. This set includes the following factors:

- physical (health and fitness, endurance, disease resistance and regenerative abilities);
- emotional (the ability to realistically perceive the world, coping with stress, flexibility and a compromise attitude to conflict);
- social (family and environmental relationships, social skills and cultural sensitivity);
- intellectual (personal development, perceptual, analytical, decision-making and rational skills);
- spiritual (religiosity and/or value system, relations to the living world, empathy and an active attitude towards society and other people);
- professional (career, finances, quality of life, professional and non-professional interactions);
- environmental (relations with the natural environment and with the material and metaphysical surroundings in general).

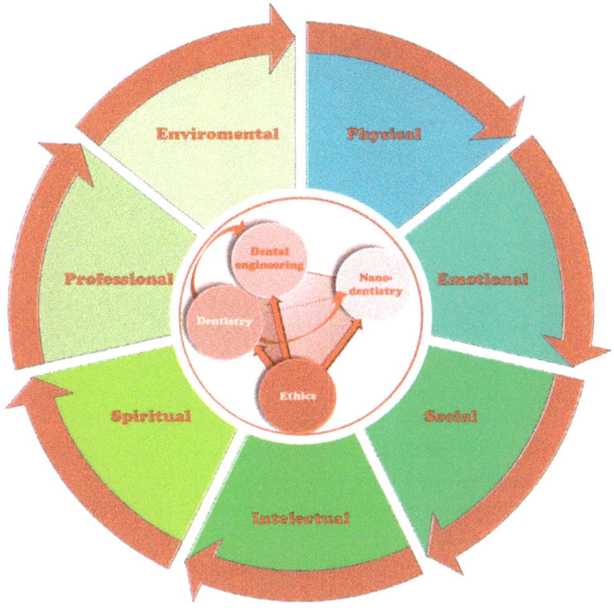

Figure 38. The scheme of the ethical aspects of the personal dimensions of sustainable oral cavity health.

Thus far, the ethical responsibilities of dentists and dental engineers involved in the design and manufacturing of prosthetic restorations have been presented in relation to the individual expectations of each patient. The ethical aspects of the entire community of dentists and dental engineers include patients and societies in general, and in this context the social function of dentistry is determined. In the literature on the development prospects of dentistry on a global scale and in individual countries, two opposing views have emerged. In fact, the matter has an ethical aspect, because the supporters of one of these views contrast with the latter by acting to the detriment of the social mission of dentistry and even acting out of low motives to achieve unjustified material benefits. The authors of this paper have already been critical regarding the professional context of formulating the goal of dentists' activities in many countries as "drilling, filling and billing", as expressed in [397], treating this as an unethical and contemptuous act towards most dentist's honestly and ethically performing their dentist tasks at the service of patients and for their well-being. It cannot be ruled out that, in order to meet the demands of wealthy patients trying to satisfy their vanity without the need for dental treatment, some dentists take such measures, but it is certainly an insignificant margin of the entire professional community. Even if medical procedures are accompanied by exaggerated cosmetic procedures, the active prevention of many systemic diseases is extremely important in the oral cavity. However, the works [6,7] present the radical view that a comprehensive change is needed in the approach to the treatment of oral diseases and a departure from the so-called Western model of dental care, the culmination of which is the Dentistry 4.0 stage, characterized by more and more technologically advanced specialist treatment, which, in the opinion of the authors of this paper, is the pinnacle of achievements in this field. The professional and ethical dilemma can be solved [6,7] by the proper organization and implementation of dental prophylactic procedures for the entire world's population [403,404]. Such an approach to the problem bears the hallmarks of an ethical delict, as it is obvious that such a theoretical assumption is unfeasible in practice, for which one can easily find a large amount of evidence in clinical practice and scientific research studies. Paradoxically, the evidence for this seemingly modern approach to solving the problem is the archaeological clinical evidence from over 14,000 years ago presented in Figure 1. Therefore, it is difficult to agree with such an opinion, which may be considered unethical, as it contradicts the level of current knowledge on this subject, and the ethical obligations of both dentists and dental engineers include continuous training and current knowledge of all scientific achievements in the field of the practiced profession. This does not mean that prevention is unnecessary and is not supported by the authors of this paper. The aim of this paper is to present a sustainable model of dentistry development, in which prophylaxis is prominent as one of the three important elements, but not excluding the other two. It is not true that the treatment of the effects of the disease is unnecessary and therefore unethical; on the contrary, it is another element of this model. As long as any of these actions are not ruled out, the ethical problem is entirely unapparent and essentially non-existent. This has been proven throughout the content of this paper. It should be clearly emphasized that it is the moral duty of every physician, including dentists, to provide medical assistance to each affected patient. It is obvious that the best method of treatment is disease prevention, which fully justifies the widespread development of methods and a prophylactic system on a global scale, but this cannot be treated as an alternative to dental treatment. Should that happen, it would be a serious breach of ethical principles. This is because it is not possible to prevent the disease in a fully effective manner in every case, for every patient, and for every clinical situation. As presented in the previous chapters of this paper, ethics requires the implementation of subsequent stages of treatment strategies resulting from the previously presented adaptations of Deming's circle [9], both in the field of caries treatment and periodontal treatment, which, unfortunately, in many cases even leads to implant-prosthetic treatment as the only available option to improve the patient's health. It is important, also from the ethical point of view of the dentist community, that proper dental care covers all, including the poorest social strata in many countries. In many cases, as was shown earlier in this paper, this depends to a small extent on the professional community of dentists, because the decisive factor is the assessments and actions of politicians in this area, subject to social evaluation in the electoral procedure and the economic condition of individual countries.

It should be clearly emphasized that interventional treatment according to the so-called Western model of dental care, the culmination of which is the Dentistry 4.0 stage, cannot constitute an antinomy of preventive actions, as suggested in the works [6,7]. Since interventional treatment, as repeatedly shown in this paper, is a necessity, ethical obligation requires the best possible use of modern scientific and technical achievements in this field. This applies to the use of cone beam computed tomography (CBCT) with intraoral and extraoral scans by dentists in dental diagnostics and imaging, as well as the use of computer-aided design and manufacturing (CAD/CAM) by dental engineers, as well as additive manufacturing methods of prosthetic restorations and implants in computerized and robotic technological processes. It should also be noted that nanodentistry is becoming an important area of modern dentistry. Nanodentistry solutions are increasingly available in clinical practice and in dental engineering. Many avant-garde solutions, or maybe just ideas, are likely to become available over time, including the design, manufacture, and implementation of dental nanobots. While professional ethics requires that patients be presented with proven innovative engineering solutions, ethical requirements make it impossible to present overly attractive visions that are practically unrealistic or that have not been properly tested in the existing situation.

On the other hand, any omission or conscious avoidance of using these achievements in the treatment of a patient, in which case they could or should be used, should be treated as a serious ethical delict. This applies to the non-use of modern materials and technologies, as well as the use of improperly selected technological conditions in the process of making prosthetic restorations [435]. Such a strict ethical assessment is also deserved by unintentional or even unknowingly committed errors related to the lack of specialist engineering knowledge in this field. The ethical duties of both dental engineers and especially the dentists cooperating with them and directly applying given solutions to patients, include continuing education and practical application. The patient, in his confidence in the dentist, should be sure that, in accordance with the first ethical rule binding the dentist, he/she obtains the optimal treatment strategy and therapeutic solution, including engineering solutions with optimal properties due to the entire set of required performance properties. From an ethical point of view, it is unacceptable to use stereotypes or unverified information in such cases, as such an approach directly harms patients [435].

A separate ethical problem is the dissemination in the scientific and scientific-technical press of unverified or underdeveloped information about allegedly advantageous technological or material solutions. In the name of patients' well-understood interests, it is necessary to point out substantive irregularities of such articles in order to eliminate them from conceptual circulation in the dentist community. Several such negative examples can be given, e.g., about doubts about the introduction of computerization in dentistry and dental engineering [254], as well as in the dissemination of the information that both casting and additive manufacturing result in identical product properties [332,442], which it is obviously contrary to the objective truth. This may eliminate or weaken dentists' interest in modern technologies in defiance of patients' health interests. Such actions of the authors of such articles must be treated as an ethical delict, because such incomplete and unverified information spreads among many dentists and harms many patients. It is not ethical and even not lawful to do this.

One more ethical aspect remains that is related to answering the question of under what circumstances some or all dental procedures may be discontinued—when and if the dentist may refuse to provide dental services at all, if even at all. This problem has gained particular practical importance with the announcement on March 20, 2020, by the Word Health Organization of the COVID-19 pandemic related to the transmission of the SARS-CoV-2 coronavirus, which is a direct threat to life. At the same time, the governments of many countries are following the recommendation of the International Dentist Federation (FDI) to stay at home and limit the provided services only to cases of so-called pain relief. There is no doubt that the dentist must be safe when doing their job. This is an obvious part of the ethical canon. On the other hand, the refusal to help patients, regardless of the reasons, must always be perceived pejoratively. In most cases, this must be treated as an ethical delict. The innovative accessory proposed by the authors of this paper to prevent the spread of the SARS-CoV-2 coronavirus (Figure 36) and the concept of the sustainable development model of dentistry presented in this paper meet the ethical requirements in this area and enable the uninterrupted work of dentists, even in the face of the special conditions of a pandemic.

The current SARS-CoV-2 coronavirus pandemic will force the stabilization of the systems to prevent droplet infections, similar to the system that once was established for the disease caused by the human immunodeficiency virus (HIV), which is transmitted through blood contamination. Both of these strategies will remain in the canon of pragmatic anti-epidemiological actions for good, which is also an obvious ethical imperative, due to the health wellbeing of dentists and patients.

These considerations, transferred from the works [434,435], indicate that contemporary activities related to dentistry and dental engineering—including the standards of Dentistry 4.0, and nanodentistry widely described in [435], and the systems protecting dentists and patients against contamination by harmful droplets—are strictly determined by the principles of ethics. The achievement of therapeutic goals and the actual improvement of the health of the oral cavity, manifested by the sustainable health personal dimensions (SHPD) in this regard, which constitute the individual feeling and assessment of each patient, are determined by the principles of general human ethics, especially medical and dental ethics, together with engineering ethics, if they are actually respected by dental clinics and dental restoration design and manufacturing centers. FollowingProf. Władysław Bartoszewski, the former Polish Minister of Foreign Affairs, it is worth repeating constantly and daily applying in clinical dental practice the phrase, "It's worth being decent!" [443]. This is absolutely necessary, unless somebody does not want to take own job, profession, and duties seriously.

7. Final Remarks

This paper discusses the general disease state in the world and the scale of oral diseases, mainly caries and periodontitis, which currently affect 3–5 billion people.

The etiology; course; and, most importantly, strategies for the prevention and treatment of both of these diseases were analyzed in detail using appropriate modifications of the Deming circle and models based on the idea of the fives' rule. This chapter presents also the development of the medical imaging state and treatment methodology and the achievement of the current computerized, and, in the most avant-garde cases, robotic stage of Dentistry 4.0, for which the use of computer-aided design and the production of CAD/CAM methods as well as additive methods of manufacturing prosthetic restorations and implants are also appropriate. Diagnostics at this stage is largely based on medical imaging using cone-beam computed tomography (CBCT) and intraoral and extraoral scanning. There are undoubtedly great achievements that make the production centers of prosthetic restorations as intelligent factories fit into the idea of Industry 4.0, the most advanced stage of development of the modern industry.

The considerations in this paper were inspired by two papers published in *The Lancet* [6,7], or rather a series of two papers or two parts of one long paper, the publishing of which was undoubtedly a highlighted event. It was impossible not to notice it, but also impossible to agree with it. The main theses of this joint paper are therefore discussed in this chapter. In general, the authors of these papers accuse the current developments in the field of dentistry of fundamental errors and demand a radical change in approach. They expect a move away from the interventionist treatment philosophy and the associated computerization and technical domination, which in fact contradicts all the achievements of dentistry. They believe that, instead, prevention should be developed very broadly, and that dental care should reach the poorest sections of society in many countries. While some of these postulates concern the dentist community and there is no doubt as to their validity, the widely discussed needs of including broadly understood aspects of dentistry into the mainstream of social security and the general development of medicine in general concern mainly politicians and national governments. For this reason, it is highly unlikely that such a demand could be implemented on a global scale. These theses, however, deserve full support and implementation in terms of an egalitarian approach to prophylaxis and dental care. However, this view cannot be an antinomy of the achievements of dentistry in terms of treatment and the achieved stage of Dentistry 4.0.

The ethical aspects of dentists' work and their moral obligations towards patients were also considered. These are closely related to the cooperation with dental engineers, and the holistic approach also takes into account the ethical obligations of dental engineers. Particular attention was paid also to the obligation to proclaim the truth as a fifth ethical rule for dentists. This applies to the dentist's intimate relations with each patient, but also to general issues of taking up challenges resulting from the avant-garde state of current technologies and the complex Industry 4.0 instrumentation in relation to dentistry defined as Dentistry 4.0. Of great ethical and practical importance is the publication of modern, but at the same time reliable and proven research results on modern materials and technologies used in dentistry, as well as methods of diagnosis, including medical imaging. The dissemination of stereotypes, the multiplication of unjustified doubts, and the dissemination of false and/or unverified news causes enormous damage and must be considered a serious ethical delict of the authors and publishers of such information. Some negative examples in this regard are given. An ordinary dentist reading such papers has the right to trust the author and publisher, and, without the slightest doubt, to apply them to his/her own patients, thinking that he/she is helping while he/she is actually causing harming without realizing. He/she cannot be burdened with ethical responsibility, because he/she acted in good faith, while the entire ethical responsibility rests with the author of a poorly prepared article that appears to be scientific which may harm thousands of patients in various places around the world. This is an extremely serious ethical aspect addressed in this paper.

An ethical delict towards the entire dentist community is the call presented at a world conference to discredit this environment and criticize its behavior as being dominated by "drilling, filling & billing". This might sound impressive were it not for the fact that the authors seem to forget about the enormous amount of work conducted by dentists in preventing the worldwide development of the huge number of systemic complications of oral diseases and toothlessness resulting from these diseases.

The trivialization of the role of dentistry is the formulation of its goals as cosmetic activities consisting of extorting money from millions of patients. This is clearly not true, even if it is possible to find such negative examples of dentists who also do or did this, because they are in the vast minority and in this sense constitute a small, negative margin of this community.

Both ethical and professional duties come down to preventing the development of caries and periodontal diseases, and, if this turns out to be impossible, then stopping these diseases at the existing stage. If these actions are not successful, one or more teeth will be extracted, or even all teeth will be lost, and then it is necessary to undertake implantological and prosthetic therapy in order to avoid the negative systemic complications of toothlessness. It is a completely ethical behavior. It should also be noted that despite the designation of care for the general health welfare of people in the world, as it belongs to the main Sustainable Development Goals designated by the United Nations, to what extent it will be implemented in individual countries depends on the economic condition and political situation of each country. The impact of the professional approach of the dentists' community is much less possible. Therefore, false premises are based on the assumption that dentists are the community that should change the general approach to oral diseases, i.e. that instead of practicing interventionist dentistry according to the so-called Western approach should focus on preventing oral diseases, starting with changing the basics of education to achieve the desired effect.

Quoting the words of Pope Francis, although in a completely different context, "Who am I, to judge someone . . . ", it is not the task of the authors of this paper to assess the ethical aspect of the matter at hand, but it is difficult to agree substantially with the approach presented in the papers [6,7]. For this reason, in the name of honesty in reporting the truth, this paper was prepared.

We recall here a mechanical pendulum clock based on an idea by Christiaan Huygens in 1657 on the basis of the principles formulated by Galileo and modified by George Graham in 1715. If it can be stated with an analogy with such a clock that the pendulum in terms of the methods and techniques of dental treatment is fully tilted, this does not mean that in currently following the suggestion of the authors of the paper from *The Lancet* [6,7] it should be tilted the other way. This would mean that, alternatively, only one of these solutions is possible, were it not for the fact that the pendulum is in constant motion. It is therefore possible for both of these approaches to coexist. In keeping with this analogy, it is necessary to maintain, continue to develop, and, above all, disseminate the achievements of "Dentistry 4.0". At the same time, symmetrically on the other side, an equally advanced level should be achieved in the field of prevention and egalitarianism in dentistry.

The two trends in the development of dentistry outlined in this way require a third fulcrum in order to stabilize them, which was illustrated in the introduction by the reference to the old Chinese censer.

The third direction of the development of dentistry, discussed in the next section of this paper, and ensuring sustainable development, is the widely understood safety of dentists and dental staff. This aspect was significantly emphasized during the COVID-19 disease pandemic caused by the transmission of the SARS-CoV-2 coronavirus. The dentist performs all dental procedures in the patient's respiratory tract, and any infection of this patient with pathogenic microorganisms exposes the dentist to direct infection; this is especially important in the case of the SARS-CoV-2 coronavirus, which is a direct threat to life. The dentist when doing his job must be safe and, more importantly, must feel safe. Its is the main determinant of the calm and responsible performance of all professional tasks. It is a wide range of activities related to dentistry, which must be developed with great intensity, as the overall level in this area is far from the one set by Dentistry 4.0.

Figure 39 shows the model of the development of modern dentistry starting from 2020 and the diagram of the relationship between the above-mentioned elements discussed in this paper.

In concluding these considerations, we will once again return to the model of stability symbolized by the old Chinese censer. It should be remembered that stability is ensured if the legs have different heights until the center of gravity of the entire object tilts beyond the base projection. Should this happen, it will lose its stability and the object will collapse. In caring for the sustainable development of dentistry, it is therefore necessary to strive for the even development of each of the components of this model—i.e., advanced interventionist Dentistry 4.0, global dental prevention, and a dentistry safety system.

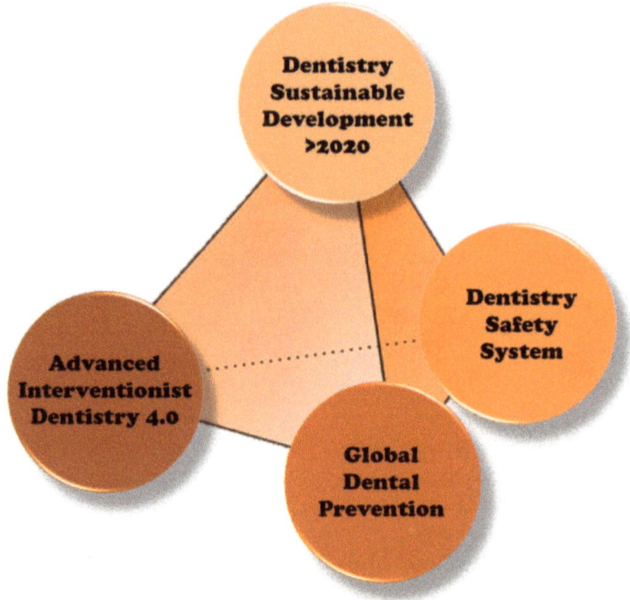

Figure 39. Model of the sustainable development of modern dentistry from 2020.

Author Contributions: Conceptualization, L=literature review, presentation—L.A.D. and L.B.D.; design, resources, data curation, software, formal analysis, practical verification, writing—original draft preparation, visualization—L.A.D., L.B.D., A.D.D.-D., J.D.; writing—review and editing, supervision, project administration, funding acquisition—L.A.D. and L.B.D. All authors have read and agreed to the published version of the manuscript.

Funding: This research was not directly financed by external funding.

Notice: The paper is prepared due to the sending of the projects applications no.: POIR.01.01.01-00-0637/20 cleanDENTworkspace—a medical aerosol absorption system generated during medical procedures in the patient's mouth during dental, ENT, and general medical procedures in order to effectively protect personnel and patients against viruses (including SARS-CoV-2) and other microorganisms, 04.06.2020, and no: POIR.01.01.01-00-0694/20 ASKLEPIOS-IFIS/ Implantological final immediate reconstruction—a system based on implants, in particular individual ones, manufactured on the basis of own innovative design and production technology using 3D printing, 17.06.2020, and due to actually implement Project POIR.01.01-00-0485/16-00 on "IMSKA-MAT Innovative dental and maxillofacial implants manufactured using the innovative additive technology supported by computer-aided materials design ADD-MAT" realized by the Medical and Dental Engineering Center for Research, Design, and Production ASKLEPIOS in Gliwice, Poland. The project was implemented in 2017–2021 and is co-financed by the Operational Program Intelligent Development of the European Union.

References

1. Kulik, R. Czym jest równowaga? *Dzikie Życie* **2008**, *3*, 165. Available online: https://dzikiezycie.pl/archiwum/2008/marzec-2008/czym-jest-rownowaga (accessed on 6 September 2020).
2. Von Carlowitz, H.C. *Sylvicultura Oeconomica Oder Haußwirthliche Nachricht und Naturmäßige Anweisung zur Wilden Baum-Zucht (Deutsch)*; Hamberger, J., Ed.; Oekom Verlag: München, Germany, 2013.

3. Brundtland, G.H. (Ed.) *Report of the World Commission on Environment and Development (WCED): Our Common Future (also known as the Brundtland Report)*; United Nations: New York, NY, USA, 1987; Available online: https://sustainabledevelopment.un.org/content/documents/5987our-common-future.pdf (accessed on 6 September 2020).
4. Meadows, D.H.; Meadows, D.L.; Randers, J.; Behrens, W.W., III. *The Limits to Growth: A Report for the Club of Rome's Project on the Predicament of Mankind*; A Potomac Associates Book: New York, NY, USA, 1972; Available online: https://collections.dartmouth.edu/teitexts/meadows/diplomatic/meadows_ltg-diplomatic.html (accessed on 6 September 2020).
5. Holcombe, R.G. A theory of the theory of public goods. *Rev. Austrian Econ.* **1997**, *10*, 1–22. [CrossRef]
6. Peres, M.A.; Macpherson, L.M.D.; Weyant, R.J.; Daly, B.; Venturelli, R.; Mathur, M.R.; Listl, S.; Celeste, R.K.; Guarnizo-Herreño, C.C.; Kearns, C.; et al. Oral diseases: A global public health challenge. *Lancet* **2019**, *394*, 249–260. [CrossRef]
7. Watt, R.G.; Daly, B.; Allison, P.; Macpherson, L.M.D.; Venturelli, R.; Listl, S.; Weyant, R.J.; Mathur, M.R.; Guarnizo-Herreño, C.C.; Celeste, R.K.; et al. Ending the neglect of global oral health: Time for radical action. *Lancet* **2019**, *394*, 261–272. [CrossRef]
8. Clarkson, J.E.; Young, L.; Ramsay, C.R.; Bonner, B.C.; Bonetti, D. How to influence patient oral hygiene behavior effectively. *J. Dent. Res.* **2009**, *88*, 933–937. [CrossRef] [PubMed]
9. Edwards Deming, W. *Out of the Crisis*; Massachusetts Institute of Technology, Center for Advanced Engineering Study: Cambridge, MA, USA, 1986.
10. Pitts, N.B.; Zero, D.T. White Paper on Dental Caries Prevention and Management. FDI World Dental Federation. Available online: http://www.fdiworlddental.org/sites/default/files/media/documents/2016-fdi_cpp-white_paper.pdf (accessed on 8 November 2020).
11. Kassebaum, N.J.; Bernabé, E.; Dahiya, M.; Bhandari, B.; Murray, C.J.; Marcenes, W. Global burden of untreated caries: A systematic review and metaregression. *J. Dent. Res.* **2015**, *94*, 650–658. [CrossRef] [PubMed]
12. Pitts, N.B.; Zero, D.T.; Marsh, P.D.; Ekstrand, K.; Weintraub, J.A.; Ramos-Gomez, F.; Tagami, J.; Twetman, S.; Tsakos, G.; Ismail, A. Dental caries. *Nat. Rev. Dis. Primers* **2017**, *3*, 17030. [CrossRef]
13. Mejàre, I.; Axelsson, S.; Dahlén, G.; Espelid, I.; Norlund, A.; Tranæus, S.; Twetman, S. Caries risk assessment. A systematic review. *Acta Odontol. Scand.* **2014**, *72*, 81–91. [CrossRef] [PubMed]
14. Twetman, S.; Banerjee, A. Caries risk assessment. In *Risk Assessment in Oral Health*; Chapple, I., Papapanou, P., Eds.; Springer: Cham, Switzerland, 2020. [CrossRef]
15. Yadav, K.; Prakash, S. Dental caries: A microbiological approach. *J. Clin. Infect. Dis. Pr.* **2017**, *2*, 1000118. [CrossRef]
16. Selwitz, R.H.; Ismail, A.I.; Pitts, N.B. Dental caries. *Lancet* **2007**, *369*, 51–59. [CrossRef]
17. Van Houte, J. Role of micro-organisms in caries etiology. *J. Dent. Res.* **1994**, *73*, 672–681. [CrossRef] [PubMed]
18. Bolin, A.K.; Bolin, A.; Jansson, L.; Calltorp, J. Children's dental health in Europe. *Swed. Dent. J.* **1997**, *21*, 25–40. [PubMed]
19. Fejerskov, O. Concepts of dental caries and their consequences for understanding the disease. *Community Dent. Oral. Epidemiol.* **1997**, *25*, 5–12. [CrossRef] [PubMed]
20. Fejerskov, O. Changing paradigms in concepts on dental caries: Consequences for oral health care. *Caries Res.* **2004**, *38*, 182–191. [CrossRef] [PubMed]
21. Pitts, N. "ICDAS"—An international system for caries detection and assessment being developed to facilitate caries epidemiology, research and appropriate clinical management. *Community Dent. Health* **2004**, *21*, 193–198. [PubMed]
22. Zero, D.T.; Zandona, A.F.; Vail, M.M.; Spolnik, K.J. Dental caries and pulpal disease. *Dent. Clin. North Am.* **2011**, *55*, 29–46. [CrossRef]
23. Zero, D.T. Dental caries process. *Dent. Clin. North Am.* **1999**, *43*, 635–664.
24. Thomson, W.M. Epidemiology of oral health conditions in older people. *Gerodontology* **2014**, *31* (Suppl. 1), 9–16. [CrossRef]
25. Psoter, W.J.; Reid, B.C.; Katz, R.V. Malnutrition and dental caries: A review of the literature. *Caries Res.* **2005**, *39*, 441–447. [CrossRef]
26. López, N.J.; Smith, P.C.; Gutierrez, J. Periodontal therapy reduce the risk of preterm low birth weight in women with periodontal disease: A randomized controlled trial. *J. Periodontol.* **2002**, *73*, 911–924. [CrossRef]

27. Al-Nawas, B.; Maeurer, M. Severe versus local odontogenic bacterial infections: Comparison of microbial isolates. *Eur. Surg. Res.* **2008**, *40*, 220–224. [CrossRef] [PubMed]
28. Biguzzi, E.; Dougall, A.; Romero-Lux, O. Non-gynaecological issues in women with bleeding disorders. *J. Haemophil. Pr.* **2019**, *6*, 39–43. [CrossRef]
29. Chapple, I.L.C.; Genco, R.; Working Group 2 of the Joint EFP/AAP Workshop. Diabetes and periodontal diseases: Consensus report of the Joint EFP/AAP Workshop on periodontitis and systemic diseases. *J. Periodontol.* **2013**, *84*, S106–S112. [CrossRef] [PubMed]
30. Fuller, E.; Steele, J.; Watt, R.; Nuttal, N. 1: Oral Health and Function—A Report from the Adult Dental Health Survey 2009; NHS Health and Social Care Information Centre, 2011. Available online: https://files.digital.nhs.uk/publicationimport/pub01xxx/pub01086/adul-dent-heal-surv-summ-them-the1-2009-rep3.pdf (accessed on 28 August 2020).
31. Albandar, J.M.; Rams, T.E. Global epidemiology of periodontal diseases: An overview. *Periodontol. 2000* **2002**, *29*, 7–10. [CrossRef] [PubMed]
32. Sheiham, A.; Netuveli, G.S. Periodontal diseases in Europe. *Periodontol. 2000* **2002**, *29*, 104–121. [CrossRef]
33. Papapanou, P.N. Commentary: Advances in periodontal disease epidemiology: A retrospective commentary. *J. Periodontol.* **2014**, *85*, 877–879. [CrossRef]
34. Albandar, J.M.; Tinoco, E.M. Global epidemiology of periodontal diseases in children and young persons. *Periodontol. 2000* **2002**, *29*, 153–176. [CrossRef]
35. Dobrzański, L.A.; Dobrzański, L.B.; Dobrzańska-Danikiewicz, A.D.; Dobrzańska, J.; Rudziarczyk, K.; Achtelik-Franczak, A. Non-antagonistic contradictoriness of the progress of advanced digitized production with SARS-CoV-2 virus transmission in the area of dental engineering. *Processes* **2020**, *8*, 1097. [CrossRef]
36. Scannapieco, F.A. Role of oral bacteria in respiratory infection. *J. Periodontol.* **1999**, *70*, 793–802. [CrossRef]
37. Scannapieco, F.A.; Bush, R.B.; Paju, S. Associations between periodontal disease and risk for nosocomial bacterial pneumonia and chronic obstructive pulmonary disease. A systemic review. *Ann. Periodontol.* **2003**, *8*, 54–69. [CrossRef]
38. Mueller, A.A.; Saldami, B.; Stübinger, S.; Walter, C.; Flückiger, U.; Merlo, A.; Schwenzer-Zimmerer, K.; Zeilhofer, H.F.; Zimmerer, S. Oral bacterial cultures in nontraumatic brain abscesses: Results of a first line study. *Oral Surg. Oral Med. Oral Pathol. Oral Radiol. Endod.* **2009**, *107*, 469–476. [CrossRef] [PubMed]
39. Li, X.; Tornstad, L.; Olsen, I. Brain abscesses caused by oral infection. *Dent. Traumatol.* **1999**, *15*, 95–101. [CrossRef] [PubMed]
40. Aleksander, M.; Krishnan, B.; Shenoy, N. Diabetes mellitus and odontogenic infections-an exaggerated risk? *Oral Maxillofac. Surg.* **2008**, *12*, 129–130. [CrossRef] [PubMed]
41. Isola, G.; Matarese, G.; Ramaglia, L.; Pedullà, E.; Rapisarda, E.; Iorio-Siciliano, V. Association between periodontitis and glycosylated haemoglobin before diabetes onset: A cross-sectional study. *Clin. Oral Investig.* **2020**, *24*, 2799–2808. [CrossRef]
42. Gocke, C.; Holtfreter, B.; Meisel, P.; Grotevendt, A.; Jablonowski, L.; Jablonowski, L.; Markus, M.R.; Kocher, T. Abdominal obesity modifies long-term associations between periodontitis and markers of systemic inflammation. *Atherosclerosis* **2014**, *235*, 351–357. [CrossRef]
43. Isola, G.; Polizzi, A.; Santonocito, S.; Alibrandi, A.; Ferlito, S. Expression of salivary and serum malondialdehyde and lipid profile of patients with periodontitis and coronary heart disease. *Int. J. Mol. Sci.* **2019**, *20*, 6061. [CrossRef]
44. Friedewald, V.E.; Kornman, K.S.; Beck, J.D.; Genco, R.; Goldfine, A.; Libby, P.; Offenbacher, S.; Ridker, P.M.; Van Dyke, T.E.; Roberts, W.C. The American journal of cardiology and journal of periodontology editors' consensus: Periodontitis and atherosclerotic cardiovascular disease. *J. Periodontol.* **2009**, *80*, 1021–1032. [CrossRef]
45. Isola, G.; Alibrandi, A.; Currò, M.; Matarese, M.; Ricca, S.; Matarese, G.; Ientile, R.; Kocher, T. Evaluation of salivary and serum ADMA levels in patients with periodontal and cardiovascular disease as subclinical marker of cardiovascular risk. *J. Periodontol.* **2020**, *91*, 1076–1084. [CrossRef]
46. Vidal, F.; Figueredo, C.M.; Cordovil, I.; Fischer, R.G. Periodontal therapy reduces plasma levels of interleukin-6, C-reactive protein, and fib-rinogen in patients with severe periodontitis and refractory arterial hypertension. *J. Periodontol.* **2009**, *80*, 786–791. [CrossRef]

47. Isola, G.; Polizzi, A.; Muraglie, S.; Leonardi, R.M.; Lo Giudice, A. Assessment of vitamin C and antioxidants profiles In Saliva and serum in patients with periodontitis and ischemic heart disease. *Nutrients* **2019**, *11*, 2956. [CrossRef]

48. Zhou, X.; Cao, F.; Lin, Z. Updated evidence of association between periodontal disease and incident erectile dysfunction. *J. Sex. Med.* **2019**, *16*, 61–69. [CrossRef] [PubMed]

49. Shamloul, R.; Ghanem, H. Erectile dysfunction. *Lancet* **2013**, *381*, 153–165. [CrossRef]

50. Tsao, C.W.; Liu, C.Y.; Cha, T.L.; Wu, S.T.; Chen, S.C.; Hsu, C.Y. Exploration of the association between chronic periodontal disease and erectile dysfunction from a population-based view point. *Andrologia* **2015**, *47*, 513–518. [CrossRef] [PubMed]

51. Rosen, R.C.; Cappelleri, J.C.; Gendrano, N., III. The international index of erectile function (IIEF): A state-of-the-science review. *Int. J. Impot. Res.* **2002**, *14*, 226–244. [CrossRef]

52. Rosen, R.C.; Cappelleri, J.C.; Smith, M.D.; Lipsky, J.; Peña, B.M. Development and evaluation of an abridged, 5-item version of the international index of erectile function (IIEF-5) as a diagnostic tool for erectile dysfunction. *Int. J. Impot. Res.* **1999**, *11*, 319–326. [CrossRef] [PubMed]

53. Sharma, A.; Pradeep, A.R.; Raju, P.A. Association between chronic periodontitis and vasculogenic erectile dysfunction. *J. Periodontol.* **2011**, *82*, 1665–1669. [CrossRef] [PubMed]

54. Matsumoto, S.; Matsuda, M.; Takekawa, M.; Okada, M.; Hashizume, K.; Wada, N.; Hori, J.; Tamaki, G.; Kita, M.; Iwata, T.; et al. Association of ED with chronic periodontal disease. *Int. J. Impot. Res.* **2014**, *26*, 13–15. [CrossRef]

55. Eltas, A.; Oguz, F.; Uslu, M.O.; Akdemir, E. The effect of periodontal treatment in improving erectile dysfunction: A randomized controlled trial. *J. Clin. Periodontol.* **2013**, *40*, 148–154. [CrossRef]

56. Lee, J.H.; Choi, J.K.; Kim, S.H.; Cho, K.H.; Kim, Y.T.; Choi, S.H.; Jung, U.W. Association between periodontal flap surgery for periodontitis and vasculogenic erectile dysfunction in Koreans. *J. Periodontal. Implant. Sci.* **2017**, *47*, 96–105. [CrossRef]

57. Martín, A.; Bravo, M.; Arrabal, M.; Magán-Fernández, A.; Mesa, F. Chronic periodontitis is associated with erectile dysfunction. A case-control study in european population. *J. Clin. Periodontol.* **2018**, *45*, 791–798. [CrossRef]

58. Oguz, F.; Eltas, A.; Beytur, A.; Akdemir, E.; Uslu, M.O.; Gunes, A. Is there a relationship between chronic periodontitis and erectile dysfunction? *J. Sex. Med.* **2013**, *10*, 838–843. [CrossRef] [PubMed]

59. Keller, J.J.; Chung, S.D.; Lin, H.C. A nationwide population-based study on the association between chronic periodontitis and erectile dysfunction. *J. Clin. Periodontol.* **2012**, *39*, 507–512. [CrossRef] [PubMed]

60. Zadik, Y.; Bechor, R.; Galor, S.; Justo, D.; Heruti, R.J. Erectile dysfunction might be associated with chronic periodontal disease: Two ends of the cardiovascular spectrum. *J. Sex. Med.* **2009**, *6*, 1111–1116. [CrossRef] [PubMed]

61. Liu, L.H.; Li, E.M.; Zhong, S.L.; Li, Y.Q.; Yang, Z.Y.; Kang, R.; Zhao, S.K.; Li, F.T.; Wan, S.P.; Zhao, Z.G. Chronic periodontitis and the risk of erectile dysfunction: A systematic review and meta-analysis. *Int. J. Impot. Res.* **2017**, *29*, 43–48. [CrossRef] [PubMed]

62. Wang, Q.; Kang, J.; Cai, X.; Wu, Y.; Zhao, L. The association between chronic periodontitis and vasculogenic erectile dysfunction: A systematic review and meta-analysis. *J. Clin. Periodontol.* **2016**, *43*, 206–215. [CrossRef] [PubMed]

63. Singh, V.P.; Nettemu, S.K.; Nettem, S.; Hosadurga, R.; Nayak, S.U. Oral health and erectile dysfunction. *J. Hum. Reprod. Sci.* **2017**, *10*, 162–166. [CrossRef]

64. Gandaglia, G.; Briganti, A.; Jackson, G.; Kloner, R.A.; Montorsi, F.; Montorsi, P.; Vlachopoulos, C. A systematic review of the association between erectile dysfunction and cardiovascular disease. *Eur. Urol.* **2014**, *65*, 968–978. [CrossRef] [PubMed]

65. Eastham, J.; Seymour, R. Is oral health a risk factor for sexual health? *Dent. Update* **2015**, *42*, 160–162. [CrossRef]

66. Peng, S.; Zhang, D.X. Chronic periodontal disease may be a sign for erectile dysfunction in men. *Med. Hypotheses* **2009**, *73*, 859–860. [CrossRef]

67. Gurav, A.N. The implication of periodontitis in vascular endothelial dysfunction. *Eur. J. Clin. Invest.* **2014**, *44*, 1000–1009. [CrossRef]

68. Brito, L.C.; DalBó, S.; Striechen, T.M.; Farias, J.M.; Olchanheski, L.R., Jr.; Mendes, R.T.; Vellosa, J.C.; Fávero, G.M.; Sordi, R.; Assreuy, J.; et al. Experimental periodontitis promotes transient vascular inflammation and endothelial dysfunction. *Arch. Oral Biol.* **2013**, *58*, 1187–1198. [CrossRef] [PubMed]

69. Zuo, Z.; Jiang, J.; Jiang, R.; Chen, F.; Liu, J.; Yang, H.; Cheng, Y. Effect of periodontitis on erectile function and its possible mechanism. *J. Sex. Med.* **2011**, *8*, 2598–2605. [CrossRef] [PubMed]

70. Kellesarian, S.V.; Malmstrom, H.; Abduljabbar, T.; Vohra, F.; Kellesarian, T.V.; Javed, F.; Romanos, G.E. Low testosterone levels in body fluids are associated with chronic periodontitis. *Am. J. Mens. Health* **2017**, *11*, 443–453. [CrossRef] [PubMed]

71. Steffens, J.P.; Wang, X.; Starr, J.R.; Spolidorio, L.C.; Van Dyke, T.E.; Kantarci, A. Associations between sex hormone levels and periodontitis in men: Results from NHANES III. *J. Periodontol.* **2015**, *86*, 1116–1125. [CrossRef]

72. Singh, B.P.; Makker, A.; Tripathi, A.; Singh, M.M.; Gupta, V. Association of testosterone and bone mineral density with tooth loss in men with chronic periodontitis. *J. Oral Sci.* **2011**, *53*, 333–339. [CrossRef]

73. Daltaban, O.; Saygun, I.; Bolu, E. Periodontal status in men with hypergonadotropic hypogonadism: Effects of testosterone deficiency. *J. Periodontol.* **2006**, *77*, 1179–1183. [CrossRef]

74. Bobjer, J.; Katrinaki, M.; Tsatsanis, C.; Lundberg Giwercman, Y.; Giwercman, A. Negative association between testosterone concentration and inflammatory markers in young men: A nested cross-sectional study. *PLoS ONE* **2013**, *8*, e61466. [CrossRef]

75. Machado, V.; Lopes, J.; Patrão, M.; Botelho, J.; Proença, L.; Mendes, J.J. Validity of the association between periodontitis and female infertility conditions: A concise review. *Reproduction* **2020**, *160*, R41–R54. [CrossRef]

76. Gajendra, S.; Kumar, J.V. Oral health and pregnancy: A review. *NY State Dent. J.* **2004**, *70*, 40–44.

77. Pallasch, T.J.; Wahl, M.J. Focal infection: New age or ancient history? *Endodon. Top.* **2003**, *4*, 32–45. [CrossRef]

78. Burzyńska, B.; Mierzwińska-Nastalska, E. Rehabilitacja protetyczna pacjentów bezzębnych. *Nowa Stomatol.* **2011**, *4*, 167–199.

79. Felton, D.A. Edentualism and comorbid factors. *J. Prosthodont.* **2009**, *18*, 88–96. [CrossRef] [PubMed]

80. Felton, D.A. Complete edentulism and comorbid diseases: An update. *J. Prosthodont.* **2016**, *25*, 5–20. [CrossRef] [PubMed]

81. Holmlund, A.; Holm, G.; Lind, L. Number of teeth as a predictor of cardiovascular mortality in a cohort of 7674 subjects followed for 12 years. *J. Periodontol.* **2010**, *81*, 870–876. [CrossRef]

82. Volzke, H.; Schwahn, C.; Hummel, A.; Wolff, B.; Kleine, V.; Robinson, D.M.; Dahm, J.B.; Felix, S.B.; John, U.; Kocher, T. Tooth loss is independently associated with the risk of acquired aortic valve sclerosis. *Am. Heart J.* **2005**, *150*, 1198–1203. [CrossRef] [PubMed]

83. Takata, Y.; Ansai, T.; Matsumura, K.; Awano, S.; Hamasaki, T.; Sonoki, K.; Kusaba, A.; Akifusa, S.; Takehara, T. Relationship between tooth loss and electrocardiographic abnormalities in octogenarians. *J. Dent. Res.* **2001**, *80*, 1648–1652. [CrossRef] [PubMed]

84. De Pablo, P.; Dietrich, T.; McAlindon, T.E. Association of periodontal disease and tooth loss with rheumatoid arthritis in the US population. *J. Rheumatol.* **2008**, *35*, 70–76.

85. Sierpinska, T.; Golebiewska, M.; Dlugosz, J.W.; Kemona, A.; Laszewicz, W. Connection between masticatory efficiency and pathomorphologic changes in gastric mucosa. *Quintessence Int.* **2007**, *38*, 31–37.

86. Abnet, C.C.; Qiao, Y.L.; Dawsey, S.M.; Dong, Z.W.; Taylor, P.R.; Mark, S.D. Tooth loss is associated with increased risk of total death and death from upper gastrointestinal cancer, heart disease, and stroke in a Chinese population-based cohort. *Int. J. Epidemiol.* **2005**, *34*, 467–474. [CrossRef]

87. Bagchi, S.; Tripathi, A.; Tripathi, S.; Kar, S.; Tiwari, S.C.; Singh, J. Obstructive sleep apnea and neurocognitive dysfunction in edentulous patients. *J. Prosthodont.* **2019**, *28*, e837–e842. [CrossRef]

88. Buset, S.L.; Walter, C.; Friedmann, A.; Weiger, R.; Borgnakke, W.S.; Zitzmann, N.U. Are periodontal diseases really silent? A systematic review of their effect on quality of life. *J. Clin. Periodontol.* **2016**, *43*, 333–344. [CrossRef] [PubMed]

89. Bui, F.Q.; Almeida-da-Silva, C.L.C.; Huynh, B.; Trinh, A.; Liu, J.; Woodward, J.; Asadi, H.; Ojcius, D.M. Association between periodontal pathogens and systemic disease. *Biomed. J.* **2019**, *42*, 27–35. [CrossRef] [PubMed]

90. Nagpal, R.; Yamashiro, Y.; Izumi, Y. The two-way association of periodontal infection with systemic disorders: An overview. *Mediat. Inflamm.* **2015**, *2015*, 793898. [CrossRef] [PubMed]

91. Henke, K. A model for memory systems based on processing modes rather than consciousness. *Nat. Rev. Neurosci.* **2010**, *11*, 523–532. [CrossRef] [PubMed]

92. Kawahata, M.; Ono, Y.; Ohno, A.; Kawamoto, S.; Kimoto, K.; Onozuka, M. Loss of molars early in life develops behavioral lateralization and impairs hippocampus-dependent recognition memory. *BMC Neurosci.* **2014**, *15*, 4. [CrossRef] [PubMed]

93. Hirano, Y.; Obata, T.; Takahashi, H.; Tachibana, A.; Kuroiwa, D.; Takahashi, T.; Ikehira, H.; Onozuka, M. Effects of chewing on cognitive processing speed. *Brain Cogn.* **2013**, *81*, 376–381. [CrossRef] [PubMed]

94. Chen, H.; Iinuma, M.; Onozuka, M.; Kubo, K.-Y. Chewing maintains hippocampus-dependent cognitive. *Int. J. Med. Sci.* **2015**, *12*, 502–509. [CrossRef]

95. Lexomboon, D.; Trulsson, M.; Wårdh, I.; Parker, W.G. Chewing ability and tooth loss: Association with cognitive impairment in an elderly population study. *J. Am. Geriatr. Soc.* **2012**, *60*, 1951–1956. [CrossRef]

96. Stein, P.S.; Desrosiers, M.; Donegan, S.J.; Yepes, J.F.; Kryscio, R.J. Tooth loss, dementia and neuropathology in the nun study. *J. Am. Dent. Assoc.* **2007**, *138*, 1314–1322. [CrossRef]

97. Onishi, M.; Iinuma, M.; Tamura, Y.; Kubo, K.Y. Learning deficits and suppression of the cell proliferation in the hippocampal dentate gyrus of offspring are attenuated by maternal chewing during prenatal stress. *Neurosci. Lett.* **2014**, *560*, 77–80. [CrossRef] [PubMed]

98. Durán, D. Ignoring the life course: GES oral health for adults of 60 years. *J. Oral Res.* **2019**, *8*, 272–274. [CrossRef]

99. Murray, C.J.L.; Lopez, A.D. *The Global Burden of Disease: A Comprehensive Assessment of Mortality and Disability from Diseases, Injuries, and Risk Factors in 1990 and Projected to 2020*; University Press on behalf of the World Health Organization and The World Bank: Boston, MA, USA; Harvard: Cambridge, MA, USA, 1996; Available online: https://apps.who.int/iris/handle/10665/41864 (accessed on 30 September 2020).

100. Available online: https://vizhub.healthdata.org/gbd-compare/ (accessed on 28 August 2020).

101. Kassebaum, N.J.; Bernabé, E.; Dahiya, M.; Bhandari, B.; Murray, C.J.L.; Marcenes, W. Global burden of severe tooth loss: A systematic review and meta-analysis. *J. Dent. Res.* **2014**, *93* (Suppl. 7), 20S–28S. [CrossRef]

102. Tan, H.; Peres, K.G.; Peres, M.A. Retention of teeth and oral health-related quality of life. *J. Dent. Res.* **2016**, *95*, 1350–1357. [CrossRef] [PubMed]

103. Ricci, S.; Capecchi, G.; Boschin, F.; Arrighi, S.; Ronchitelli, A.; Condemi, S. Toothpick use among epigravettian humans from Grotta Paglicci (Italy). *Int. J. Osteoarchaeo.* **2016**, *26*, 281–289. [CrossRef]

104. Lozano, M.; Subirà, M.; Aparicio, J.; Lorenzo, C.; Gomez-Merino, G. Toothpicking and periodontal disease in a neanderthal specimen from Cova Foradà Site (Valencia, Spain). *PLoS ONE* **2013**, *8*, e76852. [CrossRef] [PubMed]

105. Coppa, A.; Bondioli, L.; Cucina, A.; Frayer, D.W.; Jarrige, C.; Jarrige, J.-F.; Quivron, G.; Rossi, M.; Vidale, M.; Macchiarelli, R. Early neolithic tradition of dentistry. *Nature* **2006**, *440*, 755–756. [CrossRef] [PubMed]

106. Estalrrich, A.; Alarcon, J.A.; Rosas, A. Evidence of toothpick groove formation in neandertal anterior and posterior teeth. *Am. J. Physic. Anthr.* **2017**, *162*, 747–756. [CrossRef] [PubMed]

107. Mantini, S.; Marini, M.; del Monte, S.; Mazza, D.; Primicerio, P.; Brea, M.B.; Salvadei, L. Early dentistry practice in Italian neolithic site of Gaione (Parma). *Origini* **2007**, *29*, 221–225.

108. Dobrzański, L.A.; Dobrzański, L.B. Approach to the design and manufacturing of prosthetic dental restorations according to the rules of industry 4.0. *Mater. Perform. Charact.* **2020**, *9*, 394–476. [CrossRef]

109. Oxilia, G.; Peresani, M.; Romandini, M.; Matteucci, C.; Debono Spiteri, C.; Henry, A.G.; Schulz, D.; Archer, W.; Crezzini, J.; Boschin, F.; et al. Earliest evidence of dental caries manipulation in the Late Upper Palaeolithic. *Sci. Rep.* **2015**, *5*, 1–10. [CrossRef]

110. Allen, J.P. *The Art of Medicine in Ancient Egypt*; The Metropolitan Museum of Art: New York, NY, USA, 2005.

111. Frayer, D.W.; Gatti, J.; Monge, J.; Radovčic, D. Prehistoric dentistry? P4 rotation, partial M3 impaction, toothpick grooves and other signs of manipulation in krapina dental person 20. *Bull. Int. Assoc. Paleodont.* **2017**, *11*, 1–10.

112. Bernardini, F.; Tuniz, C.; Coppa, A.; Mancini, L.; Dreossi, D.; Eichert, D.; Turco, G.; Biasotto, M.; Terrasi, F.; De Cesare, N.; et al. Beeswax as dental filling on a neolithic human tooth. *PLoS ONE* **2012**, *7*, e44904. [CrossRef] [PubMed]

113. Leek, F.F. The practice of dentistry in Ancient Egypt. *J. Egypt. Archeol.* **1967**, *53*, 51–58. [CrossRef]

114. Wilwerding, T. History of Dentistry. The Free Information Society 2001. Available online: http://www.freeinfosociety.com/media/pdf/4551.pdf (accessed on 30 September 2020).

115. Townend, B.R. Story of the tooth-worm. *Bull. Hist. Med.* **1944**, *15*, 37–58.

116. Suddick, R.P.; Harris, N.O. Historical perspectives of oral biology: A series. *Crit. Rev. Oral Biol. Med.* **1990**, *1*, 135–151. [CrossRef] [PubMed]

117. Dawson, W.R. Book review: The papyrus ebers; the greatest Egyptian medical document. *J. Egypt. Archeol.* **1938**, *24*, 250–251. [CrossRef]

118. Seguin-Orlando, A.; Korneliussen, T.S.; Sikora, M.; Malaspinas, A.-S.; Manica, A.; Moltke, I.; Albrechtsen, A.; Ko, A.; Margaryan, A.; Moiseyev, V.; et al. Genomic structure in europeans dating back at least 36,200 years. *Science* **2014**, *346*, 1113–1118. [CrossRef]

119. Reisine, S.; Litt, M. Social and psychological theories and their use for dental practice. *Int. Dent. J.* **1993**, *43* (Suppl. 1), 279–287.

120. Featherstone, J.D. The continuum of dental caries—Evidence for a dynamic disease process. *J. Dent. Res.* **2004**, *83*, C39–C42. [CrossRef]

121. Pitts, N.B. A review of the current knowledge of the progress of approximal lesions. In Proceedings of the Scientific Proceedings of the 10th Asian Pacific Dental Congress, Singapore, 26–31 March 1981; Singapore Dental Association: Singapore, 1983; pp. 31–36.

122. Amaechi, B.T. Remineralization therapies for initial caries lesions. *Curr. Oral Health Rep.* **2015**, *2*, 95–101. [CrossRef]

123. Zero, D.T. Dentifrices, mouthwashes, and remineralization/caries arrestment strategies. *BMC Oral Health* **2006**, *6*, S9. [CrossRef] [PubMed]

124. Holmgren, C.; Gaucher, N.; Decerle, N.; Doméjean, S. Minimal intervention dentistry II: Part 3. Management of non-cavitated (initial) occlusal caries lesions—Non-invasive approaches through remineralization and therapeutic sealants. *Br. Dent. J.* **2014**, *216*, 237–243. [CrossRef] [PubMed]

125. Takahashi, N.; Nyvad, B. Caries ecology revisited: Microbial dynamics and the caries process. *Caries Res.* **2008**, *42*, 409–418. [CrossRef] [PubMed]

126. Margolis, H.C.; Moreno, E.C. Composition and cariogenic potential of dental plaque fluid. *Crit. Rev. Oral Biol. Med.* **1994**, *5*, 1–25. [CrossRef] [PubMed]

127. Margolis, H.C.; Moreno, E.C. Kinetics of hydroxyapatite dissolution in acetic, lactic, and phosphoric acid solutions. *Calcif. Tissue Int.* **1992**, *50*, 137–143. [CrossRef] [PubMed]

128. Vogel, G.L.; Carey, C.M.; Chow, L.C.; Gregory, T.M.; Brown, W.E. Micro-analysis of mineral saturation within enamel during lactic acid demineralization. *J. Dent. Res.* **1988**, *67*, 1172–1180. [CrossRef]

129. Ten Cate, J.M.; Featherstone, J.D. Mechanistic aspects of the interactions between fluoride and dental enamel. *Crit. Rev. Oral Biol. Med.* **1991**, *2*, 283–296. [CrossRef]

130. Lee, Y.E.; Baek, H.J.; Choi, Y.H.; Jeong, S.H.; Park, Y.D.; Song, K.B. Comparison of remineralization effect of three topical fluoride regimens on enamel initial carious lesions. *J. Dent.* **2010**, *38*, 166–171. [CrossRef]

131. Amaechi, B.T.; van Loveren, C. Fluorides and non-fluoride remineralization systems. *Monogr. Oral Sci.* **2013**, *23*, 15–26. [CrossRef] [PubMed]

132. Featherstone, J.D. Prevention and reversal of dental caries: Role of low level fluoride. *Community Dent. Oral Epidemiol.* **1999**, *27*, 31–40. [CrossRef]

133. Iheozor-Ejiofor, Z.; Worthington, H.V.; Walsh, T.; O'Malley, L.; Clarkson, J.E.; Macey, R.; Alam, R.; Tugwell, P.; Welch, V.; Glenny, A.M. Water fluoridation for the prevention of dental caries. *Cochrane Database Syst. Rev.* **2015**, *2015*, CD010856. [CrossRef] [PubMed]

134. Reynolds, E.C.; Cai, F.; Shen, P.; Walker, G.D. Retention in plaque and remineralization of enamel lesions by various forms of calcium in a mouthrinse or sugar-free chewing gum. *J. Dent. Res.* **2003**, *82*, 206–211. [CrossRef] [PubMed]

135. Kitasako, Y.; Sadr, A.; Hamba, H.; Ikeda, M.; Tagami, J. Gum containing calcium fluoride reinforces enamel subsurface lesions in situ. *J. Dent. Res.* **2012**, *91*, 370–375. [CrossRef] [PubMed]

136. Hamba, H.; Nikaido, T.; Inoue, G.; Sadr, A.; Tagami, J. Effects of CPP-ACP with sodium fluoride on inhibition of bovine enamel demineralization: A quantitative assessment using micro-computed tomography. *J. Dent.* **2011**, *39*, 405–413. [CrossRef]

137. Kidd, E.A.M.; Fejerskov, O. What constitutes dental caries? Histopathology of carious enamel and dentin related to the action of cariogenic biofilms. *J. Dent. Res.* **2004**, *83*, C35–C38. [CrossRef]

138. Mandel, I.D. The functions of saliva. *J. Dent. Res.* **1987**, *66*, 623–627. [CrossRef]

139. Hara, A.T.; Zero, D.T. The caries environment: Saliva, pellicle, diet, and hard tissue ultrastructure. *Dent. Clin. North Am.* **2010**, *54*, 455–467. [CrossRef]

140. Zero, D.T. Sugars—The arch criminal? *Caries Res.* **2004**, *38*, 277–285. [CrossRef]

141. Sheiham, A.; James, W.P.T. Diet and dental caries: The pivotal role of free sugars reemphasized. *J. Dent. Res.* **2015**, *94*, 1341–1347. [CrossRef]

142. Marsh, P.D. Microbial ecology of dental plaque and its significance in health and disease. *Adv. Dent. Res.* **1994**, *8*, 263–271. [CrossRef]

143. Zero, D.T. Adaptations in dental plaque. In *Cariology for the Nineties*; Bowen, W.H., Tabak, L., Eds.; University of Rochester Press: Rochester, NY, USA, 1993; pp. 333–350.

144. Aas, J.A.; Griffen, A.L.; Dardis, S.R.; Lee, A.M.; Olsen, I.; Dewhirst, F.E.; Leys, E.J.; Paster, B.J. Bacteria of dental caries in primary and permanent teeth in children and young adults. *J. Clin. Microbiol.* **2008**, *46*, 1407–1417. [CrossRef] [PubMed]

145. Kleinberg, I. A mixed-bacteria ecological approach to understanding the role of the oral bacteria in dental caries causation: An alternative to Streptococcus mutans and the specific-plaque hypothesis. *Crit. Rev. Oral Biol. Med.* **2002**, *13*, 108–125. [CrossRef] [PubMed]

146. Arif, N.; Sheehy, E.C.; Do, T.; Beighton, D. Diversity of Veillonella spp. from sound and carious sites in children. *J. Dent. Res.* **2008**, *87*, 278–282. [CrossRef] [PubMed]

147. Moynihan, P.J.; Kelly, S.A. Effect on caries of restricting sugars intake: Systematic review to inform WHO guidelines. *J. Dent. Res.* **2014**, *93*, 8–18. [CrossRef] [PubMed]

148. Sheiham, A.; James, W.P. A reappraisal of the quantitative relationship between sugar intake and dental caries: The need for new criteria for developing goals for sugar intake. *BMC Public Health* **2014**, *14*, 863. [CrossRef] [PubMed]

149. Liu, Y.-L.; Nascimento, M.; Burne, R.A. Progress toward understanding the contribution of alkali generation in dental biofilms to inhibition of dental caries. *Int. J. Oral Sci.* **2012**, *4*, 135–140. [CrossRef]

150. Pollick, H.F. Salt fluoridation: A review. *J. Calif. Dent. Assoc.* **2013**, *41*, 395–404.

151. Marinho, V.C.; Higgins, J.P.; Logan, S.; Sheiham, A. Topical fluoride (toothpastes, mouthrinses, gels or varnishes) for preventing dental caries in children and adolescents. *Cochrane Database Syst. Rev.* **2003**, *4*, CD002782. [CrossRef]

152. Walsh, T.; Worthington, H.V.; Glenny, A.M.; Appelbe, P.; Marinho, V.C.; Shi, X. Fluoride toothpastes of different concentrations for preventing dental caries in children and adolescents. *Cochrane Database Syst. Rev.* **2010**, *1*, CD007868, Update in: *Cochrane Database Syst Rev* **2019**, *3*, CD007868. [CrossRef]

153. Zero, D.T. The role of dietary control. In *Dental Caries: The Disease and its Clinical Management*, 2nd ed.; Fejerskov, O., Kidd, E., Eds.; Blackwell Munksgaard: Oxford, UK, 2008; pp. 329–352.

154. Pollard, M.A.; Imfeld, T.; Higham, S.M.; Agalamanyi, E.A.; Curzon, M.E.; Edgar, W.M.; Borgia, S. Acidogenic potential and total salivary carbohydrate content of expectorants following the consumption of some cereal-based foods and fruits. *Caries Res.* **1996**, *30*, 132–137. [CrossRef] [PubMed]

155. Paes Leme, A.F.; Koo, H.; Bellato, C.M.; Bedi, G.; Cury, J.A. The role of sucrose in cariogenic dental biofilm formation—New insight. *J. Dent. Res.* **2006**, *85*, 878–887. [CrossRef] [PubMed]

156. Cury, J.A.; Rebelo, M.A.; Del Bel Cury, A.A.; Derbyshire, M.T.; Tabchoury, C.P. Biochemical composition and cariogenicity of dental plaque formed in the presence of sucrose or glucose and fructose. *Caries Res.* **2000**, *34*, 491–497. [CrossRef] [PubMed]

157. Zero, D.T.; Fontana, M.; Martínez-Mier, E.A.; Ferreira-Zandoná, A.; Ando, M.; González-Cabezas, C.; Bayne, S. The biology, prevention, diagnosis and treatment of dental caries: Scientific advances in the United States. *J. Am. Dent. Assoc.* **2009**, *140* (Suppl. 1), 25S–34S. [CrossRef] [PubMed]

158. Hall, H.D.; Schneyer, C.A. Salivary gland atrophy in rat induced by liquid diet. *Proc. Soc. Exp. Biol. Med. Soc. Exp. Biol. Med.* **1964**, *117*, 789–793. [CrossRef] [PubMed]

159. Wierichs, R.J.; Meyer-Lueckel, H. Systematic review on noninvasive treatment of root caries lesions. *J. Dent. Res.* **2015**, *94*, 261–271. [CrossRef] [PubMed]

160. Clarkson, B.H.; Exterkate, R.A.M. Noninvasive dentistry: A dream or reality? *Caries Res.* **2015**, *49* (Suppl. 1), 11–17. [CrossRef]

161. Marsh, P.D.; Head, D.A.; Devine, D.A. Prospects of oral disease control in the future—An opinion. *J. Oral Microbiol.* **2014**, *6*, 261–276. [CrossRef]

162. Dirks, O.B.; van Amerongen, J.; Winkler, K.C. A reproducible method for caries evaluation. *J. Dent. Res.* **1951**, *30*, 346–359. [CrossRef]

163. Marthaler, T.M. A standardized system of recording dental conditions. *Helv. Odontol. Acta* **1966**, *10*, 1–18.

164. Black, G.V. *A Work on Operative Dentistry: The Technical Procedures in Filling Teeth*; Medico-Dental Publishing: Chicago, IL, USA, 1917; p. 5.

165. Ismail, A.I.; Sohn, W.; Tellez, M.; Amaya, A.; Sen, A.; Hasson, H.; Pitts, N.B. The international caries detection and assessment system (ICDAS): An integrated system for measuring dental caries. *Community Dent. Oral Epidemiol.* **2007**, *35*, 170–178. [CrossRef]

166. ICDAS Website. Available online: https://www.icdas.org/. (accessed on 20 November 2020).

167. Fejerskov, O.; Kidd, E. (Eds.) *Dental Caries: The Disease and its Clinical Management*, 2nd ed.; Blackwell Munksgaard: Oxford, UK, 2008.

168. Fisher, J.; Glick, M. The FDI World Dental Federation. A new model for caries classification and management: The FDI World Dental Federation caries matrix. *J. Am. Dent. Assoc.* **2012**, *143*, 546–551. [CrossRef] [PubMed]

169. Pitts, N.B.; Ekstrand, K.R. International caries detection and assessment system (ICDAS) and its international caries classification and management system (ICCMS)—Methods for staging of the caries process and enabling dentists to manage caries. *Community Dent. Oral Epidemiol.* **2013**, *41*, e41–e52. [CrossRef] [PubMed]

170. Ismail, A.; Pitts, N.B.; Tellez, M. The international caries classification and management system (ICCMS^{TM}) an example of a caries management pathway. *BMC Oral Health* **2015**, *15*, S9. [CrossRef] [PubMed]

171. Ismail, A.; Tellez, M.; Pitts, N.B.; Ekstrand, K.R.; Ricketts, D.; Longbottom, C.; Eggertsson, H.; Deery, C.; Fisher, J.; Young, D.A.; et al. Caries management pathways preserve dental tissues and promote oral health. *Community Dent. Oral Epidemiol.* **2013**, *41*, e12–e40. [CrossRef] [PubMed]

172. Ormond, C.; Douglas, G.; Pitts, N. The use of the international caries detection and assessment system (ICDAS) in a national health service general dental practice as part of an oral health assessment. *Prim. Dent. Care* **2010**, *17*, 153–159. [CrossRef] [PubMed]

173. Pitts, N.B. Discovering dental public health: From fisher to the future. *Community Dent. Health* **1994**, *11*, 172–178.

174. Pitts, N.B. Modern concepts of caries measurement. *J. Dent. Res.* **2004**, *83*, 43–47. [CrossRef]

175. Pitts, N.B. How the detection, assessment, diagnosis and monitoring of caries integrate with personalized caries management. *Monogr. Oral Sci.* **2009**, *21*, 1–14. [CrossRef]

176. Pitts, N.B.; Fyffe, H.E. The effect of varying diagnostic thresholds upon clinical caries data for a low prevalence group. *J. Dent. Res.* **1988**, *67*, 592–596. [CrossRef]

177. Pitts, N.; Melo, P.; Martignon, S.; Ekstrand, K.; Ismail, A. Caries risk assessment, diagnosis and synthesis in the context of a European core curriculum in cariology. *Eur. J. Dent. Educ.* **2011**, *15* (Suppl. 1), 23–31. [CrossRef]

178. Zandoná, A.F.; Zero, D.T. Diagnostic tools for early caries detection. *J. Am. Dent. Assoc.* **2006**, *137*, 1675–1684. [CrossRef] [PubMed]

179. Ricketts, D.; Lamont, T.; Innes, N.P.; Kidd, E.; Clarkson, J.E. Operative caries management in adults and children. *Cochrane Database Syst. Rev.* **2013**, *3*, CD003808, Update in: *Cochrane Database Syst. Rev.* **2019**, *7*, CD003808. [CrossRef] [PubMed]

180. *ISO24234:2015 Dentistry — Dental amalgam*; ISO: Geneva, Switzerland, 2015.

181. FDA, About Dental Amalgam Fillings. Available online: https://www.fda.gov/medical-devices/dental-amalgam/about-dental-amalgam-fillings (accessed on 28 August 2020).

182. SCENIHR (Scientific Committee on Emerging and Newly-Identified Health Risks). Opinion on the Safety of Dental Amalgam and Alternative Dental Restoration Materials for Patients and Users. European Union, 29 April 2015. Available online: https://ec.europa.eu/health/sites/health/files/scientific_committees/emerging/docs/scenihr_o_046.pdf (accessed on 28 August 2020).

183. Council Decision (EU) 2017/939 of 11 May 2017 on the conclusion on behalf of the European Union of the Minamata Convention on Mercury. *Off. J Eu Oj L* **2017**, *142*, 4–39. Available online: http://data.europa.eu/eli/dec/2017/939/oj (accessed on 28 August 2020).

184. Europejski Trybunał Praw Człowieka. Konwencja o Ochronie Praw Człowieka i Podstawowych Wolności sporządzona w Rzymie dnia 4 listopada 1950 r., zmieniona następnie Protokołami nr 3, 5 i 8 oraz uzupełniona Protokołem nr 2. *Dz. U.* **1993**, *61*, 284. Available online: https://www.echr.coe.int/Documents/Convention_POL.pdf (accessed on 28 August 2020).

185. Schwartz, R.S.; Robbins, J.W. Post placement and restoration of endodontically treated teeth: A literature review. *J. Endodon.* **2004**, *30*, 289–301. [CrossRef] [PubMed]

186. Peroz, I.; Blankenstein, F.; Lange, K.-P.; Naumann, M. Restoring endodontically treated teeth with post and cores—A review. *Quintessence Int.* **2005**, *36*, 737–746.

187. Shutzky-Goldberg, I.; Shutzky, H.; Gorfil, C.; Smidt, A. Restoration of endodontically treated teeth review and treatment recommendations. *Int. J. Dent.* **2009**, *2009*, 150251. [CrossRef]

188. Schilder, H.; Goodman, A.; Winthrop, A. The termomechanical properties of gutta-percha. Determination of phase transition temperatures for gutta-percha. *Oral Surg. Oral Med. Oral Pathol.* **1974**, *38*, 109–114. [CrossRef]

189. Combe, E.C.; Cohen, B.D.; Cumming, K. Alpha- and beta-forms of gutta-percha in products for root canal filling. *Int. Endodon. J.* **2001**, *34*, 447–451. [CrossRef]

190. Ferreira, C.M.; Silva, J.B.A., Jr.; Monteiro de Paula, R.C.; Andrade Feitosa, J.P.; Negreiros Cortez, D.G.; Zaia, A.A.; de Souza-Filho, F.J. Brazilian gutta-percha points. Part I: Chemical composition and X-ray diffraction analysis. *Braz. Oral Res.* **2005**, *19*, 193–197. [CrossRef]

191. Ferreira, C.M.; Gurgel-Filho, E.D.; Silva, J.B.A., Jr.; Monteiro de Paula, R.C.; Pessoa Andrade Feitosa, J.; Figueiredo de Almeida Gomes, B.P.; de Souza-Filho, F.J. Brazilian gutta-percha points. Part II: Thermal properties. *Braz. Oral Res.* **2007**, *21*, 29–34. [CrossRef] [PubMed]

192. Dobrzańska, J.; Gołombek, K.; Dobrzański, L.B. Polymer materials used in endodontic treatment—In vitro testing. *AMSE* **2012**, *58*, 110–115.

193. Dobrzańska, J. Analiza Szczelności Wypełnień Kanałów Korzeniowych. Ph.D. Thesis, Śląski Uniwersytet Medyczny w Katowicach, Zabrze, Poland, 2011.

194. Sunay, H.; Tanalp, J.; Dikbas, I.; Bayirli, G. Cross-sectional evaluation of the periapical status and quality of root canal treatment in a selected population of urban Turkish adults. *Int. Endodon. J.* **2007**, *40*, 139–145. [CrossRef]

195. Meuwissen, R.; Eschen, S. Twenty years of endodontic treatment. *J. Endodon.* **1983**, *9*, 390–393. [CrossRef]

196. Lazarski, M.P.; Walker, W.A.; Flores, C.M.; Schindler, W.G.; Hargreaves, K.M. Epidemiological evaluation of the outcomes of nonsurgical root canal treatment in a large cohort of insured dental patients. *J. Endodon.* **2001**, *27*, 791–796. [CrossRef] [PubMed]

197. Kirkevang, L.L.; Horsted-Bindslev, P.; Orstavik, D.; Wenzel, A. Frequency and distribution of endodontically treated teeth and apical periodontitis in an urban Danish population. *Int. Endodon. J.* **2001**, *34*, 198–205. [CrossRef] [PubMed]

198. Tsuneishi, M.; Yamamoto, T.; Yamanaka, R.; Tamaki, N.; Sakamoto, T.; Tsuji, K.; Watanabe, T. Radiographic evaluation of periapical status and prevalence of endodontic treatment in an adult Japanese population. *Oral Surg. Oral Med. Oral Pathol. Oral Radiol. Endod.* **2005**, *100*, 631–635. [CrossRef]

199. Chen, S.C.; Chuech, L.H.; Hsiao, C.K.; Tsai, M.Y.; Ho, S.C.; Chiang, C.P. An epidemiologic study of tooth retention after nonsurgical endodontic treatment in a large population in Taiwan. *J. Endodon.* **2007**, *33*, 226–229. [CrossRef]

200. Salehrabi, R.; Rotstein, I. Endodontic treatment outcomes in a large patient population in the USA: An epidemiological study. *J. Endodon.* **2004**, *30*, 846–850. [CrossRef]

201. Genco, R.J.; Borgnakke, W.S. Risk factors for periodontal disease. *Periodontol. 2000* **2013**, *62*, 59–94. [CrossRef]

202. Martinez-Canut, P.; Llobell, A.; Romero, A. Predictors of long-term outcomes in patients undergoing periodontal maintenance. *J. Clin. Periodontol.* **2017**, *44*, 620–631. [CrossRef] [PubMed]

203. Van Dyke, T.E.; Dave, S. Risk factors for periodontitis. *J. Int. Acad. Periodontol.* **2005**, *7*, 3–7. [PubMed]

204. Hasan, A.; Palmer, R.M. A clinical guide to periodontology: Pathology of periodontal disease. *Br. Dent. J.* **2014**, *216*, 457–461. [CrossRef] [PubMed]

205. Isola, G.; Polizzi, A.; Iorio-Siciliano, V.; Alibrandi, A.; Ramaglia, L.; Leonardi, R. Effectiveness of a nutraceutical agent in the non-surgical periodontal therapy: A randomized, controlled clinical trial. *Clin. Oral Invest.* **2020**. (Published online 17 June 2020). [CrossRef] [PubMed]

206. Neiva, R.F.; Steigenga, J.; Al-Shammari, K.F.; Wang, H.-L. Effects of specific nutrients on periodontal disease onset, progression and treatment. *J. Clin. Periodontol.* **2003**, *30*, 579–589. [CrossRef]

207. Marsh, P.D.; Zaura, E. Dental biofilm: Ecological interactions in health and disease. *J. Clin. Periodontol.* **2017**, *44* (Suppl. 18), S12–S22. [CrossRef]

208. Highfield, J. Diagnosis and classification of periodontal disease. *Aust. Dent. J.* **2009**, *54*, S11–S26. [CrossRef]

209. Lee, J.-H.; Oh, J.-Y.; Choi, J.-K.; Kim, Y.-T.; Park, Y.-S.; Jeong, S.-N.; Choi, S.-H. Trends in the incidence of tooth extraction due to periodontal disease: Results of a 12-year longitudinal cohort study in South Korea. *J. Periodont. Implant. Sci.* **2017**, *47*, 264–272. [CrossRef]

210. McCaul, L.K.; Jenkins, W.M.M.; Kay, E.J. The reasons for extraction of permanent teeth in Scotland: A 15-year follow-up study. *Br. Dent. J.* **2001**, *190*, 658–662. [CrossRef]

211. Carnevale, G.; Pontoriero, R.; Lindhe, J. Treatment of furcation-involved teeth. In *Clinical Periodontology and Impant Dentistry*; Lindhe, J., Karring, T., Lang, N.P., Eds.; Munksgaard: Copenhagen, Denmark, 1997; pp. 682–710.

212. Papapanou, P.N.; Toneti, M.S. Diagnosis and epidemiology of periodontal osseous lesions. *Periodontol. 2000* **2000**, *22*, 8–21. [CrossRef]

213. Nordland, P.; Garrett, S.; Kiger, R.; Vanooteghem, R.; Hutchens, L.H.; Egelberg, J. The effect of plaque control and root debridement in molar teeth. *J. Clin. Periodontol.* **1987**, *14*, 231–236. [CrossRef] [PubMed]

214. Hirschfeld, L.; Wasserman, B.A. Long-term survey of tooth loss in 600 treated periodontal patients. *J. Periodontol.* **1978**, *49*, 225–237. [CrossRef] [PubMed]

215. McGuire, M.K.; Nunn, M.E. Prognosis versus actual outcome. III. The effectiveness of clinical parametrs in accurately predicting tooth survival. *J. Periodontol.* **1966**, *67*, 666–674. [CrossRef] [PubMed]

216. De Vries, K. Primary care: Gingivitis. *Aust. J. Pharm.* **2015**, *96*, 64.

217. Azaripour, A.; Weusmann, J.; Eschig, C.; Schmidtmann, I.; Van Noorden, C.J.F.; Willershausen, B. Efficacy of an aluminium triformate mouthrinse during the maintenance phase in periodontal patients: A pilot double blind randomized placebo-controlled clinical trial. *BMC Oral Health* **2016**, *16*, 57. [CrossRef] [PubMed]

218. Pradeep, K.; Rajababu, P.; Satyanarayana, D.; Sagarm, V. Gingival recession: Review and strategies in treatment of recession. *Case Rep. Dent.* **2012**, *2012*, 563421. [CrossRef] [PubMed]

219. Scottish Dental Clinical Effectiveness Programme. Prevention and Treatment of Periodontal Diseases in Primary Care, Dental Clinical Guidance. *Dundee Dental Education Centre: Dundee, 2014.* Available online: https://www.sdcep.org.uk/wp-content/uploads/2015/01/SDCEP+Periodontal+Disease+Full+Guidance.pdf (accessed on 30 September 2020).

220. Roos-Jansåker, A.-M.; Egelberg, S.R.J. Treatment of peri-implant infections: A literature review. *J. Clin. Periodontol.* **2003**, *30*, 467–485. [CrossRef]

221. Wiebe, C.B.; Putnins, E.E. The periodontal disease classification system of the American academy of periodontology—An update. *J. Can. Den. Assoc.* **2000**, *66*, 594–597.

222. Caton, J.G.; Armitage, G.; Berglundh, T.; Chapple, I.L.C.; Jepsen, S.; Kornman, K.S.; Mealey, B.L.; Papapanou, P.N.; Sanz, M.; Tonetti, M.S. A new classification scheme for periodontal and peri-implant diseases and conditions—Introduction and key changes from the 1999 classification. *J. Periodontol.* **2018**, *89*, S1–S8. [CrossRef]

223. Scannapieco, F.A.; Gershovich, E. The prevention of periodontal disease—An overview. *Periodontol. 2000* **2020**, *84*, 9–13. [CrossRef]

224. Armitage, G.C. Periodontal diagnoses and classification of periodontal diseases. *Periodontol. 2000* **2004**, *34*, 9–21. [CrossRef] [PubMed]

225. Preshaw, P.M. Detection and diagnosis of periodontal conditions amenable to prevention. *BMC Oral Health* **2015**, *15*, S1–S5. [CrossRef] [PubMed]

226. Walter, C.; Schmidt, J.C.; Dula, K.; Sculean, A. Cone beam computed tomography (CBCT) for diagnosis and treatment planning in periodontology: A systematic review. *Quintessence Int.* **2016**, *47*, 26–37. [CrossRef]

227. Woelber, J.P.; Fleiner, J.; Rau, J.; Ratka-Krüger, P.; Hannig, C. Accuracy and usefulness of CBCT in periodontology: A systematic review of the literature. *Int. J. Periodontics. Restor. Dent.* **2018**, *38*, 289–297. [CrossRef] [PubMed]

228. Walter, C.; Schmidt, J.C.; Rinne, C.A.; Mendes, S.; Dula, K.; Sculean, A. Cone beam computed tomography (CBCT) for diagnosis and treatment planning in periodontology: Systematic review update. *Clin. Oral Investig.* **2020**, *24*, 2943–2958. [CrossRef]

229. Acar, B.; Kamburoğlu, K. Use of cone beam computed tomography in periodontology. *World J. Radiol.* **2014**, *6*, 139–147. [CrossRef]

230. Du Bois, A.H.; Kardachi, B.; Bartold, P.M. Is there a role for the use of volumetric cone beam computed tomography in periodontics? *Aust. Dent. J.* **2012**, *57* (Suppl. 1), 103–108. [CrossRef]

231. Basic Periodontal Examination (BPE). The British Society of Periodontology. 2011. Available online: https://www.bsperio.org.uk/assets/downloads/BPE_Guidelines_2011.pdf (accessed on 20 November 2020).

232. Vandana, K.L.; Haneet, R.K. Cementoenamel junction: An insight. *J. Indian Soc. Periodontol.* **2014**, *18*, 549–554. [CrossRef]

233. Dufty, J.; Gkranias, N.; Donos, N. Necrotising ulcerative gingivitis: A literature review. *Oral Health Prev. Dent.* **2017**, *15*, 321–327. [CrossRef]

234. Furuta, M.; Fukai, K.; Aida, J.; Shimazaki, Y.; Ando, Y.; Miyazaki, H.; Kambara, M.; Yamashita, Y. Periodontal status and self-reported systemic health of periodontal patients regularly visiting dental clinics in the 8020 promotion foundation study of Japanese dental patients. *J. Oral Sci.* **2019**, *61*, 238–245. [CrossRef]

235. Lang, N.; Bartold, P.M.; Cullinan, M.; Jeffcoat, M.; Mombelli, A.; Murakami, S.; Page, R.; Papapanou, P.; Tonetti, M.; Van Dyke, T. Consensus report: Aggressive periodontitis. *Ann. Periodontol.* **1999**, *4*, 53. [CrossRef]

236. Kinane, D.F.; Marshall, G.J. Peridonatal manifestations of systemic disease. *Aust. Dent. J.* **2008**, *46*, 2–12. [CrossRef] [PubMed]

237. Mohd-Dom, T.; Ayob, R.; Mohd-Nur, A.; Abdul-Manaf, M.R.; Ishak, N.; Abdul-Muttalib, K.; Aljunid, S.M.; Ahmad-Yaziz, Y.; Abdul-Aziz, H.; Kasan, N.; et al. Cost analysis of periodontitis management in public sector specialist dental clinics. *BMC Oral Health* **2014**, *14*, 56. [CrossRef] [PubMed]

238. Brånemark, P.-I.; Adell, R.; Breine, U.; Hansson, B.O.; Lindström, J.; Ohlsson, A. Intraosseous anchorage of dental prosthesis. I Experimental studies. *Scand. J. Plast. Reconstr. Surg.* **1969**, *3*, 81–100. [CrossRef] [PubMed]

239. Brånemark, P.-I.; Hansson, B.O.; Adell, R.; Breine, U.; Lindström, J.; Hallén, O.; Ohman, A. Osseointegrated implants in the treatment of the edentulous jaw. Experience from a 10-year period. *Scand. J. Plast. Reconstr. Surg. Suppl.* **1977**, *16*, 1–132.

240. Brånemark, P.-I.; Hansson, B.O.; Adell, R.; Breine, U.; Lindström, J.; Hallén, O.; Ohman, A. *Osseointegrated Implants in the Treatment of the Edentulous Jaw*; Almqvist and Wiksell: Stockholm, Sweden, 1977.

241. Brånemark, P.-I. Osseointegration and its experimental studies. *J. Prosthet. Dent.* **1983**, *50*, 399. [CrossRef]

242. Brånemark, P.-I.; Rydevik, B.L.; Skalak, R. (Eds.) *Osseointegration in Skeletal Reconstruction and Joint Replacement*; Quintessence Publishing Co.: Carol Stream, IL, USA, 1997.

243. Kagermann, H.; Wahlster, W.; Helbig, J. *Recommendations for Implementing the Strategic Initiative Industrie 4.0: Final Report of the Industrie 4.0 Working Group*; Federal Ministry of Education and Research: Bonn, Germany, 2013.

244. Rüßmann, M.; Lorenz, M.; Gerbert, P.; Waldner, M.; Justus, J.; Engel, P.; Harnisch, M. *Industry 4.0: The Future of Productivity and Growth in Manufacturing Industries*; Boston Consulting Group: Boston, MA, USA, 2015.

245. Hermann, M.; Pentek, T.; Otto, B. Design principles for industrie 4.0 scenarios. In Proceedings of the 2016 49th Hawaii International Conference on System Sciences (HICSS), Koloa, HI, USA, 5–8 January 2016; pp. 3928–3937. [CrossRef]

246. Jose, R.; Ramakrishna, S. Materials 4.0: Materials big data enabled materials discovery. *Appl. Mater. Today* **2018**, *10*, 127–132. [CrossRef]

247. Dobrzański, L.A.; Dobrzańska-Danikiewicz, A.D. Why are carbon-based materials important in civilization progress and especially in the industry 4.0 stage of the industrial revolution. *Mater. Perform. Charact.* **2019**, *8*, 337–370. [CrossRef]

248. Dobrzański, L.A.; Dobrzański, L.B. Dentistry 4.0 concept in the design and manufacturing of prosthetic dental restorations. *Processes* **2020**, *8*, 525. [CrossRef]

249. Beuer, F.; Schweiger, J.; Edelhoff, D. Digital dentistry: An overview of recent developments for CAD/CAM generated restorations. *Br. Dent. J.* **2008**, *204*, 505–511. [CrossRef]

250. Nejatian, T.; Almassi, S.; Shamsabadi, A.F.; Vasudeva, G.; Hancox, Z.; Singh Dhillon, A.; Sefat, F. Digital dentistry. In *Advanced Dental Biomaterials*; Khurshid, Z., Najeeb, S., Zafar, M.S., Sefat, F., Eds.; Woodhead Publishing: London, UK, 2019; pp. 507–540. [CrossRef]

251. Blatz, M.B.; Conejo, J. The current state of chairside digital dentistry and materials. *Dent. Clin. North Am.* **2019**, *63*, 175–179. [CrossRef] [PubMed]

252. Sulaiman, T.A. Materials in digital dentistry—A review. *J. Esthet. Restor. Dent.* **2020**, *32*, 171–181. [CrossRef] [PubMed]

253. Valizadeh, S.; Valilai, O.F.; Houshmand, M.; Vasegh, Z. A novel digital dentistry platform based on cloud manufacturing paradigm. *Int. J. Comp. Integr. Manufac.* **2019**, *32*, 1024–1042. [CrossRef]

254. Rekow, E.D. Digital dentistry: The new state of the art—Is it disruptive or destructive? *Dent. Mater.* **2020**, *36*, 9–24. [CrossRef] [PubMed]

255. U533; Kaye, G. Does Digital Dentist Mean Anything Today? 6 December 2017. Available online: https://www.dentaleconomics.com/science-tech/article/16389518/does-digital-dentist-mean-anything-today (accessed on 9 October 2020).

256. Dobrzański, L.A. Role of materials design in maintenance engineering in the context of industry 4.0 idea. *JAMME* **2019**, *96*, 12–49. [CrossRef]

257. Dobrzański, L.A. (Ed.) Metale i ich stopy. In *Open Access Library*; VII, 2; International OCSCO World Press: Gliwice, Poland, 2017.

258. Dobrzański, L.A. Significance of materials science for the future development of societies. *J. Mater. Process. Technol.* **2006**, *175*, 1–3, 133–148. [CrossRef]

259. Wilson, C.E.; Dhert, W.J.A.; van Blitterswijk, C.A.; Verbout, A.J.; de Bruijn, J.D. Evaluating 3D bone tissue engineered constructs with different seeding densities using the alamarBlue™ assay and the effect on in vivo bone formation. *J. Mater. Sci. Mater. Med.* **2002**, *13*, 1265–1269. [CrossRef] [PubMed]

260. Folkman, J. Tumor angiogenesis: Therapeutic implications. *N. Eng. J. Med.* **1971**, *285*, 1182–1186. [CrossRef]

261. Noga, M.; Pawlak, A.; Dybala, B.; Dabrowski, B.; Swieszkowski, W.; Lewandowska-Szumiel, M. *Biological Evaluation of Porous Titanium Scaffolds (Ti-6Al-7Nb) with HAp/Ca-P Surface Seeded with Human Adipose Derived Stem Cells*; E-MRS Fall Meeting: Warszawa, Poland, 2013.

262. Rouwkema, J.; Rivron, N.C.; van Blitterswijk, C.A. Vascularization in tissue engineering. *Trends Biotechnol.* **2008**, *26*, 434–441. [CrossRef]

263. Bramfeldt, H.; Sabra, G.; Centis, V.; Vermette, P. Scaffold vascularization: A challenge for three-dimensional tissue engineering. *Curr. Med. Chem.* **2010**, *17*, 3944–3967. [CrossRef]

264. Bose, S.; Roy, M.; Bandyopadhyay, A. Recent advances in bone tissue engineering scaffolds. *Trends Biotechnol.* **2012**, *30*, 546–554. [CrossRef] [PubMed]

265. Jain, R.K.; Au, P.; Tam, J.; Duda, D.G.; Fukumura, D. Engineering vascularized tissue. *Nat. Biotechnol.* **2005**, *23*, 821–823. [CrossRef] [PubMed]

266. Prendergast, P.; Huiskes, R. Microdamage and osteocyte-lacuna strain in bone: A microstructural finite element analysis. *J. Biomech. Eng.* **1996**, *118*, 240–246. [CrossRef] [PubMed]

267. Bettinger, C.J.; Borenstein, J.T.; Langer, R. Microfabrication techniques in scaffold development. In *Nanotechnology and Tissue Engineering: The Scaffold*; Laurencin, C.T., Nair, L.S., Eds.; CRC Press Taylor & Francis Group: Boca Raton, FL, USA, 2008.

268. Bentolila, V.; Boyce, T.M.; Fyhrie, D.P.; Drumb, R.; Skerry, T.M.; Schaffler, M.B. Intracortical remodeling in adult rat long bones after fatigue loading. *Bone* **1998**, *23*, 275–281. [CrossRef]

269. Ramirez, J.M.; Hurt, W.C. Bone remodeling in periodontal lesions. *J. Periodontol.* **1977**, *48*, 74–77. [CrossRef]

270. Wang, X.; Ni, Q. Determination of cortical bone porosity and pore size distribution using a low field pulsed NMR approach. *J. Orthop. Res.* **2003**, *21*, 312–319. [CrossRef]

271. Kufahl, R.H.; Saha, S.A. A theoretical model for stress-generated fluid flow in the canaliculi-lacunae network in bone tissue. *J. Biomech.* **1990**, *23*, 171–180. [CrossRef]

272. Sietsema, W.K. Animal models of cortical porosity. *Bone* **1995**, *17* (Suppl. 4), S297–S305. [CrossRef]

273. Dobrzański, L.A.; Dobrzańska-Danikiewicz, A.D.; Czuba, Z.P.; Dobrzański, L.B.; Achtelik-Franczak, A.; Malara, P.; Szindler, M.; Kroll, L. The new generation of the biological-engineering materials for applications in medical and dental implant-scaffolds. *AMSE* **2018**, *91*, 56–85. [CrossRef]

274. Dobrzański, L.A.; Dobrzańska-Danikiewicz, A.D.; Czuba, Z.P.; Dobrzański, L.B.; Achtelik-Franczak, A.; Malara, P.; Szindler, M.; Kroll, L. Metallic skeletons as reinforcement of new composite materials applied in orthopaedics and dentistry. *AMSE* **2018**, *92*, 53–85. [CrossRef]

275. Dikova, T.; Dzhendov, D.; Simov, M. Microstructure and hardness of fixed dental prostheses manufactured by additive technologies. *JAMME* **2015**, *71*, 60–69.

276. Dobrzański, L.A. (Ed.) *Biomaterials in Regenerative Medicine*; IntechOpen: Rijeka, Croatia, 2018. [CrossRef]

277. Dobrzański, L.A. (Ed.) *Powder Metallurgy—Fundamentals and Case Studies*; InTech: Rijeka, Croatia, 2017.

278. Dobrzański, L.A.; Dobrzańska-Danikiewicz, A.D. (Eds.) *Metalowe Materiały Mikroporowate I Lite Do Zastosowań Medycznych I Stomatologicznych*; International OCSCO World Press: Gliwice, Poland, 2017.

279. Dobrzański, L.A.; Dobrzańska-Danikiewicz, A.D.; Achtelik-Franczak, A.; Dobrzański, L.B. Comparative analysis of mechanical properties of scaffolds sintered from Ti and Ti6Al4V powders. *AMSE* **2015**, *73*, 69–81.

280. Brandt, M. (Ed.) *Laser Additive Manufacturing; Materials, Design, Technologies and Applications*; Woodhead Publishing: Sawston, UK, 2017.

281. Dikova, T. Properties of Co-Cr dental alloys fabricated using additive technologies. In *Biomaterials in Regenerative Medicine*; Dobrzański, L.A., Ed.; IntechOpen: Rijeka, Croatia, 2018; pp. 141–159. [CrossRef]

282. Dobrzański, L.A.; Dobrzańska-Danikiewicz, A.D.; Gaweł, T.G.; Achtelik-Franczak, A. Selective laser sintering and melting of pristine titanium and titanium Ti6Al4V alloy powders and selection of chemical environment for etching of such materials. *Arch. Met. Mater.* **2015**, *60*, 2039–2045. [CrossRef]

283. Dobrzański, L.A.; Dobrzańska-Danikiewicz, A.D.; Achtelik-Franczak, A.; Dobrzański, L.B.; Szindler, M.; Gaweł, T.G. Porous selective laser melted Ti and Ti6Al4V materials for medical applications. In *Powder Metallurgy—Fundamentals and Case Studies*; Dobrzański, L.A., Ed.; IntechOpen: Rijeka, Croatia, 2017; pp. 161–181. [CrossRef]

284. Dobrzański, L.A.; Matula, G.; Dobrzańska-Danikiewicz, A.D.; Malara, P.; Kremzer, M.; Tomiczek, B.; Kujawa, M.; Hajduczek, E.; Achtelik-Franczak, A.; Dobrzański, L.B.; et al. Composite materials infiltrated by aluminium alloys based on porous skeletons from alumina, mullite and titanium produced by powder metallurgy techniques. In *Powder Metallurgy—Fundamentals and Case Studies*; Dobrzański, L.A., Ed.; IntechOpen: Rijeka, Croatia, 2017; pp. 95–137. [CrossRef]

285. Dobrzański, L.A.; Dobrzańska-Danikiewicz, A.D.; Achtelik-Franczak, A.; Dobrzański, L.B.; Hajduczek, E.; Matula, G. Fabrication technologies of the sintered materials including materials for medical and dental application. In *Powder Metallurgy—Fundamentals and Case Studies*; Dobrzański, L.A., Ed.; IntechOpen: Rijeka, Croatia, 2017; pp. 17–52. [CrossRef]

286. Majkowska, B.; Jażdżewska, M.; Wołowiec, E.; Piekoszewski, W.; Klimek, L.; Zieliński, A. The possibility of use of laser-modified Ti6Al4V alloy in friction pairs in endoprostheses. *Arch. Met. Mater.* **2015**, *60*, 755–758. [CrossRef]

287. Das, S.; Wohlert, M.; Beaman, J.J.; Bourell, D.L. Producing metal parts with selective laser sintering/hot isostatic pressing. *JOM* **1998**, *50*, 17–20. [CrossRef]

288. Duan, B.; Wang, M.; Zhou, W.Y.; Cheung, W.L.; Li, Z.Y.; Lu, W.W. Three-dimensional nanocomposite scaffolds fabricated via selective laser sintering for bone tissue engineering. *Acta Biomater.* **2010**, *6*, 4495–4505. [CrossRef]

289. Herzog, D.; Seyda, V.; Wycisk, E.; Emmelmann, C. Additive manufacturing of metals. *Acta Mater.* **2016**, *117*, 371–392. [CrossRef]

290. Frazier, W.E. Metal additive manufacturing: A review. *J. Mater. Eng. Perform.* **2014**, *23*, 1917–1928. [CrossRef]

291. Malara, P.; Dobrzański, L.B.; Dobrzańska, J. Computer-aided designing and manufacturing of partial removable dentures. *JAMME* **2015**, *73*, 157–164.

292. Żmudzki, J.; Chladek, G.; Kasperski, J. Biomechanical factors related to occlusal load transfer in removable complete dentures. *Biomech. Model Mechanobiol.* **2015**, *14*, 679–691. [CrossRef]

293. Dobrzański, L.A.; Dobrzańska-Danikiewicz, A.D.; Achtelik-Franczak, A.; Szindler, M. Structure and properties of the skeleton microporous materials with coatings inside the pores for medical and dental applications. In *Frontiers in Materials Processing, Applications, Research and Technology*; Muruganant, M., Chirazi, A., Raj, B., Eds.; Springer: Singapore, 2018; pp. 297–320. [CrossRef]

294. Dikova, T.; Dzhendov, D.; Simov, M.; Katreva-Bozukova, I.; Angelova, S.; Pavlova, D.; Abadzhiev, M.; Tonchev, T. Modern trends in the development of the technologies for production of dental constructions. *J. IMAB Ann. Proc. (Sci. Pap.)* **2015**, *21*, 974–981. [CrossRef]

295. Gupta, S.; Kumar, S. Lasers in dentistry—An overview. *Trends Biomateratif. Organs.* **2011**, *25*, 119–123.

296. Olmos, L.; Cabezas-Villa, J.L.; Bouvard, D.; Lemus-Ruiz, J.; Jiménez, O.; Falcón-Franco, L.A. Synthesis and characterisation of Ti6Al4V/xTa alloy processed by solid state sintering. *Powder Met.* **2020**, *63*, 64–74. [CrossRef]

297. Esen, Z.; Bor, S. Processing of titanium foams using magnesium spacer particles. *Scr. Mater.* **2007**, *56*, 341–344. [CrossRef]

298. Dobrzański, L.B. Mechanical properties comparison of engineering materials produced by additive and substractive technologies for dental prosthetic restoration application. In *Biomaterials in Regenerative Medicine*; Dobrzański, L.A., Ed.; IntechOpen: Rijeka, Croatia, 2018; pp. 111–140. [CrossRef]

299. Dobrzański, L.A. The concept of biologically active microporous engineering materials and composite biological-engineering materials for regenerative medicine and dentistry. *AMSE* **2016**, *80*, 64–85. [CrossRef]

300. Dobrzański, L.A.; Dobrzańska-Danikiewicz, A.D.; Gaweł, T.G. Individual implants of a loss of palate fragments fabricated using SLM equipment. *JAMME* **2016**, *77*, 24–30. [CrossRef]

301. Bram, M.; Schiefer, H.; Bogdanski, D.; Köller, M.; Buchkremer, H.P.; Stöver, D. Implant surgery: How bone bonds to PM titanium. *Met. Pow. Rep.* **2006**, *61*, 26–28, 30–31. [CrossRef]

302. Bram, M.; Stiller, C.; Buchkremer, H.P.; Stover, D.; Bauer, H. High-porosity titanium, stainless steel, and superalloy parts. *Adv. Eng. Mat.* **2000**, *2*, 196–199. [CrossRef]

303. Reis de Vasconcellos, L.M.; Varella de Oliveira, M.; Lima de Alencastro Graça, M.; Oliveira de Vasconcellos, L.G.; Carvalho, Y.R.; Cairo, C.A.A. Porous titanium scaffolds produced by powder metallurgy for biomedical applications. *Mat. Res.* **2008**, *11*, 275–280. [CrossRef]

304. Ryan, G.; Pandit, A.; Apatsidis, D.P. Fabrication methods of porous metals for use in orthopaedic applications. *Biomaterials* **2006**, *27*, 2651–2670. [CrossRef]

305. Dobrzański, L.B.; Achtelik-Franczak, A.; Dobrzańska, J.; Pietrucha, P. Application of polymer impression masses for the obtaining of dental working models for the stereolithographic 3D printing. *AMSE* **2019**, *95*, 31–40. [CrossRef]

306. Dobrzański, L.A.; Dobrzański, L.B.; Dobrzańska-Danikiewicz, A.D. Additive and hybrid technologies for products manufacturing using powders of metals, their alloys and ceramics. *AMSE* **2020**, *102*, 59–85. [CrossRef]

307. Dobrzańska-Danikiewicz, A.D. The state of the art analysis and methodological assumptions of evaluation and development prediction for materials surface technologies. *JAMME* **2011**, *49*, 121–141.

308. Dobrzańska-Danikiewicz, A.D. Metodologia komputerowo zintegrowanego prognozowania rozwoju inżynierii powierzchni materiałów. In *Open Access Library*; 1, 7; Dobrzański, L.A., Ed.; International OCSCO World Press: Gliwice, Poland, 2012.

309. Dobrzańska-Danikiewicz, A.D. *Księga Technologii Krytycznych Kształtowania Struktury I Własności Powierzchni Materiałów Inżynierskich*; International OCSCO World Press: Gliwice, Poland, 2013.

310. Dobrzańska-Danikiewicz, A.D. Foresight of material surface engineering as a tool building a knowledge based economy. *Mater. Sci. Forum* **2012**, *706–709*, 2511–2516. [CrossRef]

311. Harrison, T. The Emergence of the Fourth Industrial Revolution in Dentistry, 25 April 2018. Available online: https://www.dentistryiq.com/practice-management/industry/article/16366877/the-emergence-of-the-fourth-industrial-revolution-in-dentistry (accessed on 9 October 2020).

312. Dobrzański, L.B.; Achtelik-Franczak, A.; Dobrzańska, J.; Dobrzański, L.A. The digitisation for the immediate dental implantation of incisors with immediate individual prosthetic restoration. *JAMME* **2019**, *97*, 57–68. [CrossRef]

313. Galanis, C.C.; Sfantsikopoulos, M.M.; Koidis, P.T.; Kafantaris, N.M.; Mpikos, P.G. Computer methods for automating preoperative dental implant planning: Implant positioning and size assignment. *Comp. Methods Programs Biomed.* **2007**, *86*, 30–38. [CrossRef]

314. Miyazaki, T.; Hotta, Y. CAD/CAM systems available for the fabrication of crown and bridge restorations. *Aust. Dent. J.* **2011**, *56*, 97–106. [CrossRef]

315. Malara, P.; Dobrzański, L.B. Computer-aided design and manufacturing of dental surgical guides based on cone beam computed tomography. *AMSE* **2015**, *76*, 140–149.

316. Malara, P.; Dobrzański, L.B. Designing and manufacturing of implantoprosthetic fixed suprastructures in edentulous patients on the basis of digital impressions. *AMSE* **2015**, *76*, 163–171.

317. Ender, A.; Zimmermann, M.; Attin, T.; Mehl, A. In vivo precision of conventional and digital methods for obtaining quadrant dental impressions. *Clin. Oral Investig.* **2016**, *20*, 1495–1504. [CrossRef]

318. Van Noort, R. The future of dental devices is digital. *Dent. Mater.* **2012**, *28*, 3–12. [CrossRef] [PubMed]

319. Dobrzańska-Danikiewicz, A.; Dobrzański, L.A.; Gaweł, T.G. Scaffolds applicable as implants of a loss palate fragments. In *International Conference on Processing & Manufacturing of Advanced Materials. Processing, Fabrication, Properties, Applications*; THERMEC'2016: Graz, Austria, 2016; p. 193.

320. Liacouras, P.; Garnes, J.; Roman, N.; Petrich, A.; Grant, G.T. Designing and manufacturing an auricular prosthesis using computed tomography, 3-dimensional photographic imaging, and additive manufacturing: A clinical report. *J. Prosthet. Dent.* **2011**, *105*, 78–82. [CrossRef]

321. Maroulakos, M.; Kamperos, G.; Tayebi, L.; Halazonetis, D.; Ren, Y. Applications of 3D printing on craniofacial bone repair: A systematic review. *J. Dent.* **2019**, *80*, 1–14. [CrossRef] [PubMed]

322. Warnke, P.H.; Douglas, T.; Wollny, P.; Sherry, E.; Steiner, M.; Galonska, S.; Becker, S.T.; Springer, I.N.; Wiltfang, J.; Sivananthan, S. Rapid prototyping: Porous titanium alloy scaffolds produced by selective laser melting for bone tissue engineering. *Tissue Eng. Part C Methods* **2009**, *15*, 115–124. [CrossRef]
323. Kruth, J.-P.; Vandenbroucke, B.; Van Vaerenbergh, J.; Naert, I. Digital manufacturing of biocompatible metal frameworks for complex dental prostheses by means of SLS/SLM. In *International Conference on Advanced Research and Rapid Prototyping (VRAP) 2005*; da Silva Bartolo, P.J., Ed.; Taylor & Francis: Leiden, The Netherlands; Balkema Publishers: Leiria, Portugal, 2005; pp. 139–145.
324. Traini, T.; Mangano, C.; Sammons, R.L.; Mangano, F.; Macchi, A.; Piatelli, A. Direct laser metal sintering as a new approach to fabrication of an isoelastic functionally graded material for manufacture of porous titanium dental implants. *Dent. Mater.* **2004**, *24*, 1525–1533. [CrossRef]
325. Revilla-Leon, M.; Ozcan, M. Additive manufacturing technologies used for 3D metal printing in dentistry. *Curr. Oral Health Rep.* **2017**, *4*, 201–208. [CrossRef]
326. Moraru, E.; Dontu, O.; Petre, A.; Vairenau, D.; Constantinescu, F.; Besnea, D. Some technological particularities on the execution of dental prostheses realized by selective laser deposition. *J. Optoelect. Adv. Mater.* **2018**, *20*, 208–213.
327. Mangano, F.; Chambrone, L.; van Noort, R.; Miller, C.; Hatton, P.; Mangano, C. Direct metal laser sintering titanium dental implants: A review of the current literature. *Int. J. Biomater.* **2014**, *2014*, 461534. [CrossRef]
328. Barazanchi, A.; Li, K.C.; Al-Amleh, B.; Lyons, K.; Waddell, N. Additive technology: Update on current materials and applications in dentistry. *J. Prosthodont.* **2017**, *26*, 153–163. [CrossRef]
329. Dobrzański, L.A.; Dobrzański, L.B.; Achtelik-Franczak, A.; Dobrzańska, J. Application solid laser-sintered or machined Ti6Al4V alloy in manufacturing of dental implants and dental prosthetic restorations according to dentistry 4.0 concept. *Processes* **2020**, *8*, 664. [CrossRef]
330. Dobrzański, L.B.; Achtelik-Franczak, A.; Dobrzańska, J.; Dobrzański, L.A. Comparison of the structure and properties of the solid Co-Cr-W-Mo-Si alloys used for dental restorations CNC machined or selective laser-sintered. *Mater. Perform. Charact.* **2020**, *9*, 556–578. [CrossRef]
331. Puskar, T.; Jevremovic, D.; Williams, R.J.; Eggbeer, D.; Vukelic, D.; Budak, I. A comparative analysis of the corrosive effect of artificial saliva of variable pH on DMLS and Cast Co-Cr-Mo dental alloy. *Materials* **2014**, *7*, 6486–6501. [CrossRef] [PubMed]
332. Kim, K.-B.; Kim, W.-C.; Kim, H.-Y.; Kim, J.-H. An evaluation of marginal fit of three-unit fixed dental prostheses fabricated by direct metal laser sintering system. *Dent. Mater.* **2013**, *29*, e91–e96. [CrossRef] [PubMed]
333. Lim, J.-H.; Park, J.-M.; Kim, M.; Heo, S.-J.; Myung, J.-Y. Comparison of digital intraoral scanner reproducibility and image trueness considering repetitive experience. *J. Prosthet. Dent.* **2018**, *119*, 225–232. [CrossRef] [PubMed]
334. Su, T.S.; Sun, J. Comparison of repeatability between intraoral digital scanner and extraoral digital scanner: An in-vitro study. *J. Prosthodont. Res.* **2015**, *59*, 236–242. [CrossRef] [PubMed]
335. Persson, A.; Andersson, M.; Odén, A.; Englund, G.S. A three-dimensional evaluation of a laser scanner and a touch-probe scanner. *J. Prosthet. Dent.* **2006**, *95*, 194–200. [CrossRef] [PubMed]
336. Mandelli, F.; Gherlone, E.; Gastaldi, G.; Ferrari, M. Evaluation of the accuracy of extraoral laboratory scanners with a single-tooth abutment model: A 3d analysis. *J. Prosthodont. Res.* **2017**, *61*, 363–370. [CrossRef] [PubMed]
337. Renne, W.; Ludlow, M.; Fryml, J.; Schurch, Z.; Mennito, A.; Kessler, R.; Lauer, A. Evaluation of the accuracy of 7 digital scanners: An in vitro analysis based on 3-dimensional comparisons. *J. Prosthet. Dent.* **2017**, *118*, 36–42. [CrossRef]
338. Ender, A.; Attin, R.; Mehl, A. In vivo precision of conventional and digital methods of obtaining complete-arch dental impressions. *J. Prosthet. Dent.* **2016**, *115*, 313–320. [CrossRef]
339. Fasbinder, D.J. Digital dentistry: Innovation for restorative treatment. *Compend. Contin. Educ. Dent.* **2010**, *31*, 2–12.
340. Lenguas Silva, A.L.; Ortega Aranegui, R.; Samara Shukeirm, G.; Lopez Bermejo, M.A. Tomografía computerizada de haz conico. Aplicaciones clínicas en odontología; comparacion con otras técnicas. *Cient. Dent.* **2010**, *7*, 147–159.
341. Garg, V.; Bagaria, A.; Bhat, S.S.; Bhardwaj, S.; Hedau, L.R. Application of cone beam computed tomography in dentistry—A review. *J. Adv. Med. Dent. Sci. Res.* **2019**, *7*, 73–76.

342. Bornstein, M.M.; Scarfe, W.C.; Vaughn, V.M.; Jacobs, R. Cone beam computed tomography in implant dentistry: A systematic review focusing on guidelines, indications, and radiation dose risks. *J. Oral Maxillofoc. Implant.* **2014**, *29*, 55–77. [CrossRef] [PubMed]

343. Ho, M.H.; Lee, S.Y.; Chen, H.H.; Lee, M.C. Three-dimensional finite element analysis of the effect of post on stress distribution in dentin. *J. Prosth. Dent.* **1994**, *72*, 367–372. [CrossRef]

344. Reimann, Ł.; Żmudzki, J.; Dobrzański, L.A. Strength analysis of a three-unit dental bridge framework with the finite element method. *Acta Bioeng. Biomech.* **2015**, *17*, 51–59. [CrossRef]

345. Tanaka, M.; Naito, T.; Yokota, M.; Kohno, M. Finite element analysis of the possible mechanism of cervical lesion formation by occlusal force. *J. Oral Rehab.* **2003**, *30*, 60–67. [CrossRef]

346. Dobrzański, L.A.; Dobrzańska-Danikiewicz, A.D.; Malara, P.; Achtelik-Franczak, A.; Dobrzański, L.B.; Gaweł, T.G. Implanto-Skafold Lub Proteza Elementów Anatomicznych Układu Stomatognatycznego Oraz Twarzoczaszki. Patent nr PL229148, 9 January 2018.

347. Dobrzański, L.A.; Dobrzańska-Danikiewicz, A.D.; Malara, P.; Achtelik-Franczak, A.; Dobrzański, L.B.; Gaweł, T.G. Implanto-Skafold Kostny. Patent nr PL229149, 9 January 2018.

348. Dobrzański, L.B.; Dobrzański, L.A.; Dobrzańska, J.; Achtelik-Franczak, A. Wysokorozwinięty Powierzchniowo Implant Stomatologiczny. Zgłoszenie patentowe P.434312, 15 June 2020.

349. Dobrzański, L.B.; Dobrzański, L.A.; Dobrzańska, J.; Achtelik-Franczak, A. Wysokorozwinięty Powierzchniowo Implant Kostny W Tym Twarzoczaszki. Zgłoszenie patentowe P.434314, 15 June 2020.

350. Dobrzański, L.A.; Dobrzańska-Danikiewicz, A.D. Foresight of the surface technology in manufacturing. In *Handbook of Manufacturing Engineering and Technology*; Nee, A.Y.C., Ed.; Springer-Verlag: London, UK, 2015; pp. 2587–2637.

351. Dobrzański, L.A.; Dobrzańska-Danikiewicz, A.D.; Szindler, M.; Achtelik-Franczak, A.; Pakieła, W. Atomic layer deposition of TiO_2 onto porous biomaterials. *AMSE* **2015**, *75*, 5–11.

352. Dobrzański, L.A.; Dobrzańska-Danikiewicz, A.D. *Inżynieria Powierzchni Materiałów: Kompendium Wiedzy I Podręcznik Akademicki*; International OCSCO World Press: Gliwice, Poland, 2018.

353. Vujicic, M. Our dental care system is stuck. *J. Am. Dent. Assoc.* **2018**, *149*, 167–169. [CrossRef]

354. Kassebaum, N.J.; Smith, A.G.C.; Bernabé, E.; Fleming, T.D.; Reynolds, A.E.; Vos, T.; Murray, C.J.L.; Marcenes, W. Global, regional, and national prevalence, incidence, and disability-adjusted life years for oral conditions for 195 countries, 1990–2015: A systematic analysis for the global burden of diseases, injuries, and risk factors. *J. Prosth. Res.* **2017**, *96*, 380–387. [CrossRef]

355. Cohen, L.; Dahlen, G.; Escobar, A.; Fejerskov, O.; Johnson, N.; Manji, F. Dentistry in crisis: Time to change, la cascada declaration. *Aust. Dent. J.* **2017**, *62*, 258–260. [CrossRef]

356. OECD. *Health at a Glance 2019: OECD Indicators*; OECD Publishing: Paris, France, 2019. [CrossRef]

357. Banerjee, A.; Doméjean, S. The contemporary approach to tooth preservation: Minimum intervention (MI) caries management in general practice. *Prim. Dent. J.* **2013**, *2*, 30–37. [CrossRef] [PubMed]

358. Brantley, C.F.; Bader, J.D.; Shugars, D.A.; Nesbit, S.P. Does the cycle of rerestoration lead to larger restorations? *J. Am. Dent. Assoc.* **1995**, *126*, 1407–1413. [CrossRef] [PubMed]

359. Fejerskov, O.; Escobar, G.; Jøssing, M.; Baelum, V. A functional natural dentition for all—and for life? The oral healthcare system needs revision. *J. Oral Rehab.* **2013**, *40*, 707–722. [CrossRef] [PubMed]

360. Elderton, R.J. Clinical studies concerning re-restoration of teeth. *Adv. Dent. Res.* **1990**, *4*, 4–9. [CrossRef]

361. Ten Cate, J.M. Review on fluoride, with special emphasis on calcium fluoride mechanisms in caries prevention. *Eur. J. Oral Sci.* **1997**, *105*, 461–465. [CrossRef] [PubMed]

362. Twetman, S.; Axelsson, S.; Dahlgren, H.; Holm, A.-K.; Källestål, C.; Lagerlöf, F.; Lingström, P.; Mejàre, I.; Nordenram, G.; Norlund, A.; et al. Caries-preventive effect of fluoride toothpaste: A systematic review. *Acta Odontol. Scand.* **2003**, *61*, 347–355. [CrossRef]

363. Ismail, A.I.; Hasson, H. Fluoride supplements, dental caries and fluorosis; A systematic review. *J. Am. Dent. Assoc.* **2008**, *139*, 1457–1468. [CrossRef]

364. WHO. *Guideline: Sugars Intake for Adults and Children*; World Health Organization: Geneva, Switzerland, 2015.

365. Lawrence, H.P.; Sheiham, A. Caries progression in 12- to 16-year-old schoolchildren in fluoridated and fluoride-deficient areas in Brazil. *Community Dent. Oral Epidemiol.* **1997**, *25*, 402–411. [CrossRef]

366. Slade, G.D.; Sanders, A.E.; Do, L.; Roberts-Thomson, K.; Spencer, A.J. Effects of fluoridated drinking water on dental caries in Australian adults. *J. Dent. Res.* **2013**, *92*, 376–382. [CrossRef]

367. Kunzel, W.; Fischer, T. Rise and fall of caries prevalence in German towns with different F concentrations in drinking water. *Caries Res.* **1997**, *31*, 166–173. [CrossRef]

368. Marthaler, T.M. Changes in the prevalence of dental caries: How much can be attributed to changes in diet? *Caries Res.* **1990**, *24* (Suppl. 1), 3–15. [CrossRef]

369. Leite, T.A.; de Paula, M.S.; Ribeiro, R.A.; Leite, I.C.G. Dental caries and sugar consumption in a group of public nursery school children. *Rev. Odontol. Univ. São Paulo* **1999**, *13*, 13–18. (In Portuguese) [CrossRef]

370. Sivaneswaran, S.; Barnard, P.D. Changes in the pattern of sugar (sucrose) consumption in Australia 1958–1988. *Comm. Dent. Health* **1993**, *10*, 353–363.

371. Holt, R.D. Foods and drinks at four daily time intervals in a group of young children. *Br. Dent. J.* **1991**, *170*, 137–143. [CrossRef] [PubMed]

372. Burt, B.A.; Eklund, S.A.; Morgan, K.J.; Larkin, F.E.; Guire, K.E.; Brown, L.O.; Weintraub, J.A. The effects of sugars intake and frequency of ingestion on dental caries increment in a three-year longitudinal study. *J. Dent. Res.* **1988**, *67*, 1422–1429. [CrossRef]

373. Rodrigues, C.S.; Sheiham, A. The relationships between dietary guidelines, sugar intake and caries in primary teeth in low income Brazilian 3-year-olds: A longitudinal study. *Int. J. Paediatr. Dent.* **2000**, *10*, 47–55. [CrossRef]

374. Masson, L.F.; Blackburn, A.; Sheehy, C.; Craig, L.C.; Macdiarmid, J.I.; Holmes, B.A.; McNeill, G. Sugar intake and dental decay: Results from a national survey of children in Scotland. *Br. J. Nutr.* **2010**, *104*, 1555–1564. [CrossRef]

375. Arnadottir, I.B.; Rozier, R.G.; Saemundsson, S.R.; Sigurjons, H.; Holbrook, W.P. Approximal caries and sugar consumption in Icelandic teenagers. *Comm. Dent. Oral Epidemiol.* **1998**, *26*, 115–121. [CrossRef]

376. Ruottinen, S.; Karjalainen, S.; Pienihakkinen, K.; Lagstrom, H.; Niinikoski, H.; Salminen, M.; Ronnemaa, T.; Simell, O. Sucrose intake since infancy and dental health in 10-year-old children. *Caries Res.* **2004**, *38*, 142–148. [CrossRef]

377. Rugg-Gunn, A.J.; Hackett, A.F.; Appleton, D.R.; Jenkins, G.N.; Eastoe, J.E. Relationship between dietary habits and caries increment assessed over two years in 405 English adolescent school children. *Arch. Oral Biol.* **1984**, *29*, 983–992. [CrossRef]

378. Van Loveren, C. Sugar restriction for caries prevention: Amount and frequency. Which is more important? *Caries Res.* **2019**, *53*, 168–175. [CrossRef] [PubMed]

379. Te Morenga, L.; Mallard, S.; Mann, J. Dietary sugars and body weight: Systematic review and meta-analyses of randomised controlled trials and cohort studies. *BMJ* **2012**, *346*, e7492. [CrossRef] [PubMed]

380. Newens, K.J.; Walton, J. A review of sugar consumption from nationally representative dietary surveys across the world. *J. Hum. Nutr. Diet.* **2016**, *29*, 225–240. [CrossRef] [PubMed]

381. James, W.P.T. Taking action on sugar. *Lancet Diabet. Endocrinol.* **2016**, *4*, 92–94. [CrossRef]

382. Sheiham, A.; James, W.P.T. A new understanding of the relationship between sugars, dental caries and fluoride use: Implications for limits on sugars consumption. *Pub. Health Nutr.* **2014**, *17*, 2176–2184. [CrossRef]

383. Broadbent, J.M.; Thomson, W.M.; Poulton, R. Trajectory patterns of dental caries experience in the permanent dentition to the fourth decade of life. *J. Dent. Res.* **2008**, *87*, 69–72. [CrossRef]

384. Broadbent, J.M.; Foster Page, L.A.; Thomson, W.M.; Poulton, R. Permanent dentition caries through the first half of life. *Br. Dent. J.* **2013**, *215*, E12. [CrossRef]

385. Lobstein, T. Sugar: A shove to industry rather than a nudge to consumers? *Lancet Diabet. Endocrinol.* **2016**, *4*, 86–87. [CrossRef]

386. McKee, M.; Stuckler, D. Realising an election manifesto for public health in the UK. *Lancet* **2015**, *385*, 665–666. [CrossRef]

387. Moodie, R.; Stuckler, D.; Monteiro, C.; Sheron, N.; Neal, B.; Thamarangsi, T.; Lincoln, P.; Casswell, S. Profits and pandemics: Prevention of harmful effects of tobacco, alcohol, and ultra-processed food and drink industries. *Lancet* **2013**, *381*, 670–679. [CrossRef]

388. Burt, B.A.; Pai, S. Is sugar consumption still a major determinant of dental caries? a systematic review. In *NIH Consensus Development Conference on Diagnosis and Management of Dental Caries Throughout Life*; Horowitz, A.M., Elliott, J.M., Eds.; National Institutes of Health: Bethesda, MD, USA, 2001; pp. 71–78.

389. Grammatikaki, E.; Wollgast, J.; Caldeira, S. *Feeding Infants and Young Children. A Compilation of National Food-Based Dietary Guidelines and Specific Products Available in the EU Market*; Joint Research Centre: Ispra, Italy, 2019; Available online: https://ec.europa.eu/jrc/sites/jrcsh/files/processed_cereal_baby_food_online.pdf (accessed on 28 August 2020).

390. Birch, S.; Listl, S. *The Economics of Oral Health and Health Care*; Max Planck Institute for Social Law and Social Policy Discussion Paper: Munich, Germany, 2015. [CrossRef]

391. Ramos-Gomez, F.; van Loveren, C. Caries prevention through life course approach, evidence-based caries prevention. In *Evidence-Based Caries Prevention*; Eden, E., Ed.; Springer International Publishing: Cham, Switzerland, 2016; pp. 143–161. [CrossRef]

392. Calzón Fernández, S.; Fernández Ajuria, A.; Martín, J.J.; Murphy, M.J. The impact of the economic crisis on unmet dental care needs in Spain. *J. Epidemiol. Comm. Health* **2015**, *69*, 880–885. [CrossRef] [PubMed]

393. Spencer, A.J. What options do we have for organising, providing and funding better public dental care? *Aust. Health Policy Inst. Comm. Pap. Ser.* **2001**, *02*, 1–80.

394. Baelum, V.; van Palestein Helderman, V.W.H.; Hugoson, A.; Yee, R.; Fejerskov, O. A global perspective on changes in the burden of caries and periodontitis: Implications for dentistry. *J. Oral Rehab.* **2007**, *34*, 872–906. [CrossRef] [PubMed]

395. Holst, D.; Sheiham, A.; Petersen, P. Regulating entrepreneurial behaviour in oral care health services. In *European Observatory on Health Care Systems Series: Regulating Entrepreneurial Behaviour in European Care Health Systems*; Saltman, R., Busse, R., Mossialos, E., Eds.; Open University Press: Buckingham, Philadelphia, PA, USA, 2002; pp. 215–231.

396. Bader, J.D.; Shugars, D.A. What do we know about how dentists make caries-related treatment decisions? *Comm. Dent. Oral Epidemiol.* **1997**, *25*, 97–103. [CrossRef] [PubMed]

397. Hayashi, M.; Haapasalo, M.; Imazato, S.; Il Lee, J.; Momoi, Y.; Murakami, S.; Whelton, H.; Wilson, N. Dentistry in the 21st century: Challenges of a globalising world. *Int. Dent. J.* **2014**, *64*, 333–342. [CrossRef] [PubMed]

398. WMA Declaration of Geneva; Adopted by the 2nd General Assembly of the World Medical Association, Geneva, Switzerland, September 1948 and amended by the 22nd World Medical Assembly, Sydney, Australia, August 1968 and the 35th World Medical Assembly, Venice, Italy, October 1983 and the 46th WMA General Assembly, Stockholm, Sweden, September 1994 and editorially revised by the 170th WMA Council Session, Divonne-les-Bains, France, May 2005 and the 173rd WMA Council Session, Divonne-les-Bains, France, May 2006 and amended by the 68th WMA General Assembly, Chicago, United States, October 2017. Available online: https://www.wma.net/policies-post/wma-declarationof-geneva/ (accessed on 30 September 2020).

399. Yee, R.; Sheiham, A. The burden of restorative dental treatment for children in third world countries. *Int. Dent. J.* **2002**, *52*, 1–9. [CrossRef]

400. Fejerskov, O.; Baelum, V. Changes in oral health profiles and demography necessitate a revision of the structure and organisation of the oral healthcare system. *Gerodontol* **2014**, *31*, 1–3. [CrossRef]

401. Bernabé, E.; Masood, M.; Vujicic, M. The impact of out-of-pocket payments for dental care on household finances in low and middle income countries. *BMC Public Health* **2017**, *17*, 109. [CrossRef]

402. FDI World Dental Federation. Available online: https://www.fdiworlddental.org/about-fdi (accessed on 2 September 2020).

403. Practising Dentists. Available online: https://ec.europa.eu/eurostat/databrowser/view/tps00045/default/map?lang=en (accessed on 2 September 2020).

404. Statista 2020. Available online: https://www.statista.com/statistics/186273/number-of-active-dentists-in-the-us-since-1993/ (accessed on 30 September 2020).

405. Novel Coronavirus (COVID-19) Infection Map. Available online: https://hgis.uw.edu/virus/ (accessed on 25 November 2020).

406. Van Doremalen, N.; Morris, D.H.; Holbrook, M.G.; Gamble, A.; Williamson, B.N.; Tamin, A.; Harcourt, J.L.; Thornburg, N.N.J.; Gerber, S.I.; Lloyd-Smith, J.O.; et al. Aerosol and surface stability of SARS-CoV-2 as compared with SARS-CoV-1. *N. Eng. J. Med.* **2020**, *382*, 1564–1567. [CrossRef]

407. Meng, L.; Hua, F.; Bian, Z. Coronavirus disease 2019 (COVID-19): Emerging and future challenges for dental and oral medicine. *J. Dent. Res.* **2020**, *99*, 481–487. [CrossRef]

408. Raport Amnesty International MJM. Pracownicy Medyczni Na Całym Świecie Są Uciszani, Narażani I Atakowani. Available online: https://amnesty.org.pl/pracownicy-medyczni-na-calym-swiecie-sa-uciszani-narazani-i-atakowani/ (accessed on 14 July 2020).

409. Fallahi, H.R.; Keyhan, S.O.; Zandian, D.; Kim, S.G.; Cheshmi, B. Being a front-line dentist during the Covid-19 pandemic: A literature review. *Maxillofac. Plast. Reconstr. Surg.* **2020**, *42*, 12. [CrossRef] [PubMed]

410. Guidance for Dental Settings, Interim Infection Prevention and Control Guidance for Dental Settings During the Coronavirus Disease 2019 (COVID-19) Pandemic. Available online: https://www.cdc.gov/coronavirus/2019-ncov/hcp/dental-settings.html (accessed on 28 August 2020).

411. Management of Acute Dental Problems During COVID-19 Pandemic. Available online: https://www.sdcep.org.uk/published-guidance/acute-dental-problems-covid-19/ (accessed on 28 August 2020).

412. Dental care and coronavirus (COVID-19). Available online: https://www.dentalhealth.org/Pages/FAQs/Category/coronavirus (accessed on 28 August 2020).

413. Guidance on COVID-19, Guidance on Managing Infection Related Risks in Dental Services, V1.1 15.05.2020. Available online: https://www.fdiworlddental.org/sites/default/files/media/documents/guidance_on_managing_infection_related_risks_in_dental_services_0.pdf (accessed on 28 August 2020).

414. COVID-19 and Dental Practice; What Has Been Done in China? Available online: https://www.fdiworlddental.org/sites/default/files/media/documents/covid-19_and_dental_practice_what_has_been_done_in_china.pdf (accessed on 28 August 2020).

415. Khader, Y.; Al Nsour, M.; Al-Batayneh, O.B.; Saadeh, R.; Bashier, H.; Alfaqih, M.; Al-Azzam, S.; AlShurman, B.A. Dentists' awareness, perception, and attitude regarding covid-19 and infection control: Cross-sectional study among jordanian dentists. *JMIR Public Health Surveill.* **2020**, *6*, e18798. [CrossRef] [PubMed]

416. Dobrzański, L.B.; Dobrzański, L.A.; Dobrzańska, J.; Rudziarczyk, K.; Achtelik-Franczak, A. Akcesorium Do Ochrony Osobistej Personelu Dentystycznego Przed Koronawirusem SARS-CoV-2 I Innymi Drobnoustrojami Chorobotwórczymi. Zgłoszenie patentowe P.434391, 19 June 2020.

417. Dobrzański, L.B.; Dobrzański, L.A.; Dobrzańska, J. Clean DENT work space—A medical aerosol absorption system generated during medical procedures in the patient's mouth during dental, ENT and general medical procedures in order to effectively protect personnel and patients against viruses (including SARS-cov-2) and other microorganisms. Wniosek o dofinansowanie Projektu nr POIR.01.01.01-00-0637/20, 4 June 2020.

418. Dobrzański, L.B.; Dobrzański, L.A.; Dobrzańska, J.; Rudziarczyk, K.; Achtelik-Franczak, A. Accessory for personal protection of dental staff against SARS-CoV-2 coronavirus and other pathogenic microorganisms. ICAN—International Invention Innovation Competition in Canada: Overall Selection of the Top 20 Best Inventions in iCAN 2020. Available online: https://www.tisias.org/uploads/6/9/5/1/69513309/the_finals_award_winners_-_list.pdf (accessed on 2 September 2020).

419. Dobrzański, L.B.; Dobrzański, L.A.; Dobrzańska, J.; Rudziarczyk, K.; Achtelik-Franczak, A. Accessory for Personal Protection of Dental Staff Against SARS-CoV-2 Coronavirus And Other Pathogenic Microorganisms. Sahasak Nimavum 2020 International Competition for Invention and Innovation: Gold Medal. Available online: http://slicexpo.com/sn-2020-results/ (accessed on 2 September 2020).

420. Steinke, H. Der Hippokratische eid: Ein schwieriges erbe. Horizonte medizingeschichte, schweizerische ärztezeitung. *Bull. Médecins Suisses—Boll. Med. Svizz.* **2016**, *97*, 1699–1701.

421. Murgic, L.; Hébert, P.C.; Sovic, S.; Pavlekovic, G. Paternalism and autonomy: Views of patients and providers in a transitional (post-communist) country. *BMC Med. Ethics* **2015**, *29*, 65. [CrossRef] [PubMed]

422. Will, J.F. A brief historical and theoretical perspective on patient autonomy and medical decision making: Part II: The autonomy model. *Chest* **2011**, *139*, 1491–1497. [CrossRef]

423. Baumann, A.; Audibert, G.; Guibet Lafaye, C.; Puybasset, L.; Mertes, P.M.; Claudot, F. Elective non-therapeutic intensive care and the four principles of medical ethics. *J. Med. Ethics* **2013**, *39*, 139–142, Erratum in **2013**, *39*, 409. [CrossRef] [PubMed]

424. Jotterand, F. The hippocratic oath and contemporary medicine: Dialectic between past ideals and present reality? *J. Med. Philos.* **2005**, *30*, 107–128. [CrossRef] [PubMed]

425. Gillon, R. Medical ethics: Four principles plus attention to scope. *BMJ* **1994**, *309*, 184. [CrossRef] [PubMed]

426. Beauchamp, T.L.; Childress, J.F. *Principles of Biomedical Ethics*, 8th ed.; Oxford University Press: New York, NY, USA, 2019.

427. Holm, S. Not just autonomy—The principles of American biomedical ethics. *J. Med. Ethics* **1995**, *21*, 6. [CrossRef]

428. Beauchamp, T.L.; Childress, J.F. *Principles of Biomedical Ethics*, 1st ed.; Oxford University Press: New York, NY, USA, 1979.

429. Veatch, R.M. The impossibility of a morality internal to medicine. *J. Med. Philos. A Forum Bioeth. Philos. Med.* **2001**, *26*, 621–642. [CrossRef]

430. Nash, D.A. Ethics in dentistry: Review and critique of principles of ethics and code of professional conduct. *J. Am. Dent. Assoc.* **1984**, *109*, 597–603. [CrossRef] [PubMed]

431. Seear, J.E.; Walters, L. *Law and Ethics in Dentistry Paperback*, 3rd ed.; Butterworth-Heinemann Ltd.: Oxford, UK, 1991.

432. Jessri, M.; Fatemitabar, S.A. Implication of ethical principles in chair-side dentistry. *Iran. J. Allergy Asthma Immunol.* **2007**, *6* (Suppl. 5), 53–59.

433. Nash, D.A. On ethics in the profession of dentistry and dental education. *Eur. J. Dent. Educ.* **2007**, *11*, 64–74. [CrossRef] [PubMed]

434. Dobrzański, L.A.; Dobrzański, L.B.; Dobrzańska-Danikiewicz, A.D.; Dobrzańska, J.; Kraszewska, M. The synergistic ethics interaction with nanoengineering, dentistry and dental engineering. In *Ethics in Nanotechnology*; Jeswani, G., Van De Voorde, M., Eds.; De Gruyter: Berlin, Germany, 2020. (prepared for printing).

435. Dobrzański, L.A.; Dobrzański, L.B.; Dobrzańska-Danikiewicz, A.D.; Dobrzańska, J.; Kraszewska, M. Development prospects of the dentistry 4.0, dental engineering and nanotechnology triad towards ethical imperatives. In *Ethics in Nanotechnology*; Jeswani, G., Van De Voorde, M., Eds.; De Gruyter: Berlin, Germany, 2020. (prepared for printing).

436. Davis, M. *Thinking Like an Engineer: Studies in the Ethics of a Profession*; Oxford University Press: New York, NY, USA; Oxford, UK, 1998.

437. Davis, M. *Engineering Ethics*; Routledge: London, UK, 2005. [CrossRef]

438. Martin, M.W.; Schinzinger, R. *Ethics in Engineering*, 4th ed.; McGraw-Hill: Boston, MA, USA, 2005.

439. Harris, C.E.; Pritchard, M.S.; Rabins, M.J. *Engineering Ethics: Concepts and Cases*, 4th ed.; Wadsworth: Belmont, CA, USA, 2008.

440. Hansson, S.O. Ethical criteria of risk acceptance. *Erkenntnis* **2003**, *59*, 291–309. [CrossRef]

441. Hahn, D.; Payne, W.; Lucas, E. *Focus on Health*, 11th ed.; McGrow Hill Humanities: New York, NY, USA, 2013; pp. 125–153.

442. Barro, Ó.; Arias-González, F.; Lusquiños, F.; Comesaña, R.; del Val, J.; Riveiro, A.; Badaoui, A.; Gómez-Baño, F.; Pou, J. Effect of four manufacturing techniques (casting, laser directed energy deposition, milling and selective laser melting) on microstructural, mechanical and electrochemical properties of co-CR dental alloys, before and after PFM firing process. *Metals* **2020**, *10*, 1291. [CrossRef]

443. Bartoszewski, W. *Warto Być Przyzwoitym*; Wydawnictwo W drodze: Poznań, Poland, 2019. (In Polish)

Publisher's Note: MDPI stays neutral with regard to jurisdictional claims in published maps and institutional affiliations.

 *processes*

Article

Non-Antagonistic Contradictoriness of the Progress of Advanced Digitized Production with SARS-CoV-2 Virus Transmission in the Area of Dental Engineering

Leszek A. Dobrzański [1,*], Lech B. Dobrzański [1], Anna D. Dobrzańska-Danikiewicz [2], Joanna Dobrzańska [1], Karolina Rudziarczyk [1] and Anna Achtelik-Franczak [1]

[1] Medical and Dental Engineering Centre for Research, Design and Production ASKLEPIOS, 13 D Krolowej Bony St., 44-100 Gliwice, Poland; dobrzanski@centrumasklepios.pl (L.B.D.); joanna.dobrzanska@centrumasklepios.pl (J.D.); Karolina.rudziarczyk@centrumasklepios.pl (K.R.); anna.achtelik-franczak@centrumasklepios.pl (A.A.-F.)

[2] Department of Mechanical Engineering, University of Zielona Góra, 4 Prof. Z. Szafrana St., 65-516 Zielona Góra, Poland; anna.dobrzanska.danikiewicz@gmail.com

* Correspondence: leszek.dobrzanski@centrumasklepios.pl; Tel.: +48-606-33-55-66

Received: 13 August 2020; Accepted: 2 September 2020; Published: 4 September 2020

Abstract: The general goals of advanced digitized production in the Industry 4.0 stage of the industrial revolution were presented along with the extended holistic model of Industry 4.0, introduced by the authors, indicating the importance of material design and the selection of appropriate manufacturing technology. The effect of the global lockdown caused by the SARS-CoV-2 virus transmission pandemic was a drastic decrease in production, resulting in a significant decrease in the gross domestic product GDP in all countries, and gigantic problems in health care, including dentistry. Dentists belong to the highest risk group because the doctor works in the patient's respiratory tract. This paper presents a breakthrough authors solution, implemented by the active SPEC strategy, and aims to eliminate clinical aerosol at the source by negative pressure aspirating bioaerosol at the patient's mouth line. The comparative benchmarking analysis and its results show that only the proprietary solution with a set of devices eliminates the threat at the source, while the remaining known methods do not meet the expectations. The details of this solution are described. Photopolymer materials and additive Digital Light Printing (DLP) technology were used.

Keywords: industry 4.0; dentistry 4.0; SARS-CoV-2 pandemic; SPEC strategy; elimination clinical aerosol at the source; dendrological matrix; photopolymer materials; additive digital light printing

1. General Goals of Advanced Digitized Production in the Industry 4.0 Stage of the Industrial Revolution

Since the end of the 18th century, the industrial revolution has progressed. Nowadays, the world has reached the Industry 4.0 stage related to the dissemination of cyber-physical systems CPS [1–5]. They provide communication between machines and smart products and create a virtual copy of the physical world via digital twins that can be subjected to virtual experiments, much cheaper and faster than the experimental verification of products in industrial practice. Apart from the cyber-IT module, the real progress is determined by the development of advanced engineering materials, multifaceted and various technological processes, including additive ones as part of them, and machines, and technological devices [6–8]. The augmented holistic model of Industry 4.0, introduced by the authors in the form of an octahedron, includes in its technological platform all four components that require systematic development (Figure 1) [7,8].

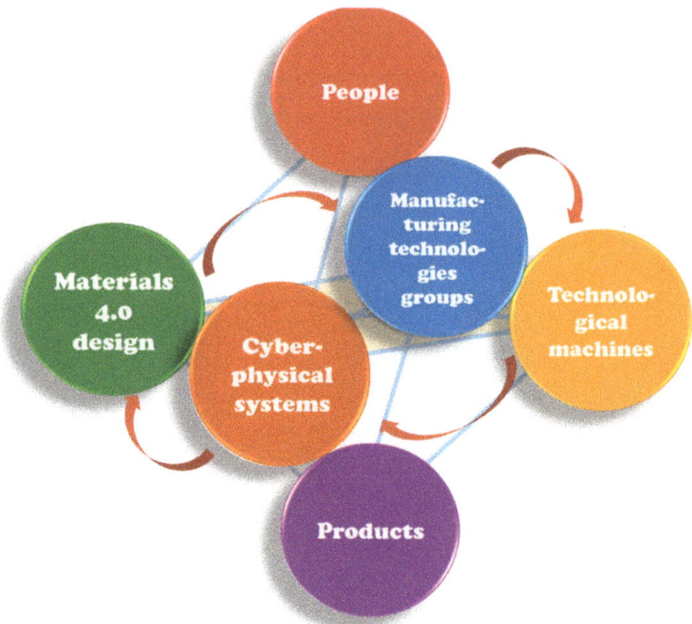

Figure 1. Diagram of the augmented holistic Industry 4.0 model in the form of an octahedron.

New technologies result in Integrated Sustainable Industrial Development (ISID) by bringing new goods to the market and improving production efficiency. By combining hardware, software, and connectivity, advanced digital production technologies (ADP) are spreading (Figure 2) [9].

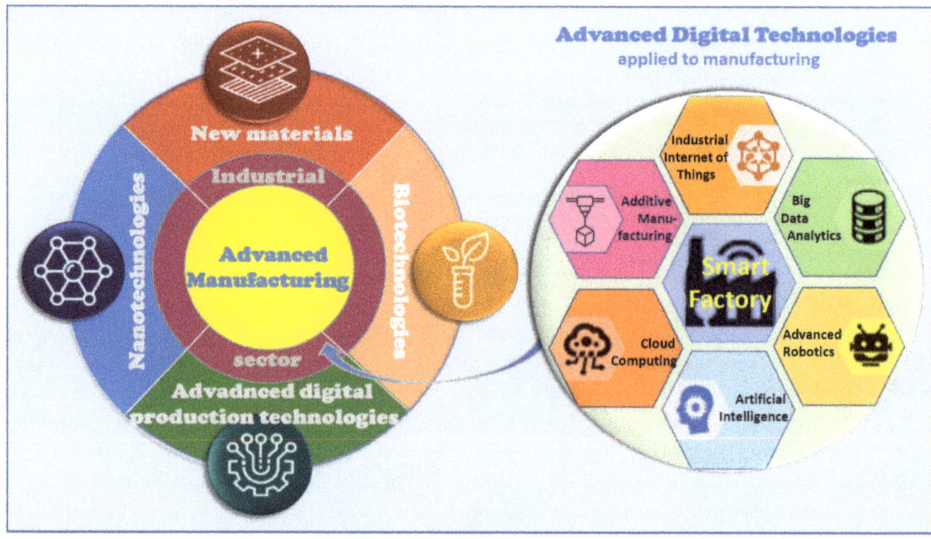

Figure 2. Diagram of advanced digital production and advanced technologies applied to manufacturing

For obvious reasons, it is in the general interest that these advanced digital production technologies be shared by the widest possible group of subjects in the world. It is increasingly difficult to imagine the design and manufacture of any product without intensive computer-aided manufacturing and manufacturing requiring significant IT support. Until recently, it seemed that the cause and effect chain of progress was clearly defined and that it could not be disrupted in any way. Unfortunately, the experimental life verification in the last few months has not confirmed this seemingly obvious view.

2. Multifaceted Implications of the COVID-19 Disease Pandemic Caused by SARS-CoV-2 Virus Transmission

On 20 March 2020, the World Health Organization declared a pandemic of the severe and acute respiratory syndrome and other severe diseases of the human body defined as the COVID-19 coronavirus 2019 disease. It is caused by the transmission of the severe acute respiratory syndrome coronavirus SARS-CoV-2 virus, which is a small nano-sized natural object with a diameter of 60–140 nm and a spherical shape, although somewhat pleomorphic, surrounded by bright spines 9–12 nm long, making it similar to a corona (Figure 3) [10].

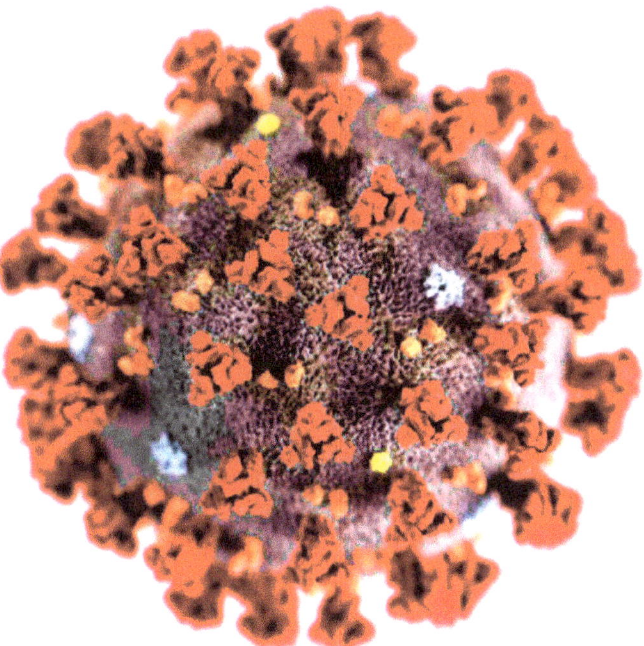

Figure 3. Visualization of SARS-CoV-2 virus.

The genome of this enveloped virus is positive polarity single-stranded ribonucleic acid. The size of this virus is 100 trillion times smaller than the size of the globe. With its arrival, virtually everything in the world has changed. The scale of the pandemic spanned the entire world, and in nearly 20 weeks of its announcement globally, as of 3:36 pm Central European Summer Time (CEST), 3 August 2020, there have been 17,918,582 confirmed cases of COVID-19, including 686,703 deaths, reported to World Health Organization WHO [11] (Figure 4).

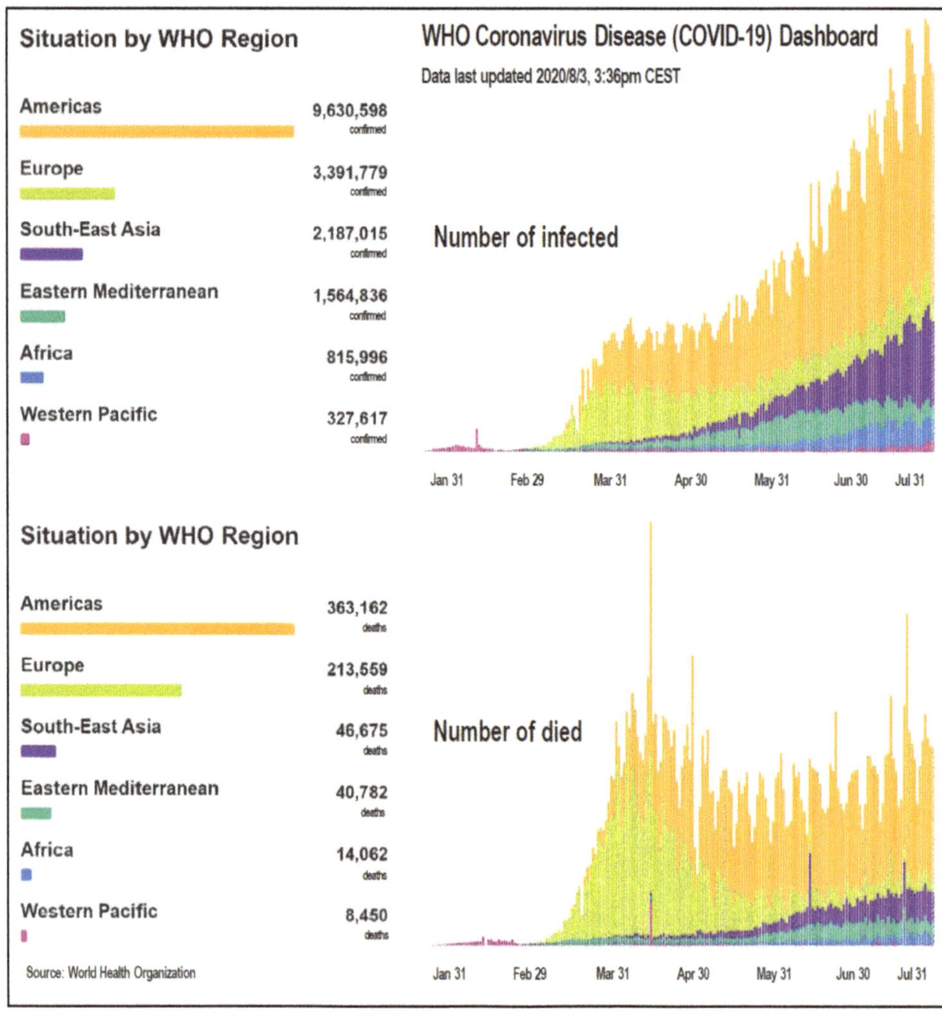

Figure 4. Graphs of changes in the number of infected and dead during the COVID-19 pandemic.

Statistically, these are huge numbers, comparable to the victims of road accidents, in which, according to WHO data, over 1.2 million people lose their lives annually in the world. The data of the European Agency for Safety and Health at Work (EU-OSHA) show that, for example, in 2017, 3362 people died in Europe, and 380,500 in the world, which is much less than the number of victims of coronavirus in the period of approx. four months.

Since the start of the COVID-19 pandemic, various national governments have taken numerous restrictive measures. The effect of the global lockdown was a drastic decrease in production, and the closure, and even many times liquidation of many industrial enterprises, as well as limiting or extinguishing activity in many other industries, resulting in a significant decrease in the gross domestic product (GDP) in Q2 2020 in all countries (Figure 5) [12] and enormous organizational problems in health care units and hospitals all over the world.

Of course, it applies mainly to infectious hospitals but causes significant perturbations in many other branches of medicine, ranging from limiting the number and scope of medical services to the complete abandonment of their provision. It even leads to deaths for reasons not directly related to COVID-19 disease, but indirectly due to it, due to gross negligence in other areas of treatment. It is the case in dentistry, where, due to the high probability of SARS-CoV-2 virus transmission in dental clinics, the US, European and Brazilian governments recommended that these clinics be closed for the first many weeks. Several dental associations around the world require dentists to postpone scheduled procedures. For example, in Poland, on 15 July 2020, by the decision of the Ministry of Health, the prohibition of normal dental practice in full force was renewed from the beginning of the pandemic, in favor of working in the emergency mode. It would be irresponsible to continue to practice in a dentist's clinic even with personal protective equipment PPE under these conditions. One of the reasons is the uncertainty about the incubation time of the virus, which is estimated at 5 to 6 days, with a range of up to 14 days [13], and even its reappearance in cured patients. More than 3000 medical workers in 79 countries around the world have died due to the disease caused by SARS-CoV-2 coronavirus [14]. According to WHO data, dentists are among the doctors at the highest risk. As a result of natural respiration processes, during dental procedures performed in the patient's open mouth, the bioaerosol exhaled by him, in the area of the nasal and oral cavity, in addition to air, also includes viruses, bacteria, and fungi, including pathogenic, abundantly absorbed by the patient while breathing. Due to the necessity of bringing the physician's face close to the patient's mouth within a clear vision distance (up to 20 cm), considering that even 5–10 cm of this distance is in the patient's mouth, the doctor inhales most of the patient's exhaled air along with pathogenic microorganisms.

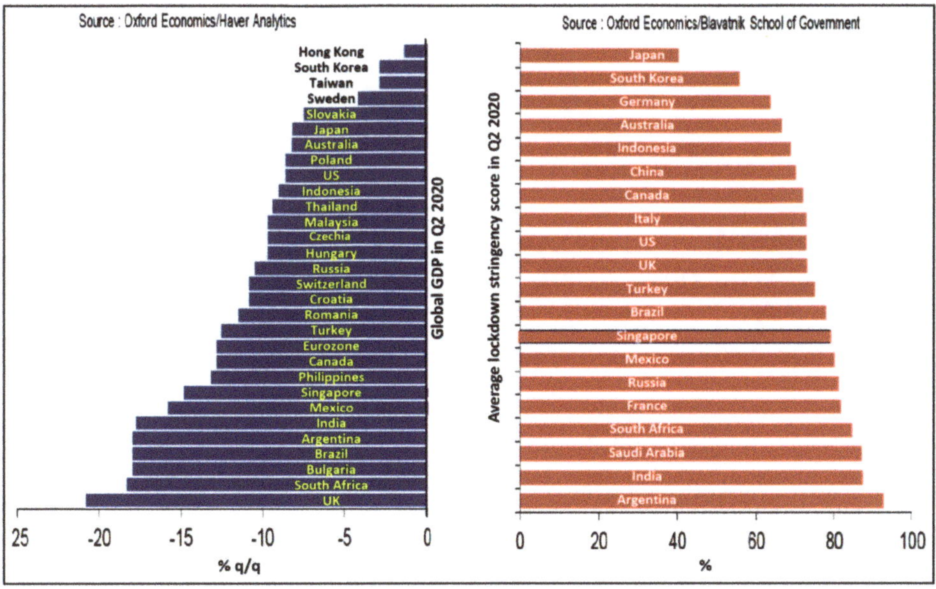

Figure 5. Fall in the gross domestic product (GDP) in the second quarter of 2020 in various countries and the lockdown stringency score.

In the case of any infection of the patient's respiratory tract, the doctor is exposed to the direct transmission of various microorganisms multiplied in the upper and lower respiratory tract of the patient, including the SARS-CoV-2 coronavirus, contained in the patient's exhaled air, as well as colloidal particles of secretion sprayed by him when coughing or sneezing, which are colloidal particles of bioaerosol. In particular, the possibility of SARS-CoV-2 coronavirus transmission in the form of bioaerosols exists even with normal breathing [15]. The problem is all the greater as the use of a turbine, or even a high-speed motor, to prepare the teeth requires intensive cooling, and the cooling jet mixed with the infected saliva and blood of the patient is intensely atomized, and the clinical aerosol is mixed with the respiratory bioaerosol, disproportionately increasing the dentist and dental staff risk. Usually, all treatments require an assistant and the so-called work for four hands. Considerable pollution of the environment by SARS-CoV-2 patients by droplets from the respiratory tract and excretion of saliva mixed with blood suggest the environment as a potential medium of transmission and the necessity to strictly observe the hygiene of the environment and hands [16]. This poses a great risk of directly infecting dentists and medical assistants in particular. In addition, some asymptomatic or subclinically infected patients with mild symptoms are also contagious, leading to the possibility of the virus spreading unintentionally in dental clinics. The mechanism is presented in Figure 6, and the transmission of infection during a week of asymptomatic infection by a dentist may reach a thousand or more people. In the last 20 weeks, every dentist practicing despite the pandemic has come across such cases.

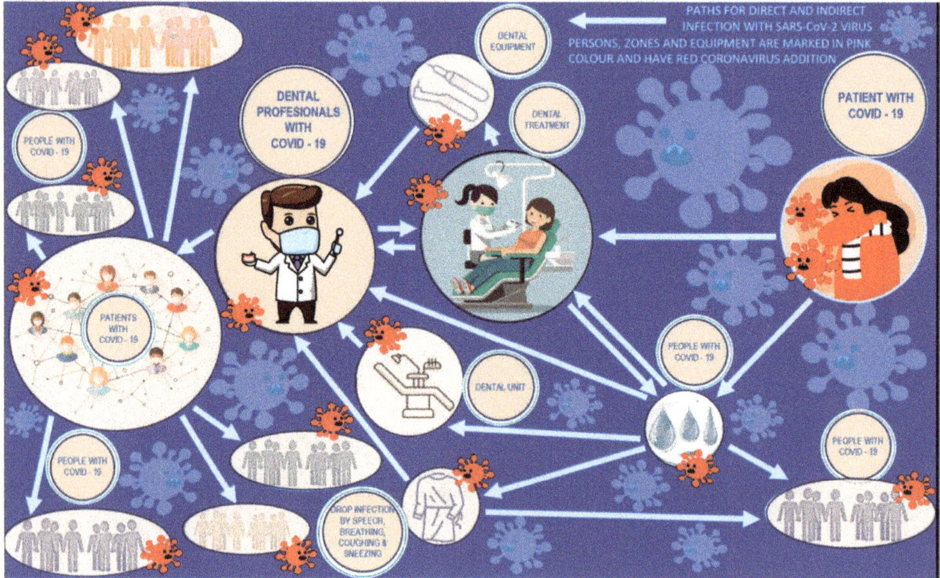

Figure 6. The mechanism of transmission of coronavirus infection by a dentist during a week of asymptomatic infection.

In view of the global efforts to promote social distancing, practical strategies to block the transmission of SARS-CoV-2 virus during dental diagnosis and treatment include a complete patient health assessment, stringent hand hygiene, personal protective equipment for dentists, rinsing the patient's mouth prior to dental procedures, isolation by a cofferdam, surgery disinfection, a surgical mask and eye protection with solid side shields or a face shield protecting the mucous membranes of the eyes, nose, and mouth during procedures, proper management of medical waste [17]. Thus, the STOP strategy is common, and in fact keeps up to date with the intensification of personal protective equipment (PPE), which is a solution contested by dentists. In fact, the society of doctors and the dental staff was deprived of the chance to return to normal working conditions, and the medical administration responsible for it did not present any reasonable proposals, except the suggestion addressed to doctors to avoid most procedures outside the emergency ones mainly with patients' pain [18–22]. In essence, it is an expectation that dentists not fulfill their obligations resulting from the Hippocratic oath, most commonly known and used nowadays after numerous modifications as the World Medical Association WMA Declaration of Geneva [23], which is made by every doctor who takes up employment in his profession. Even a basic rule was not respected—first, don't harm. The doctor, despite the fact that he risks his life and health, in accordance with the regulations and recommendations in force in many, perhaps even all countries, is obliged to refrain from providing most of the dental services [18–22]. It applies without exception to all or almost all dental offices, with full awareness of nonchalance in exposing the health of doctors and dental staff to a lethal threat at each subsequent visit of a patient, who may be asymptomatic or may have COVID-19 within 14 days of incubation and unconscious emission of infection.

In the opinion of the authors of this paper, it is an unacceptable situation from every possible point of view: ethical, legal, protection of employee interests, protection of the patient's interest, violation of constitutional rights to health protection, violation of the European Union's priorities in terms of improving the comfort and extension of human life, protection, or rather systematic violation of human rights in general [24]. In this situation, it is obvious that solutions should be sought, among other technical, which will prevent such unfavorable state of affairs.

3. The Concept of Elimination of the Mixture of Bioaerosol with Pathogenic Microorganisms and Clinical Aerosol in the Conditions of Performing Dental Procedures

This paper presents a ground-breaking authors solution that allows to solve this problem systematically and enable dentists to perform their daily work and provide all services related to the performance of a full range of dental procedures without restrictions, but with all necessary precautions. The solution is protected by a patent [18]. Works related to the serial production of the device on the market were commenced [19]. The analysis of the novelty of this solution indicates a complete change in the approach to solving the problem and a change in the way of thinking about it, making it fully unrivaled and ground-breaking in terms of its approach and essence.

A completely new SPEC strategy was applied, which was developed by the authors of this paper (Figure 7). It replaces the commonly used STOP approach in the world to solve problems of protecting medical staff in the event of an epidemiological threat. In the STOP strategy successively, more and more radical approaches are implemented, ranging from system solutions, through technical solutions, organizational solutions to the implementation of personal protection equipment. As a result, doctors and nurses are completely isolated from the environment in which are present the patients who are infected and emit pathogenic pathogens to this environment or only are suspected of being infected. By the way, the medical staff, in this case, cannot move freely, they sweat, and they are not able to perform many precise activities necessary for the needs of medical procedures. It is a passive way of preventing the effects of a pandemic or epidemic.

SPEC strategy combines a simultaneous and synergistic approach to actively solving the problem of protection against pathogenic pathogens. It combines a systemic approach, prevention, efficiency, and, most importantly, ensures the elimination of the cause and, in addition, at the source of the threat. As a result, doctors and nurses can effectively perform their medical tasks in an environment that is not contaminated by an infected patient. Their movements are not restrained, and there are no barriers to a good vision by them of the area in which medical procedures are performed.

One should hit the cause, not deal with the effect, and the causative factor should be eliminated, not tolerated, or at best weakened. While the STOP strategy is defensive and conservative, and therefore it could not lead to any solution to the problem, especially the new one, the original SPEC strategy is the essence of activity and action and that is why it led to an innovative solution of the problem.

This solution is a collection of many comprehensive devices. Its task is to eliminate the clinical aerosol at its source from the space of the dental office by collecting it directly into an epidemiologically safe container and its complete deactivation and decontamination, and not indirect isolation of medical personnel from pathogenic aerosol and tolerating its presence in the entire space of the dental clinic.

Figure 8 shows the modular structure of the dental staff protection system against the SARS-CoV-2 coronavirus and other pathogenic microorganisms. An important part of the accessory is the face cap (1), and its successive variants provide an ever wider range of accessory functions, causing negative pressure around the patient's mouth and nose and allowing the patient to aspirate bioaerosol.

The accessory also contains an extractor (2) which aspirates the patient's saliva and blood to the residues settlers (5) and (6), in which the aerosol exhaled by the patient condenses, and in both sedimentation tanks bacteria, viruses and other pathogenic microorganisms and their spores are destroyed by chemical disinfection under appropriate conditions and taking into account the elimination of risks for personnel resulting from the toxicity of the applied. Then the aerosol, pre-cleaned of microorganisms, is subjected to radiation (7), e.g., using ultraviolet radiation, filtering (8) e.g., with activated carbon, and is sucked by a suction pump (9) providing a negative pressure throughout the system to direct the purified air to the next filtration (10), and then after leaving the system directly through the pipe (11) to the atmosphere outside from laboratory rooms or to a central ventilation system. In this variant, the accessory is equipped with a nasal cannula (14) suitably attached (15) to the face cap (1) connected to an external breathing aid unit (16) in order to balance the pressure in the patient's airway against the negative pressure in the oro-space the patient's nasal passages. The system of (12) devices (5)–(10) and oxygen cylinders are mounted in an integrated housing with lockable wheels (13), allowing free movement of the entire accessory running gear in the room.

One of the elements of this device is a completely originally designed face cap, which protects the entire space of the dental office and all people present in it. It is not a filter, and this is a major novelty. However, due to the size of the coronavirus from 40 ÷ 150 nm, any available mask worn by medical personnel does not provide any protection, because the minimum size of the pores in the filters of the most effective masks is at least 2 ÷ 7.5 times larger, therefore no mask meets its requirements role. Figure 9 shows a range of face caps, each with more and more integrated functions.

Figure 7. Comparison of the basic assumptions of the conservative STOP strategy and the offensive SPEC strategy.

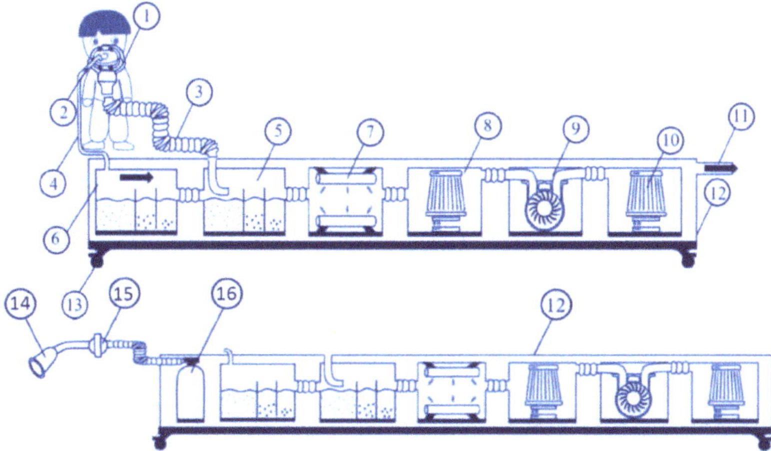

Figure 8. The modular structure of the dental personnel protection system against the SARS-CoV-2 coronavirus and other pathogenic microorganisms.

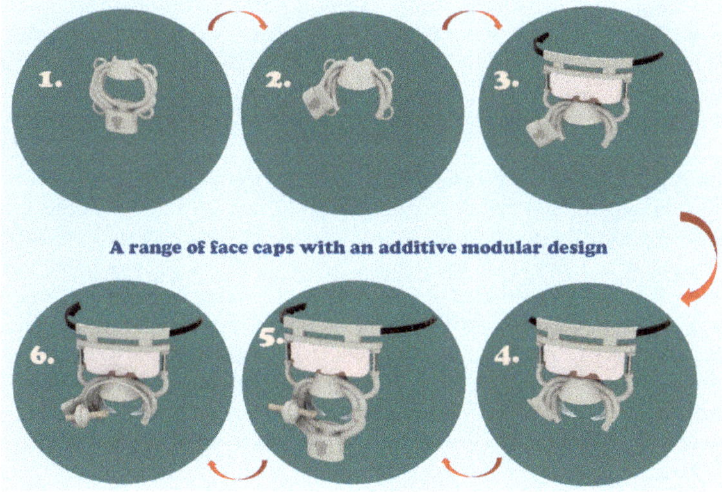

Figure 9. A range of face caps, each of which performs more and more integrated functions.

Figure 10 shows one of the most advanced versions of the face cap, and the symbology of road signs explains how it works. In its basic version, the face cap has an oval through shape and, as in any other functional variant, has an integrated structure and is composed of four tubes of approximately equal length, which requires insertion in the shortest spiral section. The sections of tubing surrounding the oval through-shape have a partially open "C"-shaped section with the obtuse portion facing the interior of the oval shape, in order to suck the bioaerosol around the patient's mouth during operation. The tubes are incorporated into a centrally located manifold, preferably in a rectangular cross-section with rounded tops, most often smoothly circular in section, allowing for dismountable connection to the flexible tubing. The face cap is placed with the flat side over the patient's face in such a way as to ensure full access to the patient's mouth by the doctors and the assistants. In order to stabilize the position of the face cap in relation to the patient's face, a flexible rubber band is mounted in the ears with the possibility of adjusting the length and pressure force, placed around the patient's head.

When connected to a system with a suction pump, a negative pressure is generated around the inner circumference of the oval through-shape, with a value regulated by the pump power, as a result of which the bioaerosol exhaled by the patient, mixed with the clinical aerosol, is sucked, creating a zone free from pathogenic or highly limited microorganisms ambient air concentration above the outer face of the face-shield. In more advanced versions, the face cup includes a shaped nose cup with the lower oval shape exposed. The collector can be centrally located, and in the versions with an exposed lower part of the oval shape, it has a collector located on the left or on the right, which significantly improves the ergonomic features of the solution, giving the doctor various chances of choice. The shaped nose piece allows the aerosol to escape from the naso-facial zone. In order to stabilize their position in relation to the patient's face, these variants of face caps have two sets of lugs for mounting elastic rubber bands with adjustable length and pressure, placed around the patient's head. Another integrated set of face caps has two bows via two distance adjustment turnbuckles connected to a slightly curved mounting bar, to which are attached two shaped spring-loaded latches aligned at respective recesses through an eccentric with a shaped bar. The cross-section of this strip has the shape of a slotted rectangle through which an elastic band is threaded with a set of holes at one end of the band and a set of hooks with heads enabling the band to be fastened around the patient's head and to adjust its length to the circumference of the head. The strip is integrated with a transparent, limited-width eye curtain, which protects, apart from the patient's eyes, also medical personnel against accidental infection. The hinges with spring-loaded latches allow two positions of the face-cap, i.e., the working position when pressed against the face and the position after a vertical rotation of 180°, where the face-piece is raised and inside out towards the environment. During the procedures, the face cap is always placed in the working position and is connected to the system of hoisting and winding devices. In addition, the integrated set, analogous to the previous one, is additionally equipped with two-wing tilting and appropriately profiled mouthpieces enabling the opening of the patient's lips during treatments performed in the oral cavity. Mouthpieces by means of snaps with ball heads are pressed into the appropriate openings of the face cap with the internal shape of a ball-less a segment of the ball. The mouthpieces are attached as needed by pressing into the integrated outer layer of the face cap. The integrated kit can furthermore have two lugs for fixing the ball joint in a seat in the shape of a central press-fit layer mounted in the lugs integral with the outer layer of the face shield by a latch. In the axis of the ball joint situated in the socket in the shape of the central layer of the sphere, a tube is inserted in a way that allows axial movement due to the elastic features of the tube with a suitably profiled flattened tip of the saliva's extractor. The described system of connections allows the superposition of the longitudinal movements of the tube and its rotation in two planes so that the tip of the mammal can, in any situation, be located in the patient's oral cavity required for the needs of the dental procedure. The integrated set (Figure 10), in addition to the double-winged mouthpieces and the tip of the suction device, is additionally equipped with a headband mounted on the outer surface opposite to the face of the face cap over which the clamp is moved by the end of the suction cap and the nasal cannula connected to the external breathing support unit of the patient to balance the positive pressure in the patient's airways in relation to the negative pressure in the oronasal space of the patient, which is part of the system of suction devices.

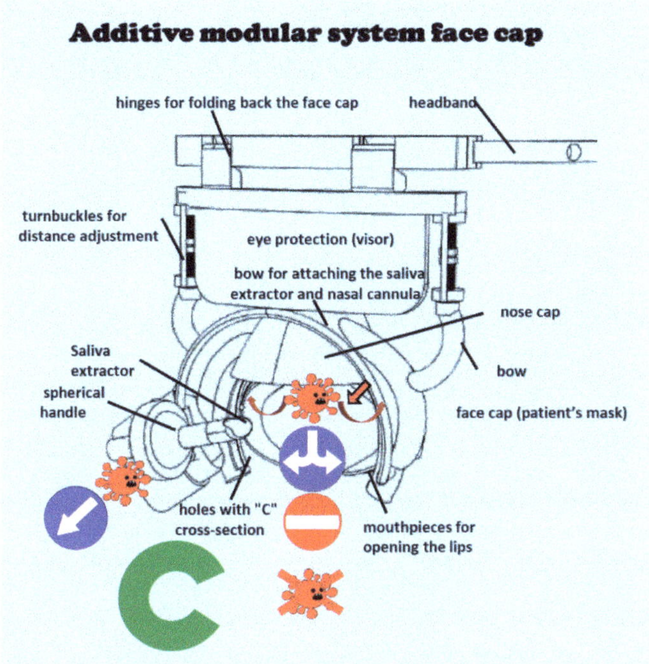

Figure 10. Diagram of the operation of the advanced face cap.

4. Comparative Assessment of the Authors Solution against the Available Methods of Protection against the Effects of the SARS-CoV-2 Coronavirus

In order to compare the methods of protection against the effects of the SARS-CoV-2 coronavirus, the authors used the scientific comparative method of procedural benchmarking in the field of knowledge engineering, developed by the authors and repeatedly tested (although in other subject areas), described in [20]. Each of these groups of methods has different attributes, so it is difficult to consider them similar, although by definition all of them are intended to improve epidemiological safety and, therefore, they are included in the considerations.

The procedural benchmarking method is based on a comparative analysis of these methods in terms of their potential and attractiveness. The point of view of medical personnel was adopted, which is faced with the choice of a method of effective protection against pathogenic virus infection, assuming that all procedures related to the widely understood clinical practice in the field of dentistry can be performed. As part of the expert assessment of the potential of the methods used, five basic criteria and five subsequent criteria were taken into account in assessing the attractiveness of the method, as given in Table 1. The methods were assessed in terms of the above-mentioned criteria of potential and attractiveness using the universal scale of relative states. It is a unipolar positive scale with no zero, with 1 being the weakest and 10 being the best. Moreover, each of the analyzed criteria, which are listed below, was given a weight which determines the importance of a given criterion for the overall evaluation of a given method. The higher the weight, the more the criterion influences the final weighted average. All groups of methods currently encountered, classified in the lower part of Table 1 in six groups, were taken into account, and the results of expert evaluation expressed numerically (from 1 to 10), calculated with the use of the authors computer program, were given in the appropriate columns.

Table 1. Criteria for comparative evaluation of groups of methods of protection against the effects of SARS-CoV-2 coronavirus.

No	Criterion	Weight	No	Criterion	Weight
	Potential			Atttractiveness	
P1	Effectiveness of eliminating the risk of coronavirus	0.50	A1	The sense of security of medical personnel	0.50
P2	Ergonomic factor and 4-hand work	0.25	A2	Patients' sense of safety	0.20
P3	Comprehensive approach	0.10	A3	Difficulties in applying the method	0.20
P4	Safety of cleaning elements	0.10	A4	Importance for the profitability of the practice	0.05
P5	Installation cost	0.05	A5	Impact on the clinics' development prospects	0.05

The results of the calculations are presented in Table 2 and graphically with the use of a dendrological matrix, presented in Figure 11. This matrix consists of four fields related to trees, intuitively indicating the potential and attractiveness as well as development prospects of the analyzed groups of methods. The methods with high potential and attractiveness include only the authors solution (6), which is located in the "Wide-stretching oak" quarter and is characterized by the greatest amount of novelty, presented in an objective manner. The field "Rooted dwarf mountain pine" includes proven, mature methods (1,2), which have high potential, but may be of complementary importance, in cooperation with the one discussed first (6). "Soaring Cypress" is a quarter containing highly attractive methods (3,4), with opportunities for the future, but requiring consolidation, otherwise they will not be successful in their application. The weakest quadrant is "Quaking Aspen". In this field, there are methods with little potential and attractiveness (5), i.e., methods that will be replaced by other, better solutions. The results of the assessments were plotted on a dendrological matrix. Among all the analyzed groups of methods, in terms of novelty compared to other methods, as well as potential and attractiveness, the authors solution (6) eliminating the threat at the source with a set of virus elimination devices clearly stands out.

Table 2. The results of calculating the potential and attractiveness of groups of methods of protection against the effects of SARS-CoV-2 coronavirus.

No	Type of Method Known in the Literature, State-Of-The-Art Including Own Solution	Potential	Attracti Veness
1	Installation of air filtering devices in the treatment room	6.7	4.0
2	Equipment operating with HEPA or carbon filters or with UV radiation	5.95	4.0
3	Dental saliva extractor attachments	3.3	5.8
4	"Extract from the treatment area" devices	3.7	6.6
5	Patient enclosures in the form of glass or PMMA boxes	1.45	3.9
6	Original solution eliminating the threat at the source with the set of devices for virus elimination	9.0	8.5

The comparative analysis and its results indicate that only the authors solution (6) with a set of devices for virus elimination eliminates the threat at the source, while the other groups of methods available on the market and known from literature and company reports are relatively much more attractive (subjective impression of the recipient) than their potential (objective technical analysis). It means that it is commonly felt their role is overestimated despite their less factual meaning. None of the current solutions exceeded both potential and attractiveness (fairly low to moderate) in any of the ratings, which proves the poor usefulness and real ineffectiveness of these methods in removing sources of infection. The author's solution receives these scores as very high, respectively 9 (potential) and 8.5 (attractiveness—there are difficulties at work). This proves the unrivaled importance of this solution and its groundbreaking importance.

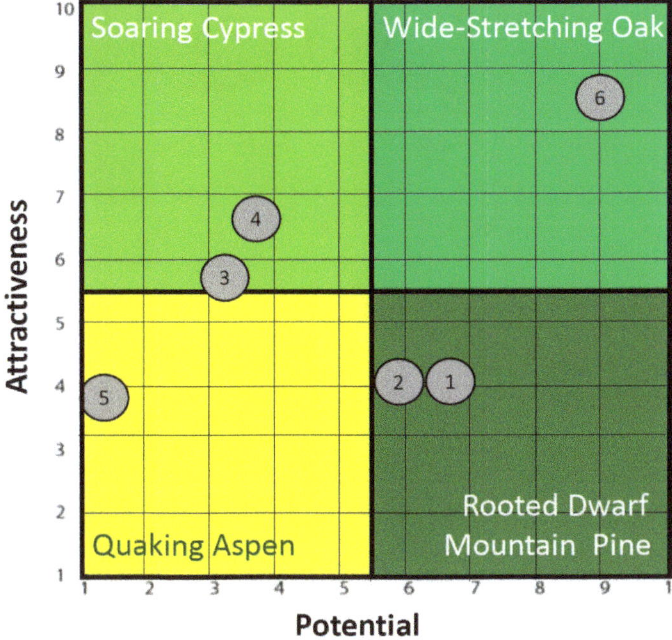

Figure 11. Dendrological matrix of methods of protection against the effects of SARS-CoV-2 coronavirus.

The essential element determining this high comparative rating is, of course, the virus capture efficiency (Figure 12). Our own solution (6) is effective due to the location of the virtual plane of the pressure difference as close as possible to the patient's mouth, and the elimination of bioaerosol takes place practically at the source of its production, preventing its diffusion beyond the patient's face. As a result, the expiratory the patient's bioaerosol mixed with the clinical aerosol is captured, and the so-called fit factor, which is the ratio of the captured aerosol to its total volume, is at least 99%. The sequence of devices and the negative pressure in it protect the entire clinic space against the diffusion of any pathogenic microorganisms. Each coronavirus captured as a result is eliminated at the next stages of the device's operation, which gives 100% efficiency in the elimination of coronaviruses that go to the elimination system. The result is a real radical reduction in the concentration of viruses and harmful microorganisms in the space of the dental clinic. The solution applies to four-handed work for all types of dental procedures. All competing solutions in the best variant most often partially reduce the concentration of coronaviruses in the area of the mouth and nose of the patient but cause them to blow out at the end of the system in the same or adjacent room, which in turn does not protect the medical team from potential infection. One of such examples representing the group marked as (3) presented recently on the Internet is not an effective solution due to the relatively large, several centimeters distance of the mouth of the saliva ejector suction nozzle adapted to suck the aerosol from the area of the patient's face (Figure 12), which, however, allows for the spreading of bioaerosol by diffusion in the entire volume of the treatment room, sucking only a part of it, and because it does not contain decontamination system of bioaerosol.

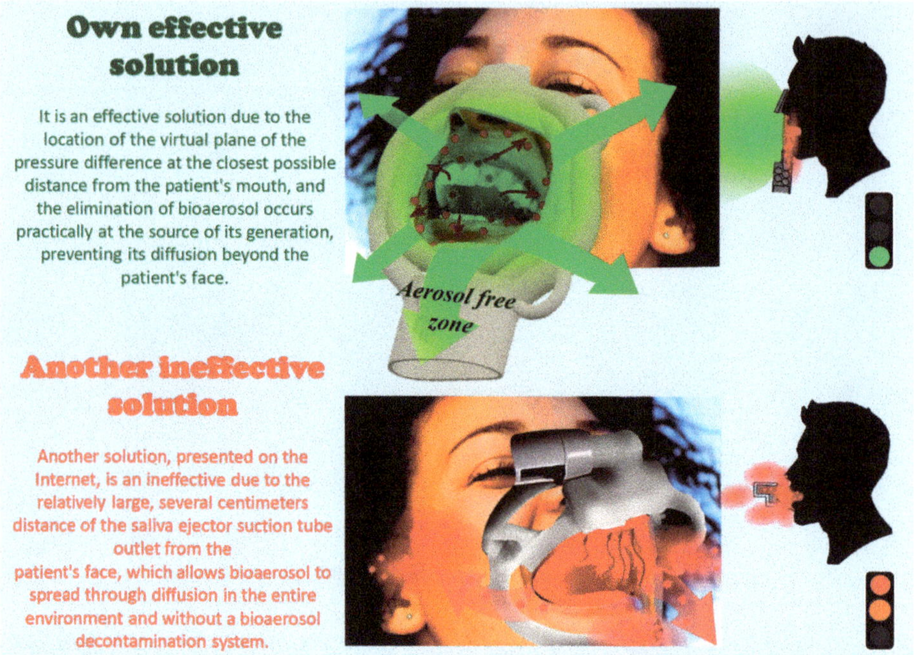

Figure 12. Comparison of the effectiveness of SARS-CoV-2 virus capture by a proprietary face cap and a solution available on the Internet.

5. Engineering Aspects of Digitized Design and Manufacturing of an Authors Device for Dental Purposes that Protect the Doctor and Medical Staff against the SARS-CoV-2 Coronavirus

In accordance with the concept of "Dentistry 4.0" [7,8], in the design and optimization of the device structure, the results of own work and information obtained by means of intraoral scanning and Cone Beam Computed Tomography (CBCT), on which own work has been carried out for about 10 years, are used. CBCT images are taken on a Gendex GXDP-700 laser device, with the SmartLogic technology applied. These data are used to develop typical 3D models of the morphology, structure, and dimensions of human faces. This is the basis for the development of a design and dimensional series of face caps with the greatest possible adaptation to the anatomical features of patients. Thus, the number of solutions, due to the anatomical dimensions of the patients, is finite and divided into quartiles. Data in the form of a set of photos DCM (DICOM Digital Imaging and Communications in Medicine image shorter DCM) taken at a certain thickness, collected during the CBCT examination are converted into data, three-dimensional models using specialized Materialize Mimics software. Using additional modules, CAD (computer-aided design) operations related to the design of face caps based on image data are performed, as well as virtual optimization using the so-called virtual twin. This allows for the proper simulation of the anatomy of the face and classification into dimensional quartiles. The implementation of virtual models according to the described methodology enables the transfer of appropriate data in the form of a three-dimensional model in STL (stereolithography or Standard Tessellation Language) format to the appropriate CAD software, in which geometric features of individual types of face caps are designed. A digital model of the oral cavity with a generated aerosol is being developed with the possibility of simulating the performance of various medical procedures generating a different amount of aerosol, including scaling and sandblasting of the side teeth, front teeth at the top or bottom, conservative treatment of teeth in various locations in the oral cavity, performing surgical procedures, as well as ear, nose, and throat (ENT) examination,

ENT procedures. The model includes models of various aerosol sources, including a dental turbine, a dental blower, a dental sandblaster, and devices. The model also enables the simulation of aerosol generation depending on the conditions necessary for the various configurations of the mouth-aerosol source, and the dimensional category of the face cap and different suction force. The developed models are the basis for the design of each element of the device settings and for determining the critical places of the entire system due to the adhesion of viruses and other microorganisms, and by successive iterations, it is possible to minimize their harmful significance. In addition, digitized engineering design includes material design and selection of appropriate materials in order to minimize the impact of critical points (e.g., pipeline kinks), designing zones throughout the suction line of the device, where it is necessary to make appropriate hydrophobic nanostructured layers Al_2O_3 and/or TiO_2 to prevent the adhesion of viruses and bacteria by the atomic layer deposition (ALD) method [21] and coatings with ZnO_2/ZnO quantum dots with a diameter of approx. $3 \div 5$ nm, functionalized hydrocarbon chains, e.g., $C_{10}H_{22}$ dispersed in an acrylate carrier, and virtual verification of selected materials, applied by immersion.

The described manufacturing method was developed on the basis of computer-processed data obtained with the use of CAD technology and Computer-Aided Manufacturing (CAM) using the AM (additive manufacturing) technology. The use of machining in the Computerized Numerical Control (CNC) Robodril CNC milling center by Fanuc and the ladle polymerization techniques of photopolymers, especially the stereolithographic SLS method, were analyzed, as well as others, including cDLM (Continuous Digital Light Manufacturing), DPP (Daylight Polymer Printing), DLP (Digital Light Printing) generally referred to as DLP (Figure 13, Table 3).

The material consumption was analyzed by comparing the manufacturing of device elements by machining from polymer discs with a diameter of 98.2 mm, because such holders are available in the machine tool and SLA manufacturing (Figure 13). It has been shown that the material savings ratio in the case of using additive methods is 1: $16 \div 20$.

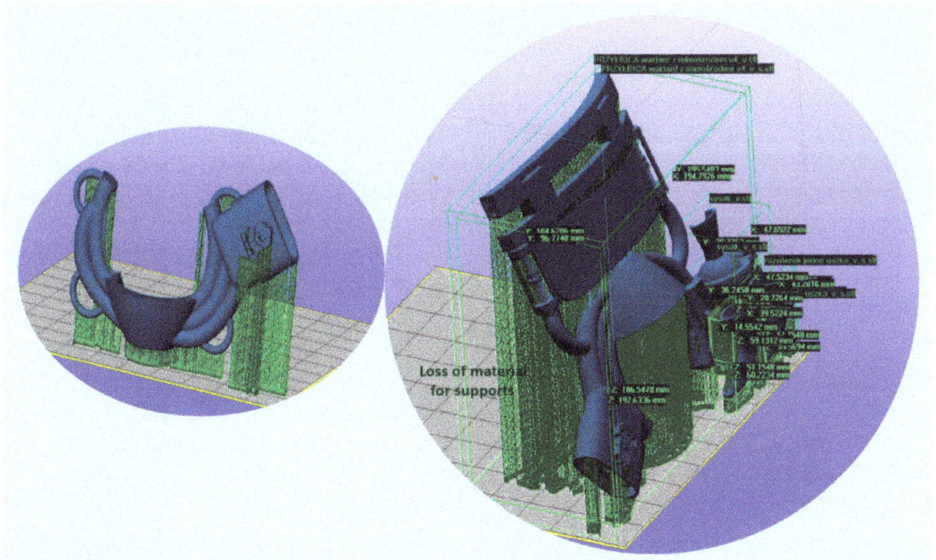

Figure 13. Diagram of the production of basic and advanced face caps using the stereolithography method.

Manufacturing with the additive method requires the creation of a virtual model showing the size, shape, and geometric features of individual elements in the STL format using appropriate CAD software, which is the basis for the implementation of a virtual technological model of these elements using the method of repeating base cells. The prepared model of all elements of the device is successively transferred to the CAM software, which is the basis for controlling the technological machine, where each element of the device is sequentially manufactured. Depending on the type of the element and the material used for this element, the SLA stereolithography method, and more preferably DLP, is used, since no manufacturing precision is required, and the production time can be significantly reduced by the DLP method. SLS (M) Selective Laser Sintering (Melting) or Electron Beam Melting (EBM) is used for a small number of metal components, and holes and threads are produced by machining in a CNC milling center when required CNC with the use of innovative additive technology aided by computer design of materials ADD-MAT [22]. After completion of the manufacturing stage, etching procedures are carried out, shaping the desired structure (texture) of the surface of the elements, and also applied surface treatment, e.g., by applying the ALD method, and in some places by immersion of coatings with quantum dots [21], followed by a sterilization procedure. The components of the device require sterilization prior to treatment and can be used repeatedly for various treatments, provided that each time is sterilized before the procedure.

Table 3. Analysis of the consumption of polymer/photopolymer materials in the case of the production of device elements by machining from polymer discs and the SLA method from light-cured polymers.

Version	Model Volume [cm^3]	SLA Approximate Material Loss (Supports) [cm^3]	Number of Discs Used for Details Manufacturing	Volume of the Working Space of the Disc [cm^3]	Total Volume of Disks [cm^3]	CNC Approximate Material Loss [cm^3]
basic	72.15	1.97	9	177.12	1594.04	1521.89
expanded	150.35	48.4	19		3365.2	3214.84

6. Recapitulation

The world is in a very strange place. We could find a few songs that describe such a situation, although none of their authors imagined that they could correspond to such a global scale and such complex problems. One can mention the English version of Czesław Niemen's song "Strange Is This World" written in Polish in 1967 and in English in 1972. Suddenly, within a matter of days, normal development was interrupted in many areas and in 218 countries and territories. This applies to private matters of each of the 7,802,780,000 inhabitants of the Earth (this is exactly how many of us were on 5 August 2020, at 4:17 AM) [23], but also to entire societies and many areas of life: culture, education, science, economy, including industry … A man should learn humility towards the majesty of Nature, even if his mistakes synergistically enhanced the highly negative effect.

Recapitulating activities identified by the subject of this paper and the thoughts contained in it can be presented on several points as follows:

I.   In the case that is dear to the authors, i.e., dentistry and dental engineering, the principle of "stay at home" quickly turned into another "do nothing", because contacting the patient you can contract a deadly disease [24–28]. The Hippocratic Oath and ethics …? [29]. We quickly developed the authors' SPEC strategy, which involves activity, not passive waiting, and hitting the heart of the problem to remove the cause that makes our life difficult. Of course, we know that the solution at the root of this problem is to invent the right vaccine. On 14 August 2020, the European Union concluded an agreement with AstraZeneca, cooperating with the University of Oxford, to purchase 300 million doses of the vaccine. The AstraZeneca vaccine is already in Phase II/III of large-scale clinical trials. Talks are also being held with other vaccine manufacturers, i.e., Sanofi-GSK, Johnson & Johnson, CureVac, and Moderna [30]. U.S. government placed a similar agreement with Pfizer and Biontech and on 22 July 2020, and submitted an initial order of 100 million doses and can acquire up to 500 million additional doses [31]. But we still do not know when these vaccines will actually be delivered to recipients. We have to wait until this process ends positively.

II. We want to help people by treating their sick teeth and removing the effects of these diseases, which can have a very complex impact on the health of many people. There is a high risk of complications, not only local but also systemic, resulting from pulp diseases [32]. Bacterial endocarditis, coronary artery disease, cerebral abscess, stroke, rheumatoid arthritis, osteoporosis, nephritis, pneumonia, and even preterm labor can be caused by an oral disease primary lesion [33–39]. In a favorable situation, caries and periodontal diseases may directly contribute to the patient's death.

III. It is obvious that the lack of teeth causes serious diseases [40–50]. It is possible to include chronic inflammatory changes gastric mucosa, upper gastrointestinal and pancreatic cancer, and an increased incidence of duodenal ulcers apart from aesthetic considerations. Tooth loss causes malocclusion, which affects the metabolism of the cerebrospinal fluid that nourishes the brain. The consequence may be headaches, migraines, pains in the cervical spine, and mood disorders, and also obstructive sleep apnea and neurocognitive dysfunction. Completely edentulous patients are more prone to electrocardiographic abnormalities, hypertension, heart failure, ischemic heart disease, stroke, aortic sclerosis, and coronary plaque formation. Toothlessness is a predictor of cardiovascular mortality. It may also cause rheumatoid arthritis, chronic kidney disease, non-insulin-dependent diabetes mellitus, and some cancers. A higher number of illnesses comorbid on the toothlessness are existing.

IV. The consequence of caries and periodontal diseases is the necessity to insert implants and prosthetic restorations, including bridges and crowns. It allows us to alleviate the effects of these diseases, and even effectively eliminate some of them. Improving the bite may have a positive effect on reducing the symptoms of multiple sclerosis. The restoring the dentition in the elderly has a positive effect on the hippocampus in the temporal lobe of the cerebral forebrain, responsible, among others, for spatial memory and for creating and recreating episodic memories. The systemic effect of masticatory disorder is an epidemiological risk factor for dementia. Toothlessness is often the direct cause of shortening human life. Chewing dysfunction causes morphological and functional changes in the hippocampus while chewing helps to maintain the proper functions of the hippocampus [51–57].

V.   The coming of the SARS-CoV-2 virus blocked almost all dental treatment activities. Most dental procedures in almost the entire world have been abandoned for many months [18–22], although starting from June/July 2020 adequately, the initial restrictions have been slightly relaxed in various countries, but not entirely waived. The result is an obvious deterioration in dental welfare and an exacerbation of the negative effects of all untreated oral diseases and the systemic comorbidities and diseases that are caused thereby, as described above. This is the conflict that is signaled in the title of this paper. On the other hand, it was its occurrence that forced the authors to be active in this area. Most likely, had it not been for this, they would never have addressed the matter. That is why it was decided that, despite the fact that this conflict arose, it is non-antagonistic because it mobilized the authors to act positively. The sooner the reasons underlying the decision to treat only emergency cases in dentistry are resolved, the sooner proper dental care will be provided again. The authors are deeply convinced that the solution proposed in this paper serves this matter well.

VI. This paper presents a breakthrough authors solution, implemented by the active SPEC strategy, and aims to eliminate clinical aerosol at the source by negative pressure aspirating bioaerosol at the patient's mouth line. The comparative benchmarking analysis and its results show that only the proprietary solution with a set of devices eliminates the threat at the source, while the remaining known methods do not meet the expectations in this aspect. It was only possible to solve this problem thanks to progress in advanced digitized production. Please note that the essence of the problem concerns the correct selection of materials to be used in the manufacture of this device and materials processing technologies with the use of additive technologies and surface engineering.

VII. So the question arises whether, when this virus disappears and its pandemic will finally end, will our solution still be needed. That's it. This pandemic has taught the people something, Nature has let the people know about its enormous power and helplessness of people, although they consider themselves the masters of the situation on Earth. As we can see, it is not so. Viruses live with us and next to us. Every now and then, some epidemic breaks out, i.e., Spanish flu, avian flu, swine flu, cholera, Ebola, SARS 1 and 2, and flu every year [58,59]. They are diseases that take thousands of lives. Each disease transmitted by airborne droplets requires protection, e.g., the way is presented in this paper. This has to be done routinely because we never know who is infected.

VIII. Despite the tragic events that inspired this action, the general conclusion is optimistic. The authors have found a solution to the problem that seriously threatens us.

Author Contributions: Conceptualization, Literature review, Presentation way—L.A.D. and L.B.D., Design, Resources, Data Curation, Software, Formal Analysis, Practical verification, Writing—Original Draft Preparation, Visualization—L.A.D., L.B.D., A.D.D.-D., J.D., K.R., A.A.-F., Writing—Review & Editing, Supervision, Project Administration, Funding Acquisition—L.A.D. and L.B.D. All authors have read and agreed to the published version of the manuscript.

Funding: The paper is prepared due to the sending of the Project application no.: POIR.01.01.01-00-0637/20 cleanDENTworkspace—a medical aerosol absorption system generated during medical procedures in the patient's mouth during dental, ENT and general medical procedures in order to effectively protect personnel and patients against viruses (including SARS-CoV-2) and other microorganisms, 04.06.2020 and actually implemented Project POIR.01.01-00-0485/16-00 on 'IMSKA-MAT Innovative dental and maxillofacial implants manufactured using the innovative additive technology supported by computer-aided materials design ADD-MAT 'realized by the Medical and Dental Engineering Center for Research, Design and Production ASKLEPIOS in Gliwice, Poland. The project is implemented in 2017–2021 and is co-financed by the Operational Program Intelligent Development of the European Union.

Conflicts of Interest: The authors declare no conflict of interest.

Intention: Authors Leszek A. Dobrzański, Lech B. Dobrzański, and Anna Dobrzańska-Danikiewicz dedicated their participation in the preparation of this work to the memory of the Beloved Mother and Grandmother of the late Krystyna Stefania Dobrzańska née Kuczyńska on the hundredth anniversary of her birth on 27 September 1920.

References

1. Kagermann, H.; Wahlster, W.; Helbig, J. *Recommendations for Implementing the Strategic Initiative INDUSTRIE 4.0: Final Report of the Industrie 4.0 Working Group*; Federal Ministry of Education and Research: Bonn, Germany, 2013.

2. Rüßmann, M.; Lorenz, M.; Gerbert, P.; Waldner, M.; Justus, J.; Engel, P.; Harnisch, M. *Industry 4.0: The Future of Productivity and Growth in Manufacturing Industries*; Boston Consulting Group: Boston, MA, USA, 2015.

3. European Commision. *Commission Sets out Path to Digitise European Industry*; European Commission: Brussels, Belgium, 2016; Available online: http://web.archive.org/web/20200403145542/https://ec.europa.eu/commission/presscorner/detail/en/IP_16_1407 (accessed on 17 August 2020).

4. Dobrzański, L.A.; Dobrzańska-Danikiewicz, A.D. Why Are Carbon-Based Materials Important in Civilization Progress and Especially in the Industry 4.0 Stage of the Industrial Revolution. *Mater. Perform. Charact.* **2019**, *8*, 337–370. [CrossRef]

5. Dobrzański, L.A.; Dobrzańska-Danikiewicz, A.D. Applications of Laser Processing of Materials in Surface Engineering in the Industry 4.0 Stage of the Industrial Revolution. *Mater. Perform. Charact.* **2019**, *8*, 1091–1129. [CrossRef]

6. Dobrzański, L.A. Effect of Heat and Surface Treatment on the Structure and Properties of the Mg-Al-Zn-Mn Casting Alloys. In *Magnesium and Its Alloys*; CRC Press: Boca Raton, FL, USA, 2019; pp. 91–202.

7. Dobrzański, L.A.; Dobrzański, L.B. Approach to the Design and Manufacturing of Prosthetic Dental Restorations According to the Rules of Industry 4.0. *Mater. Perform. Charact.* **2020**, *9*, 394–476. [CrossRef]

8. Dobrzański, L.A.; Dobrzański, L.B. Dentistry 4.0 Concept in the Design and Manufacturing of Prosthetic Dental Restorations. *Processes* **2020**, *8*, 525. [CrossRef]

9. United Nations Industrial Development Organization. Industrial Development Report 2020. Industrializing in the Digital Age. 2019. Available online: https://www.unido.org/sites/default/files/files/2019-11/UNIDO_IDR2020-MainReport_overview.pdf (accessed on 17 August 2020).

10. CDC; Eckert, A.; Higgins, D. Available online: https://phil.cdc.gov/Details.aspx?pid=23312 (accessed on 17 August 2020).

11. WHO. Coronavirus Disease (COVID-19) Dashboard Data. Available online: https://covid19.who.int/ (accessed on 17 August 2020).

12. May, B. Research Briefing/Global Coronavirus Watch: Cautious pessimism warranted (Oxford Economic, 20 July 2020). Available online: http://blog.oxfordeconomics.com/coronavirus/cautious-pessimism-warranted (accessed on 17 August 2020).

13. Meng, L.; Hua, F.; Bian, Z. Coronavirus Disease 2019 (COVID-19): Emerging and Future Challenges for Dental and Oral Medicine. *J. Dent. Res.* **2020**, *99*, 481–487. [CrossRef]

14. Raport Amnesty International MJM, Pracownicy Medyczni na Całym Świecie są Uciszani, Narażani i Atakowani. Available online: https://amnesty.org.pl/pracownicy-medyczni-na-calym-swiecie-sa-uciszani-narazani-i-atakowani/ (accessed on 17 August 2020).

15. Van Doremalen, N.; Bushmaker, T.; Morris, D.H.; Holbrook, M.G.; Gamble, A.; Williamson, B.N.; Tamin, A.; Harcourt, J.L.; Thornburg, N.J.; Gerber, S.I.; et al. Aerosol and Surface Stability of SARS-CoV-2 as Compared with SARS-CoV-1. *N. Engl. J. Med.* **2020**, *382*. [CrossRef] [PubMed]

16. Fallahi, H.R.; Keyhan, S.O.; Zandian, D.; Kim, S.-G.; Cheshmi, B. Being a front-line dentist during the Covid-19 pandemic: a literature review. *Maxillofac. Plast. Reconstr. Surg.* **2020**, *42*, 1–9. [CrossRef] [PubMed]

17. Yonis, O.B.; Alyahya, M.; Khader, Y.S.; Al Nsour, M.; Al-Batayneh, O.B.; Saadeh, R.; Bashier, H.; Alfaqih, M.; Al-Azzam, S.; Alshurman, B.A. Dentists' Awareness, Perception, and Attitude Regarding COVID-19 and Infection Control: Cross-Sectional Study Among Jordanian Dentists. *JMIR Public Health Surveill.* **2020**, *6*, e18798. [CrossRef]

18. Dobrzański, L.B.; Dobrzański, L.A.; Dobrzańska, J.; Rudziarczyk, K.; Achtelik-Franczak, A. Akcesorium do ochrony osobistej personelu dentystycznego przed koronawirusem SARS-CoV-2 i innymi drobnoustrojami chorobotwórczymi. Zgłoszenie patentowe P.434391, Poland, data zgłoszenia: 19.06.2020 r.

19. Dobrzański, L.B.; Dobrzański, L.A.; Dobrzańska, J. CleanDENTworkspace—A medical aerosol absorption system generated during medical procedures in the patient's mouth during dental, ENT and general medical procedures in order to effectively protect personnel and patients against viruses (including SARS-cov-2) and other microorganisms (wniosek o dofinansowanie Projektu nr: POIR.01.01.01-00-0637/20, 04.06.2020 złożony do NCBiR, Warsaw, Poland).

20. Dobrzańska-Danikiewicz, A.D. *Księga Technologii Krytycznych Kształtowania Struktury i Własności Powierzchni Materiałów Inżynierskich*; International OCSCO World Press: Gliwice, Poland, 2013; pp. 1–823.

21. Dobrzański, L.A.; Dobrzańska-Danikiewicz, A.D. *Inżynieria Powierzchni Materiałów: Kompendium Wiedzy i Podręcznik Akademicki*; International OCSCO World Press: Gliwice, Poland, 2018; pp. 1–1138.

22. Dobrzański, L.A.; Dobrzański, L.B. *Innovative Dental and Maxillo-Facial Implant-Scaffold Manufactured Using the Innovative Technology and Additive Computer-Aided Materials Design. Medical and Dental Engineering Centre for Research*; Design, and Production: Gliwice, Poland, 2017–2021.

23. Available online: https://www.worldometers.info/watch/world-population/ (accessed on 5 August 2020).

24. Guidance for Dental Settings, Interim Infection Prevention and Control Guidance for Dental Settings during the Coronavirus Disease 2019 (COVID-19) Pandemic. Available online: https://www.cdc.gov/coronavirus/2019-ncov/hcp/dental-settings.html (accessed on 28 August 2020).

25. Management of Acute Dental Problems during COVID-19 Pandemic. Available online: https://www.sdcep.org.uk/published-guidance/acute-dental-problems-covid-19/ (accessed on 28 August 2020).

26. Dental Care and Coronavirus (COVID-19). Available online: https://www.dentalhealth.org/Pages/FAQs/Category/coronavirus (accessed on 28 August 2020).

27. Guidance on COVID-19, Guidance on Managing Infection Related Risks in Dental Services. Available online: https://www.fdiworlddental.org/sites/default/files/media/documents/guidance_on_managing_infection_related_risks_in_dental_services_0.pdf (accessed on 28 August 2020).

28. COVID-19 and Dental Practice; What Has Been Done in China? Available online: https://www.fdiworlddental.org/sites/default/files/media/documents/covid-19_and_dental_practice_what_has_been_done_in_china.pdf (accessed on 28 August 2020).

29. WMA Declaration of Geneva (adopted by the 2nd General Assembly of the World Medical Association, Geneva, Switzerland, September 1948 and amended by the 22nd World Medical Assembly, Sydney, Australia, August 1968 and the 35th World Medical Assembly, Venice, Italy, October 1983 and the 46th WMA General Assembly, Stockholm, Sweden, September 1994 and editorially revised by the 170th WMA Council Session, Divonne-les-Bains, France, May 2005 and the 173rd WMA Council Session, Divonne-les-Bains, France, May 2006 and amended by the 68th WMA General Assembly, Chicago, United States, October 2017). Available online: https://www.wma.net/policies-post/wma-declaration-of-geneva/ (accessed on 28 August 2020).

30. Convention for the Protection of Human Rights and Fundamental Freedoms, Rome. *Refug. Surv. Q.* **2005**, *24*, 147–148. [CrossRef]

31. Coronavirus: Commission Reaches First Agreement on a Potential Vaccine. Available online: https://ec.europa.eu/commission/presscorner/detail/en/ip_20_1438 (accessed on 28 August 2020).

32. Pfizer and Biontech Announce an Agreement with U.S. Government for up to 600 Million Doses of Mrna-Based Vaccine Candidate against sars-cov-2. Available online: https://www.pfizer.com/news/press-release/press-release-detail/pfizer-and-biontech-announce-agreement-us-government-600 (accessed on 28 August 2020).

33. Al-Nawas, B.; Maeurer, M. Severe versus Local Odontogenic Bacterial Infections: Comparison of Microbial Isolates. *Eur. Surg. Res.* **2007**, *40*, 220–224. [CrossRef]

34. Mueller, A.; Saldamli, B.; Stübinger, S.; Walter, C.; Flückiger, U.; Merlo, A.; Schwenzer-Zimmerer, K.; Zeilhofer, H.-F.; Zimmerer, S. Oral bacterial cultures in nontraumatic brain abscesses: Results of a first-line study. *Oral Surg. Oral Med. Oral Pathol. Oral Radiol. Endodontol.* **2009**, *107*, 469–476. [CrossRef] [PubMed]

35. Li, X.; Tronstad, L.; Olsen, I. Brain abscesses caused by oral infection. *Dent. Traumatol.* **1999**, *15*, 95–101. [CrossRef] [PubMed]

36. Scannapieco, F.A. Role of Oral Bacteria in Respiratory Infection. *J. Periodontol.* **1999**, *70*, 793–802. [CrossRef]

37. Scannapieco, F.A.; Bush, R.B.; Paju, S. Associations between Periodontal Disease and Risk for Nosocomial Bacterial Pneumonia and Chronic Obstructive Pulmonary Disease. A Systematic Review. *Ann. Periodontol.* **2003**, *8*, 54–69. [CrossRef]

38. Alexander, M.; Krishnan, B.; Shenoy, N. Diabetes mellitus and odontogenic infections—an exaggerated risk? *Oral Maxillofac. Surg.* **2008**, *12*, 129–130. [CrossRef] [PubMed]

39. López, N.J.; Smith, P.C.; Gutiérrez, J. Periodontal Therapy May Reduce the Risk of Preterm Low Birth Weight in Women with Peridotal Disease: A randomized Controlled Trial. *J. Periodontol.* **2002**, *73*, 911–924. [CrossRef] [PubMed]

40. Pallasch, T.J.; Wahl, M.J. Focal infection: New age or ancient history? *Endod. Top.* **2003**, *4*, 32–45. [CrossRef]

41. Burzyńska, B.; Mierzwińska-Nastalska, E. Rehabilitacja protetyczna pacjentów bezzębnych. *Nowa Stomatol.* **2011**, *4*, 167–169.

42. Felton, D.A. Edentulism and Comorbid Factors. *J. Prosthodont.* **2009**, *18*, 88–96. [CrossRef]

43. Felton, D.A. Complete Edentulism and Comorbid Diseases: An Update. *J. Prosthodont.* **2015**, *25*, 5–20. [CrossRef]

44. Abnet, C.C.; Qiao, Y.-L.; Dawsey, S.M.; Dong, Z.-W.; Taylor, P.R.; Mark, S.D. Tooth loss is associated with increased risk of total death and death from upper gastrointestinal cancer, heart disease, and stroke in a Chinese population-based cohort. *Int. J. Epidemiol.* **2005**, *34*, 467–474. [CrossRef] [PubMed]

45. Sierpinska, T.; Golebiewska, M.; Dlugosz, J.; Kemona, A.; Laszewicz, W. Connection between masticatory efficiency and pathomorphologic changes in gastric mucosa. *Quintessence Int.* **2007**, *38*, 31–37. [PubMed]

46. Bagchi, S.; Tripathi, A.; Tripathi, S.; Kar, S.; Tiwari, S.C.; Singh, J. Obstructive Sleep Apnea and Neurocognitive Dysfunction in Edentulous Patients. *J. Prosthodont.* **2018**, *28*, e837–e842. [CrossRef] [PubMed]

47. Völzke, H.; Schwahn, C.; Hummel, A.; Wolff, B.; Kleine, V.; Robinson, D.M.; Dahm, J.B.; Felix, S.B.; John, U.; Kocher, T. Tooth loss is independently associated with the risk of acquired aortic valve sclerosis. *Am. Hear. J.* **2005**, *150*, 1198–1203. [CrossRef]

48. Takata, Y.; Ansai, T.; Matsumura, K.; Awano, S.; Hamasaki, T.; Sonoki, K.; Kusaba, A.; Akifusa, S.; Takehara, T. Relationship between tooth loss and electrocardiographic abnormalities in octogenarians. *J. Dent. Res.* **2001**, *80*, 1648–1652. [CrossRef]

49. Holmlund, A.; Holm, G.; Lind, L. Number of Teeth as a Predictor of Cardiovascular Mortality in a Cohort of 7674 Subjects Followed for 12 Years. *J. Periodontol.* **2010**, *81*, 870–876. [CrossRef]

50. De Pablo, P.; Dietrich, T.; McAlindon, T.E. Association of periodontal disease and tooth loss with rheumatoid arthritis in the US population. *J. Rheumatol.* **2007**, *35*, 70–76.
51. Chen, H.; Iinuma, M.; Onozuka, M.; Kubo, K.-Y. Chewing Maintains Hippocampus-Dependent Cognitive Function. *Int. J. Med Sci.* **2015**, *12*, 502–509. [CrossRef]
52. Lexomboon, D.; Trulsson, M.; Wårdh, I.; Parker, M.G. Chewing Ability and Tooth Loss: Association with Cognitive Impairment in an Elderly Population Study. *J. Am. Geriatr. Soc.* **2012**, *60*, 1951–1956. [CrossRef]
53. Stein, P.S.; Desrosiers, M.; Donegan, S.J.; Yepes, J.F.; Kryscio, R.J. Tooth loss, dementia and neuropathology in the Nun Study. *J. Am. Dent. Assoc.* **2007**, *138*, 1314–1322. [CrossRef]
54. Henke, K. A model for memory systems based on processing modes rather than consciousness. *Nat. Rev. Neurosci.* **2010**, *11*, 523–532. [CrossRef] [PubMed]
55. Kawahata, M.; Ono, Y.; Ohno, A.; Kawamoto, S.; Kimoto, K.; Onozuka, M. Loss of molars early in life develops behavioral lateralization and impairs hippocampus-dependent recognition memory. *BMC Neurosci.* **2014**, *15*, 4. [CrossRef] [PubMed]
56. Onishi, M.; Iinuma, M.; Tamura, Y.; Kubo, K.-Y. Learning deficits and suppression of the cell proliferation in the hippocampal dentate gyrus of offspring are attenuated by maternal chewing during prenatal stress. *Neurosci. Lett.* **2014**, *560*, 77–80. [CrossRef] [PubMed]
57. Hirano, Y.; Obata, T.; Takahashi, H.; Tachibana, A.; Kuroiwa, D.; Takahahi, T.; Ikehira, H.; Onozuka, M. Effects of chewing on cognitive processing speed. *Brain Cogn.* **2013**, *81*, 376–381. [CrossRef] [PubMed]
58. Procyk-Lewandowska, I. Historia Pandemii na Świecie—Koronawirus SARS-CoV-2 na tle Innych Pandemii. Available online: https://www.medicover.pl/o-zdrowiu/historia-pandemii-na-swiecie-koronawirus-sars-cov-2-na-tle-innych-pandemii,6788,n,168 (accessed on 28 August 2020).
59. Hays, J.N. *Epidemics and Pandemics: Their Impacts on Human History*; ABC-CLIO: Santa Barbara, CA, USA; Denver, CO, USA; London, UK, 2005.

Review

Dentistry 4.0 Concept in the Design and Manufacturing of Prosthetic Dental Restorations

Leszek A. Dobrzański * and Lech B. Dobrzański *

Medical and Dental Engineering Centre for Research, Design and Production ASKLEPIOS,
13 D Krolowej Bony St., 44-100 Gliwice, Poland
* Correspondence: leszek.dobrzanski@centrumasklepios.pl (L.A.D.); dobrzanski@centrumasklepios.pl (L.B.D.);
Tel.: +48-606-335-566 (L.A.D.); +48-570-470-000 (L.B.D.)

Received: 18 March 2020; Accepted: 17 April 2020; Published: 29 April 2020

Abstract: The paper is a comprehensive but compact review of the literature on the state of illnesses of the human stomatognathic system, related consequences in the form of dental deficiencies, and the resulting need for prosthetic treatment. Types of prosthetic restorations, including implants, as well as new classes of implantable devices called implant-scaffolds with a porous part integrated with a solid core, as well as biological engineering materials with the use of living cells, have been characterized. A review of works on current trends in the technical development of dental prosthetics aiding, called Dentistry 4.0, analogous to the concept of the highest stage of Industry 4.0 of the industrial revolution, has been presented. Authors' own augmented holistic model of Industry 4.0 has been developed and presented. The studies on the significance of cone-beam computed tomography (CBCT) in planning prosthetic treatment, as well as in the design and manufacture of prosthetic restorations, have been described. The presented and fully digital approach is a radical turnaround in both clinical procedures and the technologies of implant preparation using computer-aided design and manufacturing methods (CAD/CAM) and additive manufacturing (AM) technologies, including selective laser sintering (SLS). The authors' research illustrates the practical application of the Dentistry 4.0 approach for several types of prosthetic restorations. The development process of the modern approach is being observed all over the world. The use of the principles of the augmented holistic model of Industry 4.0 in advanced dental engineering indicates a change in the traditional relationship between a dentist and a dental engineer. The overall conclusion demonstrates that it is inevitable and extremely beneficial to implement the idea of Dentistry 4.0 following the assumptions of the authors' own, holistic Industry 4.0 model.

Keywords: stomatognathic system; prosthetic restorations; surgical guide; dental prosthesis restoration manufacturing center; CBCT tomography; dental implants; implant-scaffolds; hybrid multilayer biological-engineering composites biomaterials; CAD/CAM methods; additive manufacturing technologies; selective laser sintering; stereolithography; Dentistry 4.0; Industry 4.0

---

1. Introduction

Modern medicine uses numerous engineering solutions strictly [1–3]. It applies to many aspects, and very often, the production of devices replaces the body's natural functions and even some of its components. Biomaterials, also called biomedical or bioengineering materials, are used for this. An interest in the group of materials and the possibilities of their use in dentistry is the subject of this paper. Biomaterials are used to produce implants and other prosthetic restorations, for a long time, staying in direct contact with the tissues of the human body. The function of implants is to replace the hard and soft tissues of the body and reconstruct their natural functions (Table 1). They have artificially manufactured medical devices placed entirely or partially under the epithelial surface and remaining in the body for a more extended period to perform the mentioned functions [4–12]. Implants are made of one or more biomaterials. A separate group of biomaterials is surface layers on other materials used to manufacture implantable devices for the human body. Biocompatibility and non-hemolysis tendency, if the predictions of the implantable device with blood, are the most important properties of biomaterials. Such requirements are met by various materials, including different metal alloys, ceramic, and polymeric materials [4–14].

Table 1. The general classification of implants due to applications in various organs.

No.	Type of Implant or Function	Therapeutic Functions
1.	Implants of the stomatognathic system	to replace, improve, or increase any soft or hard tissue in the mouth of a jaw, mandible, or temporomandibular joint
2.	Dental implants	replacement of lost teeth and restoration of the possibility of prosthetic reconstruction in this area
3.	Craniofacial implants	to improve or replace hard tissues in the craniofacial area except for the brain, eyes, and inner ear
4.	Orthopedic implants	to support, substitute, or supplement permanently or temporarily bones, cartilage, ligaments, tendons, or related tissues
5.	Implants of the cardiovascular system	in the treatment of disorders of the heart, its valves, and the rest of the circulatory system
6.	Sensory and neurological implants	used in disorders affecting the primary senses and the brain, in the treatment of visual impairment, with hearing loss and diseases, neurological diseases
7.	Electrical implants	used to relieve pain and suffering from rheumatoid arthritis
8.	Cosmetic implants	restoring some body parts to an acceptable asthetic standard
9.	Contraceptive implants	used to prevent unintended pregnancy and to treat conditions, such as non-pathological bleeding
10.	Other implants	removal of dysfunctions of different organs of the body, including those in the digestive, respiratory, and urological systems

A dental prosthetic is a part of dentistry related to the restoration of the physiological shape and function of teeth along with maintaining appropriate occlusal conditions, enabling the appropriate grinding of food. Prosthetic restorations can refer to both prosthetic restorations based solely on own teeth, not restoring lost teeth (crowns, veneers, inlay/onlay), as well as restoring lost teeth (prosthetic bridges, skeletal, and acrylate prostheses). It is also possible to restore lost teeth as a result of extraction by performing prosthetic implant reconstruction based only on implants without disturbing the neighboring teeth. A unique case is a toothless reconstruction, which requires reconstruction of the correct relation between the lower and upper arch before proceeding with the prosthetic restoration. Modern dental engineering allows the use of advanced, accurate, and repeatable technologies, such as subtractive manufacturing, including machining with the use of numerically controlled centers, additive manufacturing, including selective laser sintering (SLS) for the manufacturing of prosthetic restorations designed using software of computer-aided design/manufacturing (CAD/CAM).

To ensure the highest quality of services are also provided for patients who, without the use of the latest technologies, are condemned to basic prosthetic solutions, such as settling dentures, the normal practice of dentists is in the area of interest of which dental prosthetics is daily in close cooperation with dental engineers that are having access to the latest achievements in engineering in prosthetics and dental implantology. A broad scope of engineering aiding in contemporary dentistry is associated with the use of the achievements of material engineering, manufacturing engineering, and tissue engineering, as part of the current stage of the Industrial 4.0 of the industrial revolution [1,15–82]. The industrial revolution began with the invention of the steam machine in the Industry 1.0 stage, followed by electricity and a production line, then automation and computerization, and now cyber-physical systems have been introduced. It is inextricably linked to the introduction of the analogous approach of Materials 4.0 [1,20,22,24,37,78–90]. Idea Industry 4.0 is a clamp connecting those issues, although the currently developed model has met with criticism as incomplete. Therefore, the authors of this paper have developed the augmented holistic Industry 4.0 model. In the technological plane of the model, the current one appears after the necessary additions as one of the four complementary components [1,20,22,37,78–81,91,92]. As part of the approach, digitization and computerization in dentistry, named as Dentistry 4.0, are becoming more common [1,93,94]. This paper addresses those issues and their practical application. It presents selected results of the authors' own studies and technological and clinical research, as well as practical examples of the use of biomaterials and advanced digitized technologies to solve various clinical problems [1,20,22,37,78–81,91,92].

The purpose of the paper is to indicate the premises and evidence for the inevitability of the application of the Dentistry 4.0 model in dental engineering and clinical practice following the extended holistic Industry 4.0 model corresponding to the current stage of development. It is expected to ensure very high-quality standards and tight dimensional tolerances of the prosthetic restorations. It is achieved by appropriate material design and, at the same time, correct technological design with the right structural design of the analyzed prosthetic restorations and implants, as well as the consistent implementation of technical assumptions in practice. It is the main focus of this paper.

2. General Characteristics of Diseases in the Human Stomatognathic System and Its Prosthetic Treatment

This paper deals with the design and production of prosthetic restorations under the Dentistry 4.0 approach, the adaptation of the augmented holistic Industry 4.0 model in the conditions of a micro-enterprise cooperating with a network of dental clinics, usually private. Therefore, an analysis of the condition in the human stomatognathic system, leading to partial or complete loss of natural dentition and the possibility of prosthetic treatment, has been made.

The stomatognathic system includes soft and hard tissues of the mouth and surrounding areas. Prosthetics of the system involves the introduction of artificial devices and substances to replace the mentioned tissues with dental prosthetic restorations. It ensures that the system will continue to function, while not only improving the patient's appearance but also his comfort and health. The scope of dental prosthetics includes not only the restoration of the proper functioning of the stomatognathic system but also prophylaxis [4].

Prosthetic treatment may be necessary for several reasons. It is most often associated with damage or loss of teeth. It may be a consequence of the tooth tissue abrasion as a result of bruxism, the need to strengthen the crown on previously treated root canal teeth in which degrading lesions, periodontal diseases, as well as caries and changes caused by it, require the implementation of restoration a prosthetic tooth crown, both in the form of a prosthetic crown and partial restorations, such as onlay, inlay, or veneer, and ultimately also single or multiple missing teeth or completely toothless. Dental deficiencies can be congenital, including due to a lack of tooth buds. However, the causes of missing teeth are most often acquired. The first case is extractions of teeth, which are directly caused by diseases in the oral cavity, such as periodontal disease. However, the leading cause of dental extractions is caries. The results of epidemiological studies illustrating examples of the extent of the disease and the causes of dental extractions caused by it are collated in publications [1,3] and are presented in Tables 2–5.

Table 2. Selected epidemiological data on the causes of dental extractions.

Country	Number of Teeth Examined	Reason for Dental Extraction, %			Literature
		Caries	Periodontal Disease	Remaining	
Brazil	404	70.3	15.1	14.6	[95]
Italy	1056	33.4	33.1	33.5	[13]
Japan	11175	55.4	38	6.6	[96]
Kuwait	2783	43.7	37.4	18.9	[97]
Scotland	917	51	21	28	[98]

Caries is a bacterial infectious disease of the hard tissues of a tooth. Bacteria that cause the development of dentin disease include anaerobes, Gram-negative bacteria, and streptococci [99]. As a result of tooth tissue degradation and reaching the pulp, it causes an inflammatory process and, finally, necrosis. Low oxidoreductive potential and low oxygen levels promote colonization by subsequent anaerobic bacteria [100], intensifying the severity of the disease. Those bacteria produce acids in the plaque, which is the result of intra- and extracorporeal metabolism of sugars. It is a reason for the demineralization of inorganic substances and the subsequent proteolysis of organic substances. The course of such a pathological process affects the successive decalcification and breakdown of the hard tissues of the tooth, which results in inflammatory changes and irreversible destruction of the periodontal tissue, leading to loss of dentition [97,101,102]. At present, caries is a human disease in many countries around the entire globe [13,95–111]. In addition to local lesions in the stomatognathic system, caries also causes numerous systemic complications in many distant organs. Caries affects most adults and 60–90% of schoolchildren [110,112] in most countries. For example, in Poland, with a high share of tooth decay [102,113], only 0.1% of adults aged 35–44 and only 19.3% of 12-year-old children are free from caries [111,112] (Table 3).

Table 3. Selected data on the occurrence of caries in the population of developmental age.

Data on Tooth Decay	Age of Children, Years				Literature
	2–3	3–4	6–7	12	
Share of caries in the population, %	35–50	56–60	85–100	90	[102,111,113]
Share in the caries-free population, %	no data	no data	14	19.3	[111,112]

The level of caries severity intensifies when the diet of a given society has an excess of purified sugars [97,102,109–111]. It is considered the most common and expensive chronic disease. Caries is a human disease, and the scale of its prevalence depends, among others, on the economic situation and the general level of culture in society and may also be the result of many years of systemic neglect in the organization of health care in a given country [97,109,111,114,115]. Minimizing the impact of caries bacteria on teeth requires strict observance of the principles of oral hygiene and proper prevention, as well as early detection of tooth decay initiation and periodontal disease [97,114,116]. Teeth with irreversible pulpitis and pulp necrosis are now effectively left in the mouth for many years after modern endodontic treatment using liquid gutta-percha [117,118]. Peri-apical lesions, such as granulomas or cysts, may develop in treated endodontic teeth [119]; however, according to statistical studies, it only applies to 3% of cases 8 years after treatment [120–123] (Table 4).

Table 4. Selected epidemiological data regarding the time of endodontic treatment of teeth in the oral cavity.

Country	Time After Which the Test Was Performed, Years	Number of Teeth Covered by the Test	Share of Teeth Retained in the Mouth, %	Share of Removed Teeth, %	Literature
USA	3.5	110,766	94.44	5.56	[121]
Taiwan	5	1,446,199	92.9	7.1	[123]
USA	8	1,462,936	97	3	[120]
Dania	17	845	55	45	[122]

Table 5, on the other hand, indicates epidemiological data on the number of teeth treated endodontically with periapical lesions revealed in pantomographic images [119–133]. Of the 3% of failures, most were revealed within the first three years of discontinuation of treatment and resulted from improperly developed and inaccurately filled root canals [119–126], which concerns 91% of teeth treated endodontically with revealed periapical lesions [119].

Table 5. Selected epidemiological data on the number of teeth treated endodontically with periapical lesions revealed in pantomographic images.

Country	Total Number of Teeth Assessed Radiologically	The Number of Teeth Treated Endodontically	The Share of Teeth with Revealed Endodontic Treatment, %	The Share of Teeth with Revealed Periapical Lesion, %	Literature
Australia	5647	499	8.84	21.43	[131]
Belgium	4617	314	6.8	40.4	[133]
Brazil	5008	553	11.04	42.5	[130]
Denmark	15,984	773	4.84	52.2	[124]
Greece	7664	680	8.87	60	[128]
Iran	28,942	1013	3.5	52	[126]
Ireland	7427	148	1.99	25	[129]
Japan	16,232	3408	21	40	[125]
Lithuania	2186	283	12.95	43.1	[132]
Spain	4453	93	2.09	64.5	[127]
Turkey	8863	470	5.3	53.5	[119]

Numerous tooth extractions are very often a necessity in the treatment of tooth decay [13,95,96,98,103,104], in particular, if the patient is negligent in prevention, in the case of cancer or as a result of injuries during an accident [134] or if there are errors in treatment, tooth fractures after endodontic treatment. Other reasons for the failure of endodontic treatment include iatrogenic causes, such as root or crown fractures [120]. The endodontic treatment of teeth is statistically more often performed in women, and the least biting teeth of the jaw are subjected to it, while most often premolars and molars of the jaw [104] and molars of the mandible [126]. Figure 1 illustrates the extent and diversity of the most common dental defects of a patient. It presents as an example of the classical classification of upper and lower jaw dental defects, as well as left and right arches.

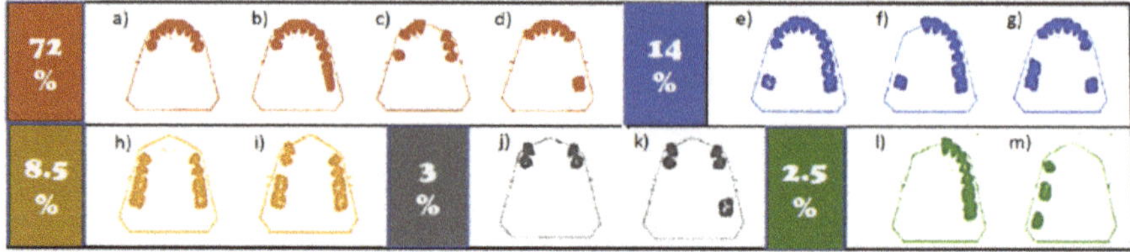

Figure 1. Classes I–IV of dental deficiencies together with the frequency of individual groups of digestive system diseases defects (on colored fields): I—two-sided areas of the toothless alveolar ridge located behind the patient's last teeth (bilateral absence of the wing), II—a one-side area of the toothless alveolar ridge located behind the patient's last teeth (unilateral lack of the arm), III—a one-sided area of the toothless alveolar ridge limited at the front and back of the patient's teeth (interdental deficiencies), IV—a single, two-sided, intersecting the median line area of the toothless alveolar ridge located forward from the patient's teeth (deficiencies front teeth); according to B—Bailyn, K—Kennedy, S—Skinner; (**a**) IK, IIB, IIIS; (**b**) IIK, IIB, IIIS; (**c**) Imed2K, IIB, IIIS; (**d**) IImed1K; (**e**) IIIK, IB, IS; (**f**) IIIK, IIIB, IS; (**g**) IIImed1K, IB, IS; (**h**) IVK, IIIB, IIS; (**i**) IIImed1K, IIIB, IIS; (**j**) Imed1K, IIB, IVS; (**k**) IImed2K, IIB, IVS; (**l**) IIK, IIB, VS; (**m**) IImed2K, IIB, VS.

The tooth as a result of conservative or endodontic treatment often requires prosthetic reconstruction due to the significant loss of hard tissues and the limitations of composite materials for direct reconstruction in the oral cavity. Teeth with living pulp are also subjected to prosthetic treatment in case of tooth wear and the need to increase the occlusion height, desire to improve the cosmetic effects, or the need to reproduce the missing tooth or teeth when creating a structure in the form of a prosthetic bridge. Tooth deficiencies from single to complete toothlessness are still prevalent and inevitable. Extractions of teeth temporarily solve problems but, at the same time, affect the formation of other issues related to the violation of the structure and balance of the stomatognathic system. It can function properly only if it ensures an even distribution of chewing forces on the teeth in all sections of the dental arches. Improving the health condition, and mainly food chewing conditions and treatment of associated digestive system diseases, elimination of frequent headaches caused by masticatory dysfunctions, as well as the desire to improve the appearance of patients' faces, requires the use of modern dental prosthetics and the use of appropriate biomaterials and appropriate technologies [3–12,79,135].

Prosthetic restorations are each time individually designed and manufactured in a way dedicated to each patient. Both alive and dead teeth after endodontic treatment are subjected to prosthetic treatment. The need for prosthetic reconstruction is often a consequence of endodontic or conservative treatment associated with significant loss of hard tissues and the limitations of tooth reconstruction using composite materials. The possibility of making a prosthetic bridge reproducing missing teeth is closely related to the presence of prosthetic pillars characterized by the required quality and correct placement. Those restorations can be based on the patient's own teeth or implants [136], but they should not be simultaneously attached to implants and their own teeth. Devices permanently attached indelibly in the mouth to stop the dysfunction of the system are generally classified as dental prostheses [6].

Out of the four groups (Table 6), they include permanent prosthetic restorations not removable from the oral cavity by the patient and forming a kind of unity with the living organism, including bridges, onlays, crown-root inlays, and crowns, as well as crowns and bridges based on implants screwed directly or indirectly through implant platforms or cemented to implantology abutments embedded in implants. The second group is mobile prosthetic restorations that can be removed by the patient from the mouth. Those include skeletal, acetal, acrylate, and nylon dentures based not only on own teeth and mucosa or exclusively on the mucosa but also on dentures based on implants on telescopes integrated with the implant connector or mounted on a beam screwed to the implants. The use of removable dentures based on teeth and mucosa or only on the mucosa is still necessary in some cases. However, it involves limited retention of those elements in the planned place during use and significant degradation of tissues affected by such a prosthesis. The financial aspect still determines its use. They are also performed for medical reasons in patients with a tooth system that is not indicated for the introduction of bridges based on their own teeth, and bone base reconstruction using a bone substitute or autograft or both of these materials is not possible at the same time, or it does not give satisfactory results. Also, during such treatment, long-term, temporary prostheses are used. The third case is combined with prosthetic restorations. Then, removable dentures and permanent restorations, which is the case with precision dentures, are connected. The connecting elements are bolts and latches.

Table 6. The general classification of dental prostheses and prosthetic restorations.

Classification Criterion	Mounting Method	Type of Prosthesis/Prosthetic Restoration
Way deposition	On implants	Solid cement
		Permanent with the option of periodic withdrawal by the patient
		Permanent with periodic download capability only by a dentist
		Movable on own teeth based on retention elements and the mucosa
	On own teeth	Permanently cemented on own teeth
		Fixed on own teeth with the possibility of periodic extraction
		Movable on own teeth based on retention elements
	On the mucosa	Mobile acrylic prostheses
	On the mucosa and own teeth	Mucosal (settling)
The method of transferring chewing force to the substrate	On implants	Implantology
		Implantology-mucosal
	On own teeth	Periodontal (non-settling)
		Periodontal-mucosa
	On the mucous	Mucosa (settling)
Period of using		Short-term use
		Long-term use

The last case is implantology hybrid solutions, combining the advantages of withdrawn prosthetic restorations with the perfect mounting of permanent prosthetic restorations. Such a type of prosthetic restoration includes prosthetic implant bridges based on telescopic crowns. The prosthetic restoration can be removed by the patient for hygienic procedures once a week or even once a month. Such a solution prevents bone base degradation, which can occur in the case of permanent prosthetic restorations and, in particular, in patients with periodontal disease. An example of such a prosthetic restoration is the design of the authors' titanium hygienic bridge, which combines the advantages of a titanium prosthetic restoration core with the option of its withdrawal.

The extent of the patient's dental defects determines the scope of prosthetic treatment and the choice of the type of prosthetic restoration (compare Figure 1). It indicates a large number of possibilities and diversity of the manufactured prosthetic restorations [1].

Implants of the stomatognathic system are used to supplement teeth defects and replace, improve, or increase any hard tissue of the jaw, mandible, or temporomandibular joint in the mouth [135]. For years, the most convenient shape of dental implants has been worked on, and their general classification is presented in Table 7.

Table 7. Classification of dental implants.

Clinical Conditions	Type	Surgical Interventions	Geometric Features
Partial lack of dentition/ Toothlessness	Intraosseous	Single-phase immediate	Individual implants
		Single-phase	Cylindrical with the smooth surface
		Two-phase immediate	
		Single-phase	Cylindrical with the helical surface
		Two-phase	
		Single-phase	Cylindrical with the helical surface with porous zones
		Two-phase	
		Single-phase immediate	Needle
			Bicortical screws
		Two-phase	Flat (razor blade)
	Through the bone	Two-phase	Yoke
Toothlessness	For periosteal	Single-phase	Individual
		Two-phase	

A correctly designed dental restoration meets biological, strength, and technological expectations. The materials used and the design features of the implant introduced to the body must, first and foremost, guarantee compliance with the bio tolerance requirements. An implant cannot initiate inflammation, nor can it promote chronic and acute immune responses and toxicological reactions, nor can it irritate adjacent tissues [13]. In such cases, various substances differing from medicines, or a mixture of various natural and synthetic substances, are classified as biomaterials, often also referred to as biomedical materials, which can replace or supplement living tissues in performing all or some of their functions [14]. Besides, depending on the intended use of implants or other dental restorations, dental biomaterials must meet different requirements [1]. Table 8 presents guidelines regarding the desired properties of implants and related requirements for the biomaterials used for them. However, in the case of a dental crown, the high compressive strength and high tribological properties are expected. Still, other expectations are concerning biomaterials used for orthopedic implants.

Table 8. General guidelines for the design of the implant.

Design Stages	Basic Conditions
Determining the geometrical features of the implant	Anatomical and physiological conditions
	Surgical techniques envisaged for usage
	Stress and displacement analysis in the implant-tissue system
	Analysis of the dimensional characteristics of the implants corresponding to the anthropometric characteristics of the patient population
Selection of implant material	Ensuring implant bio tolerance
	Obtaining the required mechanical properties of the implant
	Ensuring the desired technological properties during implant manufacturing
Development of documentation	Construction
	Technology
	Acceptance control

Implants must ensure a combination of natural body tissues, including in the mouth with artificially manufactured implants and prosthetic restorations. In addition to the requirements for the structure and mechanical properties and fulfillment of operational requirements, an essential criterion for the selection of biomaterials for individual implants and prosthetic restorations, including dental, is the diversity of technological and economic conditions, as well as the availability of clinical methods of varying levels, and crucial considerations, in this case, aesthetic. Practice indicates that the most frequently mentioned criteria in dental engineering fill foundry cobalt alloys, titanium and its alloys, and ceramics based on zirconia [83,137–147], and there are serious reasons to expand the canon with magnesium alloys.

Dental implants have gained a cylindrical shape since the 1960s and are screwed into the bone, according to the concept of P.I. Braenemark [5,7–12]. It was found out that it is the most advantageous of the possible connections. In 1965, titanium was used for dental implants, and the phenomenon of osseointegration confirmed at that time became a breakthrough and practically initiated the development of modern implantology. Osteointegration involves the fusion of implant surfaces with trabecular bone structures [5,7–12,148–156]. Starting from the 1980s, studies have shown that titanium oxide formed as a layer on the surface of titanium enables bone tissue regeneration by stimulating its growth and formation of epithelial attachment, which is a bone defensive reaction, guaranteeing implant fixation and stabilization [157–159]. The dental implantation procedure includes, in the first stage, examination, diagnostics, and preparation for implantation, in the second stage, implant placement, and the third stage consists of prosthetic loading of the implant. As part of the primary stabilization, a hole is made in the bone into which the implant is screwed, causing its immobilization with the assumed force. Ultimately, secondary stabilization by osseointegration takes place (Figure 2).

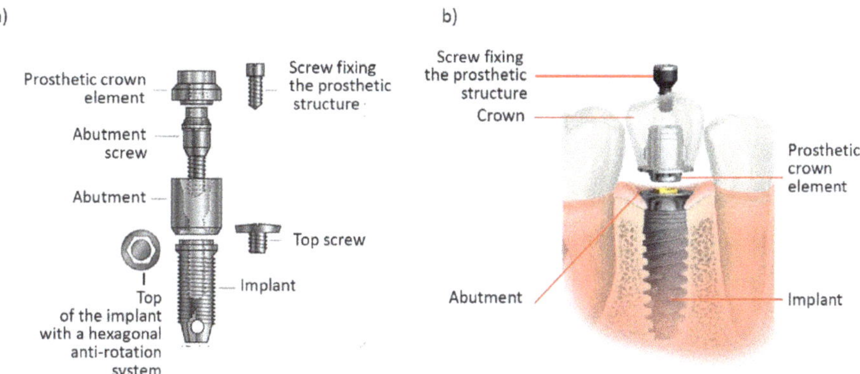

Figure 2. Examples of dental implants developed according to the Braenemark concept. (**a**) arrangement of the main elements; (**b**) view after installation in the bone.

In the first stage of the dental implantation procedure, inflammation occurs as a consequence of bone trauma as a result of surgical intervention. The second stage involves bone healing, starting from supplying blood through small blood vessels damaged during surgery to local osteonecrosis compacted to a depth of 1 mm along the entire length of the implant. The body is stimulated to differentiate stem cells into osteoblasts that stimulate bone healing and osteoid or bone matrix formation in newly created connective tissue. After about 2–3 days of the process, bone mineralization takes place over the next week, thanks to osteoblasts, so that after 4–6 weeks, there may be a repair bone, though with poor mineralization, but already connected to the patient's bone. The prosthetic restoration can take place after at least three months after the implant's placement. A temporary immediate reconstruction should be performed. It is also currently accepted that patients who meet stringent requirements can have direct implantation involving the removal of a tooth that does not have large periapical lesions and must be removed (e.g., as a result of root fracture, which often occurs after root canal treatment) and its replacement with an implant. A temporary reconstruction will be placed in the remaining alveolus, which, due to the risk of non-acceptance of the implant, must remain excluded from the occlusion.

Regardless of the widespread use of cylindrical implants with a helical surface, currently, several works are being carried out, including the own work of the authors of this article [81,91,92,160]. They deal with implants for whose manufacturing, it is necessary to use the cone-beam computed tomography (CBCT) method, to assess the state of deficiencies to patients' teeth and individual design of adequate individual prosthetic restorations. The widespread use of CBCT tomography allows planning prosthetic implant treatment and individual design of appropriate prosthetic restoration [91,92]. The technology of producing prosthetic restorations and engineering materials used for the purpose is systematically progressing. Regardless of the development of the digitization of dental diagnostics through the use of CBCT tomography, in the last decade, the role of computer-aided design/manufacturing CAD/CAM in dental engineering has been increasing [91,92,161–168].

Metal foundations made of cobalt–chromium alloys, titanium alloys, zirconium oxides, and aluminum oxides, as well as poly (methyl methacrylate) (PMMA), can be manufactured by milling in numerically controlled machining centers (CNC). Frameworks made of metal alloys are increasingly manufactured by additive methods. The digital prosthetic model is individually designed with the use of computer-aided design (CAD) and scan results of an intraoral scanner, under the supervision of the attending dentist. Such an approach allows planning the placement of the implants, ensuring entirely aesthetic reconstruction, with optimal use of the patient's bone base. Implantology abutments should not fit in the interdental spaces but the lumen of the prosthetic crown. It is necessary to use the surgical guide, developed by authors [92]. It allows the dentist to place the implants in places strictly designated for such a purpose. The authors' procedure was developed as a result of cooperation between a dentist and a dental engineer. According to the procedure, the implantation area requires a combination of a three-dimensional bone base model made with the use of a CBCT tomogram with a three-dimensional scan directly from the oral cavity or an impression scan used for a three-dimensional model of the dentition and soft tissues in the oral cavity. On this basis, the prosthetic restoration design is made, and the placement of the implants is planned. It involves planning technological holes to guide the pilot drill in such a way that holes are made in the bone in the right direction, which is achieved by creating a surgical guide using the stereolithographic (SLA) method. The guide is made of PMMA, and the pilot tunnels are filled with a titanium sleeve, produced by the selective laser sintering (SLS) method. It is a radical turnaround in the CAD/CAM design of the implant device system and in clinical procedures. On the other hand, the finite element method (FEM) is commonly used as a computational simulation tool for prosthetic restorations and enables the prediction of their behavior in application conditions [169–192].

Completion of the design of the prosthetic restoration by choosing the shape and material design by choosing, e.g., cobalt or titanium alloys, the numerical CNC milling machine performs machining of the appropriately selected material disc, when, in the case of choosing ceramic materials, e.g., ZrO_2, coloring paints and sintering are also applied using resistance or microwave ovens, followed by surface application and firing of facing ceramics, regardless of the material [91,92].

Computer-aided technologies for both design and manufacturing (CAD/CAM) are widely used in the field of additive manufacturing (AM) technologies. Among the full range of known additive technologies, mainly selective laser sintering (SLS) and stereolithography (SLA) are used in dental prosthetics [2,3,80,160,162,166,193–223]. The use of additive manufacturing technologies is unrivaled for the design and manufacturing of prosthetic restorations from solid and microporous materials compared to casting technologies, powder metallurgy, and the production of metallic foams. Selective laser sintering (SLS) is the only one that allows for a full mapping of the designed shape, which positively affects both the shape of the framework and then the entire prosthetic restoration, as well as its mass, which is crucial for the patient, especially in the case of extensive prosthetic restorations. Additionally, only by means of selective laser sintering, the porous skeletal structures of implants can be produced, e.g., teeth, but also bones from metal alloys with the required biocompatible properties can be produced [208].

Specifically, the relevant material properties include the absence of allergic and cytotoxic reactions in contact with living tissues, osteoconductivity, and osteoinductivity when the pore diameter is 50–500 µm, ensuring bone ingrowth, mechanical stabilization, the possibility of sterilization and sterility as elements of biological safety, resorptive and tissue substitution capacity with bone reconstruction, and the ability to fill the defect space [3,224,225].

The results of procedural benchmarking [85,226–229] and analysis of technology value using the weighted scoring method and dendrological matrix using a 10-stage universal unipolar positive scale without zero relative states create the conditions for the comparative assessment of individual technologies in the process of manufacturing porous materials [85,226–229]. The analysis perfectly objectifies the made estimations. Detailed attribution of weights to individual criteria of technology assessment in terms of their potential as a measure of the objective value and attractiveness indicated by customers and [230] has enabled to accurately compare those technologies in the interesting area of prosthetic restorations. The potential and attractiveness of technologies of additive manufacturing (TAM) with successively given coordinates of potential and attractiveness (8.6; 6.6) are much higher than other analyzed technologies. The determined coordinates ensure the location of the technology in the quarter-spreading oak of the said dendrological matrix, which means very favorable development prospects for those technologies. Other obtained technologies lower coordinates, that is why their development prospects in the analyzed area of application are not as desirable as technologies of powder metallurgy (TPM) (3.5; 5.0), technologies of casting (TC) (4.9; 4.4), and the technology of metallic foams (TMF) manufacturing (6.5; 4.3). The results of foresight studies [231,232] also confirm the very high development prospects of selective laser sintering (SLS) of metal and ceramic powders, which still belongs to embryonic, experimental, or prototype technologies [209,232–234] with very high potential and attractiveness. It justifies the excellent research interest in this issue concerning dental prosthetics.

It should be noted that, as in any case of sintering, among others, in powder metallurgy, sintering occurs in the liquid phase in 90% of the commercial value of sintered products [163] because mass transport in the liquid phase is hundreds of times faster than in the case of sintering in solids [163,235–240]. The liquid phase is formed under those conditions, and the liquid phase wets the particles of the remaining undissolved powders [209,223], inducing capillary forces [220,228], which affects the thickening and, as a result, the melting of a solid with high surface energy, even in relatively distant areas, and then their solidification [127]. Liquid phase sintering (LPS) includes coexisting liquid and solid during part or all of the sintering process [127]. The structure of sintered materials with a liquid phase consists of evenly distributed solid phase particles in the matrix of solidified liquid, including during SLS (Figure 3).

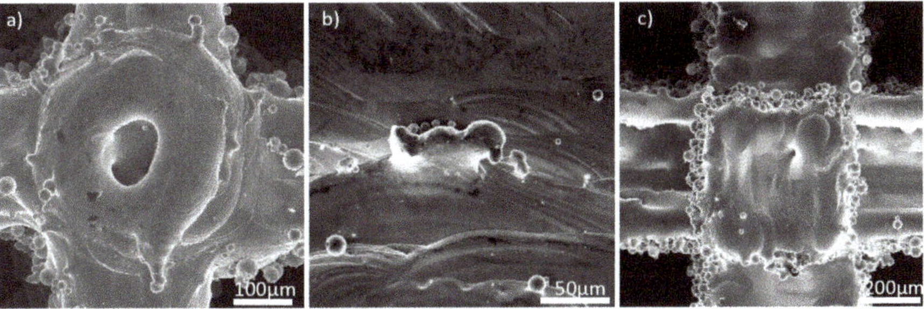

Figure 3. Structure of melted powders during selective laser sintering with the participation of the liquid phase of porous skeletons in a direction perpendicular to the working platform made of the following powders: (**a**,**b**) CoCr25W5Mo5Si alloy; (**c**) Ti6Al4V alloy (SEM).

Very modern technologies used in the manufacturing of implants and other prosthetic restorations include surface engineering technologies [211,241–243], e.g., atomic layers deposition ALD (Figure 4), and less preferably physical vapor deposition (PVD). The particular benefits of the use of those technologies in dental prosthetics include barrier prevention by applied surface layers of unfavorable greying of cover ceramics as a result of revival diffusion of titanium atoms and other metals to ceramic surfaces of crowns and bridges, and even cracking of the porcelain facing layer. In turn, inside the pores of the skeletal porous structure selectively sintered by laser, e.g., from Co–Cr alloys, titanium, or its alloys powders, thin layers, e.g., TiO_2, ZrO_2, or Al_2O_3, can be applied inside the pores (Figure 4), ensuring the proliferation and growth of living osteoblast cells. Studies using human cells from the hFOB 1.19 osteoblast culture line (Human ATCC-CRL-11372) American Type Culture Collection ATCC (USA) have been performed. In the example of the Ti6Al4V alloy, it has been indicated that coverage with ALD layers promotes the growth of viable cells, and their viability increases by up to more than 12% [244,245]. The stereolithographic (SLA) method is applicable to the implant template model.

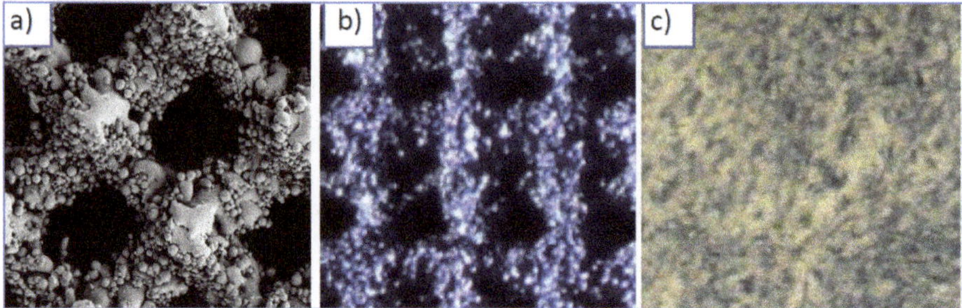

Figure 4. The structure of selectively sintered, at an angle of 45° relative to the x and y-axis of the working platform, (**a**) titanium—skeleton topography with a pore diameter ~350 μm after ultrasonic cleaning (SEM); (**b**) ultrasonically cleaned and chemically etched titanium to remove fine satellite powder particles and coated with Al_2O_3 during 1500 cycles (SEM); (**c**) layers of cultured human cells from the hFOB 1.19 osteoblast culture line (Human ATCC-CRL-11372) American Type Culture Collection ATCC (USA) on the substrate of the porous alloy Ti6Al4V selectively sintered and coated with Al_2O_3 by ALD (LM—fluorescent contrast).

A new class of implantable devices that, by all means, are desirable to use, among others, in dental implantology are implant-scaffolds. Those are new applications for which, due to the inability to depart from medical indications to classical prosthesis/implantation, it is necessary for natural cells to grow into the implanted elements and not just to generate living cells on their surface [2,3,80,244,245]. The wholly innovative and patent-protected implant-scaffolds, developed by the authors of this paper, contain a solid zone, like all implants used so far, and a hybrid porous zone is connected to it, fulfilling the function of scaffolds. The zone contains 100–600 μm micropores of various shapes.

Personalized implant-scaffolds (Figure 5) are made from powders of biomaterials, e.g., titanium and its alloys or Co–Cr alloys, through selective laser sintering (SLS). Inside the pores of implant-scaffolds, layers are applied using the atomic layers deposition (ALD) method, providing favorable conditions for the growth and proliferation of live cells after implantation. Implant-scaffolds can also be dedicated to dental implantology. The idea is developed by the author's concept of hybrid multilayer biological-engineering composites [244,245]. The structure of those hybrid multilayer composites is the artificial part, produced in the form of a porous skeleton by the SLS method with optional ALD layers inside the pores, and the biological part, consisting of living cells filling the micropores of skeletal engineering materials [244,245]. Such hybrid multilayer biological-engineering composites after completing in-vitro and in-vivo tests will find their use in dentistry and regenerative medicine, in general, to replace damaged and removed parts of the body, including teeth.

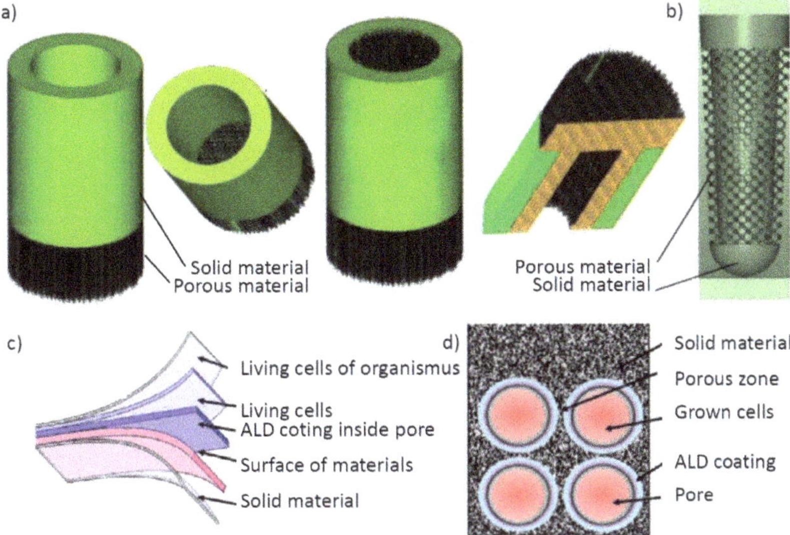

Figure 5. Scheme of composite biological-engineering; (**a**) orthopedically implant-scaffolds; (**b**) dental implant-scaffold; (**c**) the concept of multilayer engineering-biological layers inside the pores; (**d**) the structure of solid-porous engineering-biological materials.

Implantology treatment is teamwork characterized by the cooperation of a dentist, often a surgeon, prosthetist, often an orthodontist, and dental engineer, using tissue engineering methods. In each case, prosthetic restorations are individually designed and manufactured for a specific patient. The activities and operations of planning, designing, and manufacturing them constitute a complex clinical and manufacturing process. The synergy of advanced interdisciplinary knowledge in the field of imaging medicine and dentistry, tissue engineering, dental engineering, material engineering, and computer-aided design currently allows the dentist to implement complex treatment plans. The detailed design and management plan prepared in the dental engineering laboratory, resulting from the treatment plan, are the conditions for the effectiveness of clinical management. It applies, especially to the design and manufacturing of individual prosthetic components and surgical guides. Those works must be carried out in such a way that the chance factor is minimized as much as possible during the implant surgery. At the same time, it is necessary to ensure the full repeatability of medical procedures and ensure the least invasive procedure. It requires a mutual understanding by the dentist and dental engineer of the specifics and limitations of both technological and mid-surgical, creating a framework of trust and cooperation.

It should be taken into account that, as in any technical or industrial operation, it is now necessary to use the principles of the augmented holistic Industry 4.0 model in modern dental engineering. It applies to various spheres of activity, equally on both engineering design of different types of prosthetic restorations and their production. It also indicates a significant increase in the role and importance of the dental engineer, his knowledge, and experience in modern prosthetic treatment in dentistry. Thanks to this, it is not only possible to individually design prosthetic restorations, taking into account the patient's anatomical features, but the process of manufacturing these restorations is subject to all the rigors of modern industry, using the most modern technologies. In this respect, the production of prosthetic restorations does not differ in principle from the manufacturing of the most advanced products available on the market, e.g., car, aircraft, or robot components.

3. The Dentistry 4.0 Concept as a Consequence of the Current State of Civilization and Technology Development

This paper analyzes the possibilities of adapting the extended holistic Industry 4.0 model to the design and production of prosthetic restorations based on extensive literature studies. The implementation of the Industry 4.0 stage of the industrial revolution is progressing dynamically in individual industrial sectors. The process primarily involves creating so-called smart factories and, in fact, transforming existing manufacturing plants in such a way that, among others, digital technologies, data analysis, cloud computing, and additive technologies gain the practical significance. Undoubtedly, it also includes dental clinics and the processes of transformation of dental prosthesis restoration manufacturing centers into smart factories. The whole complex process is taking place in many places around the world and is termed Dentistry 4.0 [1,94] with varying degrees of advancement.

Systematic progress includes an increasing number of dental clinics, mainly centers for the production of prosthetic restorations. The progress is inevitable, and besides the technical aspect, it has a deep moral and ethical sense for those entities, which, for various reasons, but always consciously shirk active participation in the process, actually act to the detriment of patients. Patients receive less well-designed and manufactured prosthetic restorations, and the clinical procedure can be more invasive and expose them to greater pain sensations and a higher risk of complications. Those changes are evolutionary, and digital dentistry is as if absorbed successively by conventional dentistry. Over the past two decades, these changes have become increasingly dynamic and widespread [1,93]. It is undoubtedly a breakthrough, although such development depends on many factors. Those factors include the development of dental materials, manufacturing processes, and the level of equipment in technological machines. It involves ever-increasing requirements for financial commitment. It is even more challenging to overcome the barriers associated with mastering and adapting to the needs of dental prosthetics' very complex tools, machines, and cyber-IT systems being customized. It forces the cooperation of dentists with dental engineers, creating a specific supply chain. Those factors are typical attributes of the generalized Industry 4.0 holistic model. For this reason, considerations on such issues are presented later in this paper.

The value of society, as well as the increase in people's productivity and efficiency, depend on innovation and technological processes. Various strategic programs, including sustainable development goals (SDG) established by the UN, have been created in various countries. The Society 5.0 program deals with the development of the information society and human well-being. It was implemented in 2016 by the Government of Japan [22,246–252]. The Industry 4.0 program introduced in Germany [15–18] concerns the technological development of industry and IT aiding for manufacturing. The approach has been widespread in the European Union [19,21] and in many other countries of the world [20]. Currently, the most advanced developmental stage of the industrial revolution is developing, the detailed aspects of which have been extensively described in the literature [22–76]. The first stage of Industry 1.0 is associated with the invention of steam machines at the end of the 18th century [77]. Computer-aided design and production using electronic devices but also an electronic cloud, augmented reality, machine learning, and advanced robotization applies to the current stage of Industry 4.0. It is accompanied by care for the full implementation of the sustainable development principle and zeroes technology emissivity [253]. It applies to the so-called green life cycle of products and services from design to the recycling of post-production and post-operational waste. The Economy 4.0 model, taking into account the development stage of Industry 4.0, models the relationship between economic growth, concern for the climate and the human environment, and ensuring the quality of human health and life [254,255].

Due to the use of extensive information and sensory systems and technologies, with the necessary participation of people and the aiding of production automation along with the introduction of the Industry 4.0 stage of industrial development, the digital age in the industry was initiated [15,18–22,25–76]. The Internet of things (IoT), together with the Internet of people and services, integrate internal and inter-organizational networks [22,27,29,32,39,41,43,45,47,49,66], enabling connecting participants in the value chain when previously, despite the use of various other technologies' operations connecting different machines, the full visibility of processes at the factory was impossible. Currently, high computing power, metadata processing, and other cyberspace attributes, enabling the creation of social, business, biological, and cyber-physical systems (CPS), with a close connection with the technological activity, are used [22,28,33,44,46,57,58,60,63]. Those systems perform various tasks using the Internet of things, as well as sensors, radiofrequency, and identification elements, triggering devices, and even cell phones [22,27,29,32,39,41,43,45,47,66].

The Industry 4.0 concept significantly improves the manufacturing process and reduces the costs of materials, storage, and management. Smart factory systems allow improving operational flexibility with various production configurations and supply chain efficiency. The system mentioned above of sensors cooperating with the Internet of things, the use of big data sets, and results-oriented orientation ensures improvement of supply chain efficiency and accuracy of predictions in this respect, guaranteeing a sustainable increase in productivity [256]. Many elements related to the application of the Industry 4.0 concept also improve the competitiveness of not only companies but also regions. Beneficial acceleration of industrial growth, improvement of production economics, increase of productivity, and changes in the profile of labor force can be specified here [42,48].

Comparison of Industry 4.0 and Society 5.0 models indicates that the latter is much broader in terms of not only technical aspects but also includes a longer historical perspective. On the contrary to the name, the Industry 4.0 model does not even cover all cyber-physical systems. Still, in fact, it only concerns a fragment of the issue related to cyber-IT systems [22,34]. The model presented in the report [17] did not take into account engineering materials that are necessary for the manufacturing of any product or manufacturing processes, except for additive technologies. Those technologies often turn out to be impossible to use, they are rarely unrivaled, and in many cases, only alternative or complementary. There are also no technological machines in the model. In fact, the model [17] is an IT component of industrial development, but not the only one [1]. Although it is popularized by many official documents, including in the European Industry Digitization Strategy [19], through the development of so-called Digital Innovation Hubs [21] and in four flagship projects of the European Institute of Innovation and Technology (EIT) [257], the model contained in work [17] is incomplete and requires supplementation and augmentation. It is not true that by paying attention to the development of cyber-IT systems [18,258], one can ignore the need for progress in the field of materials used to manufacture any product, as well as machines and production technologies. It is also advisable to take into account the anticipated development trend of living and bioengineering machines [259]. Dynamic development of additive technologies [1,15–21,25–48,50,51,53–57,59,60,62,64,66,202,203,209,224,225,230,232–234,241,246–252,260–295] justifies their inclusion in the industrial development model, but it is necessary to take into account other technologies that make it possible to manufacture many products and which also currently require very advanced computer support [1,22,24,25,34]. Currently, various hybrid technologies are also used, combining conventional, although advanced technologies for the production of materials or products with incremental technologies of their production are commonly used. The model should also take such an aspect into account.

The authors' views on the full augmented holistic Industry 4.0 model (Figure 6) and the need to augment the current model has already been presented previously [20,37,296]. The model in the form of an octahedron contains people playing a leading role, a technological platform with four complementary components, and products that are the target of the action. Unbalance of the advanced level of the contractual level of impact of each of the individual components in the technological platform of the extended holistic model of Industry 4.0 indicates the need to increase the intensity of development of the particular component, and not only cyber-IT systems, which is allowed in the current model or any other developed in an unsustainable way.

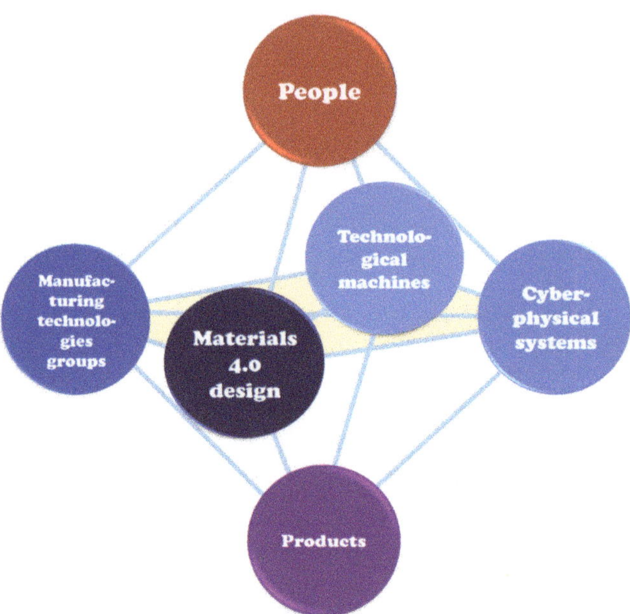

Figure 6. Octahedron of the extended holistic Industry 4.0 model containing four complementary components in the technological plane.

Such a point of view is indirectly confirmed in a recently published report because the idea of Industry 4.0 is based on a combination of modern manufacturing processes of various technology components, i.e., digital production technology, nanotechnology, biotechnology, and new materials [82]. In practice, according to the holistic augmented Industry 4.0 model presented by the authors, all technologies, machines, and materials used in practice decide on real progress. However, each of them individually determines growth to varying degrees, and no progress is possible without any of them. New technologies bring about integration and sustainable industrial development (ISID) by bringing new goods to the market and improving production efficiency. The evolutionary introduction of advanced digital production (ADP) technology of the Industry 4.0 stage of the industrial revolution consists of a combination of hardware, software, and connectivity. The dissemination of advanced digital production technologies (ADP) involves the development and use of artificial intelligence, analysis of big data sets, cloud computing, Internet of things (IoT), and advanced robotics and successively combines physical and digital production systems. Currently, only ten countries in the world introduce ADP technologies, having 90% patents and exporting up to 70%; the idea should be expected to develop in the economies of other 40 countries over time and perhaps the next ones in the future [82].

On the contrary to appearances, the augmented holistic Industry 4.0 model is used not only in large manufacturing organizations in the high technology industry, such as space, aviation, automotive, machine, defense, or electronics, but also in small and medium enterprises and smart factories created there. It is a guarantee of real progress as, e.g., as much as 99% of all enterprises in the European Union are micro and small and medium-sized enterprises (SMEs). For example, more than half of the total value added of EUR 3.9 billion in 2015 generated almost 23 million small and medium enterprises in the European Union [297]. From this point of view, the issue is significant for centers for the production of prosthetic restorations, which usually belong to the group of entrepreneurs.

The basis of manufacturing being launched and performed in engineering designing is the basis for the production of all products. Engineering designing inseparably includes structural designing (including the development of the shape and geometric form of the product or its element), material (related to shaping the structure and properties of materials corresponding to the requirements in working conditions), and technological (related to the processing of engineering materials to manufacture a product) [23,83,231,298,299].

All products require the use of engineering materials for their manufacturing, which, therefore, play a crucial role [83]. The material is solid with properties that enable it to be used by man in the manufacturing of products and can be natural or engineered when the natural resources used to require complex manufacturing processes to adapt them to technical needs. The expected functional properties of the product are ensured as a result of designing the expected material, processed in the expected manufacturing process, ensuring the expected shape and geometric features of the product, the expected material structure, ensuring the expected mechanical, physical, and/or chemical properties of the material, which guarantees the expected operating functions of products [83,299] (Figure 7). It is the 6xE principle of the six expectations, exposing the material science and material engineering paradigm.

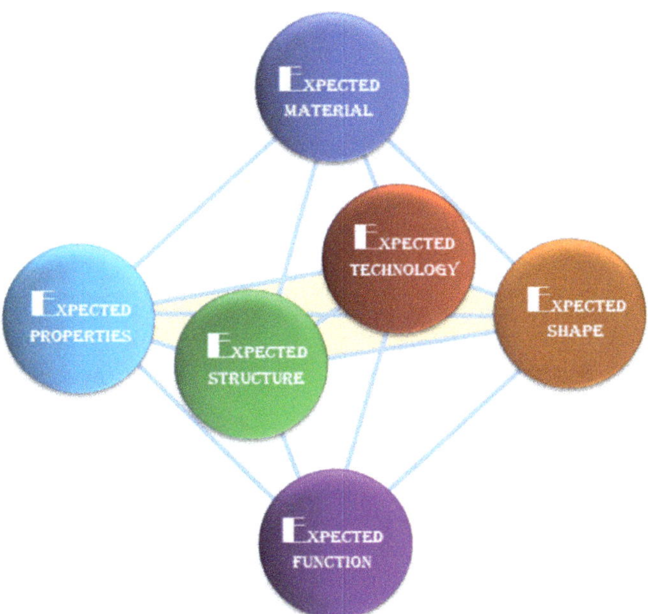

Figure 7. The 6xE principle of the six expectations, exposing the material science and material engineering paradigm.

Multi-criteria optimization, inevitably computer-aided one, enables material design and selection of one or more alternative materials for a particular product or its element from among metals, ceramics, polymers, composites, and special materials, e.g., porous or foams. The criteria are taken into account relating to the chemical composition, manufacturing conditions, operating and maintenance conditions, as well as the method of removing material waste in the post-use phase. Economic aspects and price conditions regarding its acquisition, processing into a product, and its exploitation, as well as the costs of removing post-production and post-operational waste, are also important. The modeling of processes and material properties is also included. Today, over 100,000 engineering materials are used. Initially, throughout almost the entire history of human civilization until the early twentieth century, the materials were selected by trial and error. Over time, a material selection methodology was developed, followed by the material's design in recent decades. The currently avant-garde Materials 4.0

method not only covers human systems but uses advanced cyber-physical systems containing material data, artificial intelligence tools, and intelligent machine learning algorithms (Figure 8) [24,83–89].

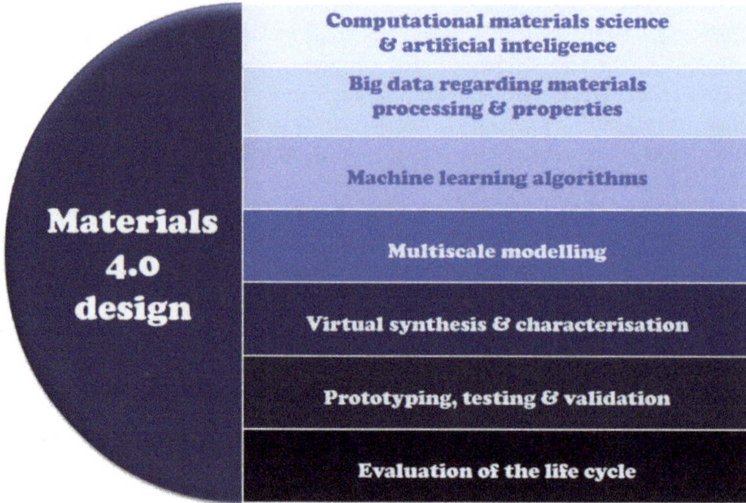

Figure 8. Materials 4.0 methodology of material design using cyber-physical systems and advanced information technologies.

Manufacturing products from raw materials requires the use of various processes, the use of many machines, and the organization of operations according to a well-prepared plan with properly used materials, energy, capital, and people. Manufacturing requires the involvement of people of different professions and to perform various tasks and the use of multiple machines, equipment, and tools. Currently, the degree of automation is usually high, and the use of computers to control the processes and products as well as computer networks and robots is significant [83,231,299]. A general diagram of technological processes is shown in Figure 9.

Figure 9. General classification scheme of product manufacturing technology.

As it was previously indicated, additive technologies play an important role [77,136,202,300–333]. They rely on joining together layer by layer of properly prepared powders or liquids, thin fibers, and sometimes in the form of a rolled role of metals and alloys, ceramics, and polymeric materials. Additive technologies are currently very popular in many applications in various industrial sectors [135,266,271,278,279,281,283,290,312] in the photovoltaic industry [285,295]. Additive technologies also find numerous applications in medicine and medical industry, among others, for planning operations by producing models of anatomical organs, for epitheses, e.g., of the eye, nose, ear, taking into account the individual anatomical features of the patient [232,284,286,288,289,293]. Additive technologies are also used to produce scaffolds with a porous structure corresponding to the anatomical shape of a natural organ [194,197,209,233,234,241,262,265,287,292,294,316]. Stereolithography (SLA) [135,202,334–337] and selective laser sintering (SLS) [2,3,80,193–202,219,224,243–245,260,262–265,267,270,274,277,280,282,293,338–347] (Figure 10) and selective laser melting (SLM) [202,225,261,264,266,268,269,272,273,275,276,291] are wrongly treated as a separate group, although it is only a variation of the previous one in the case of sintering with liquid phase used in engineering and dental prosthetics [2,3,80,193,195–201,287,294]. Selective laser sintering (SLS) is used to manufacture metal prosthetic restorations. SLA is used for the manufacturing of models and rapid prototyping and some polymeric temporary and mobile restorations.

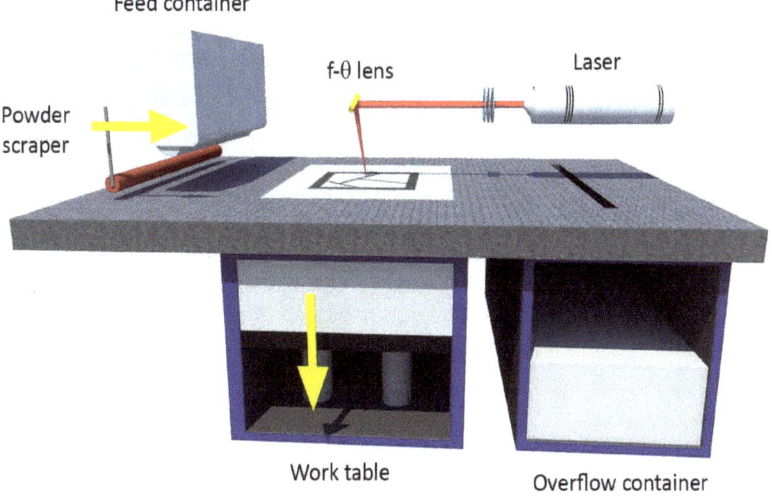

Figure 10. Scheme of selective laser sintering.

Additive technologies greatly simplify the technological process as no molds or matrices are required. The product manufacturing in other processes, e.g., casting, powder metallurgy, or plastic working or other tools in the case of machining, requires considerable financial outlays and time to prepare the production. It is particularly burdensome in unit manufacturing, and, as a result, determines that in such cases, additive technologies are extremely competitive [202,203]. However, it is necessary to make supports that attach the manufactured product to the work table. At the end of the process, the supports are cut off, and the product surface is processed by grinding or sanding.

In a large number of cases, additive technologies are complementary to other traditional technologies, so the latter still has a non-threatened position. Those technologies are considered as conventional, but now also require very advanced computer aiding and are usually highly automated and computerized. It should be noted that it is not currently and in the foreseeable future to eliminate those traditional technologies by replacing them with additive technologies. That fact must be taken into account in the current model of industrial development.

The classification of technological machines corresponds to the classification of technological processes to enable their implementation (Figure 11). Cyber-physical systems (CPS) ensure that those machines communicate with other smart components, as well as with products via the Internet of things (IoT) [22,53], and create digital twins as a virtual copy of the physical world. The existing Industry 4.0 model includes nine technologies [15,18–21,25–76,348]. The industrial Internet of things [22,27,29,32,39,41,43,45,47,66], big data sets [55,57,59], cloud computing for data processing [35,36,62] simulations, self-learning processes, and artificial intelligence [26,62] are used. The existing Industry 4.0 model also includes autonomous robots, horizontal and vertical systems integration, cybersecurity, additive production, and augmented reality [348]. However, the model should be supplemented, as previously mentioned, with engineering materials, living and bioengineering machines, and other technology groups, in addition to additive technologies (Figure 12).

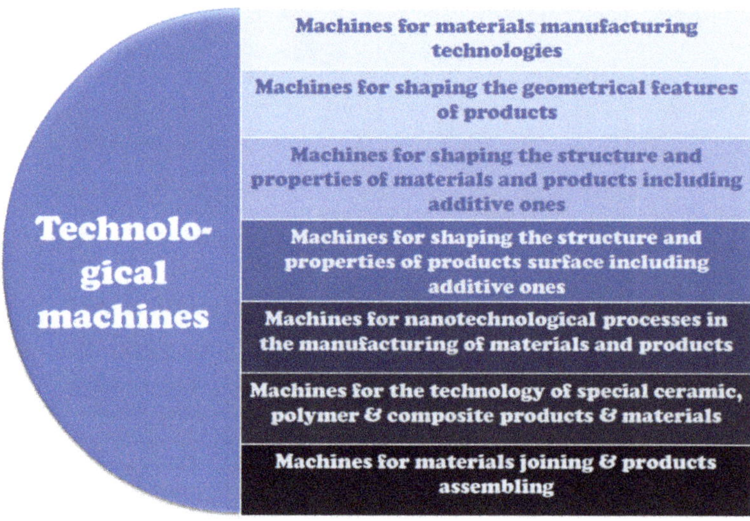

Figure 11. General scheme of the classification of technological machines.

Figure 12. The supplemented diagram of the technologies underlying the assumptions of Industry 4.0.

The Industry 4.0 system is based on smart factories [51,55] developing smart production. Smart factories use the embedded cyber-physical system and smart exchange data through smart networks connecting value creation modules with smart factories. Comprehensive engineering includes connections of stakeholders (clients, employees, or suppliers), products, and production machines, which are connected through the exchange of data from a virtual network with the participation of cloud computing [35,36,62]. Identification systems, including radiofrequency identification (RFID) chips and quick reference codes (QR), provide wireless identification and location of materials, intermediates, and products in the value chain. The actuator system using the collected data in real-time enables products, people, or product changes.

A similar situation applies to medicine and dentistry, in particular, to technical aiding centers for medicine and centers for the production of prosthetic restorations. Well-known and used tools in the area include digital documentation, online prescriptions, digital X-ray tomograms and photos, online media marketing, patient registration and treatment progress software, and the use of cloud storage [93]. The Dentistry 4.0 includes, among others, electronic medical records, digital patient records, diagnostic information, including 3D imaging using cone-beam computed tomography CBCT and intraoral and optical scanning, electronic development of treatment plans, software improved workflow tools, and increasingly available CNC milling and additive technologies, such as stereolithography and SLS. With regard to the development of Dentistry 4.0, all elements of the technological platform of the holistic augmented Industry 4.0 model regarding prosthetic and implantology dental engineering restorations are relevant, both in clinical practice and the production phase. Therefore, Dentistry 4.0 model includes digitization and computerizations of the design and manufacturing processes of prosthetic restorations for the use of cloud computing. The system contains biomaterials, as well as engineering design of prosthetic restorations and technologies for their production, including additive technologies, as well as machines and devices and supported cyber-IT systems. The issue of Dentistry 4.0 has been presented in the next subchapters of this paper.

4. General Characteristics of the Dentistry 4.0 Stage of Design and Manufacture of Prosthetic Restorations

Dentistry 4.0 is the imperative of the moment in modern dentistry and, at the same time, a unique opportunity for its development. A significant determinant of the development is the increased awareness of dentists regarding the need to use the model and the formation of a new generation of dental engineers. In addition to knowledge of the latest manufacturing technologies using appropriate machines, their competence also includes perfect IT preparation. It is not only about the ability to program and use advanced IT tools but also the use of cloud computing, the Internet of things, augmented reality, and medical imaging techniques, including cone-beam computed tomography (CBCT). It also requires knowledge of the appropriate software to connect computer networks of dental clinics and prosthetic restoration manufacturing centers. CBCT tomography (Figure 13) uses divergent cone-shaped X-rays for medical imaging. The scanner rotates around the patient's head, and, at that time, about 600 different images are recorded, each of which contains a set of volumetric data after a single rotation by another 200 degrees. Scanning software creates a digital volume composed of three-dimensional voxels of anatomical data, which are collected, reconstructed, and visualized by specialized software. The importance of this CBCT method is increasing.

Currently, the traditional relationship between a dentist and a laboratory or a center for the production of prosthetic restorations is changing. To date, the dentist has planned treatment and, according to his/her plan, commissioned the manufacturing of prosthetic restorations. The qualitative change is currently taking place. A dentist still supervises the treatment and is making the diagnosis of the patient's teeth state and registration of the situation of the teeth with the use of CBCT tomography. However, the planning, designing, and manufacturing of the prosthetic restoration are in the competence of the dental engineer cooperating with the dentist. It ensures that the highest possible quality standards and narrow dimensional tolerances are achieved. In many cases, it requires the design of a surgical guide, which, due to the diagnosis of the actual state of the bony processes of the maxilla and the mandible, allows the correct execution of holes in which implants and/or implant-scaffolds should be fixed. The idea of a digital twin gives a chance to do virtual experiments during the design and digital verification of the prosthetic restoration. On this basis, a real stereolithographic (SLA) model of rapid prototyping can be made in order to determine the correctness of short-circuit height in the articulator. In this way, the roles change because the competence of the dental engineer cooperating with the dentist undergoes the full scope of design and manufacturing of prosthetic restorations. The dentist invariably supervises the entire treatment from the medical side and assembles the manufactured prosthetic restoration in the patient's mouth.

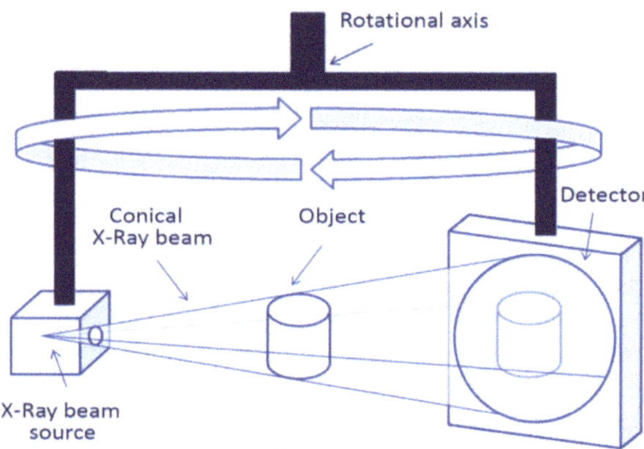

Figure 13. Diagram of a conebeam computed tomography (CBCT) scanner.

The relations according to the Dentistry 4.0 concept between the medical tasks of the dental clinic and the design and manufacturing capabilities of the manufacturing center for the design and manufacturing of dental prosthetic restorations concerning the patient's expectations are schematically shown in Figure 14. Significant progress has now been made compared to the previous historical Dentistry 3.0 stage, when, in addition to the evident improvement in conservative dentistry, X-ray evolution occurred in the diagnosis of patients' dentition [349]. In contrast, the improved level of oral care with a significant reduction in the cost of implant-prosthetic treatment and, in particular, the production of prosthetic restorations is associated with 3D imaging manifested in the use of CBCT, data processing using cloud computing, and personalization of additive technologies. The 3D printing, mainly SLS, constitutes the essence of the current very advanced stage of Dentistry 4.0. The approach appropriate for the Dentistry 4.0 stage ensures business efficiency. Digital technology affects the minimization of the cost of manufacturing personalized prosthetic restorations in some cases by more than 70% compared with conventional methods [349].

The application of the Dentistry 4.0 idea also reduces the time spent on the design and manufacturing of the prosthetic restoration. It is also associated with a reduction in the amount of time spent by the dentist and medical staff, and what is extremely important, the patient's time, logistics, and the time and expenditure required to match prosthetic restorations during necessary tests with the participation of the patient. At the same time, it is one of the reasons for reducing the total costs of the procedure, enabling the production and implementation of non-standard restorations during one visit to the dentist [349]. However, prior to the use of CBCT tomography, it requires the visualization of the state of teeth [349]. The attributes given correspond to the standards of the dental version of the smart factory. A prerequisite for the success of this modern approach and, at the same time, its opportunity is to use networks and other computer tools to significantly extend the reach and integration of the network of suppliers, cooperating dental clinics, and patients. Undoubtedly, the determinants of the possibility of implementing the Dentistry 4.0 idea are the investment limitations of the technological machines and computer software used and very high costs, incomparably greater than in the case of using conventional technologies, when the costs are even several dozen times lower. Both the horizontal and vertical technological line integration system in intelligent centers for the production of dental prosthetic restorations and the cyber-physical system ensures the creation of a value chain. Sensors in technological machines make the identification and location of the factors, creating value, things, and people, and the system for monitoring the processes of generating and collecting smart data in real-time takes into account changes in technological processes, people, and products and activates executive devices.

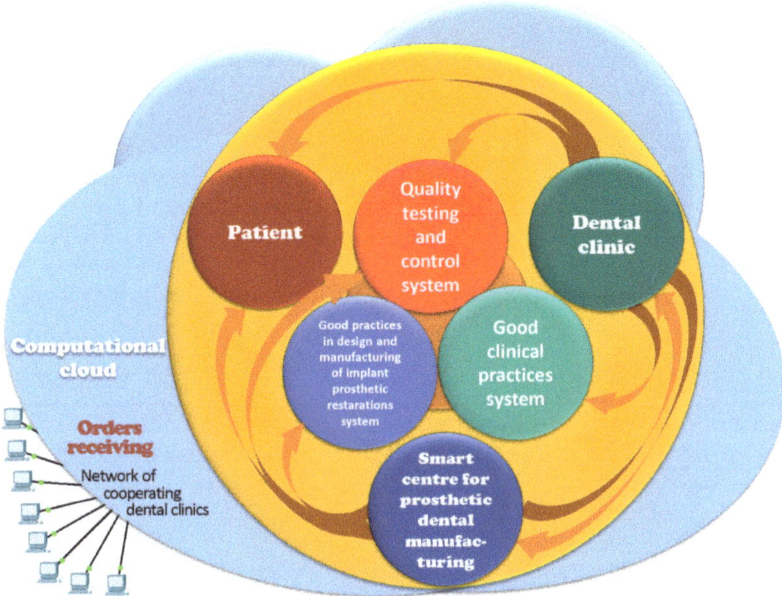

Figure 14. Diagram of the relationship between the medical tasks of the dental clinic, the design, and manufacturing capabilities of the dental prosthesis restoration center and the patient's expectations according to the Dentistry 4.0 concept.

The concept of "digital twins" of the physical world reduces the risk of design and technological errors. Data collection enables the use of artificial intelligence tools for virtual experiments and simulations. Network production integration systems in various areas are related to intelligent implant connections and prosthetic restorations as products with machines and people. In essence, those centers are connected to many dental clinics with cooperation bonds, and cloud computing can be used more and more commonly to send electronic patient data, including CBCT images. The information obtained and collected in this way creates the conditions for the design of prosthetic restorations, models, and surgical guides by means of computer-aided design (CAD) and production planning using numerically controlled CNC machines, machines for SLS additive manufacturing, and coating of internal surfaces with ALD methods. Table 9 lists materials used in dental prosthetics as an example of the authors' own experience.

The Dentistry 4.0 approach requires the dental engineer to start treatment planning with the dentist, who diagnoses the condition of the teeth based on the CBCT tomogram. The assessment of the patient's teeth condition by a dentist allows the teeth to be qualified for removal. The dental engineer and the dentist who is responsible for treatment are planning the type of prosthetic restoration and how to mount it on existing teeth or planned dental implants. In the case of restorations based on the patient's own teeth, the dental engineer can prepare wax-up or mock-up restorations in order to agree with the patient's assumed aesthetic effect. Performing those works with PMMA allows imitation of restorations faced with composite or porcelain and allows the dentist to make corrections. Preparation of the pillar teeth with the degree enables verification of the position of prosthetic restoration in the oral cavity and thus ensuring the correct occlusion position. In the case of restoration mounted on implants, the dental engineer prepares surgical guides, determining the placement of the implants in cooperation with the dentist and then, using the CAD program, determines their final positions in the mouth. In the case of individual implants, the teeth are qualified for extraction, a project of a copy of the tooth is made by reverse engineering methods using CAD software, and then the implant is made using the SLS method. An implant insertion surgical guide and customized instruments are also manufactured. In all those activities, the dental engineer works closely with the dentist. The associated clinical activity places high demands on dental engineering in all related areas from the selection of engineering materials to manufacture. It is necessary to ensure repeatability and adequacy of prosthetic restorations, including implants, to the patient's individual anatomical features. The use of conventional technologies is currently unacceptable, including related to the production of gypsum models and subsequent casting. Fully digitized technologies are highly desirable and require virtual intraoral scans. Own tests have shown that the inaccuracy of impressions is about 0.3%, dimensions differing by about ± 0.05–0.1 mm. The tested masses include polyether and polyvinyl siloxane masses—the additive silicones used for impressions reflecting the patient's teeth. The accuracy of the working model can be achieved by increasing the volume of gypsum during setting by about 0.08%. The change in the volume of mapped teeth on the working model is –0.2% to 0.4%.

Table 9. Examples of materials used in dental prosthetics in the case of the authors' own experience.

1	Type	Detailed Information
1.	Metals and their alloys	Co, Ti, and Mg alloys and their powders, including Co25Cr5W5MoSi, Ti6Al4V, Ti6Al7Nb, Ti24Nb4Zr8Sn, Ti7.5Mo and Ti40Nb, Mg8Al, Mg3Al1ZnMn, Mg9AlZnMn, Mg4Y3NdZr, Mg3Nd1GdZrZn, and pure Ti
2.	Ceramic materials	Sintered ZrO_2 and Al_2O_3
3.	Precursors for the preparation of coatings inside the pores	$TiCl_4$, $AlCl_3$, $Zr[N(CH_3)\cdot(C_2H_5)]_4$, and ZrD-04, i.e., $(C_pMe)_2Zr(OMe)Me$, where $C_p=C_5H_{5-x}$ R_x is cyclopentadienyl ligands and heteroleptic metallocenes $[Zr(C_p)_{4-n}(L)_n]$, L = Me and OMe using water and ozone as reagents and triethylphosphine (6,6,7,7,8,8,8-heptafluoro-2,2-dimethyl-3,5-octanedionate)silver(I) or Ag(fod)(PEt_3),triethylphosphine (6,6,7,7,8,8,8-heptafluoro-2,2-dimethyl-3,5-octanedionate) silver(I) connected with dimethyl amineborane ($BH_3(NHMe_2)$ or (hfac)Ag(1,5-COD) ((hexafluoroacetylace-tonato)silver(I)(1,5-cyclooctadiene)) as a metal source and tertiary butyl hydrazine (TBH) as a reactant for TiO_2, Al_2O_3, ZrO_2, and nanosilver coatings, respectively
4.	Semi-finished products for applying porcelain coatings	Pastes or powders with thinners, including, e.g., DeguDent or Willi Geller, layers connecting the metal substrate with porcelain, including bonder metallic for porcelain and paste opaque or Crea Alloy Bond and paste opaque, cervical layer Power porcelain Chrome or Opaque Dentine and subsequent layers of porcelain Flu Inside, Detine, Incisale, Incisale opalescente, Transpa with properly selected shades or Opaque Dentine, Dentine, Enamel, Clear mixed with Universal Mixing Liquide by Harvest Dental and outside Glaze with Stain dyes mixed sequentially with Ducera Liquid or Crea Color

Form: 98 mm diameter discs for milling work, laser sintering powders, liquid and gas precursors

Ensuring the required accuracy needs eliminating factors that increase the inaccuracy of the working model. It is possible to scan an impression directly from the impression tray using a prosthetic scanner or based on the results of scanning with an intraoral scanner. The accuracy of 3D models made with an intraoral scanner varies in the range of 50–70 μm and is close to the maximum accuracy of CNC numerically controlled milling machines. However, the technology still does not allow performing repetitive scans of the entire arch. In the case of casting methods, accuracy is included in a wide range and depends on many factors. As a result, the inaccuracy of the procedure of duplicating the teeth to the working model reduces the accuracy of the prosthetic restoration significantly.

To eliminate the lack of dimensional inaccuracy of manufactured prosthetic restorations, computer-aided design/computer-aided manufacturing CAD/CAM and full digitization of the manufacturing technologies [91,171,176,178,185,187,192,349–365] are used. The manufacturing of metal foundations is increasingly used by machining with milling Co–Cr alloys, titanium alloys, as well as prefabricated ceramics, including oxides zirconia and alumina, and reinforced with lithium disilicate, as well as poly (methyl methacrylate) (PMMA) and wax. Additive manufacturing technologies, including selective laser sintering, have a wide area of use, including dental engineering of computer-aided design/manufacturing technology [27,29,178,180,201,224,244,245,260,262,264,265,267,269,277,280,293,366–371].

An essential goal of using CAD technology in dental prosthetics is to increase the accuracy of the production process [349–365,370,372–377]. One of the essential criteria for the long-term success of prosthetic restorations [378–382] is the highest possible compatibility of prosthetic restorations with the patient's anatomical features. Digitization of the prosthetic restoration is carried out in one of two ways using extraoral or intraoral scanners [383–407]. The first one consists of the extraoral scanning [361,364] of a traditional classical impression. Extraoral scanners use three main methods, including using laser radiation, structured light, and direct contact. Optical scanners using laser radiation and structural light do not require physical contact, do not affect the density and quality of the surface of the scanned object [361], and usually provide a shorter scanning process than contact scanners [355]. Those scanners do not fully eliminate the impact of surface gloss, brightness, and other optical properties of the scanned object [363]. The second way to digitize a prosthetic restoration is to make a digital prosthetic model by scanning dental arches using an intraoral scanner [387,389,401]. Both types of scanners using CAD software, in turn, allow for the production of a physical model using CAM technologies using milling or additive 3D printing technologies and appropriate engineering materials, including ceramics, polymers, or metals [388,390,392–394,405].

A radical turnaround in clinical procedures and in the way of technical preparation of implantation using CAD/CAM methods is worked. They include the authors' own methods [3,206–208,210,212,223,244,245] regarding data acquisition by means of cone-beam computed tomography (CBCT) in planning and performing implant prosthetic treatment and also in endodontics and orthodontics [408–427].

Using CAM software, it is possible to manufacture precisely the prosthesis with an accuracy of up to 0.05 mm [136]. CAD/CAM techniques can also be used to prepare temporary restorations made of acrylic materials for intraoral placement. The widespread use of CAD/CAM methods and the full digitization of manufacturing technology create broad premises for the successive introduction of the assumptions of the augmented holistic Industry 4.0 model among manufacturing centers and other dental prosthetics manufacturers.

5. Examples of Digitization of the Design and Manufacture of Prosthetic Restorations and Dental Implants under the Dentistry 4.0 Model

This subchapter presents, for example, the authors' own projects regarding the application of the augmented holistic Industry 4.0 concept at the dental prosthetic restoration center according to idea Dentistry 4.0. Table 10 summarizes how the requirements of the technological platform of this model are met in the case of producing fully personalized prosthetic restorations and made on-demand.

Table 10. Examples of how to meet the requirements contained in the technological platform of the augmented holistic Industry 4.0 model in the case of the authors' own projects regarding the application of the augmented holistic Industry 4.0 concept at the dental prosthetic restoration center according to idea Dentistry 4.0.

No.	Model Module	Type of Device or Element
1.	Technological processes	Imaging of the patient's teeth
		Additive manufacturing models using stereolithography (SLA)
		Milling of metal, ceramic, and polymer discs and precision machining of elements manufactured by SLS method
		Applying atomic layers deposition (ALD) on the surface of solid bridges and crowns and inside the pores of porous structures
		Opaque and porcelain deposition and firing
2.	Machines and devices	3D progress oral scanner
		Scanner for dental models 3Shape E3, Dental Wings 7
		CNC milling machining center Fanuc Robodrill, RS-team/Kimia micro
		The machine for the additive manufacturing of 3D Systems PeoJet 6000 HD polymer elements
		Selective laser sintering machine (SLS) Orlas Creator
		Machine for applying atomic layers ALD Picosun 200
3.	Cyber-IT system	Servers, databases, set of computers, network connections, sensors, two-dimensional QR codes (quick response), and cloud computing
		Computer software for the design of geometric forms and porous structures
		Computer software for imaging and designing prosthetic restorations
		Computer software for coating design
		Software for production management systems, supply and purchasing planning, logistics, and accounting

The first of these examples concerns digitization under the Dentistry 4.0 model related to the manufacturing of dental bridges using CAD/CAM design technology and manufacturing using 3D printing in SLS technology. When designing a bridge or crown on a Co–Cr alloy or titanium framework, attention should be paid to the arrangement of the abutment teeth and ensuring adequate distribution of occlusal forces on the alveolar ridge while providing an even load on the abutment teeth. In particular, in the case of long bridges, including circular bridges, it is crucial to ensure the right mass of the manufactured element while ensuring high accuracy that will preserve the expected strength of these elements. The most crucial step in the procedure for producing a prosthetic restoration is to provide the highest possible efficiency in the reproduction of the oral situation. According to the authors' research, the best results are still achieved when performing a traditional impression using silicone impression masses and performing a 3D scan directly from the impression spoon [160].

The procedure ensures that the prosthetic elements are kept to the highest position with the minimum distance for the introduction of cement connecting tooth tissues with prosthetic restoration while maintaining the most upper marginal tightness for works in which the preparation is performed with the degree. It is crucial to ensure the highest strength of the designed element, in particular, so that the cross-sections of the connectors maintain the appropriate assumed surface area. Currently, SLS laser sintering of titanium and Co–Cr alloy powders is increasingly used in the manufacturing of prosthetic restorations. The technology is gaining recognition because it allows reducing the time necessary to manufacture the foundation by three times and, at the same time, allows to increase the accuracy of the manufactured element relative to the design, which reduces weight and post-production time (Figure 15).

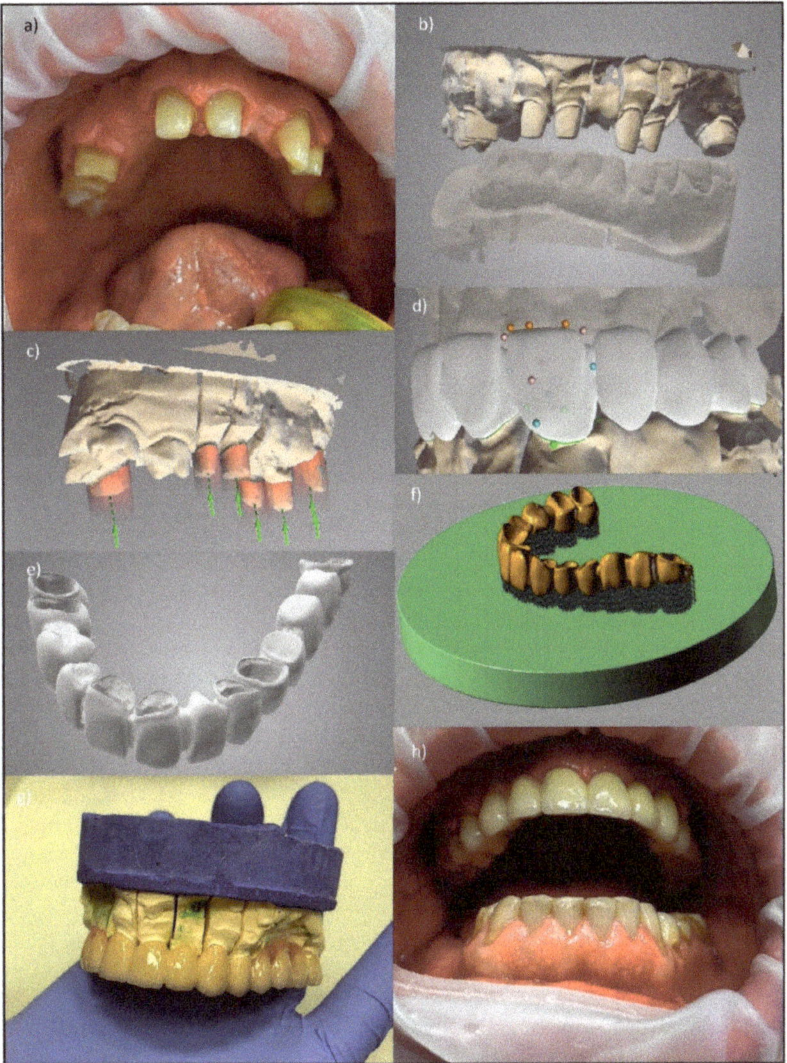

Figure 15. Description of the procedure for the production of dental bridges using the technology of selective laser sintering (SLS) from metal powders; a detailed description of the meaning of (**a**–**h**) images in the text. (**a**) preparing teeth for treatment; (**b**) virtual tooth model together with post models; (**c**) considering pillars; (**d**) prosthetic reconstruction project; (**e**) finished model; (**f**) designed supports; (**g**) tooth forming; (**h**) final result.

The dentist conducting the treatment performs the preparation of the teeth with the formation of a degree and maintaining the parallelism of all pillar teeth (a). Dental engineer using a scanner that allows making a scan of the impression spoon makes a virtual model of teeth along with models of posts (b), which will be needed to make appropriate faced porcelain. The designer must then consider pillars. Then, a prosthetic restoration project is created using libraries, which are then verified and corrected by the designer (d). The finished model (e) is transferred to the computer-aided manufacturing (CAM) software and placed on the platform where then the supports (f) are designed. After manufacturing the foundation on the SLS device, the ceramist technician performs the characterization of the foundation by forming teeth (g). The ready effect of these efforts is visible in the picture (h).

The second example concerns the digitization under the Dentistry 4.0 model related to the planning, design, and manufacturing of prosthetic elements and the surgical guide in the case of immediate implantation in place of the planned tooth extraction [81] (Figure 16). The entire design process is performed on the basis of a 3D tomogram of the patient obtained in the CBCT tomography (a) available in many implantology clinics. A digital twin of the patient's entire dentition is created. After the tomogram is converted to a 3D digital model, the bone base models and intraoral scans (b) are assembled, and the tooth (c,d) is separated, which will be removed. It is then removed from the model, and the remaining model after this operation is a virtual model of the situation after the really successful and safe removal of the tooth by the dentist (e). Then, in consultation with the attending dentist, the dental engineer plans the optimal location of the implant in the alveolus (f). After accepting the place, the implant guide (g) is designed. Then, a working model with the implant analog implemented is created. Using a typical CAD prosthetic design software, one can design an individual abutment (h), filling the extraction socket. The design of the connector is transferred to the CAM software, and then the individual abutment is made by milling using a CNC milling center. Further, it is necessary to design, also in a typical way, the design of the PMMA temporary crown (i). The crown is made in CNC technology using a milling machine.

The presented procedure is an avant-garde, authors' own method of performing the entire immediate implantation procedure, which is designed solely on the basis of data obtained in digital form. The dentist proceeds to medical procedures already having all the prosthetic elements necessary to perform the treatment (surgical guide, individual abutment, temporary crown). Thanks to the procedure, after its completion, the patient comes out with the tooth replaced with a fully useful prosthetic restoration. In addition, such a procedure may be 60% shorter compared to the same procedure performed in the traditional way using impression materials.

The third example concerns the preparation of a surgical guide for implantation of four implants, which will be the basis for the restoration of the entire arch using a permanent hybrid construction, which can be pulled out by the patient for hygiene (Figure 17). The surgical guide is made for implantation of all implants planned for placement. It is prepared for introducing first of all pilot drill, although it is also possible to replace guide sleeves with sleeves for subsequent drills. The exact surgical guide for implant must be prepared based on detailed data on the volume of bone base obtained from the 3D tomogram, which is performed before planning the procedure. A dental engineer using CAD software creates a bone base model. Throughout the entire design process, it is possible to verify the accuracy of the model, as shown in the picture (a1–a4). Then, the bone base and intraoral situation models are assembled. The correct connection of the intraoral scan with the bone base model is significant for the success of the whole procedure.

Next, the bone base and intraoral situation models are assembled. The correct connection of the intraoral scan with the bone base model is significant for the success of the whole procedure. It is necessary to make a scanning plate that is visible on the 3D tomogram pattern (reflects X rays) (b). The CBCT pattern should be performed with this plate. Using the CAD software using the 3-point folding method, the position of soft tissues relative to the bone base can be mapped. A dental engineer, in consultation and under the supervision of a dentist, plans the placement of implants (c). The thickness and distribution of the cortical layer and spongy bone base are verified, and the optimal placement of the implants is planned to take into account the anatomy of the appendix and their optimal location, taking into account the position relative to the mandibular nerve canal. Next, the surgical guide is made using 3D SLA printing technology. Using advanced 3D printing technology allows ensuring tight adherence of the surgical guide to soft tissues. It is essential because the surgical guide during the procedure is fixed only on soft tissues. The dentist only makes a small incision of soft tissue only around the holes through which the implant will be inserted. Those holes can be made in soft tissues using a dedicated drill with a diameter close to the width of the implant being inserted. Thanks to that, recovery after the procedure is mild. The assembly model of the intraoral model and bone base model is shown in the photo (d).

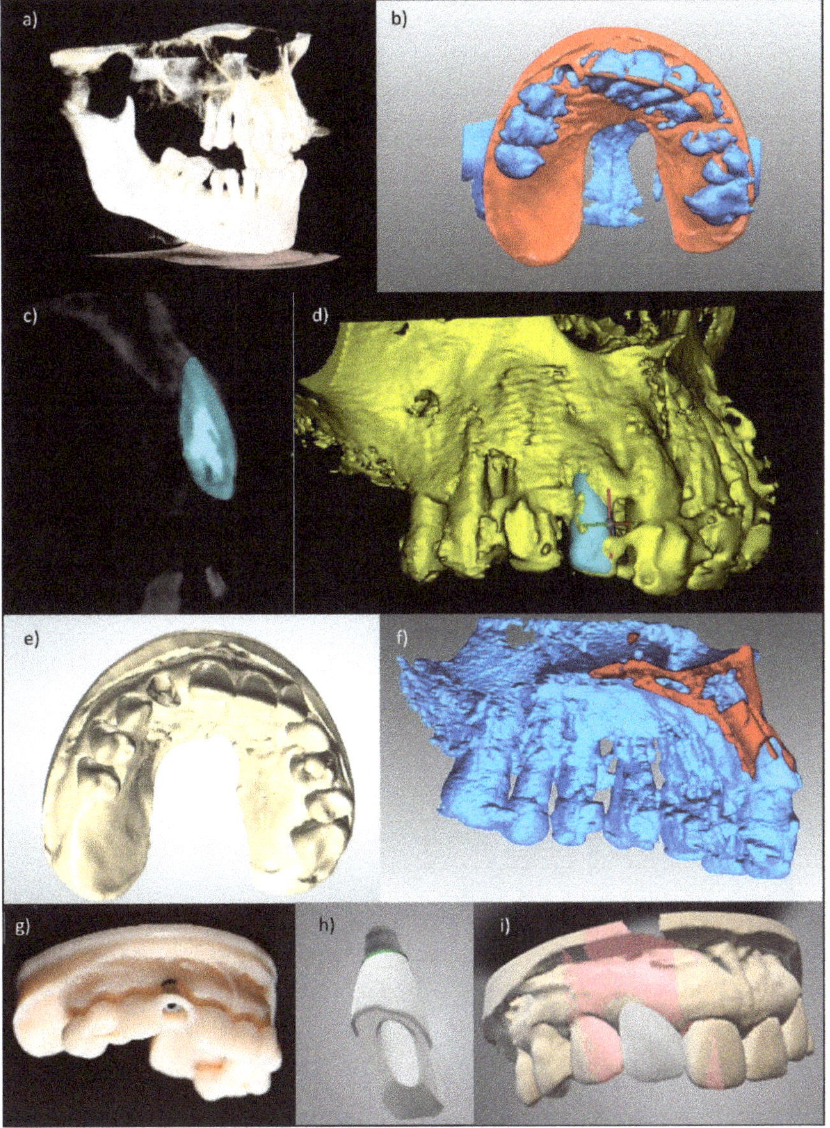

Figure 16. Description of the procedure for manufacturing prosthetic components and the surgical guide for the procedure of immediate prosthetic reconstruction using SLA technology for the production of implant and numerically controlled machining centers (CNC) for prosthetic components. (**a**) 3D tomogram of the patient obtained in the CBCT tomography; (**b**) 3D digital model, the bone base models and intraoral scans; (**c**) separated model of the tooth which will be removed; (**d**) assembled model with the tooth which will be removed; (**e**) virtual model of the situation after the removal of the tooth; (**f**) the optimal location of the implant in the alveolus; (**g**) design of the implant guide; (**h**) design of an individual abutment; (**i**) the design of the PMMA temporary crown.

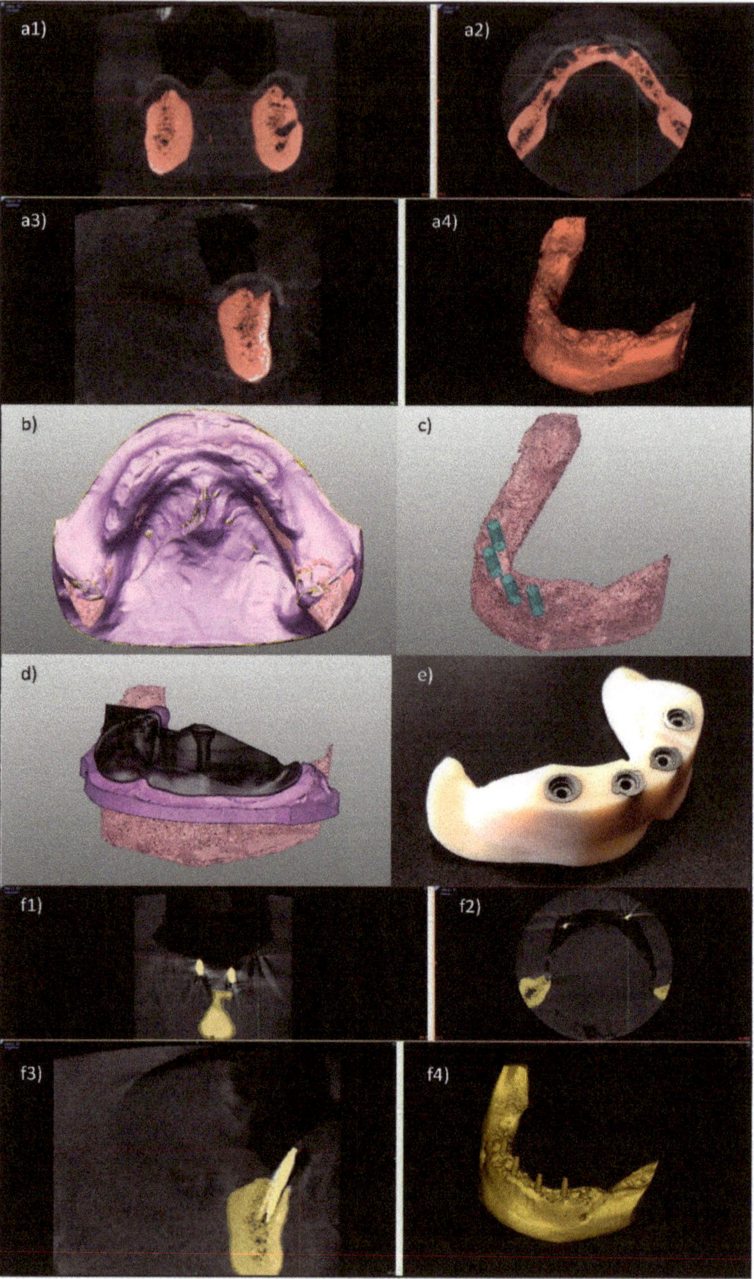

Figure 17. (On the previous page) Description of the procedure for making a template for implanting 4 implants; (**a1–a4**) verifying the accuracy of the model; (**b**) a scanning plate visible on the 3D tomogram pattern; (**c**) the plan the placement of implants; (**d**) the assembly model of the intraoral model and bone base model; (**e**) the surgical guide printed in SLA technology; (**f1–f4**) a CBCT control test with control pins.

The quick surgical guide printed in SLA technology with individually designed guide sleeves is shown in Figure 17e. After drilling the holes using the surgical guide, a CBCT control test with control pins (f1–f4) is carried out, confirming the correct placement of the holes in accordance with the plan presented in the project.

6. Summary

Based on the extensive literature review and selected results from the authors' own professional dental engineering practice, the scope of the modern idea of Dentistry 4.0 regarding technical aiding for dental prosthetics is indicated as part of the overall concept of the currently highest stage of Industry 4.0 of the industrial revolution. The progress of civilization depends on many factors associated with numerous inventions of humanity by inventing the steam machine recognized as the Industry 1.0 stage to the present stage of very advanced computerization, robotization, and automation. Since the progress is continuously being made, the UN has set 17 sustainable development goals (SDG), and several strategic programs have been developed, including the Society 5.0 program in Japan and Industry 4.0 in Germany, adopted in the European Union and widespread in many countries. While the Japanese program includes an extensive concern for the well-being of people, the German program mainly focuses on the development of industry, and both are based on very advanced computer-aiding using cyber-physical systems. The existing Industry 4.0 model is only an IT element of industrial development and, therefore, had to be augmented and supplemented. The authors of this paper have developed a full holistic model of technological development at the Industry 4.0 stage. The technology platform of the model includes four aspects, including materials, manufacturing processes, and technological machines, as well as their progress, and as the fourth element regarding cyber-information systems and the extensive use of information technology. Those issues have been generally discussed in this paper. It has been noted that this holistic augmented model has universal significance and can be used both in large manufacturing organizations in various industries, e.g., automotive, space, aviation, electronics, household appliances, and also in micro, small, and medium enterprises. Therefore, the possibility of adapting all assumptions of this universal model to the issues of designing and manufacturing prosthetic restorations as well as the cooperation of centers and laboratories for the manufacturing of those restorations with a network of generally private dental clinics has been considered in detail. An analysis of interrelationships between entities participating in the prosthetic treatment process has been made, paying attention to the relationship between patient expectations, design, and manufacturing capabilities of dental prosthesis restoration centers and the medical tasks of dental clinics. The basis of the considerations is a detailed literature analysis of diseases of the human stomatognathic system. The prevalence of caries and methods of its treatment have been pointed out, including endodontic, which allows the patient to leave his/her cured teeth for a long time in the mouth. Caries is the leading cause of dental extraction, and, in addition to periodontal disease and congenital disabilities, it mainly determines the need for prosthetic treatment. The classification of dental deficiencies has been made, and the possibilities of prosthetic treatment have been indicated along with the designation of prosthetic restorations in the event of partial or complete loss of natural dentition. Advanced prosthetic supplements offered by modern dentistry allow rebuilding all types of missing teeth. In the case of complete toothlessness, implants, permanent, and removable restorations are used for such a purpose. In the case of partial deficiencies, in particular, bridges based on their own teeth or implants are installed. It is also possible to manufacture bridges covering the full arch. In each case, dental prostheses are individually designed and manufactured for a specific patient and can be mounted alternatively on healthy or endodontically treated their own teeth or dental implants. In modern dentistry, cylindrical implants with a spiral surface are widely used. In several centers, independent work is also carried out, including own work carried out by the authors of this paper, on the introduction of individual implants produced using information obtained using cone-beam computed tomography (CBCT). Authors' own research indicates a change in the traditional relationship between a dentist and a dental engineer, which requires engineering knowledge in the field of biomaterials used, manufacturing technology, extensive knowledge in the field of IT support in design and manufacturing, as well as networks enabling internal integration and with numerous external stakeholders and with cloud computing.

The methodology in those cases is based on close cooperation between the implant dentist responsible for diagnosis and treatment and the dental engineer. His/her tasks include determining implantation areas, creating a three-dimensional model of the bone base based on a CBCT tomogram registered by the dentist and its connection with the three-dimensional model of the dentition and soft tissues in the oral cavity, prepared based on impression material, and most preferably a three-dimensional scan directly from the oral cavity. Those activities enable the design of prosthetic reconstruction and implant placement of one or several implants by planning technological holes for guiding the pilot drill to the bone based on a specific implant system and finally implementing the final design. This is necessary to design and create the surgical guide. Therefore, the dentist diagnoses the patient's condition, records the state of dental health using CBCT, and all prosthetic reconstruction is planned, designed, and manufactured by a dental engineer in consultation with the dentist while maintaining the highest possible quality standards and very narrow dimensional tolerances. Finally, the dentist assembles the manufactured prosthetic restoration in the patient's mouth in full accordance with the precise plan designed by the dental engineer using the prosthetic restorations he/she made and surgical guides. The procedure also enables virtual experiments at the design stage and digital verification of the designed prosthetic restoration, in line with the idea of a digital twin. Besides, it is possible to simultaneously make a model using rapid prototyping by stereolithographic (SLA) method and check the correctness of maintaining the correct fault height using an articulator. The presented approach is a radical turnaround in both clinical procedures and engineering techniques of implant preparation using computer-aided design and CAD/CAM manufacturing methods and AM additive manufacturing technology, including mainly selective laser sintering and stereolithography. Selective laser sintering technology also allows for the production of entirely innovative implant-scaffolds developed by the authors. They are built of a solid zone, typical of previously used implants, and a porous zone that acts as a scaffold, as hybrid connected into a uniform whole. The pores of implant-scaffolds are internally covered with bioactive thin atomic layers deposition (ALD), e.g., TiO_2 or Al_2O_3, providing favorable conditions for the proliferation and growth of living cells inside the pores of the device after implantation. The concept of hybrid multilayer engineering-biological composites developed by the authors, which is at the stage of in vitro and in vivo research, is a much more technologically advanced version of the concept. Surgical guides for the controlled making of holes in the bones are produced by methods of additive manufacturing (AM) or subtractive manufacturing by machining on a computerized numerical control (CNC) milling machine. For technological reasons, the surgical guide can be made of PMMA, in which pilot sleeves are filled with steel or titanium tubes dedicated to the selected implantation system. The authors' own designs illustrate the practical application of the augmented holistic concept of Industry 4.0 for the prosthetic restoration center, called Dentistry 4.0. The first example is the manufacturing of dental bridges using solid metal discs milling technology. Alternatively, their manufacturing is made with selective laser sintering (SLS) using metal powders hybridized with mechanical treatment using a center numerically controlled (CNC). The additive manufacturing by stereolithography (SLA) technology has been used for manufacturing the surgical guide for implantation of four implants for the assembly of a full arch bridge. Thanks to digital technology, it is possible to reduce the cost of manufacturing some prosthetic restorations compared to conventional methods. The overall conclusion of the research and analyses carried out indicates that on the one hand, unavoidable, and on the other hand, extremely beneficial, mainly due to maintaining very high-quality standards and strict dimensional tolerances, is the implementation of the idea of Dentistry 4.0 in accordance with the assumptions of the authors' own holistic Industry 4.0 model.

7. General Conclusions

The literature studies and analyses carried out take the following general conclusions:

1. The implementation of the idea of Dentistry 4.0 following the assumptions of the extended holistic model of Industry 4.0 is the imperative of the moment and the main direction of the development of dental prosthetic treatment, consisting mainly of the digitization of clinical diagnostics and the technology of designing and producing prosthetic restorations, gaining a dominant position, which is not only inevitable but also extraordinarily beneficial both from an economic and therapeutic point of view, as well as because of patient comfort.

2. The dissemination of the idea of Dentistry 4.0 poses new challenges in cooperation and mutual responsibility of dentists and dental engineers in the scope of treatment planning and implementation, which have undergone significant changes and require from dentists a wide range of self-education in IT and technical aspects with a considerable share of digitization and new challenges in creating a new generation of dental engineers and improving their competences and qualifications.

3. The engineering aspects of Dentistry 4.0 include both materials used for prosthetic restorations, manufacturing technologies using CAD/CAM, including, in particular, selective laser sintering and stereolithography, milling on center numerically controlled machines as hybrid or competitive technologies, as well as the use of atomic layers depositions in porous implants and implant-scaffolds, application of appropriate machines and devices or highly specialized computer-aided design and manufacturing software, use of computer networks, cloud computing, analysis of large data sets, augmented reality, artificial intelligence tools, and learning machines, whose range is systematically increasing in the given area.

4. Very high-quality standards and narrow dimensional tolerances of prosthetic restorations manufactured following the assumptions of Dentistry 4.0 significantly reduce the total duration of dental treatment, and patients reduce the nuisance associated with reducing the number of visits to the dental clinic in connection with the optimization of the term of implant treatment.

Author Contributions: Conceptualization, Methodology, literature review, Software, Formal Analysis, Investigation, Resources, Data Curation, Writing—Original Draft Preparation, Writing—Review & Editing, Visualization, Supervision, Project Administration, Funding Acquisition—all tasks performed jointly to half by both authors L.A.D. and L.B.D. All authors have read and agreed to the published version of the manuscript.

Funding: This research was prepared in the framework of the Project POIR.01.01-00-0485/16-00 on 'IMSKA-MAT Innovative dental and maxillofacial implants manufactured using the innovative additive technology supported by computer-aided materials design ADD-MAT' realized by the Medical and Dental Engineering Center for Research, Design, and Production ASKLEPIOS in Gliwice, Poland. The project implemented in 2017–2021 is co-financed by the Operational Programme Intelligent Development of the European Union.

Conflicts of Interest: The authors declare no conflict of interest.

References

1. Dobrzański, L.A.; Dobrzański, L.B. Approach to the design and manufacturing of prosthetic dental restorations according to the rules of the Industry 4.0 industrial revolution stage. *MPC* **2020**, prepared for printing.
2. Dobrzański, L.A. (Ed.) *Biomaterials in Regenerative Medicine*; InTech: Rijeka, Croatia, 2018.
3. Dobrzański, L.A.; Dobrzańska-Danikiewicz, A.D. (Eds.) *Metalowe Materiały Mikroporowate i lite do Zastosowań Medycznych i Stomatologicznych*; International OCSCOWorld Press: Gliwice, Poland, 2017.
4. Vieira, C.L.Z.; Caramelli, B. The history of dentistry and medicine relationship: Could the mouth finally return to the body? *Oral Dis.* **2009**, *15*, 538–546. [CrossRef] [PubMed]
5. Brånemark, P.-I.; Rydevik, B.L.; Skalak, R. (Eds.) *Osseointegration in Skeletal Reconstruction and Joint Replacement*; Quintessence Publishing Co.: Carol Stream, IL, USA, 1997.
6. Marxkors, R. Rozwój protetyki stomatologicznej w Europie Środkowej od drugiej połowy XX wieku. *Protet. Stomatol.* **2011**, *61*, 138–152. (translated from Germany: B. Płonka)

7. Brånemark, P.-I.; Hansson, B.O.; Adell, R.; Breine, U.; Lindström, J.; Hallén, O.; Ohman, A. Osseointegrated implants in the treatment of the edentulous jaw. Experience from a 10-year period. *Scand. J. Plast. Reconstr. Surg.* **1977**, *11* (Suppl. 16), 1–132.

8. Brånemark, R.; Brånemark, P.-I.; Rydevik, B.; Myers, R.R. Osseointegration in skeletal reconstruction and rehabilitation: A review. *J. Rehabil. Res. Dev.* **2001**, *38*, 175–181. [PubMed]

9. Brånemark, P.-I.; Adell, R.; Breine, U.; Hansson, B.O.; Lindström, J.; Ohlsson, A. Intra-osseous anchorage of dental prosthesis. I Experimental studies. *Scand. J. Plast. Reconstr. Surg.* **1969**, *3*, 81–100. [CrossRef] [PubMed]

10. Brånemark, P.-I. Osseointegration and its experimental studies. *J. Prosthet. Dent.* **1983**, *50*, 399–410. [CrossRef]

11. Brånemark, P.-I.; Hansson, B.O.; Adell, R.; Breine, U.; Lindström, J.; Hallén, O.; Ohman, A. *Osseointegr. Implant. Treat. Edentulous Jaw*; Almqvist and Wiksell: Stockholm, Sweden, 1977.

12. Brånemark, P.-I. Osseointegration in orthopaedic surgery. In Proceedings of the First World Congress of Osseointegration, Venice, Italy, 2 October 1994; pp. 17–18.

13. Angelillo, I.F.; Nobile, C.G.A.; Pavia, M. Survey of reasons for extraction of permanent teeth in Italy. *Community Dent. Oral Epidemiol.* **1996**, *24*, 336–340. [CrossRef]

14. Maciejewska-Szaniec, Z.; Maciejewska, B.; Piotrowski, P.; Wiskirska-Woźnica, B. Charakterystyka zaburzeń czynnościowych układu stomatognatycznego u pacjentów audiologicznych. *Fam. Med. Prim. Care Rev.* **2014**, *16*, 255–256.

15. Rüßmann, M.; Lorenz, M.; Gerbert, P.; Waldner, M.; Justus, J.; Engel, P.; Harnisch, M. *Industry 4.0: The Future of Productivity and Growth in Manufacturing Industries*; Boston Consulting Group: Boston, MA, USA, 2015. Available online: http://web.archive.org/web/20190711124617/https://www.zvw.de/media.media.72e472fb-1698-4a15-8858-344351c8902f.original.pdf (accessed on 15 December 2019).

16. Kagermann, H. Chancen von Industrie 4.0 Nutzen. In *Industrie 4.0 in Produktion, Automatisierung und Logistik*; Springer Fachmedien Wiesbaden: Wiesbaden, Germany, 2014; pp. 603–614.

17. Kagermann, H.; Wahlster, W.; Helbig, J. *Recommendations for Implementing the Strategic Initiative Industrie 4.0: Final Report of the Industrie 4.0 Working Group*; Federal Ministry of Education and Research: Bonn, Germany, 2013.

18. Hermann, M.; Pentek, T.; Otto, B. *Design Principles for Industrie 4.0 Scenarios: A Literature Review*; Technische Universität Dortmund: Dortmund, Germany, 2015.

19. European Commision. Commission sets out path to digitise European industry. Press release on 19 April 2016, Brussels. Available online: https://ec.europa.eu/commission/presscorner/detail/en/IP_16_1407 (accessed on 15 December 2019).

20. Dobrzański, L.A. Effect of Heat and Surface Treatment on the Structure and Properties of the Mg-Al-Zn-Mn Casting Alloys. In *Magnesium and Its Alloys: Technology and Applications*; Dobrzański, L.A., Totten, G.E., Bamberger, M., Eds.; CRC Press: Boca Raton, FL, USA, 2019; pp. 91–202.

21. Implementing the Digitising European Industry actions. "Digital Innovation Hubs on Smart Factories in new EU Member States," the project managed by the EC to support the European Parliament. 2017. Available online: https://ec.europa.eu/futurium/en/implementing-digitising-european-industry-actions/digital-innovation-hubs-smart-factories-new-eu (accessed on 15 December 2019).

22. Dobrzański, L.A.; Dobrzańska-Danikiewicz, A.D. Why are Carbon-Based Materials Important in Civilization Progress and Especially in the Industry 4.0 Stage of the Industrial Revolution? *Mater. Perform. Charact.* **2019**, *8*, 337–370. [CrossRef]

23. Ruehle, M.; Dosch, H.; Mittemeijer, E.J.; van de Voorde, M.H. (Eds.) *European White Book on Fundamental Research in Materials Science*; Max-Planck-Institute fuer Metallforschung: Stuttgart, Germany, 2001.

24. Jose, R.; Ramakrishna, S. Materials 4.0: Materials Big Data Enabled Materials Discovery. *Appl. Mater. Today* **2018**, *10*, 127–132. [CrossRef]

25. Lu, B.H.; Bateman, R.J.; Cheng, K. RFID Enabled Manufacturing: Fundamentals, Methodology and Applications. *Int. J. Agil. Syst. Manag.* **2006**, *1*, 73–92. [CrossRef]

26. Brettel, M.; Friederichsen, N.; Keller, M.; Rosenberg, M. How Virtualization, Decentralization and Network Building Change the Manufacturing Landscape: An Industry 4.0 Perspective. *Int. J. Mech. Aerosp. Ind. Mechatron. Manuf. Eng.* **2014**, *8*, 37–44.

27. Wan, J.; Yan, H.; Liu, Q.; Zhou, K.; Lu, R.; Li, D. Enabling Cyber-Physical Systems with Machine-to-Machine Technologies. *Int. J. Ad Hoc Ubiquitous Comput.* **2013**, *13*, 187–196. [CrossRef]

28. Zhong, R.Y.; Li, Z.; Pang, L.Y.; Pan, Y.; Qu, T.; Huang, G.Q. RFID-Enabled Real-Time Advanced Planning and Scheduling Shell for Production Decision Making. *Int. J. Comput. Integr. Manuf.* **2013**, *26*, 649–662. [CrossRef]

29. Gubbi, J.; Buyya, R.; Marusic, S.; Palaniswami, M. Internet of Things (IoT): A Vision, Architectural Elements, and Future Directions. *Future Gener. Comput. Syst.* **2013**, *29*, 1645–1660. [CrossRef]

30. Zhu, Q.; Wang, R.; Chen, Q.; Liu, Y.; Qin, W. IOT Gateway: Bridging Wireless Sensor Networks in to Internet of Things. In *2010 IEEE/IFIP International Conference on Embedded and Ubiquitous Computing*; The Institute of Electrical and Electronics Engineers: Piscataway, NJ, USA, 2010; pp. 347–352.

31. Wu, D.Z.; Greer, M.J.; Rosen, D.W.; Schaefer, D. Cloud Manufacturing: Strategic Vision and State-of-the-Art. *J. Manuf. Syst.* **2013**, *32*, 564–579. [CrossRef]

32. Buer, S.V.; Strandhagen, J.O.; Chan, T.S. The link between Industry 4.0 and lean manufacturing: Mapping current research and establishing a research agenda. *Int. J. Prod. Res.* **2018**, *56*, 2924–2940. [CrossRef]

33. Moreno-Vozmediano, R.; Montero, R.S.; Llorente, I.M. Key Challenges in Cloud Computing: Enabling the Future Internet of Services. *IEEE Internet Comput.* **2013**, *17*, 18–25. [CrossRef]

34. Giusto, D.; Iera, A.; Morabito, G.; Atzori, L. (Eds.) *The Internet of Things*; Springer: New York, NY, USA, 2010.

35. Lee, J.; Kao, H.-A.; Yang, S. Service Innovation and Smart Analytics for Industry 4.0 and Big Data Environment. *Procedia CIRP* **2014**, *16*, 3–8. [CrossRef]

36. Bi, Z.; Xu, L.D.; Wang, C. Internet of Things for Enterprise Systems of Modern Manufacturing. *IEEE Trans. Ind. Inform.* **2014**, *10*, 1537–1546. [CrossRef]

37. Dobrzański, L.A.; Dobrzańska-Danikiewicz, A.D. Applications of Laser Processing of Materials in Surface Engineering in the Industry 4.0 Stage of the Industrial Revolution. *Mater. Perform. Charact.* **2019**, *8*, 1091–1129. [CrossRef]

38. Patel, P.; Cassou, D. Enabling High-Level Application Development for the Internet of Things. *J. Syst. Softw.* **2015**, *103*, 62–84. [CrossRef]

39. Qiu, X.; Luo, H.; Xu, G.; Zhong, R.Y.; Huang, G.Q. Physical Assets and Service Sharing for IoT-enabled Supply Hub in Industrial Park (SHIP). *Int. J. Prod. Econ.* **2015**, *159*, 4–15. [CrossRef]

40. Wamba, S.F.; Akter, S.; Edwards, A.; Chopin, G.; Gnanzou, D. How 'Big Data' Can Make Big Impact: Findings from a Systematic Review and a Longitudinal Case Study. *Int. J. Prod. Econ.* **2015**, *165*, 234–246. [CrossRef]

41. Posada, J.; Toro, C.; Barandiaran, I.; Oyarzun, D.; Stricker, D.; de Amicis, R.; Pinto, E.B.; Eisert, P.; Döllner, J.; Vallarino, I. Visual Computing as a Key Enabling Technology for Industrie 4.0 and Industrial Internet. *IEEE Comput. Graph. Appl.* **2015**, *35*, 26–40. [CrossRef]

42. Zhang, Y.; Zhang, G.; Wang, J.; Sun, S.; Si, S.; Yang, T. Real-Time Information Capturing and Integration Framework of the Internet of Manufacturing Things. *Int. J. mputer Integr. Manuf.* **2015**, *28*, 811–822. [CrossRef]

43. Lee, J.; Bagheri, B.; Kao, H.-A. A Cyber-Physical Systems Architecture for Industry 4.0-Based Manufacturing Systems. *Manuf. Lett.* **2015**, *3*, 18–23. [CrossRef]

44. Georgakopoulos, D.; Jayaraman, P.P.; Fazia, M.; Villari, M.; Ranjan, R. Internet of Things and Edge Cloud Computing Roadmap for Manufacturing. *IEEE Cloud Comput.* **2016**, *3*, 66–73. [CrossRef]

45. Almada-Lobo, F. The Industry 4.0 Revolution and the Future of Manufacturing Execution Systems (MES). *J. Innov. Manag.* **2015**, *3*, 16–21. [CrossRef]

46. Gerbert, P.; Lorenz, M.; Rüßmann, M.; Waldner, M.; Justus, J.; Engel, P.; Harnisch, M. *Industry 4.0: The Future of Productivity and Growth in Manufacturing Industries*; Boston Consulting Group: Boston, MA, USA, 2015. Available online: http://web.archive.org/web/20190711125458/https://www.bcg.com/publications/2015/engineered_products_project_business_industry_4_future_productivity_growth_manufacturing_industries.aspx (accessed on 15 December 2019).

47. Farooq, M.U.; Waseem, M.; Mazhar, S.; Khairi, A.; Kamal, T. A Review on Internet of Things (IoT). *Int. J. Comput. Appl.* **2015**, *113*, 1–7. [CrossRef]

48. Yin, Y.H.; Nee, A.Y.C.; Ong, S.K.; Zhu, J.Y.; Gu, P.H.; Chen, L.J. Automating Design with Intelligent Human-Machine Integration. *CIRP Ann.* **2015**, *64*, 655–677. [CrossRef]

49. Li, B.-H.; Hou, B.-C.; Yu, W.-T.; Lu, X.-B.; Yang, C.-W. Applications of Artificial Intelligence in Intelligent Manufacturing: A Review. *Front. Inf. Technol. Ectronic Eng.* **2017**, *18*, 86–96. [CrossRef]

50. Colin, M.; Galindo, R.; Hernández, O. Information and Communication Technology as a Key Strategy for Efficient Supply Chain Management in Manufacturing SMEs. *Procedia Comput. Sci.* **2015**, *55*, 833–842. [CrossRef]

51. Hozdić, E. Smart Factory for Industry 4.0: A Review. *Int. J. Mod. Manuf. Technol.* **2015**, *7*, 28–35.

52. Kumar, K.; Zindani, D.; Davim, J.P. *Industry 4.0: Developments Towards the Fourth Industrial Revolution*; Springer Nature Singapore Pte Ltd.: Singapore, 2019.

53. Zhong, R.Y.; Huang, G.Q.; Lan, S.; Dai, Q.Y.; Zhang, T.; Xu, C. A Two-Level Advanced Production Planning and Scheduling Model for RFID-Enabled Ubiquitous Manufacturing. *Adv. Eng. Inform.* **2015**, *29*, 799–812. [CrossRef]

54. Qin, J.; Liu, Y.; Grosvenor, R. A Categorical Framework of Manufacturing for Industry 4.0 and Beyond. *Procedia CIRP* **2016**, *52*, 173–178. [CrossRef]

55. Wang, S.; Wan, J.; Zhang, D.; Li, D.; Zhang, C. Towards Smart Factory for Industry 4.0: A Self-Organized Multi-Agent System with Big Data Based Feedback and Coordination. *Comput. Netw.* **2016**, *101*, 158–168. [CrossRef]

56. DIN SPEC 91345:2016-04. *Referenzarchitekturmodell Industrie 4.0 (RAMI4.0)*; Beuth: Berlin, Germany, 2016.

57. Monostori, L.; Kádár, B.; Bauernhansl, T.; Kondoh, S.; Kumara, S.; Reinhart, G.; Sauer, O.; Schuh, G. Sihn and K. Ueda, "Cyber-Physical Systems in Manufacturing". *CIRP Ann.* **2016**, *65*, 621–641. [CrossRef]

58. Mosterman, P.J.; Zander, J. Industry 4.0 as a Cyber-Physical System Study. *Softw. Syst. Modeling* **2016**, *15*, 17–29. [CrossRef]

59. Zhong, R.Y.; Lan, S.; Xu, C.; Dai, Q.; Huang, G.Q. Visualization of RFID-Enabled Shopfloor Logistics Big Data in Cloud Manufacturing. *Int. J. Adv. Manuf. Technol.* **2016**, *84*, 5–16. [CrossRef]

60. Zhong, R.Y.; Newman, S.T.; Huang, G.Q.; Lan, S. Big Data for Supply Chain Management in the Service and Manufacturing Sectors: Challenges, Opportunities, and Future Perspectives. *Comput. Ind. Eng.* **2016**, *101*, 572–591. [CrossRef]

61. Bahrin, M.A.K.; Othman, M.F.; Azli, N.H.N.; Talib, M.F. Industry 4.0: A Review on Industrial Automation and Robotic. *J. Teknol.* **2016**, *78*, 137–143. [CrossRef]

62. Misra, G.; Kumar, V.; Agarwal, A.; Agarwal, K. Internet of Things (IoT)—A Technological Analysis and Survey on Vision, Concepts, Challenges, Innovation Directions, Technologies, and Applications (An Upcoming or Future Generation Computer Communication System Technology). *Am. J. Electr. Electron. Eng.* **2016**, *4*, 23–32. [CrossRef]

63. Harrison, R.; Vera, D.; Ahmad, B. Engineering Methods and Tools for Cyber-Physical Automation Systems. *Proc. IEEE* **2016**, *104*, 973–985. [CrossRef]

64. Schumacher, A.; Erol, S.; Sihn, W. A Maturity Model for Assessing Industry 4.0 Readiness and Maturity of Manufacturing Enterprises. *Procedia CIRP* **2016**, *52*, 161–166. [CrossRef]

65. Ardito, L.; Petruzzelli, A.; Panniello, U.; Garavelli, A. Towards Industry 4.0: Mapping digital technologies for supply chain management-marketing integration. *Bus. Process Manag. J.* **2019**, *25*, 323–346. [CrossRef]

66. Sipsas, K.; Alexopoulos, K.; Xanthakis, V.; Chryssolouris, G. Collaborative Maintenance in Flow-Line Manufacturing Environments: An Industry 4.0 Approach. *Procedia CIRP* **2016**, *55*, 236–241. [CrossRef]

67. Dobrzański, L.A.; Dobrzańska-Danikiewicz, A.D.; Malara, P.; Gaweł, T.; Dobrzański, L.B.; Achtelik-Franczak, A. Fabrication of scaffolds from Ti6Al4V powders using the computer aided laser method. *Arch. Metall. Mater.* **2015**, *60*, 1065–1070. [CrossRef]

68. Zhong, R.Y.; Xu, X.; Klotz, E.; Newman, S.T. Intelligent Manufacturing in the Context of Industry 4.0: A Review. *Engineering* **2017**, *3*, 616–630. [CrossRef]

69. Adolphs, P.; Auer, S.; Bedenbender, H.; Billmann, M.; Hankel, M.; Heidel, R.; Hoffmeister, M.; Jochem, M.; Koschnick, G.; Koziolek, H.; et al. *Structure of the Administration Shell: Continuation of the Development of the Reference Model for the Industrie 4.0 Component*; Federal Ministry for Economic Affairs and Energy: Berlin, Germany, 2016.

70. Stock, T.; Seliger, G. Opportunities of Sustainable Manufacturing in Industry 4.0. *Procedia CIRP* **2016**, *40*, 536–541. [CrossRef]

71. Pfeiffer, S. Robots, Industry 4.0 and Humans, or Why Assembly Work is More than Routine Work. *Societies* **2016**, *6*, 16. [CrossRef]

72. Vaidya, S.; Ambad, P.; Bhosle, S. Industry 4.0—A Glimpse. *Procedia Manuf.* **2018**, *20*, 233–238. [CrossRef]

73. Gorkhali, A.; Xu, L.D. Enterprise Application Integration in Industrial Integration: A Literature Review. *J. Ind. Integr. Manag.* **2016**, *1*, 1650014. [CrossRef]

74. Thoben, K.-D.; Wiesner, S.; Wuest, T. "Industrie 4.0" and Smart Manufacturing–A Review of Research Issues and Application Examples. *Int. J. Autom. Technol.* **2017**, *11*, 4–16. [CrossRef]

75. Lu, Y. Industry 4.0: A Survey on Technologies, Applications and Open Research Issues. *J. Ind. Inf. Integr.* **2017**, *6*, 1–10. [CrossRef]

76. Xu, X. Machine Tool 4.0 for the New Era of Manufacturing. *Int. J. Adv. Manuf. Technol.* **2017**, *92*, 1893–1900. [CrossRef]

77. Han, G.W.; Feng, D.; Yin, M.; Ye, W.J. Ceramic/aluminum co-continuous composite synthesized by reaction accelerated melt infiltration. *Mater. Sci. Eng. A* **1997**, *225*, 204–207. [CrossRef]

78. Dobrzański, L.A. Role of materials design in maintenance engineering in the context of industry 4.0 idea. *J. Achiev. Mater. Manuf. Eng.* **2019**, *96*, 12–49. [CrossRef]

79. Dobrzański, L.A.; Dobrzański, L.B.; Dobrzańska-Danikiewicz, A.D. Conventional and additive technologies for products manufacturing using powders of metals, their alloys and ceramics. *Mater. Perform. Charact.* **2020**, prepared for printing.

80. Dobrzański, L.A.; Dobrzański, L.B. *Innovative Dental and Maxillo-Facial Implant-Scaffold Manufactured Using the Innovative Technology and Additive Computer-Aided Materials Design ADD-MAT*; IMSKA-MAT project number POIR.01.01.00-0397/16-00; Medical and Dental Engineering Centre for Research, ASKLEPIOS: Gliwice, Poland, 2017–2021.

81. Dobrzański, L.B.; Achtelik-Franczak, A.; Dobrzańska, J.; Dobrzański, L.A. The digitisation for the immediate dental implantation of incisors with immediate individual prosthetic restoration. *J. Achiev. Mater. Manuf. Eng.* **2019**, *97*, 57–68. [CrossRef]

82. United Nations Industrial Development Organization. *Industrial Development Report 2020. Industrializing in the Digital Age. Overview*; ID/449; UNIDO: Vienna, Auatria, 2019.

83. Dobrzański, L.A. *Materiały Inżynierskie i Projektowanie Materiałowe: Podstawy Nauki o Materiałach i Metaloznawstwo*; WNT: Warsaw, Poland, 2006.

84. Ashby, M.F. *Materials Selection in Mechanical Design*; Butterworth-Heinemann: Burlington, MA, USA, 2011.

85. Dobrzańska-Danikiewicz, A.D. The State of the Art Analysis and Methodological assumptions of Evaluation and Development Prediction for Materials Surface Technologies. *J. Achiev. Mater. Manuf. Eng.* **2011**, *49*, 121–141.

86. Ashby, M.F.; Jones, D.R.H. *Engineering Materials 2: An. Introduction to Microstructures and Processing*; Butterworth-Heinemann: Waltham, MA, USA, 2013.

87. Waterman, N.A.; Ashby, M.F. *The Materials Selector, Volumes 1–3*; Chapman & Hall: London, UK; Weinheim, Germany; New York, NY, USA; Tokyo, Japan; Melbourne, Australia; Madras, India, 1997.

88. Ashby, M.F.; Jones, D.R.H. *Engineering Materials 1: An. Introduction to Properties, Applications and Design*; Butterworth-Heinemann: Waltham, MA, USA, 2012.

89. Ashby, M.F. *Materials Selection in Mechanical Design*; Pergamon Press: Oxford, UK; New York, NY, USA; Seoul, Korea; Tokyo, Japan, 1992.

90. Dobrzański, L.B.; Achtelik-Franczak, A.; Dobrzańska, J.; Dobrzański, L.A. Comparison of the structure and properties of the solid Co-Cr-W-Mo-Si alloys used for dental restorations CNC machined or selective laser-sintered. *MDPI* **2020**. prepared for printing.

91. Malara, P.; Dobrzański, L.B. Computer-aided design and manufacturing of dental surgical guides based on cone beam computed tomography. *Arch. Mater. Sci. Eng.* **2015**, *76*, 140–149.

92. Malara, P.; Dobrzański, L.B. Designing and manufacturing of implantoprosthetic fixed suprastructures in edentulous patients on the basis of digital impressions. *Arch. Mater. Sci. Eng.* **2015**, *76*, 163–171.

93. Kaye, G. Does Digital Dentist Mean Anything Today. Available online: https://www.dentaleconomics.com/science-tech/article/16389518/does-digital-dentist-mean-anything-today (accessed on 6 December 2017).

94. Harrison, T. The Emergence of the Fourth Industrial Revolution in Dentistry. Available online: https://www.dentistryiq.com/practice-management/industry/article/16366877/the-emergence-of-the-fourth-industrial-revolution-in-dentistry (accessed on 25 April 2018).

95. Caldas, A.F., Jr.; Marcenes, W.; Sheiham, A. Reason for tooth extraction in a Brazilian population. *Int. Dent. J.* **2000**, *50*, 267–273. [CrossRef]

96. Morita, M.; Kimura, T.; Kanegae, M.; Ishikawa, A.; Watanobe, T. Reasons for extraction of permanent teeth in Japan. *Community Dent. Oral Epidemiol.* **1994**, *22*, 303–306. [CrossRef]

97. Al-Shammari, K.F.; Al-Arsari, J.M.; Al-Melh, M.A.; Al-Khabhaz, A.K. Reasons for tooth extraction in Kuwait. *Med. Princ. Pract.* **2006**, *15*, 417–422. [CrossRef]

98. Chestnutt, I.G.; Binnie, V.I.; Taylor, M.M. Reason for tooth extraction in Scotland. *J. Dent.* **2000**, *28*, 295–297. [CrossRef]

99. Skawińska, A.; Błaszczak, J.; Sikorska-Jaroszyńska, M.H.J.; Mielnik-Błaszczak, M.; Borowicz, J. Profilaktyka choroby próchnicowej. *Stand. Med. Pediatr.* **2011**, *8*, 768–773.

100. Aas, J.A.; Griffen, A.L.; Dardis, S.R.; Lee, A.M.; Olsen, I.; Dewhirst, F.E.; Leys, E.J.; Paster, B.J. Bacteria of Dental Caries in Primary and Permanent Teeth in Children and Young Adults. *J. Clin. Microbiol.* **2008**, *46*, 1407–1417. [CrossRef] [PubMed]

101. Pawka, B.; Dreher, P.; Herda, J.; Szwiec, I.; Krasicka, M. Próchnica zębów u dzieci problemem społecznym. *Probl. Hig. Epidemiol.* **2010**, *91*, 5–7.

102. Wójcicka, A.; Zalewska, M.; Czerech, E.; Jabłoński, R.; Grabowska, S.Z.; Maciorkowska, E. Próchnica wieku rozwojowego chorobą cywilizacyjną. *Przegląd Epidemiol.* **2012**, *66*, 705–711.

103. Paju, S.; Scannapieco, F.A. Oral biofilms, periodontitis and pulmonary infections. *Oral Dis.* **2007**, *13*, 508–512. [CrossRef]

104. Hollandra, A.C.B.; de Alencar, A.H.G.; de Estrela, C.R.A.; Bueno, M.R.; Estrela, C. Prevalance of Endodontically Trale Teeth in a Brazilian Adult Population. *Braz. Dent. J.* **2008**, *19*, 313–317. [CrossRef]

105. Oral Health Country. *Area Profile Project CAPP*, 5th ed.; Centre for Oral Health Sciences, Malmö University: Malmö, Sweden, 2013. Available online: http://www.mah.se/CAPP/ (accessed on 15 December 2016).

106. Spratt, D.A.; Pratten, J.; Wilson, M.; Gulabivala, K. An in vitro evaluation of the antimicrobial efficacy of irrigants on biofilms of root canal isolates. *Int. Endod. J.* **2001**, *34*, 300–307. [CrossRef]

107. Petersen, P.E.; Ogawa, H. Prevention of dental caries through the use of fluoride—The WHO approach. *Community Dent. Health* **2016**, *33*, 66–68. [CrossRef]

108. Kanasi, E.; Ayilavarapu, S.; Jones, J. The aging population: Demographics and the biology of aging. *Periodontology 2000* **2016**, *72*, 13–18. [CrossRef]

109. Petersen, P.E. *Oral Health Surveys: Basic Methods*, 5th ed.; World Health Organization: Geneva, Switzerland, 2013.

110. Krzywiec, E.; Zalewska, M.; Wójcicka, A.; Jabłoński, R.; Olejnik, B.J.; Grabowska, S.Z.; Jamiołkowski, J.; Czerech, E.; Łuszcz, A.; Stepek, A.; et al. Wybrane zachowania żywieniowe a występowanie próchnicy u młodzieży. *Przegląd Epidemiol.* **2012**, *66*, 713–721.

111. Szymańska, J.; Szalewski, L. Próchnica zębów mlecznych w populacji polskich dzieci w wieku 0, 5–6 lat. *Zdr. Publiczne* **2011**, *121*, 86–89.

112. Bromblik, A.; Wierzbicka, M.; Szatko, F. Wpływ uwarunkowań środowiskowych na zapadalność i przebieg próchnicy zębów u dzieci. *Czas. Stomatol.* **2010**, *63*, 301–309.

113. Yu, C.; Abbott, P.V. An overview of the dental pulp: Its functions and responses to injury. *Aust. Dent. J.* **2007**, *52*, S4–S16. [CrossRef] [PubMed]

114. Petersen, P.E. *Changing Oral Health Profiles of Children in Central and Eastern Europe—Challenges for the 21st Century*; WHO Global Oral Health: Geneva, Switzerland, 2003.

115. Staśkiewicz, T. Analiza wpływu wybranych czynników na intensywność próchnicy wczesnej. *Ann. Acad. Med. Stetin.* **2012**, *58*, 36–39. [PubMed]

116. Miyagi, S.P.H.; Kerkis, I.; Maranduba, C.M.d.; Gomes, C.M.; Martinis, M.D.; Marques, M.M. Expression of extracellular matrix proteins in human dental pulp stem cells depends on the donor tooth conditions. *J. Endod.* **2010**, *36*, 826–831. [CrossRef]

117. Calayatud-Perez, V.; Gimbernat, J.M.; Lagunas, J.G. Parietal osteomyelitis of dental origin: A case report. *J. Cranio Maxillofac. Surg.* **1993**, *21*, 127–129. [CrossRef]

118. Dobrzańska, J.; Gołombek, K.; Dobrzański, L.B. Polymer materials used in endodontic treat-ment—In Vitro testing. *Arch. Mater. Sci. Eng.* **2012**, *58*, 110–115.

119. Sunay, H.; Tanalp, J.; Dikbas, I.; Bayirli, G. Cross-sectional evaluation of the periapical status and quality of root canal treatment in a selected population of urban Turkish adults. *Int. Endod. J.* **2007**, *40*, 139–145. [CrossRef]

120. Salehrabi, R.; Rotstein, I. Endodontic treatment outcomes in a large patient population in the USA: An epidemiological study. *J. Endod.* **2004**, *30*, 846–850. [CrossRef]

121. Lazarski, M.P.; Walker, W.A.; Flores, C.M.; Schindler, W.G.; Hargreaves, K.M. Epidemiological evaluation of the outcomes of nonsurgical root canal treatment in a large cohort of insured dental patients. *J. Endod.* **2001**, *27*, 791–796. [CrossRef]

122. Meuwissen, R.; Eschen, S. Twenty years of endodontic treatment. *J. Endod.* **1983**, *9*, 390–393. [CrossRef]

123. Chen, S.C.; Chuech, L.H.; Hsiao, C.K.; Tsai, M.Y.; Ho, S.C.; Chiang, C.P. An epidemiologic study of tooth retention after nonsurgical endodontic treatment in a large population in Taiwan. *J. Endod.* **2007**, *33*, 226–229. [CrossRef] [PubMed]

124. Kirkevang, L.; Hörsted-Bindslev, P.; Ørstavik, D.; Wenzel, A. Frequency and distribution of endodontically treated teeth and apical periodontitis in an urban Danish population. *Int. Endod. J.* **2001**, *34*, 198–205. [CrossRef] [PubMed]

125. Tsuneishi, M.; Yamamoto, T.; Yamanaka, R.; Tamaki, N.; Sakamoto, T.; Tsuji, K.; Watanabe, T. Radiographic evaluation of periapical status and prevalence of endodontic treatment in an adult Japanese population. *Oral Surg. Oral Med. Oral Pathol. Oral Radiol. Endodontol.* **2005**, *100*, 631–635. [CrossRef]

126. Asgary, S.; Shadman, B.; Ghalamkarpour, Z.; Shahravan, A.; Ghoddusi, J.; Bagherpour, A.; Baghban, A.A.; Hashemipour, M.; Ghasemian, M. Pour periapical status and quality of root canal filling and coronal restoration in Iranian population. *Iran. Endod. J.* **2010**, *5*, 74–82. [PubMed]

127. Jiménez-Pinzón, A.; Seguro-Egea, J.J.; Poyato-Ferrera, H.; Velasco-Ortega, E.; Rios-Santos, J.V. Prevalance of apical periodontitis and frequency of root fillednteeth in an adult Spanish population. *Int. Endod. J.* **2004**, *37*, 167–173. [CrossRef] [PubMed]

128. Georgopoulou, M.K.; Spanaki-Voreadi, A.P.; Pantazis, N.; Kontakiotis, E.G. Frequency and distribution of root filled teeth and apical periodontitis in a Greek population. *Int. Endod. J.* **2005**, *38*, 105–111. [CrossRef]

129. Loftus, J.J.; Keating, A.P.; McCartan, B.E. Periapical status and quality of endodontic treatment in an adult Irish population. *Int. Endod. J.* **2005**, *38*, 81–86. [CrossRef]

130. Terças, A.G.; de Oliveira, A.E.F.; Lopes, F.F.; Filho, E.M.M. Radiographic study of the prevalence of apical periodontitis and endodontic treatment in the adult population of São Luís, MA, Brazil. *J. Appl. Oral Sci.* **2006**, *14*, 183–187. [CrossRef]

131. Da Silva, K.; Lam, J.M.Y.; Wu, N.; Duckmanton, P. Cross-sectional study of endodontic treatment in an Australian population. *Aust. Endod. J.* **2009**, *35*, 140–146. [CrossRef]

132. Peciuliene, V.; Rimkuviene, J.; Maneliene, R.; Iranauskaite, D. Apical periodontitis in root filled teeth associated with the quality of root fillings. *Stomatol. Balt. Dent. Maxillofac. J.* **2006**, *8*, 122–126.

133. De Moor, R.J.G.; Hommez, G.M.G.; de Boever, J.G.; Delmé, K.I.M.; Martens, G.E.I. Periapical health related to the quality of root canal treatment in a Belgian population. *Int. Endod. J.* **2000**, *33*, 113–120. [CrossRef] [PubMed]

134. Carvalho, J.C.M.; Dias, R.B.; Mattos, B.S.C.; Andre, M. *Rehabilitacao Protetica Craniomaxilofacial*; Livraria Santos Editora: Sao Paulo, Brazil, 2013.

135. Melchels, F.P.W.; Domingos, M.A.N.; Klein, T.J.; Malda, J.; Bartolo, P.J.; Hutmacher, D.W. Additive manufacturing of tissues and organs. *Prog. Polym. Sci.* **2012**, *37*, 1079–1104. [CrossRef]

136. Allahverdi, M.; Danforth, S.C.; Jafari, M.; Safari, A. Processing of advanced electroceramic components by fused deposition technique. *J. Eur. Ceram. Soc.* **2001**, *21*, 1485–1490. [CrossRef]

137. al Jabbari, Y.S. Physico-mechanical properties and prosthodontic applications of Co-Cr dental alloys: A review of the literature. *J. Adv. Prosthodont.* **2014**, *6*, 138–145. [CrossRef] [PubMed]

138. Dejak, B.; Kacprzak, M.; Suliborski, B.; Śmielak, B. Struktura i niektóre właściwości ceramik dentystycznych stosowanych w uzupełnieniach pełnoceramicznych w świetle literatury. *Protet. Stomatol.* **2006**, *56*, 471–477.

139. Madfa, A.A.; Al-Sanabani, F.A.; Al-Qudami, N.H.; Al-Sanabani, J.S.; Amran, A.G. Use of Zirconia in Dentistry: An Overview. *Open Biomater. J.* **2014**, *5*, 1–9. [CrossRef]

140. Conrad, H.J.; Seong, W.J.; Pesun, I.J. Current ceramic materials and systems with clinical recommendations: A systematic review. *J. Prosthet. Dent.* **2007**, *98*, 389–404. [CrossRef]

141. Komine, F.; Blatz, M.B.; Matsumura, H. Current status of zirconia-based fixed restorations. *J. Oral Sci.* **2010**, *52*, 531–539. [CrossRef]

142. Vagkopoulou, T.; Koutayas, S.O.; Koidis, P.; Strub, J.R. Zirconia in dentistry: Part 1. Discovering the nature of an upcoming bioceramic. *Eur. J. Esthet. Dent.* **2009**, *4*, 130–151.

143. Bachhav, V.C.; Aras, M.A. Zirconia-based fixed partial dentures: A clinical review. *Quintessence Int.* **2011**, *42*, 173–182.

144. Craig, R.G. *Materiały Stomatologiczne*; Elsevier Urban & Partner: Wrocław, Poland, 2008.

145. Augustyn-Pieniążek, J.; Łukaszczyk, A.; Szczurek, A.; Sowińska, K. Struktura i własności stopów dentystycznych na bazie kobaltu stosowanych do wykonywania protez szkieletowych. *Inżynieria Mater.* **2013**, *34*, 116–120.

146. Miyazaki, T.; Nakamura, T.; Matsumura, H.; Ban, S.; Kobayashi, T. Current status of zirconia restoration. *J. Prosthodont. Res.* **2013**, *57*, 236–261. [CrossRef] [PubMed]

147. Malara, P.; Dobrzański, L.B. Screw-retained full arch restorations—Methodology of computer aided design and manufacturing. *Arch. Mater. Sci. Eng.* **2017**, *83*, 23–29. [CrossRef]

148. Worthington, P. Current status of osseointegration. In *International Workshop on Osseointegration in Skeletal Reconstruction and Joint Replacement*; Rydevik, B., Brånemark, P.-I., Skalak, R., Eds.; The Institute for Applied Biotechnology: Göteborg, Sweden, 1991; pp. 17–20.

149. Bjursten, L.-M. The bone-implant interface in osseointegration. In *International Workshop on Osseointegration in Skeletal Reconstruction and Joint Replacement*; Rydevik, B., Brånemark, P.-I., Skalak, R., Eds.; The Institute for Applied Biotechnology: Göteborg, Sweden, 1991; pp. 25–31.

150. Esposito, M.; Hirsch, J.M.; Lekholm, U.; Thomsen, P. Biological factors contributing to failures of osseointegrated oral implants. (I). Success criteria and epidemiology. *Eur. J. Oral Sci.* **1998**, *106*, 527–551. [CrossRef] [PubMed]

151. Skalak, R. Biomechanics of osseointegration. In *Osseointegration in Skeletal Reconstruction and Joint Replacement*; Brånemark, P.-I., Rydevik, B.L., Skalak, R., Eds.; Quintessence Publishing Co.: Carol Stream, IL, USA, 1997; pp. 45–56.

152. Skalak, R. Overview of previous development and biomechanics of osseointegration. In *International Workshop on Osseointegration in Skeletal Reconstruction and Joint Replacement*; Rydevik, B., Brånemark, P.-I., Skalak, R., Eds.; The Institute for Applied Biotechnology: Göteborg, Sweden, 1991.

153. Adell, R.; Ericksson, B.; Kekholm, U.; Branemark, P.-I.; Jemt, T. Long-term follow-up study of osseointegrated implants in the treatment of totally endentulous jaws. *Int. J. Oral Maxillofac. Implant.* **1990**, *5*, 347–359.

154. Henry, P.J. Osseointegration in dentistry. In *Proceedings of Interdisciplinary Conference on Osseointegration from Molecule to Man*; Institute for Applied Biotechnology: Goteborg, Sweden, 1999; pp. 2–12.

155. Cranin, A.N.; Lemons, J.E. Dental implantation. In *Biomaterials Science: An. Introduction to Materials in Medicine*, 2nd ed.; Ratner, B.D., Hoffman, A.S., Schoen, F.J., Lemons, J.E., Eds.; Elsevier Academic Press: Amsterdam, The Netherlands, 1997; pp. 555–573.

156. Williams, E.; Rydevik, B.; Brånemark, P.-I. (Eds.) *Proceedings of Interdisciplinary Conference on Osseointegration from Molecule to Man*; Institute for Applied Biotechnology: Goteborg, Sweden, 1999.

157. Donath, K.; Laass, M.; Günzl, H.J. The histopatology of different foreign-body reactions in oral soft tissue and bone tissue. *Virchows Archiv. A Pathol. Anat. Histopathol.* **1992**, *420*, 131–137. [CrossRef]

158. Donath, K.; Breuner, G. A method for the study of undecalcified bones and teeth with attached soft tissues. The Säge-Schliff (sawing and grinding) technique. *J. Oral Pathol. Med.* **1982**, *11*, 318–326. [CrossRef]

159. Schroeder, A.; van der Zypen, E.; Stich, H.; Sutter, F. The reactions of bone, connective tissue, and epithelium to endosteal implants with titanium-sprayed surfaces. *J. Maxillofac. Surg.* **1981**, *9*, 15–25. [CrossRef]

160. Dobrzański, L.B.; Achtelik-Franczak, A.; Dobrzańska, J.; Pietrucha, P. Application of polymer impression masses for the obtaining of dental working models for the stereolithographic 3D printing. *Arch. Mater. Sci. Eng.* **2019**, *95*, 31–40. [CrossRef]

161. Papaspyridakos, P.; White, G.S.; Lal, K. Flapless CAD/CAM-guided surgery for staged transition from failing dentition to complete arch implant rehabilitation: A 3-year clinical report. *J. Prosthet. Dent.* **2012**, *107*, 143–150. [CrossRef]

162. Malara, P.; Dobrzański, L.B.; Dobrzańska, J. Computer-aided designing and manufacturing of partial removable dentures. *J. Achiev. Mater. Manuf. Eng.* **2015**, *73*, 157–164.

163. Malkondu, Ö.; Tinastepe, N.; Akan, E.; Kazazoğlu, E. An overview of monolithic zirconia in dentistry. *Biotechnol. Biotechnol. Equip.* **2016**, *30*, 644–652. [CrossRef]

164. Malara, P.; Dobrzański, L.B. Computer aided manufacturing and design of fixed bridges restoring the lost dentition, soft tissue and the bone. *Arch. Mater. Sci. Eng.* **2016**, *81*, 68–75. [CrossRef]

165. Majewski, S. Nowe technologie wytwarzania stałych uzupełnień zębowych: Galwanoforming, technologia CAD/CAM, obróbka tytanu i współczesne systemy ceramiczne. *Protet. Stomatol.* **2007**, *57*, 124–131.

166. Galanis, C.C.; Sfantsikopoulos, M.M.; Koidis, P.T.; Kafantaris, N.M.; Mpikos, P.G. Computer methods for automating preoperative dental implant planning: Implant positioning and size assignment. *Comput. Methods Programs Biomed.* **2007**, *86*, 30–38. [CrossRef] [PubMed]

167. Majewski, S.; Pryliński, M. *Materiały i Technologie Współczesnej Protetyki Stomatologicznej*; Wydawnictwo Czelej: Lublin, Poland, 2013.

168. Miyazaki, T.; Hotta, Y. CAD/CAM systems available for the fabrication of crown and bridge restorations. *Aust. Dent. J.* **2011**, *56*, 97–106. [CrossRef] [PubMed]

169. Ho, M.H.; Lee, S.Y.; Chen, H.H.; Lee, M.C. Three-dimensional finite element analysis of the effect of post on stress distribution in dentin. *J. Prosthet. Dent.* **1994**, *72*, 367–372. [CrossRef]

170. Reimann, Ł.; Żmudzki, J.; Dobrzański, L.A. Strength analysis of a three-unit dental bridge framework with the Finite Element Method. *Acta Bioeng. Biomech.* **2015**, *17*, 51–59. [CrossRef]

171. Litonjua, L.A.; Andreana, S.; Patra, A.K.; Cohen, R.E. An assessment of stress analyses in the theory of abfraction. *Bio Med. Mater. Eng.* **2004**, *14*, 311–321.

172. Kierklo, A. Studium hipotez wytrzymałościowych tkanek twardych kości i zęba. *Zesz. Nauk. Politech. Białostockiej Bud.* **1997**, *16*, 95–106.

173. Lee, H.E.; Lin, C.L.; Wang, C.H.; Cheng, C.H.; Chang, C.H. Stresses at the cervical lesion of maxillary premolar—A finite element investigation. *J. Dent.* **2002**, *30*, 283–290. [CrossRef]

174. Żmudzki, J. *Uwarunkowania Materiałowe Wydolności Czynnościowej Całkowitych Osiadających Protez Zębowych*; International OCSCO World Press: Gliwice, Poland, 2012.

175. Kierklo, A. Wpływ ubytku klinowego na stan wytężenia twardych tkanek zęba. *Czas. Stomatol.* **1999**, *52*, 355–360.

176. Nagasao, T.; Miyamoto, J.; Kawana, H. Biomechanical Evaluation of Implant Placement in the Reconstructed Mandible. *J. Oral Maxillofocial Implant.* **2009**, *24*, 999–1005.

177. Milewski, G. *Wytrzymałościowe Aspekty Interakcji Biomechanicznej Tkanka Twarda-Implant w Stomatologii*; Wydawnictwo Politechniki Krakowskiej: Cracow, Poland, 2002.

178. Ress, J.S.; Hammadeh, M. Undermining of enamel as a mechanism of abfraction lesion formation: A finite element study. *Eur. J. Oral Sci.* **2004**, *112*, 347–352. [CrossRef]

179. Loska, S.; Paszenda, Z.; Basiaga, M.; Kiel-Jamrozik, M. Zastosowanie MES w analizie układu wkład koronowo-korzeniowy-ząb. In Proceedings of the Inżynieria Biomedyczna w Stomatologii, IBwS 2014, Wisła, Poland, 18–21 November 2014; pp. 46–47, [CD-ROM].

180. Dejak, B. Wpływ szerokości i głębokości wkładów koronowo-korzeniowych na naprężenia występujące w trzonowcach podczas cyklu żucia. *Protet. Stomatol.* **2004**, *54*, 86–92.

181. Tanaka, M.; Naito, T.; Yokota, M.; Kohno, M. Finite element analysis of the possible mechanism of cervical lesion formation by occlusal force. *J. Oral Rehabil.* **2003**, *30*, 60–67. [CrossRef]

182. Szymaniak, E.; Kierklo, A.; Tribiłło, R. Praktyczne zastosowanie metody elementów skończoych (MES) do analizy naprężeń w tkankach zęba i wypełnieniu. *Czas. Stomatol.* **1991**, *44*, 271–275.

183. Dejak, B. Wpływ anizotropowych właściwości szkliwa na naprężenia występujące w zębach trzonowych podczas żucia. *Protet. Stomatol.* **2004**, *54*, 162–167.

184. Kierklo, A.; Szymaniak, E.; Tribiłło, R.; Walendziuk, A. Zastosowanie metody elementów skończonych (MES) do oceny naprężeń w zębie z kanałem wypełnionym gutaperką. *Czas. Stomatol.* **1995**, *48*, 495–500.

185. Dobosz, A.; Panek, H.; Dobosz, K. Zastosowanie metody elementów skończonych do analizy naprężeń w twardych tkankach zębów. *Dent. Med. Probl.* **2005**, *42*, 651–655.

186. Yang, H.S.; Lang, L.A.; Guckes, A.D.; Felton, D.A. The effect of thermal change on various dowel-and-core restorative materials. *J. Prosthet. Dent.* **2001**, *86*, 74–80. [CrossRef] [PubMed]

187. Żmudzki, J.; Chladek, G. Modelowanie numeryczne biomechaniki protez całkowitych. In Proceedings of the Inżynieria Biomedyczna w Stomatologii, IBwS 2011, Ustroń, Poland, 1–3 July 2011; pp. 1–3, [CD-ROM].

188. Dejak, B.; Młotkowski, A. Badanie naprężeń w zębach w zależności od wielkości części korzeniowej wkładów. *Stomatol. Współczesna* **1995**, *2*, 410–419.

189. Ślusarski, P.; Świniarski, J.; Kozakiewicz, M.; Elgalal, M.; Bogusławski, G.; Gralewski, J.; Grądzki, R.; Ślusarska, A.; Dejak, B. Model numeryczny do analizy biomechaniki żuchwy z wykorzystaniem metody elementów skończonych (MES). *Mag. Stomatol.* **2012**, *11*, 38–41.

190. Lipski, T.; Chladek, W. Wartości sił zgryzu w zależności od wieku i płci. *Protet. Stomatol.* **1997**, *47*, 284–287.

191. Gultekin, B.A.; Gultekin, P.; Yalcin, S. Application of Finite Element Analysis in Implant Dentistry. In *Finite Element Analysis—New Trends and Developments*; Ebrahimi, F., Ed.; IntechOpen: Rijeka, Croatia, 2012. [CrossRef]

192. Habelitz, S.; Marshall, S.J.; Marschall, G.W.M., Jr.; Balooch, M. Mechanical properties of human dental enamel on the nanometre scale. *Arch. Oral Biol.* **2001**, *46*, 173–183. [CrossRef]

193. Dobrzański, L.A. (Ed.) *Powder Metallurgy—Fundamentals and Case Studies*; IntechOpen: Rijeka, Croatia, 2017.

194. Dobrzański, L.A.; Dobrzańska-Danikiewicz, A.D.; Achtelik-Franczak, A.; Dobrzański, L.B. Comparative analysis of mechanical properties of scaffolds sintered from Ti and Ti6Al4V powders. *Arch. Mater. Sci. Eng.* **2015**, *73*, 69–81.

195. Dobrzański, L.A.; Dobrzańska-Danikiewicz, A.D.; Achtelik-Franczak, A.; Dobrzański, L.B.; Hajduczek, E.; Matula, G. Fabrication Technologies of the Sintered Materials Including Materials for Medical and Dental Application. In *Powder Metallurgy—Fundamentals and Case Studies*; Dobrzański, L.A., Ed.; IntechOpen: Rijeka, Croatia, 2017; pp. 17–52. [CrossRef]

196. Dobrzański, L.A.; Dobrzańska-Danikiewicz, A.D.; Achtelik-Franczak, A.; Dobrzański, L.B.; Szindler, M.; Gaweł, T.G. Porous Selective Laser Melted Ti and Ti6Al4V Materials for Medical Applications. In *Powder Metallurgy—Fundamentals and Case Studies*; Dobrzański, L.A., Ed.; IntechOpen: Rijeka, Croatia, 2017; pp. 161–181. [CrossRef]

197. Dobrzański, L.A.; Dobrzańska-Danikiewicz, A.D.; Gaweł, T.G.; Achtelik-Franczak, A. Selective laser sintering and melting of pristine titanium and titanium Ti6Al4V alloy powders and selection of chemical environment for etching of such materials. *Arch. Metall. Mater.* **2015**, *60*, 2039–2045. [CrossRef]

198. Dobrzański, L.A.; Matula, G.; Dobrzańska-Danikiewicz, A.D.; Malara, P.; Kremzer, M.; Tomiczek, B.; Kujawa, M.; Hajduczek, E.; Achtelik-Franczak, A.; Dobrzański, L.B.; et al. Composite Materials Infiltrated by Aluminium Alloys Based on Porous Skeletons from Alumina, Mullite and Titanium Produced by Powder Metallurgy Techniques. In *Powder Metallurgy—Fundamentals and Case Studies*; Dobrzański, L.A., Ed.; IntechOpen: Rijeka, Croatia, 2017; pp. 95–137. [CrossRef]

199. Majkowska, B.; Jażdżewska, M.; Wołowiec, E.; Piekoszewski, W.; Klimek, L.; Zieliński, A. The possibility of use of laser-modified Ti6Al4V alloy in friction pairs in endoprostheses. *Arch. Metall. Mater.* **2015**, *60*, 755–758. [CrossRef]

200. Duan, B.; Wang, M.; Zhou, W.Y.; Cheung, W.L.; Li, Z.Y.; Lu, W.W. Three-dimensional nanocomposite scaffolds fabricated via selective laser sintering for bone tissue engineering. *Acta Biomater.* **2010**, *6*, 4495–4505. [CrossRef] [PubMed]

201. Das, S.; Wohlert, M.; Beaman, I.J.; Bourell, D.L. Producing metal parts with selective laser sintering/hot isostatic pressing. *JOM* **1998**, *50*, 17–20. [CrossRef]

202. Guo, N.; Leu, M.C. Additive manufacturing: Technology, applications and research needs. *Front. Mech. Eng.* **2013**, *8*, 215–243. [CrossRef]

203. Mellor, S.; Hao, L.; Zhang, D. Additive manufacturing: A framework for implementation. *Int. J. Prod. Econ.* **2014**, *149*, 194–201. [CrossRef]

204. Żmudzki, J.; Chladek, G.; Kasperski, J. Biomechanical factors related to occlusal load transfer in removable complete dentures. *Biomech. Modeling Mechanobiol.* **2015**, *14*, 679–691. [CrossRef]

205. Brandt, M. (Ed.) *Laser Additive Manufacturing; Materials, Design, Technologies and Applications*; Woodhead Publishing: Sawston, UK, 2017.

206. Dobrzański, L.A.; Dobrzańska-Danikiewicz, A.D.; Achtelik-Franczak, A.; Szindler, M. Structure and Properties of the Skeleton Microporous Materials with Coatings Inside the Pores for Medical and Dental Applications. In *Frontiers in Materials Processing, Applications, Research and Technology*; Muruganant, M., Chirazi, A., Raj, B., Eds.; Springer: Singapore, 2018; pp. 297–320. [CrossRef]

207. Dobrzański, L.B. Struktura i Własności Materiałów Inżynierskich na Uzupełnienia Protetyczne Układu Stomatognatycznego Wytwarzane Metodami Przyrostowymi i Ubytkowymi. Ph.D. Thesis, Academy of Mining and Metallurgy, Cracow, Poland, 2017.

208. Dobrzański, L.A. The concept of biologically active microporous engineering materials and composite biological-engineering materials for regenerative medicine and dentistry. *Arch. Mater. Sci. Eng.* **2016**, *80*, 64–85. [CrossRef]

209. Bandyopadhyay, A.; Espana, F.; Balla, V.K.; Bose, S.; Ohgami, Y.; Davies, N.M. Influence of porosity on mechanical properties and in vivo response of Ti6Al4 implants. *Acta Biomater.* **2010**, *6*, 1640–1648. [CrossRef] [PubMed]

210. Dobrzański, L.A.; Dobrzańska-Danikiewicz, A.D.; Gaweł, T.G. Individual implants of a loss of palate fragments fabricated using SLM equipment. *J. Achiev. Mater. Manuf. Eng.* **2016**, *77*, 24–30. [CrossRef]

211. Koeck, B.; Wagner, W. *Implantologia*; Majewski, S.W., Ed.; Urban & Partner: Wrocław, Poland, 2004.

212. Dobrzański, L.A. Applications of newly developed nanostructural and microporous materials in biomedical, tissue and mechanical engineering. *Arch. Mater. Sci. Eng.* **2015**, *76*, 53–114.

213. De Vasconcellos, L.M.R.; de Oliveira, M.V.; Graça, M.L.d.; de Vasconcellos, G.O.; Carvalho, Y.R.; Cairo, C.A.A. Porous Titanium Scaffolds Produced by Powder Metallurgy for Biomedical Applications. *Mater. Res.* **2008**, *11*, 275–280. [CrossRef]

214. Dikova, T.; Dzhendov, D.; Simov, M. Microstructure and Hardness of Fixed Dental Prostheses Manufactured by Additive Technologies. *J. Achiev. Mater. Manuf. Eng.* **2015**, *71*, 60–69.

215. Bram, M.; Schiefer, H.; Bogdanski, D.; Köller, M.; Buchkremer, H.P.; Stöver, D. Implant surgery: How bone bonds to PM titanium. *Metal. Powder Rep.* **2006**, *61*, 26–31. [CrossRef]

216. Dikova, T. Properties of Co-Cr Dental Alloys Fabricated Using Additive Technologies. In *Biomaterials in Regenerative Medicine*; Dobrzański, L.A., Ed.; IntechOpen: Rijeka, Croatia, 2018; pp. 141–159. [CrossRef]

217. Bram, M.; Stiller, C.; Buchkremer, H.P.; Stover, D.; Bauer, H. High-porosity titanium, stainless steel, and superalloy parts. *Adv. Eng. Mater.* **2000**, *2*, 196–199. [CrossRef]

218. Ryan, G.; Pandit, A.; Apatsidis, D.P. Fabrication methods of porous metals for use in orthopaedic applications. *Biomaterials* **2006**, *27*, 2651–2670. [CrossRef] [PubMed]

219. Dobrzański, L.A.; Achtelik-Franczak, A.; Król, M. Computer Aided Design in Selective Laser Sintering (SLS)—Application in medicine. *J. Achiev. Mater. Manuf. Eng.* **2013**, *60*, 66–75.

220. Esen, Z.; Bor, S. Processing of titanium foams using magnesium spacer particles. *Scr. Mater.* **2007**, *56*, 341–344. [CrossRef]

221. Dikova, T.; Dzhendov, D.; Simov, M.; Katreva-Bozukova, I.; Angelova, S.; Pavlova, D.; Abadzhiev, M.; Tonchev, T. Modern Trends in the Development of the Technologies for Production of Dental Constructions. *J. MAB Annu. Proceeding (Sci. Pap.)* **2015**, *21*, 974–981. [CrossRef]

222. Gupta, S.; Kumar, S. Lasers in Dentistry—An Overview. *Trends Biomater. Artif. Organs* **2011**, *25*, 119–123.

223. Dobrzański, L.B. Mechanical Properties Comparison of Engineering Materials Produced by Additive and Substractive Technologies for Dental Prosthetic Restoration Application. In *Biomaterials in Regenerative Medicine*, Dobrzański, L.A., Ed.; IntechOpen: Rijeka, Croatia, 2018; pp. 111–140. [CrossRef]

224. Kolan, K.C.; Doiphode, N.D.; Leu, M.C. Selective laser sintering and freeze extrusion fabrication of scaffolds for bone repair using 13–93 bioactive glass: A comparison. In *Proceedings of the Solid Freeform Fabrication Symposium*; The University of Texas at Austin: Austin, TX, USA, 2010.

225. Yu, W.H.; Sing, S.L.; Chua, C.K.; Kuo, C.N.; Tian, X.L. Particle-reinforced metal matrix nanocomposites fabricated by selective laser melting: A state of the art review. *Prog. Mater. Sci.* **2019**, *104*, 330–379. [CrossRef]

226. Dobrzański, L.A.; Dobrzańska-Danikiewicz, A.D. Foresight of the Surface Technology in Manufacturing. In *Handbook of Manufacturing Engineering and Technology*; Nee, A.Y.C., Ed.; Springer: London, UK, 2015; pp. 2587–2637.

227. Dobrzańska-Danikiewicz, A.D. *Księga Technologii Krytycznych Kształtowania Struktury i Własności Powierzchni Materiałów Inżynierskich*; International OCSCO World Press: Gliwice, Poland, 2013.

228. Dobrzannńska-Danikiewicz, A.D. *Metodologia Komputerowo Zintegrowanego Prognozowania Rozwoju Inżynierii Powierzchni Materiałów*; International OCSCO World Press: Gliwice, Poland, 2012.

229. Dobrzańska-Danikiewicz, A.D. Foresight of Material Surface Engineering as a Tool Building a Knowledge-Based Economy. *Mater. Sci. Forum* **2012**, *706–709*, 2511–2516. [CrossRef]

230. Dobrzański, L.A.; Achtelik-Franczak, A. Struktura i własności tytanowych szkieletowych materiałów mikroporowatych wytworzonych metodą selektywnego spiekania laserowego do zastosowań w implantologii oraz medycynie regeneracyjnej. In *Metalowe Materiały Mikroporowate i Lite do Zastosowań Medycznych i Stomatologicznych*; Dobrzański, L.A., Dobrzańska-Danikiewicz, A.D., Eds.; International OCSCO World Press: Gliwice, Poland, 2017; pp. 186–244.

231. Dobrzański, L.A.; Dobrzańska-Danikiewicz, A.D. *Inżynieria Powierzchni Materiałów—Kompendium Wiedzy i Podręcznik Akademicki*; International OCSCO World Press: Gliwice, Poland, 2018.

232. Ciocca, L.; Fantini, M.; de Crescenzio, F.; Corinaldesi, G.; Scott, R. Direct metal laser sintering (DMLS) of a customized titanium mesh for prosthetically guided bone regeneration of atrophic maxillary arches. *Med. Biol. Eng. Comput.* **2011**, *49*, 1347–1352. [CrossRef]

233. Sachlos, E.; Czernuszka, J.T. Making tissue engineering scaffolds work. Review: The application of solid freeform fabrication technology to the production of tissue engineering scaffolds. *Eur. Cells Mater.* **2003**, *5*, 29–40. [CrossRef]

234. Eshraghi, S.; Das, S. Mechanical and microstructural properties of polycaprolactone scaffolds with one-dimensional, two-dimensional, and three-dimensional orthogonally oriented porous architectures produced by selective laser sintering. *Acta Biomater.* **2010**, *6*, 2467–2476. [CrossRef] [PubMed]

235. German, R.M. *Sintering Theory and Practice*; John Wiley & Sons: New York, NY, USA, 1996.

236. German, R.M. Phase diagrams in liquid phase sintering treatments. *JOM* **1986**, *38*, 26–29. [CrossRef]

237. Schaffer, G.B.; Sercombe, T.B.; Lumley, R.N. Liquid phase sintering of aluminium alloys. *Mater. Chem. Phys.* **2001**, *67*, 85–91. [CrossRef]

238. Kang, S.-J.L. Basis of Liquid Phase Sintering. In *Sintering, Densification, Grain Growth and Microstructure*; Butterworth-Heinemann: Amsterdam, The Netherlands, 2005; pp. 199–203. [CrossRef]

239. Kang, S.-J.L. Liquid phase sintering. In *Sintering of Advanced Materials, Woodhead Publishing Series in Metals and Surface Engineering*; Fang, Z.Z., Ed.; Woodhead Publishing Limited: Cambridge, UK, 2010; pp. 110–129. [CrossRef]

240. Huo, S.H.; Qian, M.; Schaffer, G.B.; Crossin, E. Aluminium powder metallurgy. In *Fundamentals of Aluminium Metallurgy; Production, Processing and Applications*; Lumley, R., Ed.; Woodhead Publishing Limited: Cambridge, UK, 2011; pp. 655–701.

241. Veiseh, M.; Edmondson, D. Bone as an Open Cell Porous Material. *ME 599K Spec. Top. Cell. Solids* **2003**.

242. Dobrzański, L.A.; Dobrzańska-Danikiewicz, A.D.; Szindler, M.; Achtelik-Franczak, A.; Pakieła, W. Atomic layer deposition of TiO_2 onto porous biomaterials. *Arch. Mater. Sci. Eng.* **2015**, *75*, 5–11.

243. Dobrzański, L.A.; Dobrzańska-Danikiewicz, A.D.; Gaweł, T.G. Ti6Al4V porous elements coated by polymeric surface layer for biomedical applications. *J. Achiev. Mater. Manuf. Eng.* **2015**, *71*, 53–59.

244. Dobrzański, L.A.; Dobrzańska-Danikiewicz, A.D.; Czuba, Z.P.; Dobrzański, L.B.; Achtelik-Franczak, A.; Malara, P.; Szindler, M.; Kroll, L. Metallic skeletons as reinforcement of new composite materials applied in orthopaedics and dentistry. *Arch. Mater. Sci. Eng.* **2018**, *92*, 53–85. [CrossRef]

245. Dobrzański, L.A.; Dobrzańska-Danikiewicz, A.D.; Czuba, Z.P.; Dobrzański, L.B.; Achtelik-Franczak, A.; Malara, P.; Szindler, M.; Kroll, L. The new generation of the biological-engineering materials for applications in medical and dental implant-scaffolds. *Arch. Mater. Sci. Eng.* **2018**, *91*, 56–85. [CrossRef]

246. Japan Business Federation. Toward Realization of the New Economy and Society (Outline). Keidanren, 2016. Available online: https://www.keidanren.or.jp/en/policy/2016/029_outline.pdf (accessed on 15 December 2019).

247. Fukuyama, M. Society 5.0: Aiming for a New Human-Centered Society. 2018. Available online: https://www.jef.or.jp/journal/pdf/220th_Special_Article_02.pdf (accessed on 15 December 2019).

248. Harayama, Y. Society 5.0: Aiming for a New Human-Centered Society. *Hitachi Rev.* **2017**, *66*, 8–13.

249. Available online: https://www8.cao.go.jp/cstp/society5_0/index.html (accessed on 15 December 2019).

250. Japan Business Federation. Society 5.0: Co-Creating the Future (Excerpt). Keidanren, 2018. Available online: https://www.keidanren.or.jp/en/policy/2018/095_outline.pdf (accessed on 15 December 2019).

251. Government of Japan Cabinet Office. "Society 5.0", Cabinet Office. 2019. Available online: http://web.archive.org/web/20190710182953/ (accessed on 15 December 2019).

252. Japan Business Federation. "Japan's Initiatives–Society 5.0", Keidanren. Available online: http://www.keidanren.or.jp/en/policy/2017/010_overview.pdf (accessed on 15 December 2019).

253. Samuelson, P.A. The Pure Theory of Public Expenditure. *Rev. Econ. Stat.* **1954**, *36*, 387–389. [CrossRef]

254. Tietenberg, T.; Lewis, L. *Environmental and Natural Resource Economics*; Pearson Education: Boston, MA, USA; Columbus, OH, USA; Indianapolis, IN, USA; New York, NY, USA; San Francisco, CA, USA; Upper Saddle River, NJ, USA; Amsterdam, The Netherlands; Cape Town, South Africa; Dubai, UAE; London, UK; Madrid, Spain; Milan, Italy; Munich, Germany; Paris, France; Montreal, QC, Canada; Toronto, ON, Canada; Delhi, India; Mexico City, Mexico; Sao Paulo, Brazil; Sydney, Australia; Hong Kong, China; Seoul, Korea; Singapore; Taipei, Taiwan; Tokyo, Japan, 2003.

255. Brundtland, G.H. *Report of the World Commission on Environment and Development (WCED): Our Common Future (also known as the Brundtland Report)*; United Nations: Oslo, Norway, 20 March 1987. Available online: https://sustainabledevelopment.un.org/content/documents/5987our-common-future.pdf (accessed on 15 December 2019).

256. Rideout, J. Are Canadian Manufacturers Ready for Industry 4.0? Sadly, the Answer is no. Cisco Canada Blog. 9 June 2017. Available online: https://gblogs.cisco.com/ca/2017/06/09/are-canadian-manufacturers-ready-for-industry-4-0-sadly-the-answer-is-no/ (accessed on 15 December 2019).

257. EIT Manufacturing. Our Flagships. Available online: http://eitmanufacturing.eu/about-us/#flagships (accessed on 15 December 2019).

258. Tay, S.I.; Lee, T.C.; Hamid, N.A.A.; Ahmad, A.N.A. An Overview of Industry 4.0: Definition, Components, and Government Initiatives. *J. Adv. Res. Dyn. Control Syst.* **2018**, *10*, 1379–1387.

259. Hockfield, S. *The Age of Living Machings: How Biology Will Build. the Next Technology Revolution*; W.W.Norton & Company: New York, NY, USA, 2019.

260. Kolan, K.C.; Leu, M.C.; Hilmas, G.E.; Brown, R.F.; Velez, M. Fabrication of 13–93 bioactive glass scaffolds for bone tissue engineering using indirect selective laser sintering. *Biofabrication* **2011**, *3*, 025004. [CrossRef] [PubMed]

261. Warnke, P.H.; Douglas, T.; Wollny, P.; Sherry, E.; Steiner, M.; Galonska, S.; Becker, S.T.; Springer, I.N.; Wiltfang, J.; Sivananthan, S. Rapid prototyping: Porous titanium alloy scaffolds produced by selective laser melting for bone tissue engineering. *Tissue Eng. Part C Methods* **2009**, *15*, 115–124. [CrossRef] [PubMed]

262. Leiggener, C.; Messo, E.; Thor, A.; Zeilhofer, H.F.; Hirsch, J.M. A selective laser sintering guide for transferring a virtual plan to real time surgery in composite mandibular reconstruction with free fibula osseous flaps. *Int. J. Oral Maxillofac. Surg.* **2009**, *38*, 187–194. [CrossRef]

263. Salmoria, G.V.; Fancello, E.A.; Roesler, C.R.M.; Dabbas, F. Functional graded scaffold of HDPE/HA prepared by selective laser sintering: Microstructure and mechanical properties. *Int. J. Adv. Manuf. Technol.* **2013**, *65*, 1529–1534. [CrossRef]

264. Kruth, J.P.; Mercelis, P.; Vaerenbergh, J.V.; Froyen, L.; Rombouts, M. Binding mechanisms in selective laser sintering and selective laser melting. *Rapid Prototyp. J.* **2005**, *11*, 26–36. [CrossRef]

265. Tan, K.H.; Chua, C.K.; Leong, K.F.; Cheah, C.M.; Cheang, P.; Bakar, M.S.A.; Cha, S.W. Scaffold development using selective laser sintering of polyetheretherketone-hydroxyapatite biocomposite blends. *Biomaterials* **2003**, *24*, 3115–3123. [CrossRef]

266. Wang, Y.; Shen, Y.; Wang, Z.; Yang, J.; Liu, N.; Huang, W. Development of highly porous titanium scaffolds by selective laser melting. *Mater. Lett.* **2010**, *64*, 674–676. [CrossRef]

267. Williams, J.M.; Adewunmi, A.; Schek, R.M.; Flanagan, C.L.; Krebsbach, P.H.; Feinberg, S.E.; Hollister, S.J.; Das, S. Bone tissue engineering using polycaprolactone scaffolds fabricated via selective laser sintering. *Biomaterials* **2005**, *26*, 4817–4827. [CrossRef]

268. Peng, W.-M.; Liu, Y.-F.; Jiang, X.-F.; Dong, X.-T.; Jun, J.; Baur, D.A.; Xu, J.-J.; Pan, H.; Xu, X. Bionic mechanical design and 3D printing of novel porous Ti6Al4V implants for biomedical applications. *J. Zhejiang Univ. Sci. B* **2019**, *20*, 647–659. [CrossRef]

269. Abe, F.; Osakada, K.; Shiomi, M.; Uematsu, K.; Matsumoto, M. The manufacturing of hard tools from metallic powders by selective laser melting. *J. Mater. Process. Technol.* **2001**, *111*, 210–213. [CrossRef]

270. Shishkovsky, I.V.; Volova, L.T.; Kuznetsov, M.V.; Morozov, Y.G.; Parkind, I.P. Porous biocompatible implants and tissue scaffolds synthesized by selective laser sintering from Ti and NiTi. *J. Mater. Chem.* **2008**, *18*, 1309–1317. [CrossRef]

271. Thomas, C.L.; Gaffney, T.M.; Kaza, S.; Lee, C.H. Rapid prototyping of large scale aerospace structures. In *Proceedings of IEEE Aerospace Applications Conference*; IEEE: Aspen, CO, USA, 1996; Volume 4, pp. 219–230. [CrossRef]

272. Kruth, J.P.; Froyen, L.; van Vaerenbergh, J.; Mercelis, P.; Rombouts, M.; Lauwers, B. Selective laser melting of iron-based powder jet. *J. Mater. Process. Technol.* **2004**, *149*, 616–622. [CrossRef]

273. Gasser, A.; Backes, G.; Kelbassa, I.; Weisheit, A.; Wissenbach, K. Laser additive manufacturing: Laser metal deposition (LMD) and selective laser melting (SLM) in turbo-engine applications. *Laser Tech. J.* **2010**, *7*, 58–63. [CrossRef]

274. Pham, D.T.; Dimov, S.; Lacan, F. Selective laser sintering: Applications and technological capabilities. *Proc. Inst. Mech. Eng. Part B J. Eng. Manuf.* **1999**, *213*, 435–449. [CrossRef]

275. Zhang, J.; Song, B.; Wei, Q.; Bourell, D.; Shi, Y. A review of selective laser melting of aluminum alloys: Processing, microstructure, property and developing trends. *J. Mater. Sci. Technol.* **2019**, *35*, 270–284. [CrossRef]

276. Osakada, K.; Shiomi, M. Flexible manufacturing of metallic products by selective laser melting of powder. *Int. J. Mach. Tools Manuf.* **2006**, *46*, 1188–1193. [CrossRef]

277. Xie, F.; He, X.; Cao, S.; Qu, X. Structural and mechanical characteristics of porous 316L stainless steel fabricated by indirect selective laser sintering. *J. Mater. Process. Technol.* **2013**, *213*, 838–843. [CrossRef]

278. Cader, M.; Trojnacki, M. Analiza możliwości zastosowania technologii przyrostowych do wytwarzania elementów konstrukcji robotów mobilnych. *Pomiary Autom. Robot.* **2013**, *2*, 200–207.

279. Fukuda, A.; Takemoto, M.; Saito, T.; Fujibayashi, S.; Neo, M.; Pattanayak, D.K.; Matsushita, T.; Sasaki, K.; Nishida, N.; Kokubo, T.; et al. Osteoinduction of porous Ti implants with a channel structure fabricated by selective laser melting. *Acta Biomater.* **2011**, *7*, 2327–2336. [CrossRef]

280. Oyama, K.; Filho, L.F.S.D.; Muto, J.; de Souza, D.G.; Gun, R.; Otto, B.A.; Carrau, R.L.; Prevedello, D.M. Endoscopic endonasal cranial base surgery simulation using an artificial cranial base model created by selective laser sintering. *Neurosurg. Rev.* **2015**, *38*, 171–178. [CrossRef] [PubMed]

281. Song, Y.; Yan, Y.; Zhang, R.; Xu, D.; Wang, F. Manufacturing of the die of an automobile deck part based on rapid prototyping and rapid tooling technology. *J. Mater. Process. Technol.* **2002**, *120*, 237–242. [CrossRef]

282. Furumoto, T.; Koizumi, A.; Alkahari, M.R.; Anayama, R.; Hosokawa, A.; Tanaka, R.; Ueda, T. Permeability and strength of a porous metal structure fabricated by additive manufacturing. *J. Mater. Process. Technol.* **2015**, *219*, 10–16. [CrossRef]

283. Contreras, A.; Lopez, V.H.; Bedolla, E. Mg/TiC composites manufactured by pressureless melt infiltration. *Scr. Mater.* **2004**, *51*, 249–253. [CrossRef]

284. Choi, J.W.; Yamashita, M.; Sakakibara, J.; Kaji, Y.; Oshika, T.; Wicker, R.B. Combined micro and macro additive manufacturing of a swirling flow coaxial phacoemulsifier sleeve with internal micro-vanes. *Biomed. Microdevices* **2010**, *12*, 875–886. [CrossRef] [PubMed]

285. Dobrzański, L.A.; Drygała, A.; Musztyfaga, M.; Panek, P. Porównanie struktury i właściwości elektrycznych przednich elektrod ogniw słonecznych wypalanych w piecu taśmowym i selektywnie spiekanych laserowo. *Elektron. Konstr. Technol. Zastos.* **2011**, *52*, 50–52.

286. Klimek, M. Zastosowanie technologii SLS w wykonawstwie stałych uzupełnień protetycznych. *Twój Przegląd Stomatol.* **2012**, *12*, 47–55.

287. Nouri, A.; Hodgson, P.D.; Wen, C. Biomimetic Porous Titanium Scaffolds for Orthopedic and Dental Applications. In *Biomimetics Learning from Nature*; Mukherjee, A., Ed.; IntechOpen: London, UK, 2010; pp. 415–450. [CrossRef]

288. Mazzoli, A. Selective laser sintering in biomedical engineering. *Med. Biol. Eng. Comput.* **2013**, *51*, 245–256. [CrossRef]

289. Li, J.P.; Habibovic, P.; van den Doel, M.; Wilson, C.E.; de Wijn, J.R.; van Blitterswijk, C.A.; de Groot, K. Bone ingrowth in porous titanium implants produced by 3D fiber deposition. *Biomaterials* **2007**, *28*, 2810–2820. [CrossRef]

290. Bibb, R.; Thompson, D.; Winder, J. Computed tomography characterisation of additive manufacturing materials. *Med. Eng. Phys.* **2011**, *33*, 590–596. [CrossRef]

291. Xue, W.; Vamsi, K.B.; Bandyopadhyay, A.; Bose, S. Processing and biocompatibility evaluation of laser processed porous titanium. *Acta Biomaterialia* **2007**, *3*, 1007–1018. [CrossRef]

292. Karageorgiou, V.; Kaplan, D. Porosity of 3D biomaterial scaffolds. *Biomaterials* **2005**, *26*, 5474–5491. [CrossRef]

293. Wanibuchi, M.; Ohtaki, M.; Fukushima, T.; Friedman, A.H.; Houkin, K. Skull base training and education using an artificial skull model created by selective laser sintering. *Acta Neurochir.* **2010**, *152*, 1055–1060. [CrossRef]

294. Yoshida, K.; Saiki, Y.; Ohkubo, C. Improvement of drawability and fabrication possibility of dental implant screw made of pure titanium. *Hut. Wiadomości Hut.* **2011**, *78*, 153–156.

295. Dobrzański, L.A.; Musztyfaga, M.; Drygała, A. Selective laser sintering method of manufacturing front electrode of silicon solar cell. *J. Achiev. Mater. Manuf. Eng.* **2010**, *42*, 111–119.

296. UNO General Assembly. 70/1. Transforming our world: The 2030 Agenda for Sustainable Development. Resolution adopted by the General Assembly on 25 September 2015. Available online: https://www.un.org/ga/search/view_doc.asp?symbol=A/RES/70/1&Lang=E (accessed on 15 December 2019).

297. European Parliament. Small and Medium-Sized Enterprises. Fact Sheets on the European Union. Available online: http://www.europarl.europa.eu/factsheets/pl/sheet/63/male-i-srednie-przedsiebiorstwa (accessed on 15 December 2019).

298. Dieter, G.E. *ASM Handbook—Materials Selection and Design*; Dieter, G.E., Ed.; ASM International: Metals Park, Hong Kong, 1997; Volume 20.

299. Dobrzański, L.A. *Metale i ich Stopy*; International OCSCO World Press: Gliwice, Poland, 2017.

300. Liao, Y.S.; Li, H.C.; Chiu, Y.Y. Study of laminated object manufacturing with separately applied heating and pressing. *Int. J. Adv. Manuf. Technol.* **2006**, *27*, 703–707. [CrossRef]

301. Cormier, D.; Harrysson, O.; West, H. Characterization of H13 steel produced via electron beam melting. *Rapid Prototyp. J.* **2004**, *10*, 35–41. [CrossRef]

302. Murr, L.E.; Gaytan, S.M.; Medina, F.; Martinez, E.; Martinez, J.L.; Hernandez, D.H.; Machado, B.I.; Ramirez, D.A.; Wicker, R.B. Characterization of Ti6Al4V open cellular foams fabricated by additive manufacturing using electron beam melting. *Mater. Sci. Eng. A* **2010**, *527*, 1861–1868. [CrossRef]

303. Balla, V.K.; DeVasConCellos, P.D.; Xue, W.; Bose, S.; Bandyopadhyay, A. Fabrication of compositionally and structurally graded Ti-TiO$_2$ structures using laser engineered net shaping (LENS). *Acta Biomater.* **2009**, *5*, 1831–1837. [CrossRef]

304. Zhang, K.; Liu, W.; Shang, X. Research on the processing experiments of laser metal deposition shaping. *Opt. Laser Technol.* **2007**, *39*, 549–557. [CrossRef]

305. Crump, S.S. Fused deposition modeling (FDM): Putting rapid back into prototyping. In *The Second International Conference on Rapid Prototyping*; University of Dayton: Dayton, OH, USA, 1991; pp. 354–357.

306. Murr, L.E.; Gaytan, S.M.; Ceylan, A.; Martinez, E.; Martinez, J.L.; Hernandez, D.H.; Machado, B.I.; Ramirez, D.A.; Medina, F.; Collins, S.; et al. Characterization of titanium aluminide alloy components fabricated by additive manufacturing using electron beam melting. *Acta Mater.* **2010**, *58*, 1887–1894. [CrossRef]

307. Thomson, R.C.; Wake, M.C.; Yaszemski, M.J.; Mikos, A.G. Biodegradable polymer scaffolds to regenerate organs. In *Biopolymers II. Advances in Polymer Science*; Peppas, N.A., Langer, R.S., Eds.; Springer: Berlin, Heidelberg, 1995; Volume 122, pp. 245–274. [CrossRef]

308. Maroulakos, M.; Kamperos, G.; Tayebi, L.; Halazonetis, D.; Ren, Y. Applications of 3D printing on craniofacial bone repair: A systematic review. *J. Dent.* **2019**, *80*, 1–14. [CrossRef] [PubMed]

309. Nikzad, M.; Masood, S.H.; Sbarski, I.; Groth, A. Rheological properties of a particulate-filled polymeric composite through fused deposition process. *Mater. Sci. Forum* **2010**, *654–656*, 2471–2474. [CrossRef]

310. Bak, D. Rapid prototyping or rapid production? 3D printing processes move industry towards the latter. *Assem. Autom.* **2003**, *23*, 340–345. [CrossRef]

311. Lewis, G.K.; Schlienger, E. Practical considerations and capabilities for laser assisted direct metal deposition. *Mater. Des.* **2000**, *21*, 417–423. [CrossRef]

312. Dimitrov, D.; Schreve, K.; Beer, N. Advances in three dimensional printing—State of the art and future perspectives. *Rapid Prototyp. J.* **2006**, *12*, 136–147. [CrossRef]

313. Rännar, L.E.; Glad, A.; Gustafson, C.G. Efficient cooling with tool inserts manufactured by electron beam melting. *Rapid Prototyp. J.* **2007**, *13*, 128–135. [CrossRef]

314. Weisensel, L.; Travitzky, N.; Sieber, H.; Greil, P. Laminated object manufacturing (LOM) of SiSiC composites. *Adv. Eng. Mater.* **2004**, *6*, 899–903. [CrossRef]

315. Sachs, E.; Cima, M.; Cornie, J.; Brancazio, D.; Bredt, J.; Curodeau, A.; Fan, T.; Khanuja, S.; Lauder, A.; Lee, J.; et al. Three-dimensional printing: The physics and implications of additive manufacturing. *CIRP Ann.* **1993**, *42*, 257–260. [CrossRef]

316. Lee, M.; Dunn, J.C.Y.; Wu, B.M. Scaffold fabrication by indirect three-dimensional printing. *Biomaterials* **2005**, *26*, 4281–4289. [CrossRef]

317. Zhong, W.; Li, F.; Zhang, Z.; Song, L.; Li, Z. Short fiber reinforced composites for fused deposition modelling. *Mater. Sci. Eng. A* **2001**, *301*, 125–130. [CrossRef]

318. An, J.; Teoh, J.E.M.; Suntornnond, R.; Chua, C.K. Design and 3D Printing of Scaffolds and Tissues. *Engineering* **2015**, *1*, 261–268. [CrossRef]

319. Sachs, E.; Cima, M.; Cornie, J. Three-dimensional printing: Rapid tooling and prototypes directly from a CAD model. *CIRP Ann.* **1990**, *39*, 201–204. [CrossRef]

320. Sanjana, N.E.; Fuller, S.B. A fast flexible ink-jet printing method for patterning dissociated neurons in culture. *J. Neurosci. Methods* **2004**, *136*, 151–163. [CrossRef] [PubMed]

321. Turnbull, G.; Clarke, J.; Picard, F.; Riches, P.; Jia, L.; Han, F.; Li, B.; Shu, W. 3D bioactive composite scaffolds for bone tissue engineering. *Bioact. Mater.* **2018**, *3*, 278–314. [CrossRef] [PubMed]

322. Hue, P. Progress and Trends in Ink-jet Printing Technology. *J. Imaging Sci. Technol.* **1998**, *42*, 49–62.

323. Hofmeister, W.; Griffith, M.; Ensz, M.; Smugeresky, J. Solidification in direct metal deposition by LENS processing. *JOM* **2001**, *53*, 30–34. [CrossRef]

324. Liacouras, P.; Garnes, J.; Roman, N.; Petrich, A.; Grant, G.T. Designing and manufacturing an auricular prosthesis using computed tomography, 3-dimensional photographic imaging, and additive manufacturing: A clinical report. *J. Prosthet. Dent.* **2011**, *105*, 78–82. [CrossRef]

325. Heinl, P.; Rottmair, A.; Korner, C.; Singer, R.F. Cellular titanium by selective electron beam melting. *Adv. Eng. Mater.* **2007**, *9*, 360–364. [CrossRef]

326. Park, J.; Tari, M.J.; Hahn, H.T. Characterization of the laminated object manufacturing (LOM) process. *Rapid Prototyp. J.* **2000**, *6*, 36–50. [CrossRef]

327. Chojnowska, K. Model wirtualny wsparty wydrukiem 3D. Design News Polska. Kwiecień. 2008. Available online: http://www.designnews.pl/menu-gorne/artykul/article/model-wirtualny-wsparty-wydrukiem-3d/ (accessed on 15 December 2019).

328. Chua, C.K.; Leong, K.F.; Lim, C.S. *Rapid Prototyping: Principles and Applications*; World Scientific Publishing Company: Singapore, 2010.

329. Prechtl, M.; Otto, A.; Geiger, M. Rapid tooling by laminated object manufacturing of metal foil. *Adv. Mater. Res.* **2005**, *6*, 303–312. [CrossRef]

330. Łaskawiec, J.; Michalik, R. *Zagadnienia Teoretyczne i Aplikacyjne w Implantach*; Wydawnictwo Politechniki Śląskiej: Gliwice, Poland, 2002.

331. Agarwala, M.K.; Weeren, R.; Bandyopadhyay, A.; Whalen, P.J.; Safari, A.; Danforth, S.C. Fused deposition of ceramics and metals: An overview. In *Proceeding of Solid Freeform Fabrication Symposium*; The University of Texas at Austin: Austin, TX, USA, 1996; pp. 385–392.

332. Levy, G.N.; Schindel, R.; Kruth, J.P. Rapid manufacturing and rapid tooling with layer manufacturing (LM) technologies, state of the art and future perspectives. *CIRP Ann.* **2003**, *52*, 589–609. [CrossRef]

333. Rangarajan, S.; Qi, G.; Venkataraman, N.; Safari, A.; Danforth, S.C. Powder processing, rheology, and mechanical properties of feedstock for fused deposition of Si_3N_4 ceramics. *J. Am. Ceram. Soc.* **2000**, *83*, 1663–1669. [CrossRef]

334. Wu, H.; Li, D.; Guo, N. Fabrication of integral ceramic mold for investment casting of hollow turbine blade based on stereolithography. *Rapid Prototyp. J.* **2009**, *15*, 232–237. [CrossRef]

335. Jacobs, P.F. *Rapid Prototyping & Manufacturing: Fundamentals of Stereolithography*; SME Publication: Dearborn, MI, USA, 1992.

336. Arcaute, K.; Mann, B.K.; Wicker, R.B. Stereolithography of three-dimensional bioactive poly(ethylene glycol) constructs with encapsulated cells. *Ann. Biomed. Eng.* **2006**, *34*, 1429–1441. [CrossRef]

337. Melchels, F.P.W.; Feijen, J.; Grijpma, D.W. A review on stereolithography and its applications in biomedical engineering. *Biomaterials* **2010**, *31*, 6121–6130. [CrossRef]

338. Kumar, S. Selective Laser Sintering: A Qualitative and Objective Approach. *JOM* **2003**, *55*, 43–47. [CrossRef]

339. Wu, G.; Zhou, B.; Bi, Y.; Zhao, Y. Selective laser sintering technology for customized fabrication of facial prostheses. *J. Prosthet. Dent.* **2008**, *100*, 56–60. [CrossRef]

340. Greulich, M. Rapid prototyping and fabrication of tools and metals parts by laser sintering of metal powders. *Mater. Technol.* **1997**, *12*, 155–157. [CrossRef]

341. Dobrzańska-Danikiewicz, A.; Dobrzański, L.A.; Gaweł, T.G. Scaffolds applicable as implants of a loss palate fragments. In *International Conference on Processing & Manufacturing of Advanced Materials. Processing, Fabrication, Properties, Applications, THERMEC'2016*; TU Graz: Graz, Austria, 2016; p. 193.

342. Abe, F.; Osakada, K.; Kitamura, Y.; Matsumoto, M.; Shiomi, M. Manufacturing of titanium parts for medical purposes by selective laser melting. In *Proceedings of the Eighth International Conference on Rapid Prototyping*; University of Dayton: Tokyo, Japan, 2000; pp. 288–293.

343. Dobrzańska-Danikiewicz, A.D.; Gaweł, T.G.; Wolany, W. Ti6Al4V titanium alloy used as a modern biomimetic material. *Arch. Mater. Sci. Eng.* **2015**, *76*, 150–156.

344. Lethaus, B.; Poort, L.; Böckmann, R.; Smeets, R.; Tolba, R.; Kessler, P. Additive manufacturing for microvascular reconstruction of the mandible in 20 patients. *J. Cranio Maxillo Facial Surg.* **2012**, *40*, 43–46. [CrossRef] [PubMed]

345. Dobrzański, L.A. Overview and general ideas of the development of constructions, materials, technologies and clinical applications of scaffolds engineering for regenerative medicine. *Arch. Mater. Sci. Eng.* **2014**, *69*, 53–80.

346. Pattanayak, D.K.; Fukuda, A.; Matsushita, T.; Takemoto, M.; Fujibayashi, S.; Sasaki, K.; Nishida, N.; Nakamura, T.; Kokubo, T. Bioactive Ti metal analogous to human cancellous bone: Fabrication by selective laser melting and chemical treatments. *Acta Biomater.* **2011**, *7*, 1398–1406. [CrossRef] [PubMed]

347. Bertol, L.S.; Júnior, W.K.; da Silva, F.P.; Kopp, C.A. Medical design: Direct metal laser sintering of Ti-6Al-4V. *Mater. Des.* **2010**, *31*, 3982–3988. [CrossRef]

348. Boston Consulting Group. Embracing Industry 4.0 and Rediscovering Growth. BCG, 2019. Available online: http://web.archive.org/web/20190710182601/https://www.bcg.com/pl-pl/capabilities/operations/embracing-industry-4.0-rediscovering-growth.aspx (accessed on 15 December 2019).

349. Al-Fadda, S.A.; Zarb, G.A.; Finer, Y.A. A comparison of the accuracy of fit of 2 methods for fabricating implant-prosthodontic frameworks. *Int. J. Prosthodont.* **2007**, *20*, 125–131.

350. Redlich, M.; Weinstock, T.; Abed, Y.; Schneor, R.; Holdstein, Y.; Fischer, A. A new system for scanning, measuring and analyzing dental casts based on a 3D holographic sensor. *Orthod. Craniofacial Res.* **2008**, *11*, 90–95. [CrossRef]

351. Sachs, C.; Groesser, J.; Stadelmann, M.; Schweiger, J.; Erdelt, K.; Beuer, F. Full-arch prostheses from translucent zirconia: Accuracy of fit. *Dent. Mater.* **2014**, *30*, 817–823. [CrossRef] [PubMed]

352. Zilberman, O.; Huggare, J.A.; Parikakis, K.A. Evaluation of the validity of tooth size and arch width measurements using conventional and three-dimensional virtual orthodontic models. *Angle Orthod.* **2003**, *73*, 301–306.

353. Santoro, M.; Galkin, S.; Teredesai, M.; Nicolay, O.F.; Cangialosi, T.J. Comparison of measurements made on digital and plaster models. *Am. J. Orthod. Dentofac. Orthop.* **2003**, *124*, 101–105. [CrossRef]

354. Kusnoto, B.; Evans, C.A. Reliability of a 3D surface laser scanner for orthodontic applications. *Am. J. Orthod. Dentofac. Orthop.* **2002**, *122*, 342–348. [CrossRef]

355. Arnetzl, G.; Pongratz, D. Milling precision and fitting accuracy of Cerec Scan milled restorations. *Int. J. Comput. Dent.* **2005**, *8*, 273–281. [PubMed]

356. Luthardt, R.G.; Kuhmstedt, P.; Walter, M.H. A new method for the computer aided evaluation of three-dimensional changes in gypsum materials. *Dent. Mater.* **2003**, *19*, 19–24. [CrossRef]

357. Katsoulis, J.; Mericske-Stern, R.; Rotkina, L.; Zbären, C.; Enkling, N.; Blatz, M.B. Precision of fit of implant-supported screw-retained 10-unit computer-aided-designed and computer-aided-manufactured frameworks made from zirconium dioxide and titanium: An in vitro study. *Clin. Oral Implant. Res.* **2014**, *25*, 165–174. [CrossRef]

358. Rudolph, H.; Luthardt, R.G.; Walter, M.H. Computer-aided analysis of the influence of digitizing and surfacing on the accuracy in dental CAD/CAM technology. *Comput. Biol. Med.* **2007**, *37*, 579–587. [CrossRef]

359. Chan, D.C.; Chung, A.K.; Haines, J.; Yau, E.H.; Kuo, C.C. The accuracy of optical scanning: Influence of convergence and die preparation. *Oper. Dent.* **2011**, *36*, 486–491. [CrossRef] [PubMed]

360. Nedelcu, R.G.; Persson, A.S. Scanning accuracy and precision in 4 intraoral scanners: An in vitro comparison based on 3-dimensional analysis. *J. Prosthet. Dent.* **2014**, *112*, 1461–1471. [CrossRef] [PubMed]

361. Sousa, M.V.; Vasconcelos, E.C.; Janson, G.; Garib, D.; Pinzan, A. Accuracy and reproducibility of 3-dimensional digital model measurements. *Am. J. Orthod. Dentofac. Orthop.* **2012**, *142*, 269–273. [CrossRef]

362. Del Corso, M.; Aba, G.; Vazquez, L.; Dargaud, J.; Ehrenfest, D.M.D. Optical three-dimensional scanning acquisition of the position of osseointegrated implants: An in vitro study to determine method accuracy and operational feasibility. *Clin. Implant. Dent. Relat. Res.* **2009**, *11*, 214–221. [CrossRef]

363. Quaas, S.; Rudolph, H.; Luthardt, R.G. Direct mechanical data acquisition of dental impressions for the manufacturing of CAD/CAM restorations. *J. Dent.* **2007**, *35*, 903–908. [CrossRef]

364. Tomita, S.; Shin-ya, A.; Gomi, H.; Shin-ya, A.; Yokoyama, D. Machining accuracy of crowns by CAD/CAM system using TCP/IP: Influence of restorative material and scanning condition. *Dent. Mater. J.* **2007**, *26*, 549–560. [CrossRef]

365. Mously, H.A.; Finkelman, M.; Zandparsa, R.; Hirayama, H. Marginal and internal adaptation of ceramic crown restorations fabricated with CAD/CAM technology and the heat-press technique. *J. Prosthet. Dent.* **2014**, *112*, 249–256. [CrossRef]

366. Luthardt, R.G.; Sandkuhl, O.; Herold, V.; Walter, M.H. Accuracy of mechanical digitizing with a CAD/CAM system for fixed restorations. *Int. J. Prosthodont.* **2001**, *14*, 146–151.

367. Hayasaki, H.; Martins, R.P.; Gandini, L.G.; Saitoh, I.; Nonaka, K. A new way of analyzing occlusion 3 dimensionally. *Am. J. Orthod. Dentofac. Orthop.* **2005**, *128*, 128–132. [CrossRef]

368. Persson, A.S.; Andersson, M.; Oden, A.; Sandborgh-Englund, G. Computer aided analysis of digitized dental stone replicas by dental CAD/CAM technology. *Dent. Mater.* **2008**, *24*, 1123–1130. [CrossRef] [PubMed]

369. Kuroda, T.; Motohashi, N.; Tominaga, R.; Iwata, K. Three-dimensional dental cast analyzing system using laser scanning. *Am. J. Orthod. Dentofac. Orthop.* **1996**, *110*, 365–369. [CrossRef]

370. Jemt, T.; Hjalmarsson, L. In vitro measurements of precision of fit of implant supported frameworks. A comparison between "virtual" and "physical" assessments of fit using two different techniques of measurements. *Clin. Implant Dent. Relat. Res.* **2012**, *14*, e175–e182. [CrossRef] [PubMed]

371. Persson, M.; Andersson, M.; Bergman, B. The accuracy of a high-precision digitizer for CAD/CAM of crowns. *J. Prosthet. Dent.* **1995**, *74*, 223–229. [CrossRef]

372. Keating, A.P.; Knox, J.; Bibb, R.; Zhurov, A.I. A comparison of plaster, digital and reconstructed study model accuracy. *J. Orthod.* **2008**, *35*, 191–201. [CrossRef]

373. Tapie, L.; Lebon, N.; Mawussi, B.; Chabouis, H.F.; Duret, F.; Attal, J.P. Understanding dental CAD/CAM for restorations the digital workflow from a mechanical engineering viewpoint. *Int. J. Comput. Dent.* **2015**, *18*, 21–44.

374. Vlaar, S.T.; van der Zel, J.M. Accuracy of dental digitizers. *Int. Dent. J.* **2006**, *56*, 301–309. [CrossRef]

375. Ender, A.; Mehl, A. Accuracy of complete-arch dental impressions: A new method of measuring trueness and precision. *J. Prosthet. Dent.* **2013**, *109*, 121–128. [CrossRef]

376. Keul, C.; Stawarczyk, B.; Erdelt, K.J.; Beuer, F.; Edelhoff, D.; Güth, J.F. Fit of 4-unit FDPs made of zirconia and CoCr-alloy after chairside and labside digitalizationea laboratory study. *Dent. Mater.* **2014**, *30*, 400–407. [CrossRef] [PubMed]

377. McLean, J.W.; von Fraunhofer, J.A. The estimation of cement film thickness by an in vivo technique. *Br. Dent. J.* **1971**, *131*, 107–111. [CrossRef] [PubMed]

378. Tamim, H.; Skjerven, H.; Ekfeldt, A.; Ronold, H.J. Clinical evaluation of CAD/CAM metal-ceramic posterior crowns fabricated from intraoral digital impressions. *Int. J. Prosthodont.* **2014**, *27*, 331–337. [CrossRef]

379. Ender, A.; Zimmermann, M.; Attin, T.; Mehl, A. In vivo precision of conventional and digital methods for obtaining quadrant dental impressions. *Clin. Oral Investig.* **2016**, *20*, 1495–1504. [CrossRef] [PubMed]

380. Lim, J.-H.; Park, J.-M.; Kim, M.; Heo, S.-J.; Myung, J.-Y. Comparison of digital intraoral scanner reproducibility and image trueness considering repetitive experience. *J. Prosthet. Dent.* **2018**, *119*, 225–232. [CrossRef]

381. Persson, A.; Andersson, M.; Oden, A.; Sandborgh-Englund, G. A three-dimensional evaluation of a laser scanner and a touch-probe scanner. *J. Prosthet. Dent.* **2006**, *95*, 194–200. [CrossRef] [PubMed]

382. Ender, A.; Attin, T.; Mehl, A. In vivo precision of conventional and digital methods of obtaining complete-arch dental impressions. *J. Prosthet. Dent.* **2016**, *115*, 313–320. [CrossRef] [PubMed]

383. Mandelli, F.; Gherlone, E.; Gastaldi, G.; Ferrari, M. Evaluation of the accuracy of extraoral laboratory scanners with a single-tooth abutment model: A 3D analysis. *J. Prosthodont. Res.* **2017**, *61*, 363–370. [CrossRef]

384. Marghalani, A.; Weber, H.-P. Finkelman, M.; Kudara, Y.; El Rafie, K.; Papaspyridakos, P. Digital versus conventional implant impressions for partially edentulous arches: An evaluation of accuracy. *J. Prosthet. Dent.* **2018**, *119*, 574–579. [CrossRef]

385. Emir, F.; Ayyıldız, S. Evaluation of the trueness and precision of eight extraoral laboratory scanners with a complete-arch model: A three-dimensional analysis. *J. Prosthodont. Res.* **2019**, *63*, 434–439. [CrossRef]

386. Voegtlin, C.; Schulz, G.; Jaeger, K.; Mueller, B. Comparing the accuracy of master models based on digital intra-oral scanners with conventional plaster casts. *Phys. Med.* **2016**, *1*, 20–26. [CrossRef]

387. Anh, J.W.; Park, J.M.; Chun, Y.S.; Kim, M.; Kim, M. A comparison of the precision of three-dimensional images acquired by 2 digital intraoral scanners: Effects of tooth irregularity and scanning direction. *Korean J. Orthod.* **2016**, *46*, 3–12. [CrossRef] [PubMed]

388. Ender, A.; Mehl, A. In-vitro evaluation of the accuracy of conventional and digital methods of obtaining full-arch dental impressions. *Quintessence Int.* **2015**, *46*, 9–17. [CrossRef]

389. Renne, W.; Ludlow, M.; Fryml, J.; Schurch, Z.; Mennito, A.; Kessler, R.; Lauer, A. Evaluation of the accuracy of 7 digital scanners: An in vitro analysis based on 3-dimensional comparisons. *J. Prosthet. Dent.* **2017**, *118*, 36–42. [CrossRef] [PubMed]

390. Patzelt, S.B.; Emmanouilidi, A.; Stampf, S.; Strub, J.R.; Att, W. Accuracy of full-arch scans using intraoral scanners. *Clin. Oral Investig.* **2014**, *18*, 1687–1694. [CrossRef]

391. Bosch, G.; Ender, A.; Mehl, A. A 3-dimensional accuracy analysis of chairside CAD/CAM milling processes. *J. Prosthet. Dent.* **2014**, *112*, 1425–1431. [CrossRef] [PubMed]

392. Hayama, H.; Fueki, K.; Wadachi, J.; Wakabayashi, N. Trueness and precision of digital impressions obtained using an intraoral scanner with different head size in the partially edentulous mandible. *J. Prosthodont. Res.* **2018**, *62*, 347–352. [CrossRef]

393. Patzelt, S.B.M.; Bishti, S.; Stampf, S.; Att, W. Accuracy of computer-aided design/computer-aided manufacturing-generated dental casts based on intraoral scanner data. *JADA* **2014**, *145*, 1133–1140. [CrossRef]

394. Su, T.S.; Sun, J. Comparison of repeatability between intraoral digital scanner and extraoral digital scanner: An in-vitro study. *J. Prosthodont. Res.* **2015**, *59*, 236–242. [CrossRef]

395. Fielding, G.A.; Bandyopadhyay, A.; Bose, S. Effects of silica and zinc oxide doping on mechanical and biological properties of 3D printed tricalcium phosphate tissue engineering scaffolds. *Dent. Mater.* **2012**, *28*, 113–122. [CrossRef] [PubMed]

396. Fukazawa, S.; Odaira, C.; Kondo, H. Investigation of accuracy and reproducibility of abutment position by intraoral scanners. *J. Prosthodont. Res.* **2017**, *61*, 450–459. [CrossRef] [PubMed]

397. Kim, K.R.; Seo, K.-Y.; Kim, S. Conventional open-tray impression versus intraoral digital scan for implant-level complete-arch impression. *J. Prosthet. Dent.* **2019**, *122*, 543–549. [CrossRef]

398. Park, J.M. Comparative analysis on reproducibility among 5 intraoral scanners: Sectional analysis according to restoration type and preparation outline form. *J. Adv. Prosthodont.* **2016**, *8*, 354–362. [CrossRef]

399. Ebert, J.; Ozkol, E.; Zeichner, A.; Uibel, K.; Weiss, O.; Koops, U.; Telle, R.; Fisher, H. Direct inkjet printing of dental prostheses made of zirconia. *J. Dent. Res.* **2009**, *88*, 673–676. [CrossRef]

400. Łomżyński, Ł.G.; Mierzwińska-Nastalska, E. Cyfrowe wyciski wykonywane z zastosowaniem optycznych skanerów wewnątrzustnych. Systematyczny przegląd piśmiennictwa. Część I: Właściwości i zalety stosowania techniki wycisków cyfrowych w stomatologii oraz analiza cech pożądanych optymalnego systemu skanującego. *Protet. Stomatol.* **2019**, *69*, 217–223. [CrossRef]

401. Atieh, M.A.; Ritter, A.V.; Ko, C.C.; Duqum, I. Accuracy evaluation of intraoral optical impressions: A clinical study using a reference appliance. *J. Prosthet. Dent.* **2017**, *118*, 400–405. [CrossRef]

402. Kim, S.Y.; Kim, M.J.; Han, J.S.; Yeo, I.S.; Lim, Y.J.; Kwon, H.B. Accuracy of dies captured by an intraoral digital impression system using parallel confocal imaging. *Int. J. Prosthodont.* **2013**, *26*, 161–163. [CrossRef]

403. Fasbinder, D.J. Digital dentistry: Innovation for restorative treatment. *Compend. Contin. Educ. Dent.* **2010**, *31*, 2–12.

404. Sakornwimon, N.; Leevailoj, C. Clinical marginal fit of zirconia crowns and patients' preferences for impression techniques using intraoral digital scanner versus polyvinyl siloxane material. *J. Prosthet. Dent.* **2017**, *118*, 386–391. [CrossRef]

405. Mehl, A.; Ender, A.; Mörmann, W.; Attin, T. Accuracy testing of a new intraoral 3D camera. *Int. J. Comput. Dent.* **2009**, *12*, 11–28. [PubMed]

406. Seelbach, P.; Brueckel, C.; Wöstmann, B. Accuracy of digital and conventional impression techniques and workflow. *Clin. Oral Investig.* **2013**, *17*, 1759–1764. [CrossRef]

407. Kachata, P.R.; Geissberger, M.J. Dentistry a la carte: In office CAD/CAM technology. *J. Calif. Dent. Assoc.* **2010**, *38*, 323–330.

408. Silva, A.L.L.; Aranegui, R.O.; Shukeir, G.S.; Bermejo, M.A.L. Tomografía computerizada de haz cónico. Aplicaciones clínicas en odontología; comparación con otras técnicas. *Cient. Dent.* **2010**, *7*, 147–159.

409. Gava, M.M.; Suomalainen, A.; Vehmas, T.; Ventä, I. Did malpractice claims for failed dental implants decrease after introduction of CBCT in Finland? *Clin. Oral Investig.* **2019**, *23*, 399–404. [CrossRef] [PubMed]

410. Fakhran, S.; Alhilali, L.; Sreedher, G.; Dohatcu, A.C.; Lee, S.; Ferguson, B.; Branstetter, B.F. Comparison of simulated cone beam computed tomography to conventional helical computed tomography for imaging of rhinosinusitis. *Laryngoscope* **2014**, *124*, 2002–2006. [CrossRef]

411. Tsuchida, R.; Araki, K.; Okano, T. Evaluation of a limited cone beam volumetric imaging system: Comparison with film radiography in detecting incipient proximal caries. *Oral Surg. Oral Med. Oral Pathol. Oral Radiol. Endodontology* **2007**, *104*, 412–416. [CrossRef]

412. Rangel, F.A.; Maal, T.J.J.; de Koning, M.J.J.; Bronkhorst, E.M.; Bergé, S.J.; Kuijpers-Jagtman, A.M. Integration of digital dental casts in cone beam computed tomography scans—A clinical validation study. *Clin. Oral Investig.* **2018**, *22*, 1215–1222. [CrossRef]

413. Mischkowski, R.A.; Pulsfort, R.; Ritter, L.; Neugebauer, J.; Brochhagen, H.G.; Zöller, E.J.E. Geometric accuracy of newly developed conebeam device for maxillofacial imaging. *Oral Med. Oral Pathol. Oral Radiol. Endodontology* **2007**, *104*, 551–559. [CrossRef]

414. Bornstein, M.M.; Scarfe, W.C.; Vaughn, V.M.; Jacobs, R. Cone beam computed tomography in implant dentistry: A systematic review focusing on guidelines, indications, and radiation dose risks. *J. Oral Maxillofocial Implant.* **2014**, *29*, 55–77. [CrossRef]

415. Haiter-Neto, F.; Wenzel, A.; Gotfredsen, E. Diagnostic accuracy of cone beam tomography scans compared with intraoral image modalities for detection of caries lesions. *DentoMaxilloFacial Radiol.* **2008**, *37*, 18–22. [CrossRef] [PubMed]

416. Hicks, D.; Melkers, M.; Barna, J.; Isett, K.R.; Gilbert, G.H. Comparison of the accuracy of CBCT effective radiation dose information in peer-reviewed journals and dental media. *Gen. Dent.* **2019**, *67*, 38–46. [PubMed]

417. Scarfe, W.C.; Farman, A.G. What is cone-beam CT and how does it work? *Dent. Clin. N. Am.* **2008**, *52*, 707–730. [CrossRef] [PubMed]

418. Ludlow, J.B.; Lester, W.S.; See, M.; Bailey, L.J.; Hershey, H.G. Accuracy of measurements of mandibular anatomy in cone beam computed tomography images. *Oral Med. Oral Pathol. Oral Radiol. Endodontol.* **2007**, *103*, 534–542. [CrossRef]

419. Kim, D.M.; Bassir, S.H. When Is Cone-Beam Computed Tomography Imaging Appropriate for Diagnostic Inquiry in the Management of Inflammatory Periodontitis? An American Academy of Periodontology Best Evidence Review. *J. Periodontol.* **2017**, *88*, 978–998. [CrossRef]

420. Jacobs, R.; Salmon, B.; Codari, M.; Hassan, B.; Bornstein, M.M. Cone beam computed tomography in implant dentistry: Recommendations for clinical use. *BMC Oral Health* **2018**, *18*, 88. [CrossRef]

421. Hashimoto, K.; Arai, Y.; Iwai, K.; Araki, M.; Kawashima, S.; Terakado, M. A comparison of a new limited cone beam computed tomography machine for dental use with a multidetector row helical CT machine. *Oral Med. Oral Pathol. Oral Radiol. Endodontol.* **2003**, *95*, 371–377. [CrossRef]

422. Melkers, J.; Hicks, D.; Isett, K.R.; Kopycka-Kedzierawski, D.T.; Gilbert, G.H.; Rosenblum, S.; Burton, V.; Mungia, R.; Melkers, M.J.; Ford, G. Preferences for peer-reviewed versus other publication sources: A survey of general dentists in the National Dental PBRN. *Implement. Sci.* **2019**, *14*, 19. [CrossRef]

423. Garg, V.; Bagaria, A.; Kaur, G.; Bhardwaj, S.; Hedau, L.R. Application of Cone Beam Computed Tomography in Dentistry- A Review. *J. Adv. Med. Dent. Sci. Res.* **2019**, *7*, 73–76.

424. Tetradis, S.; Anstey, P.; Graff-Radford, S. Cone Beam Computed Tomography in the Diagnosis of Dental Disease. *J. Calif. Dent. Assoc.* **2010**, *38*, 27–32.

425. Hegde, S.; Ajila, V.; Kamath, J.S.; Babu, S.; Pillai, D.S.; Nair, S.M. Importance of cone-beam computed tomography in dentistry: An update. *SRM J. Res. Dent. Sci.* **2018**, *9*, 186–190. [CrossRef]

426. Park, J.; Baumrind, S.; Curry, S.; Carlson, S.K.; Boyd, R.L.; Oh, H. Reliability of 3D dental and skeletal landmarks on CBCT images. *Angle Orthod.* **2019**, *89*, 758–767. [CrossRef] [PubMed]

427. Drage, N. Cone Beam Computed Tomography (CBCT) in General Dental Practice. *Prim. Dent. J.* **2018**, *7*, 26–30. [CrossRef] [PubMed]

processes

Review

Virtual Approach to the Comparative Analysis of Biomaterials Used in Endodontic Treatment

Joanna Dobrzańska [1,2], Lech B. Dobrzański [1,2], Klaudiusz Gołombek [3], Leszek A. Dobrzański [1,*] and Anna D. Dobrzańska-Danikiewicz [4]

1   Medical and Dental Engineering Centre for Research, Design and Production ASKLEPIOS, 13 D Krolowej Bony St., 44-100 Gliwice, Poland; joanna.dobrzanska@centrumasklepios.pl (J.D.); dobrzanski@centrumasklepios.pl (L.B.D.)
2   Medical and Dental Centre SOBIESKI, 12/1 King Jan III Sobieski St., 44-100 Gliwice, Poland
3   Faculty of Mechanical Engineering, Silesian University of Technology, 18 A Konarski St., 44-100 Gliwice, Poland; klaudiusz.golombek@polsl.pl
4   Faculty of Mechanical Engineering, University of Zielona Góra, 4 Prof. Z. Szafrana St., 65-516 Zielona Góra, Poland; anna.dobrzanska.danikiewicz@gmail.com
*   Correspondence: leszek.dobrzanski@centrumasklepios.pl

Abstract: The importance of endodontics is presented within our own concept of Dentistry Sustainable Development (DSD) consisting of three inseparable elements; i.e., Advanced Interventionist Dentistry 4.0 (AID 4.0), Global Dental Prevention (GDP), and the Dentistry Safety System (DSS) as a polemic, with the hypothesis of the need to abandon interventionist dentistry in favour of the domination of dental prevention. In view of the numerous systemic complications of caries that affect 3−5 billion people globally, endodontic treatment effectively counteracts them. Regardless of this, the prevention of oral diseases should be developed very widely, and in many countries dental care should reach the poorest sections of society. The materials and methods of clinical management in endodontic procedures are characterized. The progress in the field of filling materials and techniques for the development and obturation of root canals is presented. The endodontics market is forecast to reach USD 2.1 billion in 2026, with a CAGR of 4.1%. The most widely used and recognized material for filling root canals is gutta-percha, recognized as the "gold standard". An alternative is a synthetic thermoplastic filler material based on polyester materials, known mainly under the trade name Resilon. There are still sceptical opinions about the need to replace gutta-percha with this synthetic material, and many dentists still believe that this material cannot compete with gutta-percha. The results of studies carried out so far do not allow for the formulation of a substantively and ethically unambiguous view that gutta-percha should be replaced with another material. There is still insufficient clinical evidence to formulate firm opinions in this regard. In essence, materials and technologies used in endodontics do not differ from other groups of materials, which justifies using material engineering methodology for their research. Therefore, a detailed methodological approach is presented to objectify the assessment of endodontic treatment. Theoretical analysis was carried out using the methods of procedural benchmarking and comparative analysis with the use of contextual matrices to virtually optimize the selection of materials, techniques for the development and obturation of root canals, and methods for assessing the effectiveness of filling, which methods are usually used, e.g., in management science, and especially in foresight research as part of knowledge management. The results of these analyses are presented in the form of appropriate context matrices. The full usefulness of the research on the effectiveness and tightness of root canal filling using scanning electron microscopy is indicated. The analysis results are a practical application of the so-called "digital twins" approach concerning the virtual comparative analysis of biomaterials used in endodontic treatment.

Keywords: dentistry; endodontics; filling materials; sealants; obturation; gutta-percha; Resilon; procedural benchmarking; comparative matrices; virtual approach; digital twin; scanning electron microscopy

Citation: Dobrzańska, J.; Dobrzański, L.B.; Gołombek, K.; Dobrzański, L.A.; Dobrzańska-Danikiewicz, A.D. Virtual Approach to the Comparative Analysis of Biomaterials Used in Endodontic Treatment. *Processes* **2021**, *9*, 926. https://doi.org/10.3390/pr9060926

Academic Editor: Tao Sun

Received: 30 April 2021
Accepted: 21 May 2021
Published: 25 May 2021

237

1. The Importance of Endodontics in the General Concept of Sustainable Dentistry Development

The authors' work [1] presents the concept of Dentistry Sustainable Development (DSD > 2020) consisting of three inseparable elements; i.e., Advanced Interventionist Dentistry 4.0 (AID 4.0), Global Dental Prevention (GDP), and the Dentistry Safety System (DSS). The concept of Dentistry Sustainable Development and the authors' view on this subject is a polemic, with the main part of the thoughts contained in the series of two papers [2,3] promoting the hypothesis regarding the need to abandon interventionist dentistry, excluding therapeutic activities undertaken by dentists for the domination of dental prevention. It is impossible to agree with the approach that one should completely abandon interventionist dentistry in favour of the universally dominant prevention, if only because in many countries, e.g., due to the socio-economic situation, widespread lack of oral hygiene or eating habits, such a postulate is highly unlikely to be implemented. Treating caries and periodontal diseases is of great importance due to the scale of these diseases of the oral cavity, which currently affect 3−5 billion people in the world [1]. Dental caries is the most common infectious disease globally. It includes interactions between bacterial biofilm and the surface of teeth, saliva, and genes; dietary carbohydrates, including sugars and starches; and the influence of behavioural, social, and psychological factors [4–6]. There are six countries with the most common caries in the 5–14 age group, including Kyrgyzstan, Uzbekistan, Romania, Bulgaria, Poland, and Ecuador [7]. The countries most affected by this disease in the entire age range also include France, Spain, Iceland, Greece, and Georgia [1].

Untreated dental caries is the direct cause of many systemic complications that place enormous pressure on health care systems and health insurance and disruptions in labour markets in all countries. The numerous complications of this type include, among others, nephritis, bacterial pneumonia, rheumatoid arthritis, endocarditis, ischemic heart disease, brain abscess, osteoporosis, premature delivery, reduction in infant birth weight, and even direct death of the patient [8–13]. Because tooth extraction caused by caries is often unavoidable, including due to the ineffectiveness of endodontic treatment, it can lead to partial or even complete toothlessness. Apart from the obvious loss of aesthetic value and eating ailments, toothlessness resulting from caries causes many systemic diseases [14–28], including even direct shortening of human life, cardiovascular diseases, heart failure, ischemic heart disease, hypertension and stroke, some cancers, diabetes, insulin independence, kidney disease and rheumatoid arthritis, gastritis, duodenal ulcer, pancreatic diseases (including tumours), neoplastic changes of the oesophagus and upper gastrointestinal tract, unfavourable course of pregnancy, mood and neurocognitive disorders, obstructive sleep apnea, cervical spine pain, migraine and headaches, disorders of cerebrospinal fluid metabolism and changes in the hippocampus in the elderly [29–35] and related spatial and episodic memory disorders, exacerbating symptoms of multiple sclerosis, and increasing the risk of dementia in general. The mentioned aspects fully justify the necessity of caries treatment.

The most important risk factors contributing to caries include excessive consumption of sugars, insufficient saliva flow, hygiene negligence, fluoride deficiency during daily tooth brushing, socio-economic degradation due to nutritional deficiencies causing deformation of the teeth, and lack of proper dental care [1]. Accurate and professional dental care reduces the risk of developing caries in every patient, which is presented schematically by the caries development pyramid (CDP), from Background Level Care (BLC) through Preventive Treatment Option (PTO) to Operative Treatment Option (OTO) (Figure 1). The 5D Caries Management Cycle Rules (CMCR), by analogy to the Deming Plan-Do-Check-Act behavioural approach, is presented in Figure 1c. Before cavitation symptoms become apparent, preventive measures should be taken to detect caries at the earliest possible stages of the disease (Figure 1d), with proper risk assessment and determination of adequate actions with an appropriate frequency of reminders [36–39].

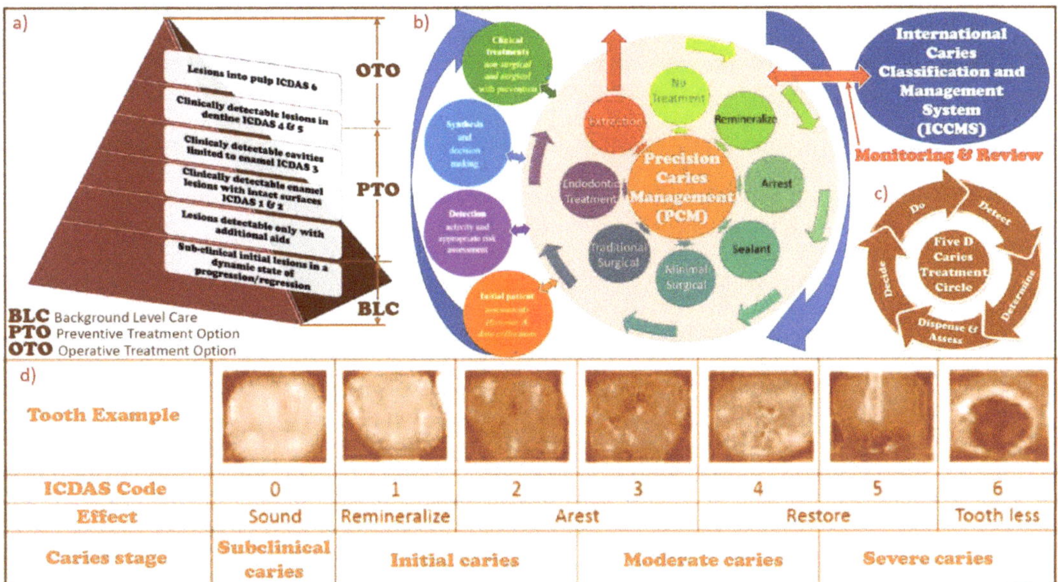

ICDAS Code	0	1	2	3	4	5	6
Effect	Sound	Remineralize	Arest		Restore		Tooth less
Caries stage	Subclinical caries	Initial caries		Moderate caries		Severe caries	

Figure 1. Diagram: (**a**) CDP caries development pyramid; (**b**) the Precision Caries Management (PCM) in cooperation with the International Caries Classification and Management System (ICCMS); (**c**) the 5D Caries Management Cycle Rules (CMCRs); (**d**) the International Caries Detection and Assessment System (ICDAS) application.

Visually, changes in enamel, dentine, and pulp allow for the separation of three primary grades of caries severity [40,41]. Work to date for over a century [42] has led to the development of the so-called "caries continuum". According to the International Caries Detection and Assessment System (ICDAS) [43–46], classification is evidence-based and oriented towards the prevention, detection, and evaluation of the stages of caries. The International Caries Classification and Management System (ICCMS) [47–50] is useful for improving long-term caries outcomes in conjunction with the Precision Caries Management (Figure 1b).

Dentists treat patients while actively preventing the spread of the previously mentioned serious systemic complications. The carefully performed endodontic treatment eliminates many of the risks as mentioned above and allows for a relatively long time to keep healed teeth in the patient's mouth, preventing complications, as well as partial or even complete toothlessness. After eight years from the end of endodontic treatment, in only 3% of cases, there are periapical changes such as cysts or granulomas [51], although in 90% of cases, these changes are caused by errors in the preparation and filling of root canals [52–56], as well as a crown or root fractures or other iatrogenic factors [57]. It is highly inappropriate to discredit the dentist community by criticizing the alleged approach in the slogan "drilling, filling, and billing" presented in the public discourse [58]. The authors remain convinced that this is a substantive manifestation of populism and trivialization of the real role of dentistry, especially endodontic treatment, or even contempt for such work, and a fundamentally inaccurate finding. Of course, the development of a root canal can be called drilling, while obturation with suitable material can be called filling. Dental services in many countries are in the private sector, so the patient has to finance all treatment costs. However, that is not a reason to mock this situation. These are serious medical activities, and treatments that are essentially cosmetological constitute a narrow range of activities of the dental community, although they most likely take place. However, they are not the motivation for the statistically overwhelming activity undertaken by dentists for the benefit of patients and in the name of well-understood ethical obligations of a dentist. Such an assessment must be surprising, especially as it comes from within the dentist community. There is every reason to believe that disseminating such false information is an ethical tort.

Although probably insufficiently developed, an important aspect of dentists' activity, especially in countries with low economic levels, is prevention. There is no doubt that prevention should be developed very widely. Dental care should cover all social strata, including the poorest and preferably in all countries, including in the original concept of the sustainable development of dentistry [1]. However, the facts seem to indicate that in many countries, especially low-income countries (LICs), this postulate is impossible to meet to a large extent or even at all.

An important part in the development of dentistry [1,10] is to ensure the safety of dentists and dental staff in the face of the COVID-19 disease pandemic caused by the transmission of the SARS-CoV-2 coronavirus. The dentist is constantly exposed to infection because they are doing their job in the patient's respiratory tract. The issue was widely discussed in the authors' work [10], which also presented the authors' technical solution serving this purpose [59]. This aspect of the endodontist's work was also highlighted in [60,61].

The basis of many therapeutic activities undertaken by dentists is close cooperation with the engineering and technical environment. In addition to the use of numerous devices, they range from a dental unit with a dental chair, through modern medical diagnosis methods using conical beam computed tomography (CBCT) and intraoral and extraoral scanning, to a wide range related to Dentistry 4.0 [62,63] as an activity within Industry 4.0. The centres for producing dental prosthetic restorations fit into this idea as smart factories [62–98].

Cooperation in the field of conservative dentistry and endodontics also requires wide participation in engineering and technical activities, including proper selection of filling materials, sealants, and appropriate techniques for the development of the root canal and obturation, requiring the use of various instruments and tools. This paper aims to develop and present the methodological assumptions of the virtual comparative analysis of biomaterials used in endodontic treatment, including the production and selection of filling and sealing materials; the selection of dental devices and tools, including specialized tools; the selection and application of the technology for the development and obturation of the root canal; and assessment effectiveness and tightness of root canal filling as a result of endodontic treatment.

2. General Characteristics of Materials and Clinical Procedures in Endodontic Treatment

Precision Caries Management (PCM) involves making the right decisions depending on the caries stage established based on the International Caries Detection and Assessment System (ICDAS) (Figure 1). Each decision is based on the evidence obtained from a thorough dental examination by a dentist. Dental pulp attacked by caries bacteria causes pain, and dental intervention becomes indispensable. As it means caries advancement above the second stage according to ICDAS, it is necessary to start endodontic treatment [99–101] because it is simply too late for any prevented treatment. This is especially true of the maxillary and mandibular incisors of the first molars, most affected by endodontic diseases [102]. Keeping the tooth functional when it is no longer possible to keep it alive is the basis of the conventional approach that prescribes endodontic treatment, which is the subject of this paper. Of course, an alternative could be extracting a diseased tooth, but dentists prefer endodontic treatment for pragmatic and ethical reasons [103,104]. The paradigm shift of this approach leads to revascularization and regenerative endodontic procedures. Regenerative endodontics aims to create and deliver tissues to replace the diseased and traumatized tooth pulp. Although it is a breakthrough approach, it should be expected to become real. It will bring tremendous benefits to patients, regardless of the scale of the costs incurred, as it offers a chance to restore the natural function of the teeth instead of surgical placement of implants or even artificial dentures. This problem is beyond the scope of these considerations and is presented in more detail in the literature review [105,106].

Endodontic treatment aims to leave the tooth in the mouth and prevent, and preferably eliminate, apical periodontitis [107]. This treatment consists of preparing the root canal by removing the pulp of the tooth and ensuring the best hermetization of the root canal after introducing replacement filling material and after prior disinfecting and shaping the canal. The selection of appropriate tools is important [108]. By introducing filling material, the tooth's hard tissues are directly restored, and the living tissue is replaced with a filling [109]. It is necessary coronally, with lateral and apical sealing of the previously infected and then prepared root canal of the tooth. It prevents the spread of bacteria with possible toxins towards the root tip [110–112] and saliva [113].

In this way, a tooth treated endodontically may continue to function in the oral cavity, even though it has been deprived of a viable intraroot structure. The filling material should meet numerous expectations: it should be biocompatible, prevent bacterial growth and preferably be bactericidal, be sterile, adapt well to the geometrical features of the prepared root canal, allow for easy removal if it is necessary to repeat endodontic treatment [114,115], and not cause adverse aesthetic effects; e.g., tooth discolouration [115–117]. Preferably, the filling material enhances the strength of the prepared tooth root [118].

Progress in filling materials and techniques for the development and obturation of root canals is made systematically, and newer materials are introduced [119–125] (Figure 2). However, the pace of these changes is not significant. Significant changes have been observed over the past half-century, and some of the methods or materials listed in Figure 2 may now be of historical importance.

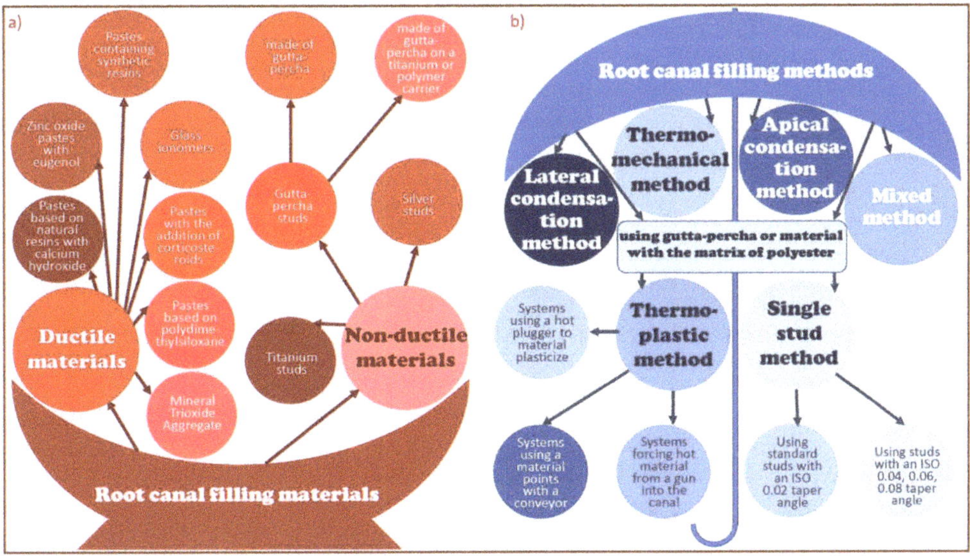

Figure 2. Scheme: (**a**) ikebana full of all types of root canal filling materials; (**b**) umbrella surrounding all types of root canal filling methods.

Apart from the substantive literature reports grouped in Figure 2 in the form of ikebana materials used in endodontics and the umbrella of obstructive techniques, an important indication for the assessment of the development of endodontics on a global scale is the results of the analysis of global markets related to this field of health protection. There are several main factors determining the growth of the endodontic market in the world. Undoubtedly, one of the most important factors is the increasing incidence of caries and other oral diseases. Together with a detailed analysis of the territorial distribution of caries and other diseases of the oral cavity, this issue was presented in the authors' study [1]. The increase in the incidence of caries and the consequent need for endodontic treatment is strongly associated with the rapid growth of the geriatric population and the ageing of societies. According to data provided by WHO, it is estimated that by 2050 the global population of elderly people will reach approximately 2 billion worldwide, when in 2010 it was already around 524 million people [126]. The rapid growth takes place in highly developed countries, including in the United States, Japan, and Europe in Italy, Germany, and the United Kingdom. According to the US Census Bureau, the US population of people aged 65 and over will increase more than twice between 2011 and 2050, from around 40 million to approximately 89 million [126]. The knowledge about oral hygiene and health in low-income groups is low, which inevitably affects the lack or inadequate level of prevention of oral diseases [127], which affects huge numbers of people worldwide. At the same time, higher- and high-income societies systematically increase awareness of oral hygiene and the need to treat oral diseases [128], which is an important driver of the growth of the endodontic market. This is accompanied by an increase in government support in many countries to promote awareness of dental disorders and diseases [128] due to the noticeable negative impact of these diseases on the macroeconomic situation in various countries and the related increasing expenditure on dental care. Such government initiatives in the last few years include, for example, the National Tobacco Control Program (NTCP) implemented in India, implementing the World Health Organization Framework Convention on Tobacco Control. In the UK, the National Health System accredits physicians to provide endodontic and oral surgery services not yet covered by the Standard General Dental Services (GDS). The growth of the endodontic market in the world is influenced by the growing number of dentists and dental offices. Technological progress and modernization of endodontic devices resulted in disseminating improved rotary equipment for the development of root canals and reducing iatrogenic errors during this treatment. The improvement of root canal development is significantly influenced by the restriction of conventional stainless-steel hand-file systems in favour of a significant increase in the use of reciprocating nickel–titanium rotary files [127].

The corollary of this is a significant improvement in root canal preparation. An overall increase in consumers' disposable income is enabling them to increase their dental care spending and increase dental tourism, especially in developing countries due to the availability of low-cost services, including transportation, accommodation, and dental care, including in India, Mexico, and Brazil [127]. There is also a huge untapped market potential in emerging countries such as South Africa and India, where the demand for dental services is growing, including motivations resulting not from health but also aesthetic reasons [128]. The opposite effect is the high cost of dental products, limited reimbursement rules, and low medical services, including dental services, especially in many low- and middle-income countries (LIC and MIC), limiting the development of this market dentistry in general [128].

The endodontics market was forecast to reach USD 1.56 billion in 2018; in 2020, it was estimated at USD 1.7 billion and to grow to USD 2.1 billion by 2026, with a CAGR of 4.1% [127], and previous forecasts indicated that the market would reach USD 1.61 billion by 2022, rising from USD 1.21 billion in 2016, with a CAGR of 5.1% in 2017 to 2022. The territorial structure of the growth of this market in this period is shown in Figure 3a, and the breakdown by end-users is shown in Figure 3b [129]. The global dental consumables market in 2016 reached USD 38.921 million, while it is forecasted that in 2023 it will reach USD 55.584 million, with a CAGR of 5.2% from 2017 to 2023 [128]. To some extent, the global dental consumables market concerns endodontics (Figure 3c) in the field of obturators and permanent endodontic sealants, files divided in terms of material into steel and Ni–Ti alloy, and type into manual and rotary, as well as filling materials and sealants and general dental materials consumables, dental sealants, dental drills, patient bibs, aspirator tubes, saliva ejectors, and other auxiliary materials.

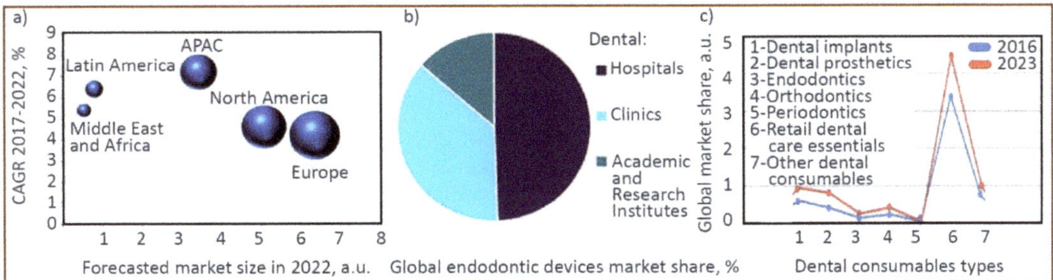

Figure 3. Forecasts for the growth of dental markets: (**a**) matrix of the forecasted size of the endodontics market (the diameter of the marker symbolizes the relative value) depending on the CAGR; (**b**) breakdown of the global endodontics market in 2018 by end-user; (**c**) an increase in the dental consumables market with the participation of endodontics in 2016−2023.

On the one hand, important elements of the global endodontic market, and most importantly, decisive for the effectiveness of endodontic treatments, are filling materials and the sealants used with them. For many years, the most commonly used and recognized material for filling root canals was gutta-percha [130–135]. The name gutta-percha refers to a stiff, naturally biologically inert, resilient, electrically nonconductive, thermoplastic latex made from the milk juice of certain plant species. The most important is the genus Palaquium from the Sapotaceae family, especially Palaquium gutta, as a tree growing in the Malay Peninsula and Indonesia, as well as Palaquium oblongifolia and Mimusops balata. However, it also occurs in Eucommia ulmoides, and in Europe in papillary and ordinary euonymus. In Australia, gutta-percha is a popular name used specifically for the euphorbia tree, Excoecaria parvifolia. The word gutta-percha comes from the name of the plant in Malay: "getah", which means "latex", while Percha or perca is the old name for the island of Sumatra. Gutta-percha is a natural polyterpene, a trans exclusively double-bonded polyisoprene polymer containing 1,4 polyisoprene, which forms a rubbery elastomer. The cis structure of polyisoprene is a typical latex elastomer. The polymer is sometimes referred to as gutta, and the name gutta-percha refers to a mixture of such a polymer with alcohol derivatives of triterpenes. While latex rubbers are amorphous with a molecular structure, the trans gutta-percha crystallizes, resulting in a more rigid material. Gutta-percha, which is less flexible than rubber, is characterized by greater chemical resistance, except for oxidizing acids, and it dissolves in aromatic and chlorinated hydrocarbons. It is not resistant to weather conditions, although it is ageing in the oxygen from the atmosphere. The product of these reactions is a brittle mass, which can be counteracted with appropriate additives. Gutta-percha is a good electrical insulator. The indigenous people of Malaysia used gutta-percha for many years, including on the handles of knives and walking sticks and for other purposes, until John Tradescant discovered it as the first European in 1656. William Montgomerie, a medical officer, was the first to use this material in 1842 for medical purposes, and even received a gold medal from the Royal Society of Arts in London in 1843 [136]. Since then, gutta-percha has been assessed as a useful practically natural thermoplastic and gained many practical and industrial uses in the second half of the 19th century. It was used as insulation for underwater telegraph cables and other electrical cables from 1851, which only changed with the invention of polyethene around 1940. It was also used to produce furniture and numerous kitchen utensils and, among others, golf balls, glues, gutta-percha paper, electroforming matrices, and stamps in the 19th century and first half of the 20th century.

Gutta-percha is a polymer that exists in three isomorphic forms: α, β, and γ, with the α form being brittle at room temperature. Forms of gutta-percha α and β go into the amorphous form and become liquid in the temperature range of 40–70 °C, while the melting point of the β form is lower than the α forms. The α form is characteristic of gutta-percha produced by trees. Both forms contain trans bonds that differ in particle size. Gutta-percha density is 0.95–1.02 g/cm^3. In dentistry, only two isomorphic forms of gutta-percha are used—β, which exists in a solid state and changes into the α form of gutta-percha at temperatures of 48.6–55.7 °C; the melting point is 64 °C. Gutta-percha in the form of studs or pellets is used in endodontics. The pure form of β gutta-percha is only 18–22%, and the rest is a mixture of zinc oxide in the proportion of 59–75%, barium and strontium sulphate in the proportion of 1.1–31.2%, as well as other polymers and wax in the proportion of 1–4.1% [1]. The addition of zinc oxide improves plasticity and reduces brittleness, while barium sulphate provides X-ray impermeability so that it is possible to identify the filling using X-rays.

The physical and chemical properties of gutta-percha, including inter alia, indifference and biocompatibility, melting point, and ductility and plasticity, determined the usefulness of gutta-percha in endodontics. Thanks to thermoplasticity, polymeric materials with a gutta-percha matrix enable three-dimensional filling of the space inside the root canal. The use of gutta-percha each time requires the combined use of sealants based on synthetic resins, both in using cold and thermoplastic condensation methods. A gutta-percha matrix material together with the sealant used ensures a tight bond with the root canal wall. The tightness of filling the root canal with the filling material and its tight connection with the canal dentin are the main factors decided in the quality of endodontic treatment. The history of the use of thermally plasticized gutta-percha and sealer slightly exceeds half a century, since its first use by Herbert Schilder in 1967 [137–139]. However, this method of filling turned out to be too time-consuming and technically difficult to master, as this technique required the use of a heat-transfer device, which was heated several times over the burner until it was red, and inserted into the channel next to the gutta-percha point and then withdrawn to leave a plasticized material inside condensed apical with a plugger.

The cycle was repeated several times using several consecutive gutta-percha points and with the heating and condensation cycles repeated several times. The dentist's ability to perform vertical condensation using hot pluggers had significantly increased since the introduction in 1982 of the Touch's Heat Electric Heat Conveyor, which was activated when the tip was inserted into the root canal, eliminating the need for a torch and preventing frequent burns to the lips. The system introduced in 1987 by Stephen Buchanan provides a constant pressure to fill the irregular space of the canal with its side branches with warm gutta-percha and sealant, thanks to the continuous supply of heat for a sufficiently long time and the condensation of the gutta-percha with a constant vertical movement in the apical direction. Buchanan pluggers, available in four sizes, are designed so that their shape closely follows the shape of the developed channel and corresponds to the expansion of nonstandard main studs. Buchanan pluggers are made of stainless steel, and their plastic deformation allows deeper condensation, especially in narrow, bent channels. The Buchanan system enables temperature control up to 600 °C. Temperatures not higher than 350 °C do not cause irreversible damage to the periapical tissues, as confirmed clinically [137–139]. However, mild periodontitis associated with overheating may be short-lived but resolve no later than 12 h. The Thermo-Hydraulic-Condensation [140–145] technique of filling root canals with the use of the Buchanan system, introduced 20 years ago, enables the lateral canals to be filled to a greater extent and consists of increasing the plasticity time of gutta-percha under the influence of temperature and pressure obtained during this procedure by using appropriate tooling (Figure 4)

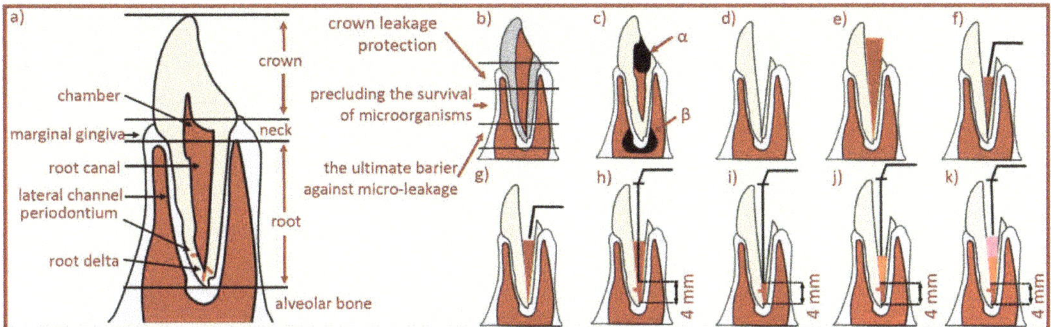

Figure 4. A diagram of a tooth cross-section showing: (**a**) the functions of the tooth pulp; (**b**) functions of filling the root canal; (**c**) dead pulp due to a carious lesion (α) with a periapical lesion (β); (**d**) evacuation of the ventricular-root pulp and conical preparation of the root canal; (**e**) fitting the main cone to the root canal; (**f**) cutting off the main stud with a heated plugger; (**g**) condensation at the mouth of the severed main cone with a cold Buchanan plugger; (**h**) plasticizing the main stud by introducing the heated plugger to a depth 4 mm less than the working length and cutting off the excess stud; (**i**) condensation of the plasticized main cone with a cold Buchanan plugger; (**j**) filling 1/3 of the central part of the canal with liquid gutta-percha plasticized outside the canal and then condensing the material with a cold Buchanan plugger as in (**i**); (**k**) final filling of the canal with liquid gutta-percha plasticized outside the canal and then condensation of the material with a cold Buchanan plugger as in point (**i**).

Gutta-percha can also be used as a liquid for filling cold canals. For this purpose, capsules are available filled with gutta-percha powder with a particle size of less than 30 μm with the addition of anti-infective nanosilver particles and sealant particles.

Despite the numerous advantages of gutta-percha, documented by multiple clinical trials, which made it even considered the "gold standard" of endodontics, there were alarming signals that in some cases, it lacked proper bonding to dentin, which could suggest inadequacies in leakage prevention [146–155]. However, obturation errors of the dentist cannot be ruled out in such cases. Undoubtedly, the result of various assessments of the effectiveness of individual filler and sealing materials is determined by the methodology of these assessments, which has been pointed out in some publications [156,157]. There are also reports of difficulties in removing gutta-percha from the root canal, if necessary, with endodontic retreatment [158,159]. It can be counteracted by selecting nonstandard sealing materials [118,160–163] and attempts to use other filling materials, among the many given in Figure 2 [164,165]. Typically, this may improve some of the expected aspects of sealing previously reported, but it almost always worsens the removal of the filling material during subsequent endodontic treatment [166].

An alternative to gutta-percha is a synthetic thermoplastic filling material based on polyester materials, consisting of an organic part constituting a matrix of methacrylate resin and inorganic fillers, including bioactive glass and barium sulfate and X-ray impermeable oxychloride [167–173], first introduced in 2004 by Resilon Research, LLC, Madison, CT, USA [168]. Hence, this filler material is mainly known by the trade name Resilon, and the sealant has an eponymous name introduced by Sybron Dental Specialities, Orange, CA, USA, although it may also be presented as RealSeal from Pentron Clinical Technologies, Wallingford, CT, USA, or AH-26/AH-plus from Dentsply Maillefer (Ballaigues, Switzerland). Endorez (Ultradent Products, South Jordan, UT, USA), for example, has the ability to dissolve in some solvents such as chloroform, or to heat soften, which is of particular importance when endodontic retreatment is required, although gutta-percha can also be removed with a suitable solvent.

This system consists of three components, including [167,174,175]:

1. A self-etching primer comprising a sulfonic acid-terminated functional monomer, HEMA 2-hydroxyelylmethacrylate, water, and a polymerization initiator;

2. A dual-curing resin sealant containing approximately 70% calcium hydroxide, bismuth oxychloride, barium glass, and silica filler;
3. A core made of synthetic thermoplastic filling material based on polyester materials, including approximately 65% of fillers with bismuth oxychloride, bioactive glass, and barium sulphate.

The core materials of this material, similarly to gutta-percha, are provided as studs and pellets, in ISO sizes in 0.02, 0.04 and 0.06 cones, and can be used as single studs or for use in the thermoplastic obturation method.

In the initial period of implementation, the filling material with a matrix of polyester materials showed leaks at the interface between the polyester–polymer material and the root canal walls much more often than gutta-percha. This was the result of, among other causes, polymerization contraction [131,132], and it was one of the reasons why many dentists refrained from using this material. Many more research results are available, focusing on identifying which of the two materials, gutta-percha or polyester matrix filling material, was more appropriate for endodontic treatment. Much less attention is paid to the remaining filling materials presented in Figure 2, which may indirectly indicate a real loss of their practical and clinical significance. Systematic research in many centres was prompted by good reports about the advantage of this new material over gutta-percha [168,176,177] and the practical importance of the newly implemented sealants [178].

Moreover, it was pointed out that the characteristics of both materials are so similar that obturation techniques, including thermoplastic ones, commonly used concerning gutta-percha, can be easily transferred to this material [168]. It was reported that the bonding of this filling material with a matrix of polyester materials with dentine forms a monoblock [179–181], which would prove the dominance of this particular material. But opinions on this subject are not uniform, and in [182], it was even stated that this monoblock concept is not true. Statements have been presented in the literature that it is a material that legitimately competes with gutta-percha and even surpasses it; e.g., it has better adhesion to the surface of the prepared root canal [183,184]. Simultaneously, in the presence of chloroform, a distinct advantage is held by gutta-percha [185], and the opposite in the case of hydration or dampness of the root canal [186]. This view does not stand up to criticism, as many papers [118,187–189] indicate that gutta-percha has undoubtedly better adhesion, and it was very often concluded that both materials behave in the same way [190], even with the use of different sealants [181,191,192]. The result of these differences is the other sealing abilities of individual materials, which have been the subject of numerous studies [168,193]. Gutta-percha provides the best filling and tightness in the root canal, which is reported in a large number of papers [160,194–207], especially after longer periods [208,209], and even better antimicrobial protection than all other filling materials, especially filling materials with a matrix made of polyester materials. In the works [193,210–229], no differences were found in ensuring the tightness between gutta-percha and filling material with a matrix of polyester materials. However, in one of these studies, the advantage of the material with a polyester matrix was noticed in the apical zone of the root canal. In several studies [223,230,231], it was found out that the material with a matrix of polyester materials is not better than gutta-percha. Finally, in the works [168,178,184,194,200,211,215,216,230,232–245], on the contrary, it is this material that provides the best tightness and reduces the number of gaps. It has also been more effective than mineral trioxide aggregate (MTA) [246,247]. The papers [225,248] also indicated that it was not found that the technique of obturation, both cold and hot, was of any importance in ensuring tightness [249], which the authors of this article find to be highly controversial and not confirmed in their clinical practice. The effect of filling root canals with various materials on the fracture strength of endodontically treated teeth was also investigated. The obtained results are as ambiguous as in the previously discussed cases. It has been found that filling with gutta-percha improves these properties to the greatest extent [250,251]. However, in the works [118,176,252–254], it is shown that it is the filling material with a matrix of polyester materials that provides better properties than gutta-percha. However, other studies [255,256] indicated that these results did not differ, that there was no effect of changing the filler material [257], and that the matrix material made of polyester materials did not improve these properties [258].

A similar dispersion of opinions and assessments concerns the importance of material selection for cytotoxicity and possible postoperative complications related to it. Gutta-percha provides better resistance to cytotoxic interactions [259,260], although the opposite also is believed [261,262], as well as that the interactions of both materials are comparable [263–265]. Moreover, these fillings do not have antibacterial activity against many bacterial strains [266], or the effect is residual [267]. The research was also carried out on removing the filling if it was necessary to repeat the procedure [268–284], especially after some time. It turns out that in several studies, the lack of any differences was indicated [268,272], both with the use of hand tools [273–275] and rotary tools [269], although the advantage of gutta-percha with and without chloroform was also indicated [272–276], and when using other solvents, such as orange or eucalyptus oil [277], and when using tetrachlorethylene [278]. On the contrary, the advantages of the filling material with a matrix of polyester materials [279–282] were distinguished in this respect, also in the case of using chloroform [268,279,280,283] and tools of the K3 type [279,282]. Thus, there are still some sceptical opinions as to the need to replace gutta-percha with another synthetic filling material based on polyester materials [284]. Many dentists still believe that this material cannot compete with gutta-percha, although many examples have shown that this may not be the case.

Nevertheless, such a significant dispersion of the results of the cited studies does not allow for a precise formulation of conclusions, especially since the commonly accepted mental abbreviation referring to the Hippocratic oath [285] dictates not to harm in the first place. These research results do not formulate a detailed and substantive, ethical view that gutta-percha should be replaced with another material, but it is no certainty that this is the truth. There is still insufficient clinical evidence to form a firm opinion in this regard, so further research is needed to determine the ranking of this filling material among all the others and, above all, its possible advantage over gutta-percha.

3. General Concept and Scope of the Work Performed

In the literature on the subject, we can find numerous reports on both clinical aspects of endodontic procedures performed by dentists, as well as publications focusing on the technical side, including the selection of materials, root canal preparation and obturation techniques, and the visualization results as given in Section 2 of this paper. Unfortunately, this last aspect is presented many times that to a small extent, and most often is not corresponding to the methodological and apparatus possibilities, which are currently at the disposal of materials science and engineering. It should be noted that, in essence, the materials and technology used in endodontics do not differ from other groups of materials, and the fact that dentists undertake such tests cannot constitute a sufficient, or actually any, justification in this matter. It should be stated, however, that works with a proper methodological level also appear. Therefore, regardless of the will and need to publish the substantive research results in this area, it was decided to present this paper on methodological issues in this area, which may be useful to subsequent authors who prepare their publications in this area.

Therefore, this paper presents a detailed methodological approach to objectivity in the assessment of endodontic treatment through in vitro studies, including a comparative analysis on the example of human incisors of the maxilla, canines, and single-canal premolars of the maxilla and mandible, removed from various patients for orthodontic, prosthetic, and periodontal indications, while respecting all ethical principles and under the supervision of the territorially competent Bioethics Committee.

The tasks presented in this paper were focused mainly on the development of research methodology. In a series of other publications [286], the detailed results of these studies in which the methodology received less attention are presented. In this paper, however, only some symptomatic research results confirmed the adopted methodological approach. In this sense, this paper is the first in a series of research results presented in this way. The analysed methodological aspects included selection of:

1. Root canal filling material during endodontic treatment, assuming that the process can be carried out both cold and warm;
2. The methodology of root canal preparation with the use of available methods and various tools to ensure three-dimensional hermetization of the root canal with replacement material;
3. Root canal obturation technology during cold and hot endodontic treatment;
4. The most advantageous visualization methods for assessing the effectiveness of endodontic treatment, including the use of methodological experiences used in materials science and engineering for the study of micro- and nanostructured objects.

To analyse these aspects, it is possible to use several basic methods of organization and management: literature review; source data analysis; issue mapping; Social, Technological, Economical, Environmental, and Political (STEEP) analysis; Strengths, Weaknesses, Opportunities, and Threats (SWOT) analysis; and auxiliary methods including benchmarking, multicriteria analysis, and statistical methods. In this paper, several of these methods were used to increase the knowledge of endodontic treatment and ensure greater clinical success. For the selection of the filling material and the most beneficial developing and obturation methods of the root canal due to the patient's health conditions, and to develop a methodology for assessing the tightness of root canal filling, with a resolving power comparable to that required in material engineering for the study of micro- or nanostructured objects, a detailed multipoint research methodology was planned, including several subsequent activities:

1. Performing a literature review and preliminary analyses to determine the scope of materials, techniques for the development and obturation of root canals, and the selection of methods for assessing the effectiveness of the filling, which will be subjected to detailed tests, with the results presented in subsequent publications in this series [286];
2. Performing a theoretical analysis using the methods of procedural benchmarking and comparative analysis, with the use of contextual matrices to optimize the selection of materials virtually, techniques for the development and obturation of root canals, methods for assessing the effectiveness of fillings, which methods are usually used among others in management science and especially in foresight research as part of knowledge management; the results of the analyses are the subject of this paper;
3. Performing a series of in vitro tests on the removed human incisors of the maxilla, canines, and single-canal premolars of the maxilla and mandible for empirical confirmation of the correctness of the methodology selected in the virtual analysis mode presented in this paper, and substantive studies of the regularities presented in the series of subsequent publications [286];
4. Development of teeth for testing with the use of rotary nickel–titanium conical drills, and comparatively, nickel–titanium hand-held drills;
5. Obturation with filling material selected in the virtual analysis mode, with the use of contextual matrices and comparative material based on polymeric polyester materials: Resilon, of the trade name RealSeal, with the use of cold and hot techniques;
6. Demineralization of parts of the prepared and filled teeth with the use of various analysed methods to prepare them for microscopic examinations;
7. Making longitudinal fractures in the parts of the prepared and filled tooth roots, after cutting the notches along the axis of the teeth, cooling them in liquid nitrogen, and making longitudinal fractures in these conditions initiated by the incisions made;
8. Performing materialographic examinations on the transverse and longitudinal sections of the prepared and filled tooth roots;
9. Performing microscopic examinations of exams, fractures, and demineralized teeth with the use of a stereoscopic light microscope, scanning electron microscope, and

confocal laser microscope to verify experimentally the developed methodology of material selection, techniques for the development and obturation of root canals, methods for assessing the effectiveness of fillings, and substantive tests of regularities presented in a series of subsequent publications [286];

10. Development of methodological procedures for the selection of materials, development and obturation techniques of root canals, assessing methods of the filling effectiveness, and the substantive results of the examination of the regularities, together with the examinations of statistical results, presented in a series of subsequent publications [286].

The implementation of the research tasks assigned in this way enables the achievement of the following basic goals:

1. Selecting the conditions for the preparation and obstruction of the root canals to ensure that the filling remains in the oral cavity of the endodontically treated teeth for the longest possible time;
2. Selecting the optimal methods for the preparation and obturation of root canals, ensuring the best tightness of the filling;
3. Development of the most appropriate and useful research methods to assess the effectiveness of endodontic treatment due to the appropriate quality of root canal fillings and the required high tightness, possibly with the minimum number of the smallest gaps on the border of the root canal wall and the filling material;
4. Explanation of the reasons for the differences in the tightness of the filling between the root canal wall and the filling material in connection with the considered methods of root canal preparation and obturation;
5. In line with the previously presented assumptions, this paper covers the first three of the presented research tasks, while the fourth task is the main content of subsequent papers from the announced series [286].

4. Description of the Methodological Concept of Materials Selection, Techniques for the Development and Obturation of Root Canals, and the Assessment of the Effectiveness of Fillings

The assumptions of the so-called "digital twins" as an approach characteristic of Industry 4.0 [64,65,67–98] and the resulting idea of Dentistry 4.0 [62,63,66] were adopted to achieve the set research goals. The concept of the "digital twin" allows many experiments to be carried out in a virtual space before the analysed product or object is physically created or made available for research, even without its physical existence. Experimental verification of the solution or variant selected virtually as the most advantageous is sufficient due to considered criteria.

Therefore, appropriate laboratory tests were preceded by a theoretical analysis with methods usually used, among others, in the science of management, especially in foresight research as part of knowledge management, which also includes procedural benchmarking and comparative analysis with the use of contextual matrices. To solve the issues constituting the essence of this paper, the methodology and research results, developed and published in own works on surface engineering technologies [69,287–304], were appropriately adapted.

Each time the issues subject to research using knowledge engineering must be considered while taking into account the background, which is the closer and further environment, it is best to use integrated methods of strategic analysis. A SWOT analysis can be directly implemented to characterize and assess the intensity of the impact of positive and negative external factors, called opportunities and threats, respectively, additionally taking into account internal factors: positive (strengths) and negative (weaknesses) (Figure 5).

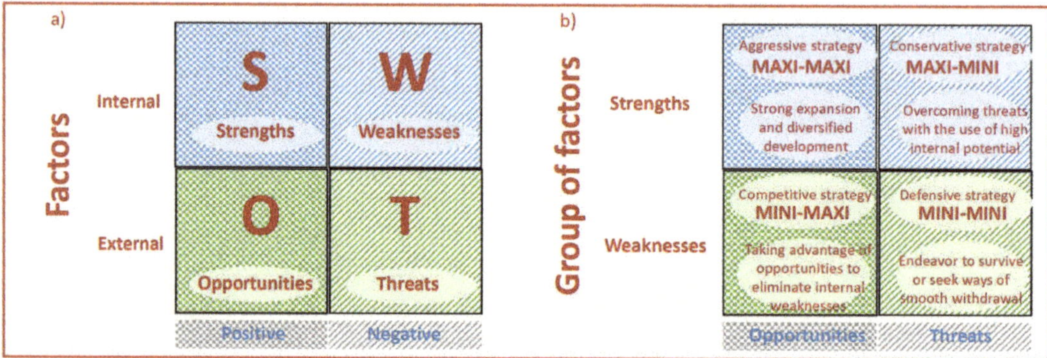

Figure 5. Scheme of the SWOT (strengths–weaknesses–opportunities–threats) analysis: (**a**) the considered factor groups; (**b**) the recommended strategies of activity concerning the analysed groups of technologies.

Multicriteria analysis can be used by defining technology assessment criteria to quantify the strengths and weaknesses of technologies, as well as opportunities and threats to them from the environment and assigning them, to sum up, weights reflecting their importance, and then assessing individual technologies according to previously adopted criteria in a specific scale. The Universal Relative State Scale (Figure 6) can be used; it is a 10-point unipolar positive interval scale without zero, where 1 is the minimum and 10 is the maximum possible score.

Numerical value	Class discriminant	Level	perfection
10	0.95	Excellent	
9	0.85	Very high	
8	0.75	High	normally
7	0.65	Quite high	
6	0.55	Moderate	
5	0.45	Medium	
4	0.35	Quite low	mediocrity
3	0.25	Low	
2	0.15	Very low	
1	0.05	Minimal	

Figure 6. The Universal Relative State Scale.

The numerical rating of the technology awarded under each of the criteria should be multiplied by its weight. Then the individual partial results are summed up, thus obtaining a weighted average that is a tool for comparative analysis, allowing us to determine the significance of a given technology against others that also were analysed. When applying multicriteria analysis of each technology, four ratings should be given, expressing numerically its strengths, weaknesses, opportunities, and threats arising from the environment. Depending on the predominance of strengths or weaknesses and opportunities or threats, each analysed group of technologies can be assigned one of four general strategies of conduct (Figure 5b). It uses such a selected methodological apparatus, allowing for a comparative analysis of technology, defining critical technologies understood as priority technologies with the best development prospects and/or key importance in the industry.

The basis for actions taken in this mode is expert knowledge; i.e., generally including the life and professional experience of experts. This assessment can be objectified by consulting a large group of experts, although the view can be obtained even after using the knowledge of one expert who is close to the issues assessed. It is about transforming inherently difficult to measure, tacit expert knowledge into explicit knowledge, available to the environment and possible to describe in the form of, e.g., a publication or a feasible procedure (Figure 7a), with the use of qualitative, semiquantitative and quantitative methods dedicated to this task. In the undertaken activities, the benchmarking method is helpful; it is based on comparing the situation of the assessed entity to the leader, on using to solve problems of a similar type, and proven and effective solutions successfully applied in other areas. The approach presented in this paper uses research methods from a completely different subject area. The material engineering research apparatus and the achievements of technological foresight of surface engineering were used to assess the quality and tightness of root canal fillings. An approach from the area of concurrent engineering was also used, the essence of which boils down to the methodology of solving important tasks, in this case concerning the methodology of endodontic treatment, by a sufficiently large group of participants, as a result of combining activities and knowledge and the obtained synergistic effect (Figure 7b).

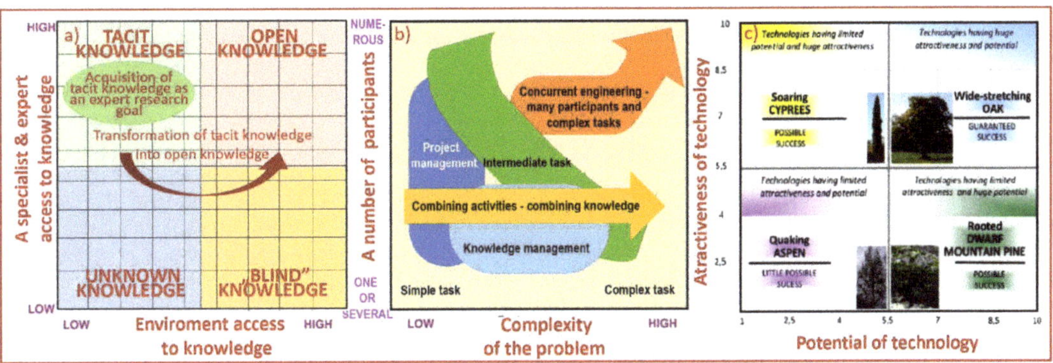

Figure 7. Schemes of context matrices: (**a**) knowledge availability; (**b**) dependence of the complexity of the problem to be solved on the number of participants in concurrent engineering; (**c**) dendrological value of technology—presentation of the approach.

Positioning of the method or procedure among those covered by the analysis and their usefulness in endodontological treatment is achieved by a set of contextual matrices, containing in particular matrices analogous to dendrological technology value matrices, adopted in the foresight analysis of surface engineering technology, where individual technologies were positioned depending on their potential and attractiveness, using original dendrological matrices (Figure 7c). These matrices are tools for comparative graphical analysis of particular methods and materials or their groups. The dendrological matrix of technology values graphically presents the assessment of individual groups of technologies in terms of their potentials, which is the real objective value of a given technology and attractiveness, reflecting the subjective perception of a given technology among its potential users.

The potential of a given group of technologies, expressed with the use of the previously described 10-degree universal scale of relative states, plotted on the horizontal axis of the dendrological matrix, is the result of a multicriteria analysis carried out based on expert assessment, taking into account in appropriate proportions the creative, application, qualitative, developmental, and technical potential expressed by an appropriately selected set of criteria that were assigned appropriate weights. On the vertical axis of the dendrological matrix, the level of attractiveness of a given group of technologies was plotted, being a weighted average of expert assessment made based on detailed criteria corresponding to economic, ecological, humanistic, natural and systemic attractiveness. Depending on the value of the potential and the level of attractiveness, which were determined as part of the expert assessment, each of the analysed technologies or, more generally, objects were placed in one of the matrix quarters, which were distinguished in the dendrological matrix of the value of technologies or other analysed objects. This approach has already been generalized and used to analyse many problems related to, among others, materials engineering, surface engineering, and technology of material processes, included in our own published works [1,64,69,132,287–314] and in the works of other authors [315–328].

Depending on the location of the analysed aspect, product, or object in a given quarter, one can conclude about its successes, respectively market or methodological, as in the discussed case concerning endodontics. Because of the generalization of the method, in this case, to the issues of endodontics, the concepts of potential and attractiveness should also be treated as generalized, and can be selected in each specific case according to actual needs.

The quarter of wide-stretching oak includes technologies or aspects characterised by the high potential in the range (5.5; 10) and high attractiveness (5.5; 10), reflecting the best possible situation for future success and expansion; e.g., market. The analysed case is, of course, a comparative analysis of the aspects characterizing individual methods and materials used in endodontics.

The quarter of the soaring cypress tree includes technologies or, more generally, objects with limited potential in the range (1; 5.5), but with high attractiveness in the range (5.5; 10), which makes these technologies possible. However, care should be taken that the apparent advantages of such an object or aspect do not obscure the actual application possibilities.

The next quarter of rooted dwarf mountain pine includes technologies and, more generally, objects of limited attractiveness in the range (1; 5.5), but with high potential in the range (5.5; 10), making their future success highly probable. However, it would require a lot of additional activity to improve the practical applicability of such locations.

The quarter of quaking aspen, on the other hand, contains the weakest technologies or objects on limited potential in the range (1; 5.5) and of limited attractiveness in the range (1; 5.5), whose future success is unlikely or impossible in such a situation; very often it is necessary to refrain from engaging in activities aimed at the practical application of such a solution.

The inspiration for developing methodological assumptions during the creation of the dendrological matrix was the portfolio methods commonly known in management sciences, serving to characterize the company's product portfolio. The methodological construction of dendrological matrices and all other portfolio matrices allows for a graphic presentation of the comparative analysis carried out, with results based on two criteria/factors placed on the horizontal (x) and vertical (y) axis of the matrix, respectively. The most famous matrix of this type was developed by the Boston Consulting Group (BCG) [329] (Figure 8a). BCG matrices owe their extraordinary popularity to their appeal to simple associations and intuitive inference. In management sciences, associations with the star as the dominant market entity and cash cow are well known, and the dog is a symbol of a failed offer while a question mark does not bode well for a given project but does not exclude it.

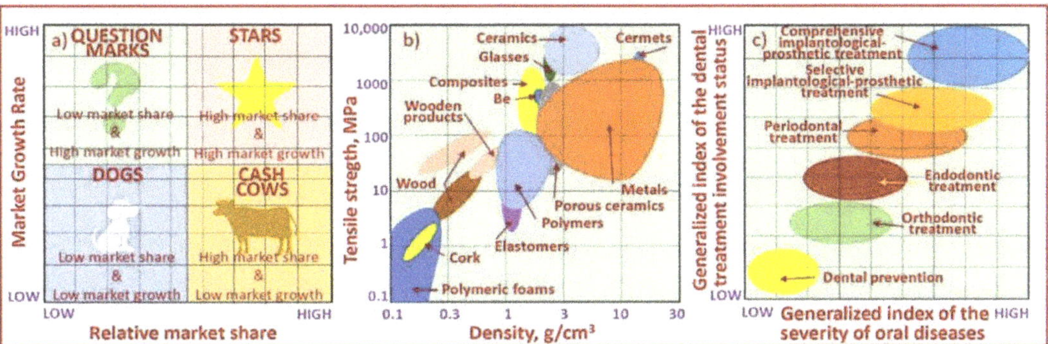

Figure 8. Contextual matrix schemas: (**a**) BCG for market assessment; (**b**) properties of various groups of engineering materials, the so-called "Maps of M.F. Ashby"; (**c**) an original concept of the matrix of advancement and treatment of oral diseases.

Even though the methodological assumptions are analogous to the previous one, such a matrix concerning the technology was called the meteorological matrix of environmental impact, concerning the graphical analysis of the influence of external factors on particular groups of technologies, or other aspects, respectively. The group of such matrices also includes the so-called "Maps of M.F. Ashby". However, their structure consists of two groups of properties of engineering materials; for example, as presented in Figure 8b. Similarly, it is possible to analyse directly the difficulties affecting negatively and opportunities positively influencing the analysed aspects in dentistry. Figure 8c schematically presents the authors' general concept of the matrix of advancement and treatment of oral diseases. In the following chapters, the authors' quantitative assessments of the impact of various factors on the effects of endodontic treatment are presented.

In the case analysed in this paper, appropriate, proprietary contextual matrices for endodontics were developed, containing strategic positions of individual methods resulting from this work. The axes are equipped with descriptors other than in the case of the surface engineering technology analysis, adequate for the analysed problem of selecting the appropriate root canal filling material, selecting the optimal method of developing and filling the root canal, and an effective method of assessing the tightness of the root canal filling, guaranteeing the effectiveness of endodontic treatment. To select the descriptors listed for research, we determined their generalized attractiveness for them and used the technique of procedural benchmarking, using the weighted scoring method for the analysis of preferences and the universal scale of relative states. The ranking and scoring methods were used to analyse tendencies, and general attractiveness in the analysed contexts was adopted as the basic criterion. This procedure was used to decide what research issues should be dealt with using the literature knowledge. First, the materials that best performed their functions when used for filling root canals were identified. Their attractiveness was analysed in terms of mechanical strength influencing a potential root fracture and the quality of the filling performed in the root canal. Next, using the effect of the first analysis, an analysis of obstruction techniques used in root canal treatment with the use of previously selected materials was carried out. The usefulness of these techniques was analysed in terms of the technique's effectiveness and the resistance of the used material to tooth root fracture. This feature is of key importance in determining the durability of the applied filling and, consequently, the effectiveness of the entire treatment. The analysis of the assessment results of obturation techniques allowed the third analysis to be carried out, which determined which techniques of root canal filling with previously selected materials and obturation techniques would meet the patient's requirements. The attractiveness of root canal filling methods was checked in terms of the quality of the performed filling and organizational aspects, including the overall cost of applying the technique and the difficulty of mastering the technique by the operator. The last analysis concerned the method selection for assessing the tightness of root canal fillings, which would be best for the analysed materials combined with organic material. The methods used on a large scale in studies dealing with the topic of tightness of root canal fillings were analysed, and methods used in other fields of knowledge could be implemented for use in this case. All the results of these analyses have been included in this paper and an exemplary description of the obtained test results. The detailed test results will be included in subsequent papers from the announced series.

The generalized attractiveness was determined for comparative evaluation, aimed at qualifying the importance of individual features in the relationship between them, using the weighted scoring method. The principle of relativization of evaluation criteria was applied, assuming differences in the significance of the criteria used, and the principle of admissibility. It used a specific group of admissibility conditions, constituting a selection filter that positively or negatively qualified a given object. The weighted scoring method allowed for the multicriteria aggregate assessment using a range scale. Detailed evaluation criteria were adopted, and their gradation was introduced. The specific weights to individual criteria were assigned based on the literature analysis on the considered criteria, interviews with dentists, and our own experience gained during dental practice. Then, the weighted values for the individual criteria were calculated. The values that were the basis for the comparative analysis were summed up to obtain finally. Table 1 presents the types of proprietary endodontic contextual matrices developed and presented in this paper.

Table 1. Types of contextual matrices related to endodontics presented in the paper.

No.	Table	Figure	Analysed Dependence on the Context Matrix	Quantity on the Horizontal Axis (Generalized Potential)	Quantity on the Vertical Axis (Generalized Attractiveness)
1.	Table 2	Figure 4	Materials selection	Generalized index of material quality for root canal filling	Generalized index of material strength in the root canal
2.	Table 3	Figure 5	Root canal development technique selection	Generalized organizational index of techniques for the development of root canals	Generalized index of the quality of filling with selected techniques
3.	Table 4	Figure 7	Techniques of obturation selection	Generalized index of material strength applied in root canal obturation techniques	Generalized index of the effectiveness of root canal obturation techniques
4.	Table 5	Figure 8	Assessing the tightness of root canal filling selection	Generalized index of organizational conditions	Generalized index of investigations effectiveness

5. Description of Authors' Contextual Matrices Concerning Materials for Filling Root Canals

The seven most common materials used to fill root canals for obturation were analysed. The suitability of individual materials depended on the material strength index in the root canal and the material quality index. To determine the strength index of the material in the root canal, we assessed to what extent the assessed material in the root canal strengthens the tooth root after time so that it maintains mechanical resistance to fracture in combination with keeping the tightness of the filling by minimizing resorption of the filling over time. The material quality index in the root canal was determined by analysing the features of individual materials according to the criteria proposed by Grossman [330]. The impact of the introduced material on the patient's body; the possibility of eliminating bacteria from the filled root canal also through the possibility of sterilization, increasing the asepsis of the treatment and minimizing the risk of reintroducing bacteria into the treatment area; the ease with which the material can be removed in the event of retreatment, as well as the possibility of correct X-ray diagnostics and control were assessed. Table 2 lists all the analysed materials used to fill the root canals and specifies the criteria of mechanical strength influencing a potential root fracture and the quality of the root canal filling.

Table 2. The set of criteria adopted for the virtual contextual analysis regarding the selection of the material filling the root canal during endodontic treatment.

General Descriptors			Criteria for Determining the Generalized Quality Index of the Root Canal Filling Material					Criteria for Determining the Generalized Strength Index of the Root Canal Filling Material	
Material	Application Form/Consistency	Method of Application to the Root Canal	The possibility of sterilization	Ease of removal from the root canal in the event of revision	Toxicity	Bactericide	X-ray transmittance	No resorption over time	Mechanical resistance of the root to fracture after using the material
Based on gutta-percha	Studs	Cold techniques							
Based on polyesters		Hot techniques							
silver studs		Applied with a sealant							
Based on synthetic resins	Pellets	Independently							
Based on zinc oxide with eugenol		On the Lentulo needle							
Based on silicones		As a sealant for studs							
Based on calcium hydroxide	Paste	As a gutta-percha sealant							
glass ionomer cements									
Weight			0.1	0.25	0.4	0.2	0.05	0.4	0.6

The performed analysis allowed for the elimination of three materials from further considerations, which stood out after the analysis, without achieving minimal results in terms of the normality of relative states. Silver studs and materials based on zinc oxide and eugenol achieved both quality and strength indexes that were below average, while materials based on calcium hydroxide achieved a quality index at a high level; however, when compared to the index of material strength in the root canal, the overall result turned out to be unsatisfactory because, despite the fulfilment of the Grossman criterion, it was not certain whether the analysed material would be able to preserve the patient's tooth several years after the treatment. The remaining four obturation methods achieved high results in the strength index of materials in the root canal. Still, none of them reached the filling quality index at the level of normality. The analysis results showed that none of the materials selected for evaluation was flawless. Still, it was possible to decide which materials stood out from the others unequivocally. Those were selected for further research, the detailed results of which are presented in the announced series. Thus, the analysis of the results made it possible to select two materials for further research: a filling material based on gutta-percha and a material based on polymeric polyester materials. The first one achieved an excellent result in terms of strength and a relatively high level of filling quality. On the other hand, the second material had a worse strength score but a relatively high-quality index. Figure 9 shows the attractiveness matrix of materials used to fill root canals, including the results of the analyses performed and described above.

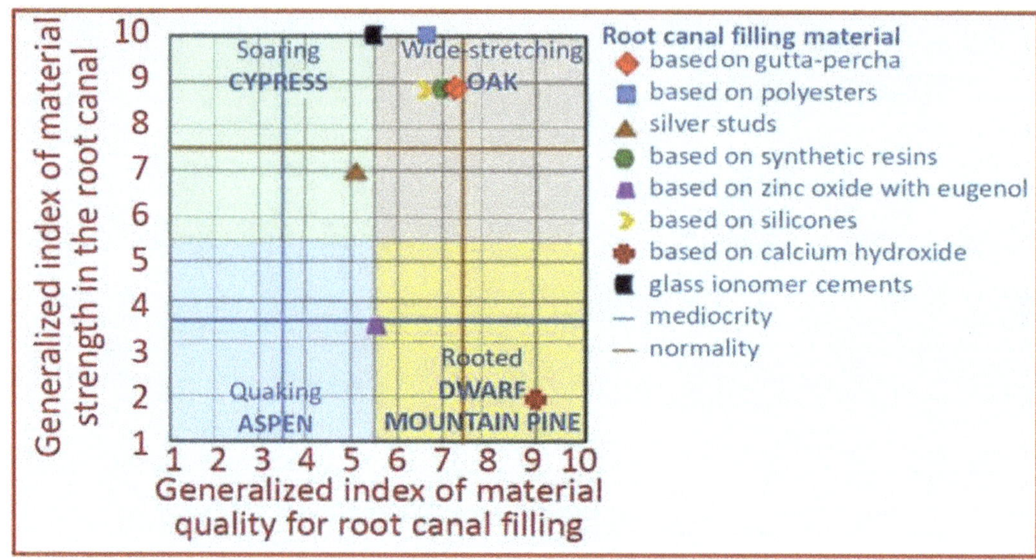

Figure 9. Contextual matrix of the attractiveness of materials used to fill root canals.

6. Description of Authors' Contextual Matrices Relating to Techniques for the Development of Root Canals

For an endodontic treatment to proceed without complications, it is extremely important to select the root canal preparation technique correctly. Therefore, all three currently used techniques of development were analysed: laser techniques, ultrasound techniques, and commonly used mechanical techniques, including the material from which the tools are manufactured. The usefulness of unique techniques for the development of the root canal and the material of the tools was dependent on the quality indicator of the performed root canal preparation. In this context, the feasibility of conical preparation and the potential for errors during the procedure, such as pushing the cut dentin beyond the periapical opening and the loss of the natural course of the canal results in the formation of niches, were assessed. Another extremely important element is the minimization of the risk of breaking the tool in the canal. Hence, the highest importance was given to this feature. The possibility of disinfecting the root canal and the risk of thermal damage to the tooth tissues were also assessed. These have a major impact on the time the treated tooth can remain in the mouth without complications, such as bacterial growth and bone changes. The second aspect in which the techniques for the development of root canals have been analysed were organizational conditions, which include the unit cost of the procedure, which should be understood as the cost of purchasing equipment, replaceable tips, materials used in a given technique, and the duration of the procedure. These factors affect the costs of the procedure and the difficulty of mastering a given technique by the dentist, taking into account access to specialist training, the duration and costs of training, and the period in which the operator will be fully operational. Table 3 summarizes all the described techniques for the development of root canals and specifies the quality criteria for developing root canals and the related organizational conditions.

Table 3. The set of criteria adopted for the virtual contextual analysis concerning the selection of techniques for the development of root canals in endodontic treatment.

General Descriptors–Techniques for the Development of Root Canals	Criteria for Determining the Generalized Quality Index of the Root Canal Filling with Selected Techniques						Criteria for Determining the Generalized Organizational Index of Techniques for the Development of Root Canals	
	No risk of thermal damage to periodontal tissues	No risk of tool breakage	No risk of losing the natural course of the root canal and creating niches	Root canal disinfection	No risk of pushing the cut dentine beyond the periapical foramen	Conical root canal preparation	Minimization of the unit cost of the procedure	Ease of mastering the technique
Laser								
Ultrasonic								
Mechanical–steel tools								
Mechanical—nickel–titanium tools								
Weight	0.3	0.2	0.1	0.1	0.2	0.1	0.5	0.5

The analysis made it possible to eliminate two techniques of root canal development from detailed studies i.e., the laser technique and the ultrasound technique. The laser processing technique achieved an average result in terms of the quality of the prepared study. The value of the indicator of organizational conditions was disqualified because it is an extremely expensive and complicated method, and the expenditure incurred does not translate into measurable benefits from its use. Therefore, it was rejected as nondevelopmental. The technique of root canal preparation using ultrasound also achieved unsatisfactory results in the attractiveness analysis. This technique was extremely average in terms of the quality of the studies performed and disqualifying in terms of organizational conditions. The costs of its use, similar to laser technology, are very high, and the quality of the study was not higher than with mechanical methods. Therefore, the analysis results allowed the choice of mechanical methods, as they have just achieved the best results due to the analysis carried out.

It should be noted, however, that mechanical methods are not without weaknesses. Their effective use also requires experience, and the risk of complications is also high. However, the cost of their use is relatively low due to the universality of their application, and the access to training materials and courses is wide, which makes it easier for the operator to acquire skills allowing for their wide application in a relatively short time. A better-quality index characterized the method using steel tools, but a lower index of organizational conditions than the technique using nickel–titanium tools. The selection between these two methods depends on the specific case and operator preference. As the results of the analyses show, there is a need to find a method of developing a breakthrough method of developing root canals that will meet the requirements of dentists to a greater extent than the ones used so far, and even more important requirements resulting from the actual needs of patients. Figure 10 shows the attractiveness matrix for techniques for the development of root canals.

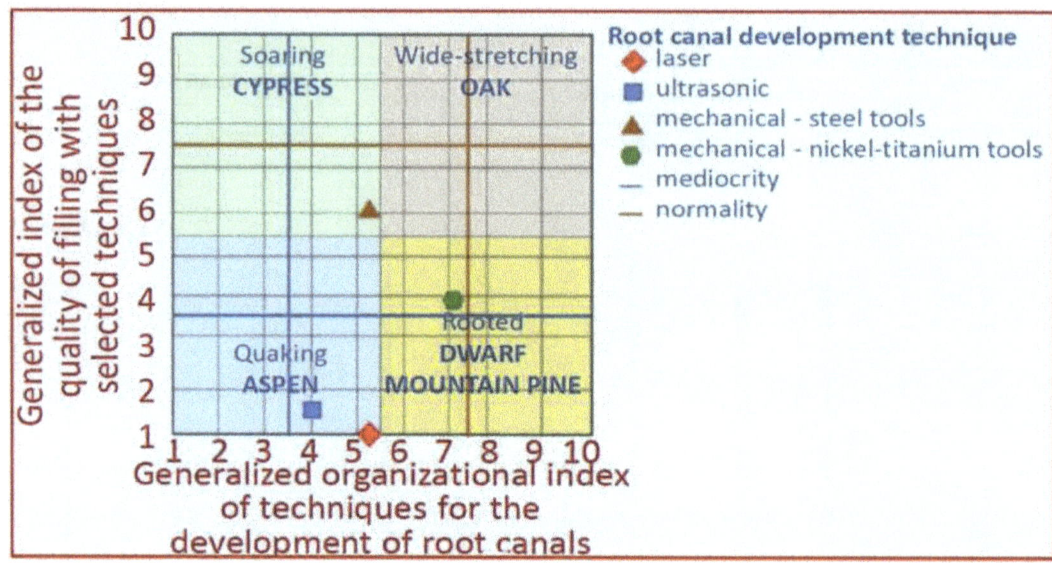

Figure 10. Contextual matrix of the attractiveness of root canal preparation techniques during endodontic treatment.

7. Description of the Authors' Contextual Matrices Concerning the Techniques of Obturation of Root Canals

The attractiveness of root canal filling methods should be analysed in terms of the quality of the filling and organizational aspects. The effectiveness of obturation methods was determined by first analysing the risk of pushing the material beyond the root apical. This is an important aspect of the period during which the treated tooth remains in the oral cavity without pain symptoms and lesions requiring its removal. An important element is the volume of the sealant used, because its high presence in the treated canal may result in the appearance of fissures in which bacteria develop over time and a reduction in the mechanical resistance of the root to fracture. The sealant is characterized by rapid resorption over time and, in fact, little mechanical strength. An additional analysed criterion for the evaluation of obturation methods is the possibility of filling additional tubules, which additionally reduces the risk of the occurrence of places where bacteria can develop. The last analysed criterion is the speed of applying the technique. It is a less-important feature, but due to the need to perform the procedure in cooperation with the patient experiencing pain, weariness, and fatigue, the obturation procedure should not take too long. For these reasons, this feature was included in the analysis, but with the lowest weight. The analysis in this aspect should be treated as control of the excessively long working time of the operator. Simultaneously, the speed of the procedure itself is not an aspect that distinguishes the method used. Table 4 summarizes all the described techniques of obturation, and lists the criteria and the weights assigned to them in detail, characterizing the analysed obturation techniques.

The analysis made it possible to unquestionably eliminate from detailed studies the method of filling the canal with paste as significantly different from the other methods, and the method of the central cone, which showed average values in terms of effectiveness. The cold lateral condensation technique achieved very good results in terms of the strength of the material used to fill the root canal, but only average results in terms of effectiveness.

Table 4. The set of criteria adopted for the virtual contextual analysis concerning the selection of assessing the tightness of root canal fillings during endodontic treatment.

Techniques of Root Canal Obstruction	Type of Material Used	Additional Devices	The amount of sealant used	Speed of proceeding using a given technique	No risk of pushing the material beyond the tip	Possibility of filling additional channels	No resorption occurs over time	Mechanical resistance of the root to fracture after using the material
	General Descriptors		**Criteria For Determining the Generalized Quality Index of the Root Canal Obturation Technique Effectiveness**				**Criteria for Determining the Generalized Index of Material Strength Applied in Root Canal Obturation Techniques**	
Filling the root canal with paste	Sealant	Lentulo needle						
Central stud	Gutta-percha + sealant							
Cold side condensation	Gutta-percha or material based on polyester polymeric materials + sealant	Cold plugger						
Hot side condensation		Hot plugger + cold plugger						
Thermo-mechanical condensation	Gutta-percha on a compactor; e.g., Quickfill + sealant	Caecum						
Gutta-percha plasticized	Gutta-percha on plastic; e.g., Thermafil + sealant	Heating furnace; e.g., TermaPrep						
Thermo-hydraulic-condensation	Gutta-percha or material based on polyester polymeric materials + sealant	Hot plugger, cold plugger, hot gutta-percha feeding device; e.g., System B + Obtura III						
Weight			0.15	0.05	0.3	0.5	0.4	0.6

The analysis shows that the lateral heat condensation of the root canal filling with material based on gutta-percha plasticized on a plasticizer achieved better results than other methods. The most attractive technique of root canal obturation turns out to be the Thermo-Hydraulic-Condensation (THC) technique, which obtained the best results both in terms of effectiveness and strength. The matrix of the attractiveness of root canal obstruction techniques is presented in Figure 11.

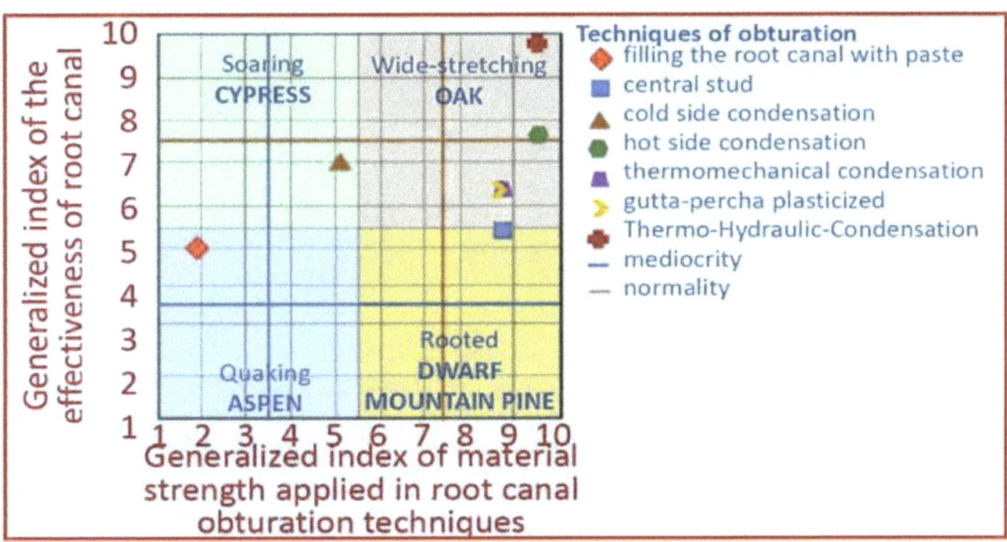

Figure 11. Contextual matrix of the attractiveness of root canal obturation techniques during endodontic treatment.

8. Description of the Authors' Contextual Matrices Concerning the Methods of Assessing the Effectiveness of Fillings

An analysis was made of which techniques worked best when using the analysed filling materials in combination with organic material. The techniques that were analysed are used on a large scale in dentistry, as well as in other fields of knowledge, mainly in material engineering, and are possible to implement in endodontics to select the best methods for assessing the tightness of root canal fillings. The visualization methods were compared with other methods described in the literature [132], taking into account the effectiveness of the tests performed and organizational conditions to assess the validity of the choice of evaluating the tightness of root canal fillings. The effectiveness of the research was analysed, which was significant from the point of view of the obtained results, depending on the incurred organizational expenses necessary to conduct the research. For this purpose, a generalized test effectiveness index was created that consisted of equal parts: the effectiveness and sensitivity of the tests performed, assessed based on the literature, and the accuracy of the determination, which should be understood as the method of evaluating the filling of the canal and the suitability of the tested test method for quantifying the leakage at the border root canal fillings and walls. The second indicator used for this analysis was the generalized indicator of organizational conditions. As part of this issue, the procedure for preparing the specimen for the test; the course of the test, including the invasiveness of the analysed method; and the costs of conducting the test, including the costs of materials, investment costs, and the costs of the entire research equipment, were analysed. The attractiveness of the methods of assessing the quality of root canal fillings was checked in terms of the quality of the filling and organizational aspects. Table 5 summarizes all the described methods of tightness testing on the border of the root canal filling and wall, and specifies the criteria for the effectiveness of the tests and organizational conditions. The graph was also supplemented with a graphic border between normality and average, which facilitated the analysis of the presented results. Simultaneously, the boundary of the excellence group has not been given, but the maximum values shown in the graph are the boundary of excellence.

Table 5. The set of criteria adopted for the virtual contextual analysis concerning the selection of techniques of obturation during endodontic treatment.

General Descriptors / Investigation Method	Criteria for Determining the Generalized Index of Organizational Conditions of Assessed Investigations Methods								Criteria for Determining the Generalized Index Investigations Effectiveness	
	Difficulty preparing the specimen	Complicating the procedure of preparing the test system	Destructive/non-destructive method	Possibility of multiple test realization	Maximum possible duration of a research experiment	Cost of materials necessary to realize the investigation	The investment cost and the purposefulness of the expenses incurred	Cost of investigation equipment	Sensitivity/resolution	Relevance of the determination
Fluent transport model										
Brightening technique										
Dye penetration										
Penetration of bacteria/metabolites										
Electrochemical										
penetration of labelled radioisotopes										
Glucose penetration										
Light microscope										
Stereoscopic microscope										
Scanning electron microscope										
Confocal laser microscope										
Weight	0.05	0.15	0.1	0.1	0.1	0.2	0.2	0.1	0.5	0.5

The weighted scoring analysis conducted allowed for the conclusion that quantitative methods obtained the lowest scores, including the liquid transport model, lightening technique, dye penetration test, test of bacterial penetration and their metabolites, glucose penetration test, and a method using an optical microscope, for which it should be said that they are of marginal importance and, as they have the little prospect and poor prospects for the future, they were not used in the further part of the research presented in the series of publications.

The analysis showed that among the methods for assessing the tightness of root canal fillings presented on the attractiveness matrix, the highest values and a clear advantage over other methods in terms of effectiveness were obtained by visualization methods using materialographic microscopes. The advantage of visualization methods is obtaining a sufficiently high resolution with which the tested preparations can be observed, which absolutely determines the observations' details and allows the measurement of the observed changes. Simultaneously, in these methods, it is possible to undoubtedly determine the type and details of the tested material, which also allows assessing how the sealant and the filling material are positioned in the channel being filled. Thanks to these methods, one obtains complete knowledge of the content of the filled channel.

All available materialographic microscopes were used in the research, including the ZEISS automated Stereo Discovery V12 stereoscopic light microscope with an HRC camera, the LSM Exciter confocal laser microscope for materialographic examinations based on the Axio Observer microscope by ZEISS (sources: HeNe 633 nm 5 mW, HeNe 543 nm 1 mW, argon 458/488/514 nm, 25 mW, V 405 nm diode), a Park Systems AFM XE-100 atomic force microscope, a SUPRA 35 high-resolution scanning electron microscope by ZEISS together with WDS, an EDS spectrometer, and an EBSD TRIDENT XM4 camera by EDAX. The assumption was made, which was confirmed experimentally, that with the use of this equipment, it is possible to test various materials, not only engineering materials, and therefore also a special group of them, which were removed human teeth filled in the root part with appropriate filling and sealing materials. The method using the light microscope obtained results on the borderline between normal and average. Separate experiments were related to the use of the confocal laser microscope and the atomic force microscope. Research with the use of these microscopes was mainly of methodological importance. Research methods using a stereoscopic light microscope, a confocal laser microscope, and a scanning electron microscope have achieved values of excellence. Therefore, they constitute the basic methods useful for the mentioned purpose in endodontics.

The best result in the analysis was obtained by using a scanning electron microscope, which has one of the highest resolutions and enables by far the best accuracy of the mapping. Using this microscope, a properly prepared specimen can be viewed with the possibility of a precise assessment of various types of materials viewed, which was of key importance in the conducted research. The sample preparation process prepared for the needs of the described research allowed for obtaining longitudinal fractures, which, without any noticeable losses, made it possible to assess the tightness of fillings in the prepared root canals.

The scanning electron microscope is useful for observing structural details on longitudinal fractures covered with a thin layer of vacuum-sputtered gold, completely invisible in the applied magnification range, ensuring electrical conductivity of the tested surface, and revealing details of the structure of the tight connection of dentin and filling material, as well as for measuring the width of the gap in in the event of its disclosure, and to identify the structure and chemical composition of sealants using the EDS (energy dispersive spectrometer) analyser.

Materialographic analysis with the use of a stereoscopic light microscope was used to observe the general view of the examined teeth after decalcification, where the course of the root canal filling was revealed, as well as two canals in the two-canal tooth, the root delta, and the course of the lateral canals. This preparation method, despite the fact that it lasted several weeks, allowed for very interesting research results. The stereoscopic light microscope was also useful for the observation of fractures, including longitudinal fractures, and was the most useful due to the adopted technique of measuring the identified gaps between the dentin and the filling material, and between the dentin and the sealant.

Materialographic examinations require proper specimen preparations to enable the observation of the examined teeth. For observation with a stereoscopic light microscope, materialographic specimens were prepared on longitudinal sections, or after the transverse cutting of the teeth filled with filling material, by embedding in a thermosetting resin and mechanical grinding successively on abrasive papers with smaller and smaller grain size, and then on diamond pastes that had a grain size of up to 3 μm. As the preliminary tests showed, due to the cracks formed during the preparation of the specimens by excessive heating, the filling material was smeared on the surface and additionally penetrated the newly formed cracks, and the abrasive from abrasive papers, and perhaps also from the diamond pastes, was arranged almost parallel on the surface of the soft elements of the structure, mainly in the filling material. They provided an example, along with other similar experiments on other preparations, that materialographic examinations are not a useful technique for preparing specimens to determine the effectiveness of root canal obturation, which practically eliminated the possibility of using this type of preparation. Therefore, the number of tests using this technique was limited, and the recording of results was limited to only methodically necessary cases. While the preparation of specimens for testing using an atomic force microscope did not require any extraordinary measures, in the case of leakage between the dentin and the seal, it posed a unique threat to the equipment. In the course of the described tests, the scanning probe blade was broken twice, jammed in the deep but narrow opening of the gap between the dentin and the filling.

The use of such specialized research equipment, especially a scanning electron microscope, requires a complex technical base and specialist knowledge necessary to operate the research equipment. The relatively unfavourable elements of the evaluation of this method included limited access to such research equipment, the purchase of which is very expensive, and the use of which requires extensive knowledge and high skills. However, we also considered that the analysis concerns the performance of research that can be carried out across multiple disciplines, and equipment of this type is found in many universities. In industrial laboratories, it can be concluded that this is not a context that disqualifies this method from research for endodontics. In the performed analysis, all organizational aspects were assessed. It can be concluded that the incurred labour and financial inputs are compensated by the high quality of the research results obtained.

The methodological studies reported in this chapter and the examples presented in it indicate the possibility of performing materialographic tests using all five analysed microscopes. The most effective way is to select the preparation of longitudinal fractures as the appropriate methodology for preparing teeth for materialographic observations, preferably using automated materialography, a stereoscopic light microscope, and a high-resolution scanning electron microscope. Figure 12 shows the attractiveness matrix for the assessment of the tightness of root canal fillings.

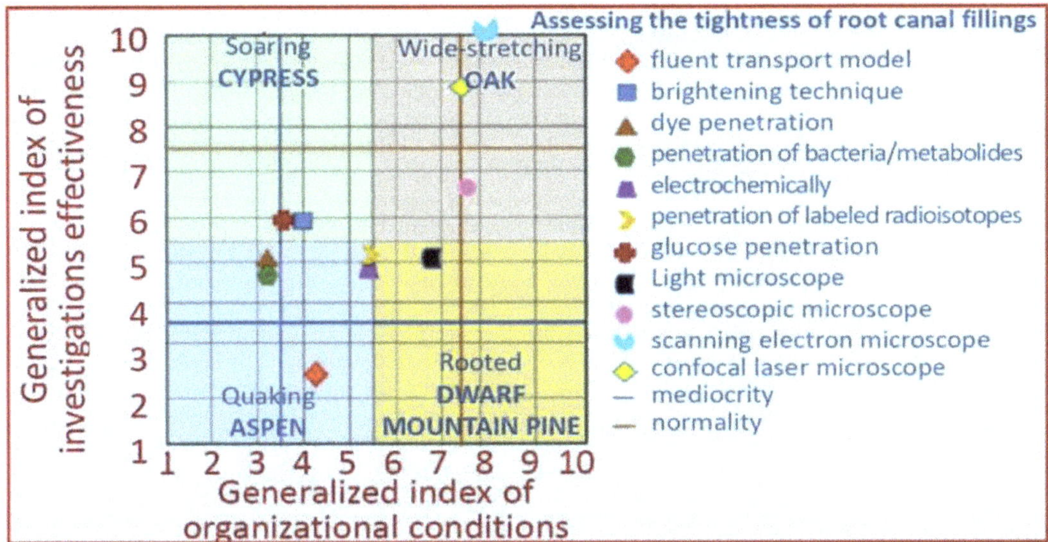

Figure 12. Matrix of attractiveness for the assessment of tightness of root canal fillings.

Measurements of the width of the gaps between the root canal wall and the filling material were made using scanning electron microscope tests in the magnification range of 1000–5000×, and the microscope software, which had a built-in mechanism for measuring the distance between two markers set by the operator. The width of the gap is a measure of the effectiveness of the treatment endodontic. Tight filling of the apical section is extremely important during endodontic treatment because it is the last barrier that protects the periapical tissues against possible bacterial infection penetrating the canal from the side of the leaky cavity in the tooth chamber. The assessment of the tightness of the root canal filling was first made by determining whether the filling leaks were present and how numerous they were in any of the three sections of the root canals: parapillary, middle, and apical. Moreover, the number of leaks in these three segments was summarized, as their sum indicated the quality of the filling in the entire tooth. In addition, it was necessary to establish:

1. The number of leaks along the entire length of the root canal;
2. A representative value characterizing the mean dimensions of such leakage.

A comparative analysis of the width of the slots was also performed, and the results of the measurements of these widths were statistically analysed. The following calculations were made:

1. Mean value;
2. Standard deviation;
3. Confidence interval with the assumed significance level of 0.95;
4. Mean difference significance test.

The "Statistica 13" program was used to perform the calculations, while the charts were prepared using the "MS Excel 365" program. The calculations results are presented in the graphs.

9. Experimental Verification and Examples of the Application of the Selected Methods to Evaluate the Effectiveness of Root Canal Filling

This section of the paper presents the results of in vitro experimental studies to confirm the possibility and correctness of selecting methods for assessing the effectiveness of root canal fillings performed in the virtual analysis mode with the use of comparative matrices.

In vitro studies included 80 single-canal incisors in the maxilla, maxillary, and mandibular canines, and maxillary and mandibular single-canal premolars removed for orthodontic, prosthetic, and periodontal indications. Among these teeth, two sets of 32 and 48 teeth, and 5 groups, each containing 16 teeth, were identified below. Immediately after extraction, the test material was placed in 0.2% chloramine solution for 24 h, then rinsed and transferred to distilled water. These teeth were selected after the rejection of approximately 25% of the preprepared single-canal teeth not meeting the established requirements. In particular, teeth with a damaged root part were excluded from the study due to fractures, cracks, or root resorption, as well as teeth from adolescent patients with incomplete development of the apical part of the root or teeth damaged during sample fractures, which was revealed only during the examination with the use of light stereoscopic microscopy methods. Correctly selected single-canal teeth were prepared and filled with the assumptions described in this paper. Before processing, each tooth was cleaned and rinsed in 0.9% NaCl solution, and then the tooth crowns were cut off at the height of the tooth neck with a diamond separator placed on the prosthetic handpiece. During the mechanical preparation of the root canals, a lubricant containing glycerin, sodium edetate, and urea peroxide RC-Prep (Premier) was used. The task of the lubricant during the preparation of the canal was to remove the smear layer covering the canal dentine and facilitate the mechanical preparation of the canal with the tools used. Between each instrument subsequently introduced into the root canal, irrigation with 2.25% sodium hypochlorite solution and 0.9% saline solution was performed alternately.

Among all prepared teeth, two sets were specified due to the preparation of root canals with hand or rotary tools. Then, five groups were completely separated due to the filling material and the method of obturation.

A 32-tooth set of 2 groups was processed with hand tools (ProTaper, Dentsply/Maillefer). Each canal was prepared in two stages. In the first stage, the peritoneal part was widened with the S1 tool, then the SX tool with the highest 19% taper was inserted into the same part of the canal, after which the length of the canal was measured with a 10 ISO Kerr file, from which 1 mm was subtracted. The working length was determined based on the length of the tooth canal minus 1 mm (as the average distance between the anatomical apex and the physiological narrowing of the root canal). The channel was subsequently developed to the full working length with the S1 and S2 tools. The next four tools, numbered F1, F2, F3, and F4, were introduced to the full working length, widening and smoothing the root canal walls. Each instrument introduced was followed by a channel recapitulation with the Kerr tool. As a result of the development and filling of root canals, the research material selected according to the previously mentioned principles was divided into five groups.

The second set of 48 teeth, consisting of 3 groups, was developed with rotary tools up to an ISO size of 40 using the X-Smart endodontic micromotor (DentSply/Maillefer). Of these teeth, 16 were processed with hand tools (ProTaper, DentSply/Maillefer), and the remaining 32 teeth were processed with K3 tools (Sybron Endo). Each channel was developed in two stages. In the first stage, the canal mouth was developed using K3 tools with the symbols 10/25 and 08/25, with a tool taper of 10 and 8%, respectively. Then, the root canal length measurement was performed by inserting a 10 ISO Kerr file and subtracting from the full length of 1 mm, determining the working length as the average distance between the anatomical apex and the physiological narrowing of the root canal. Subsequently, the canal was developed to the full working length with tools in sizes of 0.4/20, 0.4/25, 0.4/30, 0.4/35, and 0.4/40. Each instrument introduced was followed by a channel recapitulation with the Kerr tool.

In two groups, 32 teeth were filled with the method of lateral condensation. In the first group, the canals were filled with the lateral condensation method, and the filling material was made of a material based on a gutta-percha matrix. After drying the root canal, a F4 size gutta-percha point (DentSply/Maillefer) stud was fitted, corresponding to the last tool used to prepare the canal to the full working length. Before obturation, each canal was thoroughly dried with paper points. A thin layer of sealant (AH Plus type, DentSply/Maillefer) was applied to the walls with a paper filter in each canal. Each stud made of a material based on gutta-percha was decontaminated in 2.25% sodium hypochlorite solution and then placed in the root canal on the full, previously determined working length. Then, due to the large convergence of the main cone, which corresponded to the variable taper of the ProTaper tool, an expander with the size of 20 ISO, with a taper of 2%, was selected. If using a silicone stopper, a length 1 mm shorter than the length on which the main stud was inserted was determined for the plugger. After the placement prepared by the manual pusher, the canal was supplemented with studs made of gutta-percha matrix material with a convergence of 2% in the size corresponding to the pusher used. The filling of the canal was completed when it was impossible to insert the expander into the canal.

In the second group, the teeth were filled with the lateral condensation method; the filling materials were Resilon points based on polyester polymer materials (RealSeal, SybronEndo) with a convergence of 4% and 2%, respectively. Because the sodium hypochlorite solution may negatively affect the bond strength, each canal was thoroughly rinsed with 0.9% saline solution before drying to neutralize the residual hypochlorite. Before obturation, each canal was thoroughly dried with paper points. A root stud (RealSeal, SybronEndo) with a taper of 4% was fitted to each canal, corresponding to the apex of the last MAF instrument to prepare the canal. Then, the self-etching conditioner included in the kit (RealSeal, SybronEndo) was applied to the root canal walls with a paper filter. Then a thin layer of sealant (RealSeal, SybronEndo) was placed on the canal walls with a paper filter, and a previously selected main stud was inserted into the canal. A 20 ISO plugger with a taper of 2% was chosen successively.

Similar to inside condensation, a length of 1 mm shorter than the length on which the main stud was inserted was established using a silicone stopper on the pusher. After the manual plugger had prepared the site, the canal was supplemented with studs (RealSeal, SybronEndo) with a taper of 2% in the size corresponding to the plugger used. After filling the canal, the mouth of the canal was irradiated with the light of a polymerization lamp for 40 s to immediately seal the canal from the outlet side.

In the third group, the teeth were filled with the thermoplastic method using the Thermafil-type system, which began with selecting the appropriate size of this system from a material based on gutta-percha with the use of a verifier. The canals were checked with a 40 ISO verifier because the last instrument introduced during root canal preparation was 40 ISO. After selecting the appropriate Thermafil-type obturator, the working length was marked with a stopper, disinfected in 2.25% sodium hypochlorite solution, dried, and placed in a ThermaPrep Plus oven to plasticize it. A thin layer of AH Plus sealant (DentSply/Maillefer) was applied to the walls with a paper filter for each canal. After the gutta-percha was plasticized in the oven, the Thermafil-type obturator was slowly withdrawn from the oven elevator and placed in the channel to the working length previously determined, maintaining a constant pressure on the polymer support for a few seconds. When placing the obturator, care was taken that it was accurately inserted centrally into the canal to prevent and distort the material in the gutta-percha matrix at the top of the obturator. The polymeric material carrier on the gutta-percha matrix was then cut off with a Therma-Cut drill without water cooling at 300,000 rpm. The gutta-percha matrix material was condensed around the carrier with a plugger.

In the case of the next two groups, containing 32 teeth in total, they were filled with a thermoplastic method using System B and Obtura III devices (SybronEndo).

In the fourth group, the teeth were filled with studs and pellets made of material based on gutta-percha. The root canals were thoroughly dried with paper points. Then the main stud was selected from material on a gutta-percha matrix with a convergence of 4%. The selection criterion was the wedging of the stud in the periapical area after its introduction to its full working length. The Buchanan plugger (SybronEndo, size 1) was subsequently fitted to the canal's diameter, and its length was marked with a stopper, equal to the working length shortened by 4 mm. AH Plus sealant (DentSply/Maillefer) was applied with a paper filter to cover the root canal walls. During the introduction of the heated System B plugger through the gutta-percha-based stud placed in the root canal to the length marked with a stopper, the temperature was set at 200 °C. After 3 s, the heating was turned off while maintaining the pressure of the tool towards the root apex for 10 s. The plugger was then reheated to 300 °C and immediately removed along with any excess noncondensed material on the gutta-percha matrix. The apical portion of the material on the gutta-percha matrix was condensed for the next 10 s with pulsating movements using a cold, previously selected Buchanan-type plugger (SybronEndo). The canal, filled in this way at the apex, was supplemented with gutta-percha matrix material and fed into the canal using the Obtura III system (SybronEndo). The temperature was set at 160 °C. The canal was supplemented in stages by introducing small portions of liquid material on a gutta-percha matrix, each time condensing with a Buchanan-type plugger (SybronEndo). The sequence of root canal preparation and obturation using this methodology has already been presented in Figure 4.

In the fifth group, the teeth were filled with studs and pellets made of Resilon based on polyester polymeric materials (RealSeal, SybronEndo) using System B and Obtura III devices (SybronEndo). The canals were thoroughly rinsed with 0.9% saline solution to neutralize any residues. Subsequently, the main pin of the RealSeal-type (SybronEndo) with a taper of 4% was selected so that when introduced to the full working length, there was a wedging of the cone in the periapical area. Primer material was attached to the RealSeal kit (SybronEndo). A thin layer of RealSeal sealant (SybronEndo) was placed on the canal walls with a paper filter, and a previously selected stud was inserted into the canal (Figure 4).

Microscopic methods were used, adapted from materials science, to assess the tightness of the root canal fillings according to the methods described above, including a stereoscopic light microscope, a confocal laser microscope, an atomic force microscope, and a scanning electron microscope. The preparations for the tests were prepared using various available techniques; i.e., as decalcified teeth, transverse and longitudinal materialographic specimens, and longitudinal fractures. To prepare the specimens for materialographic tests, after the canals were tightly filled with each technique, the root canal orifices of the selected teeth were secured with Ketac Molar glass ionomer cement (3 M ESPE). For the next seven days, the teeth were stored in a humid environment at room temperature to bind the sealant. Each tooth was wrapped in gauze soaked in physiological saline and tightly closed in plastic containers. The prepared samples were incised longitudinally along the root to a depth of 1 mm using a diamond disc placed on a prosthetic handpiece. The research material was placed in liquid nitrogen, and then a breakthrough was made. Fifteen samples with a correctly made longitudinal fracture were selected for each of the research groups. The samples prepared in this way were subjected to preliminary examinations in the papillary, medial, and apical sections of the root canal on a Stereo Discovery V12 stereo microscope with a Zeiss AxioCam HRC digital camera. The test results were documented using digital photography at 50× magnification (Figure 13).

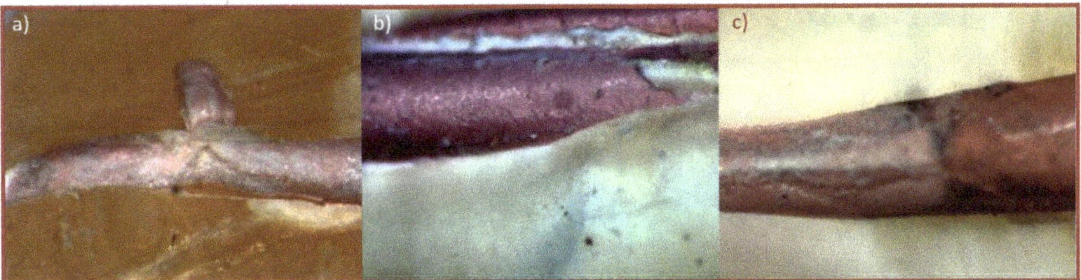

Figure 13. Results of observations in a light stereoscopic microscope of teeth after filling the root canal with filling material based on polymeric Resilon-type polyester materials, covered with a thin intermediate layer of sealant: (**a**) lateral canal filled with a thermoplastic method (50×); (**b**) joining the main stud with the complementary studs with a sealant in the middle part of the root canal after filling with the cold lateral condensation method (50×); (**c**) joining the main stud with the filling material, sealed with a thin layer of sealant in the middle part of the root canal after filling with the thermoplastic method (50×).

Five teeth filled with material based on gutta-percha with AH Plus sealant using the thermoplastic method using System B and Obtura III devices were immersed for 14 days in an aqueous solution consisting of 7% formic acid, 3% hydrochloric acid, and 8% sodium citrate. The samples were then rinsed thoroughly under running water to eliminate acid, and then immersed in 99% acetic acid for 12 h, and the samples were rinsed thoroughly with distilled water. Then, the samples were dehydrated in ethanol solutions of successively increasing concentrations of 25, 50, 70, 90, 95, and 100%, each for 30 min. Then the samples were stored in methyl salicylate and observed in the Stereo Discovery V12 stereoscopic light microscope with the AxioCam HRC digital camera by Zeiss at 8–50× magnification, to three-dimensional observation of the root canal lumen filled with a substitute material, with careful consideration of the root delta and its complicated internal system and side branches of the main canal (Figure 14).

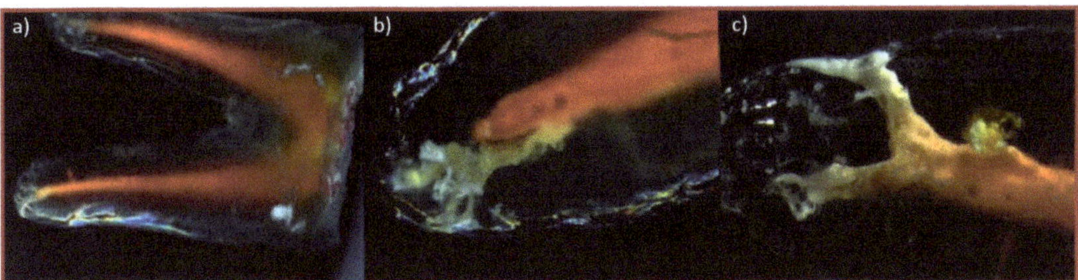

Figure 14. Results of observations in a light microscope of teeth after filling the root canal with thermoplastic gutta-percha studs prepared with the decalcification technique: (**a**) view of a two-root tooth with a root delta in each root and a lateral branch of the root canal in the lower part of the image (8×); (**b**) view of the root of the tooth with the delta-filled complex structure of the root canal filled with sealant (50×); (**c**) root canal delta filled with sealant (50×).

Each sample was sputtered with a thin layer of gold as a conductive material in an Oerlikon Balzers BAL–TEC SCD050 sputtering machine, after making longitudinal breakthroughs by cracking in liquid nitrogen. The goal of sputtering was to remove the electric charge from the surface of the sample during the scanning electron microscope test and improve the secondary electron emission factor. The longitudinal fractures sputtered with gold were subjected to quantitative analysis of the leaks between the filling material and the root canal wall using the SUPRA 35 high-resolution scanning microscope by Zeiss in the magnification range of 2000–5000×, and the results were digitally archived.

Observations were made to document the lack of discontinuities between the dentin and the layers of sealant and filling (Figure 15) and instances where such leaks appeared as local or continuous (Figure 16).

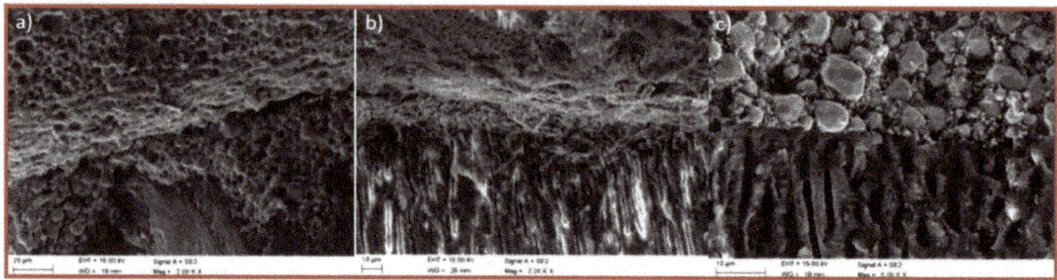

Figure 15. Scanning electron microscope view of a tooth fracture: (**a**) tight connection of three layers of filling material on a gutta-percha matrix covered with sealant, showing sealant and root dentine after filling with gutta-percha studs using the thermoplastic method (2000×); (**b**) tight connection of the root dentine with the filling material based on Resilon-type polymer materials, covered with a thin intermediate layer of the sealant, with a thermoplastic method visible after filling with the use of studs of the Resilon-type material and a visible cross-section of the dentinal tubules along their axis transversely to the axis of the root canal (2000×); (**c**) close bonding of the intermediate layer of the sealant and root canal dentin with a visible cross-section of the dentin tubules along their axis transversely to the axis of the root canal, with the method of filling and obturation as in (**a**) (5000×).

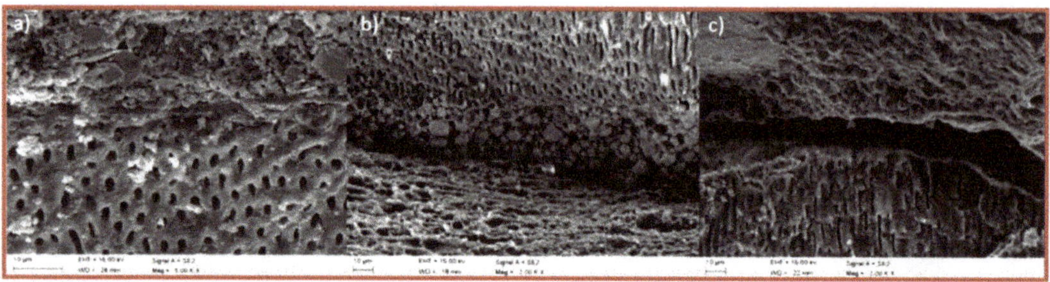

Figure 16. Scanning electron microscope view of the tooth fracture after filling the root canal by cold lateral condensation: (**a**) tight connection of the filling material based on Resilon-type polymer materials with the dentine of the root canal, showing fine particles of sealant (2000×); (**b**) border of the connection of the three layers of root canal dentin, sealant, and filling material based on gutta-percha; leakage in two places at the border of the joint between the sealant and the filling material (2000×); (**c**) leakage on the border of the sealant and the dentine of the root canal, with the method of filling and obturation as (**b**) (2000×).

Five consecutive samples—one of each group of preparations subjected to the study—were obtained by making cross-sections of these samples by cutting them on the GATAN ISOTOM device and then subjecting them to observations in the Exciter confocal laser scanning microscope by Zeiss with a 405 nm laser and electronic recording of test results, in which the samples were also observed, prepared as for observation in a scanning electron microscope, but not sputtered with a layer of gold (Figure 17).

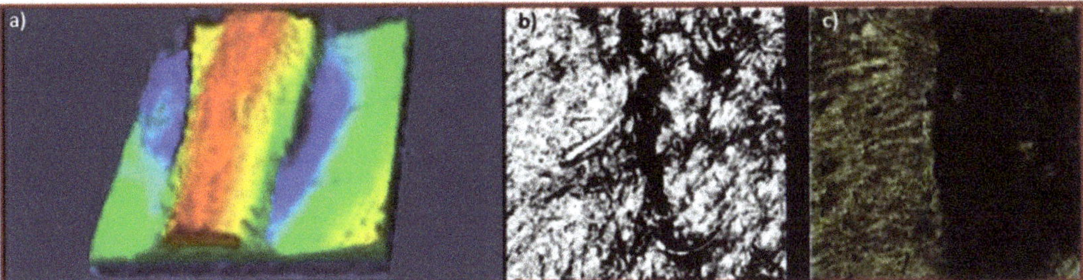

Figure 17. View in a confocal laser microscope of a tooth fracture after filling the root canal: (**a**) an exposed cylindrical stud with slight differences in geometrical features after filling with gutta-percha using the thermoplastic method (20×); (**b**) leakage of the connection between the filling using the thermoplastic method with a filling material based on polymeric Resilon-type materials with dentin of the root canal (2000×); (**c**) clear longitudinal leakage of the connection between the filling and the root dentine, with the method of filling and obturation as in (**b**) (2000×).

Observations were also made using the AFM XE-100 atomic force microscope (AFM) by Park Systems with a scanning probe, using the forces of interatomic interactions and enabling the image of the surface to be obtained with a resolving power corresponding to the size of the atom (Figure 18).

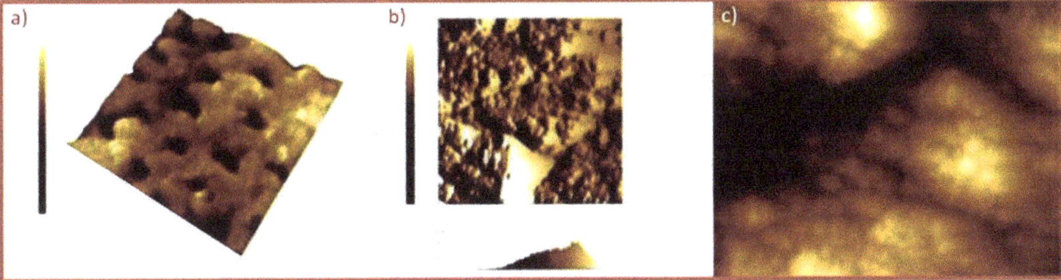

Figure 18. Microscopic surface map showing the image of the specimen surface in an atomic force microscope (**a**) with visible tubules and a visible connection in the dentin tubule with the filling (gutta-percha as a filling material using the thermoplastic method or sealant) (4000×); (**b**) with an indication of height differences and a histogram distribution of the height of the eminence and depressions of the surface after filling the root canal using the cold side condensation method with filling material based on polymeric polyester Resilon-type materials (4000×); (**c**) with visible elevations and depressions related to leaks, with the filling material and obturation method as in (**a**) (8000×).

10. Summary and Final Comments

This paper aimed to develop and present the methodological assumptions of the virtual comparative analysis of biomaterials used in endodontic treatment, including the production and selection of filling and sealing materials; the selection of dental devices and tools, including specialized tools; the selection and application of the technology for the development and obturation of the root canal; and assessment effectiveness and tightness of root canal fillings as a result of endodontic treatment. Achievements and experiences from various areas of knowledge were used synergistically to achieve this goal. The motivation of the multifaceted approach and wide-ranging activities undertaken in this work, and the synergy achieved in this way in its implementation, is referred to as the motto, the thought of W.E. Deming, the creator of the idea of continuous improvement, implemented, among others, in the Japanese automotive company the Toyota Motor Corporation [331]. This, through Kaizen [332] and Total Quality Management (TQM) [333], became the basis of its global market success and, like the German concern Volkswagen AG, sells almost

11 million passenger cars annually. The authors sought the best solution to their research and methodological problems, adopting an open and synergistic methodological approach appropriate to various and seemingly very distant scientific disciplines. The basics of quality management were used with the basic proprietary approach, contained in the slogan "Whatever you plan to do, make what is needed to do, do it right and get it done the first time". This paper dealt with clinical dentistry, especially endodontics as its branch. The knowledge and methodological experience of materials engineering, including metallography and scanning electron microscopy, was also widely used, which justifies the publication of this paper in the Special Issue on Synthesis and Characterization of Biomedical Materials. Engineering and knowledge management, including the methodology of technological foresight, as well as procedural benchmarking and comparative analysis with the use of contextual matrices, appropriately adapted methodology and research results, previously developed and published own works, concerning the analysis of many problems concerning mainly material engineering, also provided a broad view on the issues under consideration. We have referred to our own published works [1,64,69,132,287–314] and other authors [315–328].

The paper concerned one of the most common and effective methods of treating oral cavity diseases as a branch of advanced interventionist dentistry within the concept of sustainable development of dentistry [1]. Endodontic treatment ensures the maintenance of the tooth in a functional state when its vital functioning is no longer possible, undertaken when symptoms of the second stage according to ICDAS of dental pulp disease appear [99–101]. The global endodontics market reached USD 1.56 billion in 2018 and is projected to grow to USD 2.1 billion by 2026, with a CAGR of 4.1% [127]. The global dental consumables market, which also applies to endodontics to some extent, in 2016 reached USD 38.921 million, while it is forecasted to reach USD 55.584 million in 2023, with a CAGR of 5.2% [128] (Figure 3). The treatment is based on caries, as the most common infectious disease in the world, affecting 3–5 billion people [1], causing many systemic diseases [14–28], as well as personality changes leading to dementia [29–35], and if necessary the necessity to extract teeth, causing toothlessness, which may also cause many systemic complications [8–13]. However, it should be emphasized that dentists treat tooth extraction as a necessity in the event of failure to continue endodontic treatment [103,104], performed carefully to avoid complications due to iatrogenic causes [53–58]. The endodontic treatment results in sealing the previously infected and then developed root canals, preventing the spread of bacteria and their toxins to the root tip [110–112], including saliva [113].

Apart from the obturation techniques presented in the form of an umbrella (Figure 2), the effectiveness of endodontic treatment depends on the proper selection of the filling material, presented in the form of ikebana [119–125]. Apart from a good fit to the geometrical features of the prepared root canal, the filling material should be biocompatible, bactericidal, sterile, and easily removable if it is necessary to repeat endodontic treatment [114,115], and should strengthen of the root [118] and not cause tooth discolouration [115–117]. For many years, the most commonly used natural material for filling root canals was gutta-percha [130–135], even considered a "gold standard", and 20 years ago, the Thermo-Hydraulic-Condensation obturation technique [140–145] significantly improved the effectiveness of root canal fillings (Figure 4). Despite this and selecting non-standard sealing materials [118,160–163], there were suggestions of improper bonding of gutta-percha to dentin [146–155]. An alternative is a synthetic thermoplastic filler material based on polyester materials, introduced in 2004 from the company Resilon [168] with eponymous sealant [167,174,175] and amenable to all obturation techniques developed for gutta-percha [168]. The competitive advantage of this material was supposed to be a monoblock with dentin [179–181], which unfortunately did not turn out to be true [182]. There has been dispute as to whether this material is more useful, as it has been recognized in some studies [168,178,184,194,211,215,216,230,232–245], or whether gutta-percha provides the best filling and tightness in the root canal [160,194–207], especially after a longer time [208,209]. Many other evaluations and research results showed, and many dentists still believe, that Resilon cannot compete with gutta-percha. The significant dispersion of the results of the cited studies and methodological flaws of many of them did not allow for an unambiguous formulation of conclusions, and due to the lack of sufficient clinical evidence, it is impossible to firmly confirm the possible advantage of this material over gutta-percha or even the hypothesis that both materials are equivalent.

The controversies presented above regarding the choice of the more favourable filling material and the better obturation method refer to the methodological aspects of characterizing these descriptors of the adopted methodology of clinical activities, as pointed out in, among others, the works [156,157]. This paper focused on the methodological aspects of the research, as the detailed results of this research were presented in a series of other publications [286]. The detailed multipoint research methodology included the theoretical analysis carried out here using procedural benchmarking and comparative analysis using contextual matrices, in addition to reviewing the literature and preliminary analyses, the assumptions of the so-called "digital twins" as an approach characteristic of Industry 4.0 [64,65,67–98], and the resulting idea of Dentistry 4.0 [62,63,66]. The concept of the "digital twin" allows many experiments to be carried out in virtual space before the analysed product or object is physically created or made available for research, even without its physical existence. For the quantitative assessment of the analysed aspects of endodontic treatment, appropriate criteria were defined in each case, assigning them to sum up weights reflecting their significance and subsequent evaluation in each of the criteria using the universal scale of relative states (Figure 6). The positioning of the method or procedure among those covered by the analysis and their usefulness in endodontological treatment was achieved by a set of contextual matrices, containing in particular ones analogous to dendrological matrices, adopted in the foresight analysis of surface engineering technology (Figure 7), which are tools for a graphical comparative analysis of individual methods and materials or groups of them. Depending on the value of the generalized potential and the generalized level of attractiveness, which were determined as part of the expert assessment, each of the analysed objects was placed in one of the matrix quarters, which were distinguished in the dendrological value matrix, based on which it was possible to reach conclusions about its methodological success in the field of endodontics. The broad oak quarter covered the most promising aspects for future success. The soaring cypress quarter covered the objects with possible success. The rooted mountain pine quarter covered those with high probability. In contrast, the quivering aspen quarter contained the weakest objects, for which success is unlikely or impossible.

The literature analysis was based on the substantive subject of the considered criteria, interviews with dentists, and our own experience gained during dental practice. As a result of considerations, this paper presents four types of the authors' contextual matrices related to endodontics:

1. Materials selection in the coordinate system generalized index of material quality for root canal filling (x) and generalized index of material strength in the root canal (y);
2. Root canal development technique selection in the coordinate system of a generalized organizational index of techniques for the development of root canals (x) and generalized index of the quality of filling with selected techniques (y);
3. Techniques of obturation selection in the coordinate system generalized index of material strength applied in root canal obturation techniques (x) and generalized index of the effectiveness of root canal obturation techniques (y);
4. Assessing the tightness of root canal fillings selection in the coordinate system of the generalized index of organizational conditions (x) and generalized index of investigations effectiveness (y).

All the analyses performed were included in this paper, along with an exemplary description of the obtained test results. The detailed test results will be included in subsequent papers from the announced series [286]. Based on the analyses performed and the analysis of the plotted dendrological matrices, a decision was made regarding detailed research on the four aspects mentioned above. At the same time, the full usefulness of materialographic tests with the use of a scanning electron microscope was confirmed. For this purpose, 80 single-canal incisors in the maxilla, maxillary, and mandibular canines, and maxillary and mandibular single-canal premolars, removed for orthodontic, prosthetic, and periodontal indications, were included in the in vitro tests. The research was performed after obtaining the consent of the Bioethics Committee. Among these teeth, five groups were distinguished, each containing 16 teeth, tested in identical conditions of root canal preparation, using the same filling materials and using the same obturation technique. The results of good tests using various methods were also presented as an example in this work.

The analysis results confirmed the full practical usefulness of the so-called "digital twins" in a virtual comparative analysis of biomaterials used in endodontic treatment.

Author Contributions: Conceptualization, literature review, presentation, design, resources, data curation, software, formal analysis, writing, original draft preparation, visualization, J.D., L.B.D., L.A.D., and A.D.D.-D.; practical verification, J.D., L.B.D., and K.G., writing—review and editing, supervision, project administration, funding acquisition, L.A.D., L.B.D., and J.D. All authors have read and agreed to the published version of the manuscript.

Funding: This research was not directly financed by external funding.

Notice: This paper was prepared to actually implement Project POIR.01.01-00-0485/16-00 on "IMSKA-MAT Innovative dental and maxillofacial implants manufactured using the innovative additive technology supported by computer-aided materials design ADD-MAT" realized by the Medical and Dental Engineering Centre for Research, Design, and Production (ASKLEPIOS) in Gliwice, Poland. The project was implemented in 2017–2021 and was cofinanced by the Operational Program for Intelligent Development of the European Union.

Conflicts of Interest: The authors declare no conflict of interest.

References

1. Dobrzański, L.A.; Dobrzański, L.B.; Dobrzańska-Danikiewicz, A.D.; Dobrzańska, J. The Concept of Sustainable Development of Modern Dentistry. *Processes* **2020**, *8*, 1605. [CrossRef]
2. Watt, R.G.; Daly, B.; Allison, P.; Macpherson, L.M.D.; Venturelli, R.; Listl, S.; Weyant, R.J.; Mathur, M.R.; Guarnizo-Herreño, C.C.; Celeste, R.K.; et al. Ending the neglect of global oral health: Time for radical action. *Lancet* **2019**, *394*, 261–272. [CrossRef]
3. Peres, M.A.; Macpherson, L.M.D.; Weyant, R.J.; Daly, B.; Venturelli, R.; Mathur, M.R.; Listl, S.; Celeste, R.K.; Guarnizo-Herreño, C.C.; Kearns, C.; et al. Oral diseases: A global public health challenge. *Lancet* **2019**, *394*, 249–260. [CrossRef]
4. Fejerskov, O. Concepts of dental caries and their consequences for understanding the disease. *Community Dent. Oral Epidemiol.* **1997**, *25*, 5–12. [CrossRef]
5. Reisine, S.; Litt, M. Social and psychological theories and their use for dental practice. *Int. Dent. J.* **1993**, *43* (Suppl. 1), 279–287.
6. Selwitz, R.H.; Ismail, A.I.; Pitts, N.B. Dental caries. *Lancet* **2007**, *369*, 51–59. [CrossRef]
7. GBD Compare. Viz Hub. Available online: https://vizhub.healthdata.org/gbd-compare/ (accessed on 6 April 2021).
8. Li, X.; Tornstad, L.; Olsen, I. Brain abscesses caused by oral infection. *Dent. Traumatol.* **1999**, *15*, 95–101. [CrossRef] [PubMed]
9. Scannapieco, F.A.; Bush, R.B.; Paju, S. Associations between periodontal disease and risk for nosocomial bacterial pneumonia and chronic obstructive pulmonary disease. A systemic review. *Ann. Periodontol.* **2003**, *8*, 54–69. [CrossRef]
10. Dobrzański, L.A.; Dobrzański, L.B. Dentistry 4.0 Concept in the Design and Manufacturing of Prosthetic Dental Restorations. *Processes* **2020**, *8*, 525. [CrossRef]
11. Aleksander, M.; Krishnan, B.; Shenoy, N. Diabetes mellitus and odontogenic infections-an exaggerated risk? *Oral Maxillofac. Surg.* **2008**, *12*, 129–130. [CrossRef]
12. Scannapieco, F.A. Role of oral bacteria in respiratory infection. *J. Periodontol.* **1999**, *70*, 793–802. [CrossRef] [PubMed]
13. Mueller, A.A.; Saldami, B.; Stübinger, S.; Walter, C.; Flückiger, U.; Merlo, A.; Schwenzer-Zimmerer, K.; Zeilhofer, H.F.; Zimmerer, S. Oral bacterial cultures in nontraumatic brain abscesses: Results of a first line study. *Oral Surg. Oral Med. Oral Pathol. Oral Radiol. Endod.* **2009**, *107*, 469–476. [CrossRef]
14. Buset, S.L.; Walter, C.; Friedmann, A.; Weiger, R.; Borgnakke, W.S.; Zitzmann, N.U. Are periodontal diseases really silent? A systematic review of their effect on quality of life. *J. Clin. Periodontol.* **2016**, *43*, 333–344. [CrossRef] [PubMed]
15. Sierpinska, T.; Golebiewska, M.; Dlugosz, J.W.; Kemona, A.; Laszewicz, W. Connection between masticatory efficiency and pathomorphologic changes in gastric mucosa. *Quintessence Int.* **2007**, *38*, 31–37.
16. Al-Nawas, B.; Maeurer, M. Severe versus local odontogenic bacterial infections: Comparison of microbial isolates. *Eur. Surg. Res.* **2008**, *40*, 220–224. [CrossRef] [PubMed]
17. Pallasch, T.J.; Wahl, M.J. Focal infection: New age or ancient history? *Endodon. Top.* **2003**, *4*, 32–45. [CrossRef]

18. De Pablo, P.; Dietrich, T.; McAlindon, T.E. Association of periodontal disease and tooth loss with rheumatoid arthritis in the US population. *J. Rheumatol.* **2008**, *35*, 70–76. [PubMed]
19. Felton, D.A. Edentualism and comorbid factors. *J. Prosthodont.* **2009**, *18*, 88–96. [CrossRef]
20. Volzke, H.; Schwahn, C.; Hummel, A.; Wolff, B.; Kleine, V.; Robinson, D.M.; Dahm, J.B.; Felix, S.B.; John, U.; Kocher, T. Tooth loss is independently associated with the risk of acquired aortic valve sclerosis. *Am. Heart J.* **2005**, *150*, 1198–1203. [CrossRef] [PubMed]
21. Bagchi, S.; Tripathi, A.; Tripathi, S.; Kar, S.; Tiwari, S.C.; Singh, J. Obstructive sleep apnea and neurocognitive dysfunction in edentulous patients. *J. Prosthodont.* **2019**, *28*, e837–e842. [CrossRef]
22. Nagpal, R.; Yamashiro, Y.; Izumi, Y. The two-way association of periodontal infection with systemic disorders: An overview. *Mediat. Inflamm.* **2015**, *2015*, 793898. [CrossRef] [PubMed]
23. Abnet, C.C.; Qiao, Y.L.; Dawsey, S.M.; Dong, Z.W.; Taylor, P.R.; Mark, S.D. Tooth loss is associated with increased risk of total death and death from upper gastrointestinal cancer, heart disease, and stroke in a Chinese population-based cohort. *Int. J. Epidemiol.* **2005**, *34*, 467–474. [CrossRef] [PubMed]
24. Burzyńska, B.; Mierzwińska-Nastalska, E. Rehabilitacja protetyczna pacjentów bezzębnych. *Nowa Stomatol.* **2011**, *4*, 167–199.
25. Bui, F.Q.; Almeida-da-Silva, C.L.C.; Huynh, B.; Trinh, A.; Liu, J.; Woodward, J.; Asadi, H.; Ojcius, D.M. Association between periodontal pathogens and systemic disease. *Biomed. J.* **2019**, *42*, 27–35. [CrossRef]
26. Holmlund, A.; Holm, G.; Lind, L. Number of teeth as a predictor of cardiovascular mortality in a cohort of 7674 subjects followed for 12 years. *J. Periodontol.* **2010**, *81*, 870–876. [CrossRef] [PubMed]
27. Takata, Y.; Ansai, T.; Matsumura, K.; Awano, S.; Hamasaki, T.; Sonoki, K.; Kusaba, A.; Akifusa, S.; Takehara, T. Relationship between tooth loss and electrocardiographic abnormalities in octogenarians. *J. Dent. Res.* **2001**, *80*, 1648–1652. [CrossRef] [PubMed]
28. Felton, D.A. Complete edentulism and comorbid diseases: An update. *J. Prosthodont.* **2016**, *25*, 5–20. [CrossRef]
29. Chen, H.; Iinuma, M.; Onozuka, M.; Kubo, K.-Y. Chewing maintains hippocampus-dependent cognitive. *Int. J. Med. Sci.* **2015**, *12*, 502–509. [CrossRef]
30. Stein, P.S.; Desrosiers, M.; Donegan, S.J.; Yepes, J.F.; Kryscio, R.J. Tooth loss, dementia and neuropathology in the Nun study. *J. Am. Dent. Assoc.* **2007**, *138*, 1314–1322. [CrossRef]
31. Henke, K. A model for memory systems based on processing modes rather than consciousness. *Nat. Rev. Neurosci.* **2010**, *11*, 523–532. [CrossRef]
32. Lexomboon, D.; Trulsson, M.; Wårdh, I.; Parker, W.G. Chewing ability and tooth loss: Association with cognitive impairment in an elderly population study. *J. Am. Geriatr. Soc.* **2012**, *60*, 1951–1956. [CrossRef]
33. Hirano, Y.; Obata, T.; Takahashi, H.; Tachibana, A.; Kuroiwa, D.; Takahashi, T.; Ikehira, H.; Onozuka, M. Effects of chewing on cognitive processing speed. *Brain Cognit.* **2013**, *81*, 376–381. [CrossRef] [PubMed]
34. Onishi, M.; Iinuma, M.; Tamura, Y.; Kubo, K.Y. Learning deficits and suppression of the cell proliferation in the hippocampal dentate gyrus of offspring are attenuated by maternal chewing during prenatal stress. *Neurosci. Lett.* **2014**, *560*, 77–80. [CrossRef]
35. Kawahata, M.; Ono, Y.; Ohno, A.; Kawamoto, S.; Kimoto, K.; Onozuka, M. Loss of molars early in life develops behavioral lateralization and impairs hippocampus-dependent recognition memory. *BMC Neurosci.* **2014**, *15*, 4. [CrossRef] [PubMed]
36. Clarkson, B.H.; Exterkate, R.A.M. Noninvasive dentistry: A dream or reality? *Caries Res.* **2015**, *49* (Suppl. 1), 11–17. [CrossRef]
37. Pitts, N.B.; Zero, D.T. White Paper on Dental Caries Prevention and Management. FDI World Dental Federation. Available online: http://www.fdiworlddental.org/sites/default/files/media/documents/2016-fdi_cpp-white_paper.pdf (accessed on 6 April 2021).
38. Wierichs, R.J.; Meyer-Lueckel, H. Systematic review on noninvasive treatment of root caries lesions. *J. Dent. Res.* **2015**, *94*, 261–271. [CrossRef] [PubMed]
39. Marsh, P.D.; Head, D.A.; Devine, D.A. Prospects of oral disease control in the future—an opinion. *J. Oral. Microbiol.* **2014**, *6*, 261–276. [CrossRef] [PubMed]
40. Marthaler, T.M. A standardized system of recording dental conditions. *Helv. Odontol. Acta* **1966**, *10*, 1–18.
41. Dirks, O.B.; van Amerongen, J.; Winkler, K.C. A reproducible method for caries evaluation. *J. Dent. Res.* **1951**, *30*, 346–359. [CrossRef] [PubMed]
42. Black, G.V. *A Work on Operative Dentistry: The Technical Procedures in Filling Teeth*; Medico-Dental Publishing: Chicago, IL, USA, 1917; p. 5.
43. Pitts, N. "ICDAS"-an international system for caries detection and assessment being developed to facilitate caries epidemiology, research and appropriate clinical management. *Community Dent. Health* **2004**, *21*, 193–198.
44. ICDAS Website. Available online: https://www.icdas.org/ (accessed on 6 April 2021).
45. Featherstone, J.D. The continuum of dental caries—evidence for a dynamic disease process. *J. Dent. Res.* **2004**, *83*, C39–C42. [CrossRef] [PubMed]
46. Ismail, A.I.; Sohn, W.; Tellez, M.; Amaya, A.; Sen, A.; Hasson, H.; Pitts, N.B. The International Caries Detection and Assessment System (ICDAS): An integrated system for measuring dental caries. *Community Dent. Oral Epidemiol.* **2007**, *35*, 170–178. [CrossRef] [PubMed]

47. Ismail, A.; Tellez, M.; Pitts, N.B.; Ekstrand, K.R.; Ricketts, D.; Longbottom, C.; Eggertsson, H.; Deery, C.; Fisher, J.; Young, D.A.; et al. Caries management pathways preserve dental tissues and promote oral health. *Community Dent. Oral Epidemiol.* **2013**, *41*, e12–e40. [CrossRef] [PubMed]

48. Pitts, N.B.; Ekstrand, K.R. International Caries Detection and Assessment System (ICDAS) and its International Caries Classification and Management System (ICCMS)—methods for staging of the caries process and enabling dentists to manage caries. *Community Dent. Oral Epidemiol.* **2013**, *41*, e41–e52. [CrossRef]

49. Ormond, C.; Douglas, G.; Pitts, N. The use of the International Caries Detection and Assessment System (ICDAS) in a National Health Service general dental practice as part of an oral health assessment. *Prim. Dent. Care* **2010**, *17*, 153–159. [CrossRef]

50. Ismail, A.; Pitts, N.B.; Tellez, M. The international caries classification and management system (ICCMS^{TM}) an example of a caries management pathway. *BMC Oral Health* **2015**, *15*, S9. [CrossRef]

51. Sunay, H.; Tanalp, J.; Dikbas, I.; Bayirli, G. Cross-sectional evaluation of the periapical status and quality of root canal treatment in a selected population of urban Turkish adults. *Int. Endodon. J.* **2007**, *40*, 139–145. [CrossRef]

52. Tsuneishi, M.; Yamamoto, T.; Yamanaka, R.; Tamaki, N.; Sakamoto, T.; Tsuji, K.; Watanabe, T. Radiographic evaluation of periapical status and prevalence of endodontic treatment in an adult Japanese population. *Oral Surg. Oral Med. Oral Pathol. Oral Radiol. Endod.* **2005**, *100*, 631–635. [CrossRef]

53. Lazarski, M.P.; Walker, W.A.; Flores, C.M.; Schindler, W.G.; Hargreaves, K.M. Epidemiological evaluation of the outcomes of nonsurgical root canal treatment in a large cohort of insured dental patients. *J. Endodon.* **2001**, *27*, 791–796. [CrossRef]

54. Chen, S.C.; Chuech, L.H.; Hsiao, C.K.; Tsai, M.Y.; Ho, S.C.; Chiang, C.P. An epidemiologic study of tooth retention after nonsurgical endodontic treatment in a large population in Taiwan. *J. Endodon.* **2007**, *33*, 226–229. [CrossRef]

55. Meuwissen, R.; Eschen, S. Twenty years of endodontic treatment. *J. Endodon.* **1983**, *9*, 390–393. [CrossRef]

56. Kirkevang, L.L.; Horsted-Bindslev, P.; Orstavik, D.; Wenzel, A. Frequency and distribution of endodontically treated teeth and apical periodontitis in an urban Danish population. *Int. Endodon. J.* **2001**, *34*, 198–205. [CrossRef]

57. Salehrabi, R.; Rotstein, I. Endodontic treatment outcomes in a large patient population in the USA: An epidemiological study. *J. Endodon.* **2004**, *30*, 846–850. [CrossRef] [PubMed]

58. Hayashi, M.; Haapasalo, M.; Imazato, S.; Il Lee, J.; Momoi, Y.; Murakami, S.; Whelton, H.; Wilson, N. Dentistry in the 21st century: Challenges of a globalising world. *Int. Dent. J.* **2014**, *64*, 333–342. [CrossRef] [PubMed]

59. Dobrzański, L.B.; Dobrzański, L.A.; Dobrzańska, J.; Rudziarczyk, K.; Achtelik-Franczak, A. Akcesorium do ochrony osobistej personelu dentystycznego przed koronawirusem SARS-CoV-2 i innymi drobnoustrojami chorobotwórczymi. *Zgłoszenie Pat.* **2020**, 434391.

60. Galicia, J.C.; Mungia, R.; Taverna, M.V.; Mendoza, M.J.; Estrela, C.; Gaudin, A.; Zhang, C.; Vaughn, B.A.; Khan, A.A. Response by Endodontists to the SARS-CoV-2 (COVID−19) Pandemic: An International Survey. *Front. Dent. Med.* **2021**, *1*, 1–24. [CrossRef]

61. Ates, A.A.; Alomari, T.; Bhardwaj, A.; Tabnjh, A.; Gambarini, G. Differences in endodontic emergency management by endodontists and general dental practitioners in COVID-19 times. *Braz. Oral Res.* **2020**, *34*, 1–11. [CrossRef] [PubMed]

62. Dobrzański, L.A.; Dobrzański, L.B. Approach to the Design and Manufacturing of Prosthetic Dental Restorations According to the Rules of Industry 4.0. *Mater. Perform. Charact.* **2020**, *9*, 394–476. [CrossRef]

63. Dobrzański, L.A.; Dobrzańska-Danikiewicz, A.D.; Dobrzański, L.B. Effect of Biomedical Materials in the Implementation of a Long and Healthy Life Policy. *Processes* **2021**, *9*, 865. [CrossRef]

64. Dobrzański, L.A.; Dobrzański, L.B.; Dobrzańska-Danikiewicz, A.D. Overview of conventional technologies using the powders of metals, their alloys and ceramics in Industry 4.0 stage. *JAMME* **2020**, *98*, 56–85. [CrossRef]

65. Dobrzański, L.A. Role of materials design in maintenance engineering in the context of industry 4.0 idea. *JAMME* **2019**, *96*, 12–49. [CrossRef]

66. Dobrzański, L.A.; Dobrzański, L.B.; Achtelik-Franczak, A.; Dobrzańska, J. Application Solid Laser-Sintered or Machined Ti6Al4V Alloy in Manufacturing of Dental Implants and Dental Prosthetic Restorations According to Dentistry 4.0 Concept. *Processes* **2020**, *8*, 664. [CrossRef]

67. Dobrzański, L.A.; Dobrzańska-Danikiewicz, A.D. Applications of Laser Processing of Materials in Surface Engineering in the Industry 4.0 Stage of the Industrial Revolution. *Mater. Perform. Charact.* **2019**, *8*, 1091–1129. [CrossRef]

68. Dobrzański, L.A.; Dobrzańska-Danikiewicz, A.D. Why Are Carbon-Based Materials Important in Civilization Progress and Especially in the Industry 4.0 Stage of the Industrial Revolution. *Mater. Perform. Charact.* **2019**, *8*, 337–370. [CrossRef]

69. Dobrzański, L.A.; Dobrzański, L.B.; Dobrzańska-Danikiewicz, A.D.; Kraszewska, M. Manufacturing powders of metals, their alloys and ceramics and the importance of conventional and additive technologies for products manufacturing in Industry 4.0 stage. *AMSE* **2020**, *102*, 13–41. [CrossRef]

70. Hermann, M.; Pentek, T.; Otto, B. *Design Principles for Industrie 4.0 Scenarios: A Literature Review;* Technische Universität Dortmund: Dortmund, Germany, 2015.

71. Kagermann, H.; Wahlster, W.; Helbig, J. *Recommendations for Implementing the Strategic Initiative INDUSTRIE 4.0: Final Report of the Industrie 4.0 Working Group;* Federal Ministry of Education and Research: Bonn, Germany, 2013.

72. Rüßmann, M.; Lorenz, M.; Gerbert, P.; Waldner, M.; Justus, J.; Engel, P.; Harnisch, M. *Industry 4.0: The Future of Productivity and Growth in Manufacturing Industries;* Boston Consulting Group: Boston, MA, USA, 2015.

73. Kagermann, H. Chancen von Industrie 4.0 Nutzen. In *Industrie 4.0 in Produktion, Automatisierung und Logistik;* Springer Fachmedien Wiesbaden: Wiesbaden, Germany, 2014; pp. 603–614.

74. Lee, J.; Kao, H.-A.; Yang, S. Service Innovation and Smart Analytics for Industry 4.0 and Big Data Environment. *Proc. CIRP* **2014**, *16*, 3–8. [CrossRef]
75. Jose, R.; Ramakrishna, S. Materials 4.0: Materials Big Data Enabled Materials Discovery. *Appl. Mat. Today.* **2018**, *10*, 127–132. [CrossRef]
76. Buer, S.-V.; Strandhagen, J.O.; Chan, F.T.S. The Link between Industry 4.0 and Lean Manufacturing: Mapping Current Research and Establishing a Research Agenda. *Int. J. Produc. Res.* **2018**, *56*, 2924–2940. [CrossRef]
77. Brettel, M.; Friederichsen, N.; Keller, M.; Rosenberg, M. How Virtualization, Decentralization, and Network-Building Change the Manufacturing Landscape: An Industry 4.0 Perspective. *Int. J. Mech. Aerospac. Indust. Mechatron. Manuf. Eng.* **2014**, *8*, 37–44.
78. Tay, S.I.; Lee, T.C.; Hamid, N.A.A.; Ahmad, A.N.A. An Overview of Industry 4.0: Definition, Components, and Government Initiatives. *J. Adv. Res. Dynamic Control Syst.* **2018**, *10*, 1379–1387.
79. Xu, X. Machine Tool 4.0 for the New Era of Manufacturing. *Int. J. Adv. Manuf. Techn.* **2017**, *92*, 1893–1900. [CrossRef]
80. Sipsas, K.; Alexopoulos, K.; Xanthakis, V.; Chryssolouris, G. Collaborative Maintenance in Flow-Line Manufacturing Environments: An Industry 4.0 Approach. *Proc. CIRP* **2016**, *55*, 236–241. [CrossRef]
81. Posada, J.; Toro, C.; Barandiaran, I.; Oyarzun, D.; Stricker, D.; de Amicis, R.; Pinto, E.B.; Eisert, P.; Döllner, J.; Vallarino, I. Visual Computing as a Key Enabling Technology for Industrie 4.0 and Industrial Internet. *IEEE Comp. Graph Appl.* **2015**, *35*, 26–40. [CrossRef] [PubMed]
82. Hozdic, E. Smart Factory for Industry 4.0: A Review. *Int. J. Modern Manuf. Tech.* **2015**, *7*, 28–35.
83. Zhong, R.Y.; Xu, X.; Klotz, E.; Newman, S.T. Intelligent Manufacturing in the Context of Industry 4.0: A Review. *J. Eng.* **2017**, *3*, 616–630. [CrossRef]
84. Łobaziewicz, M. *Zarządzanie Inteligentnym Przedsiębiorstwem w Dobie Przemysłu 4.0*; Towarzystwo Naukowe Organizacji i Kierownictwa: Toruń, Poland, 2019.
85. Bahrin, M.A.K.; Othman, M.F.; Azli, N.H.N.; Talib, M.F. Industry 4.0: A Review on Industrial Automation and Robotic. *J. Tekno* **2016**, *78*, 137–143. [CrossRef]
86. Vaidya, S.; Ambad, P.; Bhosle, S. Industry 4.0–A Glimpse. *Proc. Manuf.* **2018**, *20*, 233–238. [CrossRef]
87. Lee, J.; Bagheri, B.; Kao, H.-A. A Cyber-Physical Systems Architecture for Industry 4.0-Based Manufacturing Systems. *Manuf. Lett.* **2015**, *3*, 18–23. [CrossRef]
88. Stock, T.; Seliger, G. Opportunities of Sustainable Manufacturing in Industry 4.0. *Proc. CIRP* **2016**, *40*, 536–541. [CrossRef]
89. Schumacher, A.; Erol, S.; Sihn, W. A Maturity Model for Assessing Industry 4.0 Readiness and Maturity of Manufacturing Enterprises. *Proc. CIRP* **2016**, *52*, 161–166. [CrossRef]
90. Kumar, K.; Zindani, D.; Davim, J.P. *Industry 4.0: Developments towards the Fourth Industrial Revolution*; Springer Nature: Singapore, 2019.
91. Pfeiffer, S. Robots, Industry 4.0 and Humans, or Why Assembly Work Is More than Routine Work. *Societies* **2016**, *6*, 16. [CrossRef]
92. Wang, S.; Wan, J.; Zhang, D.; Li, D.; Zhang, C. Towards Smart Factory for Industry 4.0: A Self-Organized Multi-Agent System with Big Data Based Feedback and Coordination. *Comp. Network* **2016**, *101*, 158–168. [CrossRef]
93. Ardito, L.; Petruzzelli, A.M.; Panniello, U.; Garavelli, A.C. Towards Industry 4.0: Mapping Digital Technologies for Supply Chain Management-Marketing Integration. *Business Proc. Manag. J.* **2019**, *25*, 323–346. [CrossRef]
94. Mosterman, P.J.; Zander, J. Industry 4.0 as a Cyber-Physical System Study. *Software Syst. Model* **2016**, *15*, 17–29. [CrossRef]
95. Almada-Lobo, F. The Industry 4.0 Revolution and the Future of Manufacturing Execution Systems (MES). *J. Innov. Manag.* **2015**, *3*, 16–21. [CrossRef]
96. Lu, Y. Industry 4.0: A Survey on Technologies, Applications and Open Research Issues. *J. Indust. Infor. Integrat.* **2017**, *6*, 1–10. [CrossRef]
97. Qin, J.; Liu, Y.; Grosvenor, R. A Categorical Framework of Manufacturing for Industry 4.0 and Beyond. *Proc. CIRP* **2016**, *52*, 173–178. [CrossRef]
98. Thoben, K.-D.; Wiesner, S.; Wuest, T. 'Industrie 4.0' and Smart Manufacturing–A Review of Research Issues and Application Examples. *Int. J. Automat. Techn.* **2017**, *11*, 4–16. [CrossRef]
99. Peroz, I.; Blankenstein, F.; Lange, K.-P.; Naumann, M. Restoring endodontically treated teeth with post and cores–A review. *Quintessence Int.* **2005**, *36*, 737–746.
100. Shutzky-Goldberg, I.; Shutzky, H.; Gorfil, C.; Smidt, A. Restoration of endodontically treated teeth review and treatment recommendations. *Int. J. Dent.* **2009**, *2009*, 150251. [CrossRef]
101. Schwartz, R.S.; Robbins, J.W. Post placement and restoration of endodontically treated teeth: A literature review. *J. Endodon.* **2004**, *5*, 289–301. [CrossRef] [PubMed]
102. De Oliveira, B.P.; Câmara, A.C.; Aguiar, C.M. Prevalence of endodontic diseases: An epidemiological evaluation in a Brazilian subpopulation. *Braz. J. Oral Sci.* **2016**, *15*, 119–123. [CrossRef]
103. Alwadani, M.; Mashyakhy, M.H.; Jali, A.; Hakami, A.O.; Areshi, A.; Daghriri, A.A.; Shaabi, F.I.; Al Moaleem, M.M. Dentists and Dental Intern's Preferences of Root Canal Treatment with Restoration Versus Extraction then Implant-Supported Crown Treatment Plan. *Open Dent. J.* **2019**, *13*, 93–100. [CrossRef]
104. Estrela, C.; Holland, R.; Estrela, C.R.; Alencar, A.H.; Sousa-Neto, M.D.; Pécora, J.D. Characterization of successful root canal treatment. *Braz. Dent. J.* **2014**, *25*, 3–11. [CrossRef]

105. Aggarwal, A.; Pandey, V.; Bansal, N. Regenerative Endodontics- Potential Approaches in Revitalizing the Tooth Pulp-A Review Article. *J. Adv. Med. Dent. Sci. Res.* **2019**, *7*, 27–32. [CrossRef]

106. Sreedev, C.P.; Karthick, K.; Mathew, S.; Raju, I. Regenerative endodontics: An overview. *J. Indian Acad. Dent. Spec. Res.* **2017**, *4*, 18–22. [CrossRef]

107. Nair, P.N.R. On the cause of persistent apical periodontitis: A review. *Int. Endod. J.* **2006**, *34*, 249–281. [CrossRef]

108. Malarvizhi, D.; Shuruthi, J.; Anuradha, B.; Subbiya, A. Etiology And Management Of Separation Of Instruments In Endodontics – An Overview. *Eur. J. Mol. Clin. Med.* **2020**, *7*, 1229–1234.

109. Drabarczyk-Nasińska, M.; Kacprzak, M. Nowoczesne leczenie endodontyczne–materiały i metody wypełniania kanału korzeniowego. *Borgis Nowa Stomatol.* **2001**, *3*, 11–14.

110. Siqueira, J.F.; Rocas, I.N.; Lopes, H.P. de Uzeda, M. Coronal leakage of two root canal sealers containing calcium hydroxide after exposure to human saliva. *J. Endod.* **1999**, *25*, 14–16. [CrossRef]

111. Hirsch, J.M.; Ahlstrom, U.; Henrikson, P.A.; Peterson, L.E. Periapical surgery. *Int. J. Oral Surg.* **1979**, *8*, 173–185. [CrossRef]

112. Sundqvist, G.; Figdor, D.; Persson, S.; Sjörgren, U. Microbiological analysis of teeth with failed endodontic treatment and the outcome of conservative re-treatment. *Oral Surg. Oral Med. Oral Pathol. Oral Radiol. Endod.* **1998**, *85*, 86–93. [CrossRef]

113. Carrotte, P. Endodontics: Part 1. The modern concept of root canal treatment. *Br. Dent. J.* **2004**, *197*, 181–183. [CrossRef] [PubMed]

114. Karamifar, K.; Tondari, A.; Saghiri, M.A. Endodontic Periapical Lesion: An Overview on the Etiology, Diagnosis and Current Treatment Modalities. *Eur. Endod. J.* **2020**, *2*, 54–67. [CrossRef]

115. Grossman, L. *Endodontic Practice*, 11th ed.; Lea & Fibiger: Phildelphia, PA, USA, 1988; p. 242.

116. Muliyar, S.; Shameem, K.A.; Thankachan, R.P.; Francis, P.G.; Jayapalan, C.S.; Hafiz, K.A. Microleakage in endodontics. *J. Int. Oral Health.* **2014**, *6*, 99–104.

117. Krastl, G.; Allgayer, N.; Lenherr, P.; Filippi, A.; Taneja, P.; Weiger, R. Tooth discoloration induced by endodontic materials: A literature review. *Dent. Traumatol.* **2013**, *29*, 2–7. [CrossRef]

118. Teixeira, F.B.; Teixeira, E.C.; Thompson, J.Y.; Trope, M. Fracture resistance of roots endodontically treated with a new resin filling material. *J. Am. Dent. Assoc.* **2004**, *135*, 646–652. [CrossRef]

119. Ahmad, P.; Dummer, P.M.H.; Noorani, T.Y.; Asif, J.A. The top 50 most-cited articles published in the International Endodontic Journal. *Int. Endod. J.* **2019**, *52*, 803–818. [CrossRef]

120. Raghavendra, S.S.; Jadhav, G.R.; Gathani, K.M.; Kotadia, P. Bioceramics in Endodontics–a review. *J. Istanbul Univ. Fac. Dent.* **2017**, *51*, S128–S137. [CrossRef]

121. Parirokh, M.; Torabinejad, M.; Dummer, P.M.H. Mineral trioxide aggregate and other bioactive endodontic cements: An updated overview - part I: Vital pulp therapy. *Int. Endod. J.* **2018**, *51*, 177–205. [CrossRef]

122. Torabinejad, M.; Parirokh, M.; Dummer, P.M.H. Mineral trioxide aggregate and other bioactive endodontic cements: An updated overview-part II: Other clinical applications and complications. *Int. Endod. J.* **2018**, *51*, 284–317. [CrossRef] [PubMed]

123. Malkondu, Ö.; Karapinar-Kazandağ, M.; Kazazoğlu, E. A review on biodentine, a contemporary dentine replacement and repair material. *Biomed. Res. Int.* **2014**, *2014*, 160951. [CrossRef]

124. Debelian, G.; Trope, M. The use of premixed bioceramic materials in endodontics. *G. Italia. Endod.* **2016**, *30*, 70–80. [CrossRef]

125. Ekambaram, M.; Yiu, C.K.Y.; Matinlinna, J.P. Bonding of adhesive resin to intraradicular dentine: A review of the literature. *Int. J. Adhes. Adhes.* **2015**, *60*, 92–103. [CrossRef]

126. Dental Endodontics Market (Product-Instruments (Endodontic Scalers & Lasers, Motors, Apex Locators, and Machine Assisted Obturation Systems) and Consumables (Obturation, Shaping and Cleaning, and Access Cavity Preparation); End User: Dental Hospitals, Dental Clinics, and Dental Academic & Research Institutes)-Global Industry Analysis, Size, Share, Growth, Trends, and Forecast 2017–2025. Dental Endodontics Market Size, Share & Trend | Industry Analysis Report, 2025. Available online: transparencymarketresearch.com (accessed on 6 April 2021).

127. Endodontic Devices Market Size, Share & Trends Analysis Report By Type (Instruments, Consumables), By End Use (Hospitals, Clinics, Dental Academic & Research Institutes), And Segment Forecasts, 2019–2026. Endodontic Devices Market Size & Share | Industry Report, 2019–2026. Available online: grandviewresearch.com (accessed on 6 April 2021).

128. Dental Consumables Market by Product [Dental Implants (Root Form Dental Implants and Plate Form Dental Implants), Dental Prosthetics (Crowns, Bridges, Dentures, Abutments, Veneers, and Inlays & Onlays), Endodontics (Endodontic Files, Obturators, and Permanent Endodontic Sealers), Orthodontics (Brackets, Archwires, Anchorage Appliances, and Ligatures), Periodontics (Dental Sutures and Dental Hemostats), Retail Dental Care Essentials (Specialized Dental Pastes, Dental Brushes, Dental Wash Solutions, Whitening Agents, and Dental Floss), and Other Dental Consumables (Dental Splints, Dental Sealants, Dental Burs, Dental Impression Materials, Dental Disposables, Bonding Agents, Patient Bibs, and Aspirator Tubes & Saliva Ejectors)]-Global Opportunity Analysis and Industry Forecast, 2017–2023. Dental Consumables Market Size, Share and Growth opportunities 2023. Available online: alliedmarketresearch.com (accessed on 6 April 2021).

129. Endodontics Market by Instruments (Scalers, Apex Locator, Motors, Handpiece, Laser), Consumables (Access Cavity Preparation, Endodontic Files, Burs, Drill, Lubricant, Obturation), End User (Clinic, Hospital)-Global Forecast to 2022. Endodontics Market by Consumables & End User | Global Forecast 2022. Available online: marketsandmarkets.com (accessed on 6 April 2021).

130. Ferreira, C.M.; Silva, J.B.A., Jr.; Monteiro de Paula, R.C.; Andrade Feitosa, J.P.; Negreiros Cortez, D.G.; Zaia, A.A.; de Souza-Filho, F.J. Brazilian gutta-percha points. Part I: Chemical composition and X-ray diffraction analysis. *Braz. Oral Res.* **2005**, *19*, 193–197. [CrossRef] [PubMed]

131. Dobrzańska, J.; Gołombek, K.; Dobrzański, L.B. Polymer materials used in endodontic treatment–in vitro testing. *AMSE* **2012**, *58*, 110–115.

132. Dobrzańska, J. Analiza Szczelności Wypełnień Kanałów Korzeniowych. Ph.D. Thesis, Śląski Uniwersytet Medyczny w Katowicach, Zabrze, Poland, 2011.

133. Schilder, H.; Goodman, A.; Winthrop, A. The termomechanical properties of gutta-percha. *Determination of phase transition temperatures for gutta-percha, Oral Surg. Oral Med. Oral Pathol.* **1974**, *38*, 109–114. [CrossRef]

134. Ferreira, C.M.; Gurgel-Filho, E.D.; Silva, J.B.A., Jr.; Monteiro de Paula, R.C.; Pessoa Andrade Feitosa, J.; Figueiredo de Almeida Gomes, B.P.; de Souza-Filho, F.J. Brazilian gutta-percha points. Part II: Thermal properties. *Braz. Oral Res.* **2007**, *21*, 29–34. [CrossRef]

135. Combe, E.C.; Cohen, B.D.; Cumming, K. Alpha- and beta-forms of gutta-percha in products for root canal filling. *Int. Endodon. J.* **2001**, *34*, 447–451. [CrossRef]

136. Tully, J. A Victorian Ecological Disaster: Imperialism, the Telegraph, and Gutta-Percha. *J. World Hist.* **2009**, *20*, 559–579. [CrossRef]

137. Schilder, H. Filling root canals in three dimensions. *Dent. Clin. North Am.* **1967**, *32*, 723–744. [CrossRef]

138. Buchanan, L.S. The continuous wave of condensation technique: A convergence of conceptual and procedural advances in obturation. *Dent. Today* **1994**, *13*, 84–85.

139. Hand, R.E.; Huget, E.F.; Tsakinis, P.J. Effects of a warm gutta-percha technique on the lateral periodontium. *Oral Surg. Oral Med. Oral Pathol.* **1976**, *42*, 395–401. [CrossRef]

140. Nahmias, Y.; Mab, T.; Dovgan, J.S. The Thermo Hydraulic Condensation Technique. *Oral Health* **2001**, *91*, 11–15.

141. Nahmias, Y.; Serota, K.S.; Watson Jr, W.R. Predictable Endodontic Success: Part II -Microstructural Replication. Available online: http://www.ecoweek.ca/issues/PrinterFriendly.asp?aid=1000156065&RType=&PC=&issue=12012003 (accessed on 6 April 2021).

142. Barattolo, R.; Santarcangelo, F. Otturazione del sistema dei canali radicolari con guttaperca termoplasticizzata: Principi, materiali e tecniche. *G. Italia. Endod.* **2011**, *25*, 112–124. [CrossRef]

143. Nahmias, Y.; Mab, T.; Dovgan, J.S. La tecnica di condensazione termoidraulica. *L'infor. Endod.* **2002**, *5*, 28–33.

144. Nahmias, Y.; Bery, P. Due radici palatine nei primi molari superiori. *L'infor. Endod.* **2007**, *10*, 48–51.

145. Carvalho-Sousa, B.; Almeida-Gomes, F.; Carvalho, P.R.; Maníglia-Ferreira, C.; Gurgel-Filho, E.D.; Albuquerque, D.S. Filling lateral canals: Evaluation of different filling techniques. *Eur. J. Dent.* **2010**, *4*, 251–256. [CrossRef] [PubMed]

146. Siqueira, J.F.; Roças, I.N.; Favieri, A.; Abad, E.C.; Castro, A.J.; Gahyva, S.M. Bacterial leakage in coronally unsealed root canals obturated with 3 different techniques. *Oral Surg. Oral Med. Oral Pathol. Oral Radiol. Endod.* **2000**, *90*, 647–650. [CrossRef]

147. Limkangwalmongkol, S.; Burtscher, P.; Abbot, P.; Sandler, A.; Bishop, B. A comparative study of the apical leakage of four root canals sealers and laterally condensed gutta percha. *J. Endod.* **1991**, *17*, 495–499. [CrossRef]

148. Swanson, K.; Madison, S. An evaluation of coronal microleakage in endodontically treated teeth. Part 1. Time periods. *J. Endod.* **1987**, *13*, 56–59. [CrossRef]

149. Friedman, S.; Torneck, C.; Komorowsji, R.; Ouzounian, Z.; Syrtash, P.; Kaufman, A. In vivo model for assessing the functional efficacy of endodontic materials and techniques. *J. Endod.* **1997**, *23*, 557–561. [CrossRef]

150. Torabinejad, M.; Ung, B.; Kettering, J. In vitro bacterial penetration of coronally unsealed endodontically treated teeth. *J. Endod.* **1990**, *16*, 566–569. [CrossRef]

151. Shipper, G.; Trope, M. In vitro microbial leakage of endodontically treated teeth using new and standard obturation techniques. *J Endod* **2004**, *30*, 154–158. [CrossRef] [PubMed]

152. Magura, M.E.; Kafrawy, A.H.; Brown, C.E.; Newton, C.W. Human saliva coronal microleakeage in obturated root canals: An in vitro study. *J. Endod.* **1991**, *17*, 324–331. [CrossRef]

153. Oliver, C.; Abbott, P. An in vitro study of apical and coronal micro leakage of laterally condensed gutta-percha with Ketac-Endo and AH-26. *Aust. Dent. J.* **1998**, *43*, 262–268. [CrossRef] [PubMed]

154. Madison, S.; Wilcox, L. An evaluation of coronal microleakge in endodontically treated teeth. Part III. In vivo study. *J. Endod.* **1988**, *14*, 455–458. [CrossRef]

155. Khayat, A.; Lee, S.J.; Torabinejad, M. Human saliva penetration of coronally unsealed obturated root canals. *J. Endod.* **1993**, *19*, 458–461. [CrossRef]

156. Jafari, F.; Jafari, S. Importance and methodologies of endodontic microleakage studies: A systematic review. *J. Clin. Exp. Dent.* **2017**, *9*, e812–e819. [CrossRef]

157. Møller, L.; Wenzel, A.; Wegge-Larsen, A.M.; Ding, M.; Væth, M.; Hirsch, E.; Kirkevang, L.-L. Comparison of images from digital intraoral receptors and cone beam computed tomography scanning for detection of voids in root canal fillings: An in vitro study using micro-computed tomography as validation. *Oral Surg. Oral Med. Oral Pathol. Oral Radiol.* **2013**, *115*, 810–818. [CrossRef]

158. Imura, N.; Kato, A.S.; Hata, G.I.; Uemura, M.; Toda, T.; Weine, F. A comparison of the relative efficacies of four hand and rotary instrumentation techniques during endodontic retreatment. *Int. Endod. J.* **2000**, *33*, 361–366. [CrossRef]

159. Ferreira, J.J.; Rhodes, J.S.; Pitt Ford, T.R. The efficacy of gutta-percha removal using ProFiles. *Int. Endod. J.* **2001**, *34*, 267–274. [CrossRef]

160. Li, G.H.; Niu, L.N.; Selem, L.C.; Eid, A.A.; Bergeron, B.E.; Chen, J.H.; Pashley, D.H.; Tay, F.R. Quality of obturation achieved by an endodontic core-carrier system with crosslinked gutta-percha carrier in single-rooted canals. *J. Dent.* **2014**, *42*, 1124–1134. [CrossRef] [PubMed]

161. Donnermeyer, D.; Bürklein, S.; Dammaschke, T.; Schäfer, E. Endodontic sealers based on calcium silicates: A systematic review. *Odontol.* **2019**, *107*, 421–436. [CrossRef]

162. Gandolfi, M.G.; Siboni, F.; Prati, C. Properties of a novel polysiloxane-guttapercha calcium silicate-bioglass-containing root canal sealer. *Dent. Mater.* **2016**, *32*, e113–e126. [CrossRef] [PubMed]

163. Zoufan, K.; Jiang, J.; Komabayashi, T.; Wang, Y.-H.; Safavi, K.E.; Zhu, Q. Cytotoxicity evaluation of Gutta Flow and Endo Sequence BC sealers. *Oral Surg. Oral Med. Oral Pathol. Oral Radiol. Endod.* **2011**, *112*, 657–661. [CrossRef] [PubMed]

164. Zmener, O. Tissue response to a new methacrylate-based root canal sealer: Preliminary observations in the subcutaneous connective tis- sues of rats. *J. Endod.* **2004**, *30*, 348–351. [CrossRef]

165. Friedman, S.; Löst, C.; Zarrabian, M.; Trope, M. Evaluation of success and failure after endodontic therapy using a glass ionomer cement sealer. *J. Endod.* **1995**, *21*, 384–390. [CrossRef]

166. Lee, K.W.; Williams, M.C.; Camps, J.J.; Pashley, D.H. Adhesion of endodontic sealers to dentin and gutta-percha. *J. Endod.* **2002**, *28*, 684–688. [CrossRef]

167. Gatwood, R.S. Endodontic materials. *Dent. Clin. North Am.* **2007**, *15*, 695–712. [CrossRef]

168. Shipper, G.; Ørstavik, D.; Teixeira, F.B.; Trope, M. An evaluation of microbial leakage in roots filled with a thermoplastic synthetic polymer-based root canal filling material (Resilon). *J. Endod.* **2004**, *30*, 342–347. [CrossRef] [PubMed]

169. Resilon™ Obturation Material-The New Standard of Care? Available online: https://www.endoexperience.com/filecabinet/Clinical%20Endodontics/Obturation/Resilon/Resilon.Fact.Sheet.pdf (accessed on 1 April 2021).

170. Lipski, M.; Woźniak, K.; Buczkowska-Radlińska, J.; Łagocka, R.; Bochińska, J.; Nowicka, A. Resilon i Epiphany nowy materiał do wypełniania kanałów korzeniowych zębów: Badania wstępne w SEM. *Mag. Stomat.* **2005**, *9*, 108–112.

171. Lotfi, M.; Ghasemi, N.; Rahimi, S.; Vosoughhosseini, S.; Saghiri, M.A.; Shahidi, A. Resilon: A comprehensive literature review. *J. Dent. Res. Dent. Clin. Dent. Prospects* **2013**, *7*, 119–130. [CrossRef] [PubMed]

172. Pawińska, M.; Kierklo, A.; Marczuk-Kolada, G. New technology in endodontics—the Resilon-Epiphany system for obturation of root canals. *Adv. Med. Sci.* **2006**, *51* (Suppl. 1), 154–157.

173. Pocket Dentistry. Gutta-percha Substitute: Resilon. Available online: https://pocketdentistry.com/33-gutta-percha-substitute-resilon/ (accessed on 1 April 2021).

174. Mohammadi, Z.; Jafarzadeh, H.; Shalavi, S.; Bhandi, S.; Kinoshita, J. Resilon: Review of a New Material for Obturation of the Canal. *J. Contemp. Dent. Pract.* **2015**, *16*, 407–414. [CrossRef]

175. Barnett, F.; Trope, M. Resilon™: A novel material to replace gutta-percha. *Contemp. Endod.* **2004**, *1*, 16–19.

176. Hammad, M.; Qualtrough, A.; Silikas, N. Effect of a new obturating material on vertical root fracture resistance of endodontically treated teeth. *J. Endod.* **2007**, *33*, 732–736. [CrossRef]

177. Teixeira, F.B.; Teixeira, E.C.N.; Thompson, J.Y.; Leinfelder, K.F.; Trope, M. Dentinal bonding reaches the root canal system. *J. Esthet. Restor. Dent.* **2004**, *16*, 348–354. [CrossRef]

178. Shipper, G.; Teixeira, F.B.; Arnold, R.R.; Trope, M. Periapical inflammation after coronal microbial inoculation of dog roots filled with gutta-percha or resilon. *J. Endod.* **2005**, *31*, 91–96. [CrossRef] [PubMed]

179. Shrestha, D.; Wei, X.; Wu, W.-C.; Ling, J.-Q. Resilon: A methacrylate resin-based obturation system. *J. Dent. Sci.* **2010**, *5*, 47–52. [CrossRef]

180. Resende, L.M.; Rached-Junior, F.J.; Versiani, M.A.; Souza-Gabriel, A.E.; Miranda, C.E.; Silva-Sousa, Y.T.; Sousa Neto, M.D. A comparative study of physicochemical properties of AH Plus, Epiphany, and Epiphany SE root canal sealers. *Int. Endod. J.* **2009**, *42*, 785–793. [CrossRef]

181. Rocha, A.W.; de Andrade, C.D.; Leitune, V.C.; Collares, F.M.; Samuel, S.M.; Grecca, F.S.; de Figueiredo, J.A.; dos Santos, R.B. Influence of endodontic irrigants on resin sealer bond strength to radicular dentin. *Bull. Tokyo Dent. Coll.* **2012**, *53*, 1–7. [CrossRef] [PubMed]

182. Strange, K.A.; Tawil, P.Z.; Phillips, C.; Walia, H.D.; Fouad, A.F. Long-term Outcomes of Endodontic Treatment Performed with Resilon/Epiphany. *J. Endod.* **2019**, *45*, 507–512. [CrossRef]

183. Pawińska, M.; Kierklo, A.; Tokajuk, G.; Sidun, J. New endodontic obturation systems and their interfacial bond strength with intraradicular dentine-ex vivo studies. *Adv. Med. Sci.* **2011**, *56*, 327–333. [CrossRef] [PubMed]

184. Kqiku, L.; Miletic, I.; Gruber, H.J.; Anic, I.; Städtler, P. Dichtigkeit von Wurzelkanalfüllungen mit GuttaFlow und Resilon im Vergleich zur lateralen Kondensation. *Wien Med. Wochenschr.* **2010**, *160*, 230–234. [CrossRef]

185. Shokouhinejad, N.; Sabeti, M.A.; Hasheminasab, M.; Shafiei, F.; Shamshiri, A.R. Push-out bond strength of Resilon/Epiphany self-etch to intraradicular dentin after retreatment: A preliminary study. *J. Endod.* **2010**, *36*, 493–496. [CrossRef]

186. De-Deus, G.; Namen, F.; Galan Jr, J.; Zehnder, M. Soft chelating irrigation protocol optimizes bonding quality of Resilon/Epiphany root fillings. *J. Endod.* **2008**, *34*, 703–705. [CrossRef] [PubMed]

187. Üreyen Kaya, B.; Keçeci, A.D.; Orhan, H.; Belli, S. Micropush-out bond strengths of gutta-percha versus thermoplastic synthetic polymer-based systems—an ex vivo study. *Int. Endod. J.* **2008**, *41*, 211–218. [CrossRef] [PubMed]

188. Fisher, M.A.; Berzins, D.W.; Bahcall, J.K. An in vitro comparison of bond strength of various obturation materials to root canal dentin using a push-out test design. *J. Endod.* **2007**, *33*, 856–858. [CrossRef]

189. Sly, M.M.; Moore, B.K.; Platt, J.A.; Brown, C.E. Push-out bond strength of a new endodontic obturation system (Resilon/Epiphany). *J. Endod.* **2007**, *33*, 160–162. [CrossRef]

190. Dumani, A.; Yoldas, O.; Isci, A.S.; Köksal, F.; Kayar, B.; Polat, E. Disinfection of artificially contaminated Resilon cones with chlorhexidine and sodium hypochlorite at different time exposures. *Oral Surg. Oral Med. Oral Pathol. Oral Radiol. Endod.* **2007**, *103*, e82–e85. [CrossRef]

191. Shokouhinejad, N.; Sharifian, M.R.; Jafari, M.; Sabeti, M.A. Push-out bond strength of Resilon/Epiphany self-etch and gutta-percha/AH26 after different irrigation protocols. *Oral Surg. Oral Med. Oral Pathol. Oral Radiol. Endod.* **2010**, *110*, e88–e92. [CrossRef] [PubMed]

192. Kumar, N.; Aggarwal, V.; Singla, M.; Gupta, R. Effect of various endodontic solutions on punch out strength of Resilon under cyclic loading. *J. Conserv. Dent.* **2011**, *14*, 366–369. [CrossRef] [PubMed]

193. Tay, F.R.; Loushine, R.J.; Weller, R.N.; Kimbrough, W.F.; Pashley, D.H.; Mak, Y.F.; Lai, C.N.; Raina, R.; Williams, M.C. Ultrastructural evaluation of the apical seal in roots filled with a polycaprolactone-based root canal filling material. *J. Endod.* **2005**, *31*, 514–519. [CrossRef] [PubMed]

194. Shin, S.J.; Jee, S.W.; Song, J.S.; Jung, I.Y.; Cha, J.H.; Kim, E. Comparison of regrowth of Enterococcus faecalis in dentinal tubules after sealing with gutta-percha or Resilon. *J. Endod.* **2008**, *34*, 445–448. [CrossRef] [PubMed]

195. Jack, R.M.; Goodell, G.G. In vitro comparison of coronal microleakage between Resilon alone and gutta-percha with a glass ionomer intraorifice barrier using a fluid filtration model. *J. Endod.* **2008**, *34*, 718–720. [CrossRef]

196. Santos, J.; Tjäderhane, L.; Ferraz, C.; Zaia, A.; Alves, M.; De Goes, M.; Carrilho, M. Long-term sealing ability of resin-based root canal fillings. *Int. Endod. J.* **2010**, *43*, 455–460. [CrossRef]

197. Shemesh, H.; Wu, M.K.; Wesselink, P.R. Leakage along the apical root fillings with and without smear layer using two different leakage models: A two month longitudinal ex vivo study. *Int. Endod. J.* **2006**, *39*, 968–976. [CrossRef]

198. De Bruyne, M.A.; De Moor, R.J. Long-term sealing ability of Resilon apical root-end fillings. *Int. Endod. J.* **2009**, *42*, 884–892. [CrossRef]

199. Deus, G.A.D.; Fábio, M.; Rocha, L.A.C.M.; Gurgel-Filho, E.D.; Maniglia, C.F.; Coutinho-Filho, T. Analysis of the film thickness of a root canal sealer following three obturation techniques. *Pesqui Odontol. Bras.* **2003**, *17*, 119–125. [CrossRef]

200. Paqué, F.; Sirtes, G. Apical sealing ability of Resilon/Epiphany versus gutta-percha/AH Plus: Immediate and 16 month leakage. *Int. Endod. J.* **2007**, *40*, 722–729. [CrossRef] [PubMed]

201. Pasqualini, D.; Scotti, N.; Mollo, L.; Berutti, E.; Angelini, E.; Migliaretti, G.; Cuffini, A.; Adlerstein, D. Microbial leakage of Gutta-Percha and Resilon root canal filling material: A comparative study using a new homogeneous assay for sequence detection. *J. Biomater. Appl.* **2008**, *22*, 337–352. [CrossRef] [PubMed]

202. Shemesh, H.; Souza, E.M.; Wu, M.K.; Wesselink, P.R. Glucose reactivity with filling materials as a limitation for using the glucose leakage model. *Int. Endod. J.* **2008**, *41*, 869–872. [CrossRef]

203. Kokorikos, I.; Kolokouris, I.; Economides, N.; Gogos, C.; Helvatjoglu-Antoniades, M. Long-term evaluation of the sealing ability of two root canal sealers in combination with self-etching bonding agents. *J. Adhes. Dent.* **2009**, *11*, 239–246. [PubMed]

204. Onay, E.O.; Ungor, M.; Orucoglu, H. An in vitro evaluation of the apical sealing ability of a new resin based root canal obturation system. *J. Endod.* **2006**, *32*, 976–978. [CrossRef] [PubMed]

205. Hirai, V.H.; da Silva Neto, U.X.; Westphalen, V.P.D.; Perin, C.P.; Carneiro, E.; Fariniuk, L.F. Comparative analysis of leakage in root canal filling performed with gutta-percha and Resilon cones with AH Plus and Epiphany sealers. *Oral Surg. Oral Med. Oral Pathol. Oral Radiol. Endod.* **2010**, *109*, 131–135. [CrossRef]

206. Saleh, I.M.; Ruyter, I.E.; Haapasalo, M.; Ørstavik, D. Bacterial penetration along different root canal filling materials in the presence or absence of smear layer. *Int. Endod. J.* **2008**, *41*, 32–40. [CrossRef]

207. De-Deus, G.; Namen, F.; Galan, J. Reduced long-term sealing ability of adhesive root fillings after water-storage stress. *J. Endod.* **2008**, *34*, 322–325. [CrossRef]

208. Pandey, P.; Aggarwal, H.; Tikku, A.P.; Singh, A.; Bains, R.; Mishra, S. Comparative evaluation of sealing ability of gutta percha and resilon as root canal filling materials- a systematic review. *J Oral Biol Craniofac Res* **2020**, *10*, 220–226. [CrossRef]

209. Barborka, B.J.; Woodmansey, K.F.; Glickman, G.N.; Schneiderman, E.; He, J. Long-term Clinical Outcome of Teeth Obturated with Resilon. *J. Endod.* **2017**, *43*, 556–560. [CrossRef]

210. Lyons, W.W.; Hartwell, G.R.; Stewart, J.T.; Reavley, B.; Appelstein, C.; Lafkowitz, S. Comparison of coronal bacterial leakage between immediate versus delayed post-space preparations in root canals filled with Resilon/Epiphany. *Int. Endod. J.* **2009**, *42*, 203–207. [CrossRef] [PubMed]

211. Silveira, F.F.; Soares, J.A.; Nunes, E.; Mordente, V.L. Negative influence of continuous wave technique on apical sealing of the root canal system with Resilon. *J. Oral Sci.* **2007**, *49*, 121–128. [CrossRef] [PubMed]

212. De-Deus, G.; Audi, C.; Murad, C.; Fidel, S.; Fidel, R.A. Sealing ability of oval-shaped canals filled using the System B heat source with either gutta-percha or Resilon: An ex vivo study using a polymicrobial leakage model. *Oral Surg. Oral Med. Oral Pathol. Oral Radiol. Endod.* **2007**, *104*, 114–119. [CrossRef] [PubMed]

213. Baumgartner, G.; Zehnder, M.; Paqué, F. Enterococcus faecalis type strain leakage through root canals filled with Gutta-Percha/AH plus or Resilon/Epiphany. *J. Endod.* **2007**, *33*, 45–47. [CrossRef] [PubMed]

214. Hollanda, A.C.B.; Estrela, C.R.; Decurcio, D.A.; Silva, J.A.; Estrela, C. Sealing ability of three commercial resin-based endodontic sealers. *Gen. Dent.* **2009**, *57*, 368–373.

215. Kocak, M.M.; Er, O.; Saglam, B.C.; Yaman, S. Apical leakage of epiphany root canal sealer combined with different master cones. *Eur. J. Dent.* **2008**, *2*, 91–95. [CrossRef]

216. Bodrumlu, E.; Tunga, U. The apical sealing ability of a new root canal filling material. *Am. J. Dent.* **2007**, *20*, 295–298.
217. Raina, R.; Loushine, R.J.; Weller, R.N.; Tay, F.R.; Pashley, D.H. Evaluation of the quality of the apical seal in Resilon/ Epiphany and gutta-percha/AH Plus-filled root canals by using a fluid filtration approach. *J. Endod.* **2007**, *33*, 944–947. [CrossRef]
218. Pitout, E.; Oberholzer, T.G.; Blignaut, E.; Molepo, J. Coronal leakage of teeth root-filled with gutta-percha or Resilon root canal filling material. *J. Endod.* **2006**, *32*, 610–615. [CrossRef]
219. Fransen, J.N.; He, J.; Glickman, G.N.; Rios, A.; Shulman, J.D.; Honeyman, A. Comparative assessment of ActiV GP/glass ionomer sealer, Resilon/Epiphany, and gutta-percha/AH plus obturation: A bacterial leakage study. *J. Endod.* **2008**, *34*, 725–727. [CrossRef]
220. Shokouhinejad, N.; Sharifian, M.R.; Aligholi, M.; Assadian, H.; Tabor, R.K.; Nekoofar, M.H. The sealing ability of Resilon and gutta-percha following different smear layer removal methods: An ex vivo study. *Oral Surg. Oral Med. Oral Pathol. Oral Radiol. Endod.* **2010**, *110*, 45–49. [CrossRef]
221. Kangarlou, A.; Dianat, O.; Esfahrood, Z.R.; Asharaf, H.; Zandi, B.; Eslami, G. Bacterial leakage of GuttaFlow-filled root canals compared with Resilon/Epiphany and Gutta-percha/AH26-filled root canals. *Aust. Endod. J.* **2012**, *38*, 10–13. [CrossRef]
222. Biggs, S.C.; Knowles, K.I.; Ibarrola, J.L.; Pashley, D.H. An in vitro assessment of the sealing ability of Resilon/Epiphany using a fluid filtration. *J. Endod.* **2006**, *32*, 759–761. [CrossRef]
223. Karapınar-Kazandağ, M.; Tanalp, J.; Bayrak, Ö.F.; Sunay, H.; Bayirh, G. Microleakage of various root filling systems by glucose filtration analysis. *Oral Surg. Oral Med. Oral Pathol. Oral Radiol. Endod.* **2010**, *109*, e96–e102. [CrossRef]
224. Oddoni, P.G.; Mello, I.; Coil, J.M.; Antoniazzi, J.H. Coronal and apical leakage analysis of two different root canal obturation systems. *Braz. Oral Res.* **2008**, *22*, 211–215. [CrossRef]
225. De Almeida-Gomes, F.; Maniglia-Ferreira, C.; de Morais Vitoriano, M.; Carvalho-Sousa, B.; Guimaraes, N.L.; dos Santos, R.A.; Gurgel-Filho, E.D.; Rocha, M.M. Ex vivo evaluation of coronal and apical microbial leakage of root canal–filled with gutta-percha or Resilon/Epiphany root canal filling material. *Indian J. Dent. Res.* **2010**, *21*, 98–103. [CrossRef]
226. Üreyen Kaya, B.; Keçeci, A.D.; Belli, S. Evaluation of the sealing ability of gutta-percha and thermoplastic synthetic polymer-based systems along the root canals through the glucose penetration model. *Oral Surg. Oral Med. Oral Pathol. Oral Radiol. Endod.* **2007**, *104*, E66–E73. [CrossRef]
227. Dultra, F.; Barroso, J.M.; Carrasco, L.D.; Capelli, A.; Guerisoli, D.M.; Pécora, J.D. Evaluation of apical microleakage of teeth sealed with four different canal sealers. *J. Appl. Oral Sci.* **2006**, *14*, 341–345. [CrossRef] [PubMed]
228. Muñoz, H.R.; Saravia-Lemus, G.A.; Florián, W.E.; Lainfiesta, J.F. Microbial leakage of Enterococcus faecalis after post space preparation in teeth filled in vivo with RealSeal versus gutta-percha. *J. Endod.* **2007**, *33*, 674–675. [CrossRef] [PubMed]
229. Tanomaru-Filho, M.; Sant'anna-Junior, A.; Bosso, R.; Guerreiro-Tanomaru, J.M. Effectiveness of gutta-percha and Resilon in filling lateral root canals using the Obtura II system. *Braz. Oral Res.* **2011**, *25*, 205–209. [CrossRef] [PubMed]
230. Bodrumlu, E.; Tunga, U. Coronal sealing ability of a new root canal filling material. *J. Can. Dent. Assoc.* **2007**, *73*, 623. [PubMed]
231. Sharifian, M.R.; Shokouhinejad, N.; Aligholi, M.; Jafari, Z. Effect of chlorhexidine on coronal microleakage from root canals obturated with Resilon/Epiphany Self-Etch. *J. Oral Sci.* **2010**, *52*, 83–87. [CrossRef] [PubMed]
232. Bodrumlu, E.; Tunga, U. Apical leakage of Resilon obturation material. *J. Contemp. Dent. Pract.* **2006**, *7*, 45–52. [CrossRef] [PubMed]
233. Verissimo, D.M.; do Vale, M.S.; Monteiro, A.J. Comparison of apical leakage between canals filled with gutta-percha/AH-plus and Resilon/Epiphany System, when submitted to two filling techniques. *J. Endod.* **2007**, *33*, 291–294. [CrossRef]
234. Moura-Netto, C.; Pinto, T.; Davidowicz, H.; de Moura, A.A.M. Apical leakage of three resin-based endodontic sealers after 810-nm-diode laser irradiation. *Photomed. Laser Surg.* **2009**, *27*, 891–894. [CrossRef]
235. Wedding, J.R.; Brown, C.E.; Legan, J.J.; Moore, B.K.; Vail, M.M. An in vitro comparison of micro leakage between Resilon and gutta-percha with a fluid filtration model. *J. Endod.* **2007**, *33*, 1447–1449. [CrossRef]
236. Ishimura, H.; Yoshioka, T.; Suda, H. Sealing ability of new adhesive root canal filling materials measured by new dye penetration method. *Dent. Mater. J.* **2007**, *26*, 290–295. [CrossRef]
237. Nawal, R.R.; Parande, M.; Sehgal, R.; Rao, N.R.; Naik, A. A comparative evaluation of 3 root canal filling systems. *Oral Surg. Oral Med. Oral Pathol. Oral Radiol. Endod.* **2011**, *111*, 387–393. [CrossRef]
238. Punia, S.K.; Nadig, P.; Punia, V. An in vitro assessment of apical microleakage in root canals obturated with guttaflow, Resilon, thermafil and lateral condensation: A stereomicroscopic study. *J. Conserv. Dent.* **2011**, *14*, 173–177. [CrossRef] [PubMed]
239. Tunga, U.; Bodrumlu, E. Assessment of the sealing ability of a new root canal obturation material. *J. Endod.* **2006**, *32*, 876–878. [CrossRef] [PubMed]
240. Shashidhar, C.; Shivanna, V.; Shivamurthy, G.; Shashidhar, J. The comparison of microbial leakage in roots filled with resilon and gutta-percha: An in vitro study. *J. Conserv. Dent.* **2011**, *14*, 21–27. [CrossRef]
241. Sagsen, B.; Er, O.; Kahraman, Y.; Orucoglu, H. Evaluation of micro leakage of roots filled with different techniques with a computerized fluid filtration technique. *J. Endod.* **2006**, *32*, 1168–1170. [CrossRef] [PubMed]
242. Kqiku, L.; Städtler, P.; Gruber, H.J.; Baraba, A.; Anic, I.; Miletic, I. Active versus passive microleakage of Resilon/Epiphany and gutta-percha/AH Plus. *Aust. Endod. J.* **2011**, *37*, 141–146. [CrossRef] [PubMed]
243. Aptekar, A.; Ginnan, K. Comparative analysis of microleakage and seal for two obturation materials: Resilon/Epiphany and gutta-percha. *J. Can. Dent. Assoc.* **2006**, *72*, 245. [PubMed]
244. Stratton, R.K.; Apicella, M.J.; Mines, P. A fluid filtration comparison of gutta-percha versus Resilon, a new soft resin endodontic obturation system. *J. Endod.* **2006**, *32*, 642–645. [CrossRef]

245. Kurtzman, G.M. Resilon Update. *Inside Dent.* **2007**, *3*, 1–5.
246. Mohammadi, Z.; Khademi, A. An evaluation of the sealing ability of MTA and Resilon: A bacterial leakage study. *Iran Endod. J.* **2007**, *2*, 43–46. [PubMed]
247. Maltezos, C.; Glickman, G.N.; Ezzo, P.; He, J. Comparison of the sealing of Resilon, Pro Root MTA, and Super-EBA as root-end filling materials: A bacterial leakage study. *J. Endod.* **2006**, *32*, 324–327. [CrossRef] [PubMed]
248. Nagas, E.; Cehreli, Z.C.; Durmaz, V.; Vallittu, P.K.; Lassila, L.V. Regional push-out bond strength and coronal microleakage of Resilon after different light-curing methods. *J. Endod.* **2007**, *33*, 1464–1468. [CrossRef] [PubMed]
249. Williamson, A.E.; Marker, K.L.; Drake, D.R.; Dawson, D.V.; Walton, R.E. Resin-based versus gutta-percha-based root canal obturation: Influence on bacterial leakage in an in vitro model system. *Oral Surg. Oral Med. Oral Pathol. Oral Radiol. Endod.* **2009**, *108*, 292–296. [CrossRef]
250. Hanada, T.; Quevedo, C.G.; Okitsu, M.; Yoshioka, T.; Iwasaki, N.; Takahashi, H.; Suda, H. Effects of new adhesive resin root canal filling materials on vertical root fractures. *Aust. Endod. J.* **2010**, *36*, 19–23. [CrossRef] [PubMed]
251. Ulusoy, Ö.İ.A.; Nayır, Y.; Darendeliler-Yaman, S. Effect of different root canal sealers on fracture strength of simulated immature roots. *Oral Surg. Oral Med. Oral Pathol. Oral Radiol. Endod.* **2011**, *112*, 544–547. [CrossRef] [PubMed]
252. Shashidhar, J.; Shashidhar, C. Gutta percha verses resilon: An in vitro comparison of fracture resistance in endodontically treated teeth. *J Indian Soc Pedod Prev Dent* **2014**, *32*, 53–57. [CrossRef]
253. Baba, S.M.; Grover, S.I.; Tyagi, V. Fracture resistance of teeth obturated with gutta percha and Resilon: An in vitro study. *J. Conserv. Dent.* **2010**, *13*, 61–64. [CrossRef] [PubMed]
254. De Temiño Morante, P.R. ¿Es el Resilon el nuevo Material de Obturación Endodóntica? Available online: https://gacetadental.com/2011/10/es-el-resilon-el-nuevo-material-de-obturacin-endodntica-25262/ (accessed on 1 April 2021).
255. Hemalatha, H.; Sandeep, M.; Kulkarni, S.; Yakub, S.S. Evaluation of fracture resistance in simulated immature teeth using Resilon and Ribbond as root reinforcements—an in vitro study. *Dent. Traumatol.* **2009**, *25*, 433–438. [CrossRef] [PubMed]
256. Sagsen, B.; Er, O.; Kahraman, Y.; Akdogan, G. Resistance to fracture of roots filled with three different techniques. *Int. Endod. J.* **2007**, *40*, 31–35. [CrossRef]
257. Grande, N.M.; Plotino, G.; Lavorgna, L.; Ioppolo, P.; Bedini, R.; Pameijer, C.H.; Somma, F. Influence of different root canal-filling materials on the mechanical properties of root canal dentin. *J. Endod.* **2007**, *33*, 859–863. [CrossRef]
258. Ribeiro, F.C.; Souza-Gabriel, A.E.; Marchesan, M.A.; Alfredo, E.; Silva-Sousa, Y.T.; Sousa-Neto, M.D. Influence of different endodontic filling materials on root fracture susceptibility. *J. Dent.* **2008**, *36*, 69–73. [CrossRef]
259. Economides, N.; Koulaouzidou, E.A.; Gogos, C.; Kolokouris, I.; Beltes, P.; Antoniades, D. Comparative study of the cytotoxic effect of Resilon against two cell lines. *Braz. Dent. J.* **2008**, *19*, 291–295. [CrossRef]
260. Donadio, M.; Jiang, J.; He, J.; Wang, Y.H.; Safavi, K.E.; Zhu, Q. Cytotoxicity evaluation of Activ GP and Resilon sealers in vitro. *Oral Surg. Oral Med. Oral Pathol. Oral Radiol. Endod.* **2009**, *107*, e74–e78. [CrossRef]
261. Susini, G.; About, I.; Tran-Hung, L.; Camps, J. Cytotoxicity of Epiphany and Resilon with a root model. *Int. Endod. J.* **2006**, *39*, 940–944. [CrossRef]
262. Leonardo, M.R.; Barnett, F.; Debelian, G.J.; de Pontes Lima, R.K.; da Silva, L.A.B. Root canal adhesive filling in dogs' teeth with or without coronal restoration: A histopathological evaluation. *J. Endod.* **2007**, *33*, 1299–1303. [CrossRef]
263. Bodrumlu, E.; Muglali, M.; Sumer, M.; Guvenc, T. The response of subcutaneous connective tissue to a new endodontic filling material. *J. Biomed. Mater. Res. B Appl. Biomater.* **2008**, *84*, 463–467. [CrossRef] [PubMed]
264. Merdad, K.; Pascon, A.E.; Kulkarni, G.; Santerre, P.; Friedman, S. Short-term cytotoxicity assessment of components of the epiphany resin-percha obturating system by indirect and direct contact millipore filter assays. *J. Endod.* **2007**, *33*, 24–27. [CrossRef]
265. Onay, E.O.; Ungor, M.; Ozdemir, B.H. In vivo evaluation of the biocompatibility of a new resin-based obturation system. *Oral Surg. Oral Med. Oral Pathol. Oral Radiol. Endod.* **2007**, *104*, e60–e66. [CrossRef]
266. Bodrumlu, E.; Alaçamm, T. The antimicrobial and antifungal activity of a root canal core material. *J. Am. Dent. Assoc.* **2007**, *138*, 1228–1232. [CrossRef] [PubMed]
267. Gomes, B.P.; Berber, V.B.; Montagner, F.; Sena, N.T.; Zaia, A.A.; Ferraz, C.C.; Souza-Filho, F.J. Residual effects and surface alterations in disinfected gutta-percha and Resilon cones. *J. Endod.* **2007**, *33*, 948–951. [CrossRef] [PubMed]
268. Cunha, R.S.; De Martin, A.S.; Barros, P.P.; da Silva, F.M.; de Castilho Jacinto, R.; da Silveira Bueno, C.E. In vitro evaluation of the cleansing working time and analysis of the amount of gutta-percha or Resilon remnants in the root canal walls after instrumentation for endodontic retreatment. *J. Endod.* **2007**, *33*, 1426–1428. [CrossRef]
269. Fenoul, G.; Meless, G.D.; Pérez, F. The efficacy of R-Endo rotary NiTi and stainless steel hand instruments to remove gutta-percha and Resilon. *Int. Endod. J.* **2010**, *43*, 135–141. [CrossRef]
270. Schirrmeister, J.F.; Meyer, K.M.; Hermanns, P.; Altenburger, M.J.; Wrbas, K.T. Effectiveness of hand and rotary instrumentation for removing a new synthetic polymer-based root canal obturation material (Epiphany) during retreatment. *Int. Endod. J.* **2006**, *39*, 150–156. [CrossRef]
271. Marfisi, K.; Mercade, M.; Plotino, G.; Duran-Sindreu, F.; Bueno, R.; Roig, M. Efficacy of three different rotary files to remove gutta-percha and resilon from root canals. *Int. Endod. J.* **2010**, *43*, 1022–1028. [CrossRef]
272. Zarei, M.; Shahrami, F.; Vatanpour, M. Comparison between gutta-percha and Resilon retreatment. *J. Oral Sci.* **2009**, *51*, 181–185. [CrossRef]

273. Somma, F.; Cammarota, G.; Plotino, G.; Grande, N.M.; Pameijer, C.H. The effectiveness of manual and mechanical instrumentation for the retreatment of three different root canal filling materials. *J. Endod.* **2008**, *34*, 466–469. [CrossRef]

274. Taşdemir, T.; Yildirim, T.; Celik, D. Comparative study of removal of current endodontic fillings. *J. Endod.* **2008**, *34*, 326–329. [CrossRef]

275. Iizuka, N.; Takenaka, S.; Shigetani, Y.; Okiji, T. Removal of resin-based root canal filling materials with K3 rotary instruments: Relative efficacy for different combinations of filling materials. *Dent. Mater. J.* **2008**, *27*, 75–80. [CrossRef]

276. Hassanloo, A.; Watson, P.; Finer, Y.; Friedman, S. Retreatment efficacy of the Epiphany soft resin obturation system. *Int. Endod. J.* **2007**, *40*, 633–643. [CrossRef]

277. Tanomaru-Filho, M.; Orlando, T.A.; Bortoluzzi, E.A.; Silva, G.F.; Tanomaru, J.M. Solvent capacity of different substances on gutta-percha and Resilon. *Braz. Dent. J.* **2010**, *21*, 46–49. [CrossRef]

278. Faria-Júnior, N.B.; Loiola, L.E.; Guerreiro-Tanomaru, J.M.; Berbert, F.L.; Tanomaru-Filho, M. Effectiveness of three solvents and two associations of solvents on gutta-percha and Resilon. *Braz. Dent. J.* **2011**, *22*, 41–44. [CrossRef] [PubMed]

279. Bodrumlu, E.; Uzun, O.; Topuz, O.; Semiz, M. Efficacy of three techniques in removing root canal filling material. *J. Can. Dent. Assoc.* **2008**, *74*, 721. [PubMed]

280. Ezzie, E.; Fleury, A.; Solomon, E.; Spears, R.; He, J. Efficacy of retreatment techniques for a resin based root canal obturation material. *J. Endod.* **2006**, *32*, 341–344. [CrossRef]

281. Ring, J.; Murray, P.E.; Namerow, K.N.; Moldauer, B.I.; Garcia-Godoy, F. Removing root canal obturation materials: A comparison of rotary file systems and re-treatment agents. *J. Am. Dent. Assoc.* **2009**, *140*, 680–688. [CrossRef] [PubMed]

282. De Oliveira, D.P.; Barbizam, J.V.; Trope, M.; Teixeira, F.B. Comparison between gutta-percha and Resilon removal using two different techniques in endodontic retreatment. *J. Endod.* **2006**, *32*, 362–364. [CrossRef]

283. Azar, M.; Khojastehpour, L.; Iranpour, N. A comparison of the effectiveness of chloroform in dissolving Resilon and gutta-percha. *J. Dent. (Tehran)* **2011**, *8*, 19–24.

284. Shanahan, D.J.; Duncan, H.F. Root canal filling using Resilon: A review. *Br. Dent. J.* **2011**, *211*, 81–88. [CrossRef]

285. Steinke, H. Der Hippokratische Eid: Ein schwieriges Erbe. *Horiz. Med. Schweiz. Ärztezeitung Bull. Médecins Suisses Boll. Med. Svizz.* **2016**, *97*, 1699–1701.

286. Dobrzańska, J.; Dobrzański, L.B.; Gołombek, K.; Dobrzański, L.A.; Dobrzańska-Danikiewicz, A.D. Results of in vitro experimental research assessing the tightness of filling materials with the pulp of extracted human teeth depending on the obturation techniques during endodontic treatment. *Processes.* (prepared for printing).

287. Dobrzański, L.A.; Dobrzański, L.B.; Dobrzańska-Danikiewicz, A.D. Manufacturing technologies thick-layer coatings on various substrates and manufacturing gradient materials using powders of metals, their alloys and ceramics. *JAMME* **2020**, *99*, 14–41. [CrossRef]

288. Dobrzańska-Danikiewicz, A.D.; Dobrzański, L.A.; Szindler, M.; Achtelik-Franczak, A.; Dobrzański, L.B. Obróbka powierzchni materiałów mikroporowatych wytworzonych metodą selektywnego spiekania laserowego w celu uefektywnienia proliferacji żywych komórek. In *Metalowe Materiały Mikroporowate i Lite do Zastosowań Medycznych i Stomatologicznych*; Dobrzański, L.A., Dobrzańska-Danikiewicz, A.D., Eds.; Open Access Library VII(1); International OCSCO World Press: Gliwice, Poland, 2017; pp. 289–375.

289. Dobrzańska-Danikiewicz, A. The methodological fundaments of development state analysis of surface engineering technologies. *JAMME* **2010**, *40*, 203–210.

290. Dobrzańska-Danikiewicz, A.D.; Hajduczek, E.; Polok-Rubiniec, M.; Przybył, M.; Adamaszek, K. Evaluation of selected steel thermochemical treatment technologies using foresight methods. *JAMME* **2011**, *46*, 115–146.

291. Dobrzańska-Danikiewicz, A.D. The development perspectives of Physical Vapour Deposition technologies. *JAMME* **2012**, *54*, 103–109.

292. Dobrzańska-Danikiewicz, A.D. *Metodologia Komputerowo Zintegrowanego Prognozowania Rozwoju Inżynierii Powierzchni Materiałów*; Dobrzański, L.A., Ed.; In Open Access Library 1(7); International OCSCO World Press: Gliwice, Poland, 2012; pp. 1–289.

293. Dobrzańska-Danikiewicz, A.D. *Księga Technologii Krytycznych Kształtowania Struktury i Własności Powierzchni Materiałów Inżynierskich*; Dobrzański, L.A., Ed.; In Open Access Library 8(26); International OCSCO World Press: Gliwice, Poland, 2013; pp. 1–823.

294. Dobrzańska-Danikiewicz, A.D. (Ed.) *Materials Surface Engineering Development Trends*; Open Access Library 6; International OCSCO World Press: Gliwice, Poland, 2011; pp. 1–594.

295. Dobrzańska-Danikiewicz, A.D.; Dobrzański, L.A.; Sękala, A. Results of Technology Foresight in the Surface Engineering Area. *AMM* **2014**, *657*, 916–920. [CrossRef]

296. Dobrzańska-Danikiewicz, A.D.; Dobrzański, L.A.; Mazurkiewicz, J.; Tomiczek, B.; Reimann, Ł. E-transfer of materials surface engineering e-foresight results. *AMSE* **2011**, *52*, 87–100.

297. Dobrzańska-Danikiewicz, A.D.; Tański, T.; Malara, S.; Domagała-Dubiel, J. Technology Foresight Results Concerning Laser Surface Treatment of Casting Magnesium Alloys. In *New Features on Magnesium Alloys*; Monteiro, W.A., Ed.; IntechOpen: Rijeka, Croatia, 2012; pp. 1–30. [CrossRef]

298. Dobrzańska-Danikiewicz, A.D. Foresight of Material Surface Engineering as a Tool Building a Knowledge-Based Economy. *MSF* **2012**, *706–709*, 2511–2516. [CrossRef]

299. Dobrzański, L.A. (Ed.) *1st Workshop on Foresight of Surface Properties Formation Leading Technologies of Engineering Materials and Biomaterials*; International OCSCO World Press: Gliwice, Poland, 2009; pp. 1–272.
300. Dobrzański, L.A. (Ed.) *2nd Workshop on Foresight of Surface Properties Formation Leading Technologies of Engineering Materials and Biomaterials*; International OCSCO World Press: Gliwice, Poland, 2009; pp. 1–324.
301. Dobrzański, L.A.; Dobrzańska-Danikiewicz, A.D. (Eds.) *3rd Workshop on Foresight of Surface Properties Formation Leading Technologies of Engineering Materials and Biomaterials; Raport z realizacji zadania 2. "Analiza Istniejącej Sytuacji w Zakresie Rozwoju Technologii oraz Uwarunkowań Społeczno-Gospodarczych w Odniesieniu do Przedmioty Foresightu pt. FORSURF-Foresight Wiodących Technologii Kształtowania Własności Powierzchni Materiałów Inżynierskich i Biomedycznych"*; International OCSCO World Press: Gliwice, Poland, 2010; pp. 1–184.
302. Dobrzański, L.A.; Achtelik-Franczak, A. Struktura i własności tytanowych szkieletowych materiałów mikroporowatych wytworzonych metodą selektywnego spiekania laserowego do zastosowań w implantologii oraz medycynie regeneracyjnej. In *Metalowe Materiały Mikroporowate i Lite do Zastosowań Medycznych i Stomatologicznych*; Dobrzański, L.A., Dobrzańska-Danikiewicz, A.D., Eds.; Open Access Library VII(1); International OCSCO World Press: Gliwice, Poland, 2017; pp. 186–244.
303. Dobrzański, L.A. Effect of heat and surface treatment on the structure and properties of the Mg-Al-Zn-Mn casting alloys. In *Magnesium and Its Alloys*; Dobrzański, L.A., Totten, G.E., Bamberger, M., Eds.; CRC Press: Boca Raton, FL, USA, 2019; pp. 91–202.
304. Dobrzański, L.A.; Dobrzańska-Danikiewicz, A.D. Perspektywy i trendy rozwojowe inżynierii powierzchni materiałów. In *Inżynieria Powierzchni Materiałów: Kompendium Wiedzy i Podręcznik Akademicki*; Dobrzański, L.A., Ed.; Open Access Library VIII(1); International OCSCO World Press: Gliwice, Poland, 2018; pp. 89–157.
305. Dobrzański, L.A.; Nieradka-Buczek, B. Transparent conductive nanocomposite layers with polymer matrix and silver nanowires reinforcement. *AMSE* **2018**, *93*, 59–84. [CrossRef]
306. Dobrzański, L.A.; Dobrzańska-Danikiewicz, A.D.; Czuba, Z.P.; Dobrzański, L.B.; Achtelik-Franczak, A.; Malara, P.; Szindler, M.; Kroll, L. Metallic skeletons as reinforcement of new composite materials applied in orthopaedics and dentistry. *AMSE* **2018**, *92*, 53–85. [CrossRef]
307. Dobrzański, L.A.; Dobrzański, L.B.; Dobrzańska-Danikiewicz, A.D.; Dobrzańska, J.; Rudziarczyk, K.; Achtelik-Franczak, A. Non-Antagonistic Contradictoriness of the Progress of Advanced Digitized Production with SARS-CoV-2 Virus Transmission in the Area of Dental Engineering. *Processes* **2020**, *8*, 1097. [CrossRef]
308. Dobrzański, L.A.; Dobrzańska-Danikiewicz, A.; Achtelik-Franczak, A. The structure and properties of aluminium alloys matrix composite materials with reinforcement made of titanium skeletons. *AMSE* **2016**, *80*, 16–30. [CrossRef]
309. Dobrzański, L.A.; Prokopiuk vel Prokopowicz, M. The influence of reduced graphene oxide on the structure of the electrodes and the properties of dye-sensitized solar cells. *AMSE* **2016**, *77*, 12–30. [CrossRef]
310. Dobrzański, L.A.; Hudecki, A.; Chladek, G.; Król, W.; Mertas, A. Biodegradable and antimicrobial polycaprolactone nanofibers with and without silver precipitates. *AMSE* **2015**, *76*, 5–26.
311. Dobrzański, L.A.; Hudecki, A. Structure, geometrical characteristics and properties of biodegradable micro- and polycaprolactone nanofibers. *AMSE* **2014**, *70*, 5–13.
312. Dobrzański, L.A.; Pawlyta, M.; Hudecki, A. Conceptual study on a new generation of the high-innovative advanced porous and composite nanostructural functional materials with nanofibers. *JAMME* **2011**, *49*, 550–565.
313. Dobrzańska-Danikiewicz, A.D.; Żmudzki, J. Development trends of mucous-borne dentures in the aspect of elastomers applications. *AMSE* **2012**, *55*, 5–13.
314. Dobrzański, L.B. Struktura i Własności Materiałów Inżynierskich na Uzupełnienia Protetyczne Układu Stomatognatycznego Wytwarzane Metodami Przyrostowymi i Ubytkowymi. Ph.D. Thesis, Akademia Górniczo-Hutnicza im. Stanisława Staszica w Krakowie, Kraków, Poland, 2017.
315. Rudziarczyk, K. Modelowanie Stanu Naprężeń w Wybranych Rodzajach Implantów Stomatologicznych Oraz w Otaczającej je Tkance Kostnej. Master's Thesis, Politechnika Śląska, Gliwice, Poland, 2020. (manuscript not punlished).
316. Berliński, M.; Suprvisory guidance Dobrzański, L.A. Modelowanie Zależności między Warunkami Wytapiania w Wielkim Piecu i Składem Chemicznym Surówki z Wykorzystaniem Sieci Neuronowych. Ph.D. Thesis, Politechnika Śląska, Gliwice, Poland, 2018.
317. Nieradka-Buczek, B.; Suprvisory Guidance Dobrzański, L.A. Zastosowanie Nanodrutów Srebra jako Wzmocnienia Transparentnych Warstw Nanokompozytowych. Ph.D. Thesis, Politechnika Śląska, Gliwice, Poland, 2018.
318. Kędzierski, K.; Suprvisory guidance Dobrzański, L.A. Charakterystyka Ekologicznych Ośrodków Polimerowych do Chłodzenia Nawęglonych Stali Niskowęglowych. Ph.D. Thesis, Politechnika Śląska, Gliwice, Poland, 2017.
319. Szindler, M.; Suprvisory Guidance Dobrzański, L.A. Polimerowe Warstwy Półprzewodnikowe zol-żel do Zastosowań Fotowoltaicznych. Ph.D. Thesis, Politechnika Śląska, Gliwice, Poland, 2017.
320. Achtelik-Franczak, A.; suprvisory guidance Dobrzański, L.A. Inżynierskie Materiały Kompozytowe O Wzmocnieniu Z Mikroporowatego Tytanu Selektywnie Spiekanego Laserowo. Ph.D. Thesis, Politechnika Śląska, Gliwice, Poland, 2016.
321. Macek, M.; Suprvisory Guidance Dobrzański, l.a. Struktura i Własności Aluminiowych Materiałów Nanokompozytowych Wzmacnianych Nanorurkami Węglowymi. Ph.D. Thesis, Politechnika Śląska, Gliwice, Poland, 2016.
322. Prokopiuk vel Prokopowicz, M.; Suprvisory Guidance Dobrzański, L.A. Wpływ zredukowanego Tlenku Grafenu na Strukturę Elektrod i Własności Barwnikowych Ogniw Fotowoltaicznych. Ph.D. Thesis, Politechnika Śląska, Gliwice, Poland, 2016.
323. Czaja, I.; Suprvisory Guidance Dobrzański, L.A. Wpływ Nanodrutów i Nanoproszków Miedzi na Strukturę i Własności Nanokompozytowych Materiałów Polimerowych. Ph.D. Thesis, Politechnika Śląska, Gliwice, Poland, 2015.

324. Hudecki, A.; Suprvisory Guidance Dobrzański, L.A. Nanowłókna kompozytowe o bioaktywnym rdzeniu i antybakteryjnej powłoce na rusztowania tkankowe. Ph.D. Thesis, Politechnika Śląska, Gliwice, Poland, 2015.

325. Mucha, A.; Suprvisory Guidance Dobrzański, L.A. Wpływ nanorurek węglowych na własności barwnikowych ogniw fotowoltaicznych. Ph.D. Thesis, Politechnika Śląska, Gliwice, Poland, 2015.

326. Górniak, M.; Suprvisory Guidance Dobrzański, L.A. Wpływ Organofilizacji Powierzchni Nanorurek Haloizytowych i Nanopłytek Montmorylonitowych na Własności Kompozytów Polimerowych. Ph.D. Thesis, Politechnika Śląska, Gliwice, Poland, 2014.

327. Szindler, M.; Suprvisory Guidance Dobrzański, L.A. Struktura i Własności Nanostrukturalnych Warstw Antyrefleksyjnych Otrzymanych Metodą ALD Oraz zol-żel na Krzemowych Ogniwach Fotowoltaicznych. Ph.D. Thesis, Politechnika Śląska, Gliwice, Poland, 2014.

328. Reimann, Ł.; Suprvisory Guidance Dobrzański, L.A. System Komputerowego Wspomagania Doboru Materiałów Na Wieloczłonowe Częściowe Stałe Protezy Stomatologiczne. Ph.D. Thesis, Politechnika Śląska, Gliwice, Poland, 2013.

329. Henderson, B. The Product Portfolio. Available online: https://www.bcg.com/publications/1970/strategy-the-product-portfolio (accessed on 8 April 2021).

330. Suresh Chandra, B.; Gopikrishna, V. (Eds.) *Grossman's Endodontic Practice*, 13th ed.; Wolters Kluwer: New Delhi, India, 2014; pp. 1–576.

331. Durey, I. Deming's Management Method and Total Quality Management. Available online: https://www.gembadesk.com/2014/10/demings-management-method-and-total-quality-management/#:~{}:text=Created%20in%201951%2C%20The%20Deming%20Prize%20was%20awarded,American%20Society%20for%20Quality%E2%80%99s%20Deming%20Medal%20in%20 (accessed on 17 April 2021).

332. Lina, L.R.; Ullah, H. The concept and implementation of Kaizen in organization. *Int. J. Bus. Adm. Manag. Res.* **2019**, *19*, 1–9.

333. Rokke, C.; Yadav, O.P. Challenges and Barriers to Total Quality Management: An Overview. *Int. J. Performability Eng.* **2012**, *8*, 653–665.

Article

Application Solid Laser-Sintered or Machined Ti6Al4V Alloy in Manufacturing of Dental Implants and Dental Prosthetic Restorations According to Dentistry 4.0 Concept

Leszek A. Dobrzański *, Lech B. Dobrzański, Anna Achtelik-Franczak and Joanna Dobrzańska

Medical and Dental Engineering Centre for Research, Design and Production ASKLEPIOS,
13 D Krolowej Bony St., 44-100 Gliwice, Poland; dobrzanski@centrumasklepios.pl (L.B.D.);
anna.achtelik-franczak@centrumasklepios.pl (A.A.-F.); joanna.dobrzanska@centrumasklepios.pl (J.D.)
* Correspondence: leszek.dobrzanski@centrumasklepios.pl

Received: 7 May 2020; Accepted: 29 May 2020; Published: 3 June 2020

Abstract: This paper presents a comparison of the impact of milling technology in the computer numerically controlled (CNC) machining centre and selective laser sintering (SLS) and on the structure and properties of solid Ti6Al4V alloy. It has been shown that even small changes in technological conditions in the SLS manufacturing variant significantly affect changes from two to nearly two and a half times in tensile and bending strengths. Both the tensile and bending strength obtained in the most favourable manufacturing variant by the SLS method is over 25% higher than in the case of cast materials subsequently processed by milling. Plug-and-play SLS conditions provide about 60% of the possibilities. Structural, tribological and electrochemical tests were carried out. In vitro biological tests using osteoblasts confirm the good tendency for the proliferation of live cells on the substrate manufactured under the most favourable SLS conditions. The use of SLS additive technology for the manufacturing of dental implants and abutments made of Ti6Al4V alloy in combination with the digitisation of dental diagnostics and computer-aided design and manufacture of computer-aided design/manufacturing (CAD/CAM) following the idea of Dentistry 4.0 is the best choice of technology for manufacturing of prosthetic and implant devices used in dentistry.

Keywords: Dentistry 4.0; additive manufacturing; selective laser sintering; dental prosthetic restorations; Ti6Al4V dental alloy; structural X-ray analysis; energy-dispersive X-ray spectroscope; metallography; tensile and bending strength; corrosion resistance; tribological tests; in-vitro tests

---

1. Introduction

Contemporary dentistry requires extensive engineering support consisting of the synergistic use of extensive knowledge in the field of material engineering, manufacturing engineering, and tissue engineering covered by the current stage of Industry 4.0 of the industrial revolution [1–7]. Stadium Industry 4.0 is associated with systematically implemented cyber-physical systems. An essential element of those activities is to achieve the Materials 4.0 stage of material design [1,8,9] as part of a complex engineering design process also involving structural and technological design. The Industry 4.0 model, resulting directly from the original reports introducing this approach [2–5], however, proved to be incomplete. Criticism of this model led to the development of the authors' extended holistic model Industry 4.0 [1,6,7,10]. This model is located in the current one appropriately extended as one of the four components of the technological plane. This technological plane also contains materials, technological machines and devices, as well as technological processes together with additive methods that cannot be considered as the only ones needed in modern industry. As part of this approach, digitisation and computerisation in dentistry are described as Dentistry 4.0 [1,10–13]. Figure 1 is a schematic diagram illustrating the synergistic interaction of the three pillars of the Dentistry 4.0 model, including dentistry, dental engineering and material engineering.

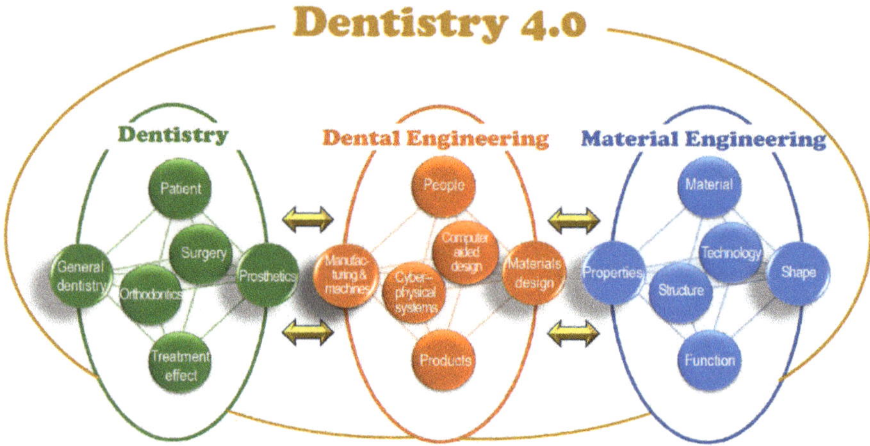

Figure 1. General diagram of the synergistic relationships between dentistry, dental engineering and materials engineering in the Dentistry 4.0 model.

In modern dentistry, as well as in medicine in general, various engineering devices are often used [1,9,12–14]. In many cases, these devices replace the components of the human body removed due to illness or, for example, during an accident and fulfill its natural functions. It applies to, among others, teeth removed due to disease or lost for other reasons, as well as those resulting from malformations. Such medical devices, including dental implants and other prosthetic restorations, are manufactured artificially and are placed under the epithelial surface in full or in part and remain in the body for a long time performing their intended functions [1,9,15–18]. They are made of biomaterials, also called bioengineering materials, and surface layers with such properties are applied to substrates from other, more common materials used on those devices. The most important properties of biomaterials include biocompatibility, and in the case of anticipated contact with blood, also resistance to haemolysis [1,12–14]. A new class of implantable devices are implant-scaffolds. They are many reasons why implant-scaffolds are desirable over other methods in dental implantology. Implant-scaffolds allow the living cells to grow into the pores of the implanted elements, not just the culturing of live cells on their surface [12–14,19–22]. Among the dentistry specialisations in the Dentistry 4.0 model, dental and maxillofacial surgery with implantology and periodontology as well as prosthetics and implant prosthetics are of particular importance. Other specialisations of general dentistry and orthodontics have a slightly smaller relationship with this concept, although they cannot be excluded.

The dental engineering model, as an important manufacturing department, fits into the general Industry 4.0 model. However, minor adjustments were made in the technological plane of the model, as manufacturing technologies and technological machines were combined into one component and engineering design was excluded as a computer-aided design constituting a separate component of this model. The role of computer-aided design/manufacturing (CAD/CAM) in dental engineering has steadily increased in the last decade [1,9,23–27]. Usually, the appropriate prosthetic restoration is individually designed [1,9] using assessments of the condition of patients' teeth damage based on diagnosis using the cone-beam computed tomography (CBCT) method. The authors participated with representatives of other centres in the work and research on the dissemination of this diagnostic method and the development of digitization of dental diagnostics [1,9,28–40]. At the same time, the concept of designing and manufacturing individual dental implants corresponds either to the shape of the root of the removed tooth, or is designed taking into account their final position in the patient's oral cavity, with a representation of the root of the removed tooth and in a standardized shape, so that it is possible to embed a prosthetic restoration. [1,9,12,21,24,28,29]. Thanks to the use of technology combining data obtained from a CBCT scanner, intraoral conditions scan and computer-aided design, it is possible to integrate individual components and produce them using additive technologies, reducing the number of potential bacterial residues, reducing mass and reducing the number of connecting elements [1,9,28,29].

For the computational simulation of implants, prosthetic restorations and their combinations, and prediction of their behaviour in application conditions, the finite element method (FEM) is commonly used [41–44]. Computer-aided design and manufacturing (CAD/CAM) methods are widely applied in the field of additive manufacturing technologies, among which selective laser sintering (SLS) and stereolithography (SLA) have gained the highest importance in dental prosthetics [1,9,45–49]. To assess the suitability of individual technologies in the process of producing materials that can be used in dentistry, the value of technology was analysed using the procedural benchmarking method and the weighted scoring method and dendrological matrix [14,50]. In this way, usefulness of technologies of additive manufacturing (TAM) with coordinates of potential and attractiveness (8.6, 6.6), which place this technology in the quarter of the dendrological matrix oak, has been confirmed. They turn out to be much larger than those corresponding to other technologies, including technologies of powder metallurgy (TPM) (6.5, 4.3), technologies of casting (TC) (4.9, 4.4) and technologies of metallic foam (TMF) manufacturing (3.3, 5.0). The results of foresight studies [51,52] also confirm the very high development prospects of selective laser sintering (SLS) of metal powders and ceramics with very high potential and attractiveness. It justifies the excellent research interest in this issue concerning dental prosthetics. Surface engineering technologies [52,53], such as atomic layers deposition (ALD), and less preferably, physical vapor deposition (PVD), are very modern technologies used in the production of implants and other prosthetic restorations. The particular benefits of using these technologies in dental prosthetics include barrier prevention of the re-diffusion of titanium and other metal atoms to the ceramic surfaces of crowns and bridges and prevention of greying and cracking of the porcelain facing layer. Inside the pores of the selective sintered laser porous skeletal structure, thin ALD layers can be applied to improve the proliferation and growth of living cells inside the pores [21].

Frameworks made of metal alloys, zirconium oxides and aluminium oxides, as well as poly (methyl methacrylate) (PMMA) can be manufactured by milling in computer numerically controlled (CNC) machining centres. Frameworks made of metal alloys are increasingly produced by additive methods. Among metals, titanium and its alloys play the most crucial role in the production of implants, including dental ones and scaffolds. It also belongs to engineering materials that are particularly suitable for use in additive technologies, including (and mainly for) selective laser sintering as a technology for the manufacture of microporous materials, like the aforementioned products and medical devices, and especially when it is necessary to colonize the manufactured pores by natural human cells.

The material intended for implants and prosthetic restorations should be characterized by: a relatively low mass, most preferably as close as possible to the mass of the element it replaces; biocompatibility understood as the lack of an allergic response of the organism to elements made of this material even in a long period of use; and strength higher than bone strength so that it can carry at least the same forces as those transmitted by the tooth/teeth it replaces. In addition, it should have flexible forming possibilities using technology, in particular additive, preferably selective laser sintering or computer numerically controlled milling, to be able to obtain shapes that match the anatomical and physiological features of the patient's mouth the best. In particular, the strength properties of the material used to manufacture implants and prosthetic restorations are essential. Firstly, the material should possess properties higher than a healthly tooth, in particular higher compressive strength and bending strength. It also needs to consider the maximum forces that the bone base can transfer when restoring the patient's teeth. Maximum voluntary bite force (MVBF) or maximum bite force

(MBF) is an essential indication of the functional state of the masticatory organ, which is taken into account when designing implants and prosthetic restorations. The methods and quality of measuring these forces significantly advanced thanks to the development of signal analysis techniques which improve the value of information obtained in this way. However, measurements remain complicated, and care should be taken when analysing the obtained research results in this respect. Regardless of the diversity of research methodology, MVBF depends on many factors. Maximum voluntary bite force (MVBF) is associated with the state of health of the masticatory system and the state of the teeth [54,55]. It also depends on the degree of jaw opening resulting from muscle length [56], on the strength of the muscles closing the jaw and on the pain threshold of the patient [57]. MVBF varies in different areas of the oral cavity, and the largest is in the first molar area [57], although there are significant differences in the first molar area [58–61]. MVBF is higher in adults with a rectangular facial morphology and deep skeletal bite than in people with a long face and open bite [62], although this is not identical with children [63]. Temporomandibular dysfunctions affect this strength in adults [64], as well as children [65]. Obtained test results and their reliability depend on factors, including ethnicity, sex, age, skull and face morphology, occlusal factors, such as the strength of jaw closure, degree of jaw opening, and thus muscle length, pain condition and threshold or pain of the temporomandibular region [57,66] including bruxism [67] and even diet [68]. It has been established that in the Western population the average maximum occlusal force usually occurs between molars and is in the range of 600–750 N [69]. However, it was indicated [67] that for men it can be about 587 N, and for women, it is less (i.e., about 425 N [67]); between incisors it is in the range of 140–200 N [70] and between the canines in the range of 120–350 N [71]. The average MBF strength depends on age, for example, those 16–18 years old with intact teeth averaged 532 N teeth or 516 N when they had small additions in the lateral teeth [54]. It was found that the maximum bite force (MBF) is not affected by oral submucous fibrosis (OSMF) because the mean MBF value is between 628 and 635 N [72]. It is also worth noting that the forces transferred by prosthetic restorations can be much higher than the forces transferred by the natural dentition of the same patient. Regardless of the number of dental implants, they increase MBF in toothless patients by an average of 64 ± 10 years, with malocclusion being higher in men, while the patient's age and type of implantation do not play a significant role [73]. In the case of all-ceramic restorations, they range between about 84 N to 1643 N, with a statistical average of approximately 430 N [74] Analogous measurements in the case of fixed partial dentures showed that the mean MBF was 596.2 ± 76.3 N at the dentate side and 580.9 ± 74.3 N at the fixed partial denture side [75]. This is because prosthetic reconstruction cannot be innervated, and nerve signals are received by the central nervous system only by the periosteum. For this reason, it is crucial to ensure the highest possible strength properties of prosthetic components. In addition, the complicated structure of both implants and prosthetic elements often necessitates the use of items with diameters below 4 mm, which means that the individual walls of those elements are a minimum of 0.3 mm long. Such technological conditions also require the use of materials having the highest possible strength properties. It is also crucial that both implants and prosthetic restorations are used in an aggressive oral environment, which also requires the use of materials resistant to it. It is preferable to use the same material to manufacture the implant, implant elements and prosthetic restorations in order to avoid the formation of galvanic cells at the joints of individual items [76], which promote the formation of inflammation and significantly reduce the life of the entire prosthetic restoration in extreme cases. Titanium and its alloys, in particular Ti6Al4V alloy, fits well with the requirements described above.

Titanium alloys with Al, V, Nb and Ta, in particular the alloy Ti6Al4V, are among the materials that meet the indicated strength requirements for use in dentistry, as well as treatment of bone fractures. Titanium and its alloys are well tolerated by the human body and increasingly used in medicine, both in prosthetics (e.g., prostheses and various implants, including intramedullary wires), as well as in dentistry and dental prosthetics. Using the above-mentioned additive technologies and materials, manufacturing of implants, abutments, crown-root inlays, bridges, crowns and skeletal dentures is possible. The biological and physicochemical properties of titanium and its alloys caused a significant breakthrough in biomedicine. It shows prosthetic materials' thermal conductivity, high hardness, mechanical strength and durability several times lower than traditional materials. In addition, it does not cause allergic reactions and is resistant to corrosion [77]. Since titanium is not ferromagnetic, patients with titanium implants can be safely examined by magnetic resonance imaging (MRI). When preparing titanium for implantation, it should be cleaned in a plasma stream, which, after the process undergoes immediate oxidation [77].

Titanium and its alloys belong to metals and alloys of relatively low density, for which additive technologies, especially selective laser sintering, have been successfully used for prosthetic restorations used in dentistry [14]. Porous Ti can also be used on non-biodegradable diving scaffolds and implant-scaffolds, including after surface treatment of pores [14]; it is primarily used because of its relatively high compressive strength and fatigue strength. A significant disadvantage is the reaction of porcelain with titanium oxide, causing bruising and darkening of the colour. However, this does not exclude the use of titanium and its alloys for the production of implants, also with integrated abutments [31]. For aesthetic reasons, it could practically eliminate such prosthetic restorations, if not for coating titanium and its alloys with thin nanostructured surface layers, most preferably in ALD processes [21,52,53]. Titanium-based alloys are corrosion resistant, which after implantation into the body do not show an allergic reaction. They are characterized by high strength and hardness and possess a thermal conductivity several times lower than traditional prosthetic materials. Titanium is a very thrombogenic material, and alloying elements added to titanium alloys improve the overall thrombotic compatibility and biocompatibility [77]. It was found that if V presents in pure form, it can be considered a potentially toxic element [78,79]. Vanadium ions bind to proteins released from its surface only when its concentration is high in the soft tissues of the body. This reaction leads to an intensification of the adverse immune response [80,81]. There are reports that V can cause aseptic abscesses, and Al can cause scars, while Ti, Zr, Nb and Ta show excellent biocompatibility [82]. Some publications [83] contain limited information on the toxic activity of V as an alloying element in the Ti6Al4V alloy. Some cellular studies point to information about the potential for cytotoxicity of the Ti6Al4V alloy [82,84–86]. It does not seem to matter if vanadium is an additive in alloys of other metals, for example, titanium. Regardless, in those reports, alloys with a dominant share of these elements or pure components are mentioned. Some publications present research results that indicate that the adverse effects of V on titanium alloys can be eliminated by replacing this element with Nb. The use of Ti24Nb4Zr8Sn alloy or other alloys above 7% Nb may be more beneficial, because they can use selective laser sintering technologies and their elastic modulus is more similar to the bone than Ti6Al4V alloy. This, in turn, prevents the implant from loosening in the body because it prevents bone resorption [87]. Although the porous Ti6Al4V was thoroughly investigated, the potential release of toxic ions led to the search for alternative Ti alloys; in addition to the Ti24Nb4Zr8Sn alloy, these include Ti7.5Mo and Ti40Nb alloys with comparable mechanical properties as their traditionally produced counterparts [87–90]. Instead of the Ti6Al4V alloy, a Ti6Al7Nb alloy with supposedly better bioavailability and better corrosion resistance can be used [77,91–93]. The differences between the properties of both alloys in a direct comparison under the same test conditions were not too significant [91], and it was even shown that the Ti6Al4V alloy showed better properties than the Ti6Al7Nb alloy [91,94–96] in the form of higher antibacterial activity and resistance to Gram-positive bacteria and thrombotic compatibility, although the opposite is true for Gram-negative bacteria [91]. It is also essential that both implants and prosthetic restorations are used in an aggressive oral environment.

Foci of corrosion, especially electrochemical corrosion associated with the formation of galvanic cells at the joints of individual elements [75], encourage the formation of inflammation and in extreme cases significantly reduce the life of the entire prosthetic restoration. Prevention of electrochemical corrosion requires the use of identical alloys for one patient, including pure titanium or its alloys for various components, including dental implants and dental prosthetic restorations. It is worth noting that due to the avoidance of sources of electrochemical corrosion, the use of pure titanium or its single-phase alloys, which are always more advantageous than multi-phase alloys due to the different electrochemical potential, different phases may locally develop corrosion [75]. Ti6Al4V alloy, especially after selective laser sintering, has a single-phase α/α' supersaturated structure [14,21,22]. This alloy is approved for medical purposes in the European Union, the USA and many other countries [14]. It is the widespread opinion among dentists that this alloy, which is known as titanium Grade 5, can be used for implants and prosthetic restorations. Despite the numerous reports of the possibility of using this alloy in dentistry, there is relatively little research evidence for the objective characterization of this alloy, and especially selectively laser sintered.

This paper presents the results of comparative studies of the structure and properties of the Ti6Al4V alloy used for dental implants and dental prosthetic restorations made alternatively using the subtractive manufacturing methods by milling from cast discs on computer numerically controlled (CNC) milling centres and the additive selective laser sintering (SLS) method for manufacturing solid elements, using techniques of computer-aided design/manufacturing (CAD/CAM). Additionally, an original technology of component manufacturing used in dental prosthetics and implantology from Ti6Al4V alloy powders through selective laser sintering (SLS) was developed.

2. Material for Research

The tests were carried out using a Ti6Al4V alloy in the form of cast discs with a diameter of 98.3 mm and height of 10 to 16 mm intended for the production of dental prosthetic restorations and powder used for the production of additive SLS. Table 1 gives the chemical composition of the materials used for testing in accordance with the manufacturer's certificates. Figure 2 shows the structure of the Ti6Al4V cast disc. For selective laser sintering, powders of spherical shape (Figure 3) and chemical composition confirmed by spectral tests using the energy dispersive spectrometry (EDS) method (Figure 4) were used. The powder has a particle diameter of 15 to 45 μm, and Figure 4 shows a cumulative screening curve and average particle size.

Table 1. Chemical composition of tested materials produced by computer-aided design/manufacturing (CAD/CAM) method by milling solid discs and by selective laser sintering from powders.

Ti6Al4V	Elements Mass Concentration (%)									
	Al	V	C	Fe	O	N	H	Other Together	Other Each	Ti
Disc	6.2	4.0						≤1.0		89.4
Powder	6.35	4.0	0.01	0.2	0.15	0.02	0.003	≤0.4	≤0.1	rest

The cast discs were milled at the FANUC CNC centre (Robodril S). The powders were subjected to selective laser sintering usually using a liquid phase sintering. The appropriate product models of the assumed shape and dimensions were designed virtually using the Power Shape Premium 2020 included in Autodesk computer software. Then the model was divided into 25 μm thick layers, the assumed number of which corresponds to the actual number of powder layers during manufacturing of the product, subjected to laser sintering. At the same time, the production conditions were selected programmatically from the laser power at a given spot diameter, laser path width, laser scanning speed and powder layer thickness to the parameter of overlapping adjacent laser paths in %. Selective laser sintering of the powder was performed in an Orleas Creator device (Table 2) in an argon atmosphere as an inert gas. Twelve variants of the sample production conditions are shown in Table 3. The workpieces are arranged in a software programme on the working platform (Figure 5).

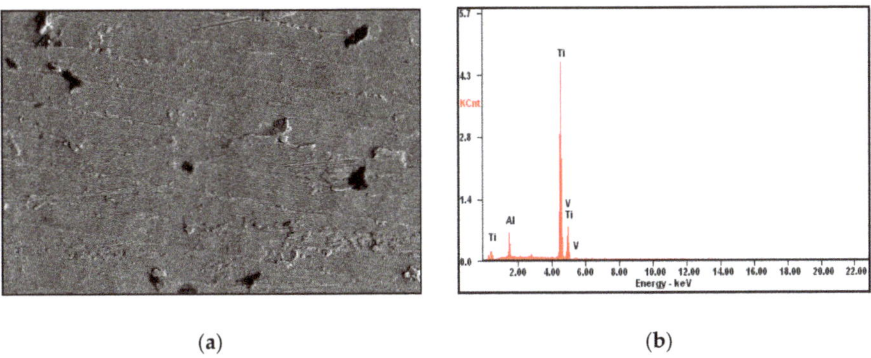

(**a**)   (**b**)

Figure 2. (**a**) Surface structure of the Ti6Al4V disc (scanning electron microscopy, SEM); (**b**) volume share of elements on the surface from (**a**).

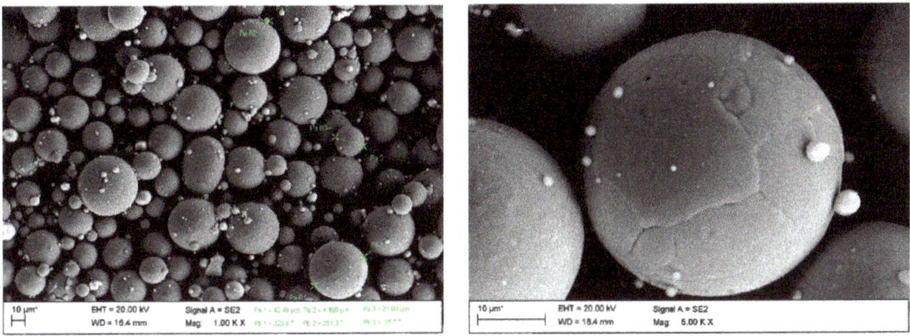

Figure 3. Ti6Al4V powder structure (SEM).

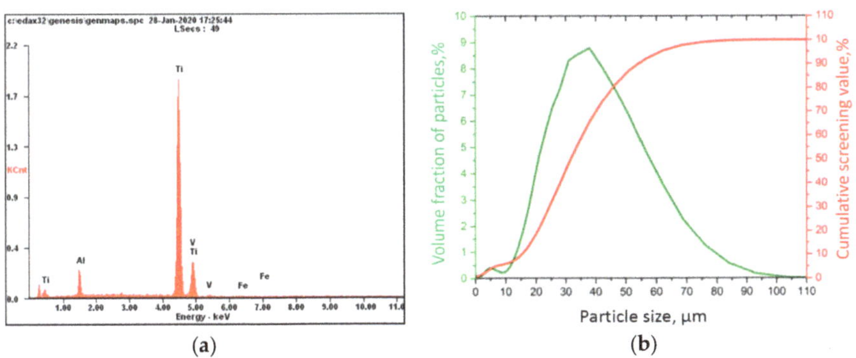

(**a**)   (**b**)

Figure 4. Test results of Ti6Al4V powder. (**a**) Volume share of elements on the particle surface from Figure 2b and (**b**) volume share and cumulative screening value depending on powder particle size.

Table 2. Operation characteristics of the Orleas Creator device.

Characteristic Values	Range of Device Operation
working space, mm	100×200
laser power, W	250 with 10%–100% power regulation
modulation frequency, Hz	50–60
laser dot diameter, μm	30–150
working environment	protective gas atmosphere: argon
powder layer thickness adjustment, μm	20–50
oxygen regulation, ppm	from 100
scanning speed, mm/s	to 10,000 mm/s
laser path width, μm	up to 200
powder characteristics	up to 45 μm, spherical and atomized powders

Table 3. Variants of conditions for preparing specimens from Ti6Al4V alloy by selective laser sintering (SLS) method.

Variant No.	Specimen Manufacturing Conditions					
	Laser Power, W	Laser Dot Diameter, μm	Scanning Speed, mm/s	Laser Path Width, μm	Allowance, %	Powder Layer Thickness, μm
1	80			80		
2	60			40		
3	100			100		
4	110			120		
5	80			120		
6	110	40	650	60	10	25
7	60			60		
8	60			120		
9	80			60		
10	100			80		
11	100			120		
12	110			80		

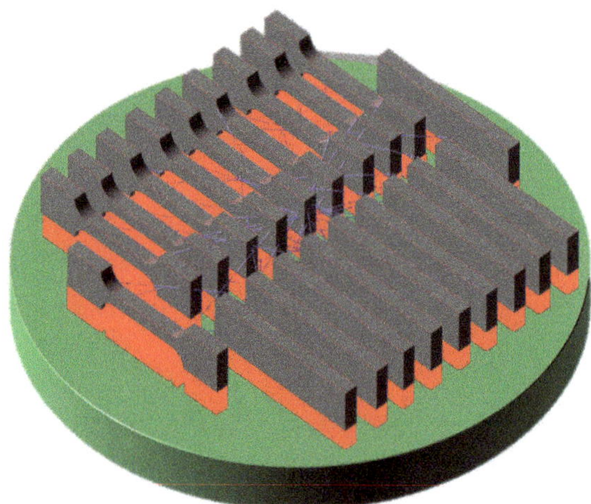

Figure 5. Example of product placement on the working platform (supports marked as red).

3. Course of Research

Specimens for metallographic, structural and physicochemical tests were in the shape of a cuboid with dimensions of $3 \times 10 \times 20$ mm. The purpose of metallographic and structural tests is to determine the impact of materials' technology and manufacturing conditions of selective laser sintering on the examined alloys' structure. In order to prepare, a line for automated grinding and polishing of Struers specimens was used, initially on an automatic grinder-polisher Tegramin-25 using SiC abrasive papers with a grain size of 120–1200 μm/mm^2, and then on polishing cloths using polycrystalline diamond suspensions with granulation of 9, 6, and 1 μm. Finishing polishing was carried out in a colloidal silica slurry with a grain size of 0.04 μm OP-S oxide.

The tests were carried out using the Nikon SMZ1270i stereoscopic and LeicaDMi8A light microscope equipped with computer image analysis systems. The Zeiss Supra 35 scanning electron microscope (SEM) was also used with the application of secondary electron (SE) detection at an accelerating voltage of 5 to 20 kV and a maximum magnification of 50,000 times. The qualitative and quantitative analysis of the chemical composition of the examined materials, including powders, was made using the EDS scattered X-ray energy spectroscopy method of the EDAX company in a scanning electron microscope (SEM).

The X'Pert PRO X-ray diffractometer from PANalytical in the Bragg–Brentano system was used for qualitative analysis of the phase composition of the tested materials using a θ/θ goniometer as well as a copper and cobalt lamp with filtered X-ray Kα1 with wavelength λ = 0.154056 nm and λ = 0.1789010 nm, at 40 kV voltage and 30 mA filament current. The angular range of the reflected radiation intensity was 5°–120° every 0.05°, and the pulse counting time was 10 s. Phase identification was performed using X'Pert Pro software with the coupled International Centre for Diffraction Data (ICCD) powder diffraction file (PDF)-4+ 2012 diffraction database.

Tests were carried out on the as-manufactured condition as well as after mechanical properties tests. which were carried out on a static universal testing machine ZwickRoell Z020 in standardized conditions for static tests with the use of individually designed micro specimens that meet internally applicable standards for testing materials produced by additive methods. Micro specimens were manufactured by milling from discs and laser sintering from powders. The specimens for tensile strength tests had the typical shape of oars with dimensions of 40 × 10 mm which were used in this test. The measuring part of the sample was 3 × 3 × 15 mm. Bending strength tests were carried out on flat micro specimens with dimensions of 35 × 10 × 3 mm. All tests were performed each time for a series of five micro specimens. Figure 6 shows the shape and dimensions of the micro specimens.

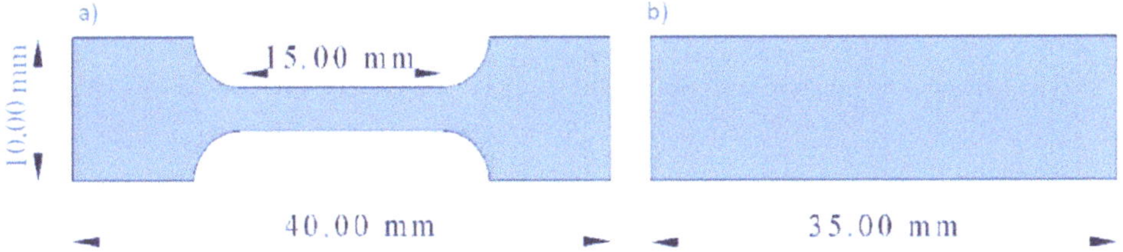

Figure 6. Dimensioned CAD model of solid specimens for strength tests: (**a**) tensile; (**b**) on bending.

The characteristic quantities, namely, tensile strength R_m, yield strength R_e and bending strength R_g were calculated in the following way applying the formulas given below.

During the tensile strength test, the tensile force was measured as a function of specimen elongation. The tensile strength R_m corresponds to the normal stress calculated as the ratio of the largest tensile force F_m to the area of the initial cross-section of the specimen S_0.

$$R_m = \frac{F_m}{S_0} \left[\frac{N}{mm^2} = MPa \right]$$

$$S_0 = a{\cdot}b \left[mm^2 \right]$$

where:

F_m—maximal tensile force;

S_0—cross-sectional area of the specimen;
R_m—tensile strength;
a—sample width.

The yield stress R_e is the value of the tensile stress in the specimen when it is reaches a clear increase in the elongation of the specimen at the constant tensile force F_e.

$$R_e = \frac{F_e}{S_0}\left[\frac{N}{mm^2} = MPa\right]$$

where:

F_e—force defining yield point.

Bending strength tests were carried out using the three-point bending method (where the force is applied to the centre of the specimen) with a preliminary force of 1 N, module speed E 1 mm/min and test speed 1 mm/min. Based on the results obtained, the bending strength was calculated as the bending strength R_g which is the basic value characterising the bending stress of the specimen and was calculated as the quotient of the bending moment M_g and the section index W_g according to the formula:

$$R_g = \frac{M_g}{W_g},$$

When the force is applied to the centre of the specimen, the bending moment was calculated according to the formula:

$$M_g = \frac{P_{kr} \cdot l_0}{4}$$

where:

P_{kr}—force applied in the center of the specimen;
l_0—distance between supports (30 mm).

While the section index for a specimen of rectangular section with width b and height h was calculated on the basis of the formula:

$$W_g = \frac{b \cdot h^2}{6}$$

where:

W_g—bending strength index for rectangular cross-section specimens.

Using the above formulas, the bending strength was calculated according to the formula:

$$R_g = \frac{P_{kr} \cdot l_0}{4 \cdot W_g}\left[\frac{N \cdot mm}{mm^3} = \frac{N}{mm^2} = MPa\right].$$

Tribological tests were carried out on samples of selected materials selectively laser sintered on both the lateral surface and the frontal surface on which the laser beam acted. The tests were performed using the ball-on-disc method, as a counter specimen using a tungsten carbide ball with a diameter of 6 mm. The atmosphere of Ringer's solution was used at 100% humidity and 25 °C. The linear speed was 15 cm/s and the measuring distance was 100 m.

Tests of corrosion resistance of the Ti6Al4V alloy were carried out on specimens gridded on paper with granulation up to 500, and then degreased in acetone. A potentiostat and a glass electrolyte cell were used for corrosion tests.

The measuring system consisted of three electrodes:

- working electrode—tested material;
- chlor-silver chloride reference electrode with potential 207 mV (at 25 °C);
- auxiliary electrode—platinum wire.

Corrosion tests were carried out in two stages:

- determination of open circuit potential (E_{ocp}) for 1 h;
- recording of anode polarization curves from the potential of $E_{start} = E_{ocp}$ - 100 mV until the potential reaches 2 V or reaches a current of 1 mA/cm^2, with a potential increase of 1 mV/s, followed by registration of the return curve in the E_{start} potential.

The results were recorded as curves. Then, using the Tafel method, the characteristic electrochemical values of materials were determined:

- i_{kor}—anode current density;
- E_{kor}—corrosion potential;
- R_{pol}—polarizing resistance.

The tests were carried out at an ambient temperature in various aquatic solutions given in Table 4:

- saline solution (artificial saliva prepared according to Fusayama's recipe);
- Ringer's solution with the addition of 30% H_2O_2;
- Ringer's solution with 1% concentrated acetic acid;
- Tyrod's solution.

Table 4. Composition of solutions for corrosion resistance tests.

Solution Component	Physiological Saline Solution (Artificial Saliva)	Ringer's Solution	Tyrod's Solution
NaCl	0.4 g/L	8.6 g/L	8.0 g/L
KCl	0.4 g/L	0.3 g/L	0.2 g/L
$NaH_2PO_4 \cdot H_2O$	0.69 g/L	-	0.05 g/L
$CaCl_2 \cdot H_2O$	0.79 g/L	0.243 g/L	0.2 g/L
Urea	1.0 g/L	-	-
Distilled water	1.0 L	1.0 L	1.0 L

The biological research was realised using human osteoblast cells from the hFOB 1.19 (Human ATCC-CRL-11372) culture line according to the authors' own work [21]. Cell culture was made in a six-well polystyrene standard plate from Thermo Scientific Nunc Cell Culture Plastics (Roskilde, Denmark). The polystyrene bottles were incubated at 33.5 °C in an atmosphere containing 95% air and 5% CO_2. The cells were etched using trypsin for their separation from the surface. The medium consisted of one part Dulbecco's modified Eagle's medium (DMEM) without phenol red and one part Hanks F12 medium. The medium was mixed with L-glutamine (2.5 mM), sodium pyruvate (0.5 mM), gentamicin sulfate (0.3 mg/mL), foetal bovine serum FBS (10%) and 4-(2-hydroxyethyl)-1-piperazineethanesulfonic acid (HEPES, 15 mM). The obtained cells were rinsed and mixed with the medium receiving working suspensions with 2.5 × 105 osteoblasts/mL. Then, 3 mL of the cell working suspension were added to one well with a flat bottom plate. Plates with cells and test materials were incubated for 72 h controlling cells growth after 24, 48 and 72 h. After 72 h of culture, 300 µL of MTT (3-(4.5-dimethylthiazol-2-yl)-2.5-diphenyltera-zolium bromide) solution (5 mg/mL) were added to the medium and incubated for 4 h. NAD(P)H-dependent cellular oxidoreductase enzymes reflected the number of viable cells present. Those enzymes reduced the tetrazolium dye MTT (3-(4.5-dimethylthiazol-2-yl)-2.5-diphenyltera-zolium bromide) to its insoluble formazan with a purple colour. The formazan was dissolved in dimethyl sulfoxide (DMSO) added to the well in a volume of 3 mL after removing culture medium. The flasks were shaken for 10 min. The portion of 150 µL for absorbance measuring was placed in the wells with a flat bottom of a 96-well microplate made of polypropylene. During the culture, photos of the bottoms of the plate at the edge of the material with the cells examined were made in the inverted microscope Olympus IX51 (Tokyo, Japan) with ColorView II Soft Imaging System Cell F (Muenster, Germany) using a magnification of 100×. Absorbance was measured at a wavelength of 550 nm using the Eon High-Performance Microplate Spectrophotometer

(BioTek Instruments, Inc., Winooski, VT, USA). Eon's monochromator-based optics allowed for flexible, filter-free 200–999 nm wavelength selection in 1 nm increments. Eon was aided by Gen5 2.0 Data Analysis Software.

From the obtained absorbance results, the value obtained for pure DMSO was subtracted and the percentage of cell viability was calculated for cells cultured without the materials tested using the following formula:

$$\% \text{ viability} = \text{absorbance of test specimen} * 100\%/\text{absorbance of control specimen}$$

The results were analysed statistically on the basis of eight specimens in each case. In each case, the standard deviation was calculated, Snedecor's F-test and Student's *t*-test were used, and the significances of differences between means were determined.

4. Test Results of Mechanical Properties Solid Laser-Sintered or Machined Ti6Al4V Alloy

The results of tests on the mechanical properties of solid laser sintered Ti6Al4V alloy, taking into account the various manufacturing conditions, are shown in Figure 7. Figure 7a illustrates the full course of the tensile curves in full range, while Figure 7b shows the initial fragment of these charts at a significant magnification.

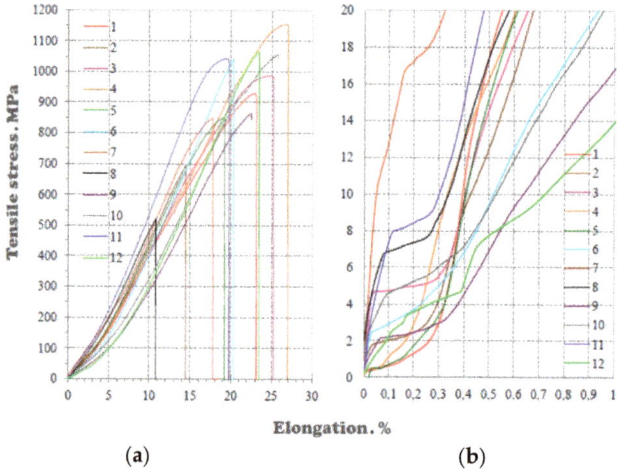

Figure 7. Charts of tensile stress versus elongation for individual micro specimens manufactured under different conditions. (**a**) Full course of tensile curves. (**b**) Initial fragment of these charts at a significant magnification.

The laser power that was used to manufacture the specimens had the highest impact on increasing the tensile strength R_m. The higher the laser power used, the higher the tensile strength results. The most significant increase in tensile strength occurs when using a 120 μm laser beam width, where an increase in power from 60 to 110 W caused more than a two-fold increase in tensile strength (Figure 7a).

For a laser beam width of 80 μm, the increase in laser power from 80 to 110 W no longer gives such a spectacular increase in tensile strength, because it is about 150 MPa (Figure 8a). Changing the laser path width with the same laser power no longer significantly affects the tensile strength. However, with the right laser beam width for each laser power it is possible to achieve the maximum tensile strength for given production conditions from 830 MPa for 60 W to over 1150 MPa for 110 W, which corresponds to an increase in tensile strength by approx. 350 MPa (Figure 8b). Table 5 summarizes the statistically averaged results of the tensile strength test depending on different manufacturing conditions, as well as compares them with the results of the bending strength test. The confidence interval in each case is not higher than 5%. The table shows the blue variant solid Ti 4 providing the highest tensile and bending strength values, while the red solid Ti 8 variant providing the lowest amounts of those strengths is given in red.

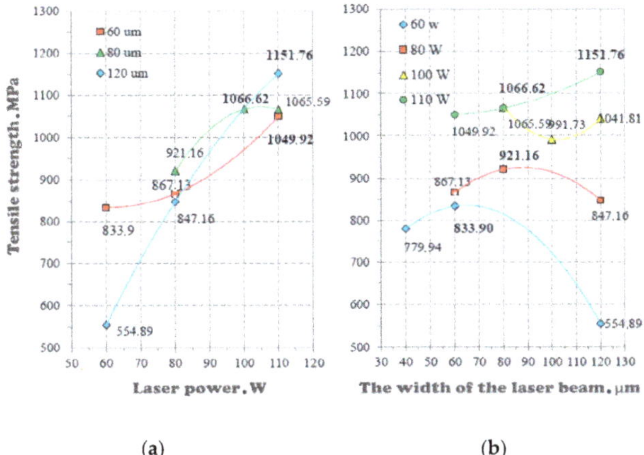

Figure 8. Relationship of tensile strength (**a**) on the laser power for different widths of the laser beam and (**b**) on the width of the laser beam for different values of the laser power.

Table 5. Summary of the results of tensile and bending strength tests of the Ti6Al4V alloy manufactured by the selective laser sintering (SLS) method.

Variant No.	Laser Power, W	Laser Beam Width, μm	Average R_m, MPa	Position for Tensile	Average R_g, MPa	Position for Bending
1	80	80	921.16	7	2194.70	4
2	60	40	779.94	11	1486.50	11
3	100	100	991.73	6	2337.22	2
4	110	120	1151.76	1	2464.03	1
5	80	120	847.16	9	1595.70	9
6	110	60	1049.92	4	2049.74	5
7	60	60	833.90	10	1546.45	10
8	60	120	554.89	12	1099.88	12
9	80	60	867.13	8	1625.21	8
10	100	80	1066.62	2	2001.85	6
11	100	120	1041.81	5	2293.33	3
12	110	80	1065.59	3	1926.89	7

The value of bending strength of Ti6Al4V micro specimens manufactured at different laser path widths of 40, 60, 80 and 120 μm is strongly dependent on the power of the laser acting on the powder (Figures 9 and 10). Figure 10a presents graphs of the dependence of bending strength on the laser power of specimens manufactured at different laser path widths of 60, 80 and 120 μm. At 60 W laser power, the bending strengths obtained are the lowest compared to other laser powers. Bending strength above 2000 MPa can be purchased for the remaining laser powers.

Increasing laser power is the most crucial factor in improving bending strength. For a laser beam width of 120 μm, increasing the laser power from 60 to 110 W increases the strength by almost 1400 MPa. Another important factor influencing the improvement of bending strength is the width of the laser beam. Within the same laser power, it is possible to select the width of the laser beam, enabling an improvement in bending strength in the range of 336–599 MPa (Figure 10b).

With a laser power of 60 W and a beam width of 120 µm, the energy emitted by the laser is not sufficient to properly sinter the powder on which the laser acts; therefore, under these conditions, the bending strength of the manufactured specimen is the lowest. At the same laser power of 60 W, narrowing the laser beam width to 60 µm generates energy with a higher energy density, which consequently increases the degree of powder remelting and increases bending strength. The use of a smaller width of the laser beam to the operating energy at a given power also means that the strength of the specimen decreases, however, not as much as when using a larger width of the laser path. For each of the selected laser powers, other manufacturing conditions (mainly the laser beam width) can be chosen in such a way to obtain a sufficiently high bending strength.

Taking into account the different conditions for producing solid specimens from Ti6Al4V alloy by SLS, it is possible to select the required set of strength properties. By changing the diameter of the laser spot, the width of the laser path, the allowance associated with the overlap of individual laser paths, and above all the laser power, it is possible to obtain a material with almost twice the strength compared to material manufactured under other conditions. These results were compared with the tensile and bending strength of the Ti6Al4V alloy machined at the FANUC CNC milling centre (Robodrill S), because such a technology is still most commonly used in implant prosthetics for the production of dental implants and dental abutments. Tensile strength values of 858 MPa were obtained for the Ti6Al4V casting alloy and 1959 MPa for bending. In the case of tensile strength, it is 300 MPa less than in the case of a selective laser sintered alloy, while in the case of bending strength it is even smaller by more than 500 MPa. Using SLS technologies and ensuring appropriate manufacturing conditions, solid material with significantly better strength properties than the casting alloy can be obtained (Figure 11). Although it dealt with the same material only manufactured differently, the characteristics of both the tensile and bending curves are definitely different. The material manufactured conventionally (i.e., casting) and processed on a CNC milling machine has the characteristics of plastic material; this type of bending and the tensile curve are characteristic for materials with plastic properties. In contrast, sintered material exhibits the properties of brittle materials. The bending stress for the sintered material in the first phase of the test increases strongly with a small deflection to drop sharply when the maximum value is reached; in the case of the tensile curve when the maximum amount is reached, the specimen breaks and the force drops to zero. In the case of plastic materials, when a certain force is reached, the force increases rapidly with a small degree of deformation of the specimen. There is a small increase in force with a large increase in deflection in the case of bending and elongation in the case of tensile.

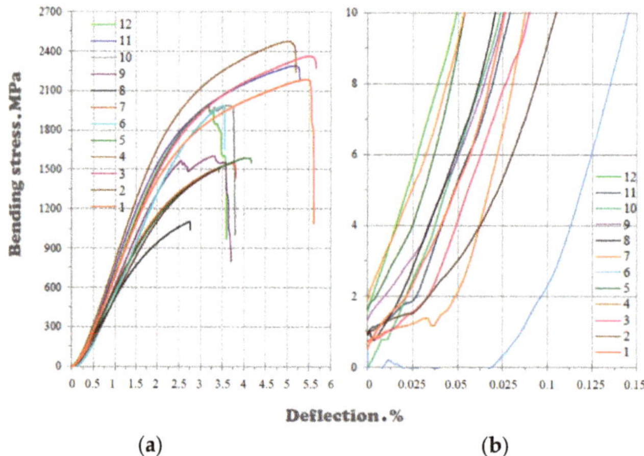

Figure 9. Charts of bending stress versus deflection for individual micro specimens manufactured in different conditions. (**a**) Full course of bending curves and (**b**) initial fragment of these charts at a significant magnification.

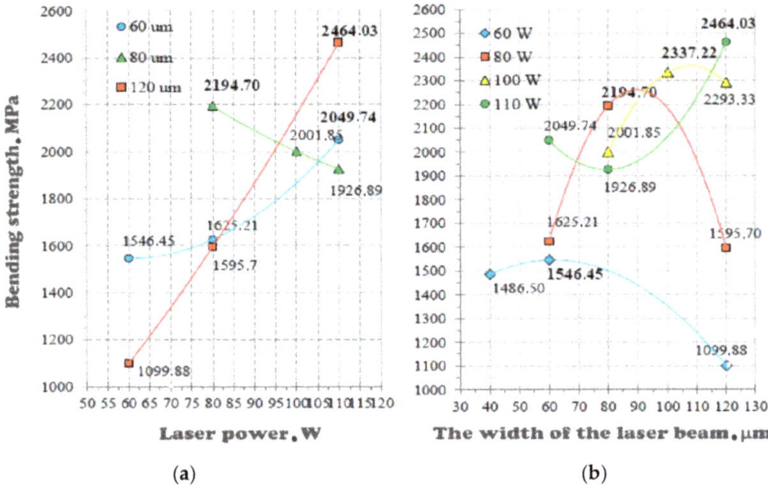

Figure 10. Relationship of bending strength (**a**) on the laser power for different widths of the laser beam (**b**) on the width of the laser beam for different values of the laser.

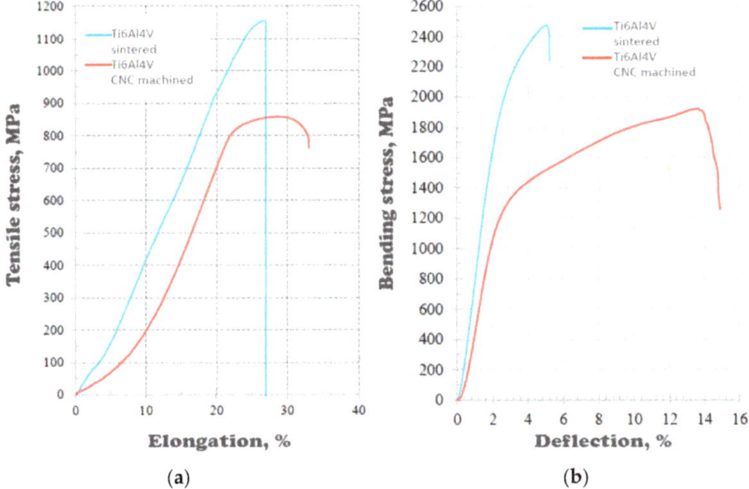

Figure 11. Comparison of tensile stress (**a**) and bending stress (**b**) for micro specimens of the Ti6Al4V alloy manufactured by SLS and the Ti6Al4V casting alloy produced by milling in a computer numerically controlled (CNC) centre.

5. The Structure of the Ti6Al4V Alloy Machined at the CNC Centre and Manufactured by the Selective Laser Sintering Method

Figure 12 shows the morphology of fractures observed in the scanning electron microscope of the Ti6Al4V casting alloy in the form of cast discs intended for milling in the CNC milling centre and sintered alloy using the SLS method after static bending tests. A breakthrough in casting alloy is plastic and sintered alloy brittle.

Furthermore, the fracture analysis does not indicate significant pores or gas bubbles resulting from the casting process. It can, therefore, be concluded that the share of pores in casting alloys is relatively small.

(a)　　　　　　　　　　　　　　　　(b)

Figure 12. Comparison of fracture morphology after static bending tests. (**a**) Ti6Al4V casting alloy after milling in a CNC milling centre. (**b**) Sintered alloy using the SLS method (Ti solid 4 variant) (SEM).

For comparison, the share of pores in the casting material was also examined (Figure 13) and compared with a selective laser sintered alloy (Figure 14). The percentage of pores on the surface of the casting alloy is 0.03%. This type of material is considered solid. Therefore, it can be stated that by using SLS technology, it is possible to manufacture solid materials that are qualitatively comparable to those manufactured by casting technologies.

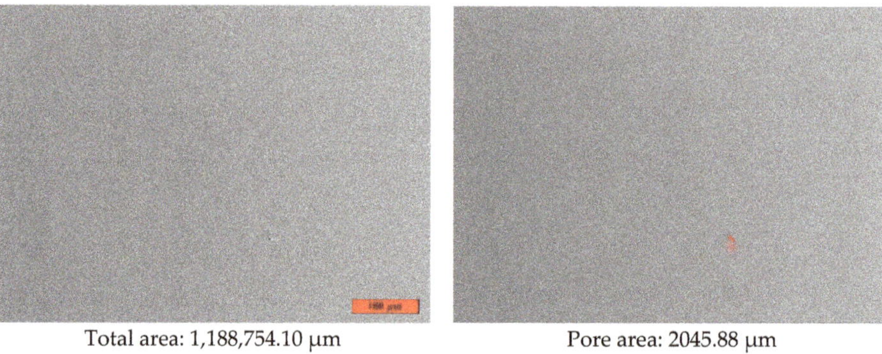

Total area: 1,188,754.10 μm　　　　　　Pore area: 2045.88 μm
Solid surface share: 99.97%　　　　　　　Pore share: 0.03%

Figure 13. Results of measuring the share of pores for specimens of the Ti6Al4V casting alloy after milling in a CNC milling centre.

Metallographic tests on non-etched specimens to assess roughness were carried out by quantitative metallographic methods using Leica's DMI8A automated light microscope software. The surface share of pores on non-etched metallographic specimens at 100× magnification corresponds to the percentage of the total area occupied by the dark spots that represent the free spaces inside the material. Performing such tests, it can be stated that by using appropriately selected production conditions by the SLS method, one can obtain a material with excellent strength properties but also with inferior properties. Figure 15 gives examples of porosity assessment in selectively sintered laser specimens according to the options given in Table 3.

The porosity of the specimen with the most strength is in the range of 0.1%, and that of the least strength is over 10%. Increasing the porosity to more than 10% caused over a two-fold decrease in both tensile and bending strength, and the surface of the specimen has evenly spaced, large-sized pores. In the case of the Ti solid 4 specimen, the porosity is very low (Figure 14).

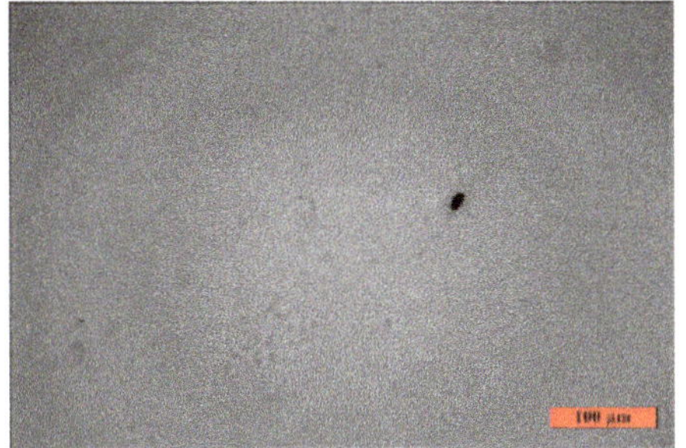

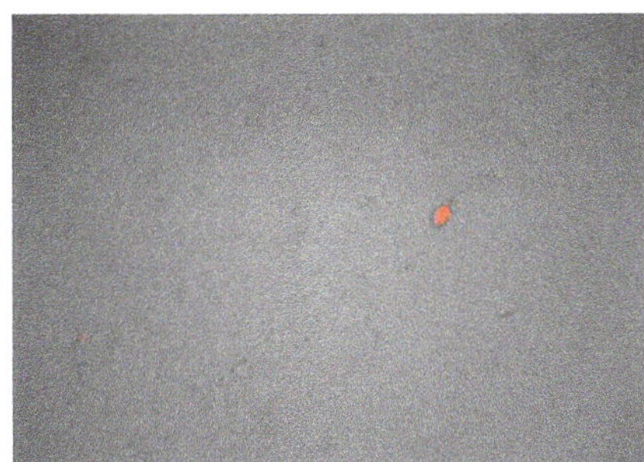

Total area: 1,188,754.10 μm   Pore area: 4093.40 μm

Solid surface share: 99.94%   Pore share: 0.06%

Ti solid 4

Figure 14. Results of measuring the share of pores for specimens of the Ti6Al4V alloy sintered by the SLS method (Ti solid 4 variant).

In the image of the surface, pores appear in a minimal amount, and the surface is almost uniformly smooth (pores in the material occur sporadically and are low). Increasing the pores' density to 0.2% of the surface area for the Ti solid 1 sample reduces the bending strength by about 250 MPa; the pores are still very small, but evenly distributed in the material. The surface of the material is no longer as homogeneous as in the case of the Ti solid 4 specimen. Along with the decrease in the strength of the specimens, much larger pores begin to appear in the material. Still, their share in relation to the volume of the entire specimen is small (Figure 15), and the surface of the specimen is relatively uniform with small places of larger pores.

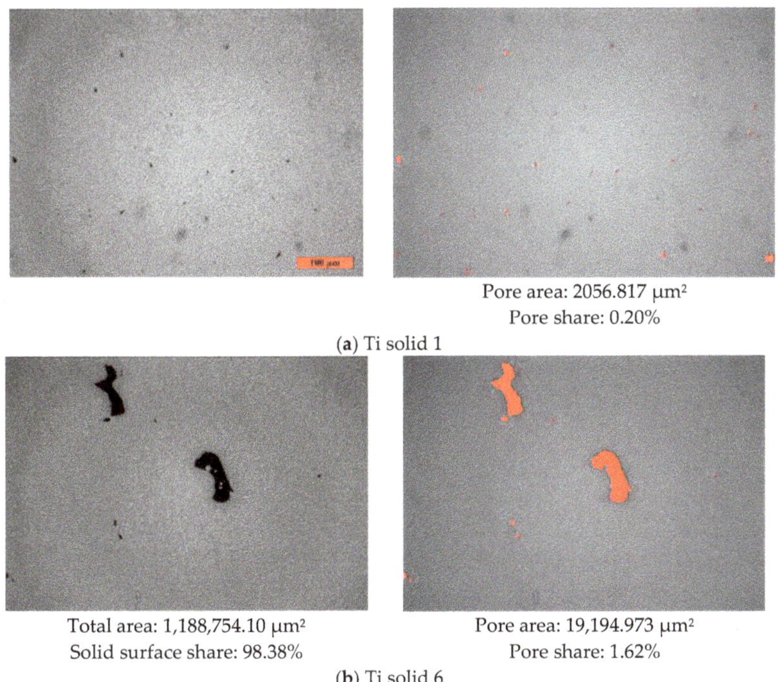

Pore area: 2056.817 μm²

Pore share: 0.20%

(a) Ti solid 1

Total area: 1,188,754.10 μm²   Pore area: 19,194.973 μm²

Solid surface share: 98.38%   Pore share: 1.62%

(b) Ti solid 6

Figure 15. *Cont.*

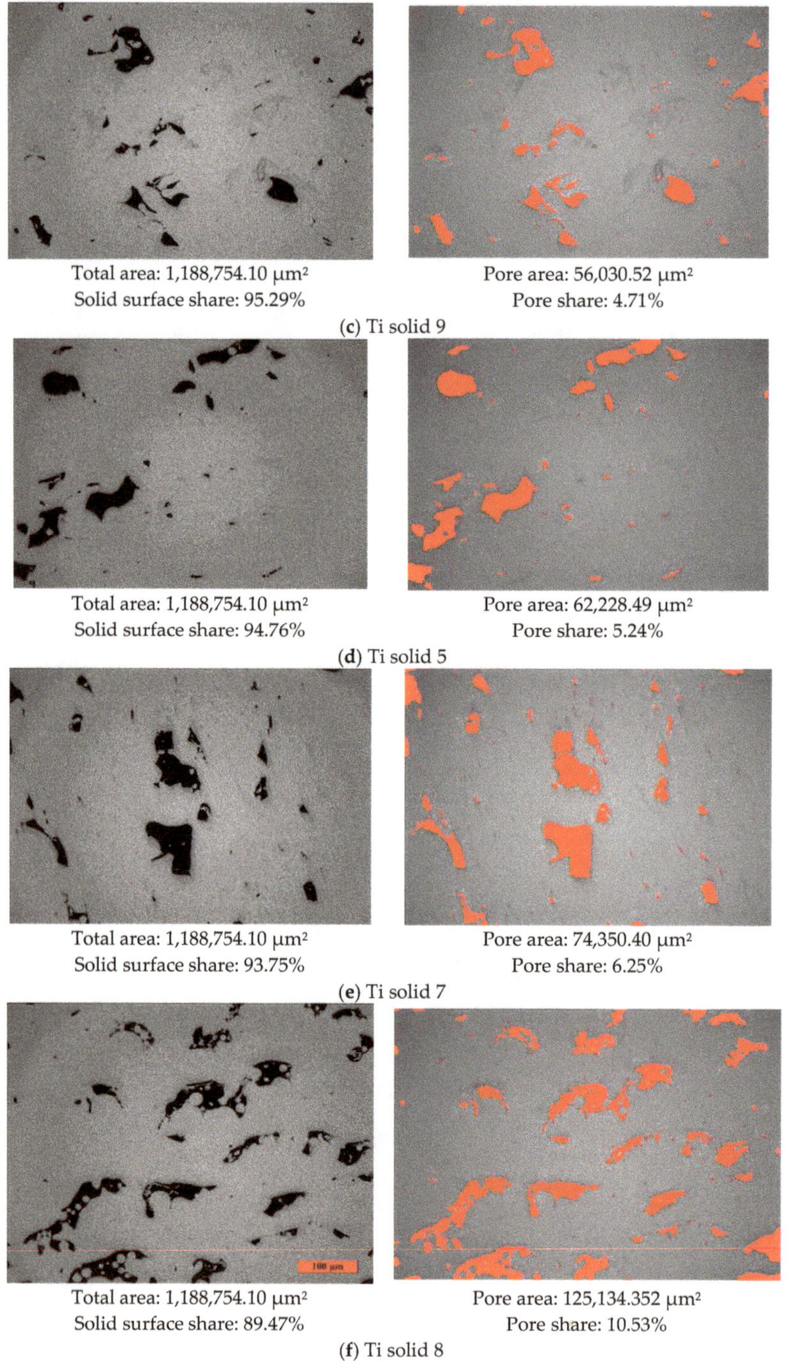

Figure 15. The share of pores in solid specimens of Ti6Al4V alloy manufactured by SLS method in various conditions.

The appearance of such pores in the material and increasing their share relative to the solid surface to a value of about 1.5% gives a reduction in strength relative to the specimen Ti solid 4 by almost 450 MPa. The measurement of porosity was made for subsequent specimens and Ti solid 5, 7, and 9 showed bending strength at a similar level; the difference in strength between these three specimens was only 80 MPa. In addition, the measured porosity does not differ much—by about 1.5%—and ranges from about 5% to over 6%. In those specimens, a much more significant number of large pores begins to appear, and the surface of the specimen is very heterogeneous, which results in a decrease in the strength of those specimens relative to those with the most strength by up to 800–900 MPa.

Detailed tests on the light microscope and in the scanning electron microscope indicate that the main reason for such significant porosity in the SLS Ti solid 8 variant, with the highest porosity in all the analysed cases, is incomplete sintering and the presence of fine non-sintered particles in individual pores (Figure 16).

It significantly affects the deterioration of the strength properties of such specimens. In other cases of porosity more significant than 4%, no such effects were found. It was confirmed that sintering takes place with the participation of the liquid phase and has the characteristics of liquid phase sintering (LPS), as in the case of 90% of sintered materials [97–100].

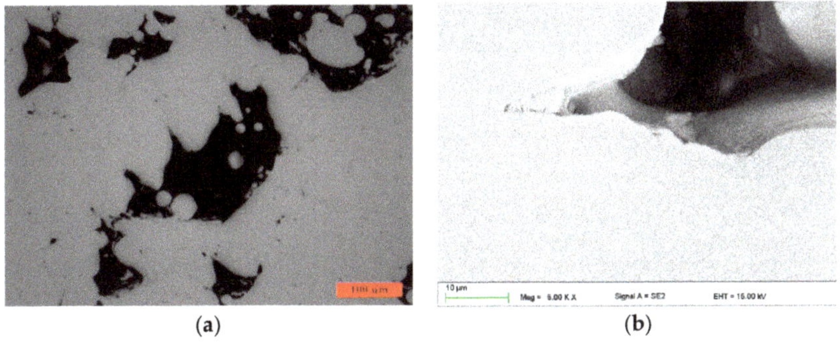

(a)   (b)

Figure 16. The structure of the Ti solid 8 specimen indicating the presence of fine sintered particles in individual pores as a result of incomplete sintering. (**a**) Light microscope; (**b**) SEM.

Materialographic tests of the selectively sintered material enabled the assessment of the quality of the specimens manufactured by assessing the additive laser transition paths manufactured on the surface and by assessing free spaces in fracture structures and the presence of loose powder particles. It turned out that the set of conditions for SLS recommended by the machine manufacturer defaults do not guarantee satisfactory results with minimal porosity, which justified the need for the detailed structural analysis described below.

Comparing the surface structure of the Ti6Al4V alloy manufactured using different SLS manufacturing conditions after passing the laser, a significant difference can be seen in the surface structure between particular specimens.

Specimens with a homogeneous surface included clearly marked laser paths; where it is possible to measure them (the paths are straight), the specimens are from sets with numbers Ti solid 1 and 4 (Figures 17 and 18). The structure described ensures the highest possible values of tensile and bending strength. Specimens' surface from sets with numbers Ti solid 8 and 5 (Figures 19 and 20) are characterized by the presence of free, empty spaces between the laser paths. The laser paths do not overlap, which is consequently reflected in the lowest properties among all tested variants.

Compared to the previously described specimens with the correct structure, this ensures significantly lower strength properties.

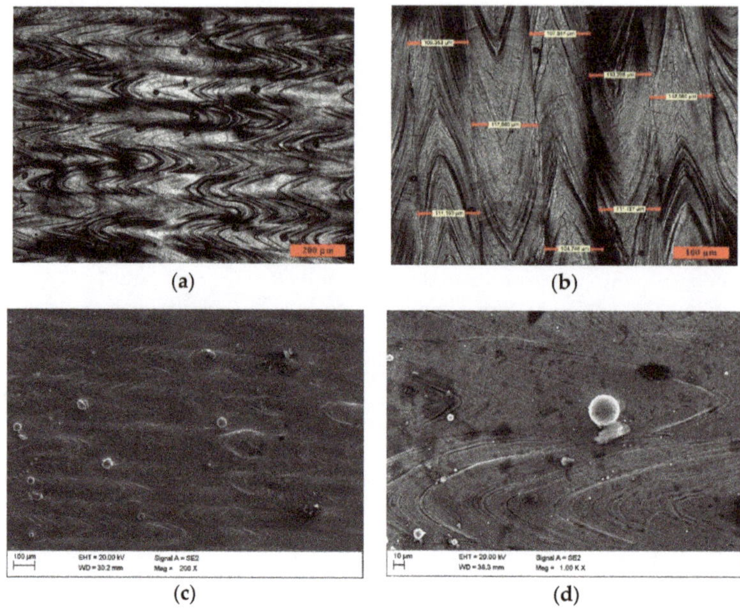

Figure 17. The surface structure of Ti6Al4V Ti solid 4 alloy specimen made (**a**,**b**) on a light microscope and (**c**,**d**) on an SEM.

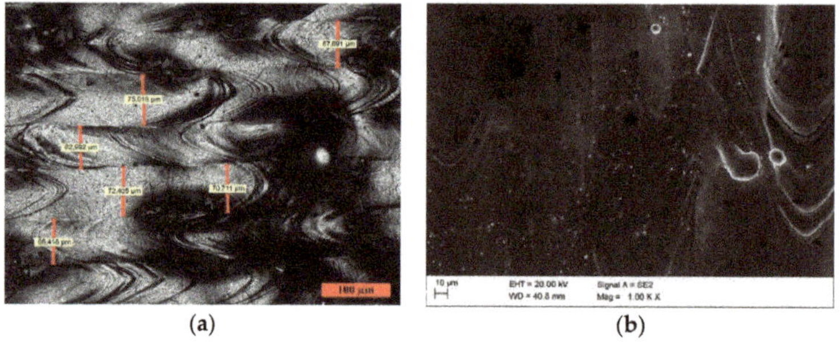

Figure 18. The surface structure of Ti6Al4V Ti solid 1 alloy specimen made (**a**) on a light microscope and (**b**) on an SEM.

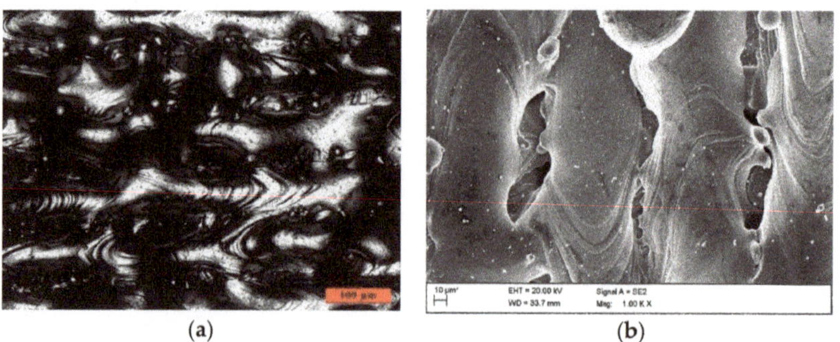

Figure 19. The surface structure of Ti6Al4V Ti solid 8 alloy specimens made (**a**) on a light microscope and (**b**) on an SEM.

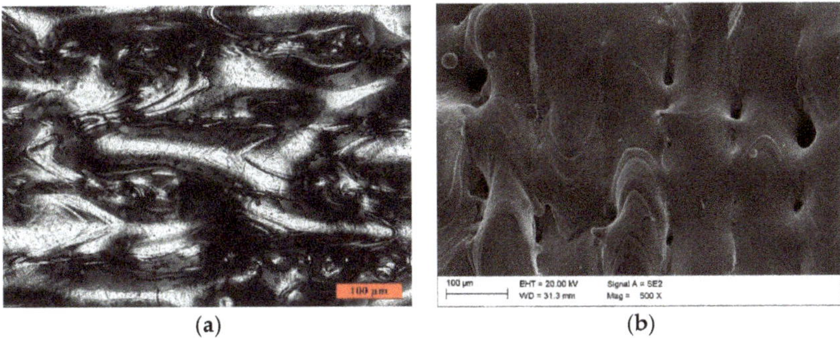

(a)　　　　(b)

Figure 20. The surface structure of Ti6Al4V Ti solid 5 alloy specimen made (**a**) on a light microscope and (**b**) on an SEM.

A detailed fractography analysis of fractures of specimens destroyed in the results of static tensile and bending tests was also carried out. Specimens with high strength are characterized by a homogeneous structure, and there are no closed pores and no visible grains of sintered powder particles (Figures 21 and 22), which of course would reduce the strength. For specimens with much lower strength, the fracture structure is heterogeneous, and between the sintered material fragments, there are free spaces that form closed pores. In many places, there are particles of powder which are not sintered (Figures 23 and 24). All this directly affects the density and strength of those samples.

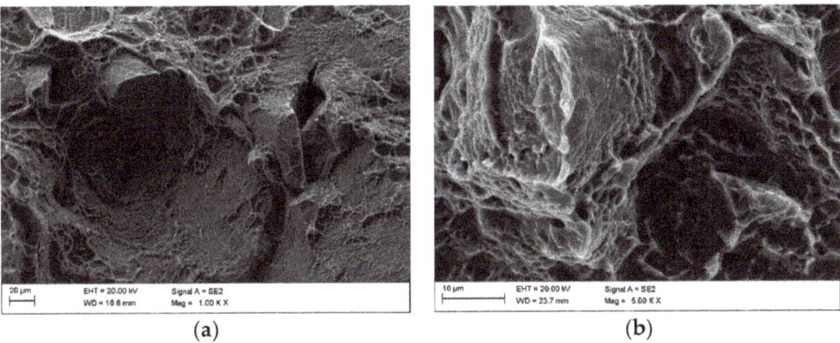

(a)　　　　(b)

Figure 21. The fracture's structure of the Ti6Al4V Ti solid 4 alloy after static testing: (**a**) tensile and (**b**) bending.

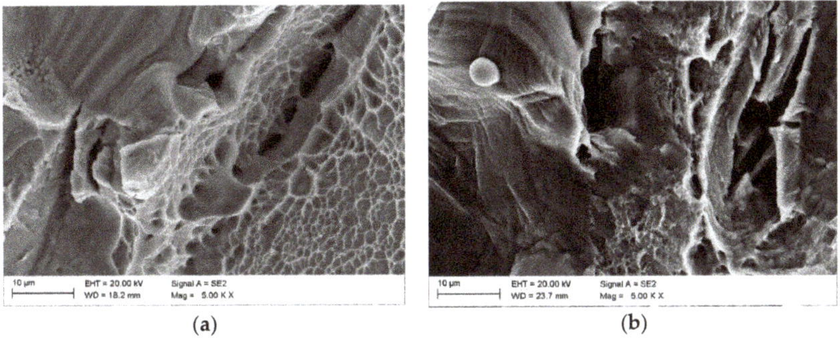

(a)　　　　(b)

Figure 22. The fracture's structure of the Ti6Al4V Ti solid 1 alloy after static testing: (**a**) tensile and (**b**) bending.

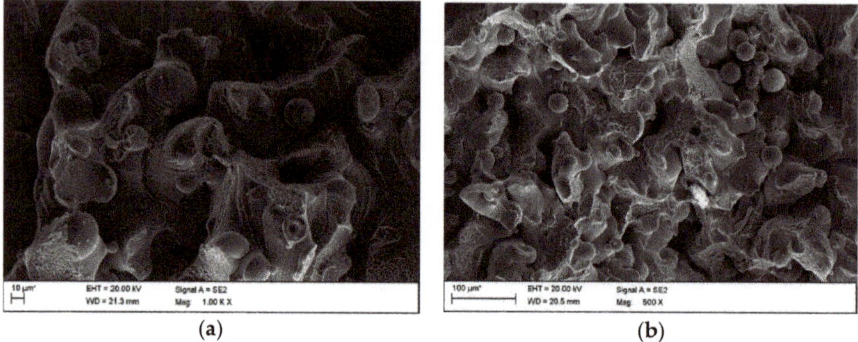

(a)   (b)

Figure 23. The fracture's structure of the Ti6Al4V Ti solid 8 alloy after static testing: (**a**) tensile and (**b**) bending.

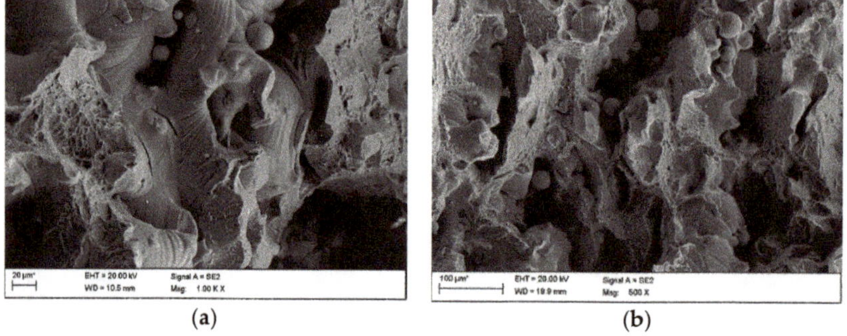

(a)   (b)

Figure 24. The fracture's structure of the Ti6Al4V Ti solid 7 alloy after static testing: (**a**) tensile and (**b**) bending.

6. Results of Tribological, Corrosive and Biological Tests of the Ti6Al4V Alloy Manufactured by the SLS Method

Tribological tests were carried out on Ti solid 4 and Ti solid 8 alloy specimens, showing the most extensive and smallest strength properties, respectively.

Both tests results, in the form of wear curves (Figure 25) and the surface structure after performing tribological tests visualized in a scanning electron microscope (Figure 26), do not indicate the fundamental differences of both materials with the highest and smallest strength properties of tribological wear, respectively.

The results of recording the curves obtained in the corrosion resistance tests are shown in Figure 27, for each material in the three corrosion solutions used. The tests were performed on Ti solid 4 and Ti solid 8 specimens that correspond to the highest and lowest mechanical properties. The comparison of test results for the both selected specimens indicates that the material with Ti solid 4 has the best corrosion resistance, which is indicated by the highest value of polarizing resistance and the lowest value of corrosive current density. In addition, it should be noted that in Ringer's solution with the addition of hydrogen peroxide, the amount of polarization resistance decreased several times, and for Ti solid 4 by 85%.

The analysis of the results also clearly indicates a higher susceptibility to corrosion damage to the tested materials in the solution in which hydrogen peroxide was added in comparison with the acid solution containing 1% acetic acid, as evidenced by the recorded values of, among others, the free and corrosive potential of materials.

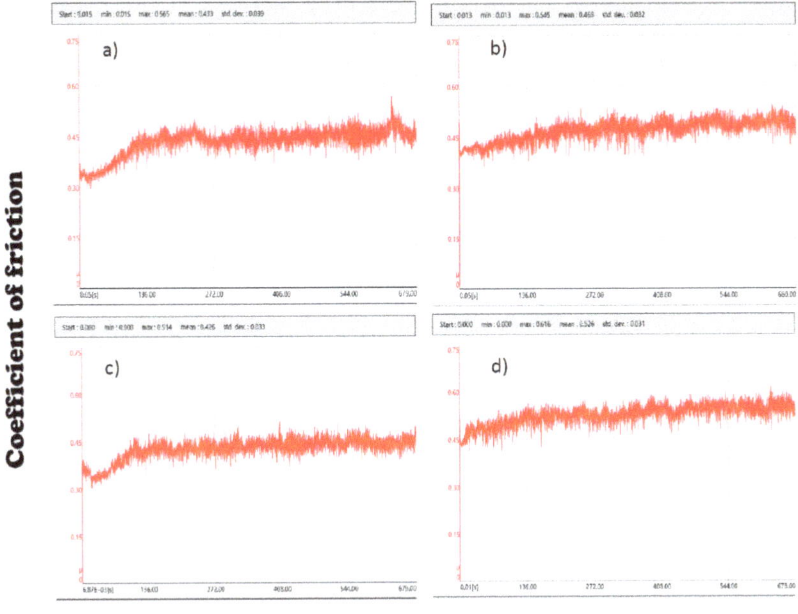

Figure 25. Wear curve for Ti solid 4 (**a**,**b**) and Ti solid 8 (**c**,**d**): (**a**,**c**) specimen side and (**b**,**d**) single-layer on specimen front.

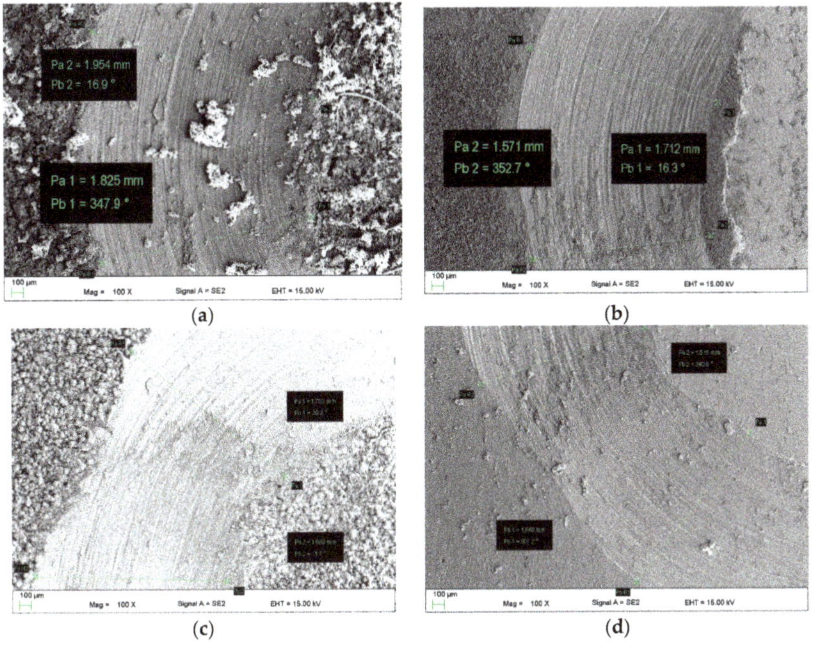

Figure 26. The surface of Ti solid 4 (**a**,**b**) and Ti solid 8 (**c**,**d**) specimens after tribological examination: (**a**,**c**) specimen side and (**b**,**d**) specimen front.

Based on the recorded anode polarization curves, it can be concluded that each of the materials after polarization above the value of the corrosion potential went into a passive state, as evidenced by a straight section of the curve parallel to the ordinate axis in Figure 26b,d.

The comparison of the polarization process of the tested materials indicates that the material designated as Ti solid 4 had the best stability compared to the others. With polarization up to 2 V, the maximum current density range was from 0.15 to 0.23 mA/cm^2 (in solution with hydrogen peroxide) for the material Ti solid 8, determined in the experiment as the limit value of current density; 1 mA/cm^2 recorded the changes of potential from 1.61 to 1.73 V.

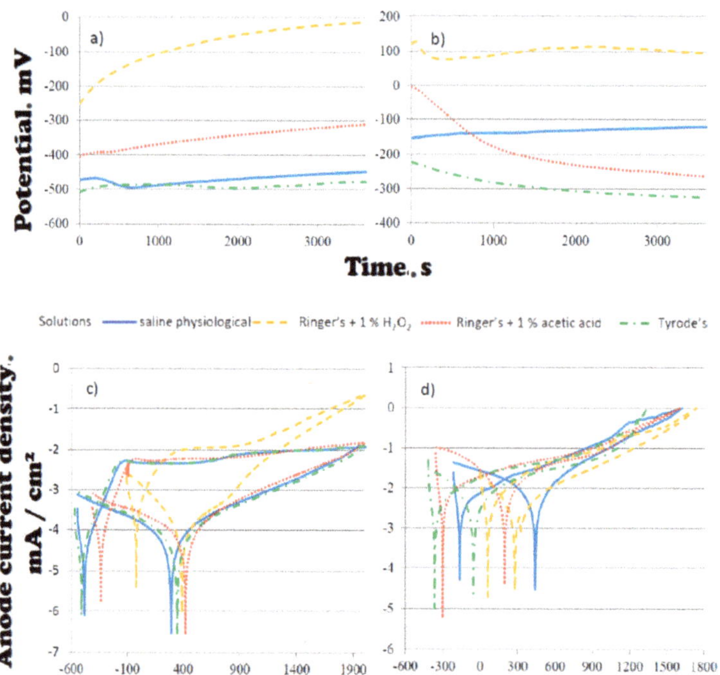

Figure 27. Curves registered in the corrosion resistance test of the Ti6Al4V alloy manufactured by the SLS method in variants Ti solid 4 (**a,c**) and Ti solid 8 (**b,d**). (**a,b**) Open circuit potential and (**c,d**) anode curves of the potentiodynamic method.

The results of the Tafel's analysis are presented in Table 6.

Table 6. Tafel's analysis results of tested materials.

Specimen	Solution	E_{ocp} (mV)	j_{kor} (µA/cm^2)	E_{kor} (mV)	R_{pol} (kΩ*cm^2)
Ti solid 4	saline	−448	0.010	−413	956
	Ringer's + 1% H$_2$O$_2$	−13	0.100	−25	140
	Ringer's + 1% acetic acid	−311	0.012	−334	549
	Tyrode's	−477	0.006	−513	1230
Tsolid 8	saline	−120	0.760	−167	13
	Ringer's + 1% H$_2$O$_2$	95	0.586	59	16
	Ringer's + 1% acetic acid	−263	1.184	−305	8
	Tyrode's	−324	0.480	−370	16

The results of biological studies indicate that the Ti6Al4V alloy correctly manufactured by the SLS method is a suitable medium for nesting and the proliferation of live cells. The survival of osteoblasts reached 97.0% when the control material in the form of neutral glass was almost identical to 96.4%, while on pure titanium grade it was lower and amounted to 88.9% with a standard deviation of less than 1.5% in each case. Figure 28 shows photos of the surface of a standard specimen and Ti6Al4V inside the pores after a 72-h culture and the addition of formazan solutions taken in DMSO.

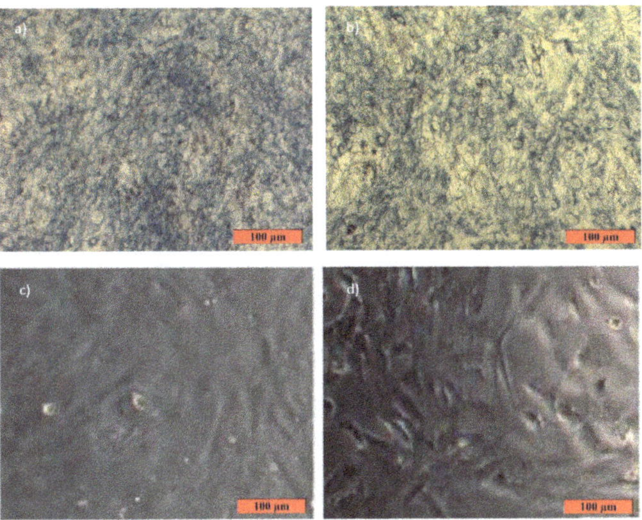

Figure 28. The surface of (**a**) a standard specimen and Ti6Al4V after a 72 h (**b**), 48 h (**c**) and 24 h (**d**) culture and the addition of formazan solutions taken in dimethyl sulfoxide (DMSO).

Figure 29 gives examples of various prosthetic restorations and implants made of Ti6Al4V alloy by the SLS method in the authors' dental practice.

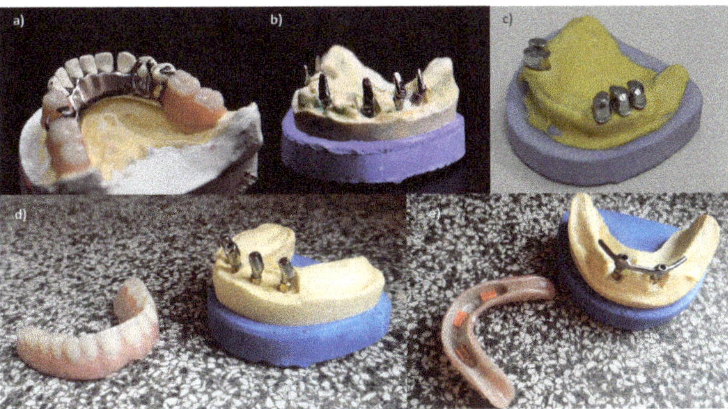

Figure 29. Examples of different prosthetic restorations and implants made of Ti6Al4V alloy by SLS: (**a**) skeletal prosthesis; (**b**) individual abutments for the round bridge; (**c**) primary telescopic crowns based on own teeth; (**d**) primary telescopic crowns integrated with individual abutments with an acrylate prosthesis; (**e**) individual bar with integrated individual abutments based on two implants with an acrylate prosthesis.

7. Final Remarks

The purpose of this paper is to compare in repeatable conditions the effects of manufacturing prosthetic restorations by the method of milling solid discs with those manufactured using SLS, as the most commonly used method in dentistry additive manufacturing technology. According to the manufacturing conditions recommended by the manufacturer of the SLS device, the so-called defaults do not ensure optimal properties of manufactured products. The authors' own technological practice also indicates that depending on the technological conditions used during the process, it is possible to obtain a very diverse range of materialographic structures of manufactured products, from entirely correct ones characterized by a practical lack of porosity to those in which the proportion of pores resulting from manufacturing errors exceeds even 10%, which significantly reduces strength properties by up to several dozen percent compared to those manufactured correctly. Positive results of these activities are aimed at convincing dentists about the lack of danger resulting from the use of modern additive technologies, and even the benefits. An unquestionable advantage is the full possibilities of making personalized prosthetic restorations adapted to the anatomical conditions and the condition of each patient's dental defects. The application of such a modern approach in conjunction with techniques for imaging the patient's dentition using cone-beam computed tomography (CBCT) and computer-aided design (CAD) is the basis for digitisation at the Dentistry 4.0 stage following the assumptions of the extended holistic Industry 4.0 model. It ensures very high standards of dental care for patients, and in the engineering aspect, maintaining very high-quality standards and tight dimensional tolerances of manufactured prosthetic restorations. It alleviates the anxiety of dentists, who are invariably responsible for the effects of treatment and relationships with patients.

Among metals, titanium and its alloys play the most crucial role in the production of implants, including dental ones and scaffolds. Titanium is particularly suitable for the production of additive methods, especially SLS. Titanium is also used in dental prosthetics, among others due to relatively low density. The disadvantage is the reaction of porcelain with titanium oxide, which causes bruising and darkening of the colour, which can be effectively counteracted by applying appropriate surface layers, most preferably using the atomic layer deposition (ALD) method. This issue is the subject of further own research and is not covered in this paper. While this problem is significant when manufacturing bridges and crowns, it does not exclude the use of titanium and its alloys for the production of implants replacing parts of the root of teeth and abutments.

Titanium alloys have the most significant practical significance, and among them, the Ti6Al4V alloy is the most commonly used. This alloy, known as titanium Grade 5, has a positive reputation among dentists due to its high usefulness in prosthetics and dental implantology. The paper presents the authors' own examples of using this alloy on various devices used in prosthetics and dental implantology. It includes comparative studies of the structure and properties of the Ti6Al4V alloy used for dental prosthetic restorations made alternatively using the loss method by milling on numerically controlled milling machines and the SLS method.

Selective laser sintering technology commonly called three-dimensional (3D) printing uses a powder with appropriate properties (i.e., grain sizes up to 45 μm, with the spherical particles form), obtained as a result of gas atomisation, as the input material. Such a powder treated with an appropriate set of technological conditions characteristic of SLS technology is transformed into a suitably designed shape. It requires the prior diagnosis of the condition of the patient's dentition using cone-beam computed tomography (CBCT) methods, as well as intraoral or less often extraoral scanning of properly made impressions and individual design of the required implant-prosthetic components. It is necessary to use highly specialised computer-aided design/manufacturing (CAD/CAM) software. The methodology of this action is described in detail in the authors' other publications.

The research results contained in this paper indicate that obtaining the right structure, especially a negligible share of pores below 0.1%, requires the selection of numerous selective laser sintering factors that determine the proper choice of technological conditions. In the case of classic sintering, it is necessary to correctly select the appropriate temperature, pressure and sintering time in quite narrow ranges for a powder with specific chemical composition and size. It ensures an optimal structure with minimal porosity providing the most favourable strength properties. In the case of selective laser sintering, the number of these factors is much more. The obtained test results confirm that sintering occurs with the participation of the liquid phase and has the characteristics of liquid phase sintering (LPS). From this statement and the theoretical premises of powder engineering, it is clear that it is not justified to distinguish selective laser sintering (SLS) and selective laser melting (SLM) as separate processes. Sintering is an essential stage of powder engineering with powder metallurgy as its part. In 90% of sintered materials, the LPS decides on the connection of individual powder particles during sintering and in this sense is the essence of sintering. Melting is, therefore, an essential element of sintering processes, and thus distinguishing it from them has no justification; it only introduces terminological confusion and information chaos. Many factors determine the provision of the desired structure and the expected mechanical and functional properties of selective laser sintered materials. The diameter of the laser spot, the width of the laser path, the allowance related to the overlap of individual laser paths, and above all the laser power, as well as the scanning speed, powder layer thickness, statistical size distribution and its bulk characteristics, determine the possibility of obtaining different structure and product properties made this way. SLS technology, while ensuring the right manufacturing conditions, allows for manufacturing solid material with strength properties significantly better than the cast alloy. Despite the identical chemical composition and significantly limited admixture concentrations, both materials behave differently undergoing plastic deformation under static tensile and bending tests. It is also confirmed by the results of fractography analysis of breakthroughs. The conventionally cast material has characteristics specific to plastic materials, with a lower maximum stress value on the tensile and bending curves, respectively, while the elongation or deflection is significantly higher. In the case of material produced by selective laser sintering, the opposite is the case; because the breakthrough is brittle, the maximum stress is significantly higher, and the maximum deformation is much lower. Tensile strength values of 858 MPa were obtained for the Ti6Al4V cast alloy and 1959 MPa for bending. In the case of tensile strength, it is 300 MPa less than in the case of a selective laser sintered alloy, while in the case of bending strength it is even lower by more than 500 MPa. The analysis of the relationship between the production conditions by selective laser sintering and porosity and morphology indicates that the proper selection of those conditions and a sharp technological regime of compliance with them throughout the whole process may lead to the manufacturing of a material with strength almost twice as high as material manufactured under extremely improper conditions. In the case of tensile strength, the maximum value for a correctly carried out process exceeds 1150 MPa, and bending is around 2465 MPa; improperly manufactured material may have a tensile strength that is over twice lower and bending strength that is almost 2.5 times lower. A similar situation is in the tensile case. The main reason is the improper sintering process, especially the local lack of LPS processes visible at the breakthroughs and the occurrence of sintered powder particles and associated local emptiness. It significantly affects the unacceptable increase in porosity, which varies almost 100 times depending on the technological variant. In the variant of obtaining a solid material, the porosity is less than 0.1%, and in the worst variants, it exceeds even 10% of the surface area of the pores. Materials manufactured by this method may ultimately differ significantly in their properties, despite the fact that in each of the discussed cases it is possible to solidify the material. Changing the laser power is crucial. If the power or actually the sintering energy density, which is related to the simultaneous selection of the diameter of the laser spot and the width of the laser path, is reduced, this leads to a situation in which the powder is either "burned" when the power and density of the laser power are too high or the powder is "unburned" when the power is too low compared to optimal standards. It is important to correctly select the technological conditions of the process so as to guarantee the best strength properties. Any change in those conditions is significant. One should be aware that failure to comply with the assumed technological regime even with respect to one or several layers of powder with thickness of 25 μm of each one may cause a discontinuity in the structure and failure to achieve the expected properties. Furthermore, it cannot be assumed that the device can operate in the plug-and-play standard using the settings indicated by the manufacturer, because the research reported in this paper, as well as the authors' own experiences obtained using other selective laser sintered materials, do not confirm this possibility. The properties of elements manufactured in this way are characterised by properties reduced by up to several dozen percent. Each time the shape and technological process of the manufactured product should be properly designed to ensure proper quality, the required tight dimensional tolerances and maintain high standards. At the same time, depending on the size of the manufactured element's desired surface structure and the range of permissible porosity, individual process conditions depend. The operator's experience and the assumptions originally adopted are decisive in this respect.

The tests carried out in the framework of this paper also confirm that it is possible to obtain high abrasion resistance and high corrosion resistance in solutions simulating body fluids, which has also been confirmed to depend on porosity. High properties characterise materials made under conditions preventing porosity higher than 0.1%. The biological properties of the selective laser sintered Ti6Al4V alloy were also checked. It is a good material for implanting osteoblast cells, better than pure titanium, which provides the possibility of using this alloy made with the use of additive technology for implants and implant-scaffolds used in dentistry.

This paper compares the results of testing the structure and tensile and bending strength of the Ti6Al4V selective laser sintered and casting machined in the CNC milling centre of the company FANUC (Robodrill S) because this technology is still most commonly used in implant prosthetics for the manufacturing of implants and dental abutments. The use of the additive manufacturing methods for the manufacturing of prosthetic elements of the Ti6Al4V alloy in combination with the digitisation of dental diagnostics and computer-aided design/manufacturing (CAD/CAM) in accordance with the idea of Dentistry 4.0 is the best choice of technology for the manufacturing of implant and prosthetic devices used in dentistry with the best strength, repeatable in shape and shape compliant with the project and requiring practically no additional post-production operations.

Author Contributions: Conceptualization, Methodology, literature review—L.A.D. and L.B.D., Investigation, Resources, Data Curation, Software, Formal Analysis, Writing—Original Draft Preparation, Visualization—L.A.D., L.B.D., A.A.-F. and J.D., Writing—Review & Editing, Supervision, Project Administration, Funding Acquisition—L.A.D. and L.B.D. All authors have read and agreed to the published version of the manuscript.

Funding: The paper is prepared in the framework of the Project POIR.01.01-00-0485/16-00 on 'IMSKA-MAT Innovative dental and maxillofacial implants manufactured using the innovative additive technology supported by computer-aided materials design ADD-MAT' realized by the Medical and Dental Engineering Centre for Research, Design and Production ASKLEPIOS in Gliwice, Poland. The project implemented in 2017–2021 is co-financed by the Operational Programme Intelligent Development of the European Union.

Conflicts of Interest: The authors declare no conflicts of interest.

References

1. Dobrzański, L.A.; Dobrzański, L.B. *Approach to the Design and Manufacturing of Prosthetic Dental Restorations According to the Rules of the Industry 4.0 Industrial Revolution Stage*; MPC, 2020. (in press)
2. Rüßmann, M.; Lorenz, M.; Gerbert, P.; Waldner, M.; Justus, J.; Engel, P.; Harnisch, M. *Industry 4.0: The Future of Productivity and Growth in Manufacturing Industries*; Boston Consulting Group: Boston, MA, USA, 2015. Available online: http://web.archive.org/web/20190711124617/https://www.zvw.de/media.media.72e472fb-1698-4a15-8858-344351c8902f.original.pdf (accessed on 15 December 2019).
3. Kagermann, H. Chancen von Industrie 4.0 nutzen. In *Industrie 4.0 in Produktion, Automatisierung und Logistik*; Springer Science and Business Media LLC: Berlin, Germany, 2014; pp. 603–614.
4. Kagermann, H.; Wahlster, W.; Helbig, J. *Recommendations for Implementing the Strategic Initiative Industrie 4.0: Final Report of the Industrie 4.0 Working Group*; Federal Ministry of Education and Research: Bonn, Germany, 2013.
5. Hermann, M.; Pentek, T.; Otto, B. *Design Principles for Industrie 4.0 Scenarios: A Literature Review*; Technische Universität Dortmund: Dortmund, Germany, 2015.

6. Dobrzański, L.A.; Dobrzańska-Danikiewicz, A.D. Why Are Carbon-Based Materials Important in Civilization Progress and Especially in the Industry 4.0 Stage of the Industrial Revolution. *Mater. Perform. Charact.* **2019**, *8*, 337–370. [CrossRef]

7. Jose, R.; Ramakrishna, S. Materials 4.0: Materials big data enabled materials discovery. *Appl. Mater. Today* **2018**, *10*, 127–132. [CrossRef]

8. Dobrzański, L. Role of materials design in maintenance engineering in the context of industry 4.0 idea. *J. Achiev. Mater. Manuf. Eng.* **2019**, *1*, 12–49. [CrossRef]

9. Dobrzański, L.A.; Dobrzański, L. Dentistry 4.0 Concept in the Design and Manufacturing of Prosthetic Dental Restorations. *Processes* **2020**, *8*, 525. [CrossRef]

10. Kaye, G. Does Digital Dentist Mean Anything -Today, 6 December 2017. Available online: https://www.dentaleconomics.com/science-tech/article/16389518/does-digital-dentist-mean-anything-today (accessed on 15 December 2019).

11. Harrison, T. The Emergence of the Fourth Industrial Revolution in Dentistry, 25 April 2018. Available online: https://www.dentistryiq.com/practice-management/industry/article/16366877/the-emergence-of-the-fourth-industrial-revolution-in-dentistry. (accessed on 15 December 2019).

12. Dobrzański, L.; Achtelik-Franczak, A.; Dobrzańska, J. The digitisation for the immediate dental implantation of incisors with immediate individual prosthetic restoration. *J. Achiev. Mater. Manuf. Eng.* **2019**, *2*, 57–68. [CrossRef]

13. Dobrzański, L.A. (Ed.) *Biomaterials in Regenerative Medicine*; InTech: Rijeka, Croatia, 2018. [CrossRef]

14. Dobrzański, L.A.; Dobrzańska-Danikiewicz, A.D. (Eds.) *Metalowe materiały Mikroporowate i Lite Do Zastosowań Medycznych i Stomatologicznych*; International OCSCOWorld Press: Gliwice, Poland, 2017.

15. Brånemark, P.-I.; Breine, U.; Adell, R.; Hansson, B.O.; Lindström, J.; Ohlsson, A. Intra-Osseous Anchorage of Dental Prostheses: I. Experimental Studies. *Scand. J. Plast. Reconstr. Surg.* **1969**, *3*, 81–100. [CrossRef]

16. Brånemark, P.-I.; Rydevik, B.L.; Skalak, R. (Eds.) *Osseointegration in Skeletal Reconstruction and Joint Replacement*; Quintessence Publishing Co.: Carol Stream, IL, USA, 1997.

17. Brånemark, P.-I.; Brånemark, B.; Rydevik, R.R. Myers, Osseointegration in skeletal recon- struction and rehabilitation: A review. *J. Rehabil. Res. Dev.* **2001**, *38*, 175–181.

18. Vieira, C.; Caramelli, B. The history of dentistry and medicine relationship: Could the mouth finally return to the body? *Oral Dis.* **2009**, *15*, 538–546. [CrossRef]

19. Dobrzański, L.A.; Dobrzańska-Danikiewicz, A.D.; Achtelik-Franczak, A.; Dobrzański, L.B. Comparative analysis of mechanical properties of scaffolds sintered from Ti and Ti6Al4V powders. *Arch. Mater. Sci. Eng.* **2015**, *73*, 69–81.

20. Dobrzański, L.; Dobrzańska-Danikiewicz, A.; Czuba, Z.; Dobrzański, L.; Achtelik-Franczak, A.; Malara, P.; Szindler, M.; Kroll, L. Metallic skeletons as reinforcement of new composite materials applied in orthopaedics and dentistry. *Arch. Mater. Sci. Eng.* **2018**, *2*, 53–85. [CrossRef]

21. Dobrzański, L.; Dobrzańska-Danikiewicz, A.; Czuba, Z.; Dobrzański, L.; Achtelik-Franczak, A.; Malara, P.; Szindler, M.; Kroll, L. The new generation of the biologicalengineering materials for applications in medical and dental implant-scaffolds. *Arch. Mater. Sci. Eng.* **2018**, *2*, 56–85. [CrossRef]

22. Dobrzański, L.B. *Struktura i Własności Materiałów Inżynierskich na Uzupełnienia Protetyczne Układu Stomatognatycznego Wytwarzane Metodami Przyrostowymi i Ubytkowymi*; Praca doktorska, Biblioteka Główna Akademii Górniczo-Hutniczej: Kraków, Poland, 2017.

23. Nouri, A.; We, C. Biomimetic Porous Titanium Scaffolds for Orthopedic and Dental Applications. *Biomim. Learn. Nat.* **2010**, 415–450. [CrossRef]

24. Malara, P.; Dobrzański, L. Computer aided manufacturing and design of fixed bridges restoring the lost dentition, soft tissue and the bone. *Arch. Mater. Sci. Eng.* **2016**, *81*, 68–75. [CrossRef]

25. Majewski, S. Nowe technologie wytwarzania stałych uzupełnień zębowych: Galwanoforming, technologia CAD/CAM, obróbka tytanu i współczesne systemy ceramiczne. *Protet. Stoma Tologiczna* **2007**, *57*, 124–131.

26. Galanis, C.C.; Sfantsikopoulos, M.M.; Koidis, P.T.; Kafantaris, N.M.; Mpikos, P.G. Computer methods for automating preoperative dental implant planning: Implant positioning and size assignment. *Comput. Methods Programs Biomed.* **2007**, *86*, 30–38. [CrossRef] [PubMed]

27. Miyazaki, T.; Hotta, Y. CAD/CAM systems available for the fabrication of crown and bridge restorations. *Aust. Dent. J.* **2011**, *56*, 97–106. [CrossRef] [PubMed]

28.   Malara, P.; Dobrzański, L.B. Computer-aided design and manufacturing of dental surgical guides based on cone beam computed tomography. *Arch. Mater. Sci. Eng.* **2015**, *76*, 140–149.

29.   Malara, P.; Dobrzański, L.B. Designing and manufacturing of implantoprosthetic fixed suprastructures in edentulous patients on the basis of digital impressions. *Arch. Mater. Sci. Eng.* **2015**, *76*, 163–171.

30.   Ender, A.; Zimmermann, M.; Attin, R.; Mehl, A. In vivo precision of conventional and digital methods for obtaining quadrant dental impressions. *Clin. Oral Investig.* **2015**, *20*, 1495–1504. [CrossRef]

31.   Lim, J.-H.; Park, J.-M.; Kim, M.; Heo, S.-J.; Myung, J.-Y. Comparison of digital intraoral scanner reproducibility and image trueness considering repetitive experience. *J. Prosthet. Dent.* **2018**, *119*, 225–232. [CrossRef]

32.   Persson, A.; Andersson, M.; Odén, A.; Englund, G.S. A three-dimensional evaluation of a laser scanner and a touch-probe scanner. *J. Prosthet. Dent.* **2006**, *95*, 194–200. [CrossRef] [PubMed]

33.   Ender, A.; Attin, R.; Mehl, A. In vivo precision of conventional and digital methods of obtaining complete-arch dental impressions. *J. Prosthet. Dent.* **2016**, *115*, 313–320. [CrossRef] [PubMed]

34.   Mandelli, F.; Gherlone, E.; Gastaldi, G.; Ferrari, M. Evaluation of the accuracy of extraoral laboratory scanners with a single-tooth abutment model: A 3D analysis. *J. Prosthodont. Res.* **2017**, *61*, 363–370. [CrossRef] [PubMed]

35.   Renne, W.; Ludlow, M.; Fryml, J.; Schurch, Z.; Mennito, A.; Kessler, R.; Lauer, A. Evaluation of the accuracy of 7 digital scanners: An in vitro analysis based on 3-dimensional comparisons. *J. Prosthet. Dent.* **2017**, *118*, 36–42. [CrossRef] [PubMed]

36.   Su, T.-S.; Sun, J. Comparison of repeatability between intraoral digital scanner and extraoral digital scanner: An in-vitro study. *J. Prosthodont. Res.* **2015**, *59*, 236–242. [CrossRef] [PubMed]

37.   Fasbinder, D.J. Digital dentistry: Innovation for restorative treatment. *Compend. Contin. Educ. Dent.* **2010**, *31*, 12.

38.   Lenguas, A.L.; Silva, R.; Ortega Aranegui, G.; Samara Shukeir, M.A.; López, B. Tomografía computerizada de haz cónico. Aplicaciones clínicas en odontología; comparación con otras técnicas. *Cient. Dent.* **2010**, *7*, 147–159.

39.   Garg, V.; Bagaria, A.; Kaur, G.; Bhardwaj, S.; Hedau, L.R. Application of Cone Beam Computed Tomography in Dentistry—A Review. *J. Adv. Med. Dent. Sci. Res.* **2019**, *7*, 73–76.

40.   Bornstein, M.M.; Scarfe, W.C.; Vaughn, V.M.; Jacobs, R. Cone beam computed tomography in implant dentistry: A systematic review focusing on guidelines, indications, and radiation dose risks. *Int. J. Oral Maxillofac. Implant.* **2014**, *29*, 55–77. [CrossRef]

41.   Ho, M.-H.; Lee, S.-Y.; Chen, H.-H.; Lee, M.-C. Three-dimensional finite element analysis of the effects of posts on stress distribution in dentin. *J. Prosthet. Dent.* **1994**, *72*, 367–372. [CrossRef]

42.   Reimann, Ł.; Żmudzki, J.; Dobrzański, L.A. Strength analysis of a three-unit dental bridge framework with the Finite Element Method. *Acta Bioeng. Biomech.* **2015**, *17*, 51–59. [PubMed]

43.   Tanaka, M.; Naito, T.; Yokota, M.; Kohno, M. Finite element analysis of the possible mechanism of cervical lesion formation by occlusal force. *J. Oral Rehabil.* **2003**, *30*, 60–67. [CrossRef] [PubMed]

44.   Lee, H.E.; Lin, C.L.; Wang, C.H.; Cheng, C.H.; Chang, C.H. Stresses at the cervical lesion of maxillary premolar—A finite element investigation. *J. Dent.* **2002**, *30*, 283–290. [CrossRef]

45.   Gupta, S.; Kumar, S. Lasers in Dentistry—An Overview. *Trends Biomater. Artif. Organs* **2011**, *25*, 119–123.

46.   Dobrzańska-Danikiewicz, A.; Dobrzański, L.A.; Gaweł, T.G. Scaffolds applicable as implants of a loss palate fragments. In Proceedings of the International Conference on Processing & Manufacturing of Advanced Materials, Fabrication, Properties, Applications, THERMEC'2016, Graz, Austria, 29 May–3 June 2016; p. 193.

47.   Liacouras, P.; Garnes, J.; Roman, N.; Petrich, A.; Grant, G.T. Designing and manufacturing an auricular prosthesis using computed tomography, 3-dimensional photographic imaging, and additive manufacturing: A clinical report. *J. Prosthet. Dent.* **2011**, *105*, 78–82. [CrossRef]

48.   Maroulakos, M.; Kamperos, G.; Tayebi, L.; Halazonetis, D.; Ren, Y. Applications of 3D printing on craniofacial bone repair: A systematic review. *J. Dent.* **2019**, *80*, 1–14. [CrossRef]

49.   Warnke, P.H.; Douglas, T.; Wollny, P.; Sherry, E.; Steiner, M.; Galonska, S.; Becker, S.T.; Springer, I.N.; Wiltfang, J.; Sivananthan, S. Rapid Prototyping: Porous Titanium Alloy Scaffolds Produced by Selective Laser Melting for Bone Tissue Engineering. *Tissue Eng. Part C Methods* **2009**, *15*, 115–124. [CrossRef]

50.   Dobrzanńska-Danikiewicz, A.D. Computer integrated development prediction methodology in materials surface engineering. *Open Access Libr.* **2012**, *1*, 1–289.

51. Dobrzański, L.A.; Dobrzańska-Danikiewicz, A.D. Foresight of the Surface Technology in Manufacturing. In *Handbook of Manufacturing Engineering and Technology*; Nee, A.Y.C., Ed.; Springer: London, UK, 2015; pp. 2587–2637.

52. Dobrzański, L.A.; Dobrzańska-Danikiewicz, A.D. Inżynieria powierzchni materiałów: Kompendium wiedzy i podręcznik akademicki. In *Materials Surface Engineering: Compendium of Knowledge and Academic Textbook*; International OCSCO World Press: Gliwice, Poland, 2018. (In Polish)

53. Dobrzański, L.A.; Dobrzańska-Danikiewicz, A.D.; Szindler, M.; Achtelik-Franczak, A.; Pakieła, W. Atomic layer deposition of TiO2 onto porous biomaterials. *Arch. Mater. Sci. Eng.* **2015**, *75*, 5–11.

54. Kampe, T.; Haraldson, T.; Hannerz, H.; Carlsson, G.E. Occlusal perception and bite force in young subjects with and without dental fillings. *Acta Odontol. Scand.* **1987**, *45*, 101–107. [CrossRef]

55. Ow, R.K.K.; Carlsson, G.E.; Jemt, T. Biting Forces in Patients with Craniomandibular Disorders. *J. CRANIO®* **1989**, *7*, 119–125. [CrossRef] [PubMed]

56. Varga, S.; Spalj, S.; Milosevic, S.A.; Mestrovic, S.; Slaj, M. Maximum voluntary molar bite force in subjects with normal occlusion. *Eur. J. Orthod.* **2010**, *33*, 427–433. [CrossRef] [PubMed]

57. Tortopidis, D.; Lyons, M.F.; Baxendale, R.H.; Gilmour, W.H. The variability of bite force measurement between sessions, in different positions within the dental arch. *J. Oral Rehabil.* **1998**, *25*, 681–686. [CrossRef] [PubMed]

58. Bates, J.F.; Stafford, G.D.; Harrison, A. Masticatory function? A review of the literature. *J. Oral Rehabil.* **1975**, *2*, 349–361. [CrossRef]

59. Proffit, W.R.; Fields, H.; Nixon, W. Occlusal Forces in Normal- and Long-face Adults. *J. Dent. Res.* **1983**, *62*, 566–570. [CrossRef]

60. Lundgren, D.; Laurell, L. Occlusal force pattern during chewing and biting in dentitions restored with fixed bridges of cross-arch extension. *J. Oral Rehabil.* **1986**, *13*, 57–71. [CrossRef]

61. Lundgren, D.; Laurell, L. Occlusal force pattern during chewing and biting in dentitions restored with fixed bridges of cross-arch extension. *J. Oral Rehabil.* **1986**, *13*, 191–203. [CrossRef]

62. Abu Alhaija, E.S.; Al Zo'Ubi, I.A.; Al Rousan, M.E.; Hammad, M.M. Maximum occlusal bite forces in Jordanian individuals with different dentofacial vertical skeletal patterns. *Eur. J. Orthod.* **2009**, *32*, 71–77. [CrossRef]

63. Sonnesen, L.; Bakke, M. Molar bite force in relation to occlusion, craniofacial dimensions, and head posture in pre-orthodontic children. *Eur. J. Orthod.* **2005**, *27*, 58–63. [CrossRef]

64. Kogawa, E.M.; Calderon, P.D.S.; Lauris, J.R.P.; Araujo, C.R.P.; Conti, P.C.R. Evaluation of maximal bite force in temporomandibular disorders patients. *J. Oral Rehabil.* **2006**, *33*, 559–565. [CrossRef]

65. Gavião, M.B.D.; Lemos, A.D.; Serra, M.D.; Gambareli, F.R.; Dos Santos, M.N. Masticatory performance and bite force in relation to signs and symptoms of temporomandibular disorders in children. *Minerva Stomatol.* **2006**, *55*, 529–539.

66. Koc, D.; Dogan, A.; Bek, B. Bite Force and Influential Factors on Bite Force Measurements: A Literature Review. *Eur. J. Dent.* **2010**, *4*, 223–232. [CrossRef] [PubMed]

67. Calderon, P.D.S.; Kogawa, E.M.; Corpas, L.D.S.; Lauris, J.R.P.; Conti, P.C.R. The influence of gender and bruxism on human minimum interdental threshold ability. *J. Appl. Oral Sci.* **2009**, *17*, 224–228. [CrossRef] [PubMed]

68. Fayad, M.; Sakr, H. Bite force and oral health impact profile in completely edentulous patients rehabilitated with two different types of denture bases. *Tanta Dent. J.* **2017**, *14*, 173. [CrossRef]

69. Hagberg, C. Assessment of bite force: A review. *J. Craniomandib. Disord. Facial Oral Pain* **1987**, *1*, 162–169.

70. Hellsing, G. On the regulation of interincisor bite force in man. *J. Oral Rehabil.* **1980**, *7*, 403–411. [CrossRef]

71. Lyons, M.; Baxendale, R. A preliminary electromyographic study of bite force and jaw-closing muscle fatigue in human subjects with advanced tooth wear. *J. Oral Rehabil.* **1990**, *17*, 311–318. [CrossRef]

72. Patil, S.R.; Maragathavalli, G.; Ramesh, D.N.S.V.; Vargheese, S.; Al-Zoubi, I.A.; Alam, M.K. Assessment of Maximum Bite Force in Oral Submucous Fibrosis Patients: A Preliminary Study. *Pesqui. Bras. Odontopediatria Clínica Integr.* **2020**, *20*, 4871. [CrossRef]

73. Bilhan, H.; Geckili, O.; Mumcu, E.; Cilingir, A.; Bozdag, E. The influence of implant number and attachment type on maximum bite force of mandibular overdentures: A retrospective study. *Gerodontology* **2010**, *29*, e116–e120. [CrossRef]

74. Van Vuuren, L.J.; Broadbent, J.M.; Duncan, W.J.; Waddell, J.N. Maximum voluntary bite force, occlusal contact points and associated stresses on posterior teeth. *J. R. Soc. N. Z.* **2019**, *50*, 132–143. [CrossRef]

75. Al-Zarea, B.K. Maximum bite force following unilateral fixed prosthetic treatment: A within-subject comparison to the dentate side. *Med. Princ. Pract.* **2015**, *24*, 142–146. [CrossRef] [PubMed]

76. Emsley, J. *Nature's Building Blocks: An A-Z Guide to the Elements*; Oxford University Press: Oxford, UK, 2001.

77. Hong, J.; Andersson, J.; Ekdahl, K.N.; Elgue, G.; Axén, N.; Larsson, R.; Nilsson, B. Titanium Is a Highly Thrombogenic Biomaterial: Possible Implications for Osteogenesis. *Thromb. Haemost.* **1999**, *82*, 58–64. [CrossRef] [PubMed]

78. Venkatarman, B.V.; Sudha, S. Vanadium Toxicity. *Asian J. Exp. Sci.* **2005**, *19*, 127–134.

79. Ngwa, H.A.; Kanthasamy, A.; Anantharam, V.; Song, C.; Witte, T.; Houk, R.; Kanthasamy, A.G. Vanadium induces dopaminergic neurotoxicity via protein kinase C delta dependent oxidative signaling mechanisms: Relevance to etiopathogenesis of Parkinson's disease. *Toxicol. Appl. Pharmacol.* **2009**, *240*, 273–285. [CrossRef] [PubMed]

80. Williams, D.F. On the mechanisms of biocompatibility. *Biomaterials* **2008**, *29*, 2941–2953. [CrossRef]

81. Steinemann, S. Metal implants and surface reactions. *Injury* **1996**, *27*, S/C16–S/C22. [CrossRef]

82. Gepreel, M.A.-H.; Niinomi, M. Biocompatibility of Ti-alloys for long-term implantation. *J. Mech. Behav. Biomed. Mater.* **2013**, *20*, 407–415. [CrossRef]

83. Okazaki, Y.; Ito, Y.; Kyo, K.; Tateishi, T. Corrosion resistance and corrosion fatigue strength of new titanium alloys for medical implants without V and Al. *Mater. Sci. Eng. A* **1996**, *213*, 138–147. [CrossRef]

84. Markwardt, J.; Friedrichs, J.; Werner, C.; Davids, A.; Weise, H.; Lesche, R.; Weber, A.; Range, U.; Meißner, H.; Lauer, G.; et al. Experimental study on the behavior of primary human osteoblasts on laser-cused pure titanium surfaces. *J. Biomed. Mater. Res. Part A* **2013**, *102*, 1422–1430. [CrossRef]

85. Pattanayak, D.K.; Fukuda, A.; Matsushita, T.; Takemoto, M.; Fujibayashi, S.; Sasaki, K.; Nishida, N.; Nakamura, T.; Kokubo, T. Bioactive Ti metal analogous to human cancellous bone: Fabrication by selective laser melting and chemical treatments. *Acta Biomater.* **2011**, *7*, 1398–1406. [CrossRef]

86. Braem, A.; Chaudhari, A.; Cardoso, M.V.; Schrooten, J.; Duyck, J.; Vleugels, J. Peri- and intra-implant bone response to microporous Ti coatings with surface modification. *Acta Biomater.* **2014**, *10*, 986–995. [CrossRef] [PubMed]

87. Zhang, L.; Klemm, D.; Eckert, J.; Hao, Y.; Sercombe, T. Manufacture by selective laser melting and mechanical behavior of a biomedical Ti–24Nb–4Zr–8Sn alloy. *Scr. Mater.* **2011**, *65*, 21–24. [CrossRef]

88. Hsu, H.-C.; Hsu, S.-K.; Hsu, S.-K.; Tsai, M.; Chang, T.-Y.; Ho, W.-F. Processing and mechanical properties of porous Ti–7.5Mo alloy. *Mater. Des.* **2013**, *47*, 21–26. [CrossRef]

89. Yao, Y.; Li, X.; Wang, Y.; Zhao, W.; Li, G.; Liu, R. Microstructural evolution and mechanical properties of Ti–Zr beta titanium alloy after laser surface remelting. *J. Alloy. Compd.* **2014**, *583*, 43–47. [CrossRef]

90. Zhuravleva, K.; Bönisch, M.; Prashanth, K.; Hempel, U.; Helth, A.; Gemming, T.; Calin, M.; Scudino, S.; Schultz, L.; Eckert, J.; et al. Production of Porous β-Type Ti–40Nb Alloy for Biomedical Applications: Comparison of Selective Laser Melting and Hot Pressing. *Materials* **2013**, *6*, 5700–5712. [CrossRef]

91. Walkowiak-Przybyło, M.; Klimek, L.; Okroj, W.; Jakubowski, W.; Chwiłka, M.; Czajka, A.; Walkowiak, B. Adhesion, activation, and aggregation of blood platelets and biofilm formation on the surfaces of titanium alloys Ti6Al4V and Ti6Al7Nb. *J. Biomed. Mater. Res. Part A* **2012**, *100*, 768–775. [CrossRef]

92. Khan, M.; Williams, R.L.; Williams, D. The corrosion behaviour of Ti–6Al–4V, Ti–6Al–7Nb and Ti–13Nb–13Zr in protein solutions. *Biomaterials* **1999**, *20*, 631–637. [CrossRef]

93. Marcu, T.; Todea, M.; Maines, L.; Leordean, D.; Berce, P.; Popa, C. Metallurgical and mechanical characterisation of titanium based materials for endosseous applications obtained by selective laser melting. *Powder Met.* **2012**, *55*, 309–314. [CrossRef]

94. Okrój, W.; Klimek, L.; Komorowski, P.; Walkowiak, B. Płytki krwi w kontakcie ze stopem tytanu Ti6Al4V i z jego zmodyfikowanymi powierzchniami, Inżynieria Biomateriałów. *Eng. Biomater.* **2005**, *8*, 13–16.

95. Tanaka, Y.; Kurashima, K.; Saito, H.; Nagai, A.; Tsutsumi, Y.; Doi, H.; Nomura, N.; Hanawa, T. In vitro short-term platelet adhesion on various metals. *J. Artif. Organs* **2009**, *12*, 182–186. [CrossRef]

96. Harris, L.; Meredith, D.O.; Eschbach, L.; Richards, R.G. Staphylococcus aureus adhesion to standard micro-rough and electropolished implant materials. *J. Mater. Sci. Mater. Electron.* **2007**, *18*, 1151–1156. [CrossRef] [PubMed]

97. German, R.M.; Dunlap, J.W. Processing of iron-titanium powder mixtures by transient liquid phase sintering. *Met. Mater. Trans. A* **1986**, *17*, 205–213. [CrossRef]

98. Bolton, J.; Grant, A. Liquid phase sintering of metal matrix composites containing solid lubricants. *J. Mater. Process. Technol.* **1996**, *56*, 136–147. [CrossRef]

99. German, R.M. *A-Z of Powder Metallurgy*; Elsevier: Oxford, UK, 2005.

100. Huppmann, W.J.; Petzow, G. The Elementary Mechanisms of Liquid Phase Sintering. *Sinter. Process.* **1980**, *70*, 189–201. [CrossRef]

Article

Corrosion Resistance of Cr–Co Alloys Subjected to Porcelain Firing Heat Treatment—In Vitro Study

Dorota Rylska [1], Bartłomiej Januszewicz [1,*], Grzegorz Sokołowski [2] and Jerzy Sokołowski [3]

[1] Institute of Materials Science and Engineering, Lodz University of Technology, 90-924 Lodz, Poland; dorota.rylska@p.lodz.pl

[2] Department of Prosthetics, Medical University of Lodz, 92-213 Lodz, Poland; grzegorz.sokolowski@umed.lodz.pl

[3] Department of General Dentistry, Medical University of Lodz, 92-213 Lodz, Poland; jerzy.sokolowski@umed.lodz.pl

* Correspondence: bartlomiej.januszewicz@p.lodz.pl; Tel.: +48-42-6313050 or +48-663-457-566

Abstract: The procedure of ceramics fusion to cobalt–chromium (Co–Cr) base dental crowns affects their corrosion behavior and biological tolerance. This study's purpose was to comparatively evaluate the effect of heat treatment (HT) applicable for dental ceramics firing on the corrosion properties among Co–Cr base alloys fabricated via different methods: casting (CST), milling soft metal and post sintering (MSM), and selective laser melting (SLM). All specimens were subjected to a heat treatment corresponding to a full firing schedule. The microstructure and elemental composition of oxidized surfaces were investigated by scanning electron microscopy and energy dispersive spectroscopy. Corrosion properties were examined by electrochemical potentiodynamic polarization tests. The values of j_{corr}, E_{corr}, R_p, and breakdown potential E_{br} were estimated. The oxide layers formed during the HT process corresponded to the composition of the original alloys' structure. Among the thermal treated alloys, SLM showed the highest corrosion resistance, followed by the MSM and CST. This may be attributed to uniform distribution of alloying elements in homogenous structure and to the reduced porosity, which enhances corrosion resistance and decreases the risk of crevice corrosion. The overall corrosion behavior was strongly influenced by the segregation of alloying elements in the microstructure, thus, is directly determined by the manufacturing method.

Keywords: Co–Cr dental alloys; corrosion; porcelain firing; SLM; MSM; CST

Citation: Rylska, D.; Januszewicz, B.; Sokołowski, G.; Sokołowski, J. Corrosion Resistance of Cr–Co Alloys Subjected to Porcelain Firing Heat Treatment—In Vitro Study. *Processes* **2021**, *9*, 636. https://doi.org/10.3390/pr9040636

Academic Editors: Lukáš Richtera and Leszek Dobrzański

Received: 17 February 2021
Accepted: 26 March 2021
Published: 5 April 2021

Publisher's Note: MDPI stays neutral with regard to jurisdictional claims in published maps and institutional affiliations.

1. Introduction

Currently, despite the growing popularity of all-ceramic restorations, a significant number of prosthetic works are made using the method of ceramics fusion on metal. Metal–ceramic restorations are used most frequently for fixed partial dentures and single crowns because of the good mechanical properties of alloys as well as the aesthetic characteristics of porcelain. Due to the high cost of noble metal alloys, base metal alloys, such as cobalt–chromium (Co–Cr), are more extensively used for the metal–ceramic restorations' construction. The low density, high hardness, and tensile strength [1] make base metal alloys superior to noble metals. The most popular technique is lost-wax casting, although there are also several dental technologies for manufacturing a final product from cobalt–chrome base alloys as alternatives to traditional casting techniques. These alternative methods include the milling of solid metal, milling in soft material, and sintering in a protective atmosphere (MSM) and selective laser melting (SLM) [2–4]. The milling alloys (1) are not likely to indicate procedural errors, such as impurities and shape deformations as a result of thermal tension and (2) show lower corrosion susceptibility than alloys manufactured by casting (CST) [5].

The effective and durable methods of improvement in corrosion resistance are still in great demand, because commonly applied dental alloys, like Co–Cr base alloys, are prone to corrosion in contact with physiological fluids. Impairment or destruction of the thin protective oxide layer on their surface, mostly composed of Cr oxide, is reported as the main reason for corrosion resistance reduction [6,7]

The metallic materials may be subjected to electrochemical pitting and crevice, fretting and fatigue corrosion, and stress corrosion cracking. The presence of chloride ions in an environment with a relatively high temperature (37 °C) is a significant cause of these alloys' corrosion. Moreover, frequently changing conditions in an oral cavity (e.g., the presence of food, pH changes, and bacteria [8]) are very harmful.

The corrosion does not only affect the metallic elements of prostheses in direct contact with saliva. It also occurs in the space of the metal inlay connection with the porcelain facing [9]. Most importantly, corrosion concerns the internal surface of metal–ceramic crowns since none of the prosthetic cements used for crown and bridge imbedding assures the connection is suitably tight. In effect, the electrolytic environment of the oral cavity penetrates and stays in contact with a metal surface (the crown's interior), causing crevice corrosion processes.

The corrosion of base alloys commences with the localized destruction of the passive layer, enabling the corrosive environment to penetrate the alloy and to cause damage to its structure. The mechanism and extent of the damage depend on the chemical composition of the alloy used. The second factor is the alloy's internal structure, conditioned by the manufacturing method.

Apart from various functional and technical advantages, the CST alloys have several drawbacks typical for casting, such as a microporosity, a tendency towards a coarse crystalline structure, dendritic microsegregation, and the coarse-grained precipitation of intermetallic phases and carbides. The foregoing detriments undoubtedly result in more critical pitting and galvanic corrosion [10–12].

The procedures of ceramics fusion on metal, which include

- surface oxidation,
- opaque ceramics fusion,
- dentin and enamel ceramics fusion, and
- glaze fusion.

Involve oxidation of the metallic surface and scale formation may affect the corrosion properties and thus the biological properties of metallic material [9,13,14].

New processing methods developed for cobalt and chrome-based alloys, not only fulfil the requirements of rapid manufacturing (RM) and rapid prototyping (RP), reduce the design lifecycle, and lower the cost of medical devices [15,16], but also help to increase the corrosion resistance of these materials, hence ensuring greater biocompatibility.

The literature includes many studies describing SLM technology, as well as comparative studies evaluating the mechanical properties, structure, and corrosion susceptibility of SLM alloys in comparison to the properties of cast elements made with the same metallic materials. The results of the studies demonstrate that the SLM elements have better properties than the ones obtained by the traditional casting methods, because of the equality of the SLM alloys' structures [17,18].

The latest technique (MSM) involves producing restorations with powder metallurgy methods [19,20]. This method seems to allow for a more homogeneous product, with a stable structure and with improved ductility compared with the traditional Co–Cr cast alloys [21].

The electrochemical characteristic of the cast and SLM-produced parts of dental Co–Cr base alloys have been extensively tested, but there are limited data concerning MSM technology. The literature on the last MSM study was considerably poorer and mainly includes studies on the fracture behavior of metal–ceramic bridges with Co–Cr-Mo frameworks obtained by powder metallurgy methods [20,22].

The studies concerning porcelain-fused-to-metal (PFM) Co–Cr dental alloys have mainly been concentrated on the evaluation of the marginal and/or internal fit of the restorations [23], whereas other studies performed an evaluation of the mechanical properties and porcelain bond strength [5,24,25], the effect of surface treatments on micro-roughness [25,26], the impact of the surface finishing process on the metal–ceramic bond strength, and the impact of different factors on the pre-oxidation process conducted before the PFM dental crown was fabricated [27–31].

In light of the results of previous research, it is not unfounded to state that the effect of the heat treatment (HT) assisting the porcelain fusion on corrosion occurrence in dental Co–Cr base alloys needs further investigation. Understanding whether and how the microstructure of metallic parts of crowns can affect their corrosion behavior after procedures of ceramics fusion in an oral cavity environment constitutes a key issue from a clinical perspective.

The aim of this study was to comparatively evaluate the effect of the thermal treatment applicable for dental ceramics firing on the surface characteristics, mainly the corrosion resistance of Co–Cr base alloys fabricated via the use of casting and two computer-aided design and manufacturing methods: milling soft metal and post-sintering (MSM) and selective laser melting (SLM). The null hypothesis was that no significant difference would be found among the alloys manufactured by different techniques after the heat treatment applied.

2. Materials and Methods

2.1. Materials Used and Processing

Five alloys (CST1, CST2, MSM, SLM1, and SLM2), each of three samples, were prepared.

In the CST group, alloys from Ceranium CC (CST1), produced by Mesa di Sala Giacomo & C. s.n.c., Italy, and Girobond NBS Co–Cr (CST2, produced by Amann Girrbach, Austria, were manufactured with the rotational method in a Formax induction-casting machine (Bego GmbH, Bremen, Germany).

The obtainment of elements produced from milled powder blocks made of Co–Cr al-loys bonded with a binding agent (Ceramill Sintron/Amann Girrbach, Dresden, Germany) and sintered in a Ceramil Argotherm furnace in an inert gas environment was followed by sinterization (MSM). The Ceramill Sintron samples were dry-milled in a process involving dust extraction using a four-axis milling CNC machine (Amann Girrbach Motion). The sintering process was performed in the Ceramill Argotherm furnace in an argon atmo-sphere at 1300 °C according to the manufacturer's instructions (sintering program duration: approximately 5 h).

The two kinds of SLM specimens were prepared from Co–Cr powders. The SLM1 alloy was produced from Modelstar S Powder 65 (Scheftner, Maintz, Germany) by using a dental laser-sintering device Realizer SLM100, (Realizer GmbH, Borhen, Germany).

The second SLM alloy (SLM2) was made from a Co–Cr alloy (similar in composition to the Starbond COS Powder (Scheftner, Maintz, Germany) using the Realizer SLM50 (Realizer GmbH, Borhen, Germany), but with an increased content of Mo and a decreased content of W (see Table 1). The composition (by weight) of all specimens is given in Table 1.

Table 1. Chemical composition (by wt %) of examined samples.

Sample	Co	Cr	W	Mo	Si	Others
CST 1	60.0	28.0	9.0	-	1.5	Mn, Fe < 1%.
CST 2	62.4	25.5	5.2	5.1	1.1	Nb, Fe, N < 1%.
MSM	66.0	28.0	-	5.0	-	Mn, Si, Fe < 1%; Organic binder: 1–2%.
SLM 1	62.7	28.8	-	6.0	0.7	C, Fe, Si, Mn < 1%.
SLM 2	58.5	25.3	7.3	6.6	0.9	Mn 0.8%; Fe < 1%.

All samples for further examinations, prepared as drilled discs with a diameter of 14 mm and a height of 7 mm, were wet grinded with abrasive papers up to P2400, and their surface was then treated with a 3-μm diamond polisher. Degreased and water-vapor-steam-cleaned specimens were exposed to a thermal treatment simulating the full firing schedule of a matched dental porcelain in a porcelain furnace Programat EP 3000 (Ivoclar Vivadent, Amherst, NY, USA).

2.2. Heat Treatment Simulating Porcelain Firing Processes

All specimens were brought under a full firing schedule (following Noritake—Super Porcelain EX3, Kuraray Noritake Inc. Japan) including two opaque, two dentine, and one final glaze simulating firings in a porcelain furnace—Ivoclar Vivadent Programat EP3000 (Ivoclar Vivadent, Amherst, NY, USA).

Particular processes of the HT corresponded to Noritake EX-3 ceramics are presented in Table 2.

Table 2. Processes of the applied heat treatment (HT).

Process	Opaque 1	Opaque 2	Dentine 1	Dentine 2	Final Glaze
B	400 °C	650 °C	600 °C	600 °C	650 °C
S	8 min	5 min	7 min	10 min	5 min
V1	400 °C	650 °C	600 °C	600 °C	-
t_1	65 °C/min	55 °C/min	45 °C/min	45 °C/min	55 °C/min
T	1000 °C	945 °C	935 °C	935 °C	910 °C
V2	1000 °C	935 °C	920 °C	925 °C	-
H	1 min	-	-	-	-

B = Initial temperature; S = Time for drying and furnace chamber closing; V1 = Temperature of vacuum activation; t_1 = Speed of temperature's increase; T = Firing temperature; V2 = Temperature of vacuum deactivation; H = Warming up time.

2.3. Microstructure and Chemical Characterization

For the microstructural characterization, the specimens of each group were examined using a SEM (scanning electron microscope) JEOL JSM-6610LV (JEOL, Akashima Tokyo, Japan) and performing EDS (Energy Dispersive X-Ray Spectroscopy) analysis.

Cross sections of oxidized samples were prepared to determine the thickness of the oxides (shown in Table 3).

Table 3. The average thickness of oxide layers covered the examined alloys.

Sample	CST1	CST2	MSM	SLM1	SLM2
Average oxide thickness [μm]	1.59	1.59	1.25	1.61	1.26

In consideration of the above data and the oxides' brittleness, surface SEM-EDS examinations were conducted. The influence of the base material could be neglected because of the following:

- The thickness of the oxide layers varied in the range of 1.16–1.78 μm.
- The average atomic mass of the elements creating the oxides is high, which directly affects the depth of penetration of the primary electron beam.

The examined surfaces were imaged under an accelerating voltage of 20 kV. The microanalyses of the chemical composition were performed with an EDS X-MAX 80 microanalyzer (Oxford Instruments, High Wycombe, UK) under an accelerating voltage of 20 kV. According to the EDS method, we performed a qualitative and quantitative analysis of the chemical composition (both the elemental constituents and oxides) of selected areas and points, and along the selected lines. The surfaces of all samples were also subjected to SEM-SE observations after electrochemical examinations. Additionally, the SEM/EDS examinations were performed for the samples subjected to the corrosion investigations.

The microstructure of the alloys before the HT was investigated by SEM-SE, and the surface distributions of the elements for all specimens before the HT were determined by the energy-dispersive X-ray spectroscopy method under the same accelerating voltage.

2.4. Corrosion Properties Examinations

The corrosion susceptibility was investigated by employing potentiodynamic polarization. The electrochemical tests were performed for all samples subjected to the HT simulating porcelain firing at room temperature in a 0.9% NaCl solution (ASTM F746-04(2014) Standard Test Method for Pitting or Crevice Corrosion of Metallic Surgical Implant Materials).

The corrosion resistance of the samples was assessed with an Atlas 0531 Electrochemical Unit & Impedance Analyzer kit. It consisted of a five-electrode accessory for measuring the chrono-volt amperometric curves of electrochemical systems consistent with AtlasCorr software (Atlas-Sollich, Rębiechowo-Gdańsk, Poland). An electrochemical cell (CEC/TH Thermostated Multipurpose Cell—Radiometer Analytical, Lyon, France) and a three-electrode measuring system were used for the corrosion tests (CT).

After being ultrasonically rinsed in double distilled water and ethanol for 10 min, respectively, three samples of each kind were placed in a special holder as a working electrode; the exposed surface area was equal to 0.95 cm^2. A platinum electrode served as a counter electrode, and a saturated calomel electrode (SCE) served as a reference electrode. The corrosion cell was a 250-mL vessel with a thermo-stating jacket. The working and platinum electrodes were mounted on the sides, facing each other; the SCE electrode was connected with the working electrode through a bridge with a Luggin capillary. The capillary tip was positioned within 2 mm of the working electrode surface to reduce the ohm drop of the solution. The specimens were exposed to the solution for 1 h to establish an open circuit potential (OCP), and potentiodynamic polarization scans were applied after establishing the OCP. The potentiodynamic polarization curves were recorded by scanning from an initial potential of -1.0 V vs. the OCP (which was measured with reference to the SCE) to a final potential of 2.0 V vs. the OCP at a scanning rate of 5 mV/s. We employed the Tafel extrapolation method, in the domain of the ± 200 mV SCE vs. the OCP, to determine the corrosion current i_{corr} and the corrosion potential E_{corr}. The values of j_{corr} and E_{corr} were obtained by means of extrapolation of the Tafel straights. The areas of potentiodynamic curves assumed as Tafel straights were chosen manually to avoid mistakes in interpretation generated by the commercial software. Analysis was performed by extrapolating the linear portions of a logarithmic current vs. a potential plot back to their intersection.

The potential (E_{br}), at which the corrosion current increases abruptly, was also estimated from the polarization curves (at potentials below the oxygen-evolution potential when the neutral saline solution was ~0.6 V).

2.5. Statistical Analysis

One-way analysis of variance (ANOVA) was used for the data of j_{corr}, E_{corr}, R_p, and E_d, followed by a Tukey post-hoc test ($\alpha = 0.05$).

3. Results

3.1. Microstructure and Chemical Characterization

Figure 1a–e presents the microstructure of all examined alloys before the thermal treatment. The SEM-SE with corresponding EDS mapping images for each alloy (before the HT) is shown in Figure 2a–f.

The results of the morphology examinations and tests, determining the chemical composition of the surface layers of the heat-treated elements, are presented below in Figures 3–7. The (a) panel of each figure shows the morphology of the sample's surface obtained by using a scanning microscope in a secondary electrons (SE) mode supplemented by the quantitative results of the average chemical composition (both elements and oxides) from the selected area. The EDS maps presenting the surface distribution of elements are shown in the (b) panel of each figure. The next panel, (c), presents the results of quantitative EDS examinations for the representative points (marked in the SEM-SE image). The (d) panel presents the EDS results of the relative concentration of elements along the selected line.

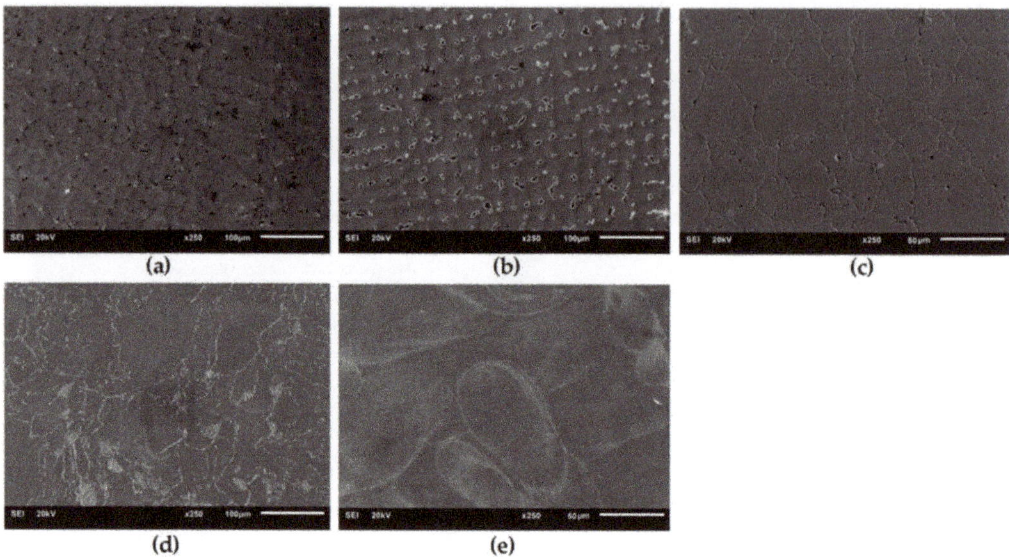

Figure 1. SEM-SE (Secondary Elctrons) images of the Co–Cr alloys before the heat treatment (HT). (**a**) CST1; (**b**) CST2; (**c**) MSM; (**d**) SLM1; (**e**) SLM2.

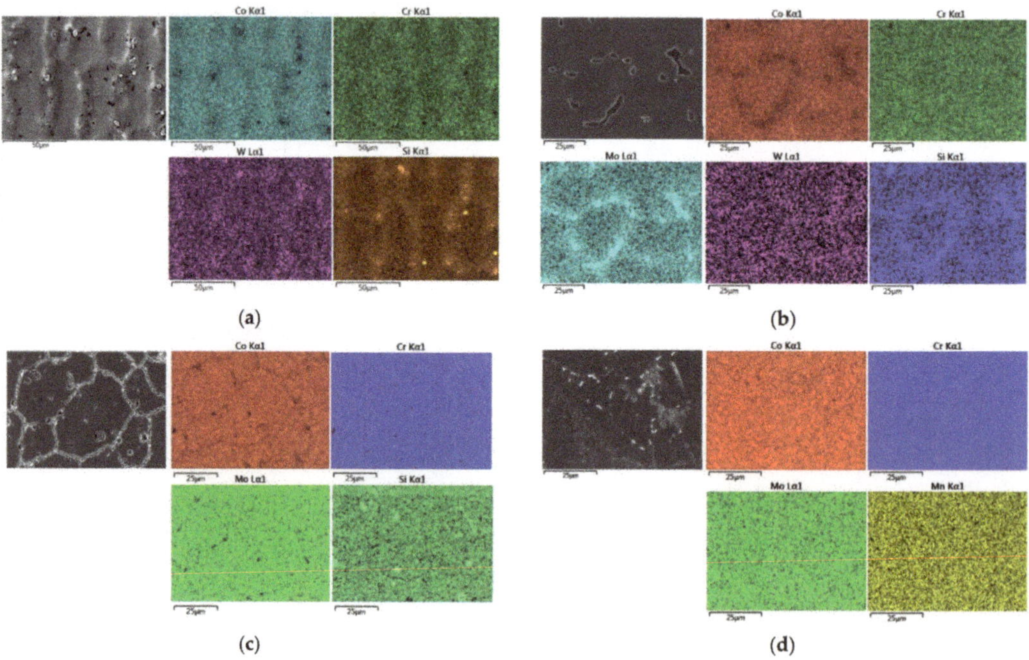

Figure 2. *Cont.*

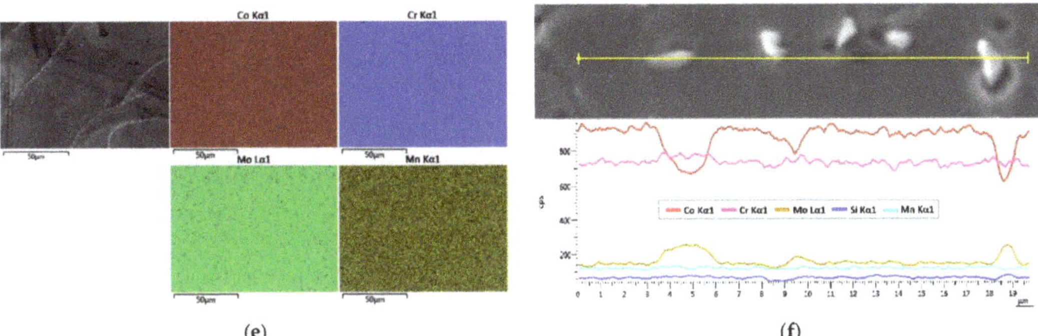

(e)

(f)

Figure 2. SEM-SE and corresponding EDS mapping images for each alloy before the HT (**a–e**): (**a**) CST1 with Co, Cr, W, and Si distributions; (**b**) CST2 with Co, Cr, Mo, W, and Si distributions; (**c**) MSM with Co, Cr, Mo, and Si distributions; (**d**) SLM1 with Co, Cr, Mo, and Mn distributions; (**e**) SLM2 with Co, Cr, Mo, and Mn distributions; (**f**) SLM1 SEM-SE micrograph with corresponding EDS line scan results from the grain boundary area with bright precipitations.

Element	wt %	Oxide	Oxide wt %
O	25.74		
Si	1.21	SiO_2	2.59
Cr	24.1	Cr_2O_3	35.23
Co	43.86	CoO	55.76
W	5.09	WO_3	6.42
Total:	100		100

(a)

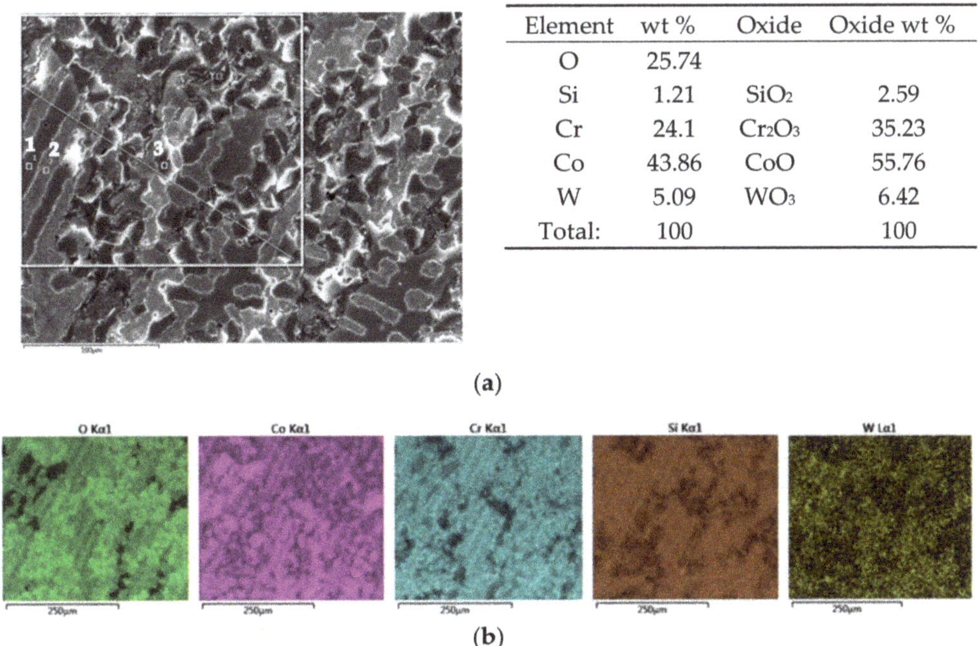

(b)

Figure 3. *Cont.*

Element [wt %]	1	2	3	Oxide [wt %]	1	2	3
O	25.85	26.29	26.33				
Si	1.02	2.58	0.85	SiO_2	2.19	5.52	1.82
Cr	25.73	21.58	29.67	Cr_2O_3	37.6	31.53	43.37
Co	41.83	43.37	38.04	CoO	53.19	55.15	48.37
W	5.56	6.18	5.11	WO_3	7.02	7.8	6.44
Total:	100	100	100		100	100	100

(**c**)

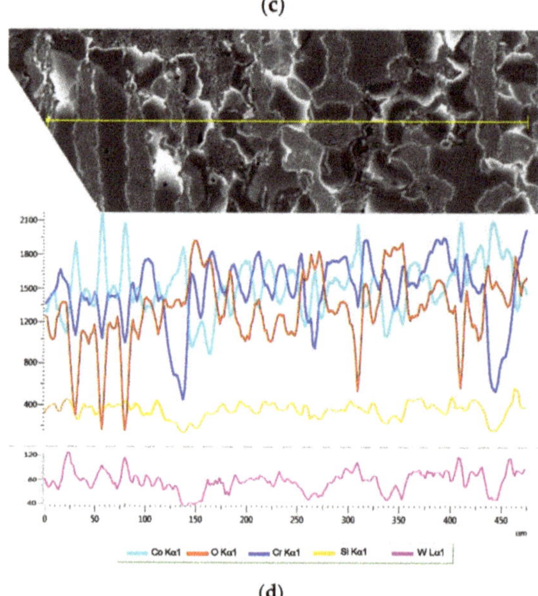

(**d**)

Figure 3. SEM EDS results for the CST1 sample: (**a**) SEM-SE morphology with average EDS results from the marked area; (**b**) maps presenting the surface distribution of O, Co, Cr, W, and Si; (**c**) EDS results for marked points; (**d**) SEM-SE micrograph with corresponding EDS line scan results.

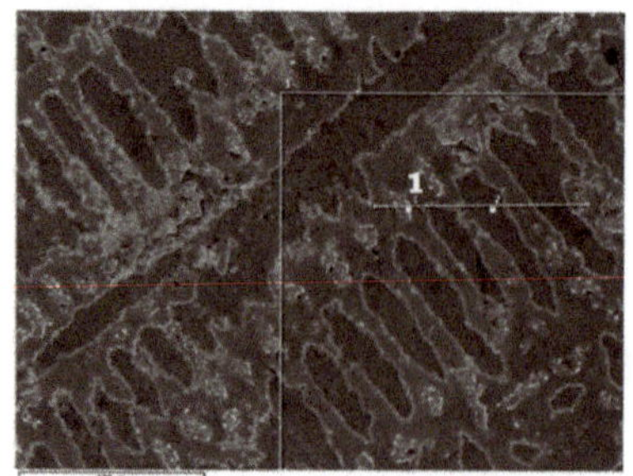

Element	wt %	Oxide	Oxide wt %
O	26.47		
Si	1.45	SiO_2	3.1
Cr	23.15	Cr_2O_3	33.83
Co	41.1	CoO	52.26
Mo	3.88	MoO_3	5.82
W	3.96	WO_3	4.99
Total:	100		100

(**a**)

Figure 4. *Cont.*

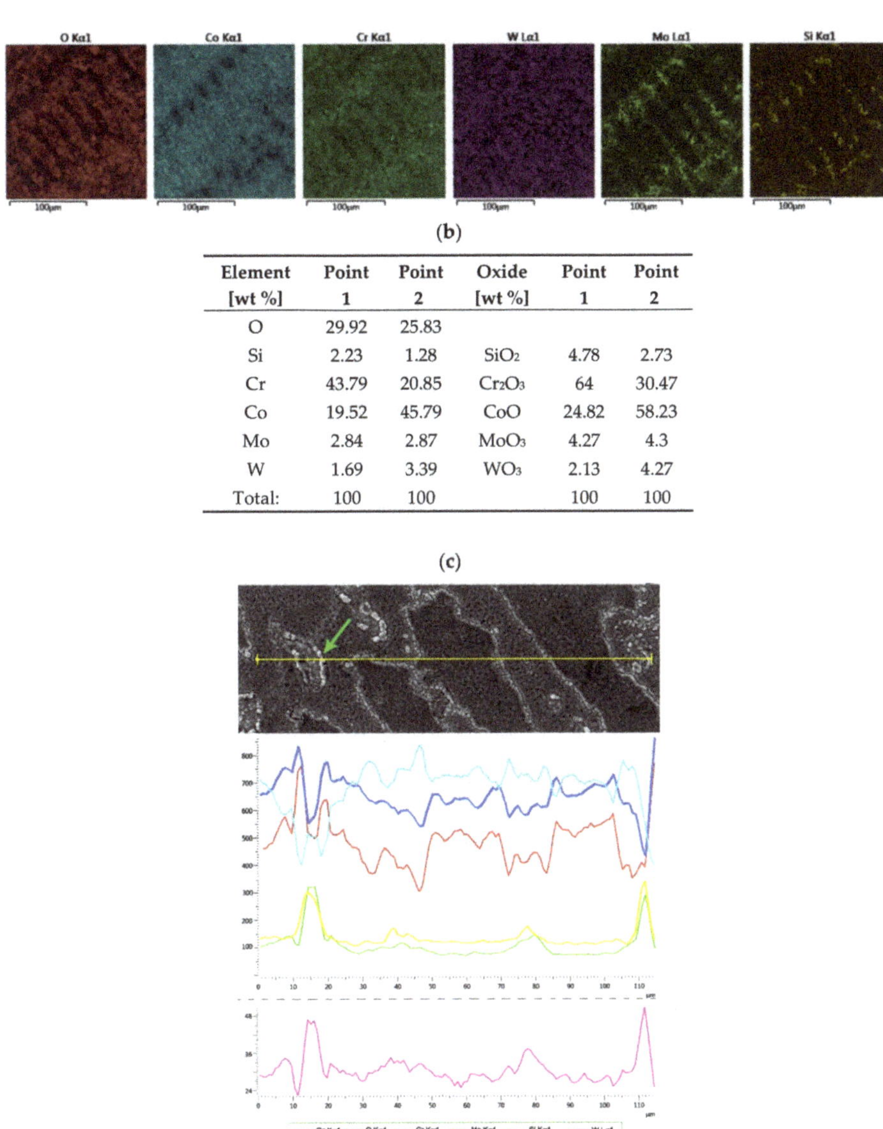

(b)

Element [wt %]	Point 1	Point 2	Oxide [wt %]	Point 1	Point 2
O	29.92	25.83			
Si	2.23	1.28	SiO_2	4.78	2.73
Cr	43.79	20.85	Cr_2O_3	64	30.47
Co	19.52	45.79	CoO	24.82	58.23
Mo	2.84	2.87	MoO_3	4.27	4.3
W	1.69	3.39	WO_3	2.13	4.27
Total:	100	100		100	100

(c)

(d)

Figure 4. SEM EDS results for the CST2 sample: (**a**) SEM-SE micrograph showing the morphology with average EDS results from the marked area; (**b**) maps presenting the surface distribution of O, Co, Cr, W, Mo, and Si; (**c**) EDS results for marked points; (**d**) SEM-SE micrograph with corresponding EDS line scan results.

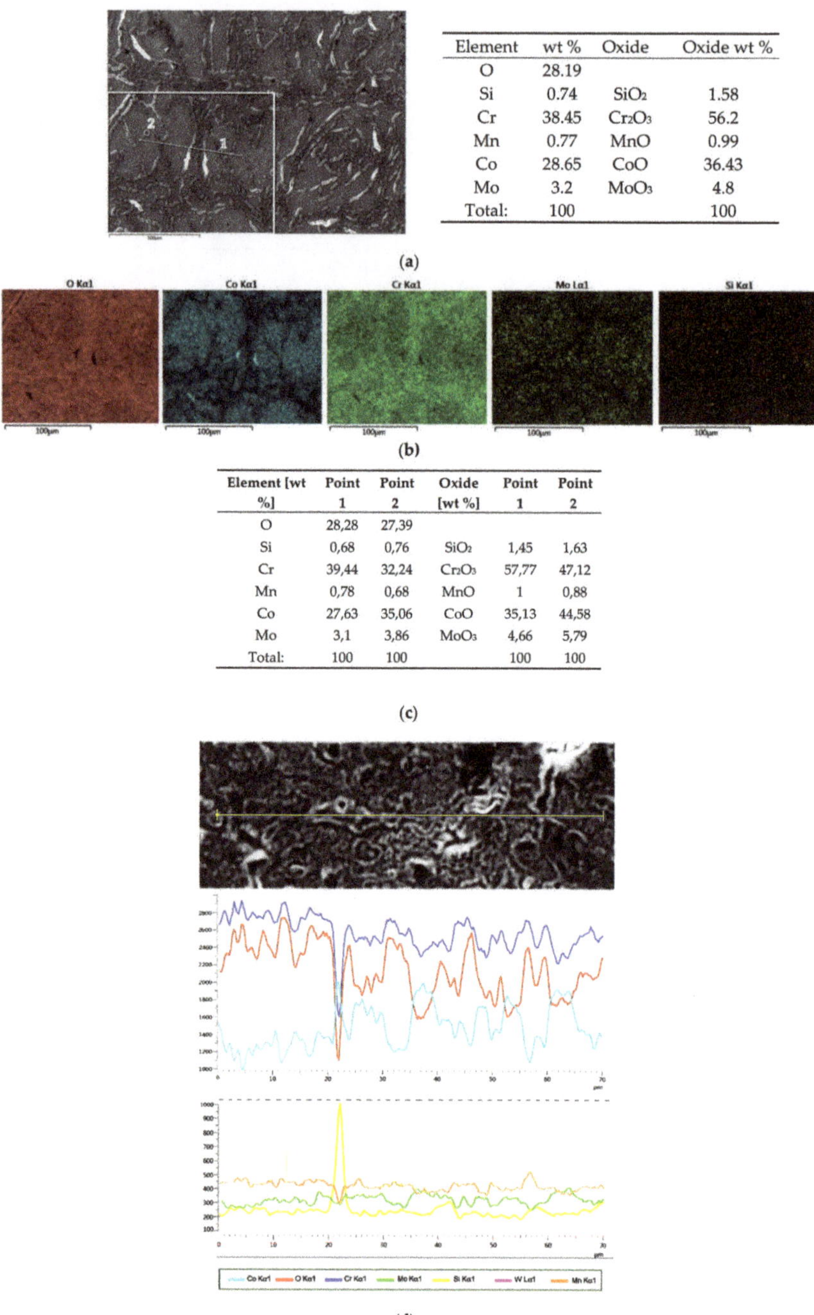

Table (a):

Element	wt %	Oxide	Oxide wt %
O	28.19		
Si	0.74	SiO_2	1.58
Cr	38.45	Cr_2O_3	56.2
Mn	0.77	MnO	0.99
Co	28.65	CoO	36.43
Mo	3.2	MoO_3	4.8
Total:	100		100

(a)

(b)

Table (c):

Element [wt %]	Point 1	Point 2	Oxide [wt %]	Point 1	Point 2
O	28,28	27,39			
Si	0,68	0,76	SiO_2	1,45	1,63
Cr	39,44	32,24	Cr_2O_3	57,77	47,12
Mn	0,78	0,68	MnO	1	0,88
Co	27,63	35,06	CoO	35,13	44,58
Mo	3,1	3,86	MoO_3	4,66	5,79
Total:	100	100		100	100

(c)

(d)

Figure 5. SEM EDS results for the MSM sample: (**a**) SEM-SE micrograph showing the morphology with average EDS results from the marked area; (**b**) maps presenting the surface distribution of O, Co, Cr, Mo, and Si; (**c**) EDS results for marked points; (**d**) SEM-SE micrograph with corresponding EDS line scan results.

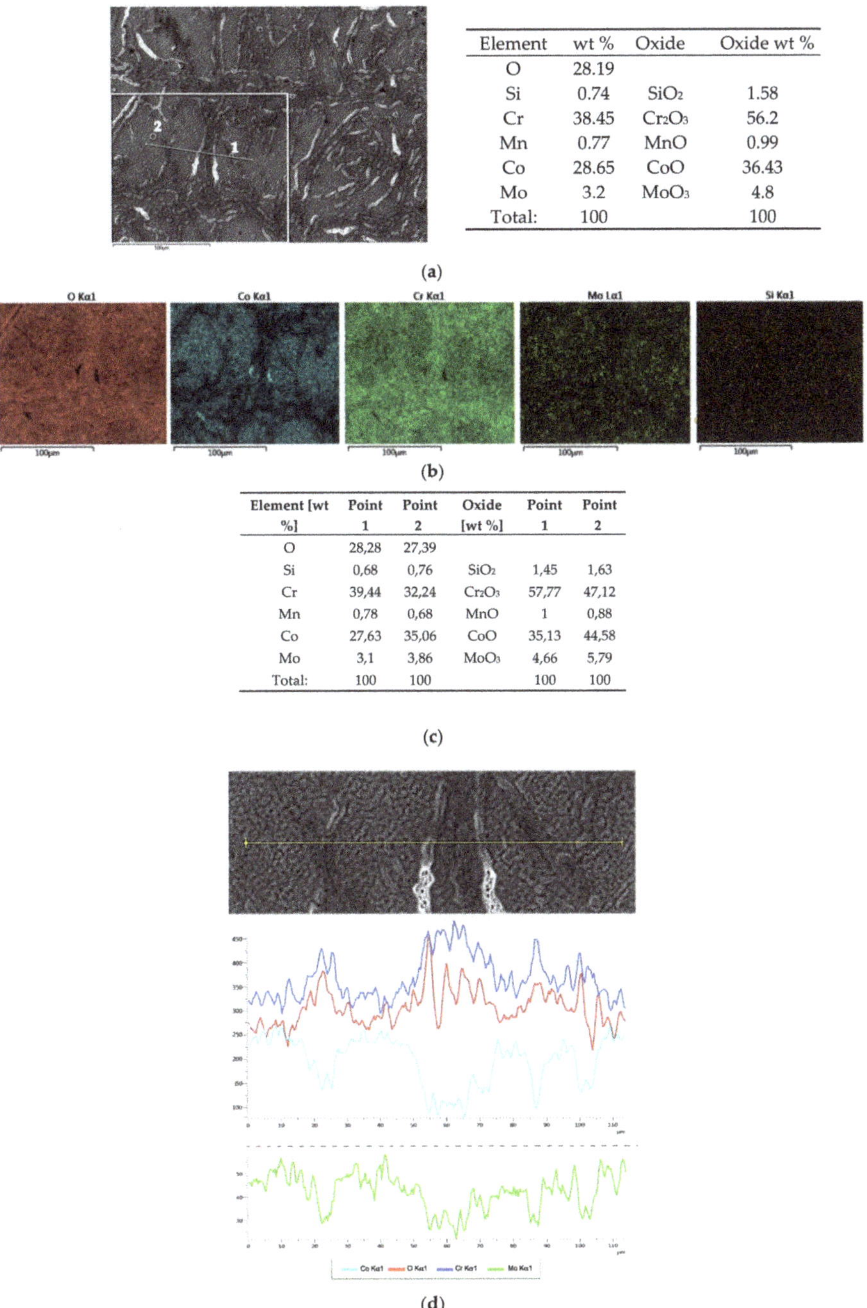

(a)

Element	wt %	Oxide	Oxide wt %
O	28.19		
Si	0.74	SiO_2	1.58
Cr	38.45	Cr_2O_3	56.2
Mn	0.77	MnO	0.99
Co	28.65	CoO	36.43
Mo	3.2	MoO_3	4.8
Total:	100		100

(b)

(c)

Element [wt %]	Point 1	Point 2	Oxide [wt %]	Point 1	Point 2
O	28,28	27,39			
Si	0,68	0,76	SiO_2	1,45	1,63
Cr	39,44	32,24	Cr_2O_3	57,77	47,12
Mn	0,78	0,68	MnO	1	0,88
Co	27,63	35,06	CoO	35,13	44,58
Mo	3,1	3,86	MoO_3	4,66	5,79
Total:	100	100		100	100

(d)

Figure 6. SEM EDS results for the SLM1 sample: (**a**) SEM-SE micrograph showing the morphology with average EDS results from the marked area; (**b**) maps presenting the surface distribution of O, Co, Cr, Mo, and Si; (**c**) EDS results for marked points; (**d**) SEM-SE micrograph with corresponding EDS line scan results.

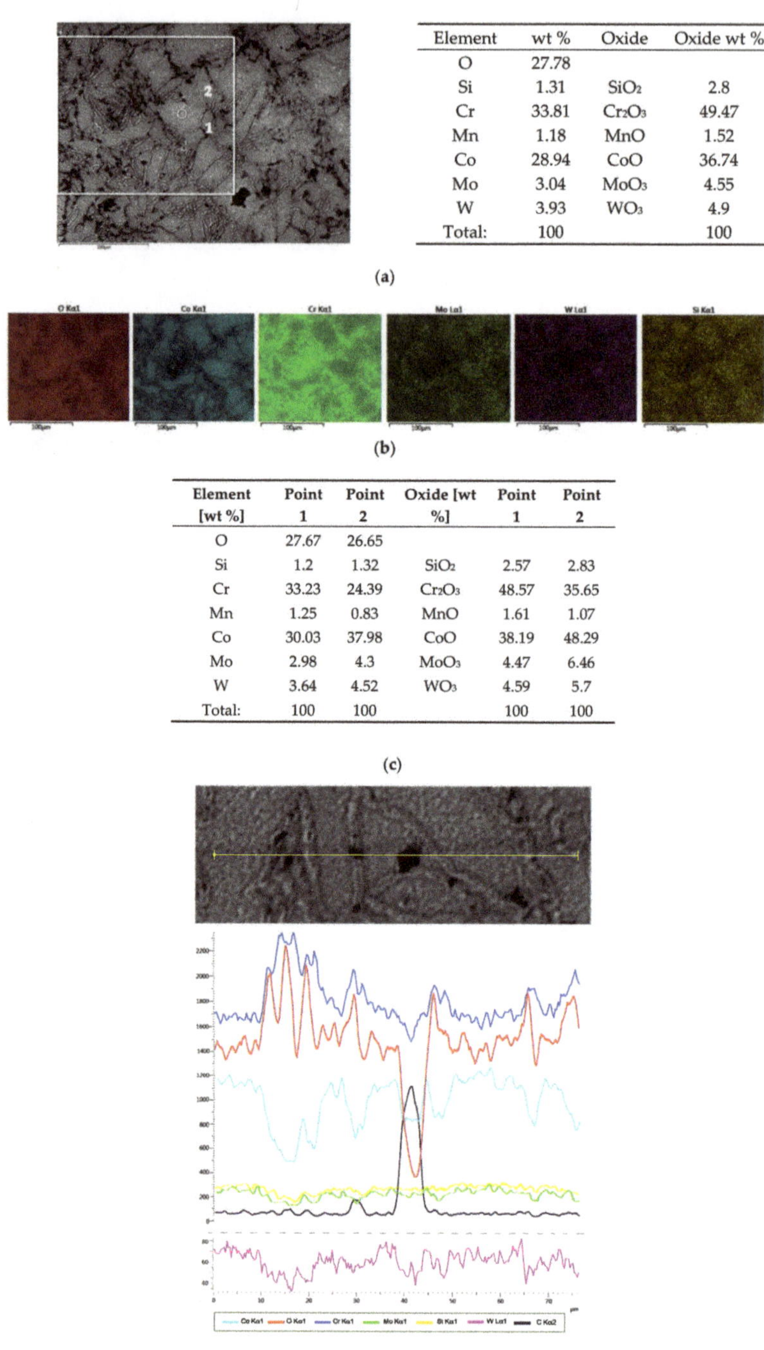

Element	wt %	Oxide	Oxide wt %
O	27.78		
Si	1.31	SiO_2	2.8
Cr	33.81	Cr_2O_3	49.47
Mn	1.18	MnO	1.52
Co	28.94	CoO	36.74
Mo	3.04	MoO_3	4.55
W	3.93	WO_3	4.9
Total:	100		100

(a)

(b)

Element [wt %]	Point 1	Point 2	Oxide [wt %]	Point 1	Point 2
O	27.67	26.65			
Si	1.2	1.32	SiO_2	2.57	2.83
Cr	33.23	24.39	Cr_2O_3	48.57	35.65
Mn	1.25	0.83	MnO	1.61	1.07
Co	30.03	37.98	CoO	38.19	48.29
Mo	2.98	4.3	MoO_3	4.47	6.46
W	3.64	4.52	WO_3	4.59	5.7
Total:	100	100		100	100

(c)

(d)

Figure 7. SEM EDS results for the SLM2 sample: (**a**) SEM-SE micrograph showing the morphology with average EDS results from the marked area; (**b**) maps presenting the surface distribution of O, Co, Cr, W, Mo, and Si; (**c**) EDS results for marked points; (**d**) SEM-SE micrograph with corresponding EDS line scan results.

Table 4. contains data summarizing the contribution of particular oxides covering the surface of thermally treated alloys (based on the quantitative EDS results concerning the average oxides contribution (panels (a) of Figures 3–7)).

Table 4. Oxide content on the alloys' surface after the HT process.

Oxide [wt %]	CST1	CST2	MSM	SLM1	SLM2
CoO	55.76	52.26	41.1	36.43	36.74
Cr_2O_3	35.23	33.83	49.6	56.2	49.47
MoO_3	-	5.82	5.24	4.8	4.55
SiO_2	2.59	3.1	2.28	1.58	2.8
WO_3	6.42	4.99	-	-	4.9
MnO	-	-	1.78	0.99	1.52

According to Table 4, one can state that the average amount of Cr_2O_3 covering surfaces after the HT was the highest using both SLM alloys, decreased using MSM samples, and reached the lowest amount using CST alloys.

The (a) panels of Figures 3–7 show the SEM-SE images of the oxidized surfaces representing all examined specimens.

For the CST samples, the distribution of elements was not uniform; consequently, the amount of different oxides formed during the HT differed at the examined points.

The surface of the first cast alloy, CST1 (Figure 3a), reflected the microstructure of the grains coated with oxides. Some parts of the areas were enriched with Cr, Si, and W and slightly lower in Co (according to the EDS results (Figure 3b,d). The same regularity occurred in the oxides' distribution—the amount of CoO, Cr_2O_3, WO_3, and SiO_2 formed during the HT differs in the examined spots (Figure 3c).

The oxide layer covering the CST2 sample replicated a typical dendritic structure featuring a dendritic matrix and interdendritic areas with some precipitations visible despite the HT effects (Figure 4a). The interdendritic areas also presented an inhomogeneous structure consisting of small, rounded, bright inclusions (an arrow in Figure 4d) and grey zones.

Unevenly distributed Co, Cr, Mo, W, and Si characterized the oxides covering the surface of CST2 according to the surface chemical composition tests (the maps presenting the surface composition, points, and linear distribution of particular elements—Figure 4b–d, respectively). Co and Cr mainly concentrated in the oxides covering the matrix space, while the oxides covering the interdendritic zones were enriched with Mo, W, and Si. The observation remains in compliance with the content of MoO_3, WO_3, SiO_2, and Cr_2O_3 measured in the marked spots (Figure 4c). The areas of rounded precipitations were highly enriched with Cr, W, and Si. The distribution of oxygen shows its increased content in the dendritic zones (Figure 4d).

The oxidized surface of the MSM sample is shown in Figure 5a. A layer consisting of the same type of oxides covers the surface. Bolder oxides, probably along the grain boundaries of the base alloy, could also be observed. Pores of diffuse distribution were also visible.

The results of chemical microanalysis showed a more uniform arrangement of Co, Cr, Mo, Si, and oxygen (Figure 5b,d) when both CST samples were compared. Additionally, the differences in the amount of oxides present at the marked points on the surface were less significant (Figure 5c).

The microstructures of the oxide layers covering both SLM specimens were much more uniform than those observed on the cast specimens' surfaces (Figures 6a and 7a). An inconsiderable amount of randomly distributed small pores was also visible.

In the specimen SLM1 after HT, the oxide layer was continuous, and bolder oxides covered the alloys' grain boundaries. The EDS studies (both presenting the surface distribution (Figure 6b) and line profile of the elements (Figure 6d)) showed a higher content of Cr and O and a lower content of Co and Mo in the oxides forming the layer over the grain boundaries. These conclusions corresponded with the results concerning the amount of CoO, Cr_2O_3, and MoO_3 in the marked spots (Figure 6c).

Similarities in the oxides' microstructure and the Cr, O, Co, and Mo distributions were evidenced by the SLM2 specimen (Figure 7b,d), where the distributions of W and Si are also similar to those of Co and Mo (Figure 7d). The amount of Cr_2O_3 was higher in the oxides covering the boundary areas, while the content values of CoO, WO_3, MoO_3, and SiO_2 remained lower, contrary to the oxides covering the grain areas (Figure 7c).

3.2. Corrosion Properties Examinations

The values of corrosion current density j_{corr}, corrosion potential E_{corr}, and polarization resistance R_p for all examined samples are presented in Table 5.

Table 5. The corrosion parameters obtained for the examined alloys in 0.9% NaCl solution with reference to the saturated calomel electrode (SCE) in potential range from -1 to 2 V.

Sample	j_{corr} [A/cm^2]	E_{corr} [V]	R_p [Ohm/cm^2]$\cdot 10^3$
CST1	$5.65 \times 10^{-7} \pm 0.19$	-0.464 ± 0.016	46.632 ± 0.055
CST2	$4.07 \times 10^{-6} \pm 0.14$	-0.524 ± 0.011	15.662 ± 0.0085
MSM	$2.6 \times 10^{-7} \pm 0.07$	-0.481 ± 0.009	99.774 ± 0.0017
SLM1	$2.05 \times 10^{-7} \pm 0.08$	-0.324 ± 0.005	168.114 ± 0.096
SLM2	$4.2 \times 10^{-8} \pm 0.2$	-0.459 ± 0.003	652.988 ± 0.138

Potentiodynamic characteristics for polarization in the range from -1.0 to 2.0 V in a semi-logarithmic arrangement are presented in Figure 8.

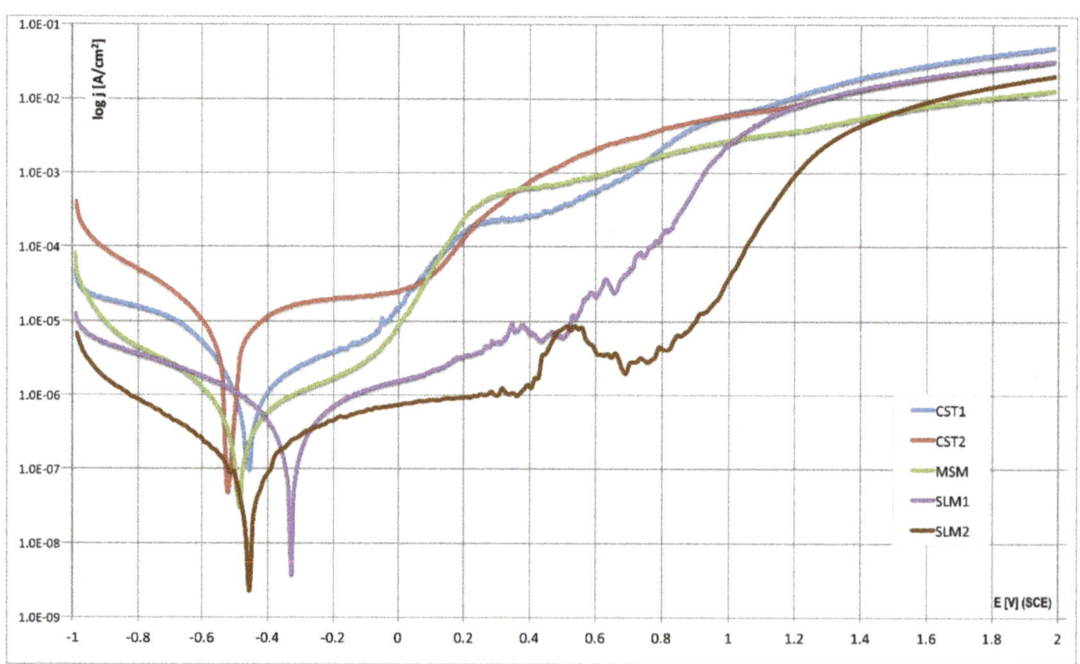

Figure 8. Potentiodynamic curves obtained for examined samples.

The values of the potential at which the corrosion current increased suddenly (the so-called breakdown potential E_{br}), as shown in Table 6.

Table 6. Values of the breakdown potential calculated from the potentiodynamic curves.

Sample	CST1	CST2	MSM	SLM1	SLM2
Ebr [V]	-0.129 ± 0.007	0.108 ± 0.006	-0.054 ± 0.003	0.320 ± 0.0015	0.428 ± 0.0025

According to Table 5, CST2 presented the highest corrosion current density (j_{corr} = $5.65 \times 10^{-7} \pm 0.19$ A/cm^2). The SLM2 sample had the best corrosion results, with a corrosion current density equal to $4.2 \times 10^{-8} \pm 0.2$ A/cm^2. The potentiodynamic curves for SLM alloys were the most shifted in the right direction. During the anodic scan, the values of the passive currents were lower in comparison to those of the CST samples and the MSM specimen.

The so-called breakdown potential's value was another parameter used for evaluating corrosion resistance. It was demonstrated by a significant increase in the current in the anodic domain. The breakdown potential was more anodic for SLM2 than for other samples ($E_{br} = 0.428 \pm 0.0025$ V) and lowest in CST1 (being equal to -0.129 ± 0.007 V (Table 6)).

For all examined data (j_{corr}, E_{corr}, R_p, and E_{br}), the p-value corresponding to the F-statistic of the one-way ANOVA was lower than 0.05, suggesting that one or more treatments were significantly different. The Tukey post-hoc test identified that, for j_{corr}, R_p, and E_{br}, all of the pairs of treatments were significantly different from each other. For E_{corr}, in CST1 and SLM2 treatments, there was no significant difference.

The microstructures of the alloys after potentiodynamic polarization tests are shown in Figure 9a–e. Additionally, Figure 10 presents the results of the EDS line profile obtained for the CST1 sample. Figures 11–14 show the point analysis results for the rest of the oxidized samples after CT (in points marked on the SEM-SE images).

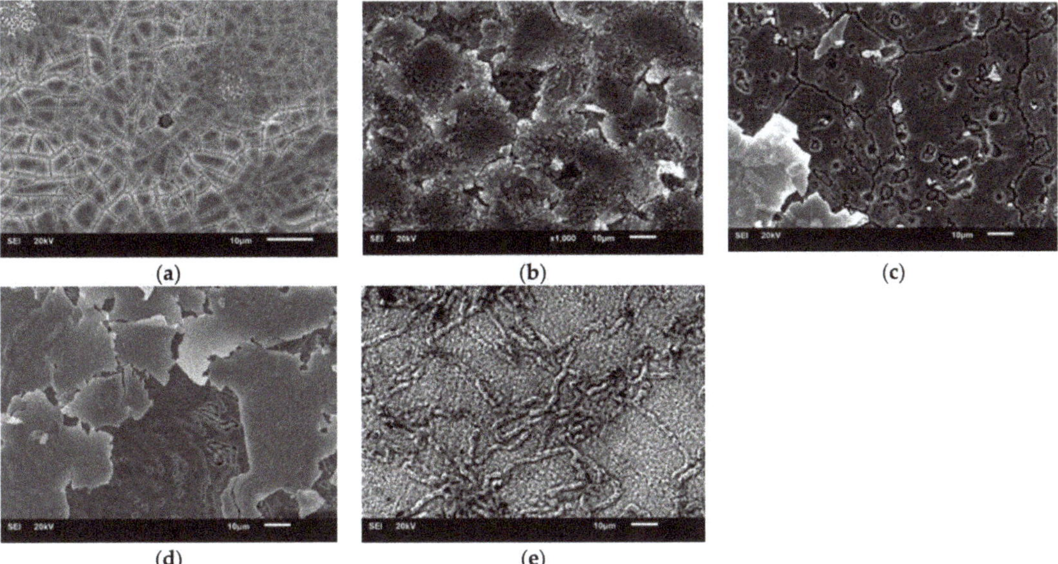

Figure 9. SEM SE images of samples' surfaces after CT: (**a**) CST1; (**b**) CST2; (**c**) MSM; (**d**) SLM1; (**e**) SLM2.

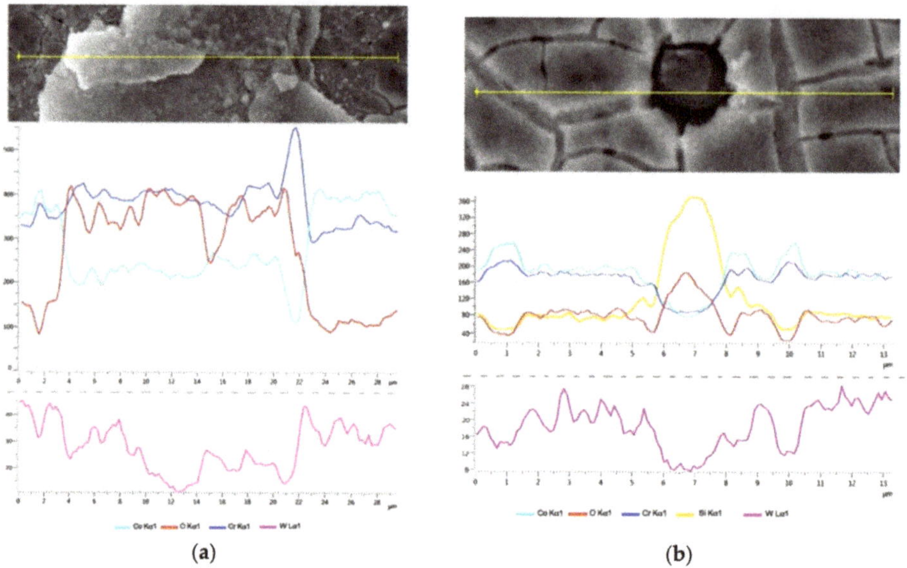

Figure 10. SEM-SE micrograph of CST1 with corresponding EDS line scan results: (**a**) along two oxide layers; (**b**) through the precipitation.

Element [wt %]	Point 1	Point 2	Oxide [wt %]	Point 1	Point 2
O	32.95	35.27			
Si	0.5	1.02	SiO_2	1.06	2.19
Cr	19.89	28,64	Cr_2O_3	33.69	46.92
Fe	0.32	0,24	FeO	0.42	0.3
Co	34.38	26.36	CoO	48.66	39,17
Mo	4.48	3.11	MoO_3	6.73	4.67
W	7.48	5.36	WO_3	9.44	6.76
Total:	100	100		100	100

Figure 11. Results of EDS point analyses for the CST2 sample after corrosion examinations with a chart comparing oxide content.

Element [wt %]	Point 1	Poin t 2	Oxide [wt%]	Point 1	Point 2
O	25.38	25.88			
Si	0.62	0.48	SiO_2	1.32	1.03
Cr	17.71	21.66	Cr_2O_3	25.88	31.66
Mn	0.33	0.58	MnO	0.43	0.75
Co	50.63	46.12	CoO	64.37	58.64
Mo	5.33	5.28	MoO_3	8.0	7.92
Total:	100	100		100	100

Figure 12. Results of EDS point analyses for MSM sample after corrosion examinations.

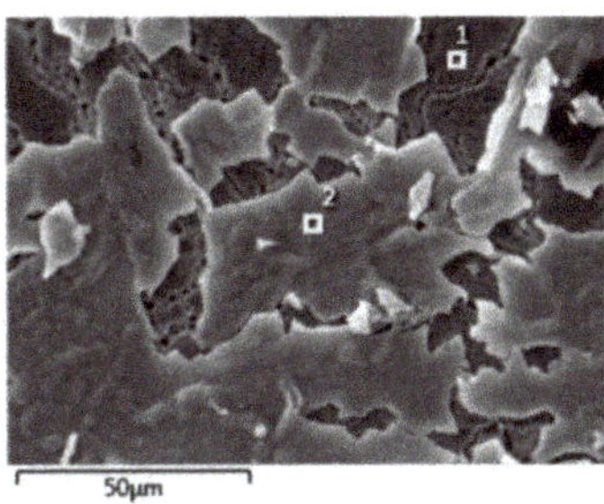

Element [wt %]	Point 1	Point 2	Oxide [wt %]	Point 1	Point 2
O	25.61	29.95			
Si	0.38	0.81	SiO_2	0.82	1.73
Cr	20.76	50.86	Cr_2O_3	30.34	74.34
Mn	0.43	1.22	MnO	0.55	1.58
Co	47.86	14.81	CoO	60.86	18.83
Mo	4.96	2.35	MoO_3	7.43	3.52
Total:	100	100		100	100

Figure 13. Results of EDS point analyses for the SLM1 sample after corrosion examinations.

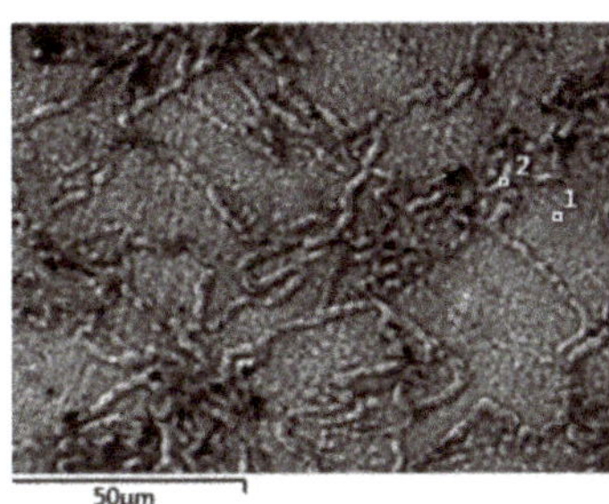

Element [wt %]	Point 1	Point 2	Oxide [wt %]	Point 1	Point 2
O	27.27	29.26			
Si	1.32	0.96	SiO_2	2.83	2.05
Cr	29.25	46.47	Cr_2O_3	42.75	67.92
Mn	1.11	2.04	MnO	1.44	2.63
Co	33.19	17.55	CoO	42.2	22.32
Mo	3.68	1.64	MoO_3	5.52	2.46
W	4.18	2.08	WO_3	5.27	2.62
Total:	100	100		100	100

Figure 14. Results of EDS point analyses for the SLM2 sample after corrosion examinations.

Figure 9a–f presents the state of the samples' surfaces after the corrosion tests (CT). The surface damage with corrosion pits could be observed only in the CST2 sample (Figure 9b).

The areas covered by a cracked oxide layer were visible on the surface of the CST1, CST2, MSM, and SLM1 samples (Figure 9a–d). Signs of slight "etching" occured around the pores present on the MSM surface. The uniform oxide layer, similar to the one created after the HT, covered the surface of the SLM2 sample (Figure 9e).

The CST1 alloy was covered by cracked oxides, with some randomly distributed areas of outer, flake-like-shaped oxides located on the inner cracked layer. The EDS line analysis (Figure 10a,b) demonstrated the presence of Cr, O, Co, and W with a fairly even distribution (taking into account different oxides' thicknesses along the analyzed line). The rounded precipitation (Figure 10b) in the central part of the analyzed line probably demonstrated an oxide rich in Si, covering the Si-rich phase that was present in the primary alloy structure.

The surface of the electrochemically tested CST2 alloy was also covered by a cracked oxide layer mainly made of CoO and Cr_2O_3 (Figure 11). The upper oxide layer had a higher amount of SiO_2 and a lower quantity of MoO_3 in comparison to the inner one.

The appearance and distribution of oxides appeared to be similar in the MSM alloy's surface after CT; however, the quantity of external oxides was decreased. Some signs of preferential dissolution at the grain boundaries could be observed in the MSM sample after corrosion tests (Figure 12), but there was no evidence of typical corrosion damage. Moreover, the type and amount of oxides present at different points on the surface remained similar (Figure 12).

The microstructure and chemical composition of the oxides examined at representative points of the SLM1 surface after CT (Figure 13) confirmed that the flake-shaped upper oxides were enriched with the Cr_2O_3 oxide and depleted of CO.

The microstructure of the SLM2 surface (Figure 14) remained almost the same as it had been before the CT. There was no evidence of a flake-shaped upper oxide layer or any corrosion signs. The composition of oxides in the bolder layer was the same as it was inside of the grains.

4. Discussion

Some studies investigated the impact of HT (being the result of porcelain firing) on the corrosion properties of dental base metal alloys obtained by different technologies. The high temperature reached during porcelain firing caused the creation of oxide layers, which in turn influenced the alloys' corrosion resistance. Different manufacturing methods leading to different alloys' morphologies may change the nature of the surface oxidation, hence influencing their corrosion performance.

The corrosion of Co–Cr dental alloys in the oral environment was inhibited by the existence of a Cr oxide layer on the surface [27,32,33]. Thus, it was crucial to determine how the structure of an alloy (which depends on the manufacturing method) influences the structure of the oxide/oxide layer formed during the HT.

In the CST1 sample, typical dendritic grains dominated before the HT (Figures 1a and 2a), and the intermetallic phases were precipitated along the grain boundaries. Cr, W, and Si were unevenly distributed in cobalt. The structure of the above cast alloy was described in the literature [34–36] as coarse-grained with solidification segregation that results in inhomogeneous material quality.

The microstructure of the oxide layer covering the CST1 (as a result of the HT) consisted of oxides presenting varied amounts of Cr, Si, W, and Co. The amount of CoO, Cr_2O_3, WO_3, and SiO_2 formed during the HT was different in the examined spots. Considering these results, we assumed that the structure of the oxide layer examined in the present study (Figure 3b–d) partly reflected the microstructure of the alloy before the HT.

An analysis of the microstructure of the untreated CST2 along with the results of the EDS tests (Figures 1b and 2b) revealed that the sample presented a matrix featuring a typical dendritic band structure, in addition to the interdendritic precipitations of irregular size. Apart from the inhomogeneous dendritic structure typical for cast elements, the CST2 also contained precipitations rich in Mo, W, and Si. Wylie et al. stated that Mo moves from the solid solution of dental alloys to the precipitations nucleating at the junction of the second phase and the matrix. Thus, their presence is considered undesirable because they deteriorate the corrosion resistance [36]. Mo is a common addition to alloys and promotes resistance to pitting and crevice corrosion. Y.S. Al Jabbari, T. Matkovic, and Karpuschewski et al. [37–39] described the presence of said kinds of precipitations and their negative effect on corrosion resistance and mechanical properties.

The oxidized CST2 specimen retained a dendritic solidification microstructure with more pronounced dendrites as well as precipitates enriched with W, Mo, and Si (Figure 4a–d). The inhomogeneous oxide layer, created as a result of the HT, affected the low corrosion resistance of the CST2 sample examined in the present paper.

This conclusion remains in accordance with results obtained by Xin et al. [40]. They investigated the characteristics of the surface and corrosion behavior of cast and SLM cobalt–chromium (Co–Cr) alloys before and after PFM firing. After the test, it was stated that both groups preserved the structure that they had before the HT.

Qiu et al. (who examined the corrosion behavior and surface properties of cast Co–Cr and two Ni–Cr dental alloys before and after simulated porcelain firing) also proved that the firing process had little effect on the microstructure of the Co–Cr alloy, since no homogenization was detected in the microstructure of the Co–Cr alloy. The examined specimens showed a noticeable dendritic structure with interdendritic areas and precipitations [14]. Ardlin, Dahl, and Tibballs investigated the structure and corrosion behavior of different dental alloys (among which were Co–Cr base alloys) before and after oxidation, simulating the temperature cycles for the application of ceramic to metal. They stated that the heating cycle did not appear to affect the Co–Cr base alloys' microstructure [41].

The MSM alloy examined in this study was produced by the use of technology based on powder metallurgy methods. These methods, compared to traditional casting, enabled us to achieve a product with a more cognate structure (also confirmed in this paper) [42]. The MSM before the HT presented a structure that was common for sintered material, i.e., with grains of different sizes and pores present both in grain boundaries and inside the grains (Figure 1c). Pores were attributed to the necking of metallic powder during sintering and the possible remainder of the organic binder; their presence was also reported in other studies [43,44]. The distribution of Co, Cr, and Mo was even, according to the EDS results (Figure 2c).

The MSM alloy sample after the HT showed no changes in the chemical composition of the oxide layers covering the surface, apart from the appearance of oxygen (following the EDS maps of element distribution and the linear and point analyses from chosen areas (Figure 5a–d)). The distribution of Co, Cr, and Mo was also more uniform in the sintered sample, which indicated a more homogeneous composition of oxides than the ones that form on CST alloys. Pores were distributed in both visible grains and grain boundaries.

Before the HT, the SLM1 alloy displayed a homogeneous microstructure consisting of grains of different size with small pores and some bright, rounded precipitations (Figure 1d). The result of EDS line scans along the line crossing this phase indicated a higher amount of Mo and Si and a decrease in Co (Figure 2f). Apart from that, the distribution of the elements was uniform (Figure 2d).

The primary microstructure of SLM2 alloy consisted of grains with arch-shaped boundaries (Figure 1e). The sample presented a uniform distribution of all examined elements (Figure 2e).

As described in the Results section, both thermally treated SLM alloys (SLM1, Figure 6a–d, and SLM2, Figure 7a–d) showed a homogeneous oxide microstructure, and a small amount of randomly distributed pores was also detected. The chemical compositions of the oxide layers proved that the main phases were Cr_2O_3 and CoO. The microstructure of both SLM samples was characterized by the oxides' swellings present in the grain boundary regions. The compositions of the oxides at the points located in the grain boundaries and grains showed no difference. Moreover, the highest percentage of Cr_2O_3 among all alloys analyzed occurred in the SLM samples (Table 4).

Considering the microstructures of the oxide layers covering the examined alloys after the HT, we can conclude that all specimens retained their structure. This statement is in accordance with the results obtained by Jabbari et al. [3], who compared the microstructure, porosity, mechanical properties, ion release, and tarnish resistance of Co–Cr alloys developed by conventional casting (CST), milling (MIL), selective laser melting (SLM), and milling soft metal (MSM). They concluded, among others, that the thermal treatment of porcelain firing did not affect the microstructural features of CST, MIL, and SLM groups. Additionally, Li et al., who investigated the bond strengths of porcelain to cobalt-chromium alloys made by casting, milling, and selective laser melting, stated that the microstructures of the cast, milled, and SLM alloys did not differ after porcelain firing [31].

The authors of the present study characterized oxide layers on the surfaces of thermally treated alloys in order to identify differences in their composition and structure to consider whether and how these diversities explain their different corrosive behavior. The investigations confirmed significant differences in the surface microstructure of the oxidized cast, sintered, and laser-sintered samples visible after HT simulating porcelain firing. The SLM technique and MSM method revealed a homogeneous oxide structure, contrary to both cast specimens. The porosity was observed in all groups of specimens, although SLM provided structures with limited internal porosity. The random slight porosity observed for the SLM samples was probably caused by an incorrect adjustment of operating conditions, and this was also reported by others [45,46].

All of the examined samples formed chromium oxide films, but other metallic elements were present in the oxides, especially the ones that covered the second phases of the CST1 and CST2 alloys.

The studies available in the literature were mostly focused on the influence of different modifications in alloys' compositions on the corrosion resistance after oxidation, rather than on the effect of the primary microstructure on their corrosion behavior after the HT. For example, K. Yamanaka et al. [28,29] described the effects of the addition of silicon and boron on oxide films formed on Co–Cr–W base dental alloys at 1273 K for 15 min in air. They proved that the addition of Si and B could make the dense Cr_2O_3 phase resistant to changes and restrict the formation of Co oxides. Y. Lu et al. investigated the oxidation behavior of Co–Cr–W alloys with Cu additions and stated that Cu content in the film structure could affect the thickness of the Cr oxide regions, accelerating the oxidation of Co–Cr–W alloys [45].

Other studies concerning the impact of production methods and various HT parameters were mainly aimed at assessing the quality of the connection with ceramics, but the conclusions on the structure of oxides could also be relevant to corrosive considerations. J. Li et al. concluded that predominantly Cr oxides were formed on the Co–Cr alloys' surface during oxidizing HT and reported that similar morphologies and similar amounts of oxides were observed for the cast and milled alloy surfaces [31]. These conclusions were partly in contrast to the ones presented in this article; however, the CNCM (computer numerical control milling) method was also based on casting technology. Serra-Prat et al. compared the shear bond strength of dental porcelain to cast, milled, and laser-sintered cobalt-chromium alloys. They reported that differences in alloy microstructures caused by different manufacturing methods might also simultaneously change the nature of surface oxidation, significantly modifying adhesion behavior as a result [46].

In the present article, conclusions concerning the microstructures of oxidized surfaces are reflected in the corrosion examination results (Tables 5 and 6; Figures 8–14).

The determination of icorr and Ecorr values was carried out in a narrow range of potentials (as described in the methodology), but the paper presents only icorr and Ecorr values obtained from the extrapolation of "Tafel regions", but no graphs (graphic results) of polarization curves considered adequate for Tafel extrapolation were published.

Considering the results of the statistical analysis, the null hypothesis could be rejected. Thick oxides, forming inner and outer cracked layers, covered all examined alloys after the CT, except for the SLM2. Hodgson et al. described similar microstructures of anodically polarized Co–Cr alloys [47]. They examined the passive and transpassive behavior of alloys and showed that the oxide film became significantly thicker, a change in its composition took place, and the dissolution rate increased in the transpassive region. In the passive region, Cr was present in the film mainly as Cr(III) oxide and with a small amount of Cr(III) hydroxide. The studies performed by the authors indicated that (at transpassive potentials) a thick and unstable oxide film (not highly protective) constantly covered the surface.

The current densities, while the anodic polarization was stopped at a potential equal to 2 V, were the same in the CST2 and SLM1 alloys and equal to 3×10^{-2} A/cm^2. While comparing the morphologies of these alloys, it was observed that, contrary to CST2, the SLM1 alloy did not exhibit any localized corrosion damage (Figure 9a,b,d).

Less evident but also comparable current densities were found for the SLM2 and MSM alloys (with values of 1.4×10^{-2} A/cm^2 and 2×10^{-2} A/cm^2) with polarizations equal to 2 V. In these cases, the corrosion morphology was different (Figure 9c,e)

The potentiodynamic polarization tests allowed us to state that the corrosion resistance of the SLM and MSM alloys after HT was improved in comparison to the cast samples (CST1 and CST2).

According to the values of the corrosion parameters obtained, corrosion resistance decreased in the following order: SLM2 > SLM1 > MSM > CST1 > CST2. The shapes of all potentiodynamic curves remained similar and typical for passive materials. The values of the passive currents for the SLM alloys were lower in comparison to the CST samples and MSM specimens, proving an improved protection efficiency not only in the corrosion potential domain, but also in the passive potential range. A transpassive area, with a significant increase in the anodic current, followed the passive range of the anodic curve. Eliaz, who described passivity and its breakdown of Co–Cr biomaterials, explained that the current increase was due to transpassive dissolution by oxidation of the passive Cr_2O_3-rich film into soluble CrO^{2-}_4 species and was not related to pitting corrosion [48].

The increase in the breakdown potential E_{br} for the SLM alloys proved that a homogeneous structure of oxides improved their corrosion resistance.

The results of the potentiodynamic polarization tests demonstrated that the corrosion potentials of the SLM alloys shifted to more positive values, and the corrosion current densities were lower in comparison to the MSM and cast alloys. The SLM1 and SLM2 alloys exhibited a similar corrosion behavior, and their higher corrosion resistance can be attributed to the homogeneous oxide layer formed during the HT process. We assumed that the last one is the result of the homogeneous structure of the alloy.

In the course of the curves for SLM1 and SLM2 samples obtained by laser sintering, temporary increases in the current values are visible in the area of potentials above ~0.35–0.45 V. The surface of the samples "at the entrance" was covered with oxides from the firing simulation process. This phenomenon may be related to the partial transformation process—the transition of Cr to a higher oxidation state (corresponding to the oxidation reactions of Cr at a higher valence level. (Dissolution by oxidation of the Cr_2O_3-rich passive film into soluble CrO^{2-}_4 species). There have been reports of this type concerning the behavior of Co–Cr–Mo alloys using potentiodynamic studies in the potential ranges of 500–700 mV in HBSS, where the authors explained the occurring increases in the value of currents by the transition of chromium to a higher degree of oxidation in oxides. Further research and immersion tests at peak potentials will allow to investigate these new products.

Additionally, the results of the potentiodynamic polarization test for the oxidized MSM alloy indicated an improved corrosion resistance in comparison to the cast specimens. The related literature included only studies evaluating the structure and corrosion resistance of non-thermally treated MSM elements. The results of these studies revealed that MSM alloys present a more improved corrosion resistance than casting alloys, due to the uniformity of the structure. H. Kim et al. [49] compared the microstructure and in vitro biocompatibility of Co–Cr alloys fabricated by casting and MSM methods. They reported that the microstructure of the MSM specimen was characterized by a fine grain size structure and showed consistently significantly smaller releases of Co ions in comparison with cast specimens. D. Rylska et al. conducted examinations of the structure and corrosion resistance of MSM Co–Cr base elements and confirmed that the sintered alloy exhibited better self-passivating behavior in comparison to cast material [50].

The thermally treated MSM alloy exhibited a lower corrosion resistance than the SLM alloys, despite exhibiting a substantial homogeneous distribution of elements in the oxide layer (Figure 5b–d). The possible reason for reduced corrosion resistance may be the porosity in the MSM alloy's structure. The existence of a direct correlation between pores and corrosion resistance was already discussed in other studies and remains in accordance with the results of the present study. For example, Buciumeanu et al. reported decreased corrosion resistance of hot-pressed Co–Cr alloys related to their increasing porosity [51]. Krasicka et al. investigated the corrosion properties of sintered compacts made of a Co–Cr–Mo alloy and concluded that the structure and porosity of the materials influenced the corrosion resistance of the investigated samples [52]. They revealed that these sintered specimens did not exhibit typical passive behavior or a current plateau in the passive region in the anodic polarization curves. They also showed that the examined compacts had higher passive current densities in the passive range than cast alloys. The authors suggested that the higher values of the passive current densities for the sintered samples were the result of their real surface areas, which were higher than the nominal areas. This conclusion may be also relevant to the present study.

Cast alloys after the HT, especially CST2, exhibited lower corrosion resistance. The oxide layer covering the CST1 surface was characterized by an uneven distribution of elements with areas enriched with Co and depleted of Cr, W, and Si (according to the EDS results; Figure 3b,d). Such distribution influenced the corrosion resistance (Tables 5 and 6; Figure 8).

The CST2 coarse-grained structure with microsegregation in dendrites and inclusions such as intermetallic phases (Figures 1b and 2b) influenced the kind and distribution of oxides formed after the HT. The presence of oxides of various compositions was responsible for the lowest corrosion resistance, which was confirmed in the results of corrosion tests. The CST2 alloy presented the highest j_{corr} and the lowest E_{br} among the examined specimens (Tables 5 and 6; Figure 8).

These results were in line with the conclusions presented by X-Z. Xin et al. [4]. They reported that, following firing, SLM samples made of Co–Cr alloys displayed significantly improved corrosion resistance compared with traditional cast specimens at pH 2.5. Microstructural differences between oxidized cast and SLM alloys and a thicker oxide layer on SLM explained the improved corrosion resistance on the surface of SLM specimens.

5. Conclusions

Within the limitations of this study, the main results can be concluded as follows:

1. The corrosion resistance after HT simulating porcelain firing was affected by the microstructure of the oxides covering surface of alloys.
2. For cast alloys, the coarse secondary phases were detrimental to the formation of protective homogeneous oxide films. The highest resistance to corrosion after the HT was observed in SLM alloys due to the microstructural homogeneity of the oxide layers.
3. The lowest corrosion resistance of cast alloys after HT was a result of their inhomogeneous oxide layers. In terms of the MSM alloy, the lower corrosion resistance may be caused by the porosity.
4. The structure of oxides forming during PFM processes reflected the morphology of the primary alloys.

Some issues require more detailed approach and attention and they are:

1. We are aware that the methods and kinds of apparatus used in the present study, e.g., scanning electron microscopy and EDS, were inadequate for detailed examinations of oxide layers.
2. Further research is needed to obtain more precise information about oxide layer structures and compositions (for example, X-ray photoelectron spectroscopy). Such research, which would include cross-section examinations, will enable us to compare the thicknesses of oxide layers on the surfaces of Co–Cr alloys and determine the possible relationship between oxide thickness and the primary alloy's structure.
3. Further investigations should also be focused on simulating the high chemical aggression of physiological fluids in the oral cavity. Many factors can change the pH of saliva, in ranges between 3.5 and 8.3, and the pH of saliva may contribute to the corrosion of dental implants.
4. The next objective would be to characterize the effect of saliva pH on the passive behavior of these alloys after PFM.
5. The authors are aware that with such a small number of tests, the test power allows only to determine that the results are not varied by accident. Increasing sample size will boost the statistical power of the tests carried. The current study should be considered as a pilot study and further investigations should be focused on the minimum number of groups (for example only SLM group with CST group as a control one) with increased sample size.

From the practical point of view, the conducted experiments provide, in the humble opinion of the authors, a valuable results on the influence of manufacturing method, which strongly affects the structure of tested materials and therefore their corrosion behavior, which can strongly affect their usefulness in dental applications.

Author Contributions: Conceptualization: J.S. and D.R.; methodology: J.S. and D.R.; validation: J.S. and D.R.; formal analysis: D.R., B.J., and G.S.; investigation: D.R. and B.J.; resources: J.S., D.R., and G.S.; writing—original draft preparation: D.R.; writing—review and editing: D.R.; visualization: D.R.; supervision: J.S. All authors have read and agreed to the published version of the manuscript.

Funding: This research received no external funding.

Institutional Review Board Statement: Not applicable.

Informed Consent Statement: Not applicable.

Data Availability Statement: Data sharing not applicable.

Conflicts of Interest: The authors declare no conflict of interest.

References

1. Wu, Y.; Moser, J.B.; Jameson, L.M.; Malone, W.F.P. The effect of oxidation heat treatment on porcelain bond strength in selected base metal alloys. *J. Prosthet. Dent.* **1991**, *66*, 439–444. [CrossRef]
2. Suleiman, S.H.; von Steyern, P.V. Fracture strength of porcelain fused to metal crowns made of cast, milled or laser-sintered cobalt-chromium. *Acta Odontol. Scand.* **2013**, *71*, 1280–1289. [CrossRef]
3. Al Jabbari, Y.S.; Barmpagadaki, X.; Psarris, I.; Zinelis, S. Microstructural, mechanical, ionic release and tarnish resistance characterization of porcelain fused to metal Co–Cr alloys manufactured via casting and three different CAD/CAM techniques. *J. Prosthodont. Res.* **2019**, *63*, 150–156. [CrossRef]
4. Kovalev, A.; Mishina, V.; Titov, V.; Moiseev, V.; Tolochko, N. Selective laser sintering of steel powders to obtain products based on SAPR models. *Metallurgist* **2000**, *4*, 206–209. [CrossRef]
5. Iseri, U.; Ozkurt, Z.; Kazazoglu, E. Shear strengths of veneering porcelain to cast, machined and laser-sintered titanium. *Dent. Mater. J.* **2011**, *30*, 274–280. [CrossRef]
6. Metikos-Hukovic, M.; Babic, R. Passivation and corrosion behaviours of cobalt and cobalt-chromium-molybdenum alloy. *Corros. Sci.* **2007**, *49*, 3570–3579. [CrossRef]
7. Manaranche, C.; Hornberger, R. A proposal for the classification of dental alloys according to their resistance to corrosion. *Dent. Mater.* **2007**, *23*, 1428–1437. [CrossRef]
8. Chang, J.-C.; Oshid, Y.; Gregory, R.L.; Andres, C.J.; Barco, T.M.; Brown, D.T. Electrochemical study on microbiology-related corrosion of metallic dental materials. *BioMed. Mater. Eng.* **2003**, *13*, 281–295. [PubMed]
9. Johnson, T.; Van Noort, R.; Stokes, C.W. Surface analysis of porcelain fused to metal systems. *Dent. Mater.* **2006**, *22*, 330–337. [CrossRef]
10. Eschler, P.Y.; Reclaru, L.; Luthy, H.; Blatter, A.; Larue, C.; Susz, C.; Bosch, J. Corrosion Testing of Cobalt-Chromium Dental Alloys doped with Precious Metals. *Eur. Cells Mater.* **2005**, *26*, 4358–4365.
11. Milosev, I. The effect of biomolecules on the behaviour of CoCrMo alloy in various simulated physiological solutions. *Electrochim. Acta* **2012**, *78*, 259–273. [CrossRef]
12. Bauer, S.; Schmuki, P.; Von der Mark, K.; Park, J. Engineering biocompatible implant surfaces. Part I: Materials and surfaces. *Prog. Mater. Sci.* **2013**, *58*, 261–326. [CrossRef]
13. Sang-Bae, L.; Ju-hye, L.; Woong-Chul, K.; Sae-Yoon, O.; Kyoung-Nam, K.; Ji-Hwan, K. Effect of different oxidation treatments on the bonding strength of new dental alloys. *Thin Solid Films* **2009**, *517*, 5370–5374.
14. Qiu, J.; Yu, W.-Q.; Zhang, F.-Q.; Smales, R.J.; Zhang, Y.-L.; Lu, C.-H. Corrosion behaviour and surface analysis of a Co-Cr and two Ni-Cr dental alloys before and after simulated porcelain firing. *Eur. J. Oral Sci.* **2011**, *119*, 93–101. [CrossRef]
15. Li, X.; He, J.; Zhang, W.; Jiang, N.; Li, D. Additive manufacturing of biomedical constructs with biomimetic structural organizations. *Materials* **2016**, *9*, 909. [CrossRef]
16. Vandenbroucke, B.; Kruth, J.P. Selective laser melting of biocompatible metals for rapid manufacturing of medical parts. *Rapid Prototyp. J.* **2007**, *13*, 196–203. [CrossRef]
17. Kim, K.B.; Kim, W.C.; Kim, H.Y.; Kim, J.H. An evaluation of marginal fit of three unit fixed dental prostheses fabricated by direct metal laser sintering system. *Dent. Mater.* **2013**, *29*, 91–96. [CrossRef]
18. Xin, X.Z.; Chen, J.; Xiang, N.; Wei, N. Surface properties and corrosion behaviour of Co-Cr alloy fabricated with selective laser melting technique. *Cell Biochem. Biophys.* **2013**, *67*, 983–990. [CrossRef]
19. Tuna, S.H.; Pekmez, N.Ö.; Kürkçüo, I. Corrosion resistance assessment of Co-Cr alloy frameworks fabricated by CAD/CAM milling, laser sintering, and casting methods. *J. Prosthet. Dent.* **2015**, *114*, 725–734. [CrossRef]
20. Krug, K.P.; Knauber, A.W.; Nothdurft, F.P. Fracture behaviour of metal-ceramic fixed dental prostheses with frameworks from cast or a newly developed sintered cobalt-chromium alloy. *Clin. Oral Investig.* **2015**, *19*, 401–411. [CrossRef]
21. Li, K.C.; Prior, D.J.; Waddell, J.N.; Swain, M.V. Comparison of the microstructure and phase stability of as-cast, CAD/CAM and powder metallurgy manufactured Co–Cr dental alloys. *Dent. Mater.* **2015**, *31*, 306–315. [CrossRef]
22. Adzali, N.M.S.; Jamaludin, S.B.; Derman, M.N. Effect of Sintering on the Physical and Mechanical Properties of Co-Cr- Mo (F-75)/HAP Composites. *Sains Malays.* **2013**, *42*, 1763–1768.

23. Quante, K.; Ludwig, K.; Kern, M. Marginal and internal fit of metal–ceramic crowns fabricated with a new laser melting technology. *Dent. Mater.* **2008**, *24*, 1311–1315. [CrossRef] [PubMed]

24. Akova, T.; Ucar, Y.; Tukay, A.; Balkaya, M.C.; Brantley, W.A. Comparison of the bond strength of laser-sintered and cast base metal dental alloys to porcelain. *Dent. Mater.* **2008**, *24*, 1400–1404. [CrossRef] [PubMed]

25. Xiang, N.; Xin, X.Z.; Chen, J.; Wei, B. Metal–ceramic bond strength of Co–Cr alloy fabricated by selective laser melting. *J. Dent.* **2012**, *40*, 453–457. [CrossRef] [PubMed]

26. Castillo-Oyague, R.; Osorio, R.; Osorio, E.; Sanchez-Aguilera, F.; Toledano, M. The effect of surface treatments on the micro roughness of laser-sintered and vacuum-cast base metal alloys for dental prosthetic frameworks. *Microsc. Res. Tech.* **2012**, *75*, 1206–1212. [CrossRef] [PubMed]

27. Koutsoukis, T.; Zinelis, S.; Eliades, G.; Al-Wazzan, K.; Rifaiy, M.A.; Al Jabbari, Y.S. Selective laser melting technique of Co–Cr dental alloys: A review of structure and properties and comparative analysis with other available techniques. *J. Prosthodont.* **2015**, *24*, 303–312. [CrossRef]

28. Yamanaka, K.; Mori, M.; Ohmura, K.; Chiba, A. Preventing high-temperature oxidation of Co–Cr-based dental alloys by boron doping. *J. Mater. Chem.* **2016**, *4*, 309–317. [CrossRef]

29. Yamanaka, K.; Mori, M.; Chiba, A. Surface characterisation of Ni-free Co–Cr–W-based dental alloys exposed to high temperatures and the effects of adding silicon. *Corros. Sci.* **2015**, *94*, 411–419. [CrossRef]

30. Li, J.; Ye, X.; Li, B.; Liao, J.; Zhuang, P.; Ye, J. Effect of oxidation heat treatment on the bond strength between a ceramic and cast and milled cobalt–chromium alloys. *Eur. J. Oral Sci.* **2015**, *123*, 297–304. [CrossRef]

31. Li, J.; Chen, C.; Liao, J.; Liu, L.; Ye, X.; Lin, S.; Ye, J. Bond strengths of porcelain to cobalt-chromium alloys made by casting, milling, and selective laser melting. *J. Prosthet. Dent.* **2017**, *118*, 69–75. [CrossRef] [PubMed]

32. Saji, V.-S.; Choe, H.-C. Electrochemical behaviour of Co–Cr and Ni–Cr dental cast alloys. *Trans. Nonferrous Met. Soc. China* **2009**, *19*, 785–790. [CrossRef]

33. Hsua, R.W.W.; Yang, C.C.; Huang, C.A.; Chen, Y.S. Electrochemical corrosion studies on Co–Cr–Mo implant alloy in biological solutions. *Mater. Chem. Phys.* **2005**, *93*, 531–538. [CrossRef]

34. Yamanaka, K.; Mori, M.; Kuramoto, K.; Chiba, A. Development of new Co–Cr–W-based biomedical alloys: Effects of microalloying and thermomechanical processing on microstructures and mechanical properties. *Mater. Des.* **2014**, *55*, 987–998. [CrossRef]

35. Yamanaka, K. Influence of carbon addition on mechanical properties and microstructures of Ni-free Co–Cr–W alloys subjected to thermomechanical processing. *J. Mech. Behav. Biomed. Mater.* **2014**, *37*, 1751–6161. [CrossRef] [PubMed]

36. Wylie, C.M.; Shelton, R.M.; Fleming, G.J.; Davenport, A.J. Corrosion of nickel-based dental casting alloys. *Dent. Mater.* **2007**, *23*, 714–723. [CrossRef]

37. Al Jabbari, Y.S.; Koutsoukis, T.; Barmpagadaki, X.; Zinelis, S. Metallurgical and interfacial characterization of PFM Co–Cr dental alloys fabricated via casting, milling or selective laser melting. *Dent. Mater.* **2014**, *30*, 79–88. [CrossRef] [PubMed]

38. Matkovic, T.; Matkovic, P.; Malina, J. Effects of Ni and Mo on the microstructure and some other properties of Co–Cr dental alloys. *J. Alloys Compd.* **2004**, *366*, 197–293. [CrossRef]

39. Karpuschewski, B.; Pieper, H.J.; Krause, M.; Doring, J. CoCr is not the same: CoCr-blanks for dental machining. In *Future Trends in Production Engineering*; Schuh, G., Neugebauer, R., Uhlmann, E., Eds.; Springer: Berlin/Heidelberg, Germany, 2013; pp. 261–274.

40. Xin, X.-Z.; Che, J.; Xiang, J.; Gong, Y.; Wei, B. Surface characteristics and corrosion properties of selective laser melted Co–Cr dental alloy after porcelain firing. *Dent. Mater.* **2014**, *30*, 263–270. [CrossRef]

41. Ardlin, B.I.; Dahl, J.E.; Tibballs, J.E. Static immersion and irritation tests of dental metal- ceramic alloys. *Eur. J. Oral Sci.* **2005**, *113*, 83–89. [CrossRef]

42. Kim, H.R.; Jang, S.H.; Kim, Y.K.; Son, J.S.; Min, B.K.; Kim, K.H. Microstructures and mechanical properties of Co–Cr dental alloys fabricated by three CAD/CAM- based processing techniques. *Materials* **2016**, *9*, 596. [CrossRef] [PubMed]

43. Takaichi, A.; Suyalatu, S.; Nakamoto, T.; Natsuka, J.; Nomura, N.; Tsutsumi, Y.; Migita, S.; Doi, H.; Kurosu, S.; Chiba, A.; et al. Microstructures and mechanical properties of Co-29Cr-6Mo alloy fabricated by selective laser melting process for dental applications. *J. Mech. Behav. Biomed. Mater.* **2013**, *21*, 67–76. [CrossRef] [PubMed]

44. Reclaru, L.; Ardelean, L.; Rusu, L.C.; Sinescu, C. Co-Cr material selection in prosthetic restoration: Laser sintering Technology. *Solid State Phenom.* **2012**, *188*, 412–415. [CrossRef]

45. Lu, Y.; Lin, W.; Xie, M.; Xu, W.; Liu, Y.; Lin, J.; Yu, C.; Tang, K.; Liu, W.; Yang, K.; et al. Examining Cu content contribution to changes in oxide layer formed on selective-laser-melted CoCrW alloys. *Appl. Surf. Sci.* **2019**, *464*, 262–272. [CrossRef]

46. Serra-Prat, J.; Cano-Batalla, J.; Cabratosa-Termes, J.; Figueras-Àlvarez, O. Adhesion of dental porcelain to cast, milled, and laser-sintered cobalt-chromium alloys: Shear bond strength and sensitivity to thermocycling. *J. Prosthet. Dent.* **2014**, *112*, 600–605. [CrossRef] [PubMed]

47. Hodgson, A.W.E.; Kurz, S.; Virtanen, S.; Fervel, V.; Olsson, C.-O.A.; Mischler, S. Passive and transpassive behaviour of CoCrMo in simulated biological solutions. *Electrochim. Acta* **2004**, *49*, 2167–2178. [CrossRef]

48. Eliaz, N. Corrosion of Metallic Biomaterials: A Review. *Materials* **2019**, *12*, 407. [CrossRef]

49. Kim, H.R.; Kim, Y.K.; Son, J.S.; Min, B.K.; Kim, K.-H.; Kwon, T.-Y. Comparison of in vitro biocompatibility of a Co–Cr dental alloy produced by new milling/post-sintering or traditional casting technique. *Mater. Lett.* **2016**, *178*, 300–303. [CrossRef]

50. Rylska, D.; Sokołowski, G.; Konieczny, B.; Sokołowski, J. The structure and corrosive properties of the CoCr-base dental alloy obtained by soft material milling followed by sinterization. *J. Achiev. Mater. Manuf. Eng.* **2016**, *2*, 60–71. [CrossRef]

51. Buciumeanu, M.; Bagheri, A.; Souza, J.C.M.; Silva, F.S.; Henriques, B. Tribocorrosion behavior of hot pressed CoCrMo alloys in artificial saliva. *Tribol. Int.* **2016**, *97*, 423–430. [CrossRef]
52. Krasicka-Cydzik, E.; Oksiuta, Z.; Dabrowski, J.R. Corrosion testing of sintered samples made of the Co-Cr-Mo alloy for surgical applications. *J. Mater. Sci. Mater. Med.* **2005**, *16*, 97–202. [CrossRef] [PubMed]

Article

Fretting Wear in Orthodontic and Prosthetic Alloys with Ti(C, N) Coatings

Katarzyna Banaszek [1], Leszek Klimek [2], Jan Ryszard Dąbrowski [3] and Wojciech Jastrzębski [4,*]

[1] Department of General Dentistry, Chair of Restorative Dentistry, Medical University of Lodz, Pomorska 251, 92-217 Lodz, Poland; katarzyna.banaszek@umed.lodz.pl

[2] Institute of Materials Science and Engineering, Lodz University of Technology, Stefanowskiego 1/15, 90-924 Lodz, Poland; leszek.klimek@umed.lodz.pl

[3] Department of Materials and Production Engineering, Bialystok University of Technology, Wiejska 45C, 15-351 Bialystok, Poland; j.dabrowski@pb.edu.pl

[4] Department of Dental Technology, Chair of Restorative Dentistry, Medical University of Lodz, Pomorska 251, 92-217 Lodz, Poland

* Correspondence: wojciechjastrzebski@tlen.pl or wojciechjastrzebski@stud.umed.lodz.pl

Received: 21 October 2019; Accepted: 18 November 2019; Published: 21 November 2019

Abstract: Fretting occurs during orthodontic treatment or wearing prosthesis. Although weight of particles is marginal, the total releasing area is more of a concern due to amount and volume of molecules. The aim of the study was to examine the fretting wear resistance of orthodontic and prosthetic alloy Ni-Cr-Mo samples coated with Ti(C, N) and to compare them with samples without any coating. Five groups of cylindrical shape samples (S1–S5) made of Ni-Cr-Mo were coated with Ti(C, N) layers with different content of C and N. The control group (S0) was without layer. The alloys underwent fretting wear resistance tests with amplitude 100 μm, at frequency 0.8 Hz with averaged unit load: 5, 10, and 15 N for 15 min. The samples were subjected to microscopic observations using scanning electron microscope and a laser scanning microscope. Samples with Ti(C, N) coatings revealed higher fretting wear resistance. The wear in each case with Ti(C, N) coatings was over twice as low. The lowest wear and thus the highest resistance was demonstrated by sample S3 (1.02 μm) whereas in control group-S0 (2.64 μm). The use of Ti(C, N)-type coatings reduces the adverse effects of fretting wear, decreasing the amount of ions released during orthodontic treatment or wearing prosthesis.

Keywords: fretting; fretting wear; Ni-Cr-Mo; dental alloys; titaniumcarbonitride; Ti(C, N) coating

---

1. Introduction

One of the types of tribological wear is fretting wear [1,2]. It includes a group of phenomena related to micro-abrasive wear in loaded and nominally immovable joints, which can be mechanical, thermal, chemical, or electrical. Its main consequence is a drastic decrease of durability and shortened operational reliability of the devices. The wear of the surface layers of elements being in contact with each other takes place as a result of oscillational micro-shifts of the contacting surfaces [3–7]. Fretting can be accompanied by four basic mechanisms: adhesion, fatigue, abrasion, and corrosion. Adhesive damage to the surface and the formation of fatigue cracks causes the creation of wear particles, followed by their oxidation and hardening [8,9].

Linked, nominally immovable surfaces are in contact only through the roughness peaks being a consequence of the roughness. Therefore, the actual contact surface is only a small fraction of the nominal surface. In the existing roughness contacts, we can distinguish between four adhesion areas, which undergo elastic deformations, as well as areas of micro-slips. A consequence of the contact surfaces' small shifts of in respect of each other as a result of the applied external loads, which are cyclic in character, can be a contact load exerted by the axial force and the tangential force. Fretting wear has also been observed to occur as a result of free vibration of the construction. Respectively to different states, Neyman [8] differs: fretting wear as a loss of mass and volume in the surface layer, fretting corrosion where oxidation of the surface layer dominates and fretting fatigue that may occur during variable load. Respectively to the mentioned different states of loads and operation, we distinguish between "fretting fatigue" and "fretting wear" [10]. The presence of areas of adhesion and micro-slips is strictly connected with the occurrence of two different wear mechanisms in them. The adhesion areas undergo cracking in the mode of contact fatigue, whereas the micro-slip areas are subjected to adhesion wear. In both cases, we observe the formation of wear products, which remain in abrasive contact. If they exhibit high hardness, they work as an abradant, accelerating the process of wear, especially in its last, catastrophic phase. Fretting almost always occurs with chemical changes of the surface except in exceptional cases of high vacuum, inert atmospheres, or precious metal contact. In active environment, where fretting corrosion takes place, wear appears sooner and is much more intensive [6,11–15]. Fretting corrosion is formed in couples working in a corrosive environment. The tensions cause an increase of the surface energy and chemical reactivity. Also, in the case of fretting corrosion, products of wear are created, which are usually metal oxides [16]. Those can be the oxides removed from the surface or formed as a result of oxidation of particles of the abraded metal. These products work as an abradant; they are refined and hardened and their amount increases until the surface is separated with a layer of oxide molecules and the wear conditions stabilize. In the contact area, we also observe processes of materials' transfer with intensive oxidation [8]. The intensity of fretting depends on the type and value of the forcing applied to the joint, the value of the stresses with which it operates, and the aggressiveness of the environment [9].

Cases of destruction through fretting have been established not only in elements of machines and constructions but also in orthopedic implants, as well as prosthetic, orthodontic, and artificial heart elements [11,13,16–23]. The fretting phenomenon refers to most biomaterials, both metallic ones and polymers and ceramics [24–27]. Fretting is especially important in the environment of the oral cavity. There are known cases of wear products of orthodontic elements (mainly ligatures) adsorbing onto the dental plaque, causing discolorations. Although weight of particles is marginal, the total releasing area is more of a concern due to amount and volume of molecules. The intensity of the process depends not only on the element but mainly on the size of contacting surfaces. A part of them is transferred into the digestive system, from which, in the form of oxides and metal ions (iron, chromium, nickel), enter the body, working toxically [28–30]. The effect of the environment is crucial for the processes of fretting and corrosive wear. Its aggressiveness is largely affected by the chemically active substances, especially chlorine, sulfur, oxygen, and phosphorus compounds [11,31]. The human saliva is necessary in the oral cavity; it has a series of protective and supplementary functions, including the lubrication function, and at the same time, it has a big influence on the processes of corrosive and fretting wear [12,31–37]. Because of adverse effects, fretting should be taken into consideration from clinical standpoint.

The non-precious alloys most commonly used in the preparation of prosthetic and orthodontic elements containing cobalt, chromium, nickel, and molybdenum demonstrate a relatively low corrosion resistance compared to precious metal alloys, which helps the elements being part of their compositions enter the body [38,39]. In order to increase their biological tolerance, various types of modification of the surface layer of the elements made of these alloys are applied, which makes it possible to obtain biocompatibility. It is worth noting that many of those layers enable an improvement of the wear resistance. Also, this resistance has been established to be higher compared to uncoated alloys [12,40]. A special focus should be put on titanium carbides and nitrides coatings. This results mainly from their high durability, corrosion resistance, and biomcompatibility [41–45].

The aim of the study was to examine the fretting wear resistance of prosthetic and orthodontic alloy Ni-Cr-Mo samples coated with Ti(C, N) layers and to compare them with samples without any coating.

2. Test Material and Methods

The test material were disks made of the Ni-Cr alloy, 8 mm in diameter and 10 mm high (Figure 1). The initial composition of the alloy determined by the X-ray fluorescence analysis technique with the use of a spectrometer SRS300 by SIEMENS is given in Table 1. The disks were divided into six groups. First group without coating (S0), others (S1–S5) differed in the amount of C and N in the deposited Ti(C, N) coating (Table 2).

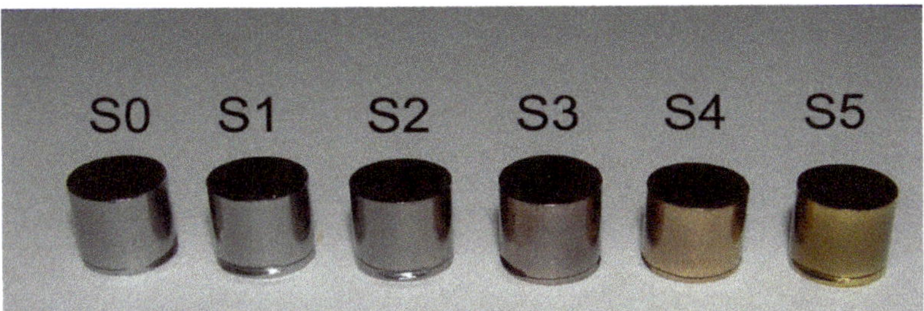

Figure 1. Test samples.

Table 1. Chemical composition of tested alloy.

Element Percentage wt %						
Cr	**Mo**	**Si**	**Fe**	**Co**	**Mn**	**Ni**
24.79	8.89	1.57	1.33	0.17	0.12	residue

Table 2. Chemical composition of tested coatings.

Coating	Element Percentage at %		
	Ti	**C**	**N**
S1	51.50	48.50	0.00
S2	52.91	33.91	13.18
S3	51.94	28.22	19.84
S4	47.78	20.05	32.17
S5	46.79	0.00	53.21
	Element Percentage wt %		
	Ti	**C**	**N**
S1	80.18	19.82	0.00
S2	79.51	13.90	6.59
S3	78.76	11.67	9.57
S4	75.26	8.61	16.13
S5	79.78	0.00	20.22

Layers were deposited using magnetron sputtering method. In order to improve adherence of Ti(C, N) layers, first adhesive sublayer of pure titanium was deposited during 120 s with argon pressure equal to 0.24 Pa and with the following work parameters of magnetron: 3 kW/approximately 4.5 A. After two minutes reactive gas was slowly introduced: nitrogen, acetylene, or their mixture. Deposition time of appropriate layer was the same for all processes and equal to 7200 s. Polarization with constant voltage during deposition was −100 V. Pressure of the process was 0.27 Pa in each case. The reactive gases and their flow is presented in Table 3. These were the only variables of the processes.

Prepared samples were examined. Layers thickness was measured and ranged from 1.25 μm to 1.62 μm, which was thoroughly described in previous article [46]. Film adhesion was also tested, in preceding article [45], according to VDI 3198 norm. The test consists in comparing the obtained imprints with the standards and determine the degree of delamination of the coating. In the impressions obtained, no cracking of the coating from the substrate was observed. The received results should be classified in the HF1 standard, with a very low density crack. Previous studies measured also the hardness and modulus of elasticity using the nanoindentation method [45]. The results for hardness ranged from 20 to 34 GPa, whereas the modulus ranged from 272 GPa to 382 GPa. Then, samples underwent fretting wear resistance tests under fretting conditions. In order to eliminate the impurities, before each test, the samples were placed in a sonifier and rinsed in ethanol for 10 min. The experiments were performed under the conditions of dry friction on a special fretting tester designed and made at the Faculty of Mechanical Engineering of Bialystok University of Technology, Poland. The schematics of the fretting test device have been shown in Figure 2.

Table 3. Reactive gas flow.

Gas	Flow Unit	Samples				
		S1	S2	S3	S4	S5
N_2	sccm	0	4	8	12	16
C_2H_2		8	6	4	2	0

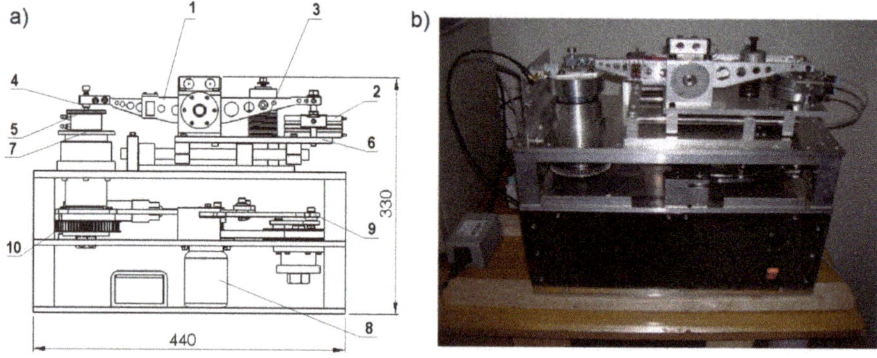

Figure 2. (**a**) Schematics of the fretting tester. 1—lever, 2—pneumatic actuator, 3—loading spring, 4—counter-sample holder, 5—sample holder, 6—slide plate, 7—turntable, 8—motor with a gear, 9—system of levers, 10—wheel activating the turntable; (**b**) Actual view.

The friction processes were realized with a small amplitude of the order of 100 μm, at the frequency of 0.8 Hz, with averaged unit load: 5, 10, and 15 N and for the predetermined time of 15 min. For the load of 15 N, additional tests were performed for the times of: 30, 60, and 120 min (Table 4).

Table 4. Testing conditions.

Load (N)	5	10	15	15	15	15
Time (min)	15	15	15	30	60	120

The movable table of the device, on which the samples were mounted, was making a reversible movement. A counter-sample was pressed onto the surface of the disk, shaped like a truncated cone, whose contact surface diameter equaled 1.3 mm (Figure 3). The counter-samples were made from the Co-Cr-Mo alloy.

After the fretting processes, the samples were subjected to microscopic observations with the use of a scanning electron microscope Hitachi S3000N and a laser scanning microscope LEXT OLS4000. The observations aimed to determine the character of the samples' wear. Under the scanning microscope, beside the assessment of the wear character, analyses of the chemical composition in the friction area were conducted. To that end, an EDS detector (with energy dispersion) by Pioneer working with a microscope was used. The test results have been presented in Figures 4–10. The observations under a laser scanning microscope with the use of the 3D option made it possible to perform tests of surface roughness and friction track profiles, which enabled the determination of the wear depth on the particular samples, thus facilitating the assessment of the fretting resistance of the examined coatings. The results of the wear measurements have been included in Table 5.

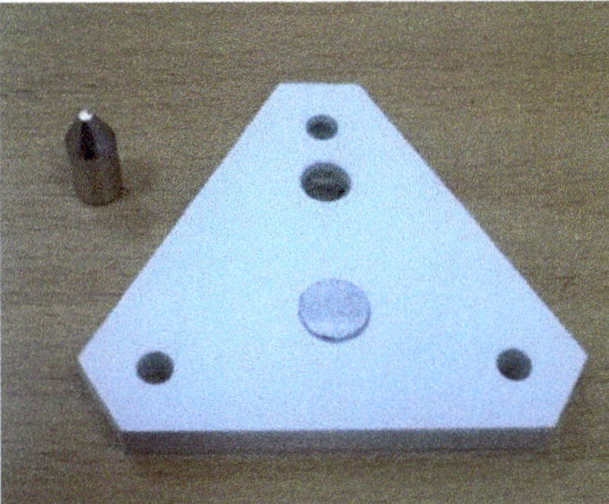

Figure 3. Sample mounted in a holder and a counter-sample used for the tests.

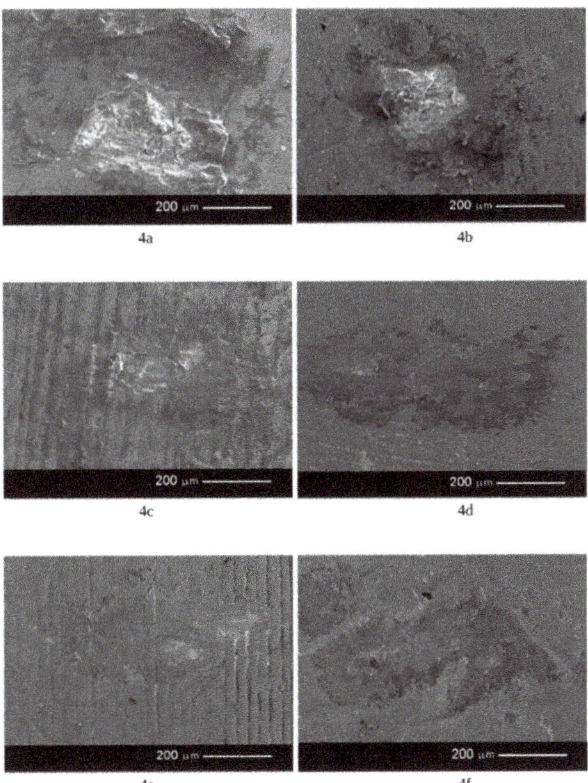

Figure 4. Microscopic images of the wear areas for sample S1 with the following test parameters: (**a**) 5 N/15 min; (**b**) 10 N/15 min; (**c**) 15 N/15 min; (**d**) 15 N/30 min; (**e**) 15 N/60 min; (**f**) 15 N/120 min.

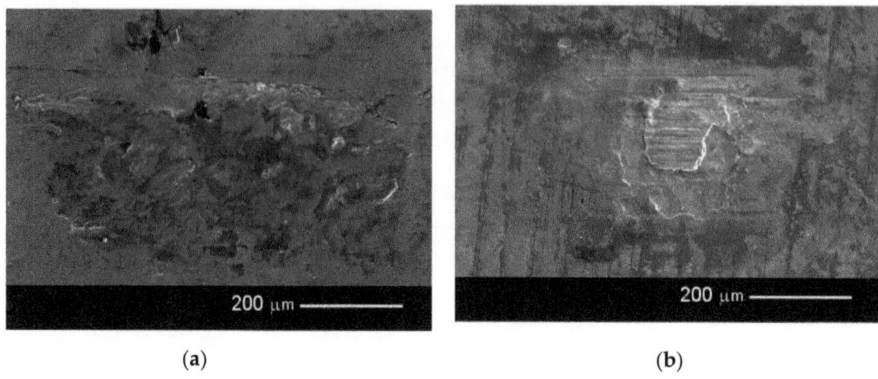

(a)   (b)

Figure 5. Images of friction tracks obtained with the load of 15 N and for the time of 120 min. (**a**) Sample S0; (**b**) Sample S5.

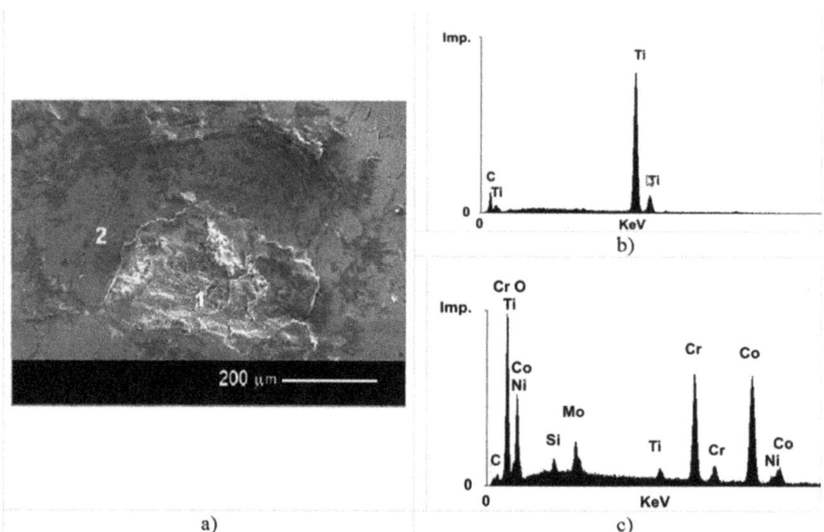

Figure 6. Area of friction tracks obtained with the load of 15 MPa and for the time of 120 min for sample S1. (**a**) Microscopic image with marked areas of EDS analyses; (**b**) EDS analysis in point 2; (**c**) EDS analysis in point 1.

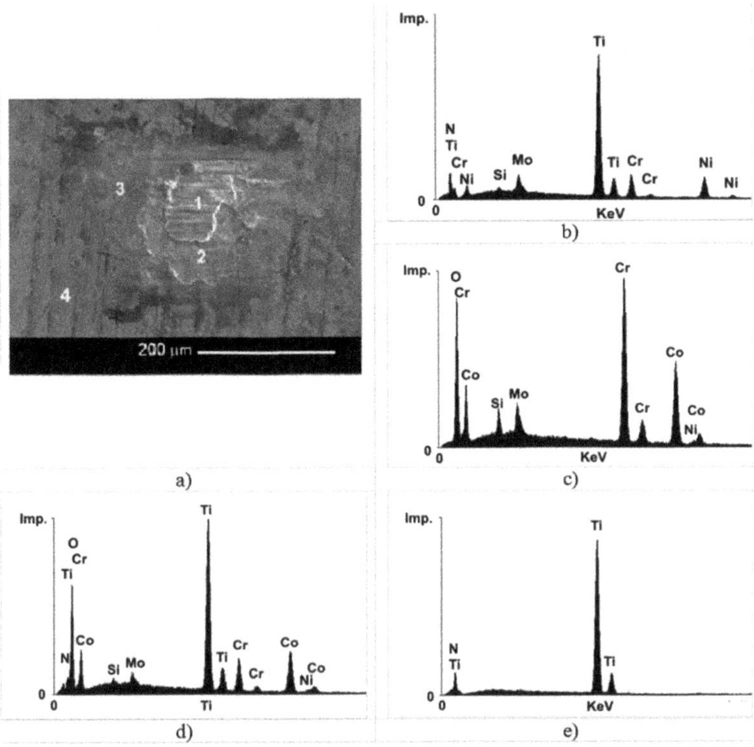

Figure 7. Area of friction tracks obtained with the load of 15 N and for the time of 120 min for sample S5. (**a**) Microscopic image with marked areas of EDS analyses; (**b**) EDS analysis in point 1; (**c**) EDS analysis in point 2; (**d**) EDS analysis in point 3; (**e**) EDS analysis in point 4.

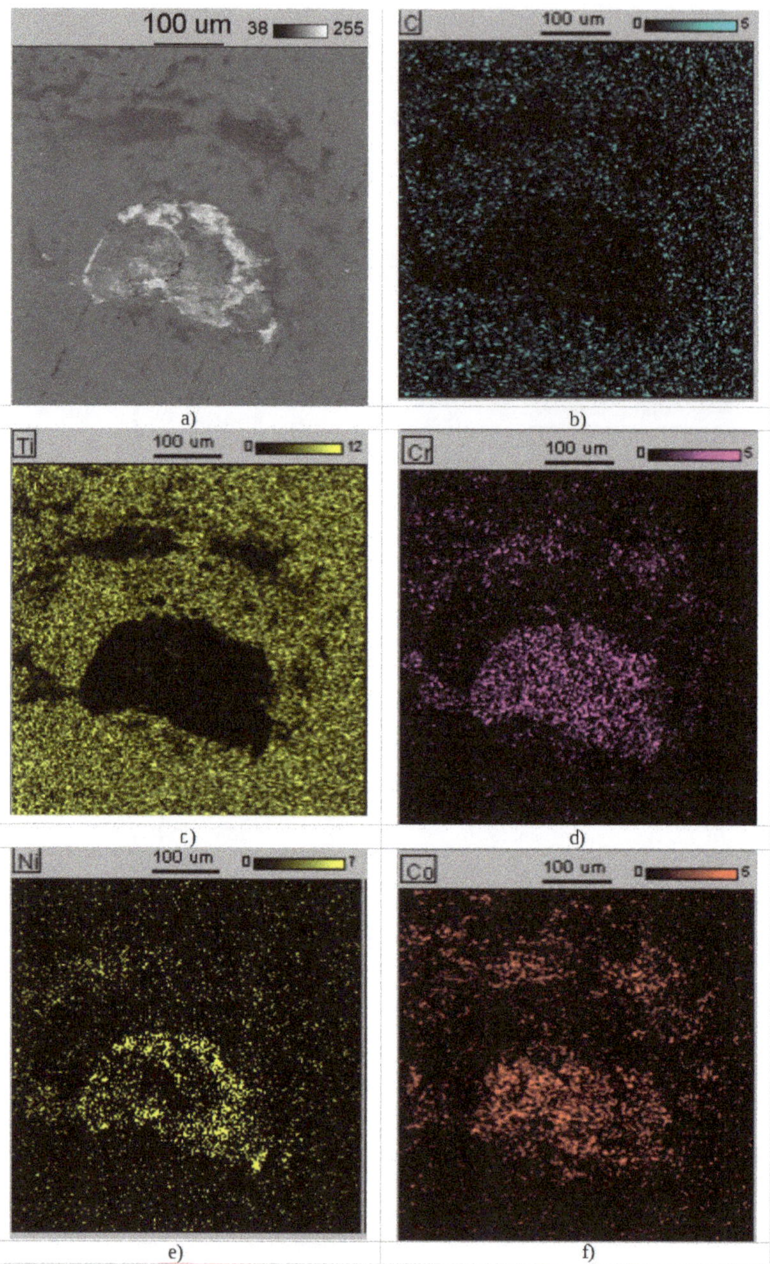

Figure 8. Microscopic image (**a**) and element distributions (**b–f**) in the friction area of sample S1. Parameters-load 15 N, time 120 min.

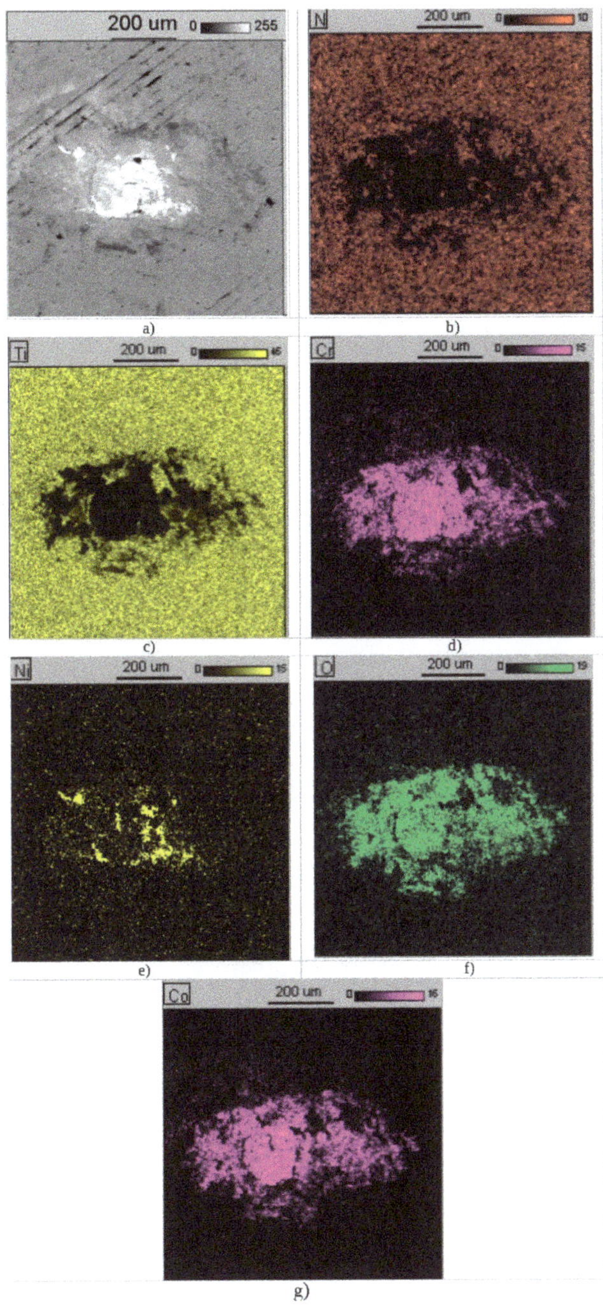

Figure 9. Microscopic image (**a**) and element distributions (**b–g**) in the friction area of sample S5. Parameters-load 15 N, time 120 min.

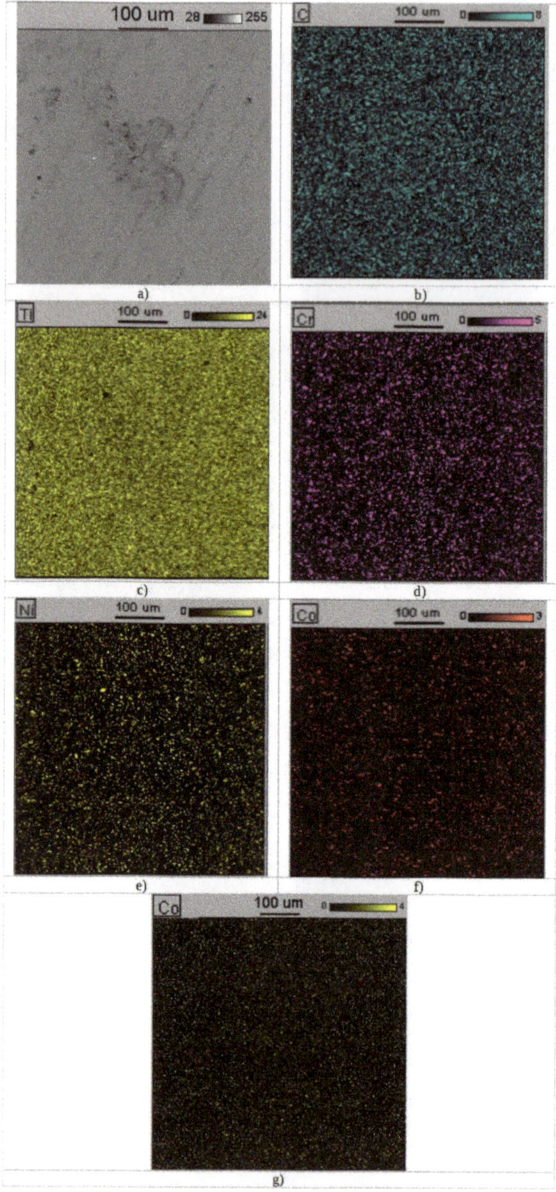

Figure 10. Microscopic image (**a**) and element distributions (**b–g**) in the friction area of sample S1. Parameters-load 15 N, time 15 min.

Table 5. Abrasion track depths for the examined samples with the load of 15 N and for the time of 15 min.

	Sample					
	S0	**S1**	**S2**	**S3**	**S4**	**S5**
Depth (µm)	2.64	1.23	1.13	1.02	1.26	1.32

3. Test Results

SEM and EDS Tests

Microscopic observations in SEM and EDS were performed on all the samples and in each of them, all the friction tracks were examined. Exemplary images of all the tracks for sample S1 has been shown in Figure 4.

On all the other samples with coatings, the friction tracks observed in SEM were analogical. Figure 5 shows friction track images obtained with the maximal loads of samples S0 and S5.

We can see in the presented microscopic images that the surface damage was diversified, depending on the test parameters. In the case of a test with higher loads and longer times, abrasion of the coating took place and the base was revealed.

In order to confirm this fact, chemical composition analyses were performed in the micro-area of the fretting wear. Exemplary results of these analyses for samples S1 and S5 have been presented in Figures 6 and 7.

As we can infer from the chemical composition analysis, in the examined micro-areas, after the abrasion of the TiC coating, the base was revealed. In the revealed area, the following elements are present: C, Ti, O, Cr, Co, and Ni. Their origins are different. Both nickel and chromium originate from the metallic base of the sample. Considering the ratio of the signal heights from chromium and nickel, it should be noted that the signal coming from chromium is significantly higher than suggested by the composition of the base alloy. The presence of cobalt proves a transfer, in the friction process, of the counter-sample material (Co-Cr-Mo alloy) into the friction area, which explains also the higher content of chromium in the examined area. The presence of titanium and carbon can be explained by the remains of the TiC coating-probably residues of the wear products. The presence of oxygen in the friction area proves the occurrence of oxidation processes during the friction. As a result, we probably observe fretting-corrosion rather than pure fretting.

Analogical results were obtained in the other samples (see Figure 7). In sample S5, the presence of nickel and chromium in point 1, in the face of a strong signal from Ti, may prove such thinning of the TiN coating that we receive a signal from the base material. Similarly to the case of the TiC coating in this sample, after the base had been revealed, there was a transfer of the counter-sample material to the friction area, which has been shown in Figure 7c,d. The presence of cobalt in point 3, with a relatively high signal from titanium and nitrogen, can be explained by the transfer of the counter-sample material onto the TiN coating.

The element distribution on the friction area is better presented by the surface element distribution maps, shown in Figures 8 and 9 for selected samples.

The next stage of investigations consisted in determining the degree of the fretting wear. As the wear degree parameter the maximal depth on the friction track was assumed, determined by means of a laser scanning microscope. In order to compare the wear resistance of the particular coatings, only those friction tracks which had not revealed the base were selected. The depth of the tracks with a revealed base is not reliable, as the base (Ni-Cr-Mo alloy) wears off in a decisively different way than the coatings. In order to determine those friction tracks which reveal the base, distribution maps of the elements present in the coatings and the base for all the examined samples and friction tracks were created. An exemplary map for the friction track with a non-revealed base for sample S1 has been shown in Figure 10, where, in the whole test area, we can see a uniform distribution of titanium and carbon. There are no visible areas with an elevated content of nickel or chromium. This proves that the base has not been revealed. It should be noted that no presence of cobalt is observed on the surface. Therefore, we can conclude that, for the TiC layer, with a short examination time, there is no transfer of the counter-sample material.

After analyzing the element distribution maps in the friction track areas for all the samples and with all the friction parameters, it was established that the determination of the abrasion depth could be performed on the samples examined with the load of 15 N and for the time of 15 min. In these samples, the coating did not undergo abrasion. For the samples tested with the load of 15 N and longer examination times, in some cases, the coating was abraded. In turn, the samples examined with lower loads, despite the fact that their coatings did not undergo abrasion, were not considered as the abrasion track was too flat, which posed the risk of a higher measurement error. Table 5 includes the measured abrasion depths for the examined samples.

Then, the roughness was measured as seen in Table 6.

As we can see, all the coatings make the fretting wear resistance higher compared to the non-coated sample. For the parameters: load 15 N and time 15 min, the wear in each case was over twice as low. The lowest wear, and thus the highest resistance, were demonstrated by sample S3.

Table 6. Mean values of roughness.

Parameter	Mean Value					
	Sample S0	Sample S1	Sample S2	Sample S3	Sample S4	Sample S5
Ra	0.33 μm	0.45 μm	0.44 μm	0.44 μm	0.38 μm	0.41 μm
Rq	0.42 μm	0.59 μm	0.55 μm	0.58 μm	0.50 μm	0.58 μm
Rz	2.13 μm	2.86 μm	2.57 μm	4.04 μm	3.49 μm	2.42 μm
Rp	0.89 μm	0.92 μm	1.04 μm	1.85 μm	1.15 μm	0.84 μm
Rv	1.32 μm	1.95 μm	1.54 μm	2.19 μm	2.34 μm	1.58 μm
Rc	1.21 μm	1.82 μm	1.57 μm	2.01 μm	1.73 μm	1.82 μm

4. Summary

Based on the performed investigations, we can establish that the use of Ti(C, N)-type coatings reduces the results of fretting wear of a Ni-Cr-Mo-type alloy. The effect of reducing wear is largely due to the significantly higher hardness of coatings compared to the alloy. Moreover, the resistance to fretting wear depends on the ratio H/E (hardness to modulus of elasticity) or H^3/E^2 -Table 7.

Table 7. Comparison of the depth [groove] and H/E ratio on samples tested at 15 N load and 15 min.

	Sample					
	S0	S1	S2	S3	S4	S5
Depth [μm]	2.64	1.23	1.13	1.02	1.26	1.32
H/E	0.016	0.089	0.093	0.084	0.083	0.074
H^3/E^2	0.00096	0.269	0.273	0.185	0.166	0.108

Analizing obtained results, confirm the mentioned relationship between depth and H/E, clearly indicating higher wear in sample without coating.

The lower the H/E (H^3/E^2), the greater the wear. However, between the groups with coatings, no strong correlation was found. No noticeable dependencies may result from small differences in the consumption of individual coatings. The highest resistance is exhibited by coatings on sample S3, containing 28.22 at % carbon and 19.84 at % nitrogen. This is not due to the hardness of the coating itself, nor its abrasion resistance, because previous studies [45] have shown that both the hardness and the modulus of elasticity and the wear resistance of coatings decrease with the content of nitrogen in it. Thus, not only the hardness of the coating, but also other surface properties e.g., the affinity of the sample materials and counter-samples determine the reduction in wear by fretting. Coatings from titanium carbides and nitrides, due to their anti-adhesive properties, are used on cutting tools to prevent sticking of the chip to the blade.

The presence of oxygen in the wear products proves the formation of oxides and thus the presence of fretting corrosion wear. It should be born in mind that the conducted research took place in the air environment. In oral conditions, with the presence of saliva, as a corrosive factor, this phenomenon may intensify. During the fretting process, a transfer of the counter-sample material to the wear products was observed. Mostly, this phenomenon was observed in the sample combination NiCr

(S0) alloy-counter-sample-Co-Cr alloy. This can be a confirmation of the participation of adhesion in fretting wear. In samples with coatings, the transfer of material from the counter-sample was clearly smaller. Moreover, the higher the nitrogen content in the coating, the stronger the phenomenon.

The obtained results make it possible to conclude that the Ti(C, N)-type coatings can be used as protective layers on prosthetic and orthodontic elements in systems where the fretting phenomenon may occur. The reduced fretting wear in respect to the alloy without a coating can significantly limit the amount of metal ions (especially nickel ions) which penetrate the oral cavity environment and, consequently, its tissues.

Due to amount of material particles, being products of fretting wear, and their very small dimensions make a relatively large real surface area. The amount of ions released in corrosive processes depends on the size of the corrosive surface. Thus, increasing the surface to be corroded increases the amount of released ions proportionately. Considering that the actual surface of the corrosion products is definitely greater than, for example, the surface of the prosthetic element, it is important to restrain the fretting wear of the prosthetic and orthodontic elements.

As the investigations included only the system: a Ni-Cr-Mo-type alloy with a Co-Cr-Mo-type alloy, these coatings can be applied in such systems where these alloys cooperate—e.g., orthodontic braces with wires and in prosthetic elements; prostheses on latches, bolts, fasteners, or locks. For other material pairs, similar tests should be performed for other associations.

The research presented in the article was of qualitative character. Its aim was not to determine the quantitative fretting wear of individual coatings, but to show the nature of this wear and to determine the value of Ti(C, N) coatings as coatings limiting the fretting wear of the Ni-Cr alloy.

5. Conclusions

Ti(C, N) coatings can be used as a protection on prosthetic and orthodontic elements, especially on contacting surfaces where fretting may occur. The use of Ti(C, N)-type coatings reduces the adverse effects of fretting wear, decreasing the amount of ions released during orthodontic treatment or wearing prosthesis.

The data used to support the findings of this study are included within the article.

Author Contributions: Conceptualization, K.B. and L.K.; Methodology, L.K. and J.R.D.; Validation, L.K. and J.R.D.; Formal Analysis, L.K. and J.R.D.; Investigation, L.K. and J.R.D.; Resources, L.K. and J.R.D.; Data Curation, K.B. and W.J.; Writing—Original Draft Preparation, K.B., L.K. and W.J.; Writing—Review and Editing, L.K. and W.J.; Visualization, K.B. and W.J.; Supervision, L.K.; Funding Acquisition, W.J.

Funding: Work carried out with funds from the Medical University of Lodz as part of the financing of research of young science workers and doctoral students. Subject no. 502-03/2-148-06/502-24-077-19.

Conflicts of Interest: The authors declare no conflict of interest.

References

1. Rocha, L.A.; Oliveira, F.; Cruz, H.V.; Sukotjo, C.; Mathew, M.T. Bio-tribocorrosion in dental applications. *Bio-Tribocorros. Biomater. Med. Implants* **2013**, *10*, 223–249.
2. Jin, Z.M.; Zheng, J.; Li, W.; Zhou, Z.R. Tribology of medical devices. *Biosurf. Biotribol.* **2016**, *4*, 173–192. [CrossRef]
3. Fallahnezhad, K.; Oskouei, R.H.; Badnava, H.; Taylor, M. An adaptive finite element simulation of fretting wear damage at the head-neck taper junction of total hip replacement: The role of taper angle mismatch. *J. Mech. Behav. Biomed. Mater.* **2017**, *75*, 58–67. [CrossRef]
4. Fallahnezhad, K.; Oskouei, R.H.; Badnava, H.; Taylor, M. The Influence of Assembly Force on the Material Loss at the Metallic Head-Neck Junction of Hip Implants Subjected to Cyclic Fretting Wear. *Metals* **2019**, *9*, 422. [CrossRef]
5. Vadiraj, A.; Kamaraj, M. Effect of surface treatments on fretting fatigue damage of biomedical titanium alloys. *Tribol. Int.* **2007**, *40*, 82–88. [CrossRef]

6. Kulesza, E.; Dąbrowski, J.R.; Siudyn, J.; Neyman, A.; Mizera, J. Fretting wear of materials—methodological aspects of research. *Acta Mech. Autom.* **2012**, *6*, 58–61.

7. Cruzado, A.; Hartelt, M.; Wäsche, R.; Urchegui, M.A.; Gómez, X. Fretting wear of thin steel wires. Part 1: Influence of contact pressure. *Wear* **2010**, *11*, 1409–1416. [CrossRef]

8. Neyman, A. *Textbook of Fretting in Machines Components*; Gdansk University of Technology: Gdansk, Poland, 2003; pp. 7–8.

9. Singh, V.; Shorez, J.P.; Mali, S.A.; Hallab, N.J.; Gilbert, J.L. Material dependent fretting corrosion in spinal fusion devices: Evaluation of onset and long-term response. *J. Biomed. Mater. Res. B Appl. Biomater.* **2018**, *106*, 2858–2868. [CrossRef]

10. Fouvry, S.; Kapsa, P.H.; Vincent, L. *Developments of Fretting Sliding Criteria to Quantify the Local Friction Coefficient Evolution under Partial Slip Condition*; Dowson, D., Taylor, C.M., Childs, T.H.C., Dalmaz, G., Berthier, Y., Flamand, L., Eds.; Elsevier Science & Technology: Oxford, UK, 1998; pp. 161–172.

11. Duisabeau, L.; Combrade, P.; Forest, B. Environmental effect on fretting of metallicmaterials for orthopaedic implants. *Wear* **2004**, *256*, 805–816. [CrossRef]

12. Dąbrowski, J.R.; Klekotka, M.; Sidun, J. Fretting and fretting corrosion of 316L implantation steel in the oral cavity environment. *Eksploat. Niezawodn.* **2014**, *16*, 3.

13. Geringer, J.; Forest, B.; Combrade, P. Fretting-corrosion of materials used as orthopaedic implants. *Wear* **2005**, *7*, 943–951. [CrossRef]

14. House, K.; Sernetz, F.; Dymock, D.; Sandy, J.R.; Ireland, A.J. Corrosion of orthodontic appliances—Should we care? *Am. J. Orthod. Dentofac. Orthop.* **2008**, *4*, 584–592. [CrossRef] [PubMed]

15. Sifakakis, I.; Eliades, T. Adverse reactions to orthodontic materials. *Aust. Dent. J.* **2017**, *62*, 20–28. [CrossRef] [PubMed]

16. Geringer, J.; Kim, K.; Boyer, B. Fretting corrosion in biomedical implants. *Tribocorros. Passiv. Met. Coat.* **2011**, 401–423.

17. Liskiewicz, T.; Fouvry, S.; Wendler, B. Impact of variable conditions on fretting wear. *Surf. Coat. Technol.* **2003**, *163*, 465–471. [CrossRef]

18. Klimek, L.; Palatyńska-Ulatowska, A. Scanning electron microscope appearances of fretting in the fixed orthodontic appliances. *Acta Bioeng. Biomech.* **2012**, *14*, 79–83.

19. Weaver, J.D.; Ramirez, L.; Sivan, S.; Di Prima, M. Characterizing fretting damage in different test media for cardiovascular device durability testing. *J. Mech. Behav. Biomed. Mater.* **2018**, *82*, 338–344. [CrossRef]

20. Halwani, D.O.; Anderson, P.G.; Brott, B.C.; Anayiotos, A.S.; Lemons, J.E. Clinical device-related article surface characterization of explanted endovascular stents: Evidence of in vivo corrosion. *J. Biomed. Mater. Res. B Appl. Biomater.* **2010**, *95*, 225–238. [CrossRef]

21. Rowan, F.E.; Wach, A.; Wright, T.M.; Padgett, D.E. The onset of fretting at the head-stem connection in hip arthroplasty is affected by head material and trunnion design under simulated corrosion conditions. *J. Orthop. Res.* **2018**, *36*, 1630–1636. [CrossRef]

22. Kumar, S.; Singh, S.; Hamsa, R.; Ahmed, S.; Prasanthma Bhatnagar, A.; Sidhu, M.; Shetty, P. Evaluation of Friction in Orthodontics Using Various Brackets and Archwire Combinations-An in Vitro Study. *J. Clin. Diagn. Res.* **2014**, *8*, ZC33–ZC36.

23. Lukina, E.; Kollerov, M.; Meswania, J.; Khon, A.; Panin, P.; Blunn, G.W. Fretting corrosion behavior of nitinol spinal rods in conjunction with titanium pedicle screws. *Mater. Sci. Eng. C* **2017**, *72*, 601–610. [CrossRef] [PubMed]

24. Bronzino, J.D. *The Biomedical Engineering HandBook*, 2nd ed.; CRC Press: Boca Raton, FL, USA, 2000.

25. He, D.; Zhang, T.; Wu, Y. Fretting and galvanic corrosion behaviors and mechanisms of Co-Cr-Mo and Ti-6Al-4V alloys. *Wear* **2002**, *249*, 883–891. [CrossRef]

26. Pellier, J.; Geringer, J.; Forest, B. Fretting-corrosion between 316L SS and PMMA: Influence of ionic strength, protein and electrochemical conditions on material wear. Application to orthopedic implants. *Wear* **2011**, *271*, 1563–1571. [CrossRef]

27. Zhu, M.H.; Yu, H.Y.; Zhou, Z.R. Radial fretting behaviours of dental ceramics. *Tribol. Int.* **2006**, *39*, 1255–1261. [CrossRef]

28. Eliades, T.; Athanasiou, A.E. In vivo aging of orthodontic Alloys: Implications for corrosion potential, nickel release and biocompatibility. *Angle Orthod.* **2002**, *72*, 222–237.

29. Hansen, D.C. Metal corrosion in the human body: The ultimate bio-corrosion scenario. *Electrochem. Soc. Interface* **2008**, *17*, 31–34.

30. Shih, C.C.; Shih, C.M.; Chen, Y.L.; Su, Y.Y.; Shih, J.S.; Kwok, C.F.; Lin, S.J. Growth inhibition of cultured smooth muscle cells by corrosion products of 316 L stainless steel wire. *J. Biomed. Mater. Res.* **2001**, *57*, 200–207. [CrossRef]

31. Lgied, M.; Liskiewicz, T.; Neville, A. Electrochemical investigation of corrosion and wear interactions under fretting conditions. *Wear* **2012**, *282*, 52–58. [CrossRef]

32. Sajewicz, E. Effect of saliva viscosity on tribological behaviour of tooth enamel. *Tribol. Int.* **2009**, *42*, 327–332. [CrossRef]

33. Schipper, R.A.; Silletti, E.; Vingerhoeds, M.H. Saliva as research material: Biochemical, physiocochemical and practical aspects. *Arch. Oral Biol.* **2007**, *52*, 1114–1135. [CrossRef]

34. Huang, H.H.; Chiu, Y.H.; Lee, T.H.; Wu, S.C.; Yang, H.W.; Su, K.H.; Hsu, C.C. Ion release from NiTi orthodontic wires in artificial saliva with various acidities. *Biomaterials* **2003**, *24*, 3585–3592. [CrossRef]

35. Klekotka, M.; Dąbrowski, J.R.; Karalus, W. Fretting—Corrosion of Co-Cr-Mo alloy in oral cavity environment. *Solid State Phenom.* **2015**, *227*, 455–458. [CrossRef]

36. Downarowicz, P.; Mikulewicz, M. Trace metal ions release from fixed orthodontic appliances and DNA damage in oral mucosa cells by in vivo studies: A literature review. *Adv. Clin. Exp. Med.* **2017**, *26*, 1155–1162. [CrossRef] [PubMed]

37. Mystkowska, J.; Łysik, D.; Klekotka, M. Effect of Saliva and Mucin-Based Saliva Substitutes on Fretting Processes of 316 Austenitic Stainless Steel. *Metals* **2019**, *9*, 178. [CrossRef]

38. Sfondrini, M.F.; Cacciafesta, V.; Maffia, E.; Scribante, A.; Alberti, G.; Biesuz, R.; Klersy, C.V. Nickel release from new conventional stainless steel, recycled, and nickel-free orthodontic brackets: An in vitro study. *Am. J. Orthod. Dentofac. Orthop.* **2010**, *137*, 809–815. [CrossRef]

39. Huang, T.H.; Ding, S.J.; Min, Y.; Kao, C.T. Metal ion release from new and recycled stainless steel brackets. *Eur. J. Orthod.* **2004**, *26*, 171–177. [CrossRef]

40. Kang, S.Y.; Huang, J.; Li, Q.H.; Diao, D.F.; Duan, Y.Z. The effects of diamond-like carbon films on fretting wear behavior of orthodontic archwire-bracket contacts. *J. Nanosci. Nanotechnol.* **2015**, *15*, 4641–4647. [CrossRef]

41. Banaszek, K.; Wiktorowska-Owczarek, A.; Kowalczyk, E.; Klimek, L. Possibilities of applying Ti (C, N) coatings on prosthetic elements—Research with the use of human endothelial cells. *Acta Bioeng. Biomech.* **2016**, *18*, 119–126.

42. Riescher, S.; Wehner, D.; Schmid, T.; Zimmermann, H.; Hartmann, B.; Schmid, C.; Lehle, K. Titaniumcarboxonitride layer increased biocompatibility of medical polyetherurethanes. *J. Biomed. Mater. Res. B Appl. Biomater.* **2014**, *102*, 141–148. [CrossRef]

43. Lehle, K.; Lohn, S.; Reinerth, G.; Schubert, T.; Preuner, J.G.; Birnbaum, D.E. Cytological evaluation of the tissue-implant reaction associated with subcutaneous implantation of polymers coated with titaniumcarboxonitride in vivo. *Biomaterials* **2004**, *25*, 5457–5466. [CrossRef]

44. Lehle, K.; Buttstaedt, J.; Birnbaum, D.E. Expression of adhesion molecules and cytokines in vitro by endothelial cells seeded on various polymer surfaces coated with titaniumcarboxonitride. *J. Biomed. Mater. Res. A* **2003**, *65*, 393–401. [CrossRef] [PubMed]

45. Banaszek, K.; Pietnicki, K.; Klimek, L. Effect of carbon and nitrogen content in Ti(C,N) coatings on selected mechanical properties. *Metal. Form.* **2015**, *XXVI*, 33–45.

46. Banaszek, K.; Klimek, L. Ti(C,N) as Barrier Coatings. *Coatings* **2019**, *9*, 432. [CrossRef]

 *processes*

Article

Evolution of Phase Composition and Antibacterial Activity of Zr–C Thin Films

Katarzyna Mydłowska [1],*, Ewa Czerwińska [1], Adam Gilewicz [1], Ewa Dobruchowska [1], Ewa Jakubczyk [2], Łukasz Szparaga [1], Przemysław Ceynowa [1] and Jerzy Ratajski [1]

[1] Faculty of Mechanical Engineering, Koszalin University of Technology (KUT), Śniadeckich 2, 75-453 Koszalin, Poland; ewa.czerwinska@tu.koszalin.pl (E.C.); adam.gilewicz@tu.koszalin.pl (A.G.); ewa.dobruchowska@tu.koszalin.pl (E.D.); lukasz.szparaga@tu.koszalin.pl (Ł.S.); przemyslaw.ceynowa@tu.koszalin.pl (P.C.); jerzy.ratajski@tu.koszalin.pl (J.R.)

[2] Centre for Engineering Materials, Mechanical Engineering Sciences, University of Surrey, Surrey, Guildford GU2 7XH, UK; e.jakubczyk@surrey.ac.uk

* Correspondence: katarzyna.mydlowska@tu.koszalin.pl; Tel.: +48-94-348-65-62

Received: 26 January 2020; Accepted: 20 February 2020; Published: 25 February 2020

Abstract: The research presented in this article concerns Zr–C coatings which were deposited on 304L steel by reactive magnetron sputtering from the Zr target in an Ar–C_2H_2 atmosphere at various acetylene flow rates, resulting in various atomic carbon concentrations in the coating. The article describes research covering the change in the antibacterial and anticorrosive properties of these coatings due to the change in their chemical and phase composition. The concentration of C in the coatings varied from 21 to 79 at.%. The coating morphology and the elemental distribution in individual coatings were characterized using field emission scanning electron microscopy with an energy-dispersive X-ray analytical system. X-ray diffraction and Raman spectroscopy were used to analyze their microstructure and phase composition. Parallel changes in the mechanical properties of the coatings were analyzed. Based on the obtained results, it was concluded that the wide possibility of shaping the mechanical properties of Zr–C coatings in combination with relatively good antibacterial properties after exceeding 50 at.% of carbon concentration in coatings and high protective potential of these coatings make them a good candidate for medical applications. In particular, corrosion tests showed the high anti-pitting potential of Zr–C coatings in the environment of artificial saliva.

Keywords: thin films; zirconium carbide; antimicrobial properties; medical implants

---

1. Introduction

Metals and alloys are widely used in the production of medical implants for use in dentistry, cardiology, orthopedic fractures, etc. Titanium alloys, austenitic stainless steel and chromium–cobalt–nickel alloys are mainly used for implants. First of all, these materials must have adequate biocompatibility, i.e., they must not be irritating, toxic, allergic, inflammatory, mutagenic or carcinogenic to surrounding tissues [1,2]. For example, in the case of dental and orthodontic implants, some studies show that biofilms caused by attachment of bacteria are responsible for about 65% of infections such as periodontal disease and peri-implant diseases [3]. To prevent these adverse effects, surface treatments are used, among which the deposition of coatings by PVD and CVD methods are one of the ways to improve their mechanical and biological properties [4–8]. In particular, the coatings should fulfill a number of functions, such as bacterial resistance, i.e., resistance to adhesion and colonization of bacteria [9,10]. They should be biocompatible [11], have adequate thrombogenicity [12,13] in cardiac surgery applications and create a suitable diffusion barrier for elements causing allergic reactions in the case of deposition of coatings on medical steels [4,14]. Simultaneously, the coatings should have pitting corrosion resistance, a low coefficient of friction and wear resistance, high hardness and low Young's modulus [15–17]. The aforementioned range of properties by which coatings should be characterized is a function of many factors, the most important of which is the chemical composition of coatings and their structure (monolayers, multilayers, nanocomposites, gradient coatings, etc. [18]). In the case of antibacterial properties, due to the fact that the interaction of the material with bacteria is adhesive, the chemical activity is important (contact angle, free surface energy, etc.) and depends primarily on the chemical composition and type of bonds in the surface layer of coatings deposited on implants. The roughness of the surface is also taken into account.

Promising candidates meeting these requirements are thin coatings made of transition metal carbides. Among them, zirconium-carbide-based coatings are interesting due to their chemical, mechanical and electrical properties [19–22]. The microstructure and phase composition of these coatings are mainly influenced by the atomic concentration of carbon as well as the parameters of the coating deposition process [23,24]. Relatively significant possibilities of controlling properties are obtained due to the formation of free carbon in the amorphous form surrounding the grains of nanocrystalline carbide [25], although comparing to coatings based on transition metal nitrides, carbide coatings with higher carbon concentrations are characterized with lower hardness, however, production of coatings with a lower coefficient of friction is possible. Hence, by appropriately changing the phase composition and using the potential of carbon to combine into various structures, there are wide possibilities of shaping the properties of coatings based on Zr–C, which also inspire research on the perspectives of their application on medical implants [4,26,27]. In particular, in the studies presented in [4], ZrC coatings were deposited at different substrate temperatures (25, 200 and 400 °C) on substrates of 316L steel. It has been shown that the surface energy, hydrophobicity and surface roughness have the greatest influence on the hemocompatibility of these coatings. The authors of [28] present research on the analysis of antibacterial activity and cell proliferation activity of Zr–C–N coating with different C concentration deposited on titanium (Ti). Coatings with the highest C concentration (22 at.%) have been shown to increase the antibacterial ability against *Staphylococcus aureus*, but also meet the requirement for HGF (human gingival fibroblasts) biocompatibility. Tests of CN_x-doped W, Ti and Zr coatings for antibacterial activity and wear are presented in [6]. It was shown that the CN_x-Zr coating, whose surface had the highest hydrophilicity, provided the best antibacterial effect. However, the CN_x-Zr coating showed lower wear resistance than the CN_x-W and CNx-Ti coatings.

The research presented in this article concerns Zr–C coatings which were deposited on 304L steel by reactive magnetron sputtering from the Zr target in an Ar-C_2H_2 atmosphere at various mass acetylene flow rates conditioning a specific carbon concentration in the coating. The evolution of phase composition of coatings was studied with increasing carbon concentration and the composition correlated with mechanical, antibacterial and corrosive properties. For comparison, these properties were also tested for 304L steel without a coating. In particular, the correspondence between chemical and phase composition of coatings and biological properties was studied. Corrosion properties and contact angle were tested in an environment that simulates the oral cavity, i.e., artificial saliva, taking into account the possible application of Zr–C coatings on orthodontic wires.

2. Materials and Methods

2.1. Coating Deposition

Zr–C coatings deposited by pulsed, reactive magnetron sputtering on high-speed steel substrates were the object of research previously described by the authors in [17]. Coatings considered in this paper were deposited on 304L medical steel substrates in the form of discs with a diameter of 32 mm and a thickness of 4 mm in the same deposition processes, methodology which has been described in detail in previous work [17]. The objects of research in this work are Zr–C coatings with different carbon concentrations obtained at various acetylene mass flow rates during the deposition process.

In addition, zirconium coatings deposited without carbon-carrier gas in the working chamber were tested for comparative purposes.

2.2. Structural and Compositional Characterisation

The surface morphology and the elemental distribution in individual coatings were characterized using field emission scanning electron microscopy (FESEM, JEOL, JSM-7100F) with an energy-dispersive X-ray analytical system (Thermo Fisher SEM-EDX).

The concentration of the individual elements in the deposited coatings was determined by wavelength-dispersive X-ray spectroscopy (WDX) using a Thermo Scientific's Magnaray system with an X-ray microprobe working in a wave mode.

The microstructure of Zr–C coatings was analyzed by means of X-ray diffraction (XRD) using the conventional symmetrical Bragg–Brentano configuration ($\theta/2\theta$) with the Co-K$_\alpha$ radiation source on a DRON device. Further data processing was performed by HighScore Plus with ICDD PDF-4+ database software.

The analysis of the structure (type of bonds) of the carbon phase in Zr–C coatings was carried out by Raman spectroscopy using Renishaw in Via with laser excitation of the 534 nm wavelength.

Chemical and phase composition of the coatings was determined by means of X-ray photoelectron spectroscopy (XPS) using a Kratos Axis Supra system with monochromatic Al Kα1,2 radiation of energy 1486.6 eV. The calibration of the photoelectron spectrum was carried out with use of carbon C 1s (284.8 eV). The deconvolution of the XPS spectrum was performed through the fitting of peaks with the GL(30) function (Gauss and Lorenz functions at a ratio 30:70) and sigmoidal asymmetric function.

2.3. Mechanical Evaluation

Hardness measurements were performed by nanoindentation using the Fisherscope HM2000 nanoindenter equipped with the diamond Berkovich indenter. The load–depth indentation curves were obtained in the mode of the linear load increase up to 50 mN. The values of the hardness and the Young's modulus were determined based on the indentation curves using the Oliver–Pharr model [29].

2.4. Microbiological Tests

Four different species of microorganisms from the American Pure Culture Collection (*Candida albicans* ATCC 2091, *Escherichia coli* ATCC 25922, *Staphylococcus aureus* ATCC 25923 and *Streptococcus mutans* ATCC 35668) were used. The antimicrobial activity of the coatings was assessed by direct method based on the (BPR-ECHA-EU) regulation criteria (SN 195920–ASTM E2922), while the evaluation of antifungal activity was performed based on the SN 195921 standard. The fungus culture was carried out on Sabouraud medium, while the bacteria culture was on TSA medium. Test microorganisms were inoculated on the surface of the medium with a suspension concentration of 10^8 cfu/mL (0.5 Mc Farland); the test samples faced with coatings were placed into the medium after 20 min and incubated at 37 °C for 48 h. After incubation, the growth inhibition zone of the tested microorganisms was evaluated. The criterion for the assessment of fungicidal and bactericidal properties was the presence and size of the zone inhibiting the growth of the tested microorganisms. The tests were carried out in three repetitions for each tested sample and each tested microorganism.

To test the susceptibility of the surface of the coatings to microbial adhesion, a bacterial suspension with a density of 0.5 McFarland was applied to the test specimens and incubated at 37 °C for 24 h. After culturing, the samples were washed three times with distilled water to remove nonadherent cells and then shaken for 15 min in distilled water. Afterwards, the coatings were dried and the 0.5 cm^3 of benzidine bis (for observation of live cells) was applied to the surface and incubated for 120 s. After rinsing in distilled water, 0.3 cm^3 of propidium iodide (dead cell observation) was placed on the surface of the coatings for 120 s and again washed with water. After drying, the surfaces of the samples were observed using a fluorescence microscope (Motic BA 410 E M, MOTIC ASIA, Hong Kong). For each coating and the observed microorganism, two series of measurements were made and the final total number of microorganisms was the average of ten 1 cm^2 fields of view calculated using the Motic Live Imaging (MOTIC ASIA, Hong Kong) module.

The results of microbial adhesion to the coatings were calculated in relation to the control sample (304L sample without coating) using the formula:

$$\eta = ((N_0 - N)/N_0) \times 100\%, \tag{1}$$

where N represents the number of living/dead cells on the sample in cfu/cm^2 and N_0 represents the number of living/dead cells on the uncoated 304L reference sample.

2.5. The Wettability of Coating Surface and Corrosion Behavior

The wettability of the coating surface was determined by measuring the static contact angles through the sessile drop method using F4 Goniometer by Rame-Hart Instrumental Co. Drops with a volume of approximately 4 μL were deposited onto the investigated surfaces using a microsyringe. Fusayama-Meyer's artificial saliva, with the composition given in Table 1, was used as a test liquid. The contact angles were measured at 3–5 different points at the surface of each sample. The images were recorded over 300 s and analyzed using the DROPimage CA (Rame-Hart Instrumental Co) program.

Table 1. Chemical composition of Fusayama-Meyer's artificial saliva, pH 5.3.

Compound	KCl	NaCl	CaCl$_2$ × 2H$_2$O	NaH$_2$PO$_4$ × 2H$_2$O	Na$_2$S × 9H$_2$O	Urea
Concentration (g/L)	0.400	0.400	0.906	0.690	0.005	1.000

The corrosion behavior of substrate/coating systems was evaluated by potentiodynamic polarization tests, which were carried out using an ATLAS 0531 Electrochemical Unit (Atlas-Sollich, Poland). Conventional electrochemical three-electrode cell was used with a sample with the active area of 0.292 cm^2 as the working electrode, saturated calomel electrode (SCE, Hg/Hg$_2$Cl$_2$/KCl) as the reference electrode and a platinum sheet as counter electrode. The electrolyte was Fusayama-Meyer artificial saliva. Polarization tests were carried out in a temperature of 25 ± 1 °C at the scan rate of 0.001 V/s. Prior to polarization, the samples were kept in contact with the electrolyte for 1 h at open circuit conditions. Measurements were repeated until obtaining three similar results (polarization curves). Corrosion potential (E_{corr}), corrosion current density (i_{corr}) and polarization resistance (R_{pol}—a value inversely proportional to i_{corr}) were estimated by the Tafel extrapolation method. Since the anodic curves were influenced by passivation reactions, only the cathodic curves were used to determine i_{corr}, according to [30].

3. Results

3.1. Chemical Composition

To determine the effect of acetylene mass flow rate on the chemical composition of the deposited coating, wavelength-dispersive X-ray spectroscopy (WDX) tests were carried out. Table 2 summarizes the results of the tests, i.e., the average concentrations of elements in individual coatings.

Table 2. Chemical composition of the Zr–C coatings.

C$_2$H$_2$ Volumetric Flow Rate, sccm	C, at.%	Zr, at.%	O, H and Others, at.%
0	~0.0	91.8	8.2
1.5	20.7	71.6	7.7
3.5	52.9	43.7	3.4
4.5	63.3	33.0	3.7
6.5	79.0	18.9	2.1

It has been proven that in coatings deposited using volumetric flow rates between 1.5–6.5 sccm, the carbon concentration increases linearly from around 20 to 80 at.%.

In order to determine the chemical composition (chemical bonds) of the surface layer of deposited coatings, important from the point of view of biological properties, an XPS test was performed.

As expected, among the atoms identified on the surface of the coatings, the main share is zirconium and carbon. As mentioned above, the samples after the deposition process were in the ambient atmosphere, and therefore a significant amount of oxygen atoms is also observed on their surface. Figure 1 shows high resolution Zr 3d, O 1s and C 1s spectra. The positions of individual peaks in the spectra obtained as a result of their deconvolution with Gaussian curves are given in Table 3. The Zr 3d spectrum for each coating (Figure 1a) can be deconvolved into four main peaks of about 180, 182, 183 and 185 eV. The first two refer to Zr-C bonds, while the others are derived from Zr-O bonds in connection with spin-orbit doublets (i.e., Zr 3d$^{5/2}$ and Zr 3d$^{3/2}$) separated by energy difference [31]. For coatings with a higher carbon concentration (63 and 79 at.%), zirconium carbide is by far the dominant phase. According to the literature [32], binding energy of metallic zirconium is typically around 178.5 eV. The location of the maximum main peak in the Zr 3d spectrum for the coating at 21 at.% C between energies typical for Zr-C bond and Zr-Zr as well as the shift of the peak assigned to Zr-C in the C 1s spectrum for this coating towards lower binding energies suggests the presence of some kind of intermediate phase, not metallic zirconium itself. At lower carbon concentrations (21 and 53 at.%), a relatively high proportion of zirconium oxide is observed. This is due to the high reactivity of zirconium, which in contact with air is oxidized to form stable, compact oxides on the surface of the material. In the O 1s (Figure 1b) spectrum, the dominant peak for all coatings is a peak of about 530.5 eV identified as Zr-O bond. Other peaks come from oxygen adsorbed on the surface and contamination by organic compounds.

In the case of spectrum C 1s (Figure 1c) for coatings from 53 at.% C inclusive, four peaks can be distinguished. The peak at about 284.8 eV is characteristic for C-C bonds. The peak at about 286.5 eV, of lower intensity, originates from C-O bonds in the alcohol and ester group (surface impurities). The ~282 eV peak is characteristic for Zr-C binding. An additional peak of 283 eV is also present for these three coatings. The presence of an additional peak between the peaks derived from Me-C and C-C bonds has been observed earlier in Zr, Ti and Nb carbide coatings [33]. It is attributed to the occurrence of a phase in which carbon atoms are partially bonded to metal and carbon simultaneously (atoms on the border of metal carbide and amorphous carbon phases) and therefore is marked with a symbor Me-C*. For the two coatings with the highest carbon concentrations (63 and 79 at.%), this peak becomes dominant over Zr-C. In addition, with increasing carbon concentration in the coating, the peak from Zr-C shifts towards higher binding energies. This may indicate an increase in the proportion of the amorphous carbon phase and that the system is transformed into the MeC/a-C nanocomposite [34].

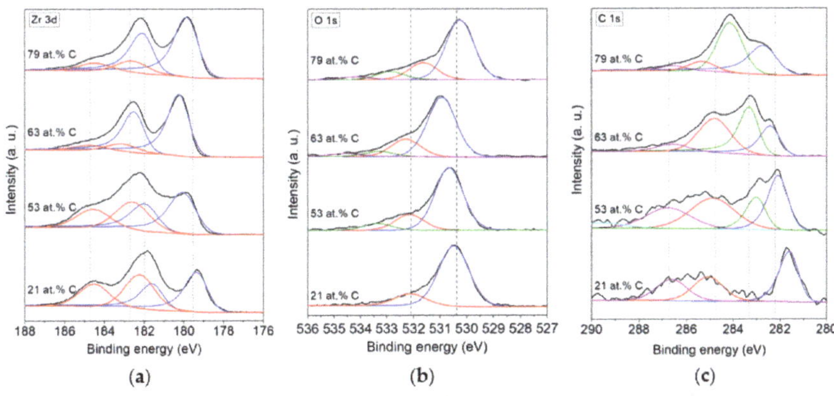

Figure 1. XPS spectra of Zr–C coatings of various carbon concentrations (a) Zr 3d, (b) O 1s, (c) C 1s.

Table 3. Binding energies calculated from the deconvolved XPS spectra.

Type of Coating	Zr (3d)			O (1s)			C (1s)		
	Phase	at.%	B.E. (eV)	Phase	at.%	B.E. (eV)	Phase	at.%	B.E. (eV)
21 at.% C	ZrC	45	$179.4\ (3d^{5/2})$	ZrO_2	83.3	530.5	ZrC	39.72	281.6
			$181.6\ (3d^{3/2})$	O ads.	16.7	532.2	C-C	32.16	285.1
	ZrO_2	55	$182.3\ (3d^{5/2})$				C-O-C, C-OH	28.12	286.7
			$184.5\ (3d^{3/2})$						
53 at.% C	ZrC	52.5	$180.1\ (3d^{3/2})$	ZrO_2	73.6	530.7	ZrC	24.64	282.1
			$182.0\ (3d^{5/2})$	O ads.	18.3	532.2	ZrC *	14.6	283
	ZrO_2	47.5	$182.6\ (3d^{3/2})$		8.1	533.4	C-C	33.18	284.9
			$184.6\ (3d^{5/2})$				C-O-C, C-OH	27.57	286.8
									289.6
63 at.% C	ZrC	84.2	$180.3\ (3d^{5/2})$	ZrO_2	70.1	531	ZrC	19.95	282.43
			$182.5\ (3d^{3/2})$	O ads.	20.2	532.3	ZrC *	31.94	283.34
	ZrO_2	15.8	$183.1\ (3d^{5/2})$		6.1	533.3	C-C	39.55	284.83
			$184.8\ (3d^{3/2})$		3.6	534.6	C-O-C, C-OH	8.56	286.61
79 at.% C	ZrC	79.2	$179.9\ (3d^{5/2})$	ZrO_2	68.5	530.3	ZrC	31.7	282.7
			$182.1\ (3d^{3/2})$	O ads.	19.2	531.7	ZrC *	53.2	284.2
	ZrO_2	20.8	$182.6\ (3d^{5/2})$		9.1	532.9	C-C	10	285.4
			$184.5\ (3d^{3/2})$		3.2	534.2	C-O-C, C-OH	5.1	286.6

3.2. Coating Morphology

The effect of C concentration on the microstructure of chosen coatings is illustrated in Figure 2a. In addition, Figure 2b–d shows the results of EDX mapping, which aimed to examine the degree of uniformity in the distribution of elements on the coating surface. It can be seen that the grain boundaries are clearly observed in the SEM micrographs (Figure 2a). Maps clearly show that Zr, C and O are evenly distributed on the surface of each coating. As mentioned in Section 3.1, the presence of oxygen is associated with the oxidation of the surface of the coatings stored in the ambient atmosphere after the deposition process or with the process conditions themselves.

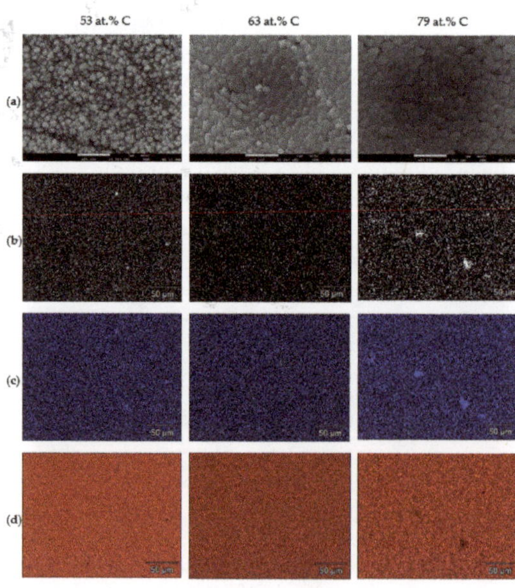

Figure 2. (a) SEM micrographs of the surfaces of coatings. (b–d) SEM-EDX mapping with a visible element distribution on the surface of samples: (b) O, (c) C, (d) Zr.

Based on the results of the EDX mapping analysis taken from the cross section of the samples, two layers can be distinguished (Figures 3 and 4). The first layer in direct contact with stainless steel (Zr adhesive layer) consistently has 300 nm thickness. The next (specific layer) is a mixture of Zr and C, in which both elements are evenly distributed in its volume.

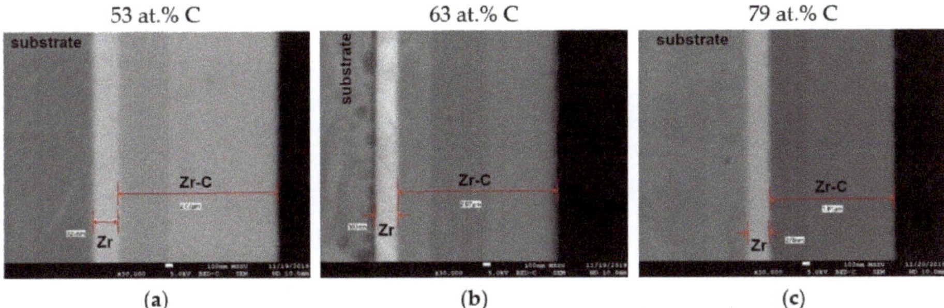

Figure 3. Backscattered electron (BSE) cross-sectional micrograph of samples (**a**) Zr–C with 53 at.% C, (**b**) Zr–C with 63 at.% C, (**c**) Zr–C with 79 at.% C.

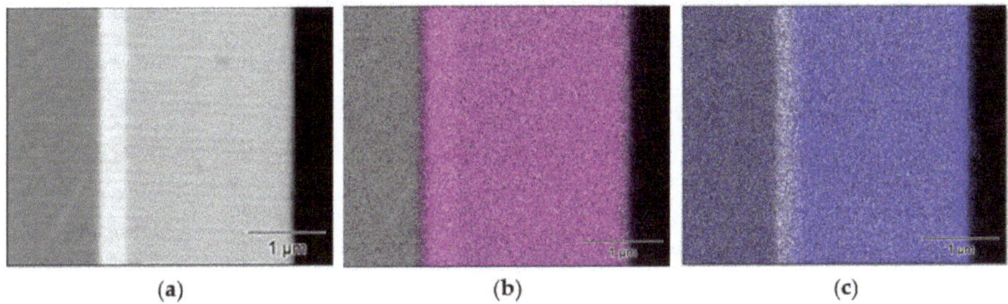

Figure 4. (**a**) Cross-sectional micrograph for sample with 53 at.% C, with (**b**) an example of mapping distribution of C and (**c**) an example of mapping distribution of Zr.

3.3. Phase Composition

3.3.1. X-ray Diffraction

Based on the analysis of X-ray diffraction patterns (Figure 5), obtained using conventional Bragg–Brentano geometry, it appears that the phase composition of the coatings changes radically with increasing carbon concentration, similarly to Ti-C coatings [35]: from the coating dominated by Zr for low carbon concentration (21 at.%) through the intermediate Zr-ZrC phases to the nanocomposite coating composed of zirconium carbide in an amorphous hydrogenated carbon matrix for the highest carbon concentration (79 at.%).

These conclusions resulting from the analysis of registered X-ray diffractograms correspond with coating tests using XPS (Section 3.1). In particular, according to the PDF-4+ ICDD database, for a coating with a carbon concentration of 21 at.%, the peak with the highest intensity comes from the ZrC phase (Ref. No. 04-019-3519) or from the $ZrC_{0.2}$ phase (Ref. No. 04-002-0154). At carbon concentrations of 63 and 79 at.%., the intensity of the main diffraction peak ZrC (111) decreases. At the same time, the width of the peak increases. This is due to the reduction of the ZrC phase content in the coating with simultaneous reduction of the ZrC crystallite size. In turn, the diffraction peaks from the Zr phase visible on the diffraction pattern, in the case of the coating with the highest carbon concentration, are derived from the Zr adhesive layer [17].

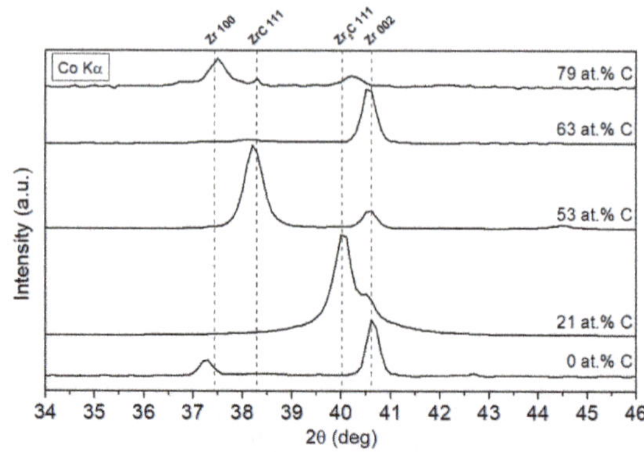

Figure 5. X-ray diffractograms of Zr–C coatings at various concentrations of carbon.

3.3.2. Raman Spectroscopy

Raman spectroscopy was used to determine the carbon structure (sp^2 and sp^3 bond ratio) in the coatings. For this purpose, Raman spectra were deconvolved with two Gaussian curves, as shown in Figure 6. Peaks were obtained with maximum positions of ~1580 cm^{-1} and ~1350 cm^{-1}, referred to as peak G and peak D, respectively [36].

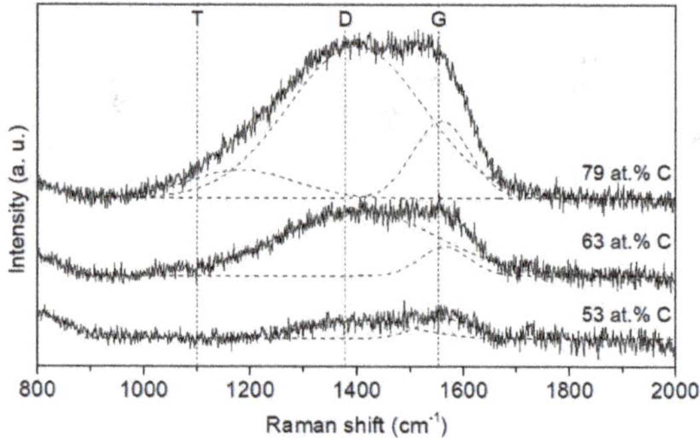

Figure 6. Raman spectra for tested coatings.

A so-called T peak (about 1100 cm^{-1}) also appears for coatings with the highest carbon concentration. According to [37], this peak corresponds to the C–C sp^3 vibration density of states (VDOS). Table 4 summarizes the calculated I(D)/I(G) ratios (integrated intensities of D and G peaks) as well as the location of the maximum of G peak for each coating. According to the model proposed in [37], it can be stated that the coatings do not have a graphite structure, as evidenced by the location of the G peak below 1580 cm^{-1}, which is typical for polycrystalline graphite. This location, together with values of I(D)/I(G) ratio above 3.0, indicate that carbon is in amorphous form with a predominant share of sp^2 hybridization bonds [17,37]. The maximum of this ratio for the coating by 63 at.% C, with the unchanged position of peak G, may indicate the largest share of sp^2 bonds in the carbon phase of this coating.

Table 4. Analysis of Raman spectra for tested coatings.

C (at.%)	I(D)/I(G)	G Peak Position (cm^{-1})
53	3.08 ± 0.13	1577.1 ± 0.42
63	4.5 ± 0.08	1575.4 ± 0.35
79	3.91 ± 0.09	1575.1 ± 0.33

3.4. Mechanical Properties

In Figure 7a, the hardness of the coatings depending on the concentration of carbon atoms is shown. It can be seen that the hardness of coatings changes dynamically as the concentration of carbon increases. At 21 at.% C, the hardness is 8 GPa, and it increases to almost 23 GPa near the stoichiometric composition. After exceeding 50 at.% C, it gradually decreases, and at a concentration of 79 at.% C it reaches 15 GPa. The Young's modulus is characterized by the analogous nature of the changes (Figure 7b). According to the interpretation in [17] regarding Ti-C coatings, the increase in hardness is caused by a higher content of the carbide phase in the hard coating and increased concentration of carbon diffused into the interstitial sites of the carbide lattice [17]. Then, based on the analysis of X-ray diffraction patterns (Figure 5), it can be seen that after exceeding the concentration of 50 at.% C in Zr–C coatings, the participation of zirconium carbide phase decreases, which corresponds to a decrease in hardness and Young's modulus.

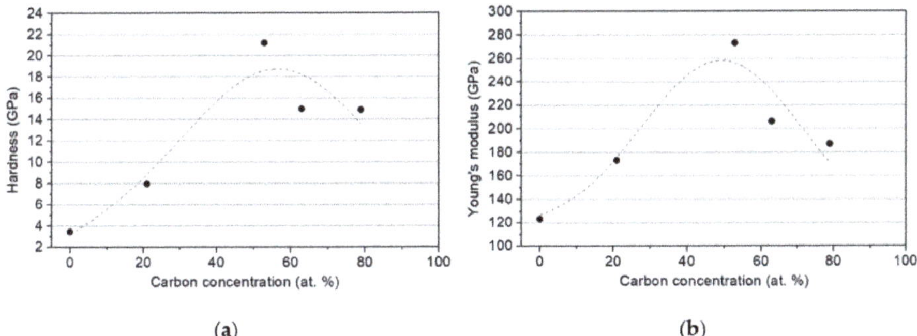

(a)   (b)

Figure 7. (a) Hardness, (b) Young's modulus of deposited coatings depending on carbon concentration.

3.5. Antibacterial Properties

3.5.1. Static Zone of Inhibition

Four species of microorganisms from the American Pure Culture Collection (*Candida albicans* ATCC 2091, *Escherichia coli* ATCC 25922, *Staphylococcus aureus* ATCC 25923 and *Streptococcus mutans* ATCC 35668) were used. The selected strains are representative of the peri-implant environment. The criterion for the assessment of fungicidal and bactericidal properties was the presence and size of the zone inhibiting the growth of microorganisms used for research.

According to the results presented in Figures 8 and 9, a 304L steel sample without coating does not create a zone of inhibition for all tested bacterial strains. This zone also does not form on samples with a Zr coating and a Zr–C coating with a carbon concentration of 21 at.% for *Candida albicans*. For all other bacteria, zones of inhibition are formed for the samples regardless of the carbon concentration in the coating.

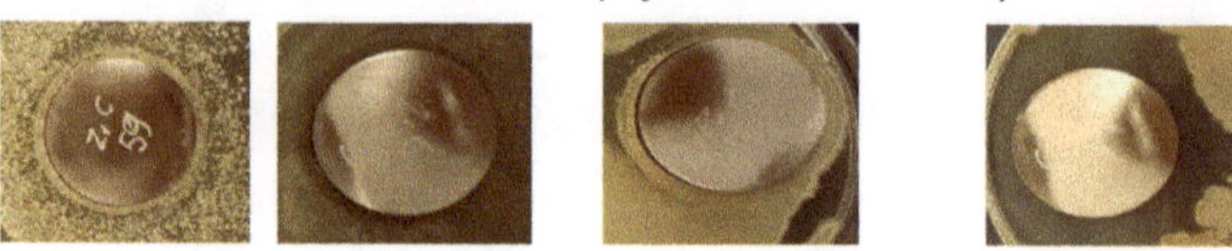

Figure 8. Antibacterial activity of the sample coated with Zr–C (63 at.% C) coating against bacterial strains based on the inhibition zone test.

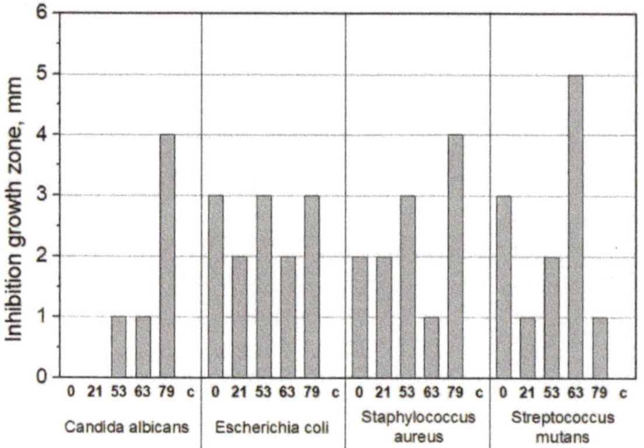

Figure 9. Growth inhibition zones of the bacterial strains in presence of the Zr–C coated samples.

3.5.2. Numbers of Live and Dead Cells

In the case of living cells (Figure 10), the number of *Candida albicans* cells on the Zr coating and the coating with 21 at.% C exceeds the amount of these cells on a 304L steel sample without a coating. The same applies to the Zr coating in the case of *Escherichia coli* bacterial cells. On coatings with a carbon concentration of 53 at.% and higher, a lower relative number of live cells (Equation (1)) is observed relative to the uncoated sample. Among the tested coatings differing in the carbon concentration, there is a coating with a concentration of ~63 at.% on which a relatively small number of living cells is observed for all bacteria used in the study (according to Equation (1), the higher the bar, the smaller the number of cells).

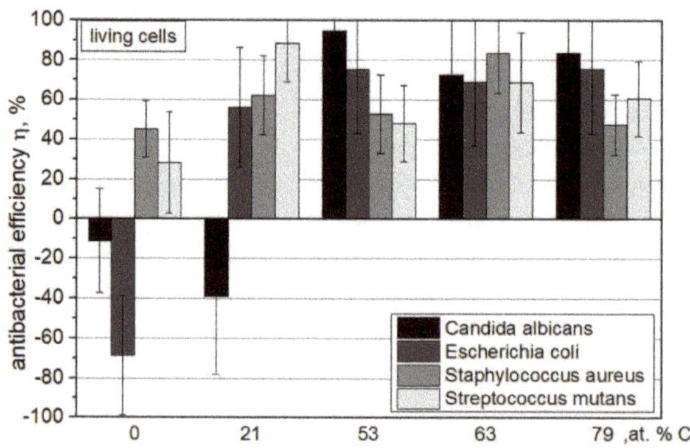

Figure 10. The relative number (cfu/cm^2) of living bacterial cells on the Zr–C coated samples relative to the number of living bacterial cells on the uncoated 304L sample.

In the case of dead cells, a similar tendency is observed regarding the number of cells as in the case of living cells (Figure 11). That is, for coatings with a carbon concentration exceeding 53 at.%, the number of dead cells for all the bacteria used in the experiment is significantly reduced compared to the uncoated sample. This effect was demonstrated by calculating the average antibacterial efficiency of the tested coatings in relation to all considered strains (Figure 12).

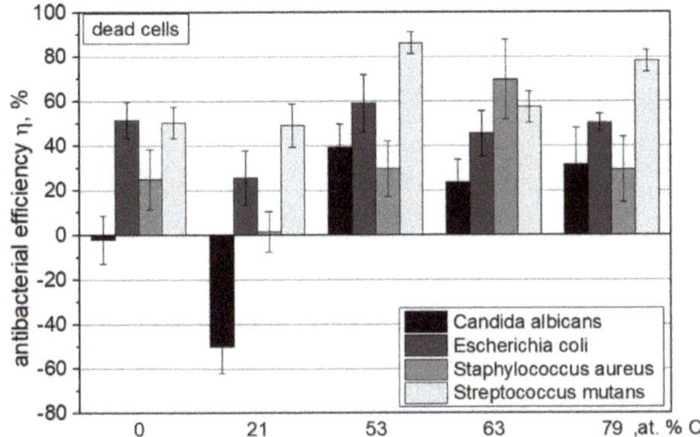

Figure 11. The relative number (cfu/cm^2) of dead bacterial cells on the Zr–C coated samples relative to the number of dead bacterial cells on the uncoated 304L sample.

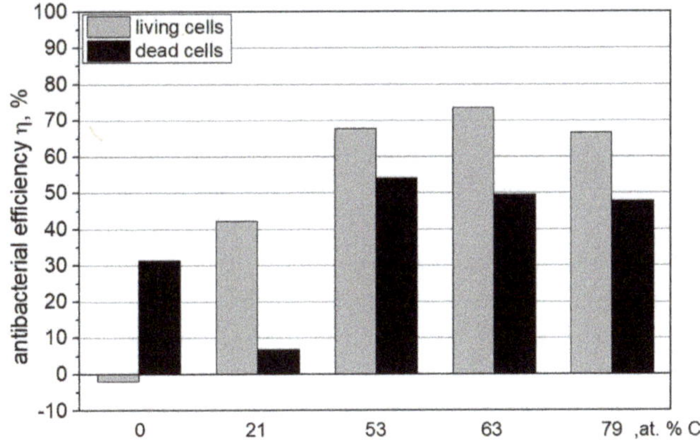

Figure 12. The relative average number (cfu/cm^2) of living and dead bacterial cells for all bacterial strains on the Zr–C coated samples relative to the number of bacterial cells on control sample.

3.5.3. Contact Angle

Microbial adhesion to the surface of the biomaterial depends on the surface properties of the microbial cell, the properties of the liquid surrounding the material and the surface properties of the material such as hydrophobicity/hydrophilicity and free surface energy. It has been shown in the literature that the formation of dental plaque in the human mouth is significantly reduced on hydrophobic surfaces relative to hydrophilic surfaces [38,39]. One of the reasons may be the fact that hydrophilic bacteria inhabiting the oral cavity tend to be surrounded by an aqueous medium such as saliva, which easily breaks away from the more hydrophobic surface of the biomaterial [39]. For this reason, an important factor in assessing the potential antibacterial properties of the coating is to study its hydrophobic properties by determining the contact angle. The results of measuring the contact angle of coatings and uncoated steel with the artificial saliva solution are shown in Table 5.

Table 5. Contact angle of 304L steel substrate, bare and coated with Zr–C, with Fusayama-Meyer's artificial saliva.

Sample	Contact Angle After 30 s	Contact Angle After 300 s
	(°)	(°)
304L	87.0 ± 0.7	80.0 ± 0.8
304L/Zr	69.1 ± 0.3	56.0 ± 1.7
304L/Zr–C-21 at.% C	73.2 ± 1.4	59.8 ± 1.3
304L/Zr–C-53 at.% C	94.3 ± 0.3	86.7 ± 0.5
304L/Zr–C-63 at.% C	99.2 ± 0.4	92.7 ± 0.8
304L/Zr–C-79 at.% C	97.1 ± 0.4	89.5 ± 1.7

Coatings deposited at a acetylene flow rate of 3.5 sccm and above showed a higher contact angle than uncoated 304L steel. Given that the contact angle is a measure of hydrophobicity, it can be concluded that surfaces coated with Zr–C coatings above 50 at.% C are more hydrophobic than surfaces of uncoated 304L steel. The highest contact angle was observed for the surface of the coating with a concentration above 60 at.% C (99.2°).

3.6. Corrosion Properties

In order to determine the protective properties of Zr–C coatings, potentiodynamic tests were carried out in the solution of artificial saliva. Potentiodynamic curves for the substrate/Zr–C coating systems with different concentrations of carbon and for the uncoated substrate are shown in Figure 13. Values of determined corrosion current density (i_{corr}), corrosion potential (E_{corr}), polarization resistance (R_{pol}), breakdown potential (R_b) and the slope of the cathodic curve ($-b_c$) are shown in Table 6.

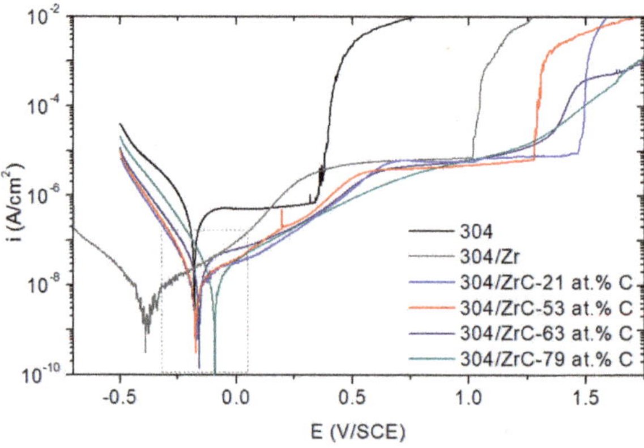

Figure 13. Polarization curves obtained for 304L steel substrate and substrate-coating systems in Fusayama-Meyer's artificial saliva.

Table 6. Electrochemical parameters characterizing corrosion processes of 304L steel substrate, bare and coated, in Fusayama-Meyer's artificial saliva.

Sample	E_{corr}	i_{corr}	$-b_c$	R_{pol}	E_b	Protective
	(V)	(A/cm²)	(V/dec)	(Wcm²)	(V)	Efficiency [%]
304L	0.191 ± 0.019	(827 ± 40) × 10⁻⁹	0.187 ± 0.012	(98 ± 3) × 10³	0.350 ± 0.022	
304L/Zr	0.430 ± 0.049	(7.80 ± 1.73 × 10⁻⁹	0.097 ± 0.011	(4.45 ± 0.86) × 10⁶	1.175 ± 0.189	99.1
304L/Zr–C-21 at.% C	0.162 ± 0.016	(27.76 ± 7.25) × 10⁻⁹	0.115 ± 0.031	(1.93 ± 0.52) × 10⁶	1.403 ± 0.033	96.6
304L/Zr–C-53 at.% C	0.161 ± 0.010	(20.41 ± 0.81) × 10⁻⁹	0.115 ± 0.004	(1.82 ± 0.55) × 10⁶	1.219 ± 0.154	97.5
304L/Zr–C-63 at.% C	0.157 ± 0.007	(41.29 ± 5.07) × 10⁻⁹	0.161 ± 0.017	(1.11 ± 0.09) × 10⁶	-	95.0
304L/Zr–C-79 at.% C	0.085 ± 0.016	(19.12 ± 2.00) × 10⁻⁹	0.137 ± 0.009	(2.18 ± 0.22) × 10⁶	-	97.7

The lowest corrosion potential among all tested samples (lower than steel) shows a sample with a zirconium coating. This is due to the greater reactivity of zirconium to alloying steel components resulting in rapid surface oxidation. The passive layer formed on the surface is very compact and provides effective protection against electrolytes (a clear passive area, the lowest corrosive current among the tested). The passive layer remains stable until reaching the breakdown potential of 1.175 V. The tests also showed active-passive behavior of the uncoated 304L substrate and the substrate with the coatings with 21 and 53 at.% C. Increasing the carbon concentration in the coating changes the shape of the polarization curves, i.e., there is no visible passive area in the anodic range. It can therefore be concluded that in the case of the coating with the ZrC nanocomposite structure in the amorphous hydrogenated carbon matrix, the oxidation of metallic system components is difficult. For coatings above 53 at.% C, on polarization curves, there is no clear pitting initiation zone (no breakdown potential was determined). This indicates a high protective potential of these coatings against pitting corrosion in an artificial saliva environment.

All of the Zr–C coatings were characterized by a reduction in the corrosion current by an order of magnitude, as well as a slight increase in corrosion potential in relation to the uncoated substrate. Corrosion potential increased with increasing concentration of carbon in the coating. Based on the value of the corrosion current density, the protective performance of the coatings was estimated, which is shown in Table 6. The protective capacity of all coatings is over 95%. The highest yield among Zr–C coatings (97.7%) is characterized by the coating with the highest carbon concentration. This property, in conjunction with the absence of a pitting initiation zone on the polarization curve, makes this coating a good candidate for anticorrosive protection of medical grade 304L in the environment of human saliva.

4. Discussion

On the basis of hydrophobicity and microbiological properties, i.e., bacteriostatic tests and resistance to bacterial adhesion and colonization, it is clear that the antibacterial properties of the Zr–C coatings studied grow in accordance with the sequence:

304L steel → 304L/Zr coating → 304L/Zr-C coating (<50 at.% C) → 304L/Zr-C coating (>50 at.% C).

As demonstrated in the assessment of the antibacterial properties of the Zr–C coatings tested, their hydrophobicity plays an important role by determining the contact angle. All tested coatings with C concentration above 50 at.% showed a higher contact angle than uncoated 304L steel. Considering the contact angle as a measure of hydrophobicity and at the same time as an assessment of potential antibacterial properties, it can be generally observed that surfaces with Zr–C coatings have better antibacterial properties than surfaces of uncoated 304L steel and that coated with Zr. In particular, a rapid increase in antibacterial properties is observed when the carbon concentration in Zr–C coatings exceeds 50 at.%, while practically no significant differences occur in the bacteriostatic tests as well as resistance to bacterial adhesion and colonization. In this context, it is interesting to compare the results of the chemical and phase composition of the coatings carried out using the XPS method and the participation in coatings sp^2 and sp^3 bonds determined by Raman spectroscopy. First of all, the results of these tests indicate low or no participation of phases with graphite and diamond carbon bonds in coatings with a carbon concentration below 50 at.%. In turn, in coatings with a carbon concentration above 50 at.%, based on the analysis of Raman spectra it can be concluded that the carbon has an amorphous form with a dominance of sp^2 bonds.

In [40], hydrogenated amorphous carbon (α-C:H) coatings were produced on PET in an acetylene atmosphere at different working pressures. Coatings differing in the share of sp^3/sp^2 bonds were obtained. On the basis of the conducted research it was shown that the adhesion of bacteria to the coatings depends on the structure of the coatings and decreases with decreasing sp^3/sp^2 ratio, i.e., an increase in the share of sp^2 bonds. Additionally, in [41] it has been proved that the sp^3/sp^2 ratio is an important factor affecting the biological response of carbon coatings. As already mentioned, in the test results presented in this article, the coatings produced above 50 at.% C have the observed advantage of sp^2 bonds over sp^3 bonds, and these relationships, i.e., the ratio of sp^3/sp^2 bonds, do not change with increasing carbon concentration; the antibacterial properties also do not change. These results correspond to the results presented in the cited papers [40,41].

In the context of the obtained results characterizing antibacterial properties, it should also be taken into account that in the tested coatings there is a specific hydrogen concentration due to the fact that Zr–C coatings are produced in the C_2H_2 atmosphere. In [5], it is stated that the presence of hydrogen in carbon coatings deteriorates their antibacterial activity compared to a coating containing no hydrogen. Bacterial adhesion is an interfacial process and therefore electrostatic and van der Waals forces are most likely responsible for the interaction between material and bacterial surfaces. Due to the presence of hydrogen in the carbon layer, polarized C-H bonds are formed, which are promoters of Lifshitz van de Waals forces between the microorganism and the polarized surface of the material [42]. Thus, on the one hand, the advantage of sp^2 bonds in the coating promotes its antibacterial effect, while on the other hand, the presence of hydrogen may slightly worsen this effect.

The antibacterial properties of the tested coatings were also examined by analyzing the number of cells attached to the coatings. Both live and dead cells were considered. In these studies, it was found that the number of adsorbed cells, both dead and living, decreases significantly after exceeding 50 at.% C compared to uncoated 304L steel and that coated with Zr, i.e., as in other tests, the antibacterial properties of coatings increase significantly after exceeding 50 at.%. In the case of dead cells, studies described in [43–45] suggest that the main mechanism for the death of bacteria is the physical interaction with carbon-based coatings. Carbon aggregates can cause irreversible destruction of the outer cell membrane of bacteria that results in the release of cell contents [43]. The smallest wettability of the coating with a carbon concentration of ~63% and the smallest roughness observed in SEM images may be the reason for the weakest adhesion, i.e., the highest antibacterial effectiveness, of this coating in relation to some bacterial strains.

In summary, due to biological properties, the tested coatings after exceeding 50 at.% C are characterized by higher antibacterial properties both compared to uncoated 304L steel and to coatings with a lower carbon concentration. Taking into account the possibility of shaping a wide range of mechanical properties of these coatings, especially for coatings with a carbon concentration above 50 at.%, it can be concluded that these coatings are a good candidate for orthodontic applications. An additional argument strengthening this conclusion is the result of corrosion tests. In particular, increasing the carbon concentration in the coating above the stoichiometry results in hindered oxidation of the metallic components of the systems. For these coatings, there is also no clear pitting initiation area on the polarization curves (no breakdown potential was determined). This indicates a high protective potential of these coatings against pitting in the environment of artificial saliva.

5. Conclusions

The research presented in the article demonstrates the impact of changing the concentration of carbon in Zr–C coatings and related structure changes on the protective potential of these coatings against bacteria and corrosion in the human saliva environment. In particular, based on the results obtained, the following conclusions were drawn:

1. It has been shown that as the concentration of carbon atoms in the coating increases, the content of the phase of carbon unbound to zirconium increases. Raman spectroscopy showed that carbon phase has an amorphous structure with a predominance of bonds with sp^2 hybridization.
2. Coatings with carbon concentrations close to 50 at.% show the highest hardness and Young's modulus. At higher and lower concentrations, there is a sharp decrease in these values.
3. Considering the contact angle as a measure of hydrophobicity and at the same time as an indirect assessment of potential antibacterial properties, it can be generally stated that surfaces coated with Zr–C coatings have better antibacterial properties than surfaces of uncoated 304L steel.

4. There is a rapid increase in antibacterial properties when the carbon concentration in Zr–C coatings exceeds 50 at.%. Above this carbon concentration in the coatings, practically no significant differences occur in the bacteriostatic tests as well as resistance to bacterial adhesion and colonization.

5. Corrosion tests showed the high anti-pitting potential of Zr–C coatings in the environment of artificial saliva.

6. The wide possibility of shaping the mechanical properties of Zr–C coatings in combination with relatively good antibacterial properties after exceeding 50 at.% C concentration in coatings and high protective potential of these coatings makes them a good candidate for orthodontic applications.

Author Contributions: Conceptualization, E.C., Ł.S., E.J., K.M. and J.R.; methodology, E.C., E.D., E.J.; validation, A.G., E.D., E.C., P.C. and K.M.; formal analysis, J.R., K.M., E.C.; investigation, E.C., E.D., A.G. and E.J.; data curation, A.G., K.M., E.C., E.D., E.J.; writing—original draft preparation, J.R., K.M., E.C.; writing—review and editing, K.M., J.R., E.J.; visualization, E.J., K.M., P.C.; supervision, J.R., Ł.S.; project administration, A.G.; funding acquisition, A.G., J.R. All authors have read and agreed to the published version of the manuscript.

Funding: This research was funded by the National Science Center, Poland [grant: Opus 11, No. 2016/21/B/ST8/01738] and grant of KUT No. 504.38.05/19.

Conflicts of Interest: The authors declare no conflict of interest.

References

1. Sharan, D. The problem of corrosion in orthopaedic implant materials. *Orthop. Update (India)* **1999**, *9*, 1–5.

2. Shih, C.C.; Shih, C.M.; Su, Y.Y.; Su, L.H.J.; Chang, M.S.; Lin, S.J. Effect of surface oxide properties on corrosion resistance of 316L stainless steel for biomedical applications. *Corros. Sci.* **2004**, *46*, 427–441. [CrossRef]

3. Becker, W.; Becker, B.E.; Newman, M.G.; Nyman, S. Clinical and microbiologic findings that may contribute to dental implant failure. *Int. J. Oral Maxillofac. Implant.* **1990**, *5*, 1–17.

4. Ding, M.H.; Zhang, H.S.; Zhang, C.; Jin, X. Characterization of ZrC coatings deposited on biomedical 316L stainless steel by magnetron sputtering method. *Surf. Coat. Tech.* **2013**, *224*, 34–41. [CrossRef]

5. Zhou, H.; Xu, L.; Ogino, A.; Nagatsu, M. Investigation into the antibacterial property of carbon films. *Diam. Relat. Mater.* **2008**, *17*, 1416–1419. [CrossRef]

6. Yao, S.H.; Su, Y.L.; Lai, Y.C. Antibacterial and tribological performance of carbonitride coatings doped with W, Ti, Zr, or Cr deposited on AISI 316L stainless steel. *Materials* **2017**, *10*, 1189. [CrossRef]

7. Größner-Schreiber, B.; Herzog, M.; Hedderich, J.; Dück, A.; Hannig, M.; Griepentrog, M. Focal adhesion contact formation by fibroblasts cultured on surface-modified dental implants: An in vitro study. *Clin. Oral Implant. Res.* **2006**, *17*, 736–745. [CrossRef]

8. Groessner-Schreiber, B.; Hannig, M.; Dück, A.; Griepentrog, M.; Wenderoth, D.F. Do different implant surfaces exposed in the oral cavity of humans show different biofilm compositions and activities? *Eur. J. Oral Sci.* **2004**, *112*, 516–522. [CrossRef]

9. Muzio, G.; Miola, M.; Perero, S.; Oraldi, M.; Maggiora, M.; Ferraris, S.; Vernè, E.; Festa, V.; Festa, F.; Canuto, R.A.; et al. Polypropylene prostheses coated with silver nanoclusters/silica coating obtained by sputtering: Biocompatibility and antibacterial properties. *Surf. Coat. Tech.* **2017**, *319*, 326–334. [CrossRef]

10. Marciano, F.R.; Bonetti, L.F.; Mangolin, J.F.; Da-Silva, N.S.; Corat, E.J.; Trava-Airoldi, V.J. Investigation into the antibacterial property and bacterial adhesion of diamond-like carbon films. *Vacuum* **2011**, *85*, 662–666. [CrossRef]

11. Lai, C.H.; Chang, Y.Y.; Huang, H.L.; Kao, H.Y. Characterization and antibacterial performance of ZrCN/amorphous carbon coatings deposited on titanium implants. *Thin Solid Films* **2011**, *520*, 1525–1531. [CrossRef]

12. Janvier, G.; Baquey, C.; Roth, C.; Benillan, N.; Belisle, S.; Hardy, J.F. Extracorporeal circulation, hemocompatibility, and biomaterials. *Ann. Thorac. Surg.* **1996**, *62*, 1926–1934. [CrossRef]

13. Huang, N.; Yang, P.; Leng, Y.X.; Chen, J.Y.; Sun, H.; Wang, J.; Wang, G.J.; Ding, P.D.; Xi, T.F.; Leng, Y. Hemocompatibility of titanium oxide films. *Biomaterials* **2003**, *24*, 2177–2187. [CrossRef]

14. Liu, C.; Lin, G.; Yang, D.; Qi, M. In vitro corrosion behavior of multilayered Ti/TiN coating on biomedical AISI 316L stainless steel. *Surf. Coat. Tech.* **2006**, *200*, 4011–4016. [CrossRef]

15. Hauert, R. An overview on the tribological behavior of diamond-like carbon in technical and medical applications. *Tribol. Int.* **2004**, *37*, 991–1003. [CrossRef]

16. Geetha, M.; Singh, A.K.; Asokamani, R.A.; Gogia, A.K. Ti based biomaterials, the ultimate choice for orthopaedic implants–a review. *Prog. Mater. Sci.* **2009**, *54*, 397–425. [CrossRef]

17. Gilewicz, A.; Mydłowska, K.; Ratajski, J.; Szparaga, Ł.; Bartosik, P.; Kochmański, P.; Jędrzejewski, R. Structural, mechanical and tribological properties of ZrC thin films deposited by magnetron sputtering. *Vacuum* **2019**, *169*, 108909. [CrossRef]

18. Szparaga, Ł.; Mydłowska, K.; Gilewicz, A.; Ratajski, J. Mechanical and anti-wear properties of multi-module Cr/CrN coatings. *Int. J. Surf. Sci. Eng.* **2019**, *13*, 37–49. [CrossRef]

19. Lewin, E.; Wilhelmsson, O.; Jansson, U. Nanocomposite nc-TiC/a-C thin films for electrical contact applications. *J. Appl. Phys.* **2006**, *100*, 054303. [CrossRef]

20. Chen, C.S.; Liu, C.P. Diffusion barrier properties of amorphous ZrCN films for copper metallization. *J. Non-Cryst. Solids* **2005**, *351*, 3725–3729. [CrossRef]

21. Sasaki, M.; Kozukue, Y.; Hashimoto, K.; Takayama, K.; Nakamura, I.; Takano, I.; Sawada, Y. Properties of carbon films with a dose of titanium or zirconium prepared by magnetron sputtering. *Surf. Coat. Tech.* **2005**, *196*, 236–240. [CrossRef]

22. Martínez-Martínez, D.; López-Cartes, C.; Fernández, A.; Sánchez-López, J.C. Influence of the microstructure on the mechanical and tribological behavior of TiC/aC nanocomposite coatings. *Thin Solid Films* **2009**, *517*, 1662–1671. [CrossRef]

23. Liu, B.; Liu, C.; Shao, Y.; Zhu, J.; Yang, B.; Tang, C. Deposition of ZrC-coated particle for HTR with ZrCl4 powder. *Nucl. Eng. Des.* **2012**, *251*, 349–353. [CrossRef]

24. Jansson, U.; Lewin, E. Sputter deposition of transition-metal carbide films—A critical review from a chemical perspective. *Thin Solid Films* **2013**, *536*, 1–24. [CrossRef]

25. Andersson, M.; Urbonaite, S.; Lewin, E.; Jansson, U. Magnetron sputtering of Zr–Si–C thin films. *Thin Solid Films* **2012**, *520*, 6375–6381. [CrossRef]

26. Zheng, Y.F.; Liu, X.L.; Zhang, H.F. Properties of Zr–ZrC–ZrC/DLC gradient films on TiNi alloy by the PIIID technique combined with PECVD. *Surf. Coat. Tech.* **2008**, *202*, 3011–3016. [CrossRef]

27. Chu, C.L.; Ji, H.L.; Yin, L.H.; Pu, Y.P.; Lin, P.H.; Chu, P.K. Fabrication, properties, and cytocompatibility of ZrC film on electropolished NiTi shape memory alloy. *Mat. Sci. Eng. C* **2011**, *31*, 423–427. [CrossRef]

28. Chang, Y.Y.; Huang, H.L.; Lai, C.H.; Hsu, J.T.; Shieh, T.M.; Wu, A.Y.J.; Chen, C.L. Analyses of antibacterial activity and cell compatibility of titanium coated with a Zr–C–N film. *PLoS ONE* **2013**, *8*, e56771. [CrossRef]

29. Oliver, W.C.; Pharr, G.M. An improved technique for determining hardness and elastic modulus using load and displacement sensing indentation experiments. *J. Mater. Res.* **1992**, *7*, 1564–1583. [CrossRef]

30. McCafferty, E. Validation of corrosion rates measured by the Tafel extrapolation method. *Corros. Sci.* **2005**, *47*, 3202–3215. [CrossRef]

31. Long, Y.; Javed, A.; Chen, J.; Chen, Z.K.; Xiong, X. Phase composition, microstructure and mechanical properties of ZrC coatings produced by chemical vapor deposition. *Ceram. Int.* **2014**, *40*, 707–713. [CrossRef]

32. Balaceanu, M.; Braic, M.; Braic, V.; Vladescu, A.; Negrila, C.C. Surface chemistry of plasma deposited ZrC hard coatings. *J. Optoelectron. Adv. Mater.* **2005**, *7*, 2557–2560.

33. Lewin, E.; Persson, P.A.; Lattemann, M.; Stüber, M.; Gorgoi, M.; Sandell, A.; Ziebert, C.; Schäfers, F.; Braun, W.; Halbritter, J.; et al. On the origin of a third spectral component of C1s XPS-spectra for nc-TiC/a-C nanocomposite thin films. *Surf. Coat. Technol.* **2008**, *202*, 3563–3570.

34. Meng, Q.N.; Wen, M.; Mao, F.; Nedfors, N.; Jansson, U.; Zheng, W.T. Deposition and characterization of reactive magnetron sputtered zirconium carbide films. *Surf. Coat. Technol.* **2013**, *232*, 876–883. [CrossRef]

35. Voevodin, A.A.; Capano, M.A.; Laube, S.J.P.; Donley, M.S.; Zabinski, J.S. Design of a Ti/TiC/DLC functionally gradient coating based on studies of structural transitions in Ti–C thin films. *Thin Solid Films* **1997**, *298*, 107–115. [CrossRef]

36. Ferrari, A.C.; Robertson, J. Interpretation of Raman spectra of disordered and amorphous carbon. *Phys. Rev. B* **2000**, *61*, 14095. [CrossRef]

37. Ferrari, A.C.; Robertson, J. Raman spectroscopy of amorphous, nanostructured, diamond–like carbon, and nanodiamond. *Philos. Trans. R. Soc. A* **2004**, *362*, 2477–2512. [CrossRef]

38. Quirynen, M.; Marechal, M.; Busscher, H.J.; Weerkamp, A.H.; Arends, J.; Darius, P.L.; van Steenberghe, D. The influence of surface free-energy on planimetric plaque growth in man. *J. Dent. Res.* **1989**, *68*, 796–799. [CrossRef]

39. Bruinsma, G.M.; Van der Mei, H.C.; Busscher, H.J. Bacterial adhesion to surface hydrophilic and hydrophobic contact lenses. *Biomaterials* **2001**, *22*, 3217–3224. [CrossRef]

40. Wang, J.; Huang, N.; Pan, C.J.; Kwok, S.C.H.; Yang, P.; Leng, Y.X.; Chen, J.Y.; Sun, H.; Wan, G.J.; Liu, Z.Y.; et al. Bacterial repellence from polyethylene terephthalate surface modified by acetylene plasma immersion ion implantation–deposition. *Surf. Coat. Tech.* **2004**, *186*, 299–304. [CrossRef]

41. Liao, T.T.; Zhang, T.F.; Li, S.S.; Deng, Q.Y.; Wu, B.J.; Zhang, Y.Z.; Zhou, Y.J.; Guo, Y.B.; Leng, Y.X.; Huang, N. Biological responses of diamond-like carbon (DLC) films with different structures in biomedical application. *Mat. Sci. Eng. C* **2016**, *69*, 751–759. [CrossRef] [PubMed]

42. Gottenbos, B.; Van der Mei, H.C.; Busscher, H.J. Models for studying initial adhesion and surface growth in biofilm formation on surfaces. *Method Enzymol.* **1999**, *310*, 523–534.

43. Kang, S.; Pinault, M.; Pfefferle, L.D.; Elimelech, M. Single-walled carbon nanotubes exhibit strong antimicrobial activity. *Langmuir* **2007**, *23*, 8670–8673. [CrossRef] [PubMed]

44. Ali, S.S.; Hardt, J.I.; Quick, K.L.; Kim-Han, J.S.; Erlanger, B.F.; Huang, T.T.; Epstein, C.J.; Dugan, L.L. A biologically effective fullerene (C60) derivative with superoxide dismutase mimetic properties. *Free Radic. Bio. Med.* **2004**, *37*, 1191–1202. [CrossRef] [PubMed]

45. Tang, Y.J.; Ashcroft, J.M.; Chen, D.; Min, G.; Kim, C.H.; Murkhejee, B.; Larabell, C.; Keasling, J.D.; Chen, F.F. Charge-associated effects of fullerene derivatives on microbial structural integrity and central metabolism. *Nano Lett.* **2007**, *7*, 754–760. [CrossRef] [PubMed]

Article

The Influence of Nitrogen Absorption on Microstructure, Properties and Cytotoxicity Assessment of 316L Stainless Steel Alloy Reinforced with Boron and Niobium

Sadaqat Ali [1,2,*], Ahmad Majdi Abdul Rani [1,*], Riaz Ahmad Mufti [2], Farooq I. Azam [2], Sri Hastuty [3], Zeeshan Baig [4], Murid Hussain [5] and Nasir Shehzad [5]

[1]   Mechanical Engineering Department, Universiti Teknologi PETRONAS (UTP), Bandar Seri Iskandar, Perak Darul Ridzuan 32610, Malaysia
[2]   School of Mechanical & Manufacturing Engineering, National University of Sciences and Technology (NUST), H-12, Islamabad 44000, Pakistan
[3]   Mechanical Engineering Department, Universitas PERTAMINA, Jakarta 12220, Indonesia
[4]   Centre of Excellence in Science & Applied Technologies, Islamabad 44000, Pakistan
[5]   Department of Chemical Engineering, COMSATS University Islamabad, Lahore Campus, Defense Road, Off Raiwind Road, Lahore 54000, Pakistan
*   Correspondence: engineersadaqat@gmail.com (S.A.); majdi@utp.edu.my (A.M.A.R.);
    Tel.: +92-333-9410290 (S.A.); +60-5368-7155 (A.M.A.R.)

Received: 23 June 2019; Accepted: 19 July 2019; Published: 2 August 2019

Abstract: In the past, 316L stainless steel (SS) has been the material of choice for implant manufacturing. However, the leaching of nickel ions from the SS matrix limits its usefulness as an implant material. In this study, an efficient approach for controlling the leaching of ions and improving its properties is presented. The composition of SS was modified with the addition of boron and niobium, which was followed by sintering in nitrogen atmosphere for 8 h. The X-ray diffraction (XRD) results showed the formation of strong nitrides, indicating the diffusion of nitrogen into the SS matrix. The X-ray photoelectron spectroscopy (XPS) analysis revealed that a nitride layer was deposited on the sample surface, thereby helping to control the leaching of metal ions. The corrosion resistance of the alloy systems in artificial saliva solution indicated minimal weight loss, indicating improved corrosion resistance. The cytotoxicity assessment of the alloy system showed that the developed modified stainless steel alloys are compatible with living cells and can be used as implant materials.

Keywords: implant; stainless steel; nickel; leaching; nitrogen; cytotoxicity

---

1. Introduction

Biomedical implants and devices are critical to enhancing the quality of life, and the nature and life span of human beings [1]. The restoration of parts of the human body using implants is an ancient field and dates back 4000 years ago when gold and iron was used for dental applications by Egyptians and Romans [2]. The advancement in producing biomedical implants is closely related to the development of biomaterials with tailored mechanical properties. This field has gained substantial attention over the last few decades, and since the first summit on the development of biomaterials was held in 1969 [3]. Since then, there has been continuous effort to develop new materials and improve the existing manufacturing methods for the betterment of humankind in terms of implantation devices. A suitable choice of biomedical material for implant manufacturing is very crucial for its long term success, and should include consideration of its wear resistance, biocompatibility, corrosion resistance, mechanical properties, economics and ease of manufacturing [4,5].

Among the various biomaterials available, austenitic 316L stainless steel has been in use since the 1930s, when it was utilized in the total hip replacement of an orthopaedic patient [6,7]. Since then, it has gained tremendous attention as a biomaterial for implant manufacturing owing to its biocompatibility, adequate mechanical properties, ease of manufacturing, decent corrosion resistance, and cheaper cost [8,9]. ASTM International also recommends the 316L type amongst the various available grades of austenitic stainless steels for its usage as an implant material. The high concentration of chromium and low carbon content improves its tendency towards chlorine bearing solutions as far as corrosion resistance is concerned [10,11]. Saline in the human body has a close resemblance to chlorine-bearing solutions and this makes 316L SS an ideal material for the human body [12]. However, there are some problems associated with the use of this medical grade alloy [13]. Several studies on retrieved 316L SS implants from patients have revealed that 90% of the implant failures were due to corrosion attacks due to pitting [14,15]. One of the most important aspects in this regard is the leaching of nickel ions from the SS matrix [13,16,17]. The release of nickel and other metal ions from implants are clearly associated with poor corrosion resistance in humans [18–22]. The leaching of metal ions and poor corrosion resistance necessitates the modification of the 316L SS matrix by alloying it with additives and improving the surface layer of the implants [23–25].

To extend the applicability of SS as an implant material, the addition of niobium in the SS matrix can help to increase the corrosion resistance of the resulting modified SS alloy. In the last few decades, niobium has attracted much attention for its potential use as a biomaterial [26–28]. Niobium has proved to be an important and promising alloying element [29] with adequate biocompatibility for biomedical applications [30,31]. It exhibits good corrosion resistance and strength for its potential use in implant applications [32,33]. Boron has proved to be a promising additive to increase the densification process [34,35] and for increasing the overall density of the resulting stainless steel alloy [36]. Both these alloying elements address the highlighted issues related to the use of SS alloy as an implant material. Sintering of SS alloy in nitrogen has been proven to improve the densification process as well as improving the mechanical properties [37]. In this study, sintering of the alloy was carried out in nitrogen atmosphere for 8 h to develop a nitride layer on the surface of the alloy.

2. Materials and Methods

In this study, gas atomized 316L SS powder was used as the matrix to prepare the samples. The amount of boron addition was limited to 0.25 wt.% for activated sintering for maximum densification, as per our previous studies [9,34,38] whereas the amount of niobium was varied from 0.5 wt.% to 2 wt.%. The SEM images and XRD analysis of all the powders used in the study are shown in Figures 1 and 2, respectively. The XRD spectrum clearly indicates the presence of chromium, iron and nickel in the stainless steel powder at their respective peaks, along with boron and niobium peaks in their respective spectrum as shown in Figure 2.

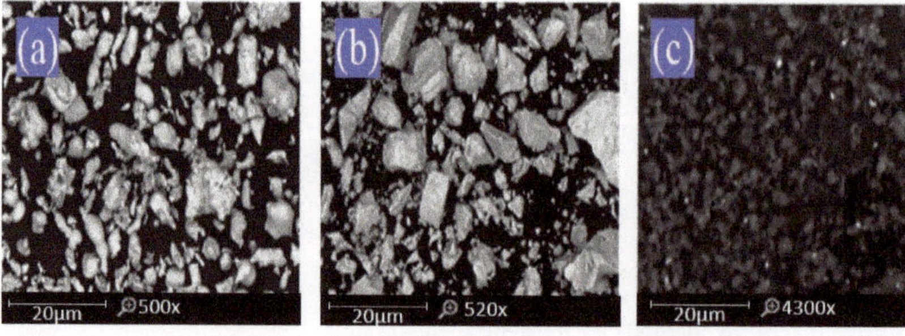

Figure 1. SEM images of (**a**) 316L SS, (**b**) niobium, and (**c**) boron powder.

A total of five composites were made in this study. These were named as S1 (pure 316L SS); S2 (316L SS + 0.25 B + 0.5 Nb); S3 (316L SS + 0.25 B + 1 Nb); S4 (316L SS + 0.25 B + 1.5 Nb); and S5 (316L SS + 0.25 B + 2 Nb). Each powder composition was prepared by blending the respective amount of each powder in a tubular mixer for 8 h. The powder from each mixture was compacted into a disc shape of 30 mm diameter and ~4 mm thick using the uniaxial cold compaction process at a pressure of 800 MPa. The compacted samples were then sintered in a tube furnace (Model: Protherm, PTF12/75/800, Alserteknik, Ankara, Turkey). Nitrogen was selected as the sintering atmosphere and a temperature of 1200 °C was used for sintering all of the samples. The dwell time was kept at 8 h for maximum absorption of nitrogen into the samples.

The green densities of the samples were calculated using geometric method, whereby the diameter and thickness of the compacted samples was measured. The green density was then calculated by dividing the mass of the sample by its volume as given by Equation (1).

$$Green\ density = \frac{Mass}{Volume} \tag{1}$$

The sintered densities for all the samples were measured using Archimedes' principle. The HR-150 AZ analytical balance manufactured by A&D Company, Limited, Tokyo, Japan was used for the density measurement. Three samples of each formulation were measured and the average value of sintered density was recorded. The relative density of each sample were calculated by dividing the sintered density by the theoretical density of pure 316L stainless steel as the reference, as shown by Equation (2). The standard deviation of green and sintered densities for all the samples were calculated and are given in Table 1.

$$Relative\ density = \frac{\rho_{experimental}}{\rho_{theoretical}} \tag{2}$$

The optical microscope (Model: Leica DM LM, Wetzlar, Germany) was used to examine the microstructure after being properly etched. The Carpenters stainless steel etchant solution was used to etch the samples, which were composed of 8.5 g Ferric Chloride ($FeCl_3$), 2.4 g Cupric Chloride ($CuCl_2$), 122 mL Hydrochloric acid (HCl), 6 mL Nitric acid (HNO_3) and 122 mL Ethanol. The samples were immersed in the solution for 30 s before being observed in the microscope.

The microhardness of the sintered samples was calculated via Vickers hardness tester (Model: Leco LM 247AT, St Joseph, MI, USA) by applying 200 gf and 15 s dwell time. At least five values were recorded for each sample from different locations and the average value was calculated.

The XRD analysis of the samples was done using the X-ray diffractometer (Model: X'pert3, Powder and Empyrean, PANalytical, B.V, Lelyweg, Almelo, The Netherlands) for the presence of possible compounds using the 10–100° scan range. The XRD peaks were then indexed and identified using High Score Plus software.

The XPS analysis was done using an X-ray photoelectron spectrometer (Model: Thermo scientific, K-alpha, East Grinstead, UK) for the analysis of elemental mass percentages of different elements on the sample surface. The XPS analysis shows all the elements that are present within the film and those bonded to it. It can be used to detect the foreign elements that might be present on the sintered sample surface. Since XPS can observe surface of nm order, the nitrogen and other elements present on the sample surface were detected by this method.

The corrosion resistance of the developed compositions was found by the weight loss method, whereby the samples were immersed in artificial saliva solution. The artificial saliva used in the testing had a composition of urea (1.0 g/L), KCl (0.4 g/L), NaCl (0.4 g/L), $Na_2S{\cdot}9H_2O$ (0.005 g/L), $NaH_2PO_4{\cdot}H_2O$ (0.69 g/L) and $CaCl_2{\cdot}2H_2O$ (0.795 g/L) with a pH of 5.5 to adjust it to the composition of natural saliva [39,40]. All the samples were first weighed and then immersed in the solution for 28 days. After a period of 28 days, each sample was cleaned and weighed again to calculate the change in weight incurred during this time period.

For the cytotoxicity testing of the samples, cell cultures were done using fibroblast cell line on all the samples. The fibroblast cell line (NIH/3T3 ATCC® CRL-1658) supplied by ATCC, Manassas, VA, USA were expanded in the media, using Dulbecco's Modified Eagle's Medium (DMEM) supplied by ThermoFisher Scientific, Waltham, MA, USA. It involved 100 µg/mL of Pen/strep (Sigma Aldrich, Life Sciences, Saint Louis, MO, USA) supplemented by 10 percent foetal bovine serum (FBS) (Sigma Aldrich, Life Sciences, USA). NIH/3T3 cells were expanded in a T75 culture flask (Corning Biosystem) at a temperature of 37 °C, and grown to 90% confluency with fresh media changes after 2 to 3 days in an incubator with 5% CO_2. After this, the cells were detached using trypsin–EDTA (Sigma Aldrich, Life Sciences, USA). On seeding day, the cells were counted with a haematocytometer and microscope. Fifty thousand cells were seeded on each sample in the 6-well plate to check the compatibility of the NIH/3T3 cells with the samples. Before seeding, the samples were diluted by adding 100 µL of the sample DMEM media into 1 mL. The cells were cultured on tissue culture plastic plates without samples, to use as control to compare the values with the cells cultured in the presence of DES samples. Before seeding of the cells, the samples were sterilized with ethanol solution for one hour and rinsed with phosphate buffer saline (PBS) for 15 min. The cells that were cultured without the samples were used as the controls. The fluorescent measurements of Alamar Blue were taken after three days to quantify the cells. The cell seeded samples were rinsed with PBS solution. They were then kept for incubation at 37 °C. Using a fluorescence plate reader (Bio-TEK, NorthStar Scientific Ltd., Bedfordshire, UK), 3–4 h absorbance was calculated at 570 nm. The cell metabolic activity took place and incorporated an oxidation-reduction indicator that changes colour as well as fluoresces. This was in response to chemical reduction that took place due to cell growth. The REDOX indicator changed colour from blue (oxidized) to red (reduced) due to growth.

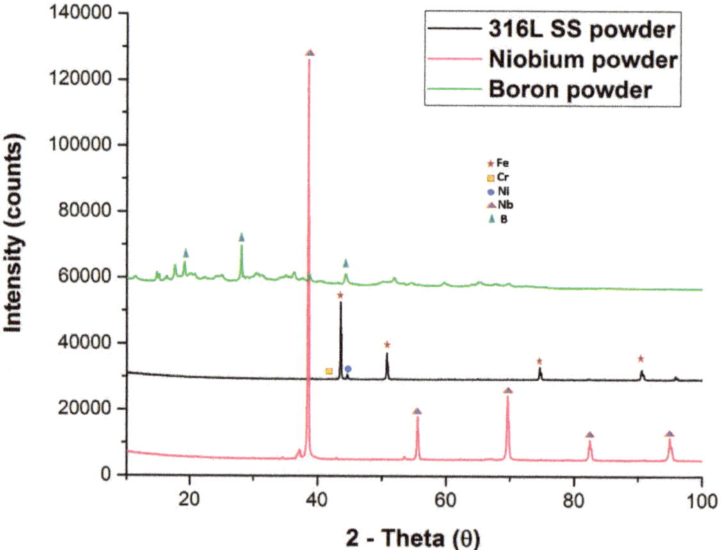

Figure 2. XRD spectra of niobium, boron and 316L SS powder.

3. Results and Discussion

The green and sintered densities of the samples that were produced are depicted in Table 1. The green density of pure 316L SS was found to be 6.5 g/cm^3. There is a slight decrease in the green density of the boron and niobium added SS samples. This is due to the fact that the boron and niobium particles tried to induce themselves in the matrix of SS. Sintering in nitrogen atmosphere resulted in improved sintered density for all the samples. Results of the sintered density show that the sintering environment and temperature promoted adequate densification. The pure 316L SS achieved a maximum sintered density of 7.575 g/cm^3. The density for boron and niobium added samples resulted in a slight decrease in sintered density. The amount of additive had a noticeable effect on the resulting properties of the alloy system. The boron addition improved the densification process and helped to maintain the density near to the sintered density of pure SS samples. Although there is a slight reduction in the sintered density of the samples with niobium and boron, their impact on the hardness, corrosion resistance and cytotoxicity were noticeable. From the analysis, the standard deviation values for green and sintered densities were within the permissible limits. This shows that the results have good repeatability.

Table 1. Densities of all the samples.

Alloy	Green Density	Standard Deviation	Sintered Density	Standard Deviation	Relative Density
S1	6.500 g/cm^3	0.112	7.575 g/cm^3	0.063	95.88%
S2	6.370 g/cm^3	0.0432	7.411 g/cm^3	0.023	93.81%
S3	6.240 g/cm^3	0.057	7.367 g/cm^3	0.051	93.25%
S4	6.160 g/cm^3	0.035	7.285 g/cm^3	0.092	92.21%
S5	6.080 g/cm^3	0.029	7.190 g/cm^3	0.028	91.01%

The microstructure of 316L samples with and without the addition of additives was observed using an optical microscope as shown in Figure 3. The sintering temperature and environment resulted in almost fully dense samples. It is evident from Figure 3 that all the samples were sintered properly with significantly low porosity. There were voids present in each sample showing that there is a very little porosity present in the samples. The nitrogen diffused into the matrix and formed strong nitrides with iron, boron and niobium. This resulted in increased microhardness of the samples and was also helpful in controlling the leaching of ions.

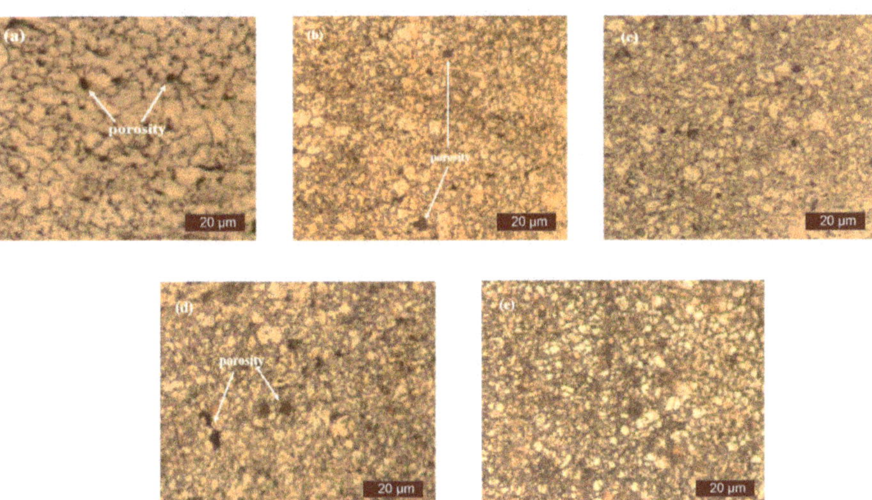

Figure 3. Microstructure of sintered samples (a) S1 (b) S2 (c) S3 (d) S4 (e) S5.

The porosity in the sintered samples was calculated from the sintered density (ρ_a) and the theoretical density (ρ_{th}) of the alloy using the formula shown in Equation (3) as per the literature [41,42].

$$Porosity = 1 - \frac{\rho_a}{\rho_{th}} \qquad (3)$$

The estimation of the pores present in the sintered samples is presented in Table 2. The results indicate that nearly all the samples have less porosity. The porosity increased with increased amounts of added titanium. The lowest porosity was observed in the pure 316L stainless steel sintered samples and this increased to 8.60 for the 2 wt.% titanium added samples.

Table 2. Porosity estimation of sintered samples.

Sample	S1	S2	S3	S4	S5
Porosity (%)	4.10	5.68	6.20	7.36	8.60

The micro hardness of the samples indicated an upsurge in micro hardness due to the infusion of nitrogen into the matrix. A micro hardness value of 235 HV was observed for 316L SS samples. The addition of boron has also had a significant role in increasing the micro hardness. The addition of niobium also increased the micro hardness of the sintered samples. The quantity of niobium addition also had an effect on the micro hardness and it increased with an increase in the quantity of niobium. A maximum of 387 HV was observed for samples with the addition of 2 wt.% niobium. The hardness of all the samples is shown in Figure 4.

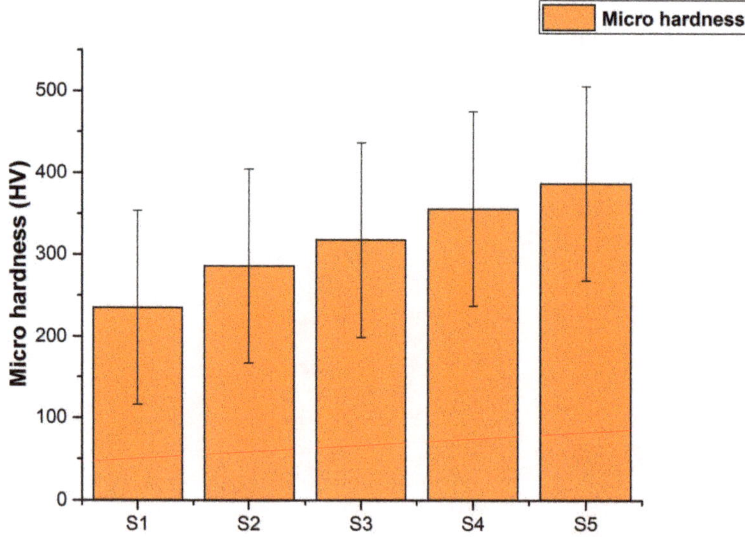

Figure 4. Micro hardness of all samples.

XRD analysis of all the sintered samples was done in order to analyse the different compounds formed during sintering of the samples. Figure 5 illustrates the XRD patterns of all the samples studied in this research. The results show the presence of austenitic structure in all the samples and γFe is also present in all the samples. The XRD patterns for pure 316L SS samples and boron and niobium added samples clearly reveal the diffusion of nitrogen into the matrix and the formation of strong nitrides. The patterns of pure 316L SS samples show the formation of $FeN_{0.324}$ at d spacing of 2.07500, C_3N_4 at d spacing of 2.51960, Ni (Cr_2O_4) at d spacing of 2.49354, and Cr_2O_3 at d spacing of 2.66348. These compounds clearly show the presence of nitrogen in the stainless steel matrix, forming nitrides with iron and carbon. For the samples of 0.5 wt.% niobium, the presence of $B_{0.47}C_{0.23}N_4$ at d spacing of 3.33850, Cr_2O_3 at d spacing of 2.66591, $B_2Fe_3Ni_3$ at d spacing of 2.07846, and NbB_2 at d spacing of 2.07838 was observed. The boron present in the matrix made a complex compound with carbon in the matrix and the diffused nitrogen. Moreover, the nitrogen also formed compounds with iron present in the SS matrix. The XRD results for 1 wt.% niobium added samples indicated the presence of $Cr_{1.3}Fe_{0.7}O_3$ at d spacing of 2.67600, $Nb_{0.76}B_2$ at d spacing of 2.08500, $FeN_{0.0499}$ at d spacing of 2.08019, and Fe_2O_3 at d spacing of 1.68000. In 1.5 wt.% niobium added samples, the following compounds were noted: B_2CN at d spacing of 2.13164, Cr_2O_3 at d spacing of 2.66439, $FeN_{0.0560}$ at d spacing of 2.08366, and $NbN_{0.9}$ at d spacing of 2.52706. The identified compounds for 2 wt.% niobium added SS samples were B_2CN at d spacing of 2.13164, $FeN_{0.0560}$ at d spacing of 2.08366, and $Cr_{1.3}Fe_{0.7}O_3$ at d spacing of 2.67600. These results indicate the formation of strong nitrides and borides that are responsible for the improvement of mechanical properties, including the hardness of the produced samples. The formation of chromium oxides favoured the creation of a passive surface oxide layer on the sample's surface. These results are similar to the results reported for the oxidation of metal surfaces [43]. These oxide layers thus increase the corrosion resistance of the produced samples.

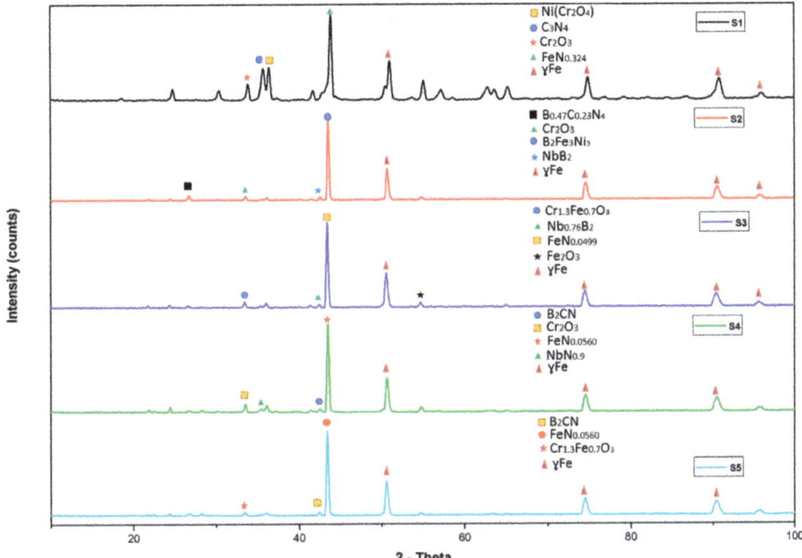

Figure 5. XRD spectra of all sintered samples.

XPS analysis was done to find the different elements present on the produced samples' surfaces. The X-ray photoelectron spectroscopy (XPS) is a quantitative surface-sensitive technique that plays an important role in determining the surface chemistry and elemental composition of a material. It involves the use of X-rays to radiate a material. The kinetic energy is measured while the number of escaped electrons is recorded. The XPS shows all the elements that are in the film and those bonded to it. It can be used to detect foreign elements that might be present on the sintered sample surface. The XPS analysis for all sintered samples is shown in Figure 6. The results indicate the presence of iron, chromium, oxygen and nitrogen, which constitute the main elements present on the sample surface. The sintered samples were covered with thin nano-size passive film consisting of chromium and iron oxide along with nitrogen. From the analysis, the amount of nitrogen present on the surface of the samples varied from 2.82% to a maximum of 3.9%. The maximum amount was observed for 2 wt.% niobium added stainless steel samples. The analysis indicated that the sintering parameters allowed the formation of a strong nitride layer along with chromium and iron oxide layers on the sintered samples. These layers serve as a coating for the samples by not allowing foreign bodies to diffuse into the matrix and they also control the leaching of ions. The presence of chromium and iron indicate the formation of surface oxide films on the surface of the sintered samples. These surface oxide films play a vital role in controlling the leaching of ions by acting as an inhibitor [44].

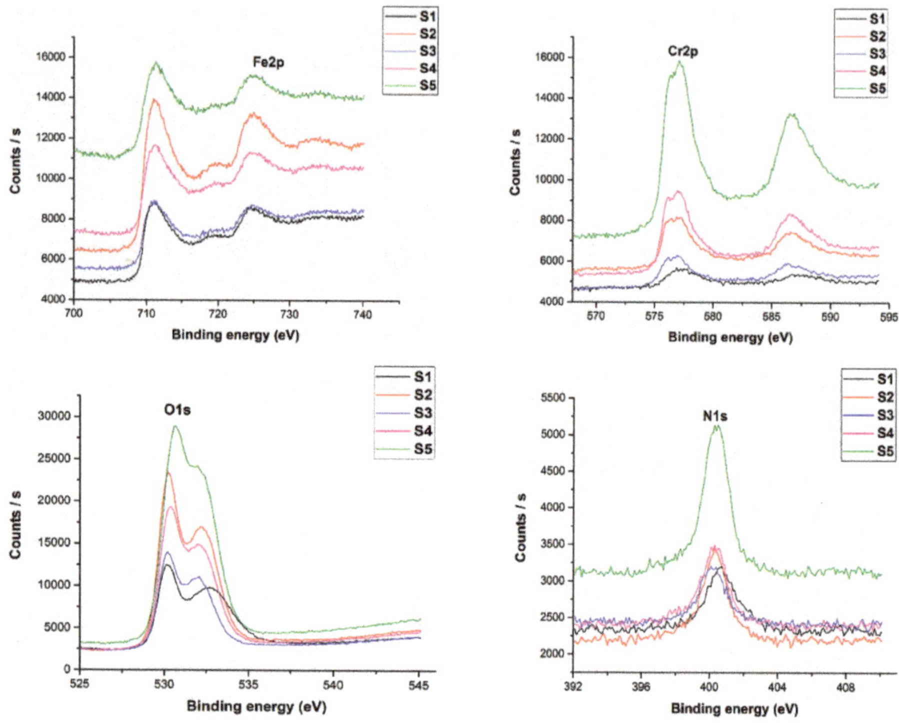

Figure 6. XPS spectra for Fe2p, Cr2p, O1s and N1s of all sintered samples.

To investigate the outcome of the alloying elements and sintering parameters, we estimated the corrosion resistance of the developed samples. The samples were cleaned and put into an artificial saliva solution and the resulting behaviour was studied. After immersing the samples in artificial saliva solution for 28 days, the minimal weight loss was observed. The weight loss measurements for all the samples are tabulated in Table 3. The results shows that the developed materials have a very good corrosion resistance to the artificial saliva solution. The pure 316L stainless steel samples showed a maximum weight loss of 0.003 g whereas the 1, 1.5 and 2 wt.% niobium added stainless steel samples showed an equal reduction in weight, even after 28 days of immersion. This indicates that these three samples were stable against the attack of solution ions. The samples showed more resistance towards corrosion compared to samples S1 and S2. The reduced weight loss also corresponds to the fact that the nitride layer along with the surface oxide layer was strong enough to retain itself throughout the process. This also relates to the fact that these layers retained the nickel and other ions present in the matrix. The proposed strategy of nitriding the surface layer proved to be efficient for the retention of nickel and other metal ions.

Table 3. Weight loss measurements for all samples in artificial saliva solution.

S.No	Sample	Weight (g) before Immersion	Weight (g) after Immersion	Δ m (g)
1	S1	17.310	17.306	0.004
2	S2	18.130	18.127	0.003
3	S3	18.190	18.188	0.002
4	S4	17.260	18.258	0.002
5	S5	18.250	18.248	0.002

In vitro cytotoxicity assessment of each of the membrane was determined by culturing NIH3T3 ATCC® CRL-1658 (fibroblast cell line) on the samples. Alamar Blue Assay was utilized to assess the cell proliferation on day 3. Microplate reader absorbance graph analysis was done to assess the cell viability of the liquid chemicals. The experiment was performed for 3 days by co-culturing with liquids and the results were then compared with the control. Figure 7 shows the comparison of the results of the control for 3 days with the results of the deep eutectic solvents for 3 days. The proliferation increased with time, i.e., up to 3 days for deep eutectic solvents which was also in contrast to the control. The increase in the absorbance rate indicates the increase of cell proliferation. All the samples showed increased absorbance as compared to the control. Sample S5 containing 2 wt.% niobium added stainless steel samples showed the best results as it shows more antibacterial properties which indicates the highest cell proliferation. The corrosion resistance of samples S3, S4 and S5 indicated that they were equally resistant to the artificial saliva solution attack and all of them showed the same weight loss. However, in terms of cytotoxicity assessment, the cell proliferation of S4 was the lowest amongst the three. The absorbance of S3 was better than S4 but less than S5. These results indicate that the material composition has a notable effect as far as the cytotoxicity assessment is concerned. From the figure, it can be noticed that all the produced samples are non-cytotoxic and compatible to living cells. All the analysis was conducted at least twice, with five readings from each of the experimental samples in order to obtain precise data. Sample S5 had the maximum increase in cell proliferation as compared to the control as well as to the other samples. This test also supports the argument for using these composites for biomedical application.

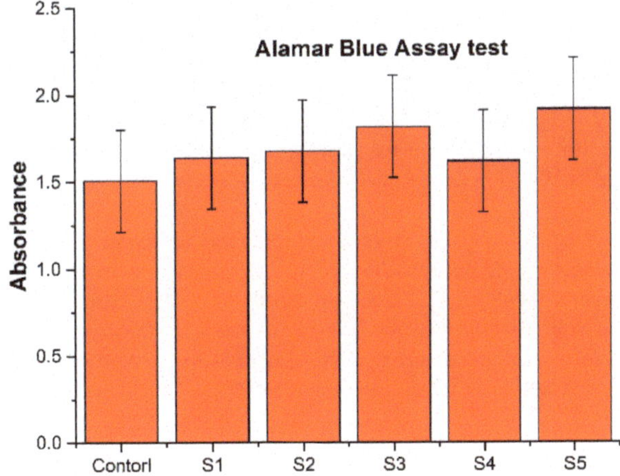

Figure 7. Cytotoxicity assessment of all the samples.

4. Conclusions

The nitrogen atmosphere for sintering austenitic stainless steel favoured the development of a strong nitride layer on the surface of the samples. An amount of 3.9% of nitrogen was found for those samples with an addition of 2 wt.% niobium. The dwell time of 8 h helped the diffusion of nitrogen into the matrix and formed strong nitrides with iron, carbon and other elements present in the stainless steel. The amount of the niobium addition was limited to 2 wt.% and this helped in the retention of the austenitic structure of the stainless steel. The samples were properly sintered at these sintering parameters and showed almost fully dense microstructures. The porosity of the samples was well within the limits and had minimal effect on the properties of the alloy systems. The 1, 1.5 and 2 wt.% niobium added samples showed maximum resistance to corrosion and had minimal weight loss. However, the cytotoxicity assessment of S5 was better compared to other samples studied in this research. All the sample compositions studied in this research are compatible with living tissues and are well suited to be used in biomedical implant manufacturing.

Author Contributions: Conceptualization, S.A.; Formal analysis, S.A.; Funding acquisition, R.A.M. and A.M.A.R.; Methodology, S.A.; Project administration, A.M.A.R. and R.A.M.; Resources, S.H., M.H. and N.S.; Supervision, A.M.A.R. and R.A.M.; Writing—original draft, S.A.; Writing—review & editing, Z.B. and F.I.A.

Funding: This research was funded by Universitas PERTAMINA—Universiti Teknologi PETRONAS (UP-UTP) International Collaborative Research Fund, cost-centers 015LB0-040 and 015ME0-081 and The APC was funded by Universiti Teknologi PETRONAS, Malaysia and National University of Sciences and Technology (NUST), Islamabad, Pakistan.

Acknowledgments: The authors would like to acknowledge Universiti Teknologi PETRONAS, Malaysia for carrying out research and other facilities.

Conflicts of Interest: The authors declare no conflict of interest.

References

1. Kang, C.-W.; Fang, F.-Z. State of the art of bioimplants manufacturing: Part I. *Adv. Manuf.* **2018**, *6*, 20–40. [CrossRef]

2. Manivasagam, G.; Dhinasekaran, D.; Rajamanickam, A. Biomedical implants: Corrosion and its prevention-a review. *Recent Patents Corros. Sci.* **2010**. [CrossRef]

3. Geetha, M.; Singh, A.K.; Asokamani, R.; Gogia, A.K. Ti based biomaterials, the ultimate choice for orthopaedic implants—A review. *Prog. Mater. Sci.* **2009**, *54*, 397–425. [CrossRef]

4. Kang, C.-W.; Fang, F.-Z. State of the art of bioimplants manufacturing: Part II. *Adv. Manuf.* **2018**, *6*, 137–154. [CrossRef]

5. Dwivedi, C.; Pandey, H.; Pandey, A.C.; Patil, S.; Ramteke, P.W.; Laux, P.; Singh, A.V. In Vivo Biocompatibility of Electrospun Biodegradable Dual Carrier (Antibiotic+ Growth Factor) in a Mouse Model—Implications for Rapid Wound Healing. *Pharmaceutics* **2019**, *11*, 180. [CrossRef]

6. Wiles, P. The surgery of the osteo-arthritic hip. *Br. J. Surg.* **1958**, *45*, 488–497. [CrossRef]

7. Wiles, P. The classic: The surgery of the osteo-arthritic hip. *Clin. Orthop. Relat. Res.* **2003**, *417*, 3–16.

8. Lo, K.H.; Shek, C.H.; Lai, J. Recent developments in stainless steels. *Mater. Sci. Eng. R Rep.* **2009**, *65*, 39–104. [CrossRef]

9. Ali, S.; Rani, A.M.A.; Altaf, K.; Baig, Z. Investigation of Boron addition and compaction pressure on the compactibility, densification and microhardness of 316L Stainless Steel. In *IOP Conference Series: Materials Science and Engineering*; IOP Publishing: Bristol, UK, 2018.

10. Park, J.B. *Biomaterials Science and Engineering*; Springer Science & Business Media: Berlin, Germany, 2012.

11. Khuenkaew, T.; Kanlayasiri, K. Resistance Spot Welding of SUS316L Austenitic/SUS425 Ferritic Stainless Steels: Weldment Characteristics, Mechanical Properties, Phase Transformation and Solidification. *Metals* **2019**, *9*, 710. [CrossRef]

12. Saini, M.; Singh, Y.; Arora, P.; Arora, V.; Jain, K. Implant. biomaterials: A comprehensive review. *World J. Clin. Cases WJCC* **2015**, *3*, 52. [CrossRef]

13. Salahinejad, E.; Hadianfard, M.J.; Macdonald, D.D.; Sharifi-Asl, S.; Mozafari, M.; Walker, K.J.; Rad, A.T.; Madihally, S.V.; Tayebi, L. In vitro electrochemical corrosion and cell viability studies on nickel-free stainless steel orthopedic implants. *PLoS ONE* **2013**, *8*, e61633. [CrossRef]

14. Sivakumar, M.; Kumar Dhanadurai, K.S.; Rajeswari, S.; Thulasiraman, V. Failures in stainless steel orthopaedic implant devices: A survey. *J. Mater. Sci. Lett.* **1995**, *14*, 351–354. [CrossRef]

15. Beddoes, J.; Bucci, K. The influence of surface condition on the localized corrosion of 316L stainless steel orthopaedic implants. *J. Mater. Sci. Mater. Med.* **1999**, *10*, 389–394. [CrossRef]

16. Asri, R.I.M.; Harun, W.S.W.; Samykano, M.; Lah, N.A.C.; Ghani, S.A.C.; Tarlochan, F.; Raza, M.R. Corrosion and surface modification on biocompatible metals: A review. *Mater. Sci. Eng. C* **2017**, *77*, 1261–1274. [CrossRef]

17. Finšgar, M.; Uzunalić, A.P.; Stergar, J.; Gradišnik, L.; Maver, U. Novel chitosan/diclofenac coatings on medical grade stainless steel for hip replacement applications. *Sci. Rep.* **2016**, *6*, 26653. [CrossRef]

18. Bayón, R.; Igartua, A.; González, J.J.; De Gopegui, U.R. Influence of the carbon content on the corrosion and tribocorrosion performance of Ti-DLC coatings for biomedical alloys. *Tribol. Int.* **2015**, *88*, 115–125. [CrossRef]

19. Fojt, J.; Joska, L.; Málek, J. Corrosion behaviour of porous Ti–39Nb alloy for biomedical applications. *Corros. Sci.* **2013**, *71*, 78–83. [CrossRef]

20. Cordeiro, J.M.; Beline, T.; Ribeiro, A.L.R.; Rangel, E.C.; da Cruz, N.C.; Landers, R.; Faverani, L.P.; Vaz, L.G.; Fais, L.M.; Vicente, F.B.; et al. Development of binary and ternary titanium alloys for dental implants. *Dent. Mater.* **2017**, *33*, 1244–1257. [CrossRef]

21. Mjöberg, B.; Hellquist, E.; Mallmin, H.; Lindh, U. Aluminum, Alzheimer's disease and bone fragility. *Acta Orthop. Scand.* **1997**, *68*, 511–514. [CrossRef]

22. Mirza, A.; King, A.; Troakes, C.; Exley, C. Aluminium in brain tissue in familial Alzheimer's disease. *J. Trace Elem. Med. Biol.* **2017**, *40*, 30–36. [CrossRef]

23. Chen, Q.; Thouas, G.A. Metallic implant biomaterials. *Mater. Sci. Eng. R Rep.* **2015**, *87*, 1–57. [CrossRef]

24. Singh, R.; Dahotre, N.B. Corrosion degradation and prevention by surface modification of biometallic materials. *J. Mater. Sci. Mater. Med.* **2007**, *18*, 725–751. [CrossRef]

25. Manam, N.S.; Harun, W.S.W.; Shri, D.N.A.; Ghani, S.A.C.; Kurniawan, T.; Ismail, M.H.; Ibrahim, M.H.I. Study of corrosion in biocompatible metals for implants: A review. *J. Alloy. Compd.* **2017**, *701*, 698–715. [CrossRef]

26. Wang, X.; Li, Y.; Xiong, J.; Hodgson, P.D.; Wen, C.E. Porous TiNbZr alloy scaffolds for biomedical applications. *Acta Biomater.* **2009**, *5*, 3616–3624. [CrossRef]

27. Ramírez, G.; Rodil, S.E.; Arzate, H.; Muhl, S.; Olaya, J.J. Niobium based coatings for dental implants. *Appl. Surf. Sci.* **2011**, *257*, 2555–2559. [CrossRef]

28. Reyes, K.M.; Kuromoto, N.K.; Claro, A.A.; Marino, C.E.B. Electrochemical stability of binary TiNb for biomedical applications. *Mater. Res. Express* **2017**, *4*, 075402. [CrossRef]

29. Kapnisis, K.; Constantinou, M.; Kyrkou, M.; Nikolaou, P.; Anayiotos, A.; Constantinides, G. Nanotribological response of aC: H coated metallic biomaterials: The cases of stainless steel, titanium, and niobium. *J. Appl. Biomater. Funct. Mater.* **2018**, *16*, 230–240.

30. Ou, K.L.; Weng, C.C.; Lin, Y.H.; Huang, M.S. A promising of alloying modified beta-type Titanium-Niobium implant for biomedical applications: Microstructural characteristics, in vitro biocompatibility and antibacterial performance. *J. Alloy. Compd.* **2017**, *697*, 231–238. [CrossRef]

31. Tavares, A.M.G.; Fernandes, B.S.; Souza, S.A.; Batista, W.W.; Cunha, F.G.C.; Landers, R.; Macedo, M.C.S.S. The addition of Si to the Ti–35Nb alloy and its effect on the corrosion resistance, when applied to biomedical materials. *J. Alloy. Compd.* **2014**, *591*, 91–99. [CrossRef]

32. Lee, C.; Ju, C.-P.; Chern Lin, J. Structure–property relationship of cast Ti–Nb alloys. *J. Oral Rehab.* **2002**, *29*, 314–322. [CrossRef]

33. Gabriel, S.B.; Panaino, J.V.P.; Santos, I.D.; Araujo, L.S.; Mei, P.R.; De Almeida, L.H.; Nunes, C.A. Characterization of a new beta titanium alloy, Ti–12Mo–3Nb, for biomedical applications. *J. Alloy. Compd.* **2012**, *536*, S208–S210. [CrossRef]

34. Ali, S.; Rani, A.M.A.; Altaf, K.; Hussain, P.; Prakash, C.; Hastuty, S.; Rao, T.V.V.L.N.; Subramaniam, K. Investigation of alloy composition and sintering parameters on the corrosion resistance and microhardness of 316L Stainless Steel alloy. In *Advances in Manufacturing II*; Gapiński, B., Szostak, M., Ivanov, V., Eds.; Springer: Berlin, Germany, 2019; pp. 532–541.

35. Bagliuk, G. Properties and structure of sintered boron containing carbon steels. In *Sintering-Methods and Products*; Shatokha, V., Ed.; IntechOpen: London, UK, 2012.

36. Bayraktaroglu, E.; Gulsoy, H.O.; Gulsoy, N.; Er, O.; Kilic, H. Effect of boron addition on injection molded 316L stainless steel: Mechanical, corrosion properties and in vitro bioactivity. *Bio Med. Mater. Eng.* **2012**, *22*, 333–349.

37. Kurgan, N. Effects of sintering atmosphere on microstructure and mechanical property of sintered powder metallurgy 316L stainless steel. *Mater. Des. (1980–2015)* **2013**, *52*, 995–998. [CrossRef]

38. Ali, S.; Rani, A.; Majdi, A.; Mufti, R.A.; Hastuty, S.; Hussain, M.; Shehzad, N.; Baig, Z.; Aliyu, A.; Azeez, A. An Efficient Approach for Nitrogen Diffusion and Surface Nitriding of Boron-Titanium Modified Stainless Steel Alloy for Biomedical Applications. *Metals* **2019**, *9*, 755. [CrossRef]

39. Chao, Z.; Yaomu, X.; Chufeng, L.; Conghua, L. The effect of mucin, fibrinogen and IgG on the corrosion behaviour of Ni–Ti alloy and stainless steel. *Biometals* **2017**, *30*, 367–377. [CrossRef]

40. Hussein, M.A.; Yilbas, B.; Kumar, A.M.; Drew, R.; Al-Aqeeli, N. Influence of Laser Nitriding on the Surface and Corrosion Properties of Ti-20Nb-13Zr Alloy in Artificial Saliva for Dental Applications. *J. Mater. Eng. Perform.* **2018**, *27*, 4655–4664. [CrossRef]

41. Nassar, A.E.; Nassar, E.E. Properties of aluminum matrix Nano composites prepared by powder metallurgy processing. *J. King Saud Univ. Eng. Sci.* **2017**, *29*, 295–299. [CrossRef]

42. Čapek, J.; Stehlíková, K.; Michalcová, A.; Msallamová, Š.; Vojtěch, D. Microstructure, mechanical and corrosion properties of biodegradable powder metallurgical Fe-2 wt % X (X= Pd, Ag and C) alloys. *Mater. Chem. Phys.* **2016**, *181*, 501–511. [CrossRef]

43. Singh, A.V.; Jahnke, T.; Xiao, Y.; Wang, S.; Yu, Y.; David, H.; Richter, G.; Laux, P.; Luch, A.; Srivastava, A.; et al. Peptide-Induced Biomineralization of Tin Oxide (SnO_2) Nanoparticles for Antibacterial Applications. *J. Nanosci. Nanotech.* **2019**, *19*, 5674–5686. [CrossRef]

44. Bauer, S.; Schmuki, P.; Von Der Mark, K.; Park, J. Engineering biocompatible implant surfaces: Part I: Materials and surfaces. *Prog. Mater. Sci.* **2013**, *58*, 261–326. [CrossRef]

Article

Optimization of Sintering Parameters of 316L Stainless Steel for In-Situ Nitrogen Absorption and Surface Nitriding Using Response Surface Methodology

Sadaqat Ali [1,*], Ahmad Majdi Abdul Rani [2,*], Riaz Ahmad Mufti [1], Syed Waqar Ahmed [3], Zeeshan Baig [4], Sri Hastuty [5], Muhammad Al'Hapis Abdul Razak [6] and Abdul Azeez Abdu Aliyu [7]

[1] School of Mechanical & Manufacturing Engineering, National University of Sciences and Technology (NUST), H-12, Islamabad 44000, Pakistan; riazmufti@smme.nust.edu.pk

[2] Mechanical Engineering Department, Universiti Teknologi PETRONAS (UTP), Seri Iskandar 32610, Malaysia

[3] Department of Mechanical Engineering, Ghulam Ishaq Khan Institute of Engineering Sciences and Technology, Khyber Pakhtunkhwa 19201, Pakistan; waqarg05@gmail.com

[4] Centre of Excellence in Science & Applied Technologies, Islamabad 44000, Pakistan; meteng81@gmail.com

[5] Mechanical Engineering Department, Universitas PERTAMINA, Jakarta 12220, Indonesia; sri.hastuty@universitaspertamina.ac.id

[6] Manufacturing Section, Universiti Kuala Lumpur Malaysian Spanish Institute, Kulim Hi-Tech Park, Kedah 09000, Malaysia; alhapis@unikl.edu.my

[7] Mechanical Engineering Department, Bayero University Kano, Kano 700241, Nigeria; garoabdul@gmail.com

* Correspondence: engineersadaqat@gmail.com (S.A.); majdi@utp.edu.my (A.M.A.R.)

Received: 11 January 2020; Accepted: 25 February 2020; Published: 5 March 2020

Abstract: This research investigates the simultaneous sintering and surface nitriding of 316L stainless steel alloy using powder metallurgy method. The influence of sintering temperature and dwell time are investigated for maximum nitrogen absorption, densification and increased microhardness using response surface methodology (RSM). In this study, 316L stainless steel powder was compacted at 800 MPa and sintered at two different temperatures of 1150 and 1200 °C with varying dwell times of 1, 3, 5 and 8 h in nitrogen atmosphere. The sintered compacts were then characterized for their microstructure, densification, microhardness and nitrogen absorption. The results revealed that increased dwell time assisted nitrogen to diffuse into stainless steel matrix along with the creation of nitride layer onto the sample surface. The microhardness and density also increased with increasing dwell time. A densification of 7.575 g/cm^3 and microhardness of 235 HV were obtained for the samples sintered at 1200 °C temperature with 8 h dwell time. The simultaneous sintering and surface nitriding technique developed in this research work can help in improving corrosion resistance of this material and controlling leaching of metal ions for its potential use in biomedical applications.

Keywords: 316L stainless steel; sintering; surface nitriding; nitrogen absorption; response surface methodology

---

1. Introduction

Among the commonly used biomaterials, 316L stainless steel is in use since ancient times when total hip replacement of a patient was carried out using 316L stainless steel for implantation [1,2]. Since then, 316L stainless steel has been gaining great attention in biomedical applications. It is available at a low cost with adequate mechanical properties, biocompatibility and corrosion resistance [3,4].

The implants and medical devices can be manufactured from 316L stainless steel at a relatively low cost as compared to its counterpart materials including cobalt chromium, titanium and other biomaterials [5–10]. Austenitic 316L stainless steel is also recommended by ASTM International for use in implant manufacturing and medical devices [11–13]. The low carbon content and high chromium concentration in this material makes it resistant to corrosive environment specially in chlorine bearing solutions [14]. The human body saline resembles the chlorine bearing solution making this material an ideal choice for usage in implant manufacturing [15].

The 316L stainless steel contains 10%–14% nickel for promoting the austenitic structure [16]. It has been reported that the implants made of biomaterials have been reported to be associated with release of metal ions and debris due to poor surface finish and corrosion resistance [17,18]. Among the released metal ions, nickel and chromium have been proven to be the cause of many genotoxic and mutagenic activities [19]. Nickel has been associated with contact allergy for patients allergic to metals causing skin diseases including dermatitis and eczematous rash [16,19]. The leaching of ions and debris due to ionic leaching have been reported to be the cause of swelling and localized pain due to inflammation [20]. The leaching of metal ions demands the surface improvement of 316L stainless steel implant material [21,22]. Various methods have been developed to improve the surface quality and corrosion resistance of stainless-steel including surface modification techniques. These include physical, chemical and mechanical surface modification techniques [23,24]. However, certain disadvantages are associated with these techniques which limit their application in biomedical field Including low coating deposition rate, crack free coatings, line of sight and expensive raw materials [25,26]. To overcome these short comings and develop an alternative solution, optimization of sintering parameters for surface nitriding of the material can address these issues [27–29]. The optimal sintering parameters with increased dwell time can help in creating a protective surface layer on the stainless-steel surface. This will help in improving the corrosion resistance of the material and act as an inhibitor for controlling the leaching of ions.

2. Materials and Methods

In this research, commercially available pure irregular shape water atomized 316L stainless steel powder was selected due to high packing density and better sinter ability than gas atomized powder. The chemical composition of the powder has been summarized in Table 1. The particle size distribution of the powder was characterized and is presented in Table 2.

Table 1. Chemical composition of 316L stainless steel powder used in the study.

Element	C	Si	O	Mn	Ni	Cr	Mo	Fe
wt.%	0.028	0.9	0.068	1.5	12.01	17.04	2.4	Balance

Table 2. Particle size distribution of 316L stainless steel powder.

Particle Distribution	D10	D50	D90
Particle size (μm)	3.98	10.27	19.61

Uniaxial cold compaction process was used to prepare the samples of size 30 mm diameter and 5 mm thickness using stainless steel die. The powder was compressed at 800 MPa pressure to prepare the green compacts followed by sintering in nitrogen atmosphere at two different temperatures of 1150 and 1200 °C. The dwell time was varied for each temperature and was set at 1, 3, 5 and 8 h. The SEM and XRD of the powder have been presented in Figures 1 and 2 respectively.

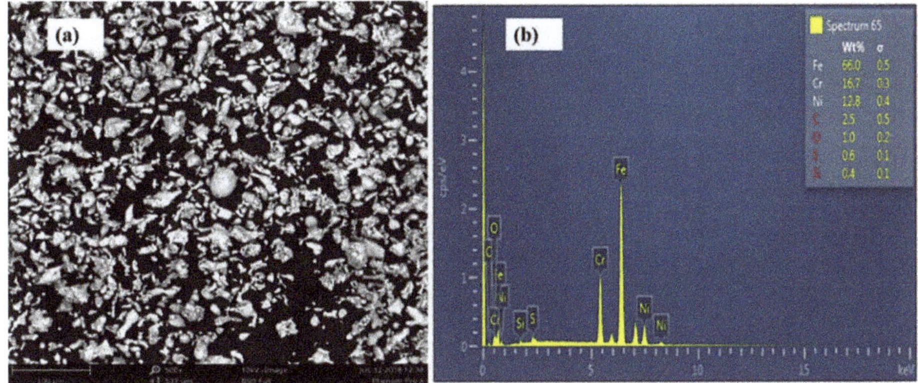

Figure 1. SEM (**a**) and EDX (**b**) of 316L stainless steel powder.

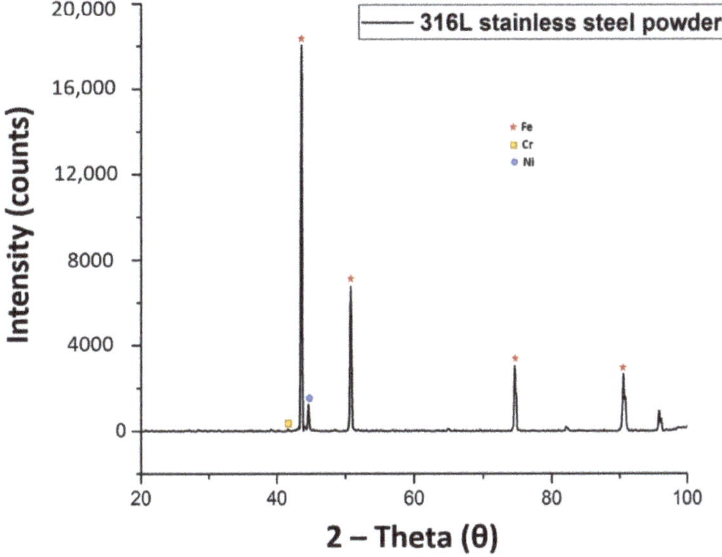

Figure 2. XRD spectrum of 316L stainless steel powder.

The most influential parameter settings, optimizing and generating a mathematical model for the selected responses was carried out by response surface methodology (RSM) (D-optimum custom design). This design generates a design that estimates the influence of the process parameters in the best possible way, particularly suited for screening studies. Considering the role of nitrogen absorption (N-abs), density (D) and microhardness (HV) in developing suitable implant material and the capability of powder metallurgy process, three major responses (N-abs, D and HV) were selected for parameters optimizations. The two important factors influencing these responses namely sintering temperature (St) and dwell time (Dt) and mixed level design were used to plan the experiment. A total of 16 experimental runs were conducted according to the procedure described above and have been presented in Table 3.

Table 3. Design of Experiments (DOE) matrix and experimental results for the responses.

Run Order	Sintering Temperature (°C)	Dwell Time (hours)	Nitrogen Absorption (%)	% of Theoretical Density	Sintered Density (g/cm^3)	Micro Hardness (HV)
1	1150	5	1.61	92.74	7.410	203
2	1200	1	1.04	92.01	7.352	185
3	1156	8	2.41	93.82	7.497	229
4	1150	5	1.61	92.74	7.410	203
5	1178	8	2.56	94.11	7.520	231
6	1158	3	1.10	92.20	7.367	196
7	1200	5	1.93	94.80	7.575	210
8	1175	3	1.20	92.30	7.375	197
9	1175	3	1.20	92.30	7.375	197
10	1178	8	2.56	94.11	7.520	231
11	1200	5	1.61	94.80	7.575	231
12	1167	1	0.98	91.73	7.330	182
13	1192	3	1.26	92.35	7.379	198
14	1175	3	1.20	92.30	7.375	197
15	1200	8	2.82	94.80	7.575	235
16	1150	1	0.96	91.58	7.318	180

The sintered samples were then characterized using X-ray diffractometer (Model: X'pert3, Powder and Empyrean, PANalytical, B.V, Lelyweg, Almelo, The Netherlands), X-ray photoelectron spectrometer (Model: Thermo Scientific, K-alpha, East Grinstead, UK), optical microscope (Model: Leica DM LM, Wetzlar, Germany) and Vickers hardness tests using tester (Model: Leco LM 247AT, St Joseph, MI, USA) to examine the resulting properties of the sintered samples.

3. Results and Discussion

The XRD evaluation was carried out to analyze the diffusion of nitrogen into the matrix of the sintered samples. The results indicated that the sintering atmosphere and dwell times helped in diffusing nitrogen to form strong nitrides with the constituents of 316L stainless steel matrix. The XRD spectrum of the sintered samples have been presented in Figure 3, indicating S1 sample sintered for 1 h dwell time, S2 sintered for 3 h dwell time, S3 sintered for 5 h dwell time and S4 sintered at dwell time of 8 h at sintering temperature of 1200 °C. The results indicate the formation of C_3N_4 at d spacing of 2.51960 Å, $FeN_{0.324}$ at d spacing of 2.07500 Å, Cr_2O_3 at d spacing of 2.66348 Å and Ni (Cr_2O_4) at d spacing of 2.49354 Å. The presence of γFe in all the samples indicated that the austenitic structure of the 316L stainless steel was retained at increased dwell times.

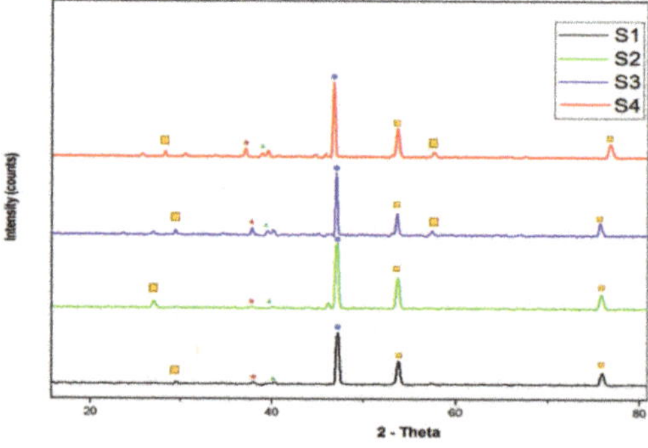

Figure 3. XRD spectra of sintered samples at different dwell times.

The XPS was done to analyze the existence of nitrogen onto the sample surface. The results indicated that the dwell time has a notable effect on the amount of nitrogen on surface of the sample. The amount of nitrogen increased with increasing dwell time and a maximum of 2.82% nitrogen was observed for samples sintered at 1200 °C temperature and 8 h dwell time. The XPS analysis has been presented in Figure 4.

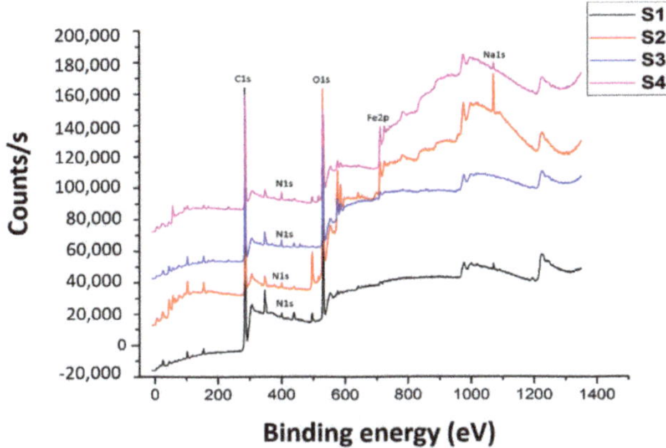

Figure 4. XPS analysis of sintered samples at different dwell times.

The microstructure of the sintered samples was observed under optical microscope after complete metallographic preparation. The micrographs of the sintered samples as viewed from optical microscope have been presented in Figure 5, indicating S1 sample sintered for 1 h dwell time, S2 sintered for 3 h dwell time, S3 sintered for 5 h dwell time and S4 sintered at dwell time of 8 h at sintering temperature of 1200 °C. It can be observed from the micrographs that the samples are sintered properly with clear grain boundaries. Moreover, increased sintering dwell time helped in creation of clear grain boundaries with reduced porosity as observed in the micrographs.

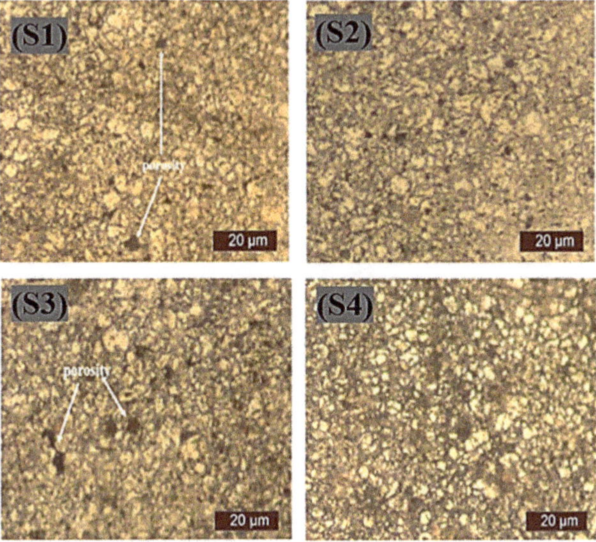

Figure 5. Microstructure of the sintered samples.

The FESEM (Field Emission Scanning Electron Microscope) of sintered samples was done to further explore the effect of sintering parameters and presence of nitrogen. The elemental mapping was held for different elements present in the stainless steel. It was also helpful in identifying the effect of prolonged sintering dwell time. The results indicated the occurrence of nitrogen in the samples indicating that nitrogen diffused into the stainless steel matrix. The elemental mapping of the selected sample has been demonstrated in the Figure 6.

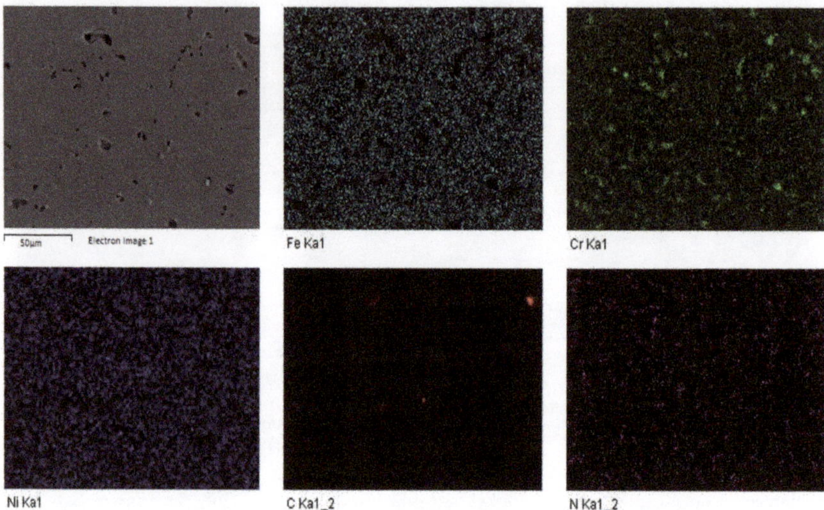

Figure 6. FESEM-EDS elemental maps for sample sintered at 8 h dwell time.

The increased dwell time of sintering created a strong nitride layer onto the surface of the sample. The parameters of sintering simultaneously sintered and surface nitride the samples. The nitride layer developed onto the sample surface improves the corrosion resistance of the material along with controlling the leaching of metal ions during exposure to corrosive environments. The nitride layer formation of selected sample has been illustrated in Figure 7.

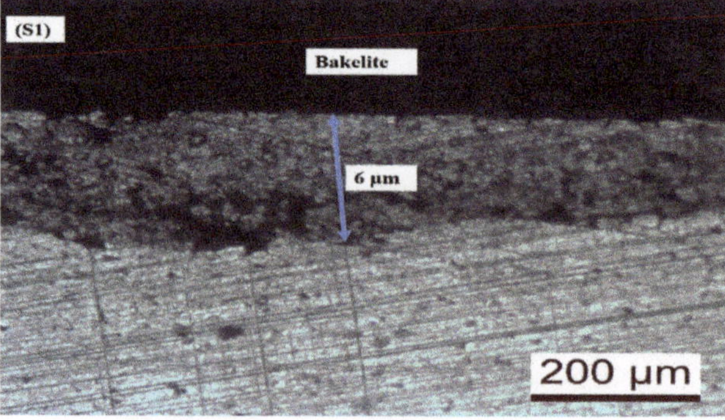

Figure 7. Surface nitride layer of sintered samples.

The microhardness of the sintered samples was measured using Vickers hardness testing whereby, 200 gf was applied on the polished surface sample for a time period of 15 s. The microhardness value of 185 HV was measured for 1 h dwell time sintered sample, whereas its value increased to 235 HV for 8 h dwell time sintered samples. The increased absorption of nitrogen at increased dwell times can be attributed to the increase of microhardness values. The microhardness values of the sintered samples have been presented in Figure 8.

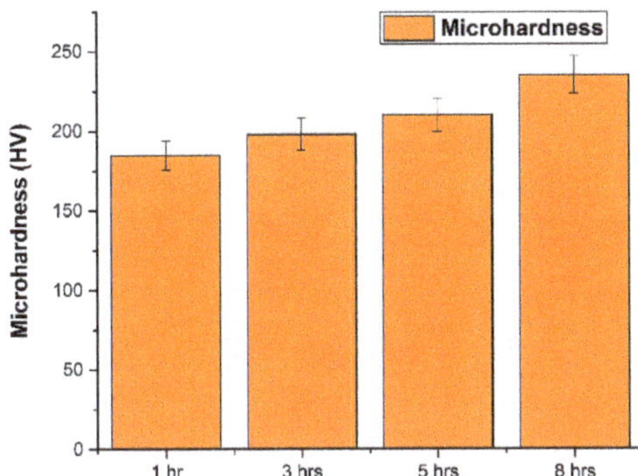

Figure 8. Microhardness values of sintered samples at varying dwell times.

3.1. Influence of Input Parameters on Nitrogen Absorption

The impact of process parameters on nitrogen absorption was determined using D-optimum design analysis. The analysis of variance (ANOVA) for the nitrogen absorption has been presented in Table 4. The regression model, significant factors, interactions and lack of fit are summarized. The model and the model terms with the values of "prob > F" less than 0.0500 are considered as significant terms. It can be noticed from the ANOVA table that the factors St and Dt are significant model terms for nitrogen absorption. The lack of fit "prob > F" value of 0.7039 employs that it is non-significant. The difference between adjusted R-square predicted R-square shows that the terms are in reasonable agreement with each other as presented in Table 5.

Table 4. ANOVA for nitrogen absorption.

Source	Sum of Squares	df	Mean Square	F Value	p-Value Prob > F	Significance
Model	2.03	5	0.41	154.42	<0.0001	significant
A-St	0.024	1	0.024	9.25	0.0125	
B-Dt	1.82	1	1.82	689.78	<0.0001	
AB	6.5×10^{-4}	1	6.5×10^{-4}	0.25	0.6298	
A^2	9.91×10^{-3}	1	9.91×10^{-3}	3.77	0.0810	
B^2	9.93×10^{-3}	1	9.93×10^{-3}	3.77	0.0808	
Residual	0.026	10	2.63×10^{-3}			
Lack of fit	9.9×10^{-3}	5	1.98×10^{-3}	0.60	0.7039	not significant
Pure Error	0.016	5	3.28×10^{-3}			
Cor Total	2.06	15				

Table 5. Predicted R-square vs. adjusted R-square.

Std. Dev.	0.051	R-squared	0.9872
Mean	0.42	Adj R-squared	0.9808
C.V.%	12.19	Pred R-squared	0.9731
PRESS	0.055	Adeq Precision	35.175
−2 Log Likelihood	−57.14		

Most of the data was found to lie along straight line as observed in the normal probability plot of residuals shown in Figure 9a. This indicated the selected model terms are only significant factors and the errors are normally distributed.

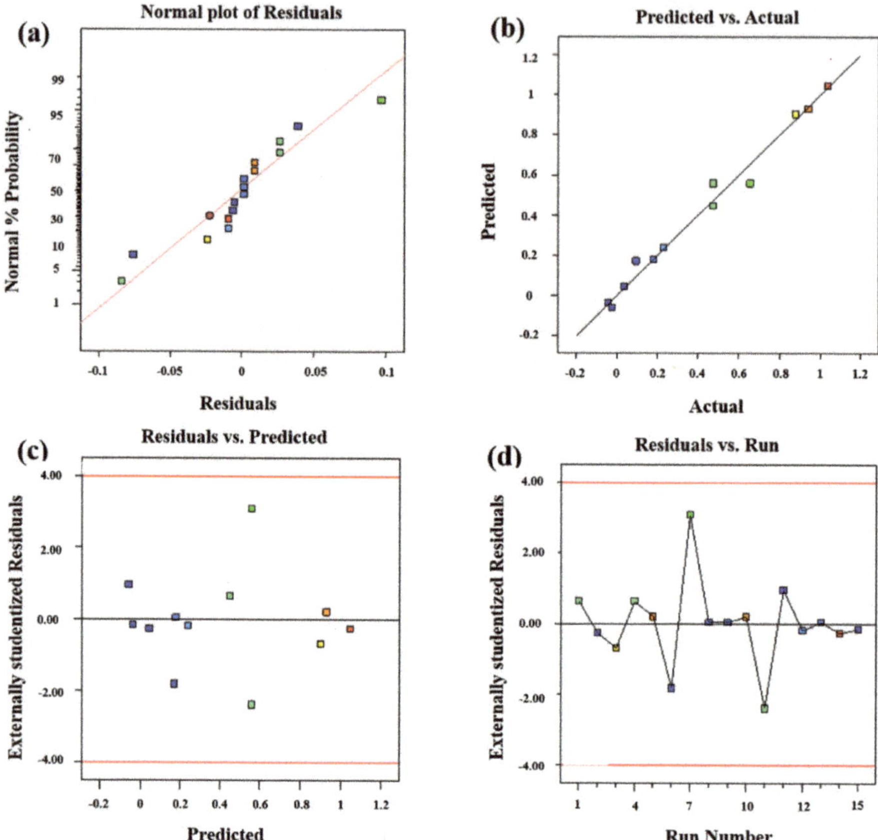

Figure 9. (**a**) Normal plot of residuals (**b**) predicted vs. actual plot (**c**) residuals vs. predicted plot (**d**) residuals vs. run plot.

A well fitted regression which scattered along straight line could be observed in the predicted (software calculated) versus actual (experimental) values plot shown in Figure 9b. This confirmed that the model is very well designated. A non-clear pattern and randomly scattered data as required was observed in the residual versus predicted values plot (Figure 9c). To measure the number of the standard deviation, an internally studentized residuals is calculated by taking the ratio between the residuals to that of the residuals of estimated standard deviation. The residual values are obtained by calculating the difference between measured and predicted values. The plot of residuals versus internally studentized residuals presented in Figure 9d confirmed that all values lie within limits and that the model can be used to predict the response.

The one factor effect plot for nitrogen absorption has been depicted in Figure 10. The one factor plot helps in predicting the behavior of each output by changing the respective input parameter. It can be noticed that increase in the dwell time from 1 h to 8 h results in increased nitrogen absorption values as shown in Figure 10d–f. The increase in sintering temperature also resulted in increased nitrogen absorption and can be observed in Figure 10a–c.

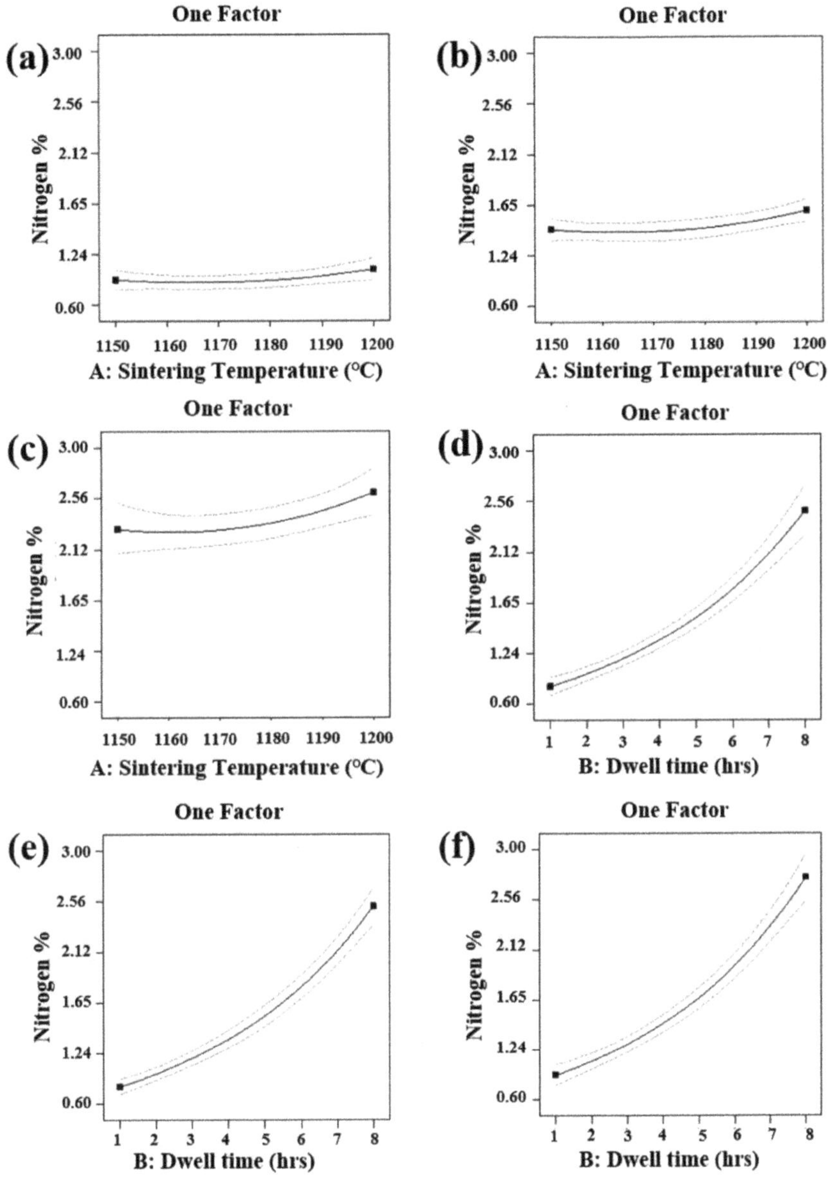

Figure 10. One factor effect plot for nitrogen absorption.

The two-dimensional representation of nitrogen absorption for the selected parameters is shown by the contour plot presented in Figure 11a. The nitrogen absorption values indicated in the middle of each contour increases linearly from the first to the last contour. The influence of sintering parameters on nitrogen absorption in the form of a three-dimensional (3D) surface plot has been presented in Figure 11b. The plot shows that nitrogen absorption increases with increase of sintering temperature and dwell time.

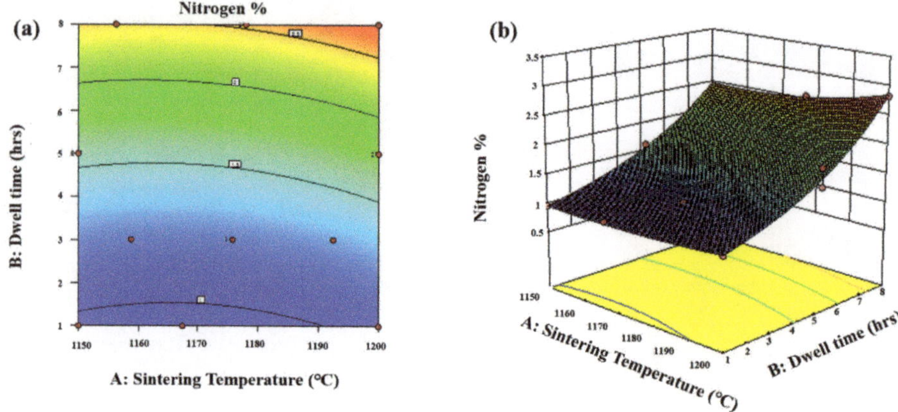

Figure 11. Contour and three-dimensional (3D) plots of nitrogen absorption.

3.2. Influence of Input Parameters on Density

The ANOVA table indicating the contribution of each factor on density is presented in Table 6. It signifies the significance of the model since "Prob > F" value is 0.0001. The table also shows that the factors St and Dt are significant model terms with "Prob > F" values <0.05.

Table 6. ANOVA for density.

Source	Sum of Squares	df	Mean Square	F Value	p-Value Prob > F	Significance
Model	2.132×10^{-3}	5	4.264×10^{-4}	23.66	<0.0001	significant
A-St	3.695×10^{-4}	1	3.695×10^{-4}	20.50	0.0011	
B-Dt	1.457×10^{-3}	1	1.457×10^{-3}	80.83	<0.0001	
AB	3.56×10^{-5}	1	3.568×10^{-5}	1.98	0.1897	
A2	6.941×10^{-5}	1	6.941×10^{-5}	3.85	0.0781	
B2	2.847×10^{-5}	1	2.847×10^{-5}	1.58	0.2374	
Residual	1.802×10^{-4}	10	1.802×10^{-5}			
Lack of fit	1.802×10^{-4}	5	3.604×10^{-5}			
Pure Error	0.000	5	0.000			
Cor Total	2.312×10^{-3}	15				

The value of 0.7315 for predicted R-square indicates that it lies within acceptable range to that of adjusted R-square as shown in Table 7. The value "Adeq precision" of 16.423 is good for a suitable signal indicating that the model can be employed to navigate the design space.

Table 7. Predicted R-square vs. adjusted R-square.

Std. Dev.	4.24×10^{-3}	R-squared	0.9221
Mean	2.01	Adj R-squared	0.8831
C.V.%	0.21	Pred R-squared	0.7315
PRESS	6.209×10^{-4}	Adeq Precision	16.423
−2 Log Likelihood	−136.90		

The normal probability plot of the residuals plot shows that nearly all the data lies along a straight line as shown in Figure 12a. A well fitted regression which scattered along straight line could be observed in the predicted (software calculated) versus actual (experimental) values plot shown in Figure 12b. This confirmed that the selected terms are only significant factors and the errors are normally distributed. The residuals versus predicted values plot shows a randomly distributed data as shown in Figure 12c. The studentized residuals of regression lines of residuals versus run plot lie within the limits as shown in Figure 12d. This confirms the capability of the model to predict the density.

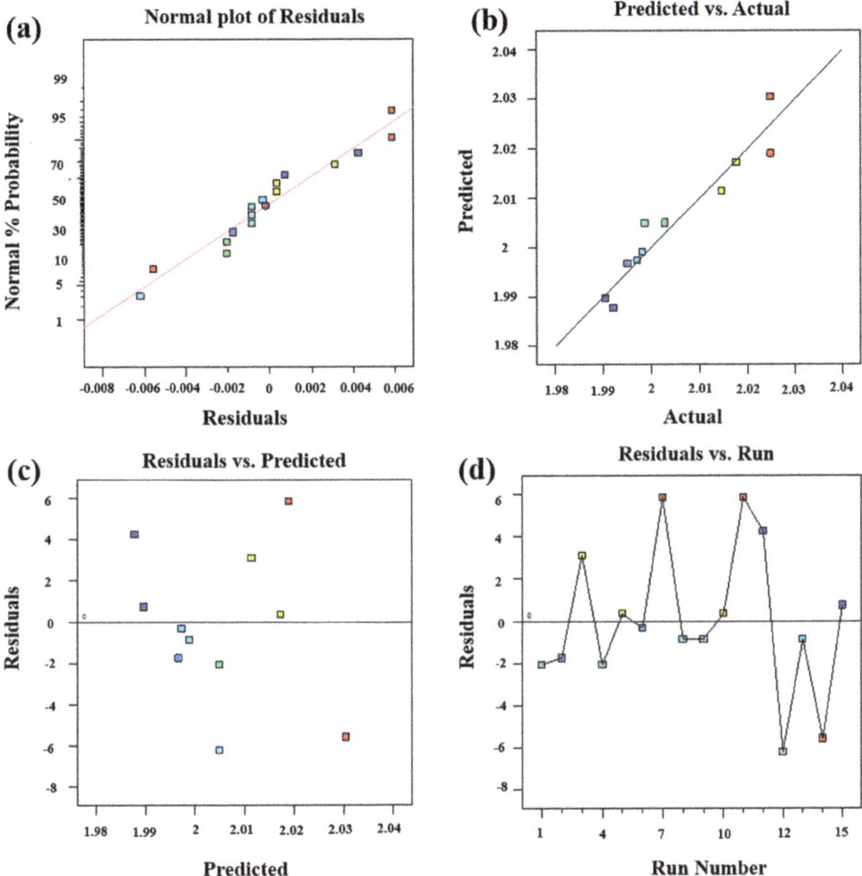

Figure 12. (**a**) Normal plot of residuals (**b**) predicted vs. actual plot (**c**) residuals vs. predicted plot (**d**) residuals vs. run plot.

The one factor effect plot for density has been presented in Figure 13. It can be observed that increased dwell time from 1 h to 8 h results in increased density value as shown in Figure 13d–f. The increase in sintering temperature also has also a notable effect on the densification as can be seen in Figure 13a–c.

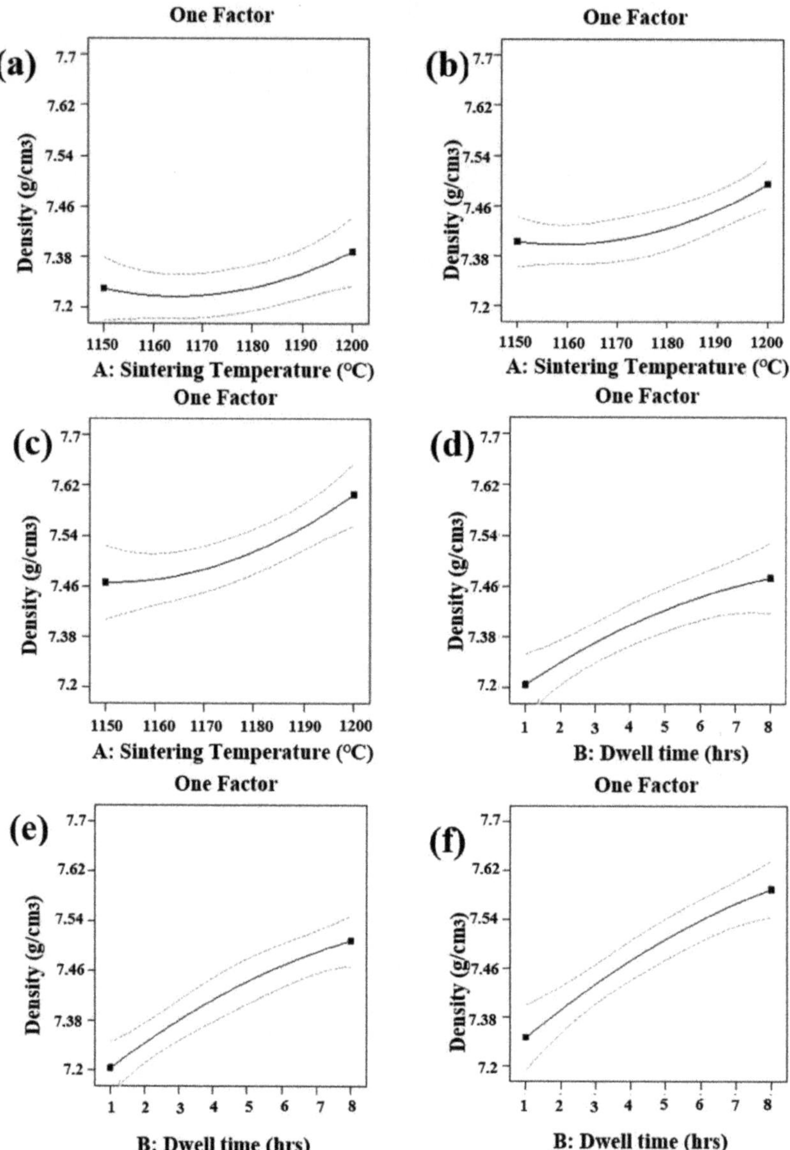

Figure 13. One factor effect plot for density.

The two-dimensional representation of density for the selected parameters is shown by the contour plot presented in Figure 14a. The influence of sintering parameters on densification in the form of 3D surface plot has been presented in Figure 14b. The plot shows that density increases with increase of sintering temperature and dwell time.

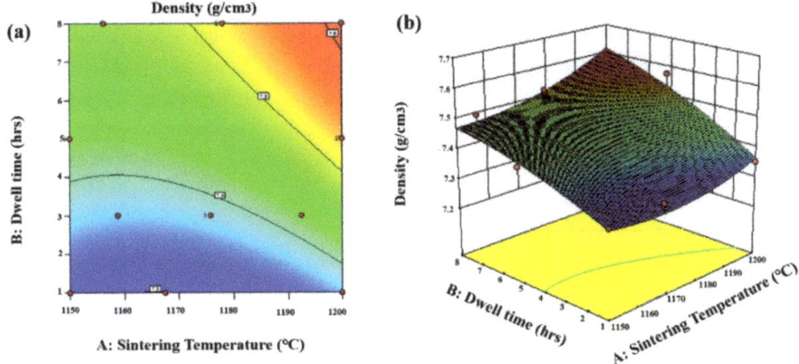

Figure 14. Contour and 3D plots of density.

3.3. Influence of Input Parameters on Microhardness

The impact of process parameters on microhardness was determined using D-optimum design analysis. The analysis of variance (ANOVA) for microhardness has been presented in Table 8. Th model and the model terms with the values of "prob > F" less than 0.0500 are considered as significant terms. It can be noticed from the ANOVA table that the factors St and Dt are significant model terms for micro hardness. The model is expected to fit, the reason being the lack of fit "prob > F" value of 0.9126 employs that it is non-significant. The difference between adjusted R-square and predicted R-square shows that the terms are in reasonable agreement with each other as seen in Table 9.

Table 8. ANOVA for microhardness.

Source	Sum of Squares	df	Mean Square	F Value	*p*-Value Prob > F	Significance
Model	0.12	3	0.039	71.38	<0.0001	significant
A-St	5.825×10^{-3}	1	5.825×10^{-3}	10.58	0.0069	
B-Dt	0.11	1	0.11	192.71	<0.0001	
AB	9.377×10^{-5}	1	9.377×10^{-5}	0.17	0.6871	
A2	6.607×10^{-3}	12	5.506×10^{-4}			
B2	2.065×10^{-3}	7	2.950×10^{-4}	0.31	0.9126	not significant
Residual	4.542×10^{-3}	5	9.084×10^{-4}			
Lack of fit	0.12	15	0.039			
Pure Error	0.12	3	5.825×10^{-3}			
Cor Total	5.825×10^{-3}	1				

Table 9. Predicted R-square vs. adjusted R-square.

Std. Dev.	0.023	R-squared	0.9469
Mean	5.33	Adj R-squared	0.9337
C.V.%	0.44	Pred R-squared	0.9055
PRESS	0.012	Adeq Precision	24.226
−2 Log Likelihood	−79.27		

Most of the data was found to lie along straight line as observed in the plot of residuals for normal probability and can be seen in Figure 15a. This indicated the selected model terms are only significant factors and the errors are normally distributed.

A well fitted regression which scattered along straight line could be observed in the predicted (software calculated) versus actual (experimental) values plot shown in Figure 15b. This confirmed that the model is very well designated. A non-clear pattern and randomly scattered data as required was observed in the residual versus predicted values plot (Figure 15c). To measure the number of the standard deviations between predicted and actual values, an internally studentized residuals is calculated by taking the ratio between the residuals to that of the residuals of estimated standard deviation. The residual values are obtained by calculating the difference between measured and predicted values. The plot of residuals versus internally studentized residuals shown in Figure 15d verified that the model can be used to predict the response, due to the absence of any outlier and all the values lie within the limits.

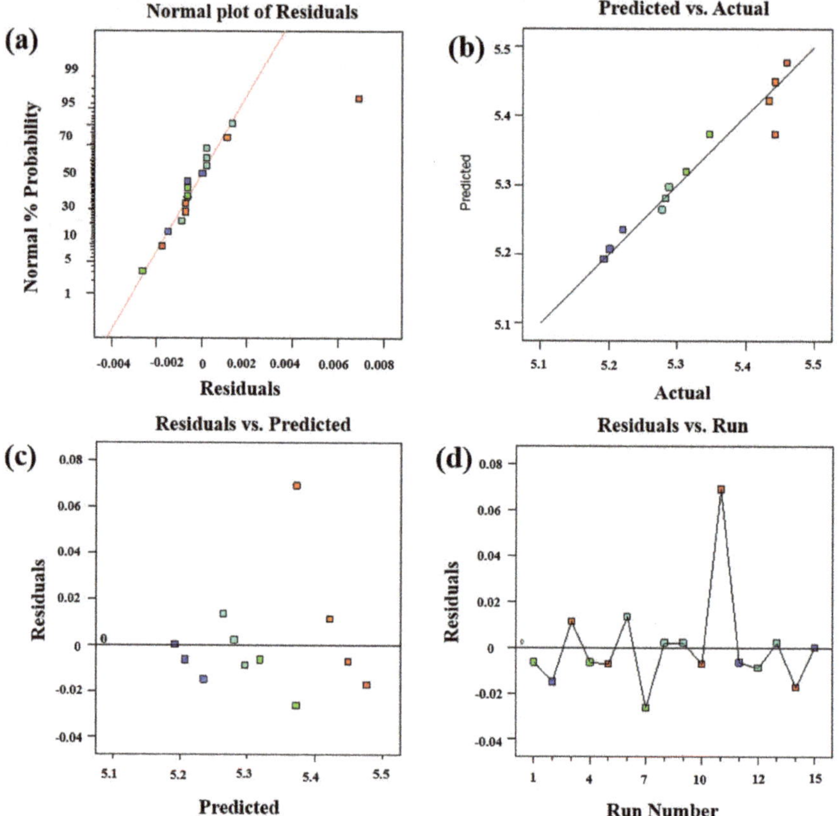

Figure 15. (**a**) Normal plot of residuals (**b**) predicted vs. actual plot (**c**) residuals vs. predicted plot (**d**) residuals vs. run plot.

The one factor effect plot for micro hardness has been depicted in Figure 16. It can be noticed that increase in the dwell time from 1 h to 8 h results in increased micro hardness values as shown in Figure 16d–f. The increase in sintering temperature also has a notable effect on the increased nitrogen absorption as can be seen in Figure 16a–c.

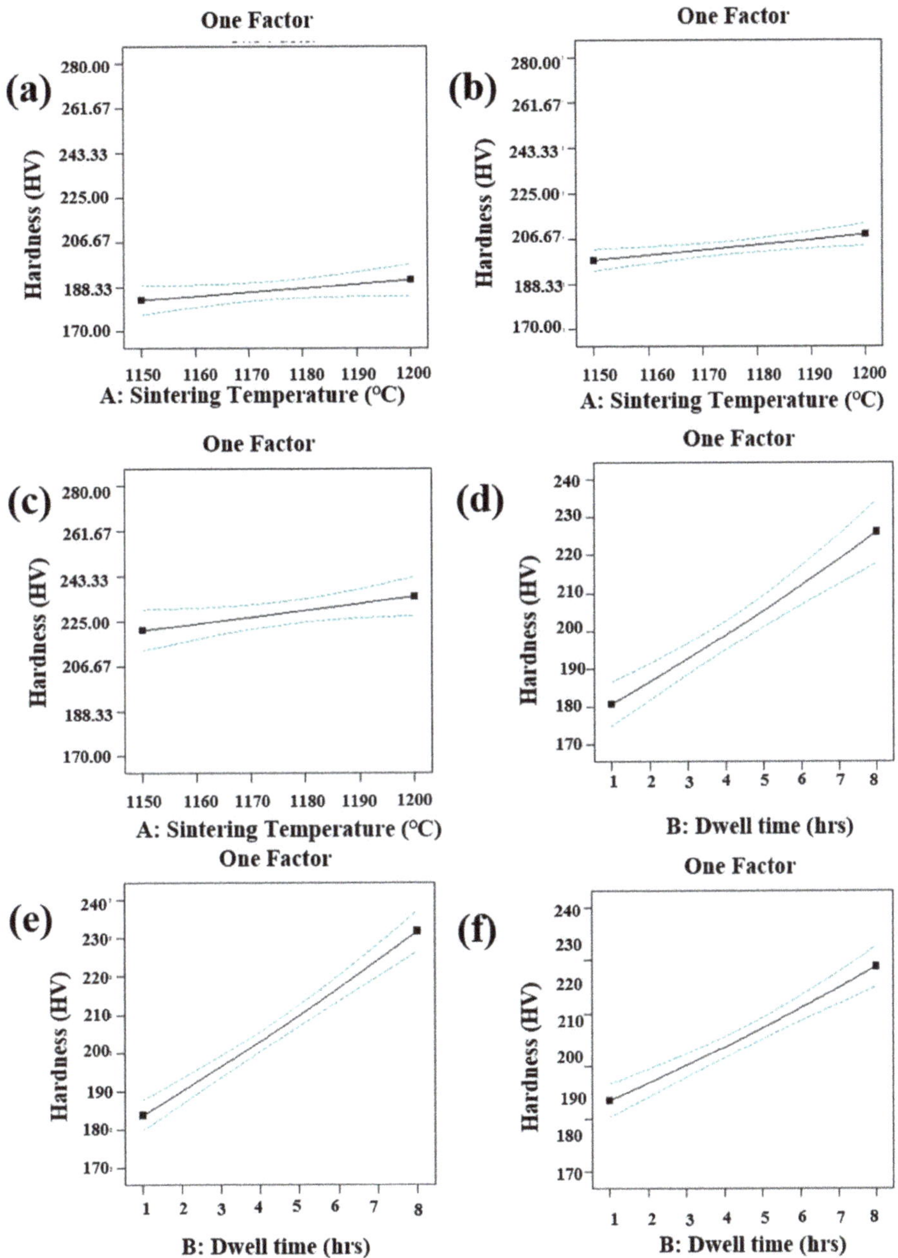

Figure 16. One factor effect plot for micro hardness.

The two-dimensional representation of micro hardness for the selected parameters is shown by the contour plot presented in Figure 17a. The nitrogen absorption values indicated in the middle of each contour increases linearly from the first to the last contour. The influence of sintering parameters on microhardness in the form of 3D surface plot has been presented in Figure 17b. The plot shows that nitrogen absorption increases with increase of sintering temperature and dwell time.

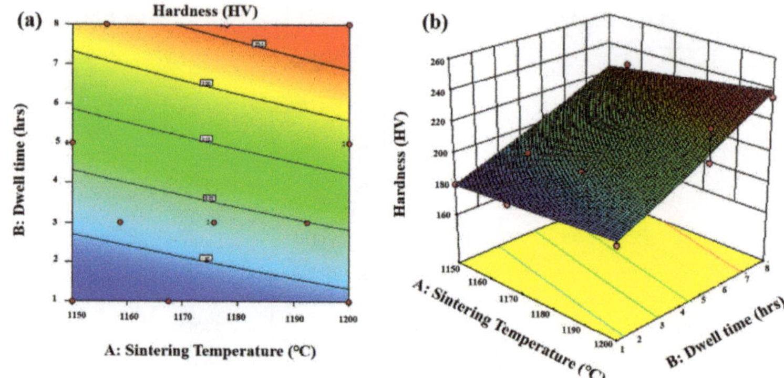

Figure 17. Contour and 3D plots of microhardness.

4. Modelling of Responses

The ANOVA tables discussed in the previous sections developed the equations which relate the responses nitrogen absorption (N-abs), density (D) and micro hardness (HV) and the input parameters. The final predicted model equations in terms of the actual factors for N-abs, D and HV are presented in the equations below.

4.1. Predicted Model Equation for Nitrogen Absorption

The predicted model equation for nitrogen absorption from the analysis has been presented in the Equation (1).

$$\text{Ln (N-abs)} = +124.27050 - 0.21326 * St - 0.078854 * \text{Dwell time} + 1.49490 \times 10^{-4}$$
$$* St * Dt + 9.1380 \times 10^{-5} * St^2 + 4.71247 \times 10^{-3} * Dt^2 \tag{1}$$

4.2. Predicted Model Equation for Density

The predicted model equation for density from the analysis has been presented in Equation (2).

$$\text{Ln (D)} = +12.40992 - 0.017856 * St - 0.034901 * \text{Dwell time} + 3.49901 \times 10^{-5}$$
$$* St * Dt + 7.64426 \times 10^{-6} * St^2 - 2.52253 \times 10^{-4} * Dt^2 \tag{2}$$

4.3. Predicted Model Equation for Microhardness

The predicted model equation for microhardness from the analysis has been presented in Equation (3).

$$\text{Ln (HV)} = +4.24692 + 7.94843 \times 10 - 4 * St - 0.032930 * \text{Dwell time} + 5.6212 \times 10^{-5} * St * Dt \tag{3}$$

5. Model Adequacy Test

The model summary statistics for each response was recorded, the details of which have been discussed below:

5.1. Model Adequacy Test for Nitrogen Absorption

The model summary statistics for nitrogen absorption have been given in Table 10. It can be observed that the linear model for nitrogen absorption has been suggested by the design expert software.

Table 10. Model summary statistics for nitrogen absorption.

Source	Std. Dev.	R-Squared	Adjusted R-Squared	Predicted R-Squared	PRESS	
Linear	0.059	0.9778	0.9743	0.9665	0.069	suggested
2FI	0.061	0.9787	0.9733	0.9584	0.086	
Quadratic	0.051	0.9872	0.9802	0.9731	0.055	
Cubic	0.053	0.9919	0.9798	0.9323	0.14	
Quartic	0.057	0.9920	0.9761		+	Aliased

5.2. Model Adequacy Test for Density

The model summary statistics for density have been given in Table 11. It can be observed that the linear and cubic models for density has been suggested by the design expert software.

Table 11. Model summary statistics for density.

Source	Std. Dev.	R-Squared	Adjusted R-Squared	Predicted R-Squared	PRESS	
Linear	5×10^{-3}	0.8590	0.8373	0.7864	4×10^{-4}	suggested
2FI	4×10^{-3}	0.8741	0.8427	0.7751	5×10^{-4}	
Quadratic	4×10^{-3}	0.9221	0.8831	0.7315	6×10^{-4}	
Cubic	6×10^{-4}	0.9988	0.9969	0.5274	1×10^{-3}	suggested
Quartic	0.000	1.0000	1.0000		+	Aliased

5.3. Model Adequacy Test for Microhardness

The model summary statistics for micro hardness have been given in Table 12. It can be observed that the linear model for micro hardness has been suggested by the software.

Table 12. Model summary statistics for microhardness.

Source	Std. Dev.	R-Squared	Adjusted R-Squared	Predicted R-Squared	PRESS	
Linear	0.023	0.9462	0.9379	0.9182	0.010	suggested
2FI	0.023	0.9469	0.9337	0.9055	0.012	
Quadratic	0.025	0.9509	0.9264	0.8740	0.016	
Cubic	0.028	0.9633	0.9083	0.7729	0.028	
Quartic	0.030	0.9635	0.8906		+	Aliased

6. Optimization of the Responses

The optimization of the responses was obtained from the design expert software by considering the response value and desirability. The software suggested the optimum parameters combination. The details of the constraints and optimized responses are shown in Tables 13 and 14 respectively.

Table 13. Constraints of the process.

Constraints						
Name	Goal	Lower Limit	Upper Limit	Lower Weight	Upper Weight	Importance
A: St	is in range	1150	1200	1	1	3
B: Dt	is in range	1	8	1	1	3
N-abs	maximize	0.96	2.82	1	1	5
D	maximize	7.318	7.575	1	1	3
HV	maximize	180	235	1	1	4

Table 14. Suggested solutions for maximum output.

Number	St	Dt	N-abs	D	HV	Desirability	
1	1199.434	7.995	2.836	7.614	238.965	1.000	Selected
2	1199.751	7.956	2.823	7.615	238.740	1.000	
3	1200.000	8.000	2.850	7.617	239.178	1.000	
4	1199.226	7.981	2.825	7.612	238.795	1.000	
5	1199.946	7.956	2.827	7.616	2.8.797	1.000	
6	1199.631	7.999	2.842	7.615	239.057	1.000	
7	1198.868	7.999	2.827	7.611	238.836	1.000	
8	1199.537	7.966	2.823	7.614	238.757	1.000	
9	1198.588	7.999	2.821	7.609	238.752	1.000	
10	1199.819	7.986	2.839	7.616	239.012	1.000	
11	1197.444	8.000	2.799	7.603	238.418	0.995	

7. Experimental Validation Study

The validation of the developed models for the responses (N-abs, D and HV) was carried out by a confirmation test using the same procedure as discussed above. Thus, it becomes possible to authenticate the developed mathematical models. The optimum parameter settings not only improve the quality but also in the industries by reducing parts production time and cost. Therefore, the development of optimum parameters settings plays a major role in this context.

The experimental results of this study were in close agreement with the predicted values of the responses between them and was less than 2% as shown in Table 15.

Table 15. Experimental validation of the responses.

Response	Predicted	Experimental	% Error
N-abs	2.84977	2.82	1.05
D	7.61741	7.575	0.55
HV	239.178	235	1.77

8. Conclusions

This study analyzed and evaluated the results on the experimental application of modifying 316L stainless steel alloy surface with increased dwell time for simultaneous sintering and in situ surface nitriding. In this research, the feasibility of sintering process to simultaneously sinter and nitride the surface layer of 316L stainless steel alloy was tested and established successfully. The experimental results revealed that a significant amount of nitride layer was formed on the surface of the sintered samples. The XRD results revealed that the sintering parameters helped in diffusion of nitrogen into the matrix of 316L stainless steel. The diffusion resulted in the formation of nitrides of iron and carbon. The XPS results indicated the presence of nitrogen onto the surface of the sintered samples in the form of nitride layer. This nitride layer was also observed under optical microscope with a thickness of 6 μm. The nitrogen gas as sintering atmosphere, sintering temperature and increased dwell time for sintering of green compacts helped in increased properties of the samples. The nitrogen uptake increased the microhardness of the sintered samples by 27%. Nitrogen as the sintering atmosphere helped in nitrogen diffusion and surface nitriding of the sintered samples. Austenitic stainless steels have low hardness and wear resistance values, therefore, this nitride layer formation not only helps in improving surface layer but also increases the corrosion and wear resistance of the material. The best results in the form of densification, microhardness and nitrogen absorption were observed at 1200 °C and 8 h dwell time.

The influence of processing parameters on densification, microhardness, nitrogen absorption and spectroscopic properties were investigated. Further analysis to set the optimum parameters setting for surface nitriding has been carried out. The mathematical model and equations for numerical prediction of densification, micro hardness and nitrogen absorption were also developed.

The newly modified approach of simultaneous sintering and surface nitriding was successful in addressing the related issues of this material.

The model and equations developed can be used to estimate the theoretical values of such responses by the industries for the production of implants and medical devices with desired properties. Moreover, the confirmation tests revealed that the experimental results are in close agreement with the predicted values of the responses with an error of less than 2%. This indicates that the developed model and equations are helpful in predicting material behavior in terms of its improved properties.

Author Contributions: Conceptualization, S.A.; formal analysis, S.A.; funding acquisition, A.M.A.R.; methodology, S.A.; project administration, A.M.A.R. and R.A.M.; resources, S.W.A. and A.A.A.A.; supervision, A.M.A.R. and R.A.M.; writing—original draft, S.A.; writing—review and editing, Z.B., S.H. and M.A.A.R. All authors have read and agreed to the published version of the manuscript.

Funding: This research was funded by YUTP-Fundamental Research grant and "The APC was funded by YUTP FRG, cost center, 015LC0-164".

Acknowledgments: The authors would like to acknowledge Universiti Teknologi PETRONAS, Malaysia and National University of Sciences and Technology (NUST), Islamabad, Pakistan for carrying out research and other facilities.

Conflicts of Interest: The authors declare no conflict of interest.

References

1. Wiles, P. The classic: The surgery of the osteo-arthritic hip. *Clin. Orthop. Relat. Res.* **2003**, *417*, 3–16.
2. Wiles, P. The surgery of the osteo-arthritic hip. *Br. J. Surg.* **1958**, *45*, 488–497. [CrossRef] [PubMed]
3. Lo, K.H.; Shek, C.H.; Lai, J. Recent developments in stainless steels. *Mater. Sci. Eng. R Rep.* **2009**, *65*, 39–104. [CrossRef]
4. Ali, S.; Rani, A.; Altaf, K.; Baig, Z. Investigation of Boron addition and compaction pressure on the compactibility, densification and microhardness of 316L Stainless Steel. *IOP Conf. Ser. Mater. Sci. Eng.* **2018**, *344*, 012023. [CrossRef]
5. Disegi, J.; Eschbach, L. Stainless steel in bone surgery. *Injury* **2000**, *31*, D2–D6. [CrossRef]
6. Dewidar, M. Influence of processing parameters and sintering atmosphere on the mechanical properties and microstructure of porous 316L stainless steel for possible hard-tissue applications. *Int. J. Mech. Mech. Eng.* **2012**, *12*, 10–24.
7. Talha, M.; Behera, C.; Sinha, O. A review on nickel-free nitrogen containing austenitic stainless steels for biomedical applications. *Mater. Sci. Eng. C* **2013**, *33*, 3563–3575. [CrossRef]
8. Alvarez, K.; Sato, K.; Hyun, S.; Nakajima, H. Fabrication and properties of Lotus-type porous nickel-free stainless steel for biomedical applications. *Mater. Sci. Eng. C* **2008**, *28*, 44–50. [CrossRef]
9. Niinomi, M.; Nakai, M.; Hieda, J. Development of new metallic alloys for biomedical applications. *Acta Biomater.* **2012**, *8*, 3888–3903. [CrossRef]
10. Muley, S.V.; Vidvans, A.N.; Chaudhari, G.P.; Udainiya, S. An assessment of ultra fine grained 316L stainless steel for implant applications. *Acta Biomater.* **2016**, *30*, 408–419. [CrossRef]
11. Wong, J.Y.; Bronzino, J.D.; Peterson, D.R. *Biomaterials: Principles and Practices*; CRC Press: Boca Raton, FL, USA, 2012.
12. Bronzino, J.D.; Peterson, D.R. *Biomedical Engineering Fundamentals*; CRC press: Boca Raton, FL, USA, 2014.
13. Bronzino, J.D. *Biomedical Engineering Handbook*; CRC press: Boca Raton, FL, USA, 1999; Volume 2.
14. Park, J.B. *Biomaterials Science and Engineering*; Springer Science & Business Media: Berlin, Germany, 2012.
15. Saini, M.; Singh, Y.; Arora, P.; Arora, V.; Jain, K. Implant biomaterials: A comprehensive review. *World J. Clin. Cases WJCC* **2015**, *3*, 52. [CrossRef] [PubMed]
16. Yang, K.; Ren, Y. Nickel-free austenitic stainless steels for medical applications. *Sci. Technol. Adv. Mater.* **2010**, *11*, 014105. [CrossRef] [PubMed]
17. Cordeiro, J.M.; Beline, T.; Ribeiro, A.L.R.; Rangel, E.C.; da Cruz, N.C.; Landers, R.; Faverani, L.P.; Vaz, L.G.; Fais, L.M.; Vicente, F.B.; et al. Development of binary and ternary titanium alloys for dental implants. *Dent. Mater.* **2017**, *33*, 1244–1257. [CrossRef] [PubMed]
18. Mirza, A.; King, A.; Troakes, C.; Exley, C. Aluminium in brain tissue in familial Alzheimer's disease. *J. Trace Elem. Med. Biol.* **2017**, *40*, 30–36. [CrossRef] [PubMed]

19. Vahter, M.; Berglund, M.; Åkesson, A.; Liden, C. Metals and women's health. *Environ. Res.* **2002**, *88*, 145–155. [CrossRef]
20. Bhola, R.; Bhola, S.M.; Mishra, B.; Olson, D.L. Corrosion in titanium dental implants/prostheses—A review. *Trends Biomater. Artif. Organs* **2011**, *25*, 34–46.
21. Manam, N.; Harun, W.; Shri, D.; Ghani, S.; Kurniawan, T.; Ismail, M.; Ibrahim, M.H. Study of corrosion in biocompatible metals for implants: A review. *J. Alloys Compd.* **2017**, *701*, 698–715. [CrossRef]
22. Chen, Q.; Thouas, G.A. Metallic implant biomaterials. *Mater. Sci. Eng. R Rep.* **2015**, *87*, 1–57. [CrossRef]
23. Singh, R.; Dahotre, N.B. Corrosion degradation and prevention by surface modification of biometallic materials. *J. Mater. Sci. Mater. Med.* **2007**, *18*, 725–751. [CrossRef]
24. Liu, X.; Chu, P.K.; Ding, C. Surface modification of titanium, titanium alloys, and related materials for biomedical applications. *Mater. Sci. Eng. R Rep.* **2004**, *47*, 49–121. [CrossRef]
25. Dayss, E.; Leps, G.; Meinhardt, J. Surface modification for improved adhesion of a polymer–metal compound. *Surf. Coat. Technol.* **1999**, *116*, 986–990. [CrossRef]
26. Huang, Y.; Qu, Y.; Yang, B.; Li, W.; Zhang, B.; Zhang, X. In vivo biological responses of plasma sprayed hydroxyapatite coatings with an electric polarized treatment in alkaline solution. *Mater. Sci. Eng. C* **2009**, *29*, 2411–2416. [CrossRef]
27. Ali, S.; Rani, A.; Majdi, A.; Mufti, R.A.; Hastuty, S.; Hussain, M.; Shehzad, N.; Baig, Z.; Aliyu, A.; Azeez, A. An Efficient Approach for Nitrogen Diffusion and Surface Nitriding of Boron-Titanium Modified Stainless Steel Alloy for Biomedical Applications. *Metals* **2019**, *9*, 755. [CrossRef]
28. Ali, S.; Rani, A.; Majdi, A.; Mufti, R.A.; Azam, F.I.; Hastuty, S.; Baig, Z.; Hussain, M.; Shehzad, N. The Influence of Nitrogen Absorption on Microstructure, Properties and Cytotoxicity Assessment of 316L Stainless Steel Alloy Reinforced with Boron and Niobium. *Processes* **2019**, *7*, 506. [CrossRef]
29. Ali, S.; Rani, A.M.; Altaf, K.; Hussain, P.; Prakash, C.; Hastuty, S.; Rao, T.V.; Subramaniam, K. Investigation of Alloy Composition and Sintering Parameters on the Corrosion Resistance and Microhardness of 316L Stainless Steel Alloy. In *Advances in Manufacturing II*; Springer: Cham, Switzerland, 2019; pp. 532–541.

processes

MDPI

Article

Optimization of Photopolymerization Process of Dental Composites

Tsanka Dikova [1,*], Jordan Maximov [2], Vladimir Todorov [2], Georgi Georgiev [1] and Vladimir Panov [1]

[1] Faculty of Dental Medicine, Medical University of Varna, 84 Tsar Osvoboditel Blvd, 9000 Varna, Bulgaria; Georgi.p.georgiev@mu-varna.bg (G.G.); vladimir.panov@mu-varna.bg (V.P.)
[2] Faculty of Mechanical Engineering, Technical University of Gabrovo, 4 Hadji Dimitar Str, 5300 Gabrovo, Bulgaria; maximov@tugab.bg (J.M.); v_p_todorov@abv.bg (V.T.)
* Correspondence: tsanka.dikova@mu-varna.bg

Abstract: The aim of this paper is to perform optimization of photopolymerization process of dental composites in order to obtain maximum hardness. Samples (5 mm diameter; 2, 3 and 4 mm thickness) were made of Universal Composite (UC), Bulk fill Composite (BC) and Flowable Composite (FC). Light curing of specimens was performed with 600, 1000 and 1500 mW/cm^2 light intensity and an irradiation time of 20, 40 and 60 s. Vickers microhardness on the top and bottom surfaces of samples was measured. Optimization was carried out via regression analysis using QStatLab software. Photopolymerization process parameters were calculated using a specially designed MatLab software-based algorithm. For all composites, regression models for hardness on top and bottom surfaces of composite layer were established. Layer thickness as well as hardness on top and bottom surfaces of each composite was calculated for 21 curing modes varying with light intensity and irradiation time. It was established that photopolymerization guidelines only of FC manufacturer guarantee the required hardness, while recommended regimes for UC and BC did not satisfy this requirement. Tables, containing recommended light curing regimes, were developed for three composite types, guaranteeing high hardness of composite restoration. They were designed to facilitate work of dentists in dental offices.

Keywords: light-cured composites; photopolymerization process; microhardness; optimization; regression analysis

Citation: Dikova, T.; Maximov, J.; Todorov, V.; Georgiev, G.; Panov, V. Optimization of Photopolymerization Process of Dental Composites. *Processes* **2021**, *9*, 779. https://doi.org/10.3390/pr9050779

Academic Editor: Leszek Adam Dobrzański

Received: 3 April 2021
Accepted: 26 April 2021
Published: 28 April 2021

1. Introduction

The introduction of light-cured resin-based composites (RBCs) is a revolutionary step in restorative dentistry because it allows clinicians to determine the beginning of the polymerization process. The reasons for their wide application in everyday practice are the increased aesthetic requirements of patients on the one hand, and on the other—the disadvantages of the amalgam such as low aesthetics, galvanic current flow and corrosion, staining of hard dental tissues, soft tissue tattoos, mercury vapor release and others [1].

Dental composites are essentially a mixture of an organic resin matrix, inorganic filler particles, a coupling agent and a photoinitiator system [2,3]. Depending on the size of the filler particles, RBCs are defined as: macrofilled, microfilled, nanofilled and hybrid [4–7]. However, several manufacturers are now incorporating nano-sized particles into their formulations, resulting in the creation of yet another category, the "nanohybrid" composites. Nanofilled composites have silicon-zirconium particles with a size between 0.005 and 0.01 μm. The small particle size allows good filling of the organic matrix—about 80%. The nanofilled composites present higher mechanical and physical properties to those of microhybrid composites, superior polishability with a gloss comparable to that of the enamel, better wear resistance and transparency [5–8]. Due to their desirable aesthetics, strength and durability, they are increasingly preferred by clinicians as a universal restorative material for both anterior and posterior fillings.

Depending on the viscosity, RBCs are defined as compactable with high viscosity, and flowable with low viscosity. In flowable composites, the percentage of inorganic

411

filler is lower and rheological modifiers (substances intended mainly to improve handling properties) have been removed from their composition. Their main advantages are high wettability of the tooth surface, providing penetration into any unevenness; ability to form layers with a minimum thickness, which significantly reduces the inclusion or retention of air; high flexibility, so they are less likely to be displaced in areas with high stress concentration; radiopacity and availability in different colors [9]. The disadvantages include a high level of polymerization shrinkage due to the smaller filler amount and lower mechanical properties. Indications for the use of flowable composites are cervical defects, very small occlusal defects as liners in class I and II cavities [10]. In the last decade, universal highly filled flowable composites have been developed, which allow for the restoration of a wide range of of anterior and posterior defects.

One of the main setbacks of RBCs as a direct restorative material is the application technique sensitivity, meaning that the success of composite restorations largely depends on the operator's skills [11]. The implementation of the adhesive protocol involves many steps and there are enough possibilities for error. In addition, the incremental technique is time consuming and contributes to even more inaccuracies. To simplify the procedure, the bulk fill composites are created.

Bulk fill composites are available in flowable and regular (nonflowable) consistency with the possibility of placing layers up to 5 mm (compared to 2 mm for conventional composites), while ensuring an adequate depth of cure. This goal is achieved in various ways, including optimization of the photoinitiator system (new photoinitiators or higher concentration of conventional ones), modification of the fillers (larger size or higher translucency of the particles) or inclusion of various chemicals in the composition [12,13]. The use of bulk fill composites for posterior restorations reduces cusp deflection [14] and polymerization stress [15] while increasing the fracture resistance of the hard tissues and the restoration itself [16]. However, flowable bulk fill composites have lower mechanical properties than non-flowable bulk fill and conventional ones, so they should not be used as a final layer, which is directly exposed to the masticatory forces [17].

The hardness of the materials determines their wear resistance, or their ability to abrade or be abraded from the opposite tooth structures [18,19]. The hardness of RBCs is influenced by several factors, the most important of which are the composition of the organic matrix, the type and amount of inorganic fillers and the degree of monomer-polymer conversion. A positive relationship has been found between the increase in hardness and the increase in the degree of conversion. The microhardness tests are most commonly used for indirect evaluation of the depth of cure and the degree of conversion of dental composites [20–22], because a small change in the degree of conversion can lead to a large hardness change [23].

The amount of light energy reaching the surface and the bottom of the composite restoration depends on many variables, such as the light intensity of the light curing unit (LCU), curing time, distance and angulation of the LCU's tip, composition, thickness, color and opacity of the composite. Therefore, the measured hardness at the top surface of the restoration cannot be accepted as an indicator of the hardness at its bottom. Less than 20% difference between the maximum hardness at the top of the composite and that found at the bottom is suggested as a guideline for adequate curing of the resin composite [24–26]. In addition, a bottom-to-top KHN ratio of 80% has been reported to correspond to a bottom-to-top degree of conversion ratio of 90% [24].

Many researchers have investigated the efficiency of different curing modes for successful polymerization of the resin-based composites expressed mainly by their hardness. Zhu S. and Platt J. [27] established that the polymerization mode and the curing tip distance had a significant effect on the hardness of the composites. Aguiar F.H.B. et al. [28] evaluated the influence of light curing modes and curing time on the microhardness of a hybrid composite. It was found that the increase of the light curing time and usage of appropriate LCU could lead to maximum hardness in polymerization of composites in deep cavities. The research of Spajic J. et al. [29] showed that in the light-cured materials, the material type had the highest effect on the microhardness, followed by the irradiation time, while the curing mode had the lowest impact. Alkhughairy F.I. [30] investigated the effect of two curing light intensities on the mechanical properties of bulk fill composites. A positive influence of the higher curing light intensity (1200 mW/cm^2) on the microhardness of the bulk fill composites was found.

Despite the daily and routine placement of composite restorations, dentists' level of knowledge about the composite properties and the main factors of the photopolymerization process—light intensity, irradiation time and layer thickness is not high [31,32]. The poor awareness can lead to incorrect curing protocol, which in turn can lead to an incomplete polymerization of the material with all the adverse consequences: reduced hardness and wear resistance, low adhesive bond strength and increased risk of restoration fracture, elution of residual monomers and faster color change. Therefore, there is a need for development of guidance for the curing modes, ensuring effective polymerization of the resin-based composites mostly used in dental practice.

In evaluation the influence of different curing modes on the polymerization process and the composites properties, most of the researchers use different methods of the statistical analysis; independent and paired sample *t*-test, one-way and two-way analysis of variance, multiple comparisons with Tukey test, full factorial ANOVA [27,29,30]. However, these methods allow only the influence and significance of the different parameters of polymerization process as well as the correlation between them to be revealed. Using the regression analysis, the optimal parameters of each process, which guarantee maximal properties of the object or material, can be calculated [33–37].

Therefore, the aim of this paper is to use regression analysis and to perform an optimization of the parameters of photopolymerization process in order to obtain maximum hardness of the dental composites. To the best of our knowledge, this kind of engineering approach is applied here for the first time in order to investigate resin-based composites. Universal nanohybrid, bulk fill and flowable light-cured composites are used in the research. Light curing modes for each composite are developed, which can serve as guidance for successful polymerization in the daily work of dentists.

2. Materials and Methods

2.1. Materials and Samples Preparation

Round samples with diameter of 5 mm and thickness of 2, 3 and 4 mm were prepared of three types of light-cured resin-based composites: Universal nanohybrid Composite (UC) Evetric (Ivoclar Vivadent, Lichtenstein, Germany), nanohybrid Bulk fill Composite (BC) for posterior restorations Filtek One Bulk Fill Restorative (3M Oral Care, St. Paul, MN, USA) and universal nanofilled Flowable Composite (FC) G-aenial Universal Flo (GC, Tokyo, Japan), indicated for restorations of all cavity classes. All composites were of A2 shade, but had a different composition and organic matrix/fillers ratio (Table 1).

The samples were prepared in three polyurethane molds with internal dimeter of 5 mm and thickness of 2, 3, and 4 mm. Glass slides and transparent celluloid strips on the top and bottom of the molds were used that guaranteed smooth surfaces and the same dimensions of all samples in the groups. The polymerization was performed with light curing unit Curing Pen (Eighteeth, Changzhou, China) with 600, 1000 and 1500 mW/cm^2 light intensity and irradiation time of 20, 40 and 60 s [29,31]. LCU's tip was placed parallel and in contact with the glass slide at a distance of 1 mm from the sample. Three specimens were prepared for each combination of parameters. They were stored in a dry dark container at room temperature for 24 h, after which the hardness measurements were performed.

Table 1. Composition of the composites used [38–42].

Composite	Composition		
	Component	Amount	Matrix/Filler Ratio, wt%
UC Evetric	Matrix: BIS-GMA (Bisphenol A glycydil dimethacrylate) UDMA (Urethane dimethacrylate) Bis-EMA (Bisphenol A polyethethylene glycol dimethacrylate) Fillers: Barium glass, Ytterbium Fluoride (YbF3), Mixed oxides and prepolymers 40 nm–3μm	3–10% 10–25% 3–10%	19–20/80–81
BC Filtek One Bulk Fill Restorative	Matrix: AUDMA (Aromatic Urethane Dimethacrylate) DDDMA (1,12-Dodecane Dimethycrylate) UDMA (Urethane dimethacrylate) Fillers: Silane Treated Ceramic, Silica, Zirconia and Ytterbium Fluoride	10–20% <10% 1–10%	23.5/76.5
FC G-aenial Universal Flo	Matrix: UDMA (Urethane dimethacrylate) Bis-EMA (Bisphenol A polyethethylene glycol dimethacrylate) Dimethacrylate component Fillers: Silicon dioxide (16 nm), Strontium glass (200 nm), pigments	10–20% 5–10% 5–10%	31/69

2.2. Hardness Measurements

The Vickers microhardness was investigated by ZHVμ-S (Zwick/Roell, Ulm-Einsingen, Germany) hardness tester with 50 gr loading for 10 s. Five measurements were performed on the top and bottom surfaces of each specimen and the mean values were recorded.

2.3. Regression Analysis

An optimization of the three parameters of the photopolymerization process, light intensity I, irradiation time t and layer thickness d was caried out. The aim was to obtain a composite layer with a certain thickness, having maximum microhardness on the top surface (HVmax) and microhardness on the bottom surface, 80% of HVmax [24–26], in given intensity values and irradiation times, specific for each LCU.

The governing factors, namely intensity I, irradiation time t and layer thickness d as well as their levels, are listed in Table 2. The factors, measured in natural physical units, are marked with $\tilde{x}_i$ and have different dimensions. In order to eliminate the experimental plan's dependence from the dimensions, the factors $\tilde{x}_i$ are transformed into a coded form x_i through dependence

$$x_i = (\tilde{x}_i - \tilde{x}_{i,0}) / |\tilde{x}_{i,max} - \tilde{x}_{i,0}| \qquad (1)$$

The objective functions are: Y_1, HV with hardness on the top surface of the composite layer and Y_2, HV hardness on the bottom surface. A planned experiment for the three investigated composites was carried out. The experimental design is shown in Table 3.

Regression analyses of the obtained experimental results for each composite were carried out through QStatLab v 6.1 software. For the objective functions Y_i, i = 1, 2, polynomials from second order were chosen since the governing factors were changed of three levels [33]:

$$Y_k(\{X\}) = a_0 + \sum_{i=1}^{m} a_i x_i + \sum_{i=1}^{m-1} \sum_{j=i-1}^{m} a_{ij} x_i x_j + \sum_{i=1}^{m} a_{ii} x_i^2, k = 1, 2, \qquad (2)$$

where $\{X\} = [x_1 x_2]^T \in \Gamma_x$ is the vector of the governing factors, Γ_x is the admissible space of the governing factors and m is their number.

For each of the composites, regression models were created for the objective functions Y_1, hardness on the top surface of the composite layer, and Y_2, hardness on the bottom surface. Using regression models for UC, *Evetric* optimizations were made in nine variations of the governing factors intensity (x_1) and irradiation time (x_2). As a condition for optimization, for each composite the average value of the microhardness on the top surface obtained in the experiment was accepted, and for the microhardness on the bottom surface 80% of the microhardness on the top surface.

Table 2. Governing factors and their levels.

	Governing Factors				
Natural $\tilde{x}_i$	**Coded x_i**	**Levels of the Factors**			
		Coded			
		For the first factor			
		-1	-0.1111		1
		For the rest factors			
		-1		0	1
		Natural			
Intensity I [mW/cm^2] $\tilde{x}_1$	x_1	600	1000		1500
Time t [s] $\tilde{x}_2$	x_2	20		40	60
Thickness d [mm] $\tilde{x}_3$	x_3	2		3	4

Table 3. Experimental design.

№	Composite Type						UC		BC		FC	
	Governing Factors						Y_1	Y_2	Y_1	Y_2	Y_1	Y_2
	Coded			Natural			HV Top	HV Bottom	HV Top	HV Bottom	HV Top	HV Bottom
	x_1	x_2	x_3	I, mW/cm^2	t, s	d, mm						
1	-1	-1	-1	600	20	2	42.0	33.5	59.1	55.8	42.4	37.9
2	1	-1	-1	1500	20	2	52.4	42.9	61.7	60.2	50.0	46.3
3	-1	1	-1	600	60	2	45.9	41.1	61.8	61.1	47.5	45.5
4	1	1	-1	1500	60	2	57.8	51.3	68.4	67.5	49.9	47.7
5	-1	-1	1	600	20	4	45.0	12.2	57.9	45.4	42.9	13.1
6	1	-1	1	1500	20	4	58.9	26.1	61.7	55.3	45.0	27.0
7	-1	1	1	600	60	4	49.3	32.7	60.3	57.5	45.3	31.7
8	1	1	1	1500	60	4	62.7	45.0	67.2	65.3	48.1	42.6
9	-0.1111	-1	-1	1000	20	2	54.1	42.2	62.2	60.3	47.7	42.3
10	-0.1111	1	1	1000	60	4	56.9	35.8	65.1	61.6	45.8	37.6
11	-1	0	-1	600	40	2	42.4	38.2	63.8	60.3	45.3	43.5
12	1	0	1	1500	40	4	61.7	38.4	65.3	62.5	45.5	36.5
13	-1	-1	0	600	20	3	44.3	22.2	59.2	51.8	46.1	29.9
14	1	1	0	1500	60	3	59.3	48.7	69.1	67.3	51.1	47.1

2.4. Calculation of the Parameters of Photopolymerization Process

The above-mentioned condition for optimization led to incorrect solutions of the equations of regression analysis for some regimes of UC *Evetric*. These incorrect solutions were referred to the photopolymerization parameters, in which the values of the microhardness on the top surface were less than the acceptable ones. Thus, the difference in the microhardness between the top and bottom surfaces was larger than 20%.

For that reason, a *MatLab* software-based algorithm was developed to calculate the microhardness on the top and bottom surfaces as well as the layer thickness, which met the requirement for maximum microhardness and 20% difference.

The regression model for each composite is used as basis of the algorithm in the newly designed program. If UC *Evetric* is taken as an example, the microhardness (at top and bottom, respectively) as a function of the governing factors is:

$$Y_1 \;=\; 56.266 + 6.402x_1 + 1.5325x_2 + 2.0025x_3 - 4.458x_1^2, \tag{3}$$

$$Y_2 \;=\; 35.936 + 6.118x_1 + 6.459x_2 - 6.991x_3 + 1.189x_1x_3 + 3.089x_2x_3. \tag{4}$$

The following equation expresses the relationship between the two objective functions:

$$0.8Y_1 - Y_2 \;=\; 0 \tag{5}$$

The Equation (5) is solved with respect to the thickness x_3:

$$x_3 \;=\; x_3(x_1, x_2) \tag{6}$$

or, taking into account (3) and (4):

$$x_3 \;=\; \frac{-9.0768 + 0.9964x_1 + 5.233x_2 + 3.5664x_1^2}{8.593 - 1.189x_1 - 3.089x_2}. \tag{7}$$

In the above expression, the coded values of intensity (x_1) and time (x_2) are replaced, and the thickness (coded values) is calculated. The coded coordinates x_1 and x_2 as well as the calculated thickness x_3 are replaced in the expressions for the objective functions Y_1 and Y_2 and the corresponding microhardnesses are calculated. It should be noted that they would be obtained in proportion $1 : 0.8$.

The thickness of the sample in physical coordinates is obtained via the transformation:

$$\tilde{x}_3 \;=\; (\tilde{x}_{3,max} - \tilde{x}_{3,0})x_3 + \tilde{x}_{3,0} \tag{8}$$

where $\tilde{x}_3$ is the thickness (mm), $\tilde{x}_{3,max}$ is the maximal thickness (mm), $\tilde{x}_{3,0}$ is the average level of the thickness (mm).

3. Results

3.1. Microhardness

The mean values of the microhardness measurements on the composite sample top and bottom surfaces are shown in Table 3. BC Filtek One Bulk Fill Restorative is characterized by a highest maximum microhardness of 65+/−4 HV, followed by UC Evetric with 56+/−4 HV, and the lowest microhardness 47+/−4 HV is obtained for FC G-aenial Universal Flo. These values are used as optimization condition in the regression analysis of the investigated composites.

3.2. Universal nanohybrid composite Evetric

The regression models for the objective functions Y_1—hardness on the top surface of the composite layer and Y_2—hardness on the bottom surface of UC *Evetric* are shown in Formulas (3) and (4). Using regression models, nine optimizations are conducted for this composite in different combinations of the parameters intensity (x_1) and irradiation time (x_2) and the results are given in Table 4. In regimes with lower intensity (600 mW/cm^2 and 1000 mW/cm^2/20 s) incorrect solutions are obtained, referring to the lower microhardness values or HV difference between the top and bottom surfaces larger than 20%. With the help of the newly designed program, maximum microhardness on the top surface, 80% microhardness on the bottom surface and the layer thickness (which guarantees it) were calculated for variations of irradiation time and LCU intensity. The results obtained are shown in Table A1 (Appendix A).

Table 4. Optimal regimes for photopolymerization of UC *Evetric* obtained by regression analysis.

№	I mW/cm^2	t s	d mm	HV Top	HV Bottom	Note
1	1500	60	3.10	60.0	48.0	
2	1500	40	3.05	58.3	41.6	
3	1500	20	2.35	55.3	41.6	Regimes that meet the requirement of max HV 56 +/−4 on the top surface and HV on the bottom surface ≥ 80%.
4	1000	60	2.85	57.0	43.0	
5	1000	40	2.15	53.9	41.6	
6	1000	20		Incorrect solution		HV difference between top and bottom surface is larger than 20%.
7	600	60		Incorrect solution		
8	600	40		Incorrect solution		HV values on the top surface are lower than the acceptable 56 +/−4.
9	600	20		Incorrect solution		

The data in Table A1 (Appendix A) can be classified into three groups. The first group includes the curing modes that do not provide the necessary microhardness on the top and bottom surfaces, which is a sign of insufficient polymerization. These are the regimes with an intensity of 600–700 mW/cm^2 for all irradiation times, and an intensity of 800 mW/cm^2 for time 20 and 40 s. The second is the boundary group with two modes: 800 mW/cm^2/60 s and 1000 mW/cm^2/20 s, which provide hardness close to the lower limit. Taking into account the microhardness increase of the composite over time [43], these regimes can be considered acceptable. The third group includes all modes with intensity 1000 mW/cm^2 for time 40 and 60 s, as well as those over 1000 mW/cm^2. They guarantee maximum microhardness on the top surface, and 80% HVmax on the bottom surface at the calculated thickness of the composite layer. In irradiating for a maximum of 60 s, the layer thickness at which 80% HVmax is achieved on the bottom surface is larger than that recommended by the manufacturer—2 mm [39]. Our results show that in this case it is possible to work with a layer thickness of up to 2.39 mm for 1200 mW/cm^2, 2.54 mm for 1300 mW/cm^2 and 3.17 mm for 1500 mW/cm^2. On the other hand, the manufacturer recommends 2 mm composite layer to be polymerized for 20 s with LCU intensities between 500 and 1000 mW/cm^2 and for 10 s with intensities above 1000 mW/cm^2. The results, obtained by us, disprove these recommendations, as observance of the specified parameters would not lead to satisfactory microhardness of the material.

3.3. Nanohybrid Bulk Fill Composite Filtek One Bulk Fill Restorative

For BC Filtek One Bulk Fill Restorative the regression models for the objective functions Y_1 and Y_2 are of the following type:

$$Y_1 \; = \; 64.55 + 2.327x_1 + 2.256x_2 - 0.704x_3 - 1.866x_2^2 + 1.003x_1x_2, \tag{9}$$

$$Y_2 \; = \; 59.113 + 3.596x_1 + 4.170x_2 - 2.740x_3 + 1.162x_1x_3 + 1.397x_2x_3. \tag{10}$$

The regression models are used for constitution of Equation (5) in order to obtain the thickness x_3:

$$x_3 \; = \; \frac{7.473 + 1.7344x_1 + 2.3652x_2 + 1.4928x_2^2 - 0.8024x_1x_2}{2.1768 - 1.162x_1 - 1.397x_2}. \tag{11}$$

The results obtained are shown in Table A2 (Appendix B). Only at $I = 600$ mW/cm^2 and $t = 20$ s there is a thickness of 3.86 mm in the range 2–4 mm, where the bottom/top microhardness ratio is equal to 0.8 and the accuracy of the calculated hardness (56.63 HV on top and 45.30 HV on bottom) is guaranteed. For all other "intensity—time" combinations in the table, the calculated limit thickness is larger than the upper limit of the range 2–4 mm. In this composite, for all combinations in the table (except for the first), the bottom/top microhardness ratio is higher or less than 0.8. For the calculated thicknesses larger than 4 mm, i.e., outside the defined range 2–4 mm, the accuracy of the calculated hardness is not guaranteed, as the regression models are valid only for the intervals in which the governing factors are changed.

Figure 1 illustrates the dependences of the microhardness on the top surface Y_1 and the bottom/top microhardness ratio Y_2/Y_1 as a function of thickness for mode 3 in Table A2 (Appendix B) (I = 600 mW/cm^2 and t = 60 s). For the first mode in Table A2 (Appendix B) the limit thickness is 3.86 mm, while for all others it is over 4 mm. Taking into account the increase of Y_2/Y_1 with thickness decrease (Figure 1), it can be assumed that for the entire thickness interval (2–4 mm) the bottom/top microhardness ratio should be higher than 0.8, which satisfies the requirement.

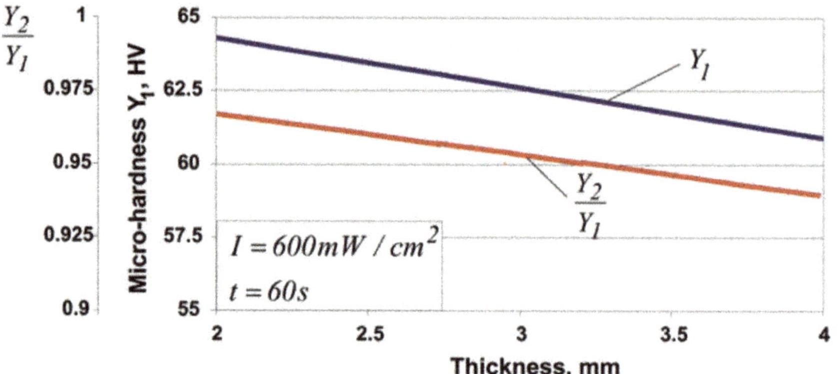

Figure 1. Dependences of the hardness on top surface Y_1 and the bottom/top hardness ratio Y_1/Y_2 as a function of thickness of BC.

The graphs in Figure 2 show the dependence of the microhardness Y_1 on the top surface on the intensity, irradiation time and thickness. It can be clearly seen that for the three thicknesses (2, 3 and 4 mm) the increase of the irradiation time over 40 s is practically unnecessary, as the microhardness on the top surface increases insignificantly. Using Figure 2 and according to the specific conditions, an optimal combination of photopolymerization parameters can be selected to ensure the minimum allowable hardness of 61 HV.

The parameters of photopolymerization of BC Filtek One Bulk Fill Restorative are summarized in Table 5. These parameters provide the minimum allowable hardness 61 HV on the top surface and hardness on the bottom surface—80% of that of the top for layer thickness between 2–4 mm. For comparison, the manufacturer recommends the irradiation time for 4 mm layer to be 40 s with intensity in the range 550–1000 mW/cm^2 and 20 s with intensity above 1000 mW/cm^2 [40,41]. The results of our study show that a satisfactory microhardness cannot be obtained by polymerization for 40 s with intensity of 550–700 mW/cm^2 and for 20 s with an intensity of 1000–1250 mW/cm^2.

Table 5. Optimal parameters of photopolymerization of BC Filtek One Bulk Fill Restorative.

№	Intensity, mW/cm^2	Time, s	Layer Thickness, mm
1	I > 1000	20	2
2	600–1500	40	2
3	I > 1000	20	3
4	600–1500	40	3
5	I > 1250	20	4
6	700–1250	40	4
7	I < 700	60	4

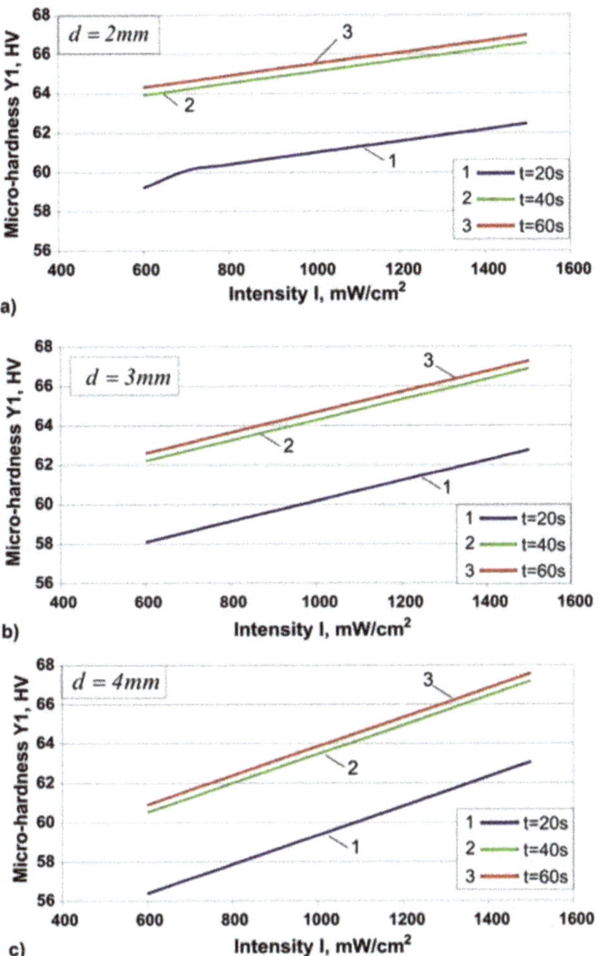

Figure 2. Dependence of the hardness on top surface on the intensity, irradiation time and layer thickness of: (**a**) 2 mm; (**b**) 3 mm and (**c**) 4 mm.

3.4. Universal Nanofilled Flowable Composite G-Aenial Universal Flo

The regression models for the objective functions Y1 and Y2 of FC G-aenial Universal Flo have the following form:

$$Y_1 = 46.762 + 1.715x_1 + 1.054x_2 - 1.326x_3 - 0.766x_1x_3 + 0.730x_1x_2x_3 \tag{12}$$

$$Y_2 = 37.091 + 4.214x_1 + 5.271x_2 - 7.769x_3 + 1.994x_1x_3 + 3.185x_2x_3 \tag{13}$$

They are used to constitute Equation (5) from which the thickness x_3 is expressed:

$$x_3 = \frac{-0.3186 + 2.842x_1 + 4.4278x_2}{6.7082 - 2.6068x_1 - 3.185x_2 + 0.584x_1x_2}. \tag{14}$$

The calculated parameters of photopolymerization for the investigated composite are shown in Table A3 (Appendix C). The data analysis shows that all light curing regimes satisfy the requirement for the surface microhardness. It is noteworthy that two modes lead to lowest hardness on the top surface: regime 1, which is characterized by the lowest parameters intensity and time, and regime 2 with the highest parameters. The other combinations of parameters provide HVmax in the range 44.46–47.50 HV.

The most probable reason for the reduced surface microhardness at high values of the irradiation time is the large layer thickness (7.63 mm at 1500 mW/cm^2 and 5.37 mm at 1300 mW/cm^2). In these two cases, the volume of the composite to be polymerized is too large and the photopolymerization process cannot proceed completely even at 60 s of irradiation. Therefore in order to ensure maximum microhardness 46–48 HV on the top surface, when working with FC G-aenial Universal Flo the thickness of the layer should not exceed 5 mm.

According to the recommendations of the manufacturer, 1.5 mm layer should be cured for 20 s with LCU intensity of 700 mW/cm^2 and for 10 s with intensity of 1200 mW/cm^2 [42]. The results of our study confirm that if these guidelines are followed, a satisfactory hardness or degree of polymerization of FC G-aenial Universal Flo could be obtained.

4. Discussion

The microhardness of resin-based composites depends mostly on the type and amount of the fillers, as well as the ratio of the monomer-polymer conversion [3,5,23,26]. The fillers of FC G-aenial Universal Flo consists of nano-powders of silicon dioxide and strontium glass with matrix/fillers ratio 31/69 wt% (Table 1). Compared to the other two composites, FC has the lowest filler amount, which defines its lowest hardness. The matrix/filler ratio of BC Filtek One Bulk Fill Restorative is 23.5/76.5 wt%, while that of UC Evetric is 19–20/80–81 wt%. As it can be seen in Table 1, the filler amount of BC is lower than the UC, but its composition is different. The fillers of BC consist only of materials with high hardness: ceramic, silicon dioxide and zirconium particles. While in the composition of the UC fillers in addition to the Ba glass particles, ytterbium fluoride and mixed oxides, prepolimers with lower hardness are included. Therefore, the comparatively high filler content and its composition define the highest hardness of BC Filtek One Bulk Fill Restorative composite.

The conducted optimization shows that for UC Evetric (Table A1, Appendix A) not all light curing modes provide the required microhardness on the top surface. Hence, in this case it is not recommended to use LCUs, operating with intensity below 800 mW/cm^2. In the case of LCU with intensity of 800 mW/cm^2, the required microhardness on the top surface is guaranteed only at irradiation of 60 s. When using LCUs with an intensity in the range of 1000–1500 mW/cm^2, it is possible to work with all irradiation times. It should be taken into account that as the duration of irradiation increases, the thickness of the composite layer and the hardness increase. When the maximum values of intensity and time (1500 mW/cm^2 and 60 s) are used, a layer with thickness of 3 mm having 60 HV microhardness on the top surface, can be successfully polymerized. The recommended light curing regimes of UC Evetric are marked with * in Table A1 (Appendix A).

The situation with BC Filtek One Bulk Fill Restorative is quite different. This composite is intended for posterior restorations [40], and therefore it is characterized by high hardness after polymerization and the ability to be applied in one layer in deep cavities. In our study, we investigated the composite layer thicknesses in the range of 2–4 mm. Regression analysis carried out ensures the accuracy of the results only in this range, i.e., where the bottom/top microhardness ratio is 0.8. According to the results shown in Table A2 (Appendix B), in regimes between mode 2 (600 mW/cm^2 and 40 s) and mode 12 (1000 mW/cm^2 and 60 s) this ratio is equal to or greater than 0.8, and for regimes above mode 12—the ratio is less than 0.8. This means that the modes up to 12th provide a microhardness on the bottom surface higher than 80% of that on the top surface, but the hardness on the top surface is below the minimum allowable 61 HV with the exception of mode 11 (1000 mW/cm^2 and 40 s). In the modes above 12th, the hardness on the bottom surface is within 0.66–0.78 (66–78%) of the hardness on the top surface, which is below the required minimum of 80% [24–26]. The graphs in Figure 2 show that increasing the time above 40 s does not lead to a significant increase in the microhardness of the top surface, therefore in the photopolymerization of BC *Filtek One Bulk Fill Restorative* it is not necessary to irradiate the obturation for 60 s. Modes that include intensities in the range 1000–1500 mW/cm^2 and irradiation times of 20 and 40 s guarantee microhardness above the minimum allowable for layer thicknesses of 4.71–12.07 mm, but with bottom/top microhardness ratio less than 0.8. Taking into account that a tooth cavity with 7 mm depth is a rare case, and then modes in which the layer thickness is above this value are excluded. The recommended regimes for photopolymerization of BC *Filtek One Bulk Fill Restorative* are marked with * in Table A2 (Appendix B). The layer thickness, greater than the upper limit of the range and the bottom/top microhardness ratio less than 0.8, are achieved in these curing regimes. In these cases, in order to ensure a higher microhardness on the bottom surface ($\geq$80% HVmax), it is necessary to work with a layer thickness less than that indicated in Table A2 (Appendix B). When working with a layer thickness in the range of 2–4 mm, the photopolymerization modes can be selected from Table 5.

Regarding FC G-aenial Universal Flo, it is found that all light curing regimes used in the study, provide the required microhardness on the top surface, but to ensure its maximum value of 46–48 HV, the layer thickness should not exceed 5 mm. Therefore, all modes in Table A3 (Appendix C) can be recommended for successful photopolymerization.

The results of the present study have shown that the instructions for light curing only of the manufacturer of FC G-aenial Universal Flo guarantee the required microhardness of the filling. The recommendations for work with the other two composites, UC Evetric and BC Filtek One Bulk Fill Restorative, do not meet the requirement for high microhardness, which means that a sufficient degree of polymerization can not be achieved. The recommended photopolymerization regimes for the three types of composites, developed in this study, guarantee high microhardness of the fillings.

5. Conclusions

In this study, optimization of the parameters of photopolymerization process of dental composites from three different groups was conducted using regression analysis. To the best of our knowledge, an engineering approach is applied here for the first time in investigation of resin-based composites. For all composites, regression models for the microhardness on the top and bottom surfaces of the composite layer were established. Both the layer thickness and the microhardness on the samples top and bottom surfaces of each composite were calculated for 21 modes of photopolymerization varying with the light intensity and irradiation time. It was established that photopolymerization guidelines only of FC manufacturer guarantee the required hardness, while recommended regimes for UC and BC do not satisfy this requirement. Tables, containing recommended light curing regimes, were developed for three composite types, guaranteeing high hardness of composite restoration. They were designed to facilitate work of dentists in dental offices.

Author Contributions: Conceptualization, T.D. and J.M.; methodology, J.M. and T.D.; software, J.M.; experiment, G.G. and V.T.; data analysis, T.D. and J.M.; writing—original draft preparation, T.D., G.G. and J.M.; writing—review and editing, J.M.; visualization, J.M.; supervision, T.D. and J.M.; project administration, V.P. All authors have read and agreed to the published version of the manuscript.

Funding: This research received no external funding.

Institutional Review Board Statement: Not applicable.

Informed Consent Statement: Not applicable.

Conflicts of Interest: The authors declare no conflict of interest.

Appendix A

Table A1. Parameters of photopolymerization of UC *Evetric*.

№	Intensity, mW/cm^2	Time, s	Layer Thickness, mm	Hardness, HV	
				Top	Bottom
1	600	20	2.09	42.05	33.64
2	600	40	2.33	44.07	35.26
3	600	60	2.81	46.56	37.25
4	700	20	1.97	45.00	36.00
5	700	40	2.19	46.97	37.58
6	700	60	2.62	49.36	39.48
7	800	20	1.88	47.57	38.05
8	800	40	2.08	49.49	39.59
9 *	800	60	2.46	51.80	41.44
10 *	1000	20	1.78	51.53	41.22
11 *	1000	40	2.95	53.40	42.72
12 *	1000	60	2.31	55.64	44.51
13 *	1200	20	1.76	53.96	43.17
14 *	1200	40	1.98	55.86	44.69
15 *	1200	60	2.39	58.21	46.57
16 *	1300	20	1.85	54.61	43.69
17 *	1300	40	2.06	56.57	45.26
18 *	1300	60	2.54	59.07	47.26
19 *	1500	20	2.07	54.82	43.85
20 *	1500	40	2.39	56.98	45.59
21 *	1500	60	3.17	60.08	48.06

Note: *—recommended regimes for successful photopolymerization.

Appendix B

Table A2. Parameters of photopolymerization of BC *Filtek One Bulk Fill Restorative*.

№	Intensity, mW/cm^2	Time, s	Layer Thickness, mm	Hardness, HV		Y_2/Y_1
				Top, Y_1	Bottom, Y_2	
1	600	20	3.86	56.63	45.30	0.80
2	600	40	4.72	59.29	48.81	0.82
3	600	60	8.36	53.47	46.27	0.86
4	700	20	4.03	57.08	46.94	0.82
5	700	40	4.99	59.78	49.07	0.82
6	700	60	9.30	53.78	46.33	0.86
7	800	20	4.23	57.58	47.06	0.82
8 *	800	40	5.31	60.35	49.31	0.82
9	800	60	10.58	54.08	46.20	0.85
10	1000	20	4.71	58.78	47.26	0.80
11 *	1000	40	6.16	61.72	49.66	0.80
12	1000	60	15.35	54.61	44.70	0.82
13 *	1200	20	5.34	60.34	47.38	0.78
14 *	1200	40	7.50	63.66	49.73	0.78
15	1200	60	32.66	54.75	36.13	0.66
16 *	1300	20	5.73	61.32	47.39	0.77
17	1300	40	8.51	65.03	49.57	0.76
18	1300	60	91.26	53.28	The results have no physical sense	
19 *	1500	20	6.79	63.89	47.27	0.74
20	1500	40	12.07	69.59	48.39	0.70
21	1500	60		The results have no physical sense		

Note: *—recommended regimes for successful photopolymerization.

Appendix C

Table A3. Parameters of photopolymerization of FC *G-aenial Universal Flo*.

№	Intensity, mW/cm^2	Time, s	Layer Thickness, mm	Hardness, HV	
				Top	Bottom
1	600	20	2.42	43.89	35.11
2	600	40	2.66	45.24	36.19
3	600	60	3.23	45.81	36.65
4	700	20	2.44	44.46	35.57
5	700	40	2.71	45.64	36.51
6	700	60	3.37	46.00	36.80
7	800	20	2.46	45.02	36.02
8	800	40	2.77	46.02	36.82
9	800	60	3.54	46.15	36.92
10	1000	20	2.51	46.09	36.87
11	1000	40	2.91	46.68	37.35
12	1000	60	4.01	46.29	37.03
13	1200	20	2.57	47.06	37.65
14	1200	40	3.11	47.16	37.73
15	1200	60	4.77	46.01	36.81
16	1300	20	2.61	47.50	38.00
17	1300	40	3.24	47.29	37.84
18	1300	60	5.37	45.58	36.46
19	1500	20	2.72	48.22	38.58
20	1500	40	3.61	47.19	37.75
21	1500	60	7.63	43.22	34.58

References

1. Georgiev, G.; Panov, V.; Dikova, T. Investigation of light intensity of wireless LED light curing units. *J. Technol. Univ. Gabrovo* **2020**, *60*, 40–45.
2. Sensi, L.G.; Strassler, H.E.; Webley, W. Direct Composite Resins. *Inside Dent.* **2007**, *3*, 76.
3. Anusavice, K.J.; Shen, C.; Rawls, H.R. *Phillips' Science of Dental Materials*; Elsevier Saunders: St. Louis, MO, USA, 2012; pp. 291–293.
4. Van Noort, R.; Barbour, R. *Introduction to Dental Materials-E-Book*; Elsevier Mosby: Edinburgh, UK, 2013; pp. 73–95.
5. Dikova, T. *Dental Materials Science: Lectures and Laboratory Classes Notes Part II*; MU-Varna: Varna, Bulgaria, 2014; p. 150.
6. Dikova, T.; Milkov, M. Nanomaterials in dental medicine. In Proceedings of the 10th Workshops "Nanoscience & Nanotechnology", Sofia, Bulgaria, 4–9 November 2009; Balabanova, E., Dragieva, I., Eds.; BAS-NCCNT: Sofia, Bulgaria, 2009; pp. 203–209.
7. Dikova, T.; Abadzhiev, M. Clinical application of the contemporary nano-materials (part 1–laboratory composites). *J. IMAB* **2009**, *15*, 67–70.
8. Mitra, S.B.; Wu, D.; Holmes, B.N. An application of nanotechnology in advanced dental materials. *J. Am. Dent. Assoc* **2003**, *134*, 1382–1390. [CrossRef]
9. Olmez, A.; Oztas, N.; Bodur, H. The effect of flowable resin composite on microleakage and internal voids in class II composite restorations. *Oper. Dent.* **2004**, *29*, 713–719. [PubMed]
10. Yacizi, A.R.; Ozgunaltay, G.; Dayangac, B. The effect of different types of flowable restorative resins on microleakage of Class V cavities. *Oper. Dent.* **2003**, *28*, 773–778.
11. Scotti, N.; Comba, A.; Gambino, A.; Manzon, E.; Breschi, L.; Paolino, D.; Pasqualini, D.; Berutti, E. Influence of operator experience on non-carious cervical lesion restorations: Clinical evaluation with different adhesive systems. *Am. J. Dent.* **2016**, *29*, 33–38.
12. Miletic, V.; Pongprueksa, P.; De Munck, J.; Brooks, N.R.; Van Meerbeek, B. Curing characteristics of flowable and sculptable bulk-fill composites. *Clin. Oral Investig.* **2017**, *21*, 1201–1212. [CrossRef] [PubMed]
13. Son, S.A.; Park, J.K.; Seo, D.G.; Ko, C.C.; Kwon, Y.H. How light attenuation and filler content affect the microhardness and polymerization shrinkage and translucency of bulk-fill composites? *Clin. Oral Investig.* **2017**, *21*, 559–565. [CrossRef]
14. Van Ende, A.; Lise, D.P.; De Munck, J.; Vanhulst, J.; Wevers, M.; Van Meerbeek, B. Strain development in bulk-filled cavities of different depths characterized using a non-destructive acoustic emission approach. *Dent. Mater.* **2007**, *33*, e165–e177. [CrossRef] [PubMed]
15. Fronza, B.M.; Rueggeberg, F.A.; Braga, R.R.; Mogilevych, B.; Soares, L.E.S.; Martin, A.A.; Giannini, M. Monomer conversion, microhardness, internal marginal adaptation, and shrinkage stress of bulk-fill resin composites. *Dent. Mater.* **2015**, *31*, 1542–1551. [CrossRef] [PubMed]
16. Rosatto, C.M.P.; Bicalho, A.A.; Veríssimo, C.; Bragança, G.F.; Rodrigues, M.P.; Tantbirojn, D.; Soares, C.J. Mechanical properties, shrinkage stress, cuspal strain and fracture resistance of molars restored with bulk-fill composites and incremental filling technique. *J. Dent.* **2015**, *43*, 1519–1528. [CrossRef] [PubMed]
17. Tomaszewska, I.M.; Kearns, J.O.; Ilie, N.; Fleming, G.J. Bulk fill restoratives: To cap or not to cap—That is the question? *J. Dent.* **2015**, *43*, 309–316. [CrossRef] [PubMed]

18. Lucey, S.; Lynch, C.D.; Ray, N.J.; Burke, F.M.; Hannigan, A. Effect of pre-heating on the viscosity and microhardness of a resin composite. *J. Oral Rehab.* **2010**, *37*, 278–282. [CrossRef]

19. Poggio, C.; Lombardini, M.; Gaviati, S.; Chiesa, M. Evaluation of Vickers hardness and depth of cure of six composite resins photo-activated with different polymerization modes. *J. Conserv. Dent.* **2012**, *15*, 237. [CrossRef] [PubMed]

20. Torres, C.R.; Caneppele, T.M.; Borges, A.B.; Torres, A.; Araújo, M.A. Influence of pre-cure temperature on Vickers microhardness of resin composite. *Int. J. Contemp. Dent.* **2011**, *2*, 41–45.

21. Saade, E.G.; Bandeca, M.C.; Rastelli, A.N.D.S.; Bagnato, V.S.; Porto-Neto, S.T. Influence of pre-heat treatment and different light-curing units on Vickers hardness of a microhybrid composite resin. *Laser Phys.* **2009**, *19*, 1276–1281. [CrossRef]

22. Osternack, F.H.R.; Caldas, D.B.D.M.; Rached, R.N.; Vieira, S.; Platt, J.A.; Almeida, J.B.D. Impact of refrigeration on the surface hardness of hybrid and microfilled composite resins. *Braz. Dent. J.* **2009**, *20*, 42–47. [CrossRef]

23. Fennis, W.M.; Ray, N.J.; Creugers, N.H.; Kreulen, C.M. Microhardness of resin composite materials light-cured through fiber reinforced composite. *Dent. Mater.* **2009**, *25*, 947–951. [CrossRef]

24. Price, R.B.; Felix, C.A.; Andreou, P. Evaluation of a second-generation LED curing light. *J. Can. Dent. Assoc.* **2003**, *69*, 666.

25. Yap, A.U.; Wong, N.Y.; Siow, K.S. Composite cure and shrinkage associated with high intensity curing light. *Oper. Dent.* **2003**, *28*, 357–364.

26. Bouschlicher, M.R.; Rueggeberg, F.A.; Wilson, B.M. Correlation of bottom-to-top surface microhardness and conversion ratios for a variety of resin composite compositions. *Oper. Dent.* **2004**, *29*, 698–704.

27. Zhu, S.; Platt, J. Curing efficiency of three different curing modes at different distances for four composites. *Oper. Dent.* **2011**, *36*, 362–371. [CrossRef]

28. Aguiark, F.H.; Braceiro, A.; Lima, D.A.N.L.; Ambrosano, G.M.B.; Lovadino, J.R. Effect of Light Curing Modes and Light Curing Time on the Microhardness of a Hybrid Composite Resin. *J. Contemp. Dent. Pract.* **2007**, *8*, 1–8.

29. Spajic, J.; Par, M.; Milat, O.; Demoli, N.; Bjelovucic, R.; Prskalo, K. Effects of curing modes on the microhardness of resin-modified glass ionomer cements. *Acta Stomatol. Croat.* **2019**, *53*, 37. [CrossRef] [PubMed]

30. Alkhudhairy, F.I. The effect of curing intensity on mechanical properties of different bulk-fill composite resins. *Clin. Cosmet. Investig. Dent.* **2017**, *9*, 1. [CrossRef] [PubMed]

31. Santini, A.; Turner, S. General dental practitioners' knowledge of polymerisation of resin-based composite restorations and light curing unit technology. *Br. Dent. J.* **2011**, *211*, E13. [CrossRef] [PubMed]

32. Georgiev, G.P. Factors associated with light curing units: A questionnaire survey. *Scr. Sci. Med. Dent.* **2019**, *5*, 37–43. [CrossRef]

33. Dikova, T.D.; Kulinich, S.A.; Iwamori, S.; Yamaguchi, S. Technological parameters optimization in picosecond laser texturing of titanium surfaces. *J. Phys. Conf. Ser.* **2021**, *1859*, 012037. [CrossRef]

34. Maximov, J.T.; Duncheva, G.V.; Anhev, A.P.; Dunchev, V.P.; Capec, J. A cost-effective optimization approach for improving the fatigue strength of diamond-burnished steel components. *J. Braz. Soc. Mech. Sci. Eng.* **2021**, *43*, 1–13. [CrossRef]

35. Maximov, J.T.; Duncheva, G.V.; Anchev, A.P.; Dunchev, V.P. Smoothing, deep or mixed diamond burnishing of low-alloy steel components—Optimization procedures. *Int. J. Adv. Manuf. Technol.* **2020**, *106*, 1917–1929. [CrossRef]

36. Malakov, I.; Zaharinov, V.; Tzenov, V. Size ranges optimization. *Procedia Eng.* **2015**, *100*, 791–800. [CrossRef]

37. Malakov, I.; Zaharinov, V. Computer Aided Determination of Criteria Priority for Structural Optimization of Technical Systems. *Procedia Eng.* **2014**, *69*, 735–744. [CrossRef]

38. Objelean, A.C.; Silaghi-Dimitrescu, L.; Furtos, G.; Badea, M.A.; Moldovan, M. The influence of organic-inorganic phase mixures on degradation behavior of some resin composites used in conservative dentistry. *J. Optoelectron. Adv. Mater.* **2016**, *18*, 567–575.

39. *Safety Data Sheet, Evetric*; Ivoclar Vivadent AG: Schaan, Liechtenstein, 2015; 6p.

40. *3M Filtek One Bulk Fill Restorative, Technical Product Profile*; 3m.com: St. Paul, MN, USA; Available online: https://multimedia.3m.com/mws/media/1317671O/3m-filtek-one-bulk-fill-restorative-technical-product-profile.pdf (accessed on 2 April 2021).

41. *3M Filtek One Bulk Fill Restorative, Safety Data Sheet*; 3m.com: St. Paul, MN, USA, 31 July 2020; Available online: https://multimedia.3m.com/mws/mediawebserver?mwsId=SSSSSuUn_zu8l00xm82Bm8_ZPv70k17zHvu9lxtD7SSSSSS-- (accessed on 2 April 2021).

42. *G-Aenial Universal Flo, Technical Manual*; GC: Leuven, Belgium; 20p, Available online: www.gceurope.com; https://cdn.gceurope.com/v1/PID/gaenialuniversalflo/manual/MAN_G-aenial_Universal_Flo_Technical_Manual_en.pdf; (accessed on 2 April 2021).

43. Yilmaz, E.Ç.; Sadeler, R.; Oner, V.; Yeşilyurt, M. Investigation of mechancal properties of restorative composites after artificial aging. In Proceedings of the International Energy and Engineering Congress, UEMK 2016 Conference Proceedings, Gaziantep University, Gaziantep, Turkey, 13–14 October 2016; pp. 739–743.

Article

Finite Element Analysis in Setting of Fillings of V-Shaped Tooth Defects Made with Glass-Ionomer Cement and Flowable Composite

Tsanka Dikova [1,*], Tihomir Vasilev [2], Vesela Hristova [1] and Vladimir Panov [1]

[1] Faculty of Dental Medicine, Medical University of Varna, 84 Tsar Osvoboditel Blvd, Varna 9000, Bulgaria; vesseladobreva@abv.bg (V.H.); vladimir.panov@mu-varna.bg (V.P.)
[2] Faculty of Engineering, Nikola Vaptsarov Naval Academy, 73 Vasil Drumev Str, Varna 9026, Bulgaria; t.vasilev@naval-acad.bg
* Correspondence: tsanka.dikova@mu-varna.bg

Received: 17 February 2020; Accepted: 16 March 2020; Published: 21 March 2020

Abstract: The aim of the present paper is to investigate the deformation–stress state of fillings of V-shaped tooth defects by finite element analysis (FEA). Two different materials are used—auto-cured resin-reinforced glass-ionomer cement (GIC) and flowable photo-cured composite (FPC). Two materials are placed into the cavity in one portion, as before the application of the composite the cavity walls are covered with a thin adhesive layer. Deformations and equivalent von Mises stresses are evaluated by FEA. Experimental study of micro-leakage is performed. It is established that there is an analogous non-homogeneous distribution of equivalent Von Mises stresses at fillings of V-shaped defects, made with GIC and FPC. Maximum stresses are generated along the boundaries of the filling on the vestibular surface of the tooth and at the bottom of the filling itself. Values of equivalent Von Mises stresses of GIC fillings are higher than that of FPC. Magnitude and character of deformation distribution at GIC and FPC fillings are similar—deformation is maximum along the vestibular surface of the filling and is 0.056 and 0.053 mm, respectively. In FPC fillings, the adhesive layer, located along the cavity/filling boundary, is characterized with greatest strain. The experimental study of micro-leakage has confirmed the adequacy of models used in FEA.

Keywords: FEA; V-shaped tooth defects; fillings; glass-ionomer cement; flowable composite

1. Introduction

V-shaped defects are non-carious cervical lesions on hard tooth tissues (HTT), which have a specific etiology, pathogenesis and clinical picture. Characteristic of this disease is that the defects are located mainly on the vestibular surface of the teeth, most often on the first premolars. Their deepening leads to the destruction of the dental crown as well as diseases of the pulp and periodontium. The anatomic shape of the teeth is disturbed. The treatment of V-shaped defects is a challenge for the dental practitioner because of the difficulties associated with holding the obturation over time, which requires a good knowledge of the restorative options and the sound choice of a suitable material.

In the case of medium and deep V-shaped defects, due to abnormalities in the anatomical shape of the teeth and deterioration of the aesthetic appearance as well as presence of hyperesthesia and functional disorders when the bottom of the lesion is close to the pulp, their restoration is required. Depending on the size of the defect and its location, a suitable material is selected. Today, plastic materials with a transparency and color similar to those of the natural enamel are used. According to the defect localization, coronary or cervical glass-ionomer cements, modified glass-ionomer cements, composite materials or compomers are recommended [1–5].

It is common practice for V-shaped defects to be filled with glass-ionomer cement (GIC) because they show a very good bond with the dentin without requiring additional adhesion. In the in vitro study of I.P. Ichim et al. [6], two techniques for filling V-shaped defects on the buccal surface were selected—insertion of the entire amount of material (GIC) at once and an incremental technique (GIC and composite material). Different investigations show that not all contemporary materials are appropriate for the treatment of all kinds non-carious lesions, because they exhibit a modulus of elasticity of 10.2 GPa. It has been found that the most suitable material for cervical restorations needs to be more flexible and with relatively low modulus of elasticity—about 1 GPa. Resin-modified glass-ionomer cements (RMGIC) were created in 1988 and since 1995 have been converted to enamel–dentin adhesives [7–9]. They show less restoration strength than the latest generation of adhesive systems, but higher than the conventional GIC.

The effectiveness of the composite material for treatment of V- shaped defects depends on the success of the bonding agents. If the adhesion of the composite to the enamel and dentin is not strong enough, a gap is formed between them, leading to sensitivity of the teeth, loss of retention and increased risk of secondary caries.

Scientific investigations of different adhesive systems—*three-step etch-and-rinse, two-step etch-and-rinse, two-step self-etch, one-step self-etch* [10,11]—show good results for adhesion. The cervical restoration, made with *etch-and-rinse* adhesive, show a better retention level than the *all-in-one* adhesive [12] with using of nano-hybrid composite. The retention rate of obturations is 82% and 75% for 6 months and 77% and 62% for 12 months. For 24 months, the retention level decreases to 69% for the *etch-and-rinse* adhesive and 49% for the *all-in-one* adhesive. There are no significant changes in color and marginal adaptation between the adhesive systems. None of the restorations have secondary caries or display any loss in anatomical shape or texture changes. Sabine May et al. [13] used two types of flowable composite to maintain V-shaped tooth defects. Fifty patients were treated with both composites and the results were monitored for 18 and 36 months. A 95.8% success rate for 36 months was found.

In recent years, simulation analysis has been used very often in the study of dental constructions, because it gives faster and relatively more reliable results than in clinical experiments when predicting complex cases of recovery [14–16]. The biomechanical behavior of different types of dental construction, such as crowns, bridges, implants, fillings, implant/bone surfaces, obturation/HTT etc., made of different materials or combinations of materials—metals, alloys, porcelain, metal-ceramics, composites and polymers—can be investigated with the help of the method of finite elements analysis (FEA) [16–18]. The laboratory simulation is mainly applied to investigate the biomechanical behavior of complex dental constructions and to understand the reasons for their deformations and failure [15,16,19,20].

Due to the great variety of direct restorative materials in dentistry nowadays, most research targets clinical investigations and in vitro experiments. The results can help one to evaluate the applicability of the particular restorative material for the specific purpose. However, the reasons, determining the behavior of dental constructions and materials can be revealed by simulations using numerical modelling.

The aim of the present paper is to study the stresses and deformations generated during the setting process of fillings of V-shaped defects made with GIC and flowable photo-cured composite (FPC) using FEA. The distribution and magnitude of the equivalent Von Mises stresses, displacements and strains in the tooth tissue and the filling itself are investigated. To validate the results, an in vitro experimental study of the micro-leakage of obturations, made with both materials, is performed.

2. Materials and Methods

2.1. Numerical Modelling by FEA

Two teeth (tooth 11 and tooth 15 according to the ISO System by the World Health Organization) were scanned with a *3Shape* laboratory scanner (*3Shape A/S, Denmark*). Virtual models in .stl file format were obtained, which were then converted to .part format of the *SolidWorks* software. On the vestibular surface of each tooth, one cavity of the size of the actual experiment was made at a distance of 1.25 mm from the cement/enamel border. As the V-shaped defects were located in the areas of the teeth with thin enamel, and the dentin was considered as a "gold standard" [21] when developing new dental composites, in the simulations the enamel was eliminated and only dentin was used as the tooth material. The cavity of tooth 11 was filled with GIC, while that of tooth 15 was filled with FPC.

In order to investigate the effect of the shrinkage of GIC and FPC on the stress-deformation state of the model, the analogy of volume reduction during the solidification process of the filling materials and during the cooling process was used. The capabilities of the *SolidWorks Simulation* software were used to analyze deformations and stresses due to temperature differences. Three different temperatures were applied: 19.99, 19.98 and 19.97 °C for shrinkage of 1%, 2% and 3%, respectively.

The virtual model of tooth 11 consisted of two parts, a GIC filling and a tooth of dentin (Figure 1a), while that of tooth 15 represented a system of three elements: tooth/30-µm adhesive layer/FPC filling (Figure 2a). The mesh, used in the simulations, is shown in Figures 1b and 2b. Standard solid mesh of high quality and tetragonal elements with 0.1 mm average size was used. The total number of the finite elements in GIC simulation was 986,861 and the number of the total nodes was 1,377,670. These parameters for FPC simulation are 1,170,403 and 1,629,798, respectively. Concerning the boundary conditions, the teeth were fixed at the bottom of the root, as shown in Figures 1a and 2a. In the numerical modeling, it was assumed that there was complete continuous contact between the tooth and the GIC filling and the three elements of the system in the case of FPC filling.

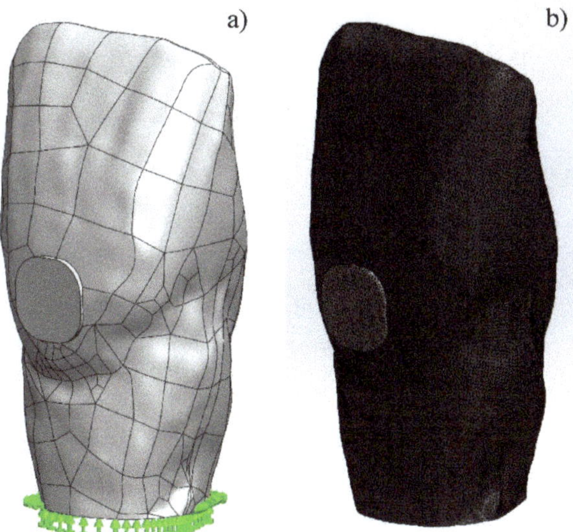

Figure 1. Virtual model of tooth 11: (**a**) fixed tooth with filling of glass-ionomer cement (GIC); (**b**) simulation mesh used.

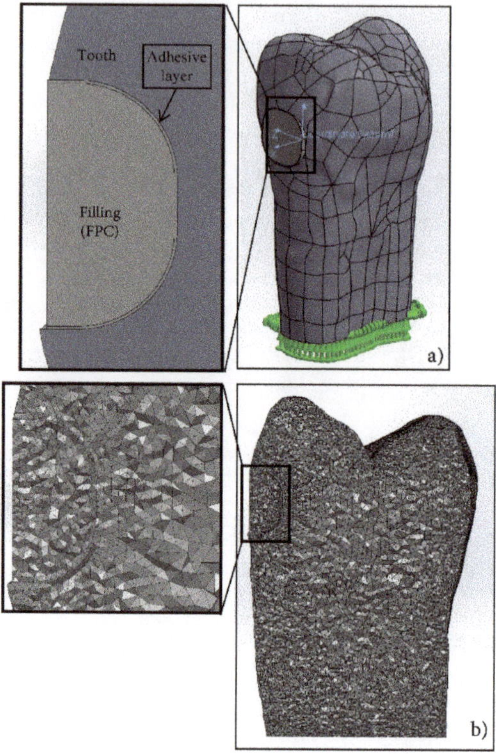

Figure 2. Virtual model of tooth 15: (**a**) fixed tooth with filling of flowable composite; (**b**) mesh used in simulation.

In the cavity, filled with GIC, the mechanical properties of the materials were set under the conditions of linear elastic isotropic strengthening. In the case of the cavity, filled with FPC, only the mechanical properties of the tooth and the adhesive layer were set under the above-mentioned conditions. Meanwhile, for the FPC filling, a model of linear elastic orthotropic strengthening was used, allowing the introduction of various physical-mechanical properties in the three directions. In this way, the shrinkage of FPC in the direction of the lamp, irradiating the filling material, was simulated. Unlike the actual photopolymerization process, which is non-homogeneous along the depth of the filling, numerical modeling was performed with the assumption of a uniform shrinkage along the X-axis due to the limitations of the software used. The values of modulus of elasticity and Poisson's ratio of the materials used are given in Table 1. The simulations were performed with the assumption of a homogeneous structure of the materials of filling and the adhesive layer, as well as of the dentin of the tooth.

Table 1. Data for materials used in the simulations [22–28].

Material/Tissue Type	Modulus of Elasticity, GPa	Poisson's Ratio	Shrinkage during Setting, %	Adhesion Strength to Dentin, MPa
Dentin	19	0.31	-	-
GIC: *FUJI VIII GP*	8.32	0.27	3	5.8
Adhesive: *Adhese Universal*	4.85	0.30	-	35
FPC: *Estelite Flow Quick - High Flow*	9.9	0.30	3	-

In both cases, a static task for linear strength analysis of the model is generated in a *SolidWorks Simulation* environment. During the simulation analysis, the Von Mises equivalent stresses, displacements and strains in the process of shrinkage of the fillings of V-shape defects made with GIC and FPC were evaluated.

2.2. In-Vitro Investigation of Micro-Leakage

Twenty-four freshly extracted human teeth (incisors, canines, premolars) were used to produce V-shaped defects on the vestibular surface of the teeth in the area of the enamel-cement border with the following shapes and average sizes: length L= 2.5 mm × width B = 2.5 mm × depth d = 1.3 mm.

The teeth were divided into two groups, with 12 teeth in each. The teeth of the first group were filled with auto-cured resin-reinforced glass-ionomer restorative cement *FUJI VIII GP (GC Corporation, Japan)*, the second group with flowable light-cured composite with lower viscosity *Estelite Flow Quick - High Flow (Tokuyama Dental Corporation inc., Japan)*. Prior to the application of the restorative material, all cavities were processed according to the manufacturer's recommendations. In the first group, after the cavity preparation and processing, the GIC was applied in one portion. The cavities of the second group were treated with 37% orthophosphoric acid for 40 s, after which they were thoroughly washed with a water-air jet. Self-etching adhesive *Adhese Universal (Ivoclar Vivadent AG, Liechtenstein)* was applied along the entire surface of the enamel and dentin of the cavity and photopolymerized for 20 s. The cavity was then filled with a flowable composite, pressed with celluloid tape and photopolymerized for 10 s.

All samples were subjected to thermal cycling in the following mode: 500 cycles, temperature 5–50 °C, holding time 15 s [29]. The teeth, with the exception of the cavities, were then isolated with contrast lacquer, placed in a dye, 2% methylene blue solution, for 8 h and thoroughly washed.

An examination of the filled cavities was done with the *Olympos SZ51* optical microscope. The degree of micro-leakage was evaluated by measuring the depth of dye penetration between the filling and the tooth walls.

3. Results

3.1. Numerical Modelling by FEA

Figure 3 shows the distribution of the equivalent Von Mises stresses on the vestibular surface of tooth 11 and tooth 15, filled with GIC and FPC, respectively. It can be seen that the equivalent stresses are generated mainly in the dentin of the tooth at the beginning of the setting process (1% shrinkage).

During the setting process of the fillings made of the two materials, the equivalent stresses increase with shrinkage increase and reach their maximum value at the end of the process. After final curing, in a maximum shrinkage of 3%, their distribution along the periphery of the filling is uneven. The highest equivalent stresses are observed at the rounded corners and the relatively lower ones in the straight sectors, which indicates the influence of the cavity shape.

The study of the equivalent stresses in depth shows that their distribution is non-homogeneous, both in the dentin of the tooth and in the filling itself (Figure 3). Regardless of the shrinkage ratio, their values are highest on the vestibular surface in the dentin at the edge of the filling. Increasing the shrinkage up to 3% leads to an increase in the equivalent stresses in the dentin at the edge of the obturation, reaching 810 and 705 MPa in the GIC and FPC, respectively, while in the dentin near the bottom of the filling they remain lower (203–270 MPa in the FPC). The stress distribution in the fillings of GIC and PFC is similar—they are smallest in the surface layer, increase with the distance to the deepest zone and reach their maximum value. At 3% shrinkage, the maximum stresses at the bottom of the GIC filling are 470–540 MPa, while at the FPC filling they are significantly lower—270–338 MPa. As the technique for preparation of the FPC filling differs from the GIC one, in this case the adhesive layer used between the dentin and FPC is clearly seen. In the adhesive layer, higher equivalent stresses in the range of 203–270 MPa appear only at 3% shrinkage. They are distributed not along the entire depth of the obturation, but only at about 30% distance from the vestibular surface.

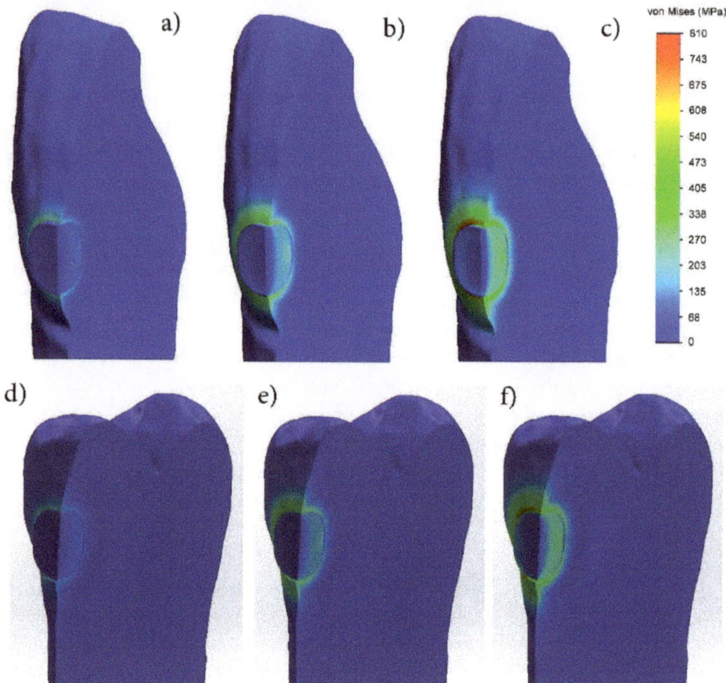

Figure 3. Distribution of equivalent Von Mises stresses during setting process of fillings made with: (**a**–**c**) glass-ionomer cement; (**d**–**f**) flowable photo-cured composite (FPC). (1% shrinkage—(**a**,**d**); 2% shrinkage—(**b**,**e**); 3% shrinkage—(**c**,**f**)).

The displacements in the processes of setting and photopolymerization of GIC and FPC fillings, respectively, are shown in Figure 4. It can be clearly seen that in both cases, the surface layer of the filling is subjected to maximum deformation, and it increases with increasing the shrinkage. In a maximum shrinkage of 3%, the displacement along the vestibular surface reaches 0.056 mm in the GIC and 0.053 mm in the FPC. In both materials, the deformation decreases with distance from the surface to the bottom of the filling.

The investigation of the strain has shown that at the GIC filling, it is located along its periphery—in the dentin of the tooth and the GIC of the filling (Figure 5a–c) and increases with shrinkage increase. At maximum shrinkage, the strain is uneven along the periphery, with larger values at the rounded corners and with smaller values in the linear sectors. In the FPC filling (Figure 5d–f), the strain is highest in the adhesive layer located along the cavity/filling boundary and increases with increasing the shrinkage of the restorative material.

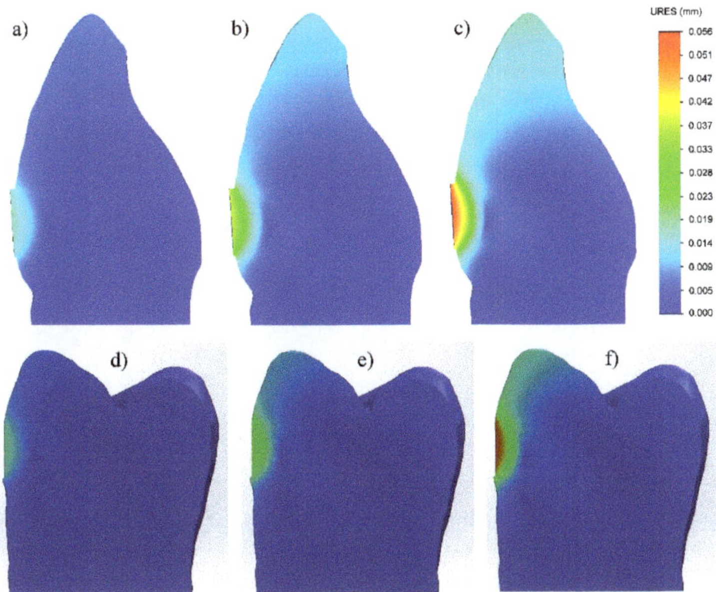

Figure 4. Displacement during setting process of fillings made with: (**a**–**c**) glass-ionomer cement; (**d**–**f**) flowable photo-cured composite. (1% shrinkage—(**a**,**d**); 2% shrinkage—(**b**,**e**); 3% shrinkage—(**c**,**f**)).

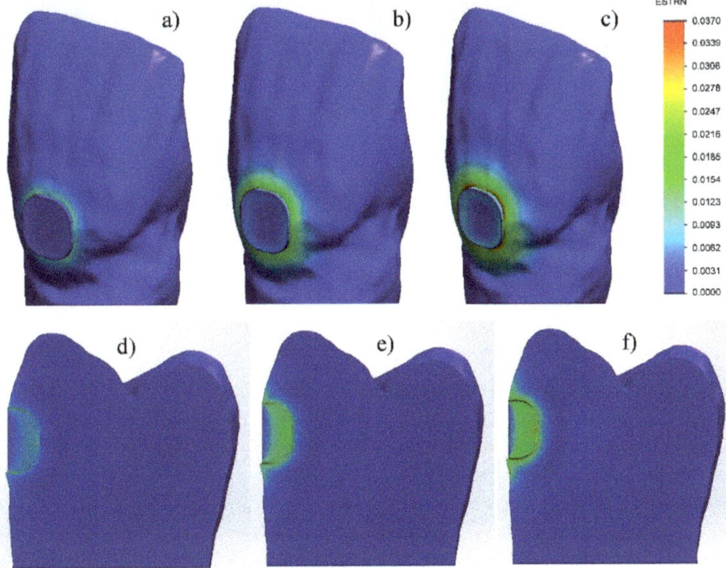

Figure 5. Strain during setting process of fillings made with: (**a**–**c**) glass-ionomer cement; (**d**–**f**) flowable photo-cured composite. (1% shrinkage—(**a**,**d**); 2% shrinkage—(**b**,**e**); 3% shrinkage—(**c**,**f**)).

3.2. In-Vitro Investigation of Micro-Leakage

During the experiment, it was found that the V-shaped defects, filled with FPC, have a lower micro-leakage (24.49%) than those filled with GIC (38.02%). In the GIC fillings, the dye had penetrated more than one third of the depth of the obturation (Figure 6); whereas, in the case of teeth filled with FPC, only a slight penetration of dye was observed, mainly in the cervical region (indicated by arrows on the picture of Figure 7).

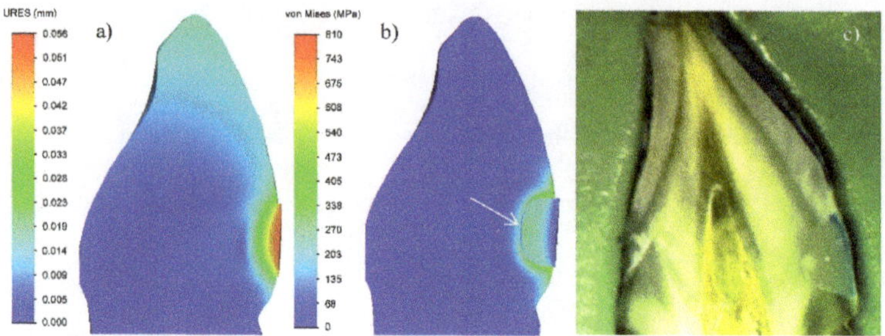

Figure 6. GIC filling of tooth cavity: (**a**) Displacement in the zone of filling; (**b**) Von Mises equivalent stresses; (**c**) Micro-leakage. (3% shrinkage).

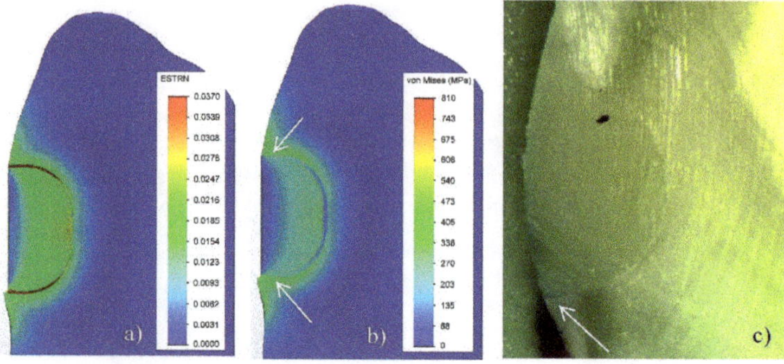

Figure 7. Filling of tooth cavity made with FPC: (**a**) Strain in the zone of filling; (**b**) Von Mises equivalent stresses; (**c**) Micro-leakage. (3% shrinkage).

4. Discussion

4.1. V-Shaped Defects Filled with GIC

The analysis of the equivalent stresses and deformations has shown that their non-homogeneous distribution in the GIC filling is due to the peculiarities of the tooth/filling system. Since the vestibular surface of the obturation does not come into contact with the tooth, and there is a complete continuous contact between the other surfaces of the tooth and the filling, in the process of cement setting, the greatest deformation will occur in the surface layer of the obturation (Figure 4a–c). This will lead to the relaxation of the stresses in this zone. The full continuous contact with the hard tooth tissues impedes the deformation of the material along the depth of the filling volume and leads to an increase in the stresses in this region during the shrinkage process (Figure 3a–c). The modulus of elasticity of the tooth dentin is about 2.5 times higher than that of the GIC used for the filling (Table 1). This determines lower deformation and, accordingly, higher equivalent stresses on the vestibular surface of the tooth along the boundary with the filling. It is clearly seen in Figure 3c that the values of the equivalent stresses in the dentin are maximal at almost half of the depth of the obturation. When these values exceed the adhesion strength of the GIC to the dentin, which is relatively low—5.8 MPa (Table 1)—then there is an opportunity for the detaching of the cement from the cavity walls. In this way, gaps between the tooth and the filling will occur, creating conditions for micro-leakage of fluids (Figure 6).

The results obtained in the experimental investigation of GIC fillings show a micro-leakage of 38% along the depth of the cavities. These data correlate very well with the stress distribution along the depth of the filling in the numerical modeling with FEA (Figure 6). In spite of the high values of the stresses resulting from the use of an analogue of the shrinkage process in the simulation analysis, the model used gives a grounded explanation of the reasons for the relatively high micro-leakage. Therefore, the experiment has confirmed the adequacy of the model used in the simulation analysis.

4.2. V-Shaped Defects Filled with FPC

In an investigation of the deformation-stress state of cavities, filled with dental composites, in addition to the features of the tooth/filling system, the characteristics of the materials used and the specificity of their setting process must be taken into account. Since in the experimental study, a thin adhesive layer is applied on the entire surface of the cavity, the virtual model for simulations consists of three components: the tooth (made of dentin), the adhesive layer, and the filling of a flowable light-cured composite. The comparison of their modulus of elasticity shows the following relation: 19–4.85–9.9 GPa, respectively. Due to the large difference in the modulus of elasticity, the highest equivalent stresses occur in the dentin along the vestibular surface at the boundary with the filling (Figure 3d–f). Since there is an adhesive layer with a low modulus of elasticity between the composite and the cavity walls, it deforms in the shrinkage process of FPC (Figure 5d–f) and leads to the relaxation of the stresses in both the dentin and the filling material. As a result, the equivalent stresses at the tooth-filling boundary are lower than those in the GIC obturation. Due to the deformation of the adhesive layer, stresses in it are observed only in the maximum of 3% shrinkage (Figure 3f).

The deformation in the FPC filling is uneven in depth (Figure 4d–f) not only because of the fact that the lack of contact on the vestibular surface facilitates the shrinkage, but also because of the specificity of the photopolymerization process. In this case, the degree of polymerization (respectively shrinkage) depends on the light intensity, which is strongest on the surface [30], leading to larger shrinkage in the surface areas and less in depth. Due to the complexity of simulation of this type of process with FEA, it is assumed that the model of the FPC filling is of a linearly elastic orthotropic type. Therefore, in the process of numerical modeling, it is assumed that the shrinkage should be performed uniformly only along the X-axis. These two reasons predetermine the non-homogeneous distribution of the stresses in the volume of the FPC filling, with the lowest along the vestibular surface and the highest at the bottom. Due to the presence of an intermediate adhesive layer with a low modulus of elasticity, which in this case serves as a buffer [2,3,31,32], the stresses at the bottom of the FPC filling are almost two times lower than those of the GIC.

Since the bond between the adhesive and the composite is chemical and that between the HTT and the adhesive layer is micromechanical, it is expected that the bond at the adhesive–dentin boundary will be broken when the stresses exceed the adhesion strength of 35 MPa (Table 1). Numerical modeling has shown that stresses above this value are obtained only in the final 3% shrinkage of FPC (Figure 3d–f), as their maximum values reach about 30% of the depth of filling. This determines the lower micro-leakage value (24.49%) compared to the GIC fillings (38.08%) and correlates well with the results obtained in the experimental study.

The model used in the simulation analysis provides an adequate explanation of the reasons for the lower micro-leakage in V-shaped defects filled with FPC. On the one hand, the elastic adhesive layer has a positive effect, which leads to the relaxation of the stresses along the cavity/composite boundary, and on the other, the specificity of the photopolymerization process and, as a result, the inhomogeneous shrinkage in depth of the filling.

5. Conclusions

The deformation–stress state of V-shaped tooth defects filled with two different materials, glass-ionomer cement and flowable composite, is investigated by numerical modeling using FEA. The results are validated by an in vitro investigation of the micro-leakage of the fillings.

It is established that there is analogous non-homogeneous distribution of Von Mises equivalent stresses at the fillings of V-shaped defects, made with GIC and FPC. The maximum stresses are generated along the boundary of the filling on the vestibular surface of the tooth and at the bottom of the filling itself. The values of Von Mises equivalent stresses of GIC fillings are higher than that of FPC.

The magnitude and character of the deformation distribution at GIC and FPC fillings are similar—the displacement is maximum along the vestibular surface of the filling and is 0.056 and 0.053 mm, respectively. The strain in FPC fillings is greatest in the adhesive layer, located along the cavity/filling boundary.

The results of the FEA provide an adequate explanation of the reasons for the lower micro-leakage of the V-shaped defects filled with flowable composite. Therefore, the experiment has confirmed the adequacy of the models used in the simulation analysis.

Author Contributions: Conceptualization, T.D.; methodology, T.D., T.V., V.H.; FEA, T.V.; in-vitro investigation, V.H.; results analysis, T.D.; writing—original draft preparation, T.D., T.V., V.H.; writing—review and editing, V.P.; supervision, V.P. All authors have read and agreed to the published version of the manuscript.

Funding: This research received no external funding.

Acknowledgments: The simulation analysis was done with the help of Assoc. Prof. Diyan Dimitrov from the Department of Mechanics and Machine Elements at Technical University of Varna.

Conflicts of Interest: The authors declare no conflict of interest.

References

1. Ichim, I.P.; Schmidlin, P.R.; Li, Q.; Kieser, J.A.; Swain, M.V. Restoration of non-carious cervical lesions: Part II. Restorative material selection to minimize fracture. *Dent. Mater.* **2007**, *23*, 1562–1569. [CrossRef] [PubMed]
2. Kemp-Scholte, C.M.; Davidson, C.L. Complete marginal seal of Class V resin composite restorations effected by increased flexibility. *J. Dent. Res.* **1990**, *69*, 1240–1243. [CrossRef] [PubMed]
3. Kemp-Scholte, C.M.; Davidson, C.L. Marginal integrity related to bond strength and strain capacity of composite resin restorative systems. *J. Prosthet. Dent.* **1990**, *64*, 658–664. [CrossRef]
4. Li, Q.; Jepsen, S.; Albers, H.K.; Eberhard, J. Flowable materials as an intermediate layer could improve the marginal and internal adaptation of composite restorations in Class-V-cavities. *Dent. Mater.* **2006**, *22*, 250–257. [CrossRef] [PubMed]
5. OäNAL, B.A.N.U.; Pamir, T. The two-year clinical performance of esthetic restorative materials in noncarious cervical lesions. *J. Am. Dent. Assoc.* **2005**, *136*, 1547–1555. [CrossRef] [PubMed]
6. Ichim, I.; Li, Q.; Loughran, J.; Swain, M.V.; Kieser, J. Restoration of non-carious cervical lesions: Part I. Modelling of restorative fracture. *Dent. Mater.* **2007**, *23*, 1553–1561. [CrossRef]
7. Van Dijken, J.W. Retention of a resin-modified glass ionomer adhesive in non-carious cervical lesions. A 6-year follow-up. *J. Dent.* **2005**, *33*, 541–547. [CrossRef]
8. Swift, E.J.; Pawlus, M.A.; Vargas, M.A. Shear bond strengths of resin-modified glass-ionomer restorative materials. *Oper. Dent.* **1995**, *20*, 138.
9. Tyas, M.J.; Burrow, M.F. Clinical evaluation of a resin-modified glass ionomer adhesive system: Results at five years. *Oper. Dent.* **2002**, *27*, 438–441.
10. Peumans, M.; Kanumilli, P.; De Munck, J.; Van Landuyt, K.; Lambrechts, P.; Van Meerbeek, B. Clinical effectiveness of contemporary adhesives: A systematic review of current clinical trials. *Dent. Mater.* **2005**, *21*, 864–881. [CrossRef]
11. Chee, B.; Rickman, L.J.; Satterthwaite, J.D. Adhesives for the restoration of non-carious cervical lesions: A systematic review. *J. Dent.* **2012**, *40*, 443–452. [CrossRef] [PubMed]
12. Tuncer, D.; Yazici, A.R.; Özgünaltay, G.; Dayangac, B. Clinical evaluation of different adhesives used in the restoration of non-carious cervical lesions: 24-month results. *Aust. Dent. J.* **2013**, *58*, 94–100. [CrossRef] [PubMed]

13. May, S.; Cieplik, F.; Hiller, K.A.; Buchalla, W.; Federlin, M.; Schmalz, G. Flowable composites for restoration of non-carious cervical lesions: Three-year results. *Dent. Mater.* **2017**, *33*, e136–e145. [CrossRef] [PubMed]

14. Dikova, T.; Dzhendov, D.; Simov, M.; Katreva-Bozukova, I.; Angelova, S.; Pavlova, D.; Abadzhiev, M.; Tonchev, T. Modern trends in the development of the technologies for production of dental constructions. *J. IMAB Annu. Proc. Sci. Pap.* **2015**, *21*, 974–981. [CrossRef]

15. Dikova, T.; Vasilev, T. Bending fracture of Co-Cr dental bridges, produced by additive technologies: Simulation analysis and test. *Eng. Fract. Mech.* **2019**, *218*, 106583. [CrossRef]

16. Dikova, T.; Vasilev, T.; Dolgov, N. Failure of ceramic coatings on cast and selective laser melted Co-Cr dental alloys under tensile test: Experiment and finite element analysis. *Eng. Fail. Anal.* **2019**, *105*, 1045–1054. [CrossRef]

17. AL-Gharrawi, H.A.; Saeed, M.W. Static stress analysis for three different types of composite materials experimentally and numerically. *Int. J. Sci. Eng. Res.* **2016**, *7*, 498–504.

18. Miletic, V.; Peric, D.; Milosevic, M.; Manojlovic, D.; Mitrovic, N. Local deformation fields and marginal integrity of sculptable bulk-fill, low-shrinkage and conventional composites. *Dent. Mater.* **2016**, *32*, 1441–1451. [CrossRef]

19. Rees, J.S.; Jacobsen, P.H. The effect of cuspal flexure on a buccal Class V restoration: A finite element study. *J. Dent.* **1998**, *26*, 361–367. [CrossRef]

20. Tanaka, M.; Naito, M.; Yokota, M.; Kohno, M. Finite element analysis of the possible mechanism of cervical lesion formation by occlusal force. *J. Oral Rehabil.* **2003**, *30*, 60–67. [CrossRef]

21. Papadogiannis, D.Y.; Lakes, R.S.; Papadogiannis, Y.; Palaghias, G.; Helvatjoglu-Antoniades, M. The effect of temperature on the viscoelastic properties of nano-hybrid composites. *Dent. Mater.* **2008**, *24*, 257–266. [CrossRef] [PubMed]

22. Aleksandrova, V. Biomechanical Problems in Direct Restoration of Class I and II Carious Defects in Patients with Bruxism [In Bulgarian]. Ph.D. Thesis, Medical University of Plovdiv, Plovdiv, Bulgaria, 2018.

23. Petrovic, B.; Markovic, D.; Kojic, S.; Peric, T.; Dubourg, G.; Drijaca, M.; Stojanovic, G. Characterization of glass ionomer cements stored in various solution. *Mater. Technol.* **2019**, *53*, 285–293. [CrossRef]

24. GC Fuji VIII GP; GC Corporation Japan; GC EUROPE N.V.; GC EEO-Bulgaria. Available online: https://cdn.gceurope.com/v1/PID/fuji8gp/leaflet/LFL_Fuji_VIII_GP_bg.pdf (accessed on 13 February 2020).

25. Scientific Documentation Adhese® Universal, Ivoclar Vivadent AG, Liechtenstein, April. 2015, p. 58. Available online: https://dspconnect.s3.amazonaws.com/Ivoclar/Adhesive/AdheseUniversalScientificDocumentation.pdf (accessed on 20 March 2020).

26. Technical Report, Estelite® Flow Quick and Estelite® Flow Quick High Flow, Tokuyama Dental Corporation Inc.: Japan. p. 23. Available online: http://www.tokuyama-dental.com/tdc/pdf/technicalreport/EFQ_and_EFQ_HF_TechnicalReport.pdf (accessed on 13 February 2020).

27. Safty, S. Elastic and Viscoelastic Properties of Resin Composites at the Macroscopic and Nano Scales. Ph.D. Thesis, University of Manchester, Manchester, UK, 2012.

28. Hirayama, S.; Iwai, H.; Tanimoto, Y. Mechanical evaluation of five flowable resin composites by the dynamic micro-indentation method. *J. Dent. Biomech.* **2014**, *5*, 1–8. [CrossRef] [PubMed]

29. Anastasova, R.; Dikova, T.; Panov, V. In Vitro study of dental composite roughness and microleakage of repaired obturations by various techniques. *J. IMAB Annu. Proc. Sci. Pap.* **2019**, *25*, 2419–2425. [CrossRef]

30. Anusavice, K.J.; Shen, C.; Rawls, H.R. *Phillips' Science of Dental Materials*, 12th ed.; Elsevier Health Sciences, Elsevier Saunders: St. Louis, MO, USA, 2012; pp. 275–307.

31. Van Meerbeek, B.; Willems, G.; Celis, J.P.; Roos, J.R.; Braem, M.; Lambrechts, P.; Vanherle, G. Assessment by nano-indentation of the hardness and elasticity of the resin-dentin bonding area. *J. Dent. Res.* **1993**, *72*, 1434–1442. [CrossRef]

32. Senawongse, P.; Pongprueksa, P.; Tagami, J. The effect of the elastic modulus of low-viscosity resins on the microleakage of Class V resin composite restorations under occlusal loading. *Dent. Mater. J.* **2010**, *29*, 324–329. [CrossRef]

Article

New Hybrid Bioactive Composites for Bone Substitution

Anna Ślósarczyk, Joanna Czechowska *, Ewelina Cichoń and Aneta Zima *

Faculty of Material Science and Ceramics, AGH University of Science and Technology, Al. Mickiewicza 30, 30-059 Krakow, Poland; aslosar@agh.edu.pl (A.Ś.); ecichon@agh.edu.pl (E.C.)

* Correspondence: jczech@agh.edu.pl (J.C.); azima@agh.edu.pl (A.Z.)

Received: 10 February 2020; Accepted: 7 March 2020; Published: 12 March 2020

Abstract: Recently, intensive efforts have been undertaken to find new, superior biomaterial solutions in the field of hybrid inorganic–organic materials. In our studies, biomicroconcretes containing hydroxyapatite (HAp)–chitosan (CTS) granules dispersed in an α tricalcium phospahate (αTCP) matrix were investigated. The influence of CTS content and the size of granules on the physicochemical properties of final bone implant materials (setting time, porosity, mechanical strength, and phase composition) were evaluated. The obtained materials were found to be promising bone substitutes for use in non-load bearing applications.

Keywords: hybrid materials; chitosan; hydroxyapatite

---

1. Introduction

Development of new bioactive, hybrid-type biomaterials may limit the application of auto- and allografts, which are nowadays commonly used in bone substitution. Hybrid materials are composed of at least two chemically bonded constituents, usually inorganic and organic, interpenetrating at the submicron level [1]. The above hybrids may provide numerous favorable physicochemical and biological properties necessary for bone tissue regeneration. These types of biomaterials, if supplemented with bone morphogenetic proteins or mesenchymal stem cells, are called biohybrids [2].

Among polymeric components of hybrid materials for medicine, chitosan (CTS), which belongs to the group of polysaccharides, deserves special attention. It is characterized by biocompatibility and fast biodegradability as well as hemostatic, antibacterial, and fungicidal properties. It also accelerates the wound healing process [3]. The combination of chitosan with calcium phosphate based bioceramics (hydroxyapatite—HAp, tricalcium phosphate—TCP) is an interesting material solution in biomaterials engineering [4–8]. Hybrids composed of calcium phosphate bone cement and chitosan belong to this group of biomaterials. The new HAp-based, self-hardening composites with high strength and resorbability may be promising for moderate stress-bearing craniofacial, dental and orthopedic repairs [9,10]. A HAp–CTS composite can serve as a delivery system for drugs and biologically active agents, including human bone morphogenetic proteins (e.g., rhBMP-2) [7,11]. The combination of bioresorbable polymer and biocompatible, bioactive ceramic material can mimic the natural function of bone. Chemical interactions between CTS and HAp generate a possibility to form coordination bonds between the –NH$_2$ of chitosan and the Ca^{2+} of hydroxyapatite. Despite many composite materials having been recently developed, to date, only a few fulfil the medical requirements. One of the reasons for this is that producing a favorable microstructure with uniformly dispersed inorganic and organic constituents still remains a challenge. Furthermore, materials often do not fulfill the requirements regarding mechanical properties, which strongly depend on the material's composition (e.g., HAp:CTS ratio) and method of manufacture [12,13].

Previous studies have considered less complicated HAp–CTS composites for the reconstruction and regeneration of bone. Different methods may be applied to the manufacture of such HAp–CTS composite implants in the form of blocks, granules, membranes, coatings, and laminates of varying porosity and pore size distribution [5,12,14]. Nowadays, scientists are focused on more sophisticated materials containing HAp–CTS hybrids. It should be noted that the existing knowledge about HAp–CTS composites is largely unordered and sometimes controversial. Biomicroconcretes based on well-established calcium phosphate bioceramics (CaPs) and calcium phosphate bone cements (CPCs), including α-TCP based biomaterials, which have been applied in medicine for years, seem to be one of the most innovative solutions in this field. Composites consisting of a CPC matrix and HAp as well as BCP (biphasic calcium phosphate) granules were obtained and examined in one study [15,16]. Furthermore, biomicroconcretes composed of α-TCP based bone cement and biodegradable calcium sulphate granules were investigated [17]. The factors influencing the properties of hybrid-type biomaterials, including biomicroconcretes, have yet not been clearly identified.

The most promising form of the potential implant materials may be biomicroconcretes containing porous hydroxyapatite–chitosan (HAp–CTS) granules (acting as the aggregate) and α-tricalcium phosphate (α-TCP)-based bone cement as a binding agent. This more complex system was the subject of our research. The combination of hybrid granules and cement in one material is supposed to ensure the handiness of the implant preparation and its good adaptation to the size and shape of a bone defect. The aim of our studies was to determine the influence of the size of hybrid HAp–CTS granules and the content of chitosan within them on the physicochemical properties of final bone implant materials, such as setting time, microstructure, porosity, mechanical strength, and phase composition.

2. Materials and Methods

2.1. Preparation of the Materials

α-Tricalcium phosphate (α-TCP) powder was synthesized by the wet chemical method [18]. Hydroxyapatite (HAp)–chitosan (CTS) hybrid materials were obtained by a method described previously [9]. The following substrates were used: Ca(OH)$_2$ ($\geq$99.5 wt.%, Merck), H$_3$PO$_4$ (85.0 wt.%, POCH), and medium-molecular-weight chitosan ($\sim$100,000 kDa, DD $\geq$ 75.0 wt.%, Sigma-Aldrich, NJ, USA). HAp–CTS hybrids were applied in the form of granules, playing the role of aggregates in the biomicroconcrete. Three different sizes of granules were prepared: 0.3–0.6 mm; 0.6–0.8 mm; 0.8–1.0 mm, with 17 wt. % of chitosan (Series A) and 23 wt. % of chitosan (Series B). The biomicroconcretes were prepared by mixing the solid phase, i.e., HAp–CTS granules and αTCP powder in a 2:3 weight ratio (ball mill MM400, Retsch, Haan, Germany). The liquid phase (1.0 wt. % of chitosan solution in 0.5 wt. % of acetic acid solution, POCH, Gliwice, Poland) was then introduced into the powder phase (Figure 1).

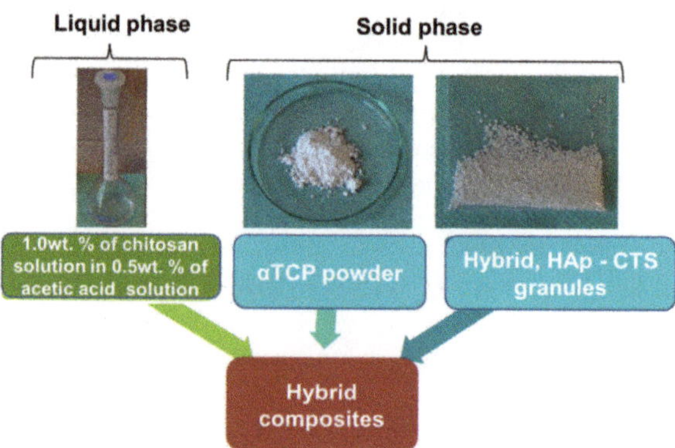

Figure 1. Schematic representation of the content of hybrid biomicroconcrete-type composites.

The liquid to powder ratio (L/P) that produced the appropriate consistency of pastes was established to be between 0.52–0.56 g/g (Series A) and 0.56–0.62 g/g (Series B) (Table 1).

Table 1. The composition of solid and liquid phases of biomicroconcretes.

Material	α-TCP: Hybrid Granules	Chitosan Content in Hybrid Granules (wt. %)	Size of Granules (mm)	Liquid Phase Composition	L/P (g/g)
Series A	3:2	17	0.3–0.6 0.6–0.8 0.8–1.0	1.0 wt. % solution of chitosan in 0.5 wt. % solution of acetic acid	0.56 0.52 0.52
Series B	3:2	23	0.3–0.6 0.6–0.8 0.8–1.0	1.0 wt. % solution of chitosan in 0.5 wt. % solution of acetic acid	0.62 0.56 0.56

The procedure applied during the preparation of biomicroconcretes is shown in Figure 2.

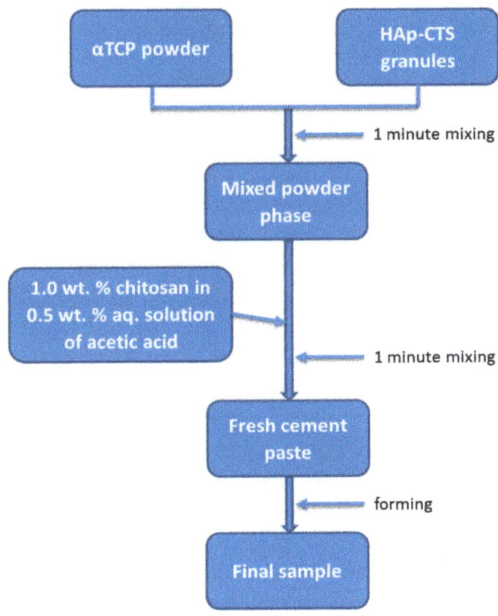

Figure 2. The flow chart of the preparation procedure.

2.2. Phase Composition

Phase composition was examined via the X-ray diffraction method (XRD). The crystalline phases of initial α-TCP powder and set biomicroconcretes were analyzed via X-ray diffraction (D-2 Phaser, Bruker, Karlsruhe, Niemcy), within the 2θ range of 10°–90° (scanning speed of 1 °/min). Before measurements were obtained, materials were crushed and sieved to below 63 μm. The crystalline phases were identified by comparing the experimental X-ray diffractograms with the PDF (powder diffraction file) of the ICDD (International Center of Diffraction Data). Phase quantification was done using the Rietveld method (Diffrac Suite Topas Rietveld Analysis Software, Bruker, Karlsruhe, Germany, 2011). Quality fits were done by refining the zero offset, scale factors, and lattice parameters and applying a pseudo-Voigt peak shape model. All measurements were made in triplicate.

2.3. Setting Times

The setting times (initial and final) were measured according to the ASTM C266-08 standard using Gillmore Needles (Syl&Ant Instruments, Poniszowice, Poland) [19]. Results are presented as the average value of six measurements ± standard deviation (SD).

2.4. Microstructure

The microstructures of developed materials was studied using a scanning electron microscope (Nova NanoSem 200, JEOL JSM 5400, Fei Europe Company, Eindhoven, The Netherlands). All samples were coated with a thin layer of carbon and then imaged using SEM. Chemical compositions in microareas were determined via energy-dispersive X-ray spectroscopy (EDX, Oxford Instruments, Oxford, UK). The accelerating voltage was equal to 18kV.

2.5. Porosity

Mercury intrusion porosimetry was applied to measure the open porosity and pore size distribution of set and hardened biomicroconcretes (AutoPore IV 9500 Porosimeter, Micromeritics, Norcross, GA, USA). The examined samples were dried before being introduced into the penetrometer of the apparatus. All measurements were made in triplicate.

2.6. Compressive Strength

Cylindrical samples of 6 mm in diameter and 12 mm height were tested using a universal testing machine Instron 3345, with a crosshead displacement rate equal to 1.0 mm min^{-1}. The compressive strength parameters were calculated by averaging 10 measurements and are presented as the mean ± standard deviation (SD).

2.7. Statistics

Data points in graph figures represent the mean, with error bars representing the standard deviation (SD). Comparative studies were performed using one-way ANOVA. Furthermore, Tukey's HSD post hoc tests were used.

3. Results and Discussion

A new generation of biomaterials combining hybrid hydroxyapatite–chitosan granules with α-TCP based bone cement was obtained. α-TCP is a degradable, self-setting phase and acted as a binder for hybrid granules. X-ray examinations of the synthesized α-TCP powder were carried out. The initial α-TCP powder was composed mainly of α-TCP phase (91 ± 2 wt. %) and a small amount of hydroxyapatite (9 ± 2 wt. %). Our previous studies showed that the hybrid HAp–CTS granules were composed of hydroxyapatite as the only crystalline phase [9]. In the case of biomicroconcretes, 7 days after setting and hardening, two crystalline phases, i.e., α-TCP and hydroxyapatite, were detected (Figure 3).

The quantitative analysis of phase composition showed that the amount of individual phases changed after setting and hardening (Table 2).

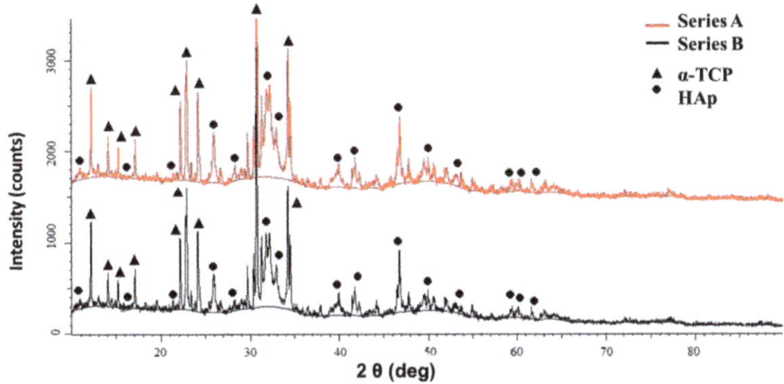

Figure 3. XRD patterns of the biomicroconcretes.

Table 2. The phase composition of biomicroconcretes (7 days after setting and hardening).

Size of Granules (mm)	Series A		Series B	
	α-TCP (wt. %)	HAp (wt. %)	α-TCP (wt. %)	HAp (wt. %)
0.3–0.6	31.7 ± 4.8	68.3 ± 4.8	49.5 ± 6.1	50.5 ± 6.0
0.6–0.8	39.7 ± 5.3	60.3 ± 5.3	51.9 ± 6.1	48.1 ± 6.1
0.8–1.0	29.9 ± 5.9	70.1 ± 5.9	47.9 ± 6.9	51.9 ± 6.6

The hydroxyapatite contained in the material comes from: (1) hybrid granules (HAp–CTS) and (2) hydrolysis of α-TCP. An increase in hydroxyapatite content and a decrease in α-TCP content, in comparison to the composition of initial powder batches (Table 1), were noticed in both series (Table 2). This indicated a high reactivity of α-TCP powder, which to a large extent hydrolyzed to hydroxyapatite (Equation (1)). This reaction plays a key role in the setting process of calcium phosphate bone cements [6,17,20].

$$3\ \alpha\text{-}Ca_3(PO_4)_2 + H_2O \rightarrow Ca_9(HPO_4)\ (PO_4)_5(OH) \tag{1}$$

$$\alpha\text{-}TCP + H_2O \rightarrow \text{calcium-deficient hydroxyapatite} \tag{2}$$

In vivo, the two-phasic α-TCP/hydroxyapatite matrix is supposed to be selectively resorbed and replaced by newly formed bone tissue. The quantitative phase composition differed for materials from Series A and B. A higher amount of hydroxyapatite was detected in the case of composites belonging to Series A (60–70 wt. %) in comparison to Series B (48–52 wt. %). These results indicate that hybrid granules HAp–CTS inhibited the hydrolysis of α-TCP to HAp. The process was more intensive for granules with a higher content of chitosan. It was probably due to the water uptake by chitosan present in the granules. Water is necessary for α-TCP hydrolysis (Equation (1)), and a lack of an appropriate amount of H_2O can inhibit the process. The absorption of water molecules by chitosan-containing composites was observed inter alia by Correlo et al. [21].

Developed biomicroconcretes, regardless of granule size, possessed initial setting times between 7 and 10 min (Table 3). The final setting times of biomaterials from both series were longer than recommended in the literature (<15 min).

Table 3. The setting times of the biomicroconcretes.

Size of Granules (mm)	Series A		Series B	
	Initial t_I (min)	Final t_F (min)	Initial t_I (min)	Final t_F (min)
0.3–0.6	9 ± 1	27 ± 1	8 ± 1	29 ± 1
0.6–0.8	8 ± 1	23 ± 1	7 ± 1	29 ± 1
0.8–1.0	9 ± 1	28 ± 1	10 ± 1	37 ± 1

Materials containing medium granules (0.6–0.8mm) and small granules (0.3–0.6) revealed final setting times between 23–29 min. These biomicroconcretes set in a period of time that was acceptable from a medical point of view, and would allow the surgeon to place material in a bone void and adjust its shape to the shape of the defect. The influence of granule size and the content of chitosan in hybrid granules was observed. The granules with larger size (0.8–1.0 mm) and higher chitosan content negatively delayed the final setting time, up to 37 min. A significant shortening of the setting times of biomicroconcretes in comparison to cement based on α-TCP without granules was noticed. According to Czechowska et al. [6], α-TCP-based bone cement possesses an initial setting time equal to 7 ± 1 min and a final setting time equal to 41 ± 3 min (L/P = 0.48 g/g). Thus, it can be concluded that the presence of hybrid granules generally reduces the final setting times of composites in comparison to a single component, α-TCP-based biomaterial. The observed shortening of the setting times may have been connected to the presence of hydroxyapatite, introduced with hybrid granules (HAp–CTS). Precipitated hydroxyapatite is often used as a crystallization nucleus, accelerating the setting process of calcium phosphate bone cements [22]. Overlong setting times may result in a continuation of the setting process after wound closure. Considering the requirements for cement-type bone substitutes (initial setting time: 3–8 min and final setting time: <15 min) [23], the results obtained for Series A, with lower chitosan content in hybrid HAp–CTS (17 wt. %), were found to be the most favorable.

SEM observations revealed that hybrid HAp–CTS granules possessed a compact, homogenous microstructure with the polymeric bridges visible on their surfaces. The presence of polymer bridges allowed granules to be easily distinguished from the calcium phosphate matrix. The granules were evenly distributed in the biomicroconcretes. The presence of chitosan ensured good cohesion of granules and final materials, preventing their disintegration. Good adhesion between the matrix and hybrid granules was observed (Figure 4).

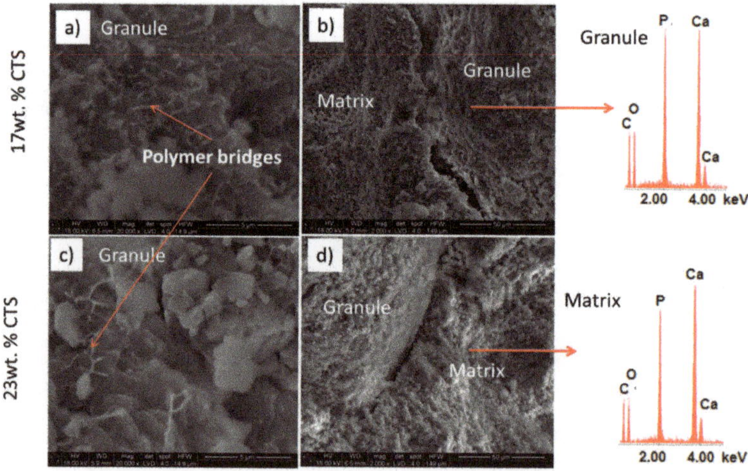

Figure 4. SEM microphotographs of cross-sections of materials from Series A (**a,b**) and B (**c,d**), and results of EDS analysis.

The total open porosity of the samples from both series was similar and equal to 45 ± 5 vol. % (Series A) and 50 ± 5 vol. % (Series B). A multimodal pore size distribution was obtained for all biomicroconcrestes, with a pore size distribution in the range of 0.005–40 µm (Figure 5).

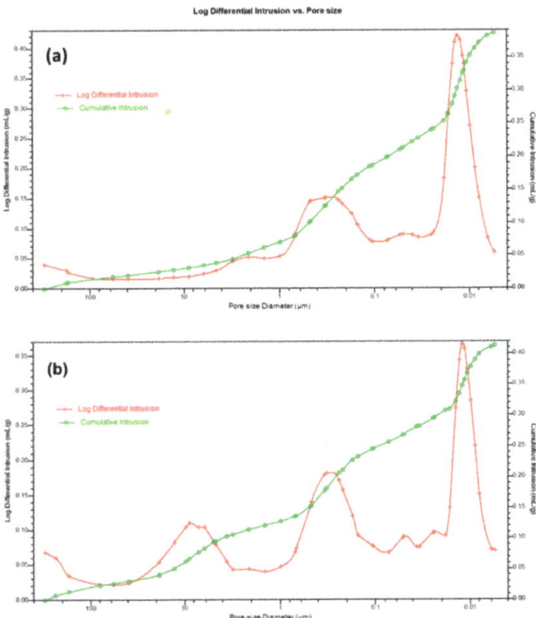

Figure 5. Pore size distribution of biomicroconcretes of materials from Series A (**a**) and Series B (**b**).

In the case of composites with lower chitosan content (17 wt. %—Series A), pores between 0.005 and 4 µm were visible. The peaks corresponding to lower pore size values were connected with the existence of pores located between CaP grains and agglomerates in the cement matrix [6]. Peaks occurring above 1 µm were connected with the presence of hybrid granules in the material. Biomicroconcretes with higher chitosan content (23 wt. %—Series B) possessed pores with diameters ranging from 0.005 µm up to ~40 µm. Thus, a higher amount of chitosan in hybrid granules contributed to the formation of pores of a larger diameter, which, in combination with higher porosity, caused a significant reduction of the mechanical strength of biomicroconcretes for this series. This was confirmed by the outcomes of the compressive strength measurements (Figure 6).

The results obtained for biomicroconcretes containing granules with 23 wt. % chitosan (Series B, 2.1 ± 0.4 MPa–2.7 ± 0.4 MPa) were even two or three times lower compared to the materials composed of granules with 17 wt. % content of chitosan (Series A). Composites from Series A possessed mechanical strength ranging from 5.4 ± 0.8 MPa to 6.2 ± 1.0 MPa. These values are similar to compressive strength of trabecular bone (4–12 MPa) [24]. The obtained results proved that the use of granules with a larger content of chitosan adversely affected the mechanical strength of the final biomicroconcretes. In addition, it was demonstrated that the size of granules only slightly affected the mechanical properties of HAp–CTS–α-TCP composites.

The developed biomaterials combining hybrid hydroxyapatite–chitosan granules and α-TCP-based bone cement, with their phase composition, microstructure, and surgical handiness, will probably meet the requirements of regenerative medicine [6,25].

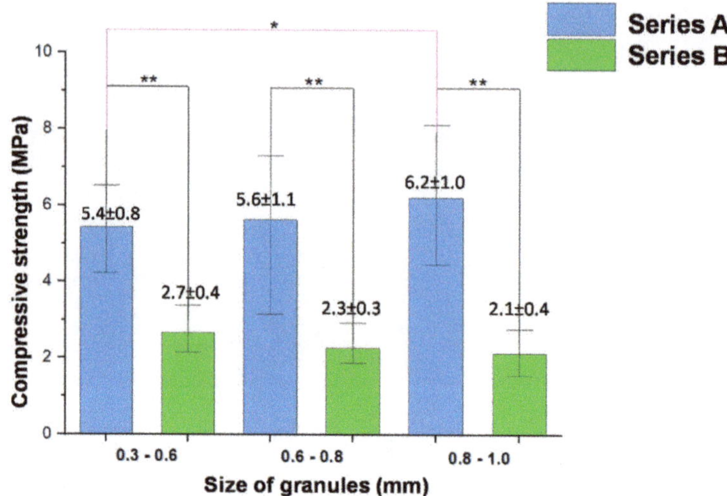

Figure 6. Compressive strength of biomicroconcretes after 7 days of setting and hardening in the air (statistically significant difference: * $p < 0.05$, ** $p < 0.0001$).

4. Summary

New microconcrete-type implant materials based on α-TCP–HAp–chitosan were obtained. The influence of chitosan content and the size of the hybrid HAp–CTS granules on the physicochemical properties of final potential bone substitutes were investigated. The obtained biomaterials, after 7 days of setting and hardening in the air, were composed of two crystalline phases: HAp (~60–70 wt. %) and α-TCP (~30–40 wt. %). Regardless of the chitosan content and the size of granules, a significant decrease in the amount of α-TCP phase compared to the initial composition was observed. This indicated the high reactivity of α-TCP powder, which hydrolyzed to non-stoichiometric hydroxyapatite (CDHAp). Our research revealed that generally, the presence of granules caused a shortening of setting times of composites, in comparison to the single component, α-TCP-based material. Biomicroconcretes containing smaller granules with 17 wt. % of chitosan possessed satisfying setting times, similar to those recommended in the literature. Such values allow for the preparation of implant material in operating room conditions and a relatively simple procedure for its introduction into the bone defect. Higher sizes of granules adversely extended the final setting times of materials. The developed biomaterials possessed high open porosity equal to ~45–50 vol. % and multimodal pore size distribution in the range of 0.005 to 40 μm. SEM observations confirmed good adhesion between matrix and hybrid granules. The compressive strength (2–6 MPa) of the obtained complex composites was close to the values corresponding to spongy bone. The amount of chitosan in hybrid HAp–CTS granules influenced the mechanical strength of studied materials. Higher compressive strength was noticed for biomicroconcretes containing granules with a lower amount of chitosan (17 wt. %). The influence of granule size in the range of 0.3 to 1.0 mm on the mechanical strength of the studied biomicroconcretes was not studied. The promising physicochemical properties of developed biomicroconcretes and their potential as novel bone substitutes need to be confirmed in biological studies.

Author Contributions: A.Ś., A.Z., J.C. conceived and planned the experiments. A.Ś. supervised the project. A.Z., J.C., E.C. carried out the experiments. All authors have read and agreed to the published version of the manuscript.

Funding: This research was supported by the National Science Centre, Poland Grant No. 2017/27/B/ST8/01173.

Conflicts of Interest: The authors declare no conflict of interest.

References

1. Kickelbick, G. Hybrid materials-past, present and future. *Hybrid Mater.* **2014**, *1*, 39–51. [CrossRef]
2. Muzzarelli, R. Chitosan composites with inorganics, morphogenetic proteins and stem cells, for bone regeneration. *Carbohydr. Polym.* **2011**, *83*, 1433–1445. [CrossRef]
3. Dorozhkin, S.V. Medical Application of Calcium Orthophosphate Bioceramics. *BIO* **2011**, *1*, 1–51. [CrossRef]
4. Venkatesan, J.; Kim, S.K. Chitosan composites for bone tissue engineering—An overview. *Mar. Drugs* **2010**, *8*, 2252–2266. [CrossRef] [PubMed]
5. Peniche, C.; Solís, Y.; Davidenko, N.; García, R. Chitosan/hydroxyapatite-based composites. *Biotecnol. Apl.* **2010**, *27*, 202–210.
6. Czechowska, J.; Zima, A.; Paszkiewicz, Z.; Lis, J.; Ślósarczyk, A. Physicochemical properties and biomimetic behaviour of α-TCP-chitosan based materials. *Ceram. Int.* **2014**, *40*, 5523–5532. [CrossRef]
7. Lai, Y.L.; Cheng, Y.M.; Yen, S.K. Doxorubicin-chitosan-hydroxyapatite composite coatings on titanium alloy for localized cancer therapy. *Mater. Sci. Eng. C* **2019**, *104*, 109953. [CrossRef]
8. Cichoń, E.; Ślósarczyk, A.; Zima, A. Influence of selected surfactants on physicochemical properties of calcium phosphate bone cements. *Langmuir* **2019**, *35*, 13656–13662. [CrossRef]
9. Zima, A. Hydroxyapatite-chitosan based bioactive hybrid biomaterials with improved mechanical strength. *Spectrochim. Acta Part A* **2018**, *193*, 175–184. [CrossRef]
10. Sun, L.; Xu, H.H.K.; Takagi, S.; Chow, I.C. Fast setting calcium phosphate cement -chitosan composite mechanical properites and sdissolution rates. *J. Biomater. Appl.* **2007**, *21*, 299–315. [CrossRef]
11. Kashiwazaki, H.; Yamaguchi, K.; Harada, N.; Akazawa, T.; Murata, M.; Iizuka, T.; Ikoma, T.; Tanaka, J.; Inoue, N. In vivo evaluation of a novel chitosan/HAp composite biomaterial as a carrier of rhBMP-2. *J. Hard Tissue Biol.* **2010**, *19*, 181–186. [CrossRef]
12. Li, H.; Zhou, C.R.; Zhu, M.Y.; Tian, J.H.; Rong, J.H. Preparation and characterization of homogeneous hydroxyapatite/chitosan composite scaffolds via in-situ hydration. *J. Biomater. Nanobiotechnol.* **2010**, *1*, 42. [CrossRef]
13. Im, K.H.; Park, J.H.; Kim, K.N.; Kim, K.M.; Choi, S.H.; Kim, C.K.; Lee, Y.K. Organic-Inorganic hybrids of hydroxyapatite with chitosan. In *Key Engineering Materials*; Trans Tech Publications Ltd.: Baech, Switzerland, 2005; Volume 284, pp. 729–732.
14. Reves, B.T.; Jennings, J.A.; Bumgardner, J.D.; Haggard, W.O. Preparation and functional assessment of composite chitosan-nano-hydroxyapatite scaffolds for bone regeneration. *J. Funct. Biomater.* **2012**, *3*, 114–130. [CrossRef] [PubMed]
15. Son, Y.J.; Chung, T.J.; Oh, K.S. Brushite Bone Cement Prepared from Granular Hydroxyapatite and β-Tricalcium Phosphate. In *Key Engineering Materials*; Trans Tech Publications Ltd.: Baech, Switzerland, 2017; Volume 720, pp. 153–156.
16. Lee, G.H.; Makkar, P.; Paul, K.; Lee, B. Incorporation of BMP-2 loaded collagen conjugated BCP granules in calcium phosphate cement based injectable bone substitutes for improved bone regeneration. *Mater. Sci. Eng. C* **2017**, *77*, 713–724. [CrossRef] [PubMed]
17. Czechowska, J.; Zima, A.; Ślósarczyk, A. Comparative study on physicochemical properties of alpha-TCP/calcium sulphate dihydrate biomicroconcretes containing chitosan, sodium alginate or methylcellulose. *Acta Bioeng. Biomech.* **2020**, in press. [CrossRef]
18. Kolmas, J.; Kaflak, A.; Zima, A.; Ślósarczyk, A. Alpha-tricalcium phosphate synthesized by two different routes: Structural and spectroscopic characterization. *Ceram. Int.* **2015**, *41*, 5727–5733. [CrossRef]
19. Standard test method for time setting of hydraulic-cement paste by gillmore needles, 19428–2959. In *ASTM Annual Book of Standards*; ASTM C266-15; American Society for Testing and Materials: West Conshohocken, PA, USA, 2015.
20. Weichhold, J.; Gbureck, U.; Goetz-Neunhoeffer, F.; Hurle, K. Setting mechanism of a CDHA forming α-TCP cement modified with sodium phytate for improved injectability. *Materials* **2019**, *12*, 2098. [CrossRef]
21. Correlo, V.M.; Pinho, E.D.; Pashkuleva, I.; Bhattacharya, M.; Neves, N.M.; Reis, R.L. Water absorption and degradation characteristics of chitosan-based polyesters and hydroxyapatite composites. *Macromol. Biosci.* **2007**, *7*, 354–363. [CrossRef]

22. Fernandez, E.; Ginebra, M.P.; Boltong, M.G.; Driessens, F.C.M.; Planell, J.A.; Ginebra, J.; De Maeyer, E.A.P.; Verbeeck, R.M.H. Kinetic study of the setting reaction of a calcium phosphate bone cement. *J. Biomed. Mater. Res.* **1996**, *32*, 367–374. [CrossRef]

23. Ginebra, M.P.; Fernández, E.; Boltong, M.G.; Bermúdez, O.; Planell, J.A.; Driessens, F.C.M. Compliance of an apatitic calcium phosphate cement with the short-term clinical requirements in bone surgery, orthopaedics and dentistry. *Clin. Mater.* **1994**, *17*, 99–104. [CrossRef]

24. Seidi, A.; Ramalingam, M.; Elloumi-Hannachi, I.; Ostrovidov, S.; Khademhosseini, A. Gradient biomaterials for soft-to-hard interface tissue engineering. *Acta Biomater.* **2011**, *7*, 1441–1451. [CrossRef] [PubMed]

25. Pilia, M.; Guda, T.; Appleford, M. Development of composite scaffolds for load-bearing segmental bone defects. *BioMed Res. Int.* **2013**, *2013*, 458253. [CrossRef] [PubMed]

Review

A Comparison of Bioactive Glass Scaffolds Fabricated by Robocasting from Powders Made by Sol–Gel and Melt-Quenching Methods

Basam A. E. Ben-Arfa * and Robert C. Pullar *

Department of Materials and Ceramic Engineering/CICECO—Aveiro Institute of Materials, University of Aveiro, 3810-193 Aveiro, Portugal
* Correspondence: basam@ua.pt (B.A.E.B.-A.); rpullar@ua.pt (R.C.P.)

Received: 26 February 2020; Accepted: 16 May 2020; Published: 21 May 2020

Abstract: Bioactive glass scaffolds are used in bone and tissue biomedical implants, and there is great interest in their fabrication by additive manufacturing/3D printing techniques, such as robocasting. Scaffolds need to be macroporous with voids ≥ 100 μm to allow cell growth and vascularization, biocompatible and bioactive, with mechanical properties matching the host tissue (cancellous bone for bone implants), and able to dissolve/resorb over time. Most bioactive glasses are based on silica to form the glass network, with calcium and phosphorous content for new bone growth, and a glass modifier such as sodium, the best known being 45S5 Bioglass®. 45S5 scaffolds were first robocast in 2013 from melt-quenched glass powder. Sol–gel-synthesized bioactive glasses have potential advantages over melt-produced glasses (e.g., greater porosity and bioactivity), but until recently were never robocast as scaffolds, due to inherent problems, until 2019 when high-silica-content sol–gel bioactive glasses (HSSGG) were robocast for the first time. In this review, we look at the sintering, porosity, bioactivity, biocompatibility, and mechanical properties of robocast sol–gel bioactive glass scaffolds and compare them to the reported results for robocast melt-quench-synthesized 45S5 Bioglass® scaffolds. The discussion includes formulation of the printing paste/ink and the effects of variations in scaffold morphology and inorganic additives/dopants.

Keywords: robocasting; bioactive glass; scaffold; sol–gel; 45S5 Bioglass®; biomaterials; biomedical implants

---

1. Introduction

Robocasting is an additive manufacturing or 3D printing technique, in which designed 3D structures are built layer-by-layer by extruding a continuous filament of a paste/printing ink from a nozzle, guided by an automated computerized system (Figure 1) [1]. It was initially developed at Sandia National Laboratories in 1996 to fabricate free-formed objects with a slurry containing low levels of binder [2,3], and was rapidly adopted to manufacture periodic 3D ceramic structures [4]. Robocasting is a particularly promising technique for bone scaffold fabrication as the scaffold can be designed to specifically mimic and replace an injured or missing part of the body as a "bespoke" implant [5], with improved mechanical properties compared to older fabrication techniques [6]. In robocasting, the ink is extruded from a nozzle to construct 3D structures. The extrusion speed and scaffold shape and size are all controlled by a computer-aided design–computer-aided manufacturing (CAD–CAM) [7] model, deposited layer-by-layer as a 3D structure [8] from an ink or paste formulated from a powder with liquid and polymeric additives.

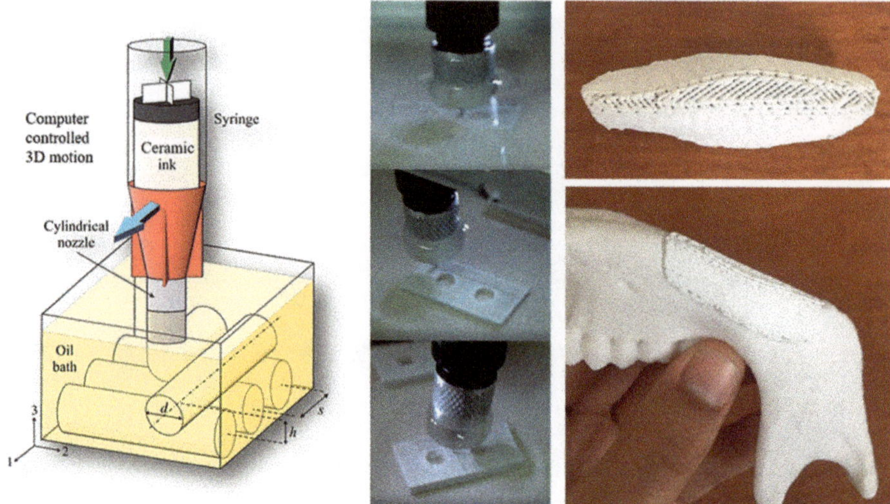

Figure 1. Schematic illustration and in-situ images of the robocasting process within an oil bath, and an example of a sintered scaffold for human mandibular defect reconstruction. [https://www.euroceram. org/en/technologies/material-extrusion/robocasting-direct-ink-writing.html, accessed on January 2020].

Robocasting has been used to produce porous scaffolds of calcium phosphate bioceramics for biomedical bone implants since the beginning of the 21st century [9,10], but SiO_2-CaO-P_2O_5-based bioactive glasses have only been robocast in the last decade [11–15]. The best known example of a bioactive glass is Bioglass® (45S5) [16], discovered by Hench and co-workers in 1969 [17]. 45S5 is known by its commercial brands Bioglass®, PerioGlas® and NovaBone® in particulate form, and NovaMin® in toothpaste, and it contains 45 wt% SiO_2, 24.5 wt% CaO, 24.5 wt% Na_2O, and 6.0 wt% P_2O_5, with a Ca/P ratio of ~5 [17]. This high sodium content (as a glass modifier) means it is soft in comparison to other glasses and can be easily machined or ground to a powder [18], but it has to be stored in a dry environment as it readily absorbs and reacts with moisture. It has also proven difficult to fabricate as porous 3D scaffolds and sinter into a dense material, as it crystallizes at a relatively low temperature, with a small sintering window [19]. Indeed, it was not robocast successfully until 2013 [11].

An alternative preparation method to melt-quenching, which has been greatly explored for the synthesis of bioactive glasses, is sol–gel [20]. Apart from the potential to avoid unwanted crystallization during synthesis, one of the great advantages of sol–gel bioactive glasses is that the glass powder can be much more porous than that produced from a melt. This should result in a greater degree of bioactivity in any resulting implant. For example, the ternary 58S glass (containing 60 wt% SiO_2, 36 wt% CaO, and 4 wt% P_2O_5), when produced by the sol–gel process, had a much greater specific surface area than melt-derived 45S5 (79 vs. 2.7 m^2/g, respectively) [20]. Hench used sol–gel about two decades after its initial discovery to prepare 45S5 glass, which enhanced both its surface morphology and bioactivity [21,22], but this sol–gel 45S5 has never been successfully robocast to date. This is because sol–gel-derived bioactive glasses have proven problematic for robocasting, as the high specific surface area and porous structure of the glass powder results in absorption of a significant portion of the dispersing liquid from the suspension, impeding paste/ink formulation and greatly reducing printability. The first robocast sol–gel-synthesized bioactive glass scaffolds were only reported in 2019 [23,24]. These were of a quaternary high-silica sol–gel glass (HSSGG) with a low sodium composition of 64.4 wt% SiO_2, 4.9 wt% Na_2O, 21.5 wt% CaO, and 9.1 wt% P_2O_5 [24]. This glass was synthesized by an innovative rapid sol–gel synthesis route, which produced a bioactive glass powder in about 1 h—more than one hundred times faster than the quickest reported standard sol–gel methods, which require lengthy drying and ageing processes of several days to weeks [25].

2. Bone Scaffolds

Scaffolds for bone regeneration must contain interconnected pores/voids on the scale of 100 µm or more to enable cell seeding, proliferation, and vascularization (formation of blood vessels), which results in the formation of new bone [26]. The scaffold must be strong enough to replace the bone temporarily while providing a suitable substrate for cell attachment and proliferation, and it should also slowly degrade (be resorbable) and be replaced as the new tissue grows [27]. Therefore, it must be made of a bioactive and biocompatible material, which allows precipitation/crystallization of the calcium phosphate minerals (typically hydroxyapatite, $Ca_{10}(PO_4)_6(OH)_2$, HAp), which enhances the bonding between the implant and the bone, ensuring good fixation and preventing micro-movements and implant failure, and it cannot be cytotoxic to the cells [28,29].

Scaffolds are usually formed in what is known as a "wood pile" or "log-cabin" structure of extruded struts, with voids between them (Figure 2). This forms an extracellular matrix in which the cells can grow and function, and this also helps define the structural properties of the resultant tissue. The scaffold serves as a temporary template to support/guide the growth of new tissue to restore normal function, but during the healing process, the scaffold should also gradually degrade to be replaced by new cells as they grow and proliferate to regenerate new tissue [30].

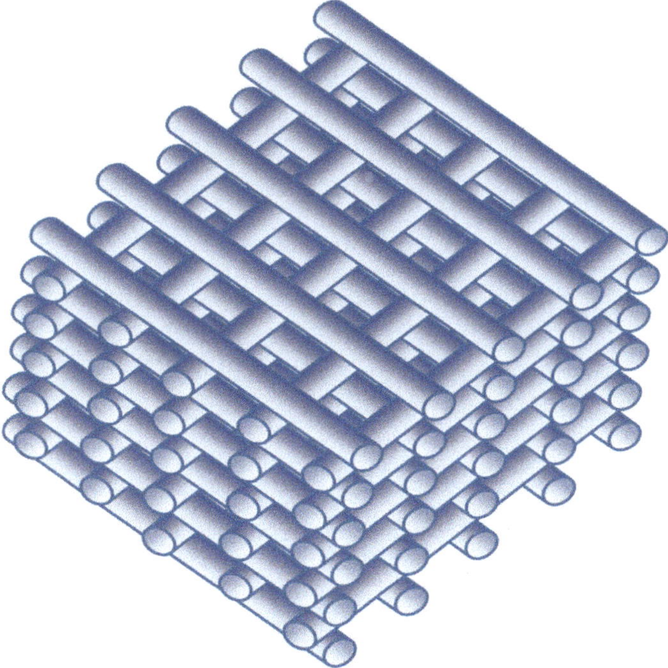

Figure 2. The typical "wood pile" or "log-cabin" scaffold structure.

As the 3D shapes are fabricated by the extrusion of a paste from a nozzle layer-by-layer, this paste, sometimes also called a printing ink, must have optimized rheological properties such as a high viscosity at zero shear and thinning behavior at applied shear. It must also have excellent shape retention capabilities, such as high elastic modulus and yield stress, after deposition and throughout the printing, drying, and sintering steps [31]. The extruded filament must retain its shape and resist the exerted loads of the subsequent layers printed above it, and casting (extrusion) can either take place in air or in a liquid medium to allow homogeneous drying. In the case of a robocast glass, the green scaffold must be dried at room temperature and then heated or sintered gradually to eliminate any organic additives to produce the final form.

To function well, a scaffold should have:

1. Biocompatibility: It must be non-cytotoxic and tissue-compatible, allowing the necessary cells to attach, differentiate, and proliferate. Any dissolved materials from the scaffold must not have any toxic/inflaμμatory influence on cells and tissues, and the body should be able to get rid of these degraded products without any side-effects.
2. Bioactivity: It must actively interact with body fluids and tissues to function, for example, crystallize new calcium phosphate minerals for bone growth. A scaffold can also be used to influence cell morphology and position, deliver growth factors to speed regeneration, or to release or trap biomolecules in a regulated manner.
3. Biodegradability: It must degrade with time (weeks/months) after implantation as it is gradually replaced by new generative tissues. It should degrade at a rate similar to that of the new tissue growth.
4. Suitable morphology and structure: It must have sufficient void/pore volume for new tissue formation and growth, needing large interconnected macropores >50–100 μm to permit cells to infiltrate easily into the scaffold core, and allow any waste to diffuse out of it. Ideally, it should also have a hierarchical degree of porosity within the scaffold struts in the size range of 2–50 μm (mesopores) to facilitate cell anchorage and proliferation, permit vascularisation, and enable nutrient/metabolite transport without compromising mechanical stability.
5. Similar mechanical properties to the host tissue: It must withstand the exerted loads, but with no stress shielding effect from a mechanical mismatch, which can cause bone resorption and implant failure. Cells have shown an ability to be sensitive to the stiffness of the scaffold, which can affect cell adhesion [32] and differentiation [33], and the scaffold needs to maintain overall mechanical integrity during the new bone formation and resorption processes [27,34].

3. Bioactive Glass

When implanted in the body, bioactive glasses such as 45S5 react with body fluid, exchanging the Na^+ and Ca^{2+} ions on their surface with H^+ ions in the fluid to hydrolyze and form Si-OH bonds, accompanied by an initial increase in pH. This creates more OH^- ions at the surface, which go on to break Si-O-Si bonds (forming more Si-OH bonds), dissolving the silica glass network. The Si-OH groups at the surface then recondense and re-polymerize, reforming new Si-O-Si bonds as a new silica–gel layer on the surface, which will now have a very low Na^+ content. At the same time, Ca^{2+} and PO_4^{3-} ions from both the bioactive glass and body fluid form a calcium phosphate layer on top of the silica layer. This also incorporates CO_3^{2-} and OH^- ions from the body fluid, crystallizing a hydroxycarbonate apatite $(Ca_{10-x-y/2}(HPO_4)_x(PO_4)_{6-x-y}(CO_3)_y(OH)_{2-x}$, HCAp) layer, which, in a few hours, begins to convert to crystalline HAp, forming bone–like material [35]. Over the following weeks, M2 macrophages (which promote wound healing) attract mesenchymal stem cells and osteoprogenitor cells, which, in turn, differentiate to become bone tissue cells such as osteoblasts. Other bone components such as collagen also form, and they combine with HAp nanocrystals to form new bone, as the silica bioactive glass implant degrades, dissolves, and resorbs [36].

3.1. Bioactive Glass Synthesis Methods

Bioactive glasses are mainly produced by two techniques: Sol–gel [25,37–39] and melt-quenching methods [39–41]. Hench was a pioneer in using both of these techniques for bioactive glass synthesis. He initially used the melt-quenching technique in 1969 [17] to prepare 45S5 and used sol–gel about twenty years later to prepare the same glass, but enhancing the surface morphology and bioactivity [22]. Both methods are now commonly adopted for producing bioactive glasses.

3.1.1. Melt-Quenching Method

In the melt-quenching method, glass is obtained by mixing, homogenizing, calcining, and fusing glass precursors such as oxides or carbonates above 1300–1500 °C. After heating the mixed precursors in a platinum crucible, the melt is quenched (rapidly cooled to avoid crystallization) in cold water to obtain a glass frit, or cast/quenched in graphite molds to obtain bulk glasses [19,40,41]. This is the method typically used to obtain Bioglass® 45S5 and other commercial bioactive glasses.

3.1.2. Sol–Gel Method

In the sol–gel method, an aqueous solution of the precursors (alkoxides and/or metal salts) is stirred for hydrolysis and condensation, resulting in a clear sol. This is then dried and aged for a long period of time, often weeks [25], followed by calcination to obtain glass granules or monolithic shapes [22,42]. Sol–gel-derived bioactive glasses have high porosity in comparison to those made from melt-quenching, which generally improves their bioactivity. The minimum ideal total porosity for good bone growth in a scaffold is generally accepted to be ~50% [43]. The long aging and drying periods can be problematic and lead to unwanted crystallization, which needs to be minimized, and this long process is very much a rate limiting step for the sol–gel method.

3.1.3. Rapid Sol–Gel Method

For the reasons given above, Ben-Arfa et al. developed an innovative rapid sol–gel synthesis route for a quaternary bioactive glass system, which can produce a bioactive glass powder in about 1 h—more than one hundred times faster than the quickest reported standard sol–gel methods [25]. This involves the sequential combination of the bioactive glass precursors in an aqueous solution, followed by rapid drying on a rotary evaporator, avoiding any ageing (Figure 3). This was shown to produce stabilized (devitrified) bioactive glasses with virtually identical properties, and enhanced bioactivity, compared to those made by conventional routes.

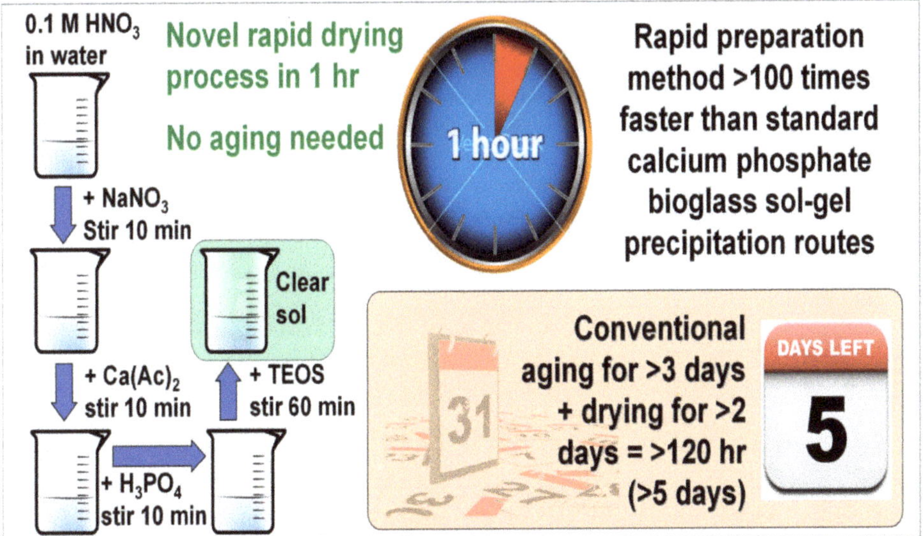

Figure 3. A depiction of the rapid sol–gel bioactive glass synthesis route developed by Ben-Arfa et al. [25].

4. Robocasting and Printing Paste/Ink Preparation

Many materials have been used to robocast scaffolds including calcium phosphate bioceramics [43–45] and polymer/ceramic composites [46,47]. However, the robocasting of bioactive glass scaffolds proved to be more difficult, even when using melt-quenched glass frit powders, which themselves exhibit less porosity than sol–gel bioactive glasses. The high porosity of sol–gel-derived bioactive glass powders created even greater difficulties in the paste preparation process, as a significant part of the dispersing liquid is absorbed and trapped in the pores, reducing the flow and stability of the paste/ink. Due to these issues, there have been far fewer reports on the robocasting of bioactive glass scaffolds compared to bioceramics, and this was limited entirely to glass compositions prepared by melt-quenching [48,49], until the successful robocasting of sol–gel-derived bioactive glass scaffolds was reported for the first time in 2019 by Ben-Arfa et al. [23,24].

Essential for the successful fabrication of scaffolds by robocasting are factors such as the simplicity of ink preparation, printability, and the ability of the glass composition to sinter [15]. The well-known 45S5 Bioglass® does not meet several of these requirements, as the high degree of leaching of Na^+ and the abrupt increase in pH hinders an efficient dispersion, and the successful preparation of robocasting pastes of melt-quenched 45S5 glasses was only quite recently achieved [15,48]. Much of the work on the robocasting of glass scaffolds was carried out using less soluble and low-alkali glasses with broader sintering windows, such as the 13–93 composition (53 wt% SiO_2, 6 wt% Na_2O, 12 wt% K_2O, 5 wt% MgO, 20 wt% CaO, 4 wt% P_2O_5) [49], but any gains in sinterability and mechanical properties were achieved at the expense of poorer bioactivity [8]. Indeed, despite over 40 years of research on bioactive glasses by many groups since Hench's invention of 45S5, no other bioactive glass composition has been found to have significantly better biological properties than the original Bioglass® 45S5 composition [19].

The formulation of inks/pastes for robocasting from bioactive sol–gel glasses is even more difficult due to their higher specific surface area and more porous structure, which, while improving bioactivity, has until recently constituted the main barrier to using them in robocasting.

A high-quality scaffold must have well-defined and interconnected pores/voids to allow good vascularization, and it must be mechanically stable and fully capable of supporting its own weight during robocasting, drying, sintering, and use without evident shape deformation or the overlapping of layers, but with strong adhesion between adjacent layers. The strength of cancellous bone (also known as trabecular bone, the spongy interior part) is between 2 and 12 MPa [50], so a bone scaffold should ideally be in this range as well. This is important, as one of the major causes of postoperative bone implant failure is a stiffness mismatch between the implanted biomaterial and surrounding bone tissue. 45S5 robocast scaffolds are typically within this range, but the mechanical properties of the denser low-sodium-glass robocast scaffolds such as 13–93 are too high for compatibility with cancellous bone. 13–93 scaffolds have been measured as having a compressive strength between 48 and 142 MPa [8,12,14], in the region of cortical (hard exterior) bone (100–150 MPa) [50], which is less useful for implants and much greater than the values for cancellous bone, resulting in a mismatch. For these reasons, this review will focus on comparing 45S5 glass robocast scaffolds with sol–gel glass robocast scaffolds.

Throughout the rest of this review, we will use the word "voids" to describe the macropores between the struts from robocasting to avoid confusion with any microporosity within the struts themselves.

4.1. Powder Milling

Milling, comminution, and pulverization are interchangeable phrases for the mechanical reduction in particle size of the glass powder. Powder milling is a very important part of the successful formulation of the paste in robocasting, particularly for bioactive glasses for robocasting, which justifies renewed investigations of the powder milling process. The powder can be subjected to wet or dry milling, although wet milling usually proves more effective and is recommended when agglomeration between particles is induced due to their high surface energy [51]. The decrease in particle size taking place during milling results from accumulated stresses induced in the particles from the applied mechanical energy. This causes cracks that propagate and break the particle [52].

Various important parameters need to be taken into account when performing wet milling, such as solids loading (liquid-to-powder ratio (LPR)), the balls-to-powder ratio (BPR) [53], and the speed and type of milling machine [54]. Elevated heat treatment temperatures (HTT) can result in the formation of hard agglomerates that are difficult to destroy with milling. The individual and mutual influences of HTT and BPR on the wet-milling performance and morphology of the powders produced have been studied and well documented [55].

Although powder milling is of key importance in the formulation of the printing ink, few details are usually given of the milling conditions/parameters in articles published on the robocasting of bioactive glasses produced by melt-quenching, often just stating the mill type/model and milling time. Some research groups employed exaggerated milling times of the glass continuously for several days, neglecting the effect of contamination from the milling balls and milling jars, as in the case of wet milling an alkali-free bioactive glass for 2 h in a planetary mill, then 6 days in a ball mill, followed by attrition milling for time periods from 2 and 12 h (158 h in total, in ethanol) [56]. Such a long milling time will almost certainly lead to contamination of the glass, and involves high costs of energy and time. On the other hand, it was demonstrated by Ben Arfa et al. [31,55] that a simple, cheap rapid milling machine was sufficient to mill glass granules for robocasting in 2 h, and that highly expensive and sophisticated milling machines (attrition mills, planetary mills) are not necessary. These authors also used ethanol as the milling medium and, employing a BPR of 10 and LPR of 1.5, reduced the milling time further to just 1 h to prepare a sol–gel glass robocasting paste [23,24].

A detailed experimental procedure for different milling conditions for 13–93 glass was fully described in [8], and these authors also compared the effects of different milling media (water/ethanol) on the subsequent mechanical and biomineralization performance of the robocast 13–93 scaffolds [57]. The results revealed that wet milling using ethanol led to scaffolds with higher bioactivity and better mechanical properties than using water as milling media. The effects of wet-milling parameters and calcination temperature on the final powder morphology of the HSSGG sol–gel glass were also studied in detail [55], and it was found that BPR has an important influence on the final particle size distribution of the powders, as shown in Figure 4, with higher BPR values generally resulting in a smaller particle size and lower size distribution. A narrow particle size distribution is generally preferable in ink formulation to achieve a good dispersion, as, with large distributions, fine particles will be dispersed while the coarser ones will tend to sediment.

On the other hand, pore volume and powder morphology were found to be mostly determined by variations in calcination temperature (Figure 5) [55]. These are important parameters, as an increase in pore volume will lead to the absorption of a larger fraction of the dispersion solution, effectively "freezing" the motion of particles and their flow. The availability of open, small-volume pores is a good attribute for the easier preparation of a glass suspension suitable for robocasting. As Figure 5 shows, heating sol–gel HSSGG to 800 °C was sufficient to obtain a glass powder with pore morphology and reduced the pore volume suitable for ink preparation for robocasting.

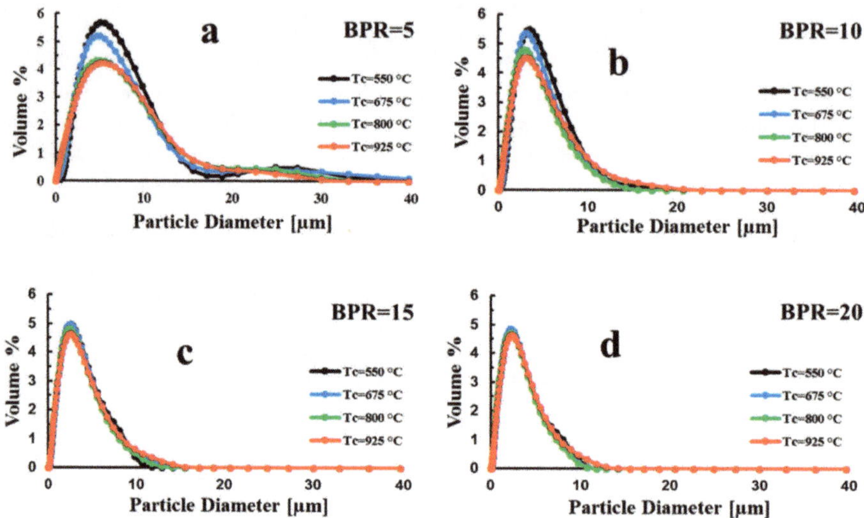

Figure 4. Particle size distributions of the bioactive sol–gel glass powders heat-treated at different temperatures followed by wet-ball milling in ethanol under given values of balls-to-powder ratio (BPR): (**a**) 5; (**b**) 10; (**c**) 15; (**d**) 20. Reproduced with permission from [55], Copyright Elsevier Ltd. and Techna Group S.r.l., 2018.

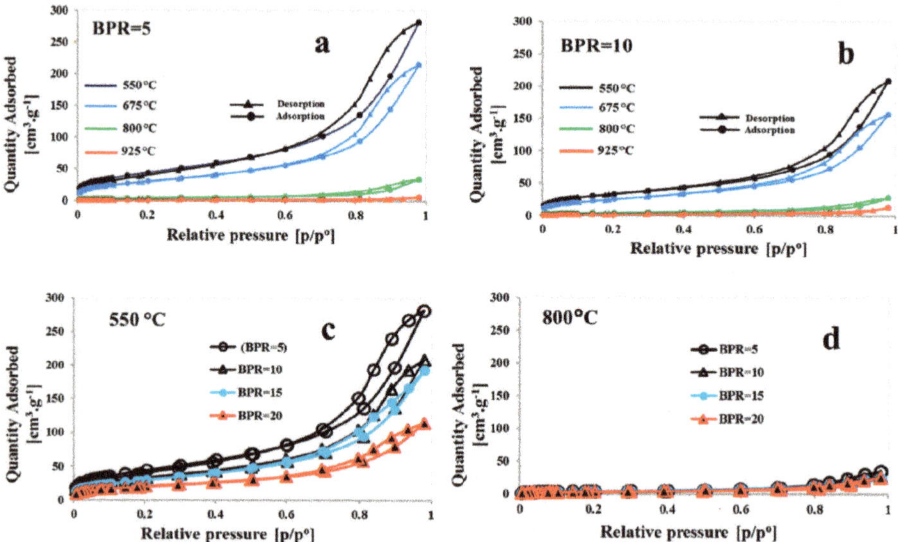

Figure 5. Effects of the processing parameters on the nitrogen sorption isotherms for the bioactive sol–gel glass powders: (**a**,**b**) Comparison of two BPRs (5 and 10) used in wet-ball milling in ethanol for samples heat-treated at different temperatures (550, 675, 800, 925 °C); (**c**,**d**) comparison of two heat treatment temperatures for powders wet-ball milling in ethanol under different BPR values (5, 10, 15, 20). Reproduced with permission from [55], Copyright Elsevier Ltd. and Techna Group S.r.l., 2018.

4.2. Paste Formulation and Preparation

High-solids loading pastes are recommended in order to reduce the shrinkage of the final sintered scaffold, and values of up to 45 vol% solids loading have been achieved in robocast bioactive glass pastes. Additives are also an important constituent in the recipe of paste formulation. However, it is preferred to use a very small amount of additive in paste preparation to facilitate their elimination in the post-treatment process. The additives play two significant roles in paste formulation, providing pseudoplastic behavior that promotes paste flowability, and aiding the shape retention and providing resistance to the loads of subsequent layers during and after robocasting. The paste must not contain air bubbles or agglomerations of powder, as the air will result in an irregular, discontinuous extrusion, whereas agglomerates tend to block the nozzle and prevent extrusion. Moreover, the paste must exhibit relatively high viscosity at zero shear stress, and thinning behavior at high shear rates, to ensure the extrusion of the paste from the nozzle [58].

Solids loading is the fractional volume of the solids phase to the total volume of powder and liquid forming the paste [59], the liquid being removed during the drying process. A high solids loading significantly increases the viscosity of the paste, as well as the mechanical properties of the green body and the final sintered scaffold [60]. The glass synthesis method used does not present a great change in the maximum solids loading, with a narrow variation in range of between 35 and 45 vol% for 45S5 [15,48] and HSSGG robocast glasses [23,24], reflecting the need for a reasonably high solids loading value for a successful robocasting glass paste/ink.

In general, the paste is formulated from a mixture of glass powder, water, and additives such as dispersants, thickeners (binder), and coagulants. Dispersants are polymers used for steric stabilization (surface active agents—surfactants). They are added to aqueous suspensions to be adsorbed on the powder particle surfaces to overcome the Van der Waals interparticle attractive forces and prevent aggregation, enabling the printability of the paste by robocasting. Binders/thickeners/plasticizers are polymers with high molecular weight and are added to increase the viscosity of the suspension to manifest flow behavior, e.g., hydroxyl propyl methylcellulose. Coagulants are also types of polymers that act as physical glue, providing resistance to the external forces exerted on the extruded struts due to the load of the subsequent layer above. Such mixtures were used in the early robocasting of bioactive glasses, but these days, a single additive is mostly used, such as carboxymethyl cellulose (CMC) [23,24,48] or Pluronic F–127 [8,12], as a combined dispersant, plasticizer, and coagulant. The use of a single additive enables a fast and easy process, especially in the case of CMC, which can be used at room temperature, whereas Pluronic F–127 needs a non-ambient temperature to prepare the suspension.

An example of the effects of BPR and calcination temperature of the glass powder on the rheological properties of a robocasting paste, in this case, for sol–gel HSSGG printing inks with 25 and 40 vol% solids loading, is shown in Figure 6 [31]. Ideally, both a reasonably high viscosity at zero shear stress followed by a gradual decrease in the viscosity (shear thinning behavior) with increased shear rate, and a stable elastic modulus of ~1 MPa throughout the viscoelastic region, are required features for an ink suitable for robocasting.

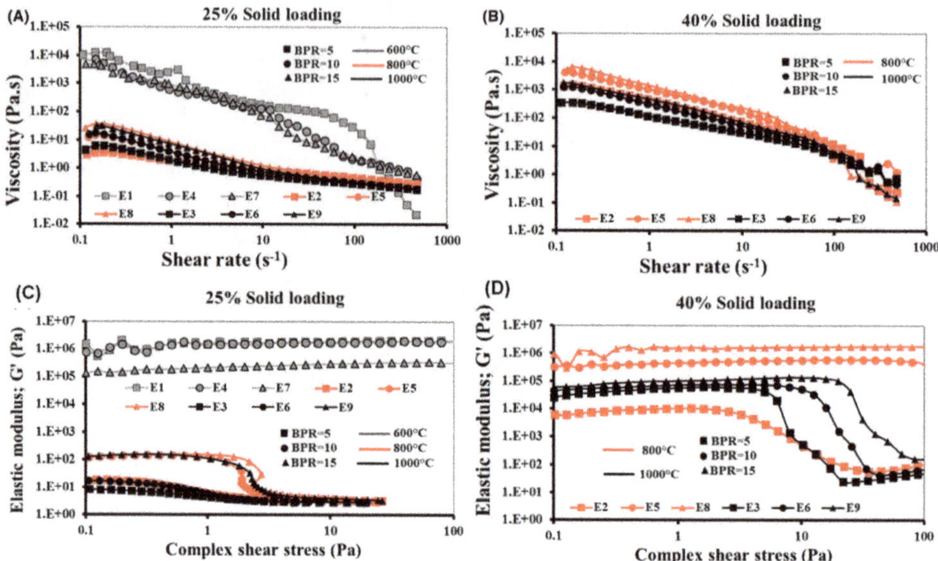

Figure 6. Effects of the processing parameters on the rheological properties of the bioactive glass suspensions. The viscosity behavior of 25% SL (**A**) and 40% SL (**B**) and the viscoelastic properties of the paste at 25% SL (**C**) and 40% SL (**D**). Reproduced with permission from [31], Copyright the American Ceramic Society, 2018.

5. Glass-Melt-Synthesized Bioglass® Robocast Scaffolds

45S5 was first robocast in 2013 by Eqtesadi et al. [15] with a solids loading of 45 vol%, and sintered at 1000 °C/1 h. The scaffolds had ~200 μm struts and ~100 × 250 μm rectangular voids, and were robocast in a variety of geometries and up to 60 layers deep. They had a total porosity of 63% (from an initial macroporosity of 52%) and compressive strength of 13 MPa, stated to be vastly superior to previously reported values for 45S5 scaffolds made by lithography, although they were less porous than scaffolds made by foam replication.

A very detailed paper on the sintering of 45S5 robocast scaffolds from a melt-derived glass was published by Eqtesadi et al. in 2014 [48]. TGA/DTA measurements on the as-fabricated scaffold showed a weight loss of ~10% between 200 and 400 °C due to water and surface OH groups being lost, followed by CMC degradation at 225 and 350 °C. They observed a second phase of weight loss of ~4% between 700 and 800 °C attributed to the loss of carbonaceous remains of the CMC, and isotropic shrinkage of ~5% between 200 and 400 °C and ~7% between 800 and 900 °C, reaching a total of ~18% shrinkage at 1000 °C (Figure 7b). They also stated that there was an equal degree of shrinkage in both the struts and the voids. A glass transition was observed at 560 °C by DTA, followed by primary crystallization at 600 °C, a second glass transition at 910 °C, and secondary crystallization at 850 °C, with the glass melting at 1100 °C, matching findings on bulk 45S5 glass [61]. A very minor Na_2CO_3 phase was detected by XRD in the otherwise amorphous glass, and this persisted to 550 °C, beyond which $Na_2CaSi_2O_6$ crystallized as the major phase at 600 °C. A secondary phase of $Na_2Ca_4(PO_4)_2Si_2O_4$ crystallized at 850 °C, with both $Na_2CaSi_2O_6$ and $Na_2Ca_4(PO_4)_2Si_2O_4$, then remaining at 1050 °C.

The 45S5 scaffold sintered at 1000 °C is shown in Figure 7a. Upon heating to 500–1100 °C/1 h to study their sintering behavior, the initial strut (400 μm) and void (200 μm) diameters began to significantly shrink above 500 °C, decreasing 300 and 150 μm, respectively, at ~1000 °C (Figure 7b) [48]. This printing paste had a solids loading of 45 vol% with an initial porosity of ~80%, 50% being macroporosity from the voids and 30% being microporosity in the struts. This remained constant to

500 °C, until a small initial loss of porosity occurred between 500 and 550 °C (the glass transition) followed by a continuous loss of porosity from ~70% at 800 °C to ~60% at 1000 °C (Figure 7c). The bulk of this porosity was lost within the struts, and not in the macroporosity of the scaffold framework. Indeed, the surface of the struts appeared sealed at 1000 °C as the intergranular phase was lost, and the 10% microporosity remaining in the struts was in the form of closed internal pores. This 80% and 70% total porosity at 550 and 800 °C, respectively, was less than those compared to 45S5 scaffolds made from other methods, but was still enough to allow bone regeneration and vascularization. The poor sinterability of 45S5 is demonstrated by the 60% porosity remaining at 1000 °C. The compressive strength of the sintered scaffolds was ~2–5 MPa at 600 °C, ~5 MPa at 800 °C, and ~10 MPa at 1000 °C (Figure 7d), within the range for cancellous bone, the authors claiming this to be up to 4000% greater than values seen in 45S5 scaffolds made by foam replication and additive lithography methods.

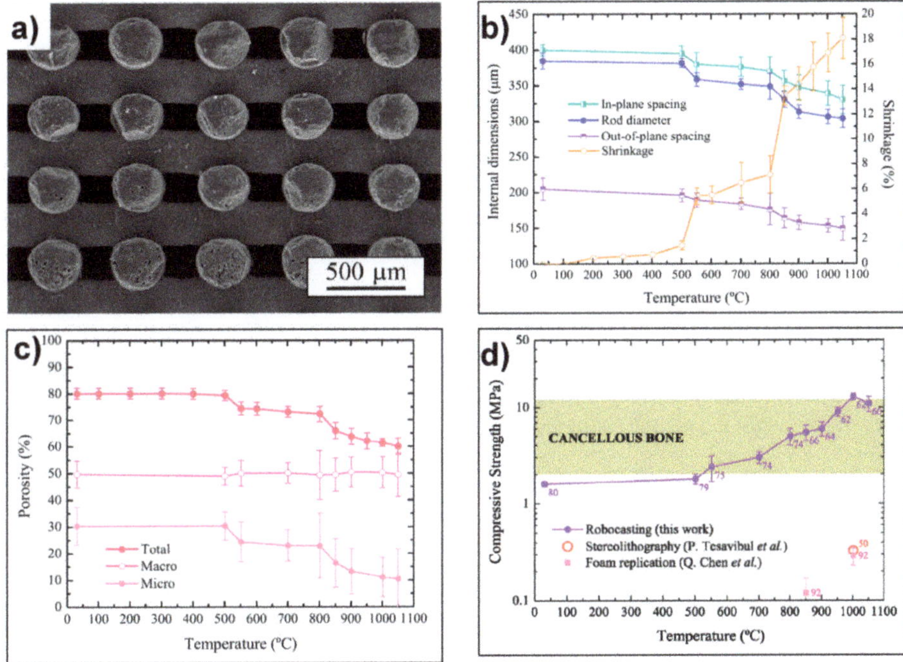

Figure 7. (**a**) Robocast 45S5 scaffold sintered at 1000 °C/1 h. (**b**) Shrinkage of the 45S5 scaffold during sintering. (**c**) Change in porosity of the scaffold with sintering temperature. (**d**) Variation in compressive strength with sintering temperature, and comparison to other 45S5 scaffolds made by alternative additive manufacturing methods. Reproduced with permission from [48], Copyright Elsevier Ltd. 2014.

Robocast 45S5 bioactive glass scaffolds were also reported in another paper by Eqtesadi et al., this time with a total of 35 vol% solids in the printing paste [62]. However, these were sintered under a flowing Ar atmosphere in a graphite furnace, first heated to 400 °C/1 h, and then to 550 or 1000 °C/1 h. Tetragonal and cylindrical lattice scaffolds with up to 50 layers high were created, and after sintering, the strut diameter was ~400 μm with rectangular ~100–150 × 250 μm voids. A wide endotherm up to 200 °C was seen as free water and surface OH groups were lost, with an exothermic peak at 225 °C as the CMC was removed in TGA/DTA measurements. The bioactive glass crystallized between 600 and 650 °C with a total weight loss of ~10% at 800 °C. Sintering in Ar had the effect of improving the density of the scaffolds, being 4%–5% more than those sintered in air. There was less bubbling and lower in-strut microporosity due to a decrease in the glass's viscosity in the Ar atmosphere, but some pores were still observed as a result of the glass bubbling.

XRD measurements of these amorphous Bioglass® scaffolds sintered in Ar at 550 °C indicated crystalline $Na_2Ca(CO_3)_2 \cdot 2H_2O$, originating from the hydration of $Na_2Ca(CO_3)_2$ crystals formed during heating in the graphite furnace and that became hydrated on standing at ambient temperature and humidity. Upon sintering at 1000 °C in Ar, $Na_2CaSi_2O_6$ was the major phase and $Na_2Ca_4(PO_4)_2Si_2O_4$ the minor phase, due to the decomposition of $Na_2Ca(CO_3)_2$. When sintered in air at 1000 °C, $Na_2CaSi_2O_6$ and $Na_2Ca_4(PO_4)_2Si_2O_4$ were both observed, without previous $Na_2Ca(CO_3)_2$ formation. On sintering in Ar, the relative density of the scaffolds was ~31% at 550 °C and ~42.5% at 1000 °C. Their compressive strength was 2.5 MPa after heating at 550 °C in Ar, at the bottom of the cancellous bone to range, and this increased to ~12 MPa when heated to 1000 °C in Ar.

It should be noted that in this same paper [62], Eqtesadi et al. also produced composite scaffolds based on 45S5 glass mixed with reduced graphene oxide additives. As these are not pure bioactive glass scaffolds, exhibited no significant improvements in mechanical properties over other 45S5 scaffolds, and had no reported data for biocompatibility or bioactivity, we have not added them to the discussion here.

6. Sol–Gel-Synthesized Bioactive Glass Robocast Scaffolds

Due to its high sodium content, the excessive leaching of Na^+ from 45S5 printing pastes, accompanied by an increase in pH, causes problems in the creation of an efficient dispersion. Although this was partially overcome by the use of CMC as a single printing additive, the sintered 45S5 scaffolds still exhibited poor densification due to the narrow sintering window between the onset of crystallization and the first glass transition in 45S5.

For this reason, a high-silica, low-sodium sol–gel-based bioactive glass (HSSGG) was developed by Ben-Arfa et al. for robocasting. Despite the potential advantages of bioactive sol–gel glasses, until recently, there had been insurmountable problems in formulating robocasting inks. Their high specific surface area and porous structure, the very features that are advantageous in terms of bioactivity, were also responsible for up-taking a significant portion of the dispersing liquid, adversely affecting printability. The first robocast sol–gel-synthesized bioactive glass scaffolds were reported by the authors in 2019 [24].

As discussed above, an in-depth investigation to understand the combined effects of porosity and particle/agglomerate size on the rheological properties of the suspensions prepared from sol–gel-derived bioactive glass powders was required [31]. Powder milling was found to be a crucial step in creating a printable sol–gel ink, with control of factors such as rotation speed and BPR, allowing the milling time of the sol–gel bioactive glass powders to be reduced to only 1 h using a simple mechanical mill. This work enabled the prediction of sol–gel glass ink printability for a complex quaternary system, and resulted in the successful printing of a scaffold from the high-silica sol–gel glass (HSSGG, composition = 64.4 wt% SiO_2, 4.9 wt% Na_2O, 21.5 wt% CaO, and 9.1 wt% P_2O_5) [24]. This was robocast from a paste with 40 vol% solids loading and the scaffolds were sintered at 800 °C/2 h [24]. Three types of scaffold were produced with as-printed void widths of 300, 400, and 500 μm, and after sintering, the strut diameter was 383 μm. The cubic voids had shrunk to ~270, ~370, and ~460 μm wide after sintering (9.5% shrinkage). The sintered HSSGG scaffold with an original void size of 300 μm is shown in Figure 8a. The total porosity of this scaffold was 46.7%, close to the optimum 50% value, and much lower than that of 45S5 scaffolds sintered at equivalent temperatures. This was achievable because, coupled with the low shrinkage, the sol–gel-derived glass was highly microporous, with an estimated microporosity of ~36 vol% in the bioactive glass struts—this being a great advantage of the sol–gel process. The majority of these micropores were also shown to be open, not closed, which is relevant as open micropores play an important role in biological activity.

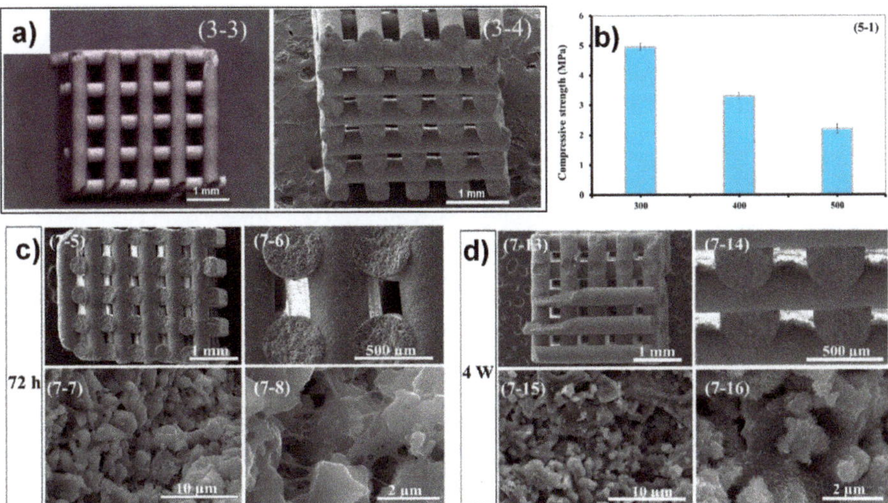

Figure 8. Robocast scaffolds of high-silica-content sol–gel bioactive glasses (HSSGG) sol–gel bioactive glass. (**a**) Optical and SEM images of scaffolds with 300 μm as-printed void size, after sintering at 800 °C/2 h. (**b**) Compressive strength of the scaffolds with different void sizes (300, 400, and 500 μm as printed) sintered at 800 °C/2 h. Structural and morphological features of the scaffolds with 300 μm printed void size after immersion in SBF for 72 h (**c**) and 4 weeks (**d**). Reproduced with permission from [24], Copyright Elsevier Ltd. 2018.

When heated to 550 °C, HSSGG was an amorphous but highly porous powder [37,38], this high degree of porosity preventing the preparation of printing pastes with over 25 vol% solids loading. To overcome this, the HSSGG powder was heat-treated at 800 °C (the same temperature was used to sinter the scaffolds), reducing the porosity of the glass powder and enabling a solids loading of 40 vol% in the paste, after milling of the powder. This higher heat treatment of the HSSGG powder resulted in the crystallization of three phases in the glass: Sodium calcium silicate, HAp, and cristobalite silica. This partial crystallization of the powder prior to the robocasting of the scaffold was also partly responsible for the high degree of microporosity remaining in the struts after sintering of the scaffold, as it limits the diffusion of the atoms.

The compressive strength of the HSSGG scaffolds is shown in Figure 8b, and the effects of void size can be clearly observed. It decreased from 5.0 MPa with a void size of 270 μm, to 3.3 and 2.2 MPa with voids of 370 and 460 μm, respectively, as the macroporosity of the scaffold structure increases (values not published). All of these compressive strength values are in the region of cancellous bone, despite the inherent microporosity of the HSSGG scaffolds. The biocompatibility of the scaffolds was tested, and they were shown not to be cytotoxic toward MG63 osteoblast cells, with no significant difference between the scaffolds with different void sizes. They were also submersed in SBF for up to 4 weeks to test their bioactivity, and showed evidence of nano-HAp crystallization on the surface after 72 h (Figure 8c), with web-like filaments forming, probably consisting of a mixture of silica gel and nanoHAp. XRD patterns indicated an increase in intensity of peaks for both cristobalite and HAp with immersion, and HAp formation was also observed in FT-IR spectra. After 4 weeks immersion, both nano-HAp platelets/needles and micron-scale HAp grains were observed (Figure 8d), along with some dissolution of the bulk scaffold structure being observed.

In order to enhance the mechanical properties of these scaffolds, HSSGG was doped with 5 mol% copper and lanthanum ions [23]. A previous sintering study utilizing Taguchi analysis had indicated that the addition of these ions could aid the sintering of HSSGG [63]. These scaffolds were also printed from pastes with 40 vol% solids loading, and sintered at 800 °C/2 h. The SEM images in

Figure 9a–d show that while the La-doped scaffold did not appear to be significantly different, the Cu-doped scaffold contained less microporosity and appeared more sintered. The total porosities of the 300 μm void scaffolds had decreased with doping, to 42.6% and 37.8% respectively for La and Cu. High-magnification SEM images showed that there was a small increase in the formation of necks between glass particles with La addition in comparison to the parent glass, and a much greater increase with Cu addition. This demonstrated that both additives behaved as sintering aids for the glass, but with a much greater effect observed with the lower-melting-point copper additive acting as a flux agent. Due to this, a small increase in the compressive strength (7%–18%) accompanied the addition of La (Figure 9e). However, a much greater improvement in the compressive strength was observed with Cu addition, it being up to 221% greater than the HSSGG scaffolds, with compressive strength values of up to 14 MPa for the scaffold with the smallest void dimensions (Figure 9e). As with the pure HSSGG glass, an increase in void size and macroporosity led to a decrease in compressive strength in all cases. This demonstrated that the compressive strength of the HSSGG scaffolds could be modified to cover the whole cancellous bone region by addition of copper as a sintering aid, without significantly impacting the printability of the ink. It was also demonstrated that the La^{3+} and Cu^{2+}-doped HSSGG were still bioactive and fully biocompatible with various cell lines, including osteoblasts [64].

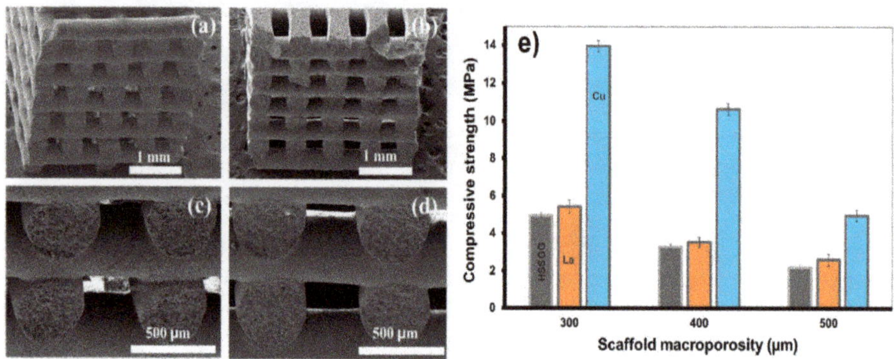

Figure 9. SEM images of robocast HSSGG sol–gel bioactive glass scaffolds, with 300 μm as-printed void size and after sintering at 800 °C/2 h, with addition of (**a,c**) 5 wt% Cu^{2+} and (**b,d**) 5 wt% La^{3+}. (**e**) Compressive strength of the scaffolds with different void sizes (300, 400, and 500 μm as printed) sintered at 800 °C/2 h. Reproduced with permission from [23], Copyright Acta Materialia Inc. 2019.

7. Comparison of Robocast Scaffolds from 45S5 Bioglass® and HSSGG Sol–Gel Bioactive Glass

The properties of the 45S5 Bioglass® robocast scaffolds and sol–gel HSSGG robocast scaffolds are compared in Table 1. These two glasses make for a good comparison as they have similar compressive strength in the unadulterated sintered scaffolds at similar temperatures, although the compressive strength of the HSSGG sol–gel glass could be improved with metallic dopants. It is immediately apparent that the range of compressive strength values covers the entire cancellous bone region (~2–12 MPa). They were also both produced with a single organic additive to the printing paste (CMC). The main difference in composition between the two glasses is in silica and sodium content. 45S5 is a high-sodium-content glass, with 45 wt% SiO_2 and 24.5 wt% Na_2O, whereas HSSGG contains 64.4 wt% SiO_2 and only 4.9 wt% Na_2O. Due to problems with sintering, 45S5 robocast scaffolds rarely achieve under 60% porosity, even when sintered at 1000 °C, and at the temperatures at which they remain amorphous (<600 °C), they can have up to 80% porosity. The lowest reported porosity is still a high 57.5%, and that was only attained when sintered in an Ar atmosphere. The sol–gel-synthesized HSSGG scaffolds had surprisingly low total porosity values, all <47%, but nevertheless contained a high level of internal microporosity, an advantage of the sol–gel method and ideal for bioactivity, and compressive strength was also in the cancellous bone region. Dopants, especially 5 wt% copper, could further reduce the total porosity by acting as a sintering aid and encouraging shrinkage of the scaffold skeleton, increasing compressive strength to the top end of the cancellous bone range in a scaffold still only sintered at 800 °C. To achieve similar values, 45S5 needs to be sintered at 1000 °C. It can also be observed in both 45S5 and HSSGG scaffolds that the void dimensions have a significant effect on compressive strength when other factors are equal, with smaller-diameter voids resulting in significantly higher values. This allows for the opportunity to tune the compressive strength by merely adjusting the CAD–CAM model for the robocasting scaffold design, without changing other parameters such as glass or printing paste composition, strut dimension, or sintering temperature.

Table 1. Comparison of sol–gel robocast HSSGG bioactive glass scaffolds and 45S5 Bioglass® scaffolds robocast from glass powders produced by melt-quenching.

Glass Synthesis	Glass Type	Sintering Temp. (°C)	Strut Diameter (µm)	Void Size (µm)	Porosity (%)	Compressive Strength (MPa)	Ref
Melt-quenching	45S5 Bioglass®	600	400	200	80	2.5	[48]
		800	360	180	72	5	
		1000	300	150	60	10	
		1000	200	100 × 250	63	13	[15]
		550 in Ar	~400	~100–150 × 250	69	2.5	[62]
		1000 in Ar	~400	~100–150 × 250	57.5	12	
Sol–gel	HSSGG	800	383	270	46.7	5.0	[24]
			383	370	-	3.3	
			383	460	-	2.2	
	HSSGG + 5 wt% La^{3+}	800	380	270	42.6	5.4	[23]
			380	370	-	3.5	
			380	460	-	2.6	
	HSSGG + 5 wt% Cu^{2+}	800	380	270	37.8	13.9	[23]
			380	370	-	10.6	
			380	460	-	4.9	

The development of the first robocastable sol–gel-synthesized glasses is an important step, and may lead to glass scaffolds with an improved, greater microporosity in the strut structure and enhanced bioactivity as a result. The rapid sol–gel synthesis route, which is 100 times quicker than the fastest conventional sol–gel ageing bioactive glass synthesis methods, has also attracted a great deal of interest, and a patent for this is currently being applied for, hopefully to develop and commercialize this technology further.

Author Contributions: Conceptualization, B.A.E.B.-A. and R.C.P.; writing—original draft preparation, B.A.E.B.-A. and R.C.P.; writing—review and editing, B.A.E.B.-A. and R.C.P.; supervision, R.C.P.; funding acquisition, R.C.P. All authors have read and agree to the published version of the manuscript.

Funding: R. C. Pullar was funded by the FCT (Fundação para a Ciência e a Tecnologia, Portugal) grant number IF/00681/2015. This research was funded by, and this work was developed within the scope of, the project CICECO–Aveiro Institute of Materials, UIDB/50011/2020 & UIDP/50011/2020, financed by national funds through the FCT/MEC and when appropriate co-financed by FEDER under the PT2020 Partnership Agreement.

Conflicts of Interest: The authors declare no conflict of interest. The funders had no role in the design of the study; in the collection, analyses, or interpretation of data; in the writing of the manuscript, or in the decision to publish the results.

References

1. Bandyopadhyay, A.; Bose, S. *Additive Manufacturing*; CRC Press: Baton Rouge, LA, USA, 2016.
2. Cesarano, J.; Calvert, P. Method for freeforming objects with low-binder slurry. U.S. Patent 6,027,326; application filed 28 October 1997, application granted 22 February 2000,
3. Cesarano, J.; Segalman, R.; Calvert, P. Robocasting Provides Moldless Fabrication from Slurry Deposition. *Ceram. Ind.* **1998**, *148*, 94–102.
4. Smay, J.E.; Iii, J.C.; Lewis, J.A. Colloidal Inks for Directed Assembly of 3-D Periodic Structures. *Langmuir* **2002**, *84*, 5429–5437. [CrossRef]
5. Cesarano, J.; Dellinger, J.G.; Saavedra, M.P.; Gill, D.D.; Jamison, R.D.; Grosser, B.A.; Sinn-Hanlon, J.M.; Goldwasser, M.S. Customization of Load-Bearing Hydroxyapatite Lattice Scaffolds. *Int. J. Appl. Ceram. Technol.* **2005**, *220*, 212–220. [CrossRef]
6. Entezari, A.; Zhang, Z.; Chen, J.; Li, Q. Optimization of bone tissue scaffolds fabricated by robocasting technique. In Proceedings of the 11th World Congress on Structural and Multidisciplinary Optimization (WCSMO-11), Sydney, Australia, 7–12 June 2015; pp. 989–994.
7. Cesarano, J., III. A review of robocasting technology. *Mater. Res. Soc.* **1999**, *542*, 133–139. [CrossRef]
8. Nommeots-Nomm, A.; Lee, P.D.; Jones, J.R. Direct ink writing of highly bioactive glasses. *J. Eur. Ceram. Soc.* **2018**, *38*, 837–844. [CrossRef]
9. Tay, B.Y.; Evans, J.R.G.; Edirisinghe, M.J.; Tay, B.Y.; Evans, J.R.G.; Edirisinghe, M.J. Solid freeform fabrication of ceramics Solid freeform fabrication of ceramics. *Int. Mater. Rev.* **2003**, *48*, 341–370. [CrossRef]
10. Miranda, P.; Saiz, E.; Gryn, K.; Tomsia, A.P. Sintering and robocasting of beta-tricalcium phosphate scaffolds for orthopaedic applications. *Acta Biomater.* **2006**, *2*, 457–466. [CrossRef]
11. Fu, Q.; Saiz, E.; Tomsia, A.P. Bioinspired Strong and Highly Porous Glass Scaffolds. *Adv. Funct. Mater.* **2011**, *21*, 1058–1063. [CrossRef]
12. Deliormanlı, A.M.; Rahaman, M.N. Direct-write assembly of silicate and borate bioactive glass scaffolds for bone repair. *J. Eur. Ceram. Soc.* **2012**, *32*, 3637–3646. [CrossRef]
13. Dorj, B.; Park, J.; Kim, H. Robocasting chitosan/nanobioactive glass dual-pore structured scaffolds for bone engineering. *Mater. Lett.* **2012**, *73*, 119–122. [CrossRef]
14. Motealleh, A.; Eqtesadi, S.; Perera, F.H.; Ortiz, A.L.; Miranda, P.; Pajares, A.; Wendelbo, R. Reinforcing 13—93 bioglass scaffolds fabricated by robocasting and pressureless spark plasma sintering with graphene oxide. *J. Mech. Behav. Biomed. Mater.* **2019**, *97*, 108–116. [CrossRef] [PubMed]
15. Eqtesadi, S.; Motealleh, A.; Miranda, P.; Lemos, A.; Rebelo, A.; Ferreira, J.M.F. A simple recipe for direct writing complex 45S5 Bioglass®3D scaffolds. *Mater. Lett.* **2013**, *93*, 68–71. [CrossRef]
16. Hench, L.L.; Roki, N.; Fenn, M.B. Bioactive glasses: Importance of structure and properties in bone regeneration. *J. Mol. Struct.* **2014**, *1073*, 24–30. [CrossRef]
17. Hench, L.L. The story of Bioglass. *J. Mater. Sci. Mater. Med.* **2006**, *17*, 967–978. [CrossRef]
18. Rahaman, M.N.; Day, D.E.; Bal, B.S.; Fu, Q.; Jung, S.B.; Bonewald, L.F.; Tomsia, A.P. Bioactive glass in tissue engineering. *Acta Biomater.* **2011**, *7*, 2355–2373. [CrossRef]
19. Jones, J.R. Review of bioactive glass: From Hench to hybrids. *Acta Biomater.* **2013**, *9*, 4457–4486. [CrossRef]
20. Zheng, K.; Boccaccini, A.R. Sol-gel processing of bioactive glass nanoparticles: A review. *Adv. Colloid Interface Sci.* **2017**, *249*, 363–373. [CrossRef]
21. Saravanapavan, P.; Jones, J.R.; Pryce, R.S.; Hench, L.L. Bioactivity of gel—Glass powders in the CaO-SiO_2 system: A comparison with ternary (CaO-P_2O_5-SiO_2) and quaternary glasses (SiO_2-CaO-P_2O_5-Na_2O). *J. Biomed. Mater. Res. A* **2003**, *66*, 110–119. [CrossRef]
22. Li, R.; Clark, A.E.; Hench, L.L. An Investigation of Bioactive Glass Powders by Sol-Gel Processing. *J. Appl. Biomater.* **1991**, *2*, 231–239. [CrossRef]
23. Ben-Arfa, B.A.E.; Neto, S.; Salvado, I.M.M.; Pullar, R.C.; Ferreira, J.M.F. Robocasting of Cu^{2+} & La^{3+} doped sol—Gel glass scaffolds with greatly enhanced mechanical properties: Compressive strength up to 14 MPa. *Acta Biomater.* **2019**, *87*, 265–272.
24. Ben-Arfa, B.A.E.; Neto, A.S.; Palamá, I.E.; Salvado, I.M.M.; Pullar, R.C.; Ferreira, J.M.F. Robocasting of ceramic glass scaffolds: Sol–gel glass, new horizons. *J. Eur. Ceram. Soc.* **2018**, *39*, 1625–1634. [CrossRef]
25. Ben-Arfa, B.A.E.; Salvado, M.M.; Ferreira, J.M.F.; Pullar, R.C. A hundred times faster: Novel, rapid sol-gel synthesis of without aging. *Int. J. Appl. Glass Sci.* **2016**, *8*, 337–343. [CrossRef]

26. Hulbert, S.F.; Morrison, S.J.; Klawitter, J.J. Tissue reaction to three ceramics of porous and non-porous structures. *J. Biomed. Mater. Res.* **1972**, *6*, 347–374. [CrossRef]

27. Houmard, M.; Fu, Q.; Saiz, E.; Tomsia, A.P.; Division, M.S.; Berkeley, L. Sol–gel method to fabricate CaP scaffolds by robocasting for tissue engineering. *J. Mater. Sci. Mater. Med.* **2012**, *23*, 921–930. [CrossRef]

28. Saiz, E.; Gremillard, L.; Menendez, G.; Miranda, P.; Gryn, K.; Tomsia, A.P. Preparation of porous hydroxyapatite scaffolds. *Mater. Sci. Eng. C* **2007**, *27*, 546–550. [CrossRef]

29. Chan, B.P.; Leong, K.W. Scaffolding in tissue engineering: General approaches and tissue-specific considerations. *Eur. Spine J.* **2008**, *17*, S467–S479. [CrossRef]

30. O'Brien, F.J.; Christenson, L.; Mikos, A.G.; Gibbons, D.F.; Picciolo, G.L.E.E. Biomaterials & scaffolds for tissue engineering. *Mater. Today* **2011**, *14*, 88–95.

31. Ben-Arfa, B.A.E.; Salvado, I.M.M.; Pullar, R.C.; Ferreira, J.M.F. Robocasting: Prediction of ink printability in solgel bioactive glass. *J. Am. Ceram. Soc.* **2019**, *102*, 1608–1618. [CrossRef]

32. Discher, D.E. Tissue cells feel and respond to the stiffness of their substrate. *Science* **2005**, *310*, 1139–1143. [CrossRef]

33. Engler, A.J.; Sen, S.; Sweeney, H.L.; Discher, D.E. Matrix Elasticity Directs Stem Cell Lineage Specification. *Cell* **2006**, *126*, 677–689. [CrossRef]

34. Karp, J.M.; Dalton, P.D.; Shoichet, M.S.; Cahn, F. Scaffolds for tissue engineering, clinical use of porous scaffolds for tissue engineering of skin. *MRS Bull.* **2003**, *28*, 301–306. [CrossRef]

35. Rabiee, S.M.; Nazparvar, N.; Azizian, M.; Vashaee, D.; Tayebi, L. Effect of ion substitution on properties of bioactive glasses: A review. *Ceram. Int.* **2015**, *41*, 7241–7251. [CrossRef]

36. Agarwal, R.; García, A.J. Biomaterial strategies for engineering implants for enhanced osseointegration and bone repair. *Adv. Drug Deliv. Rev.* **2015**, *1*, 53–62. [CrossRef]

37. Ben-Arfa, B.A.E.; Salvado, I.M.M.; Ferreira, J.M.F.; Pullar, R.C. Enhanced bioactivity of a rapidly-dried sol-gel derived quaternary bioglass. *Mater. Sci. Eng. C* **2018**, *91*, 36–43. [CrossRef]

38. Ben-Arfa, B.A.E.; Fernandes, H.R.; Salvado, I.M.M.; Ferreira, J.M.F.; Pullar, R.C. Synthesis and bioactivity assessment of high silica content quaternary glasses with Ca: P ratios of 1.5 and 1.67, made by a rapid sol-gel process. *J. Biomed. Mater. Part A* **2017**, *106A*, 510–520. [CrossRef]

39. Kaur, G.; Pickrell, G.; Sriranganathan, N.; Kumar, V.; Homa, D. Review and the state of the art: Sol–gel and melt quenched bioactive glasses for tissue engineering. *J. Biomed. Mater. Res. Part B Appl. Biomater.* **2016**, *104*, 1248–1275. [CrossRef]

40. Hench, L.L. Bioceramics. *J. Am. Ceram. Soc.* **2005**, *81*, 1705–1728. [CrossRef]

41. Hench, L.L. Bioceramics: From Concept to Clinic. *J. Am. Ceram. Soc.* **1991**, *74*, 1487–1510. [CrossRef]

42. Karageorgiou, V.; Kaplan, D. Porosity of 3D biomaterial scaffolds and osteogenesis. *Biomaterials* **2005**, *26*, 5474–5491. [CrossRef]

43. Krell, A.; Klimke, J.; Hutzler, T. Advanced spinel and sub-µm Al_2O_3 for transparent armour applications. *J. Eur. Ceram. Soc.* **2009**, *29*, 275–281. [CrossRef]

44. Miranda, P.; Pajares, A.; Saiz, E.; Tomsia, A.P.; Guiberteau, F. Mechanical properties of calcium phosphate scaffolds fabricated by robocasting. *J. Biomed. Res. Part A* **2008**, *85*, 218–227. [CrossRef]

45. Maazouz, Y.; Montufar, E.B.; Guillem-Marti, J.; Fleps, I.; Öhman, C.; Persson, C.; Ginebra, M.P. Robocasting of biomimetic hydroxyapatite scaffolds using self-setting inks. *J. Mater. Chem. B* **2014**, *2*, 5378–5386. [CrossRef]

46. Feilden, E.; Ferraro, C.; Zhang, Q.; García-Tuñón, E.; D'Elia, E.; Giuliani, F.; Vandeperre, L.; Saiz, E. 3D Printing Bioinspired Ceramic Composites. *Sci. Rep.* **2017**, *7*, 13759. [CrossRef]

47. Russias, J.; Saiz, E.; Deville, S.; Gryn, K.; Liu, G.; Nalla, R.K.; Tomsia, A.P. Fabrication and in vitro characterization of three- dimensional organic/inorganic scaffolds by robocasting. *J. Biomed. Mater. Res. Part A* **2006**, *38*, 434–445. [CrossRef]

48. Eqtesadi, S.; Motealleh, A.; Miranda, P.; Pajares, A.; Lemos, A.; Ferreira, J.M.F. Robocasting of 45S5 bioactive glass scaffolds for bone tissue engineering. *J. Eur. Ceram. Soc.* **2014**, *34*, 107–118. [CrossRef]

49. Liu, X.; Rahaman, M.N.; Hilmas, G.E.; Bal, B.S. Mechanical properties of bioactive glass (13-93) scaffolds fabricated by robotic deposition for structural bone repair. *Acta Biomater.* **2013**, *9*, 7025–7034. [CrossRef]

50. Fung, Y.C. *Biomechanics: Mechanical Properties of Living Tissues*, 2nd ed.; Springer: New York, NY, USA, 1993.

51. Oliveira, M.L.L.; Chen, K.; Ferreira, J.M.F. Influence of the deagglomeration procedure on aqueous dispersion, slip casting and sintering of Si3N4-based ceramics. *J. Eur. Ceram. Soc.* **2002**, *22*, 1601–1607. [CrossRef]

52. Loh, Z.H.; Samanta, A.K.; Heng, P.W.S. Overview of milling techniques for improving the solubility of poorly water-soluble drugs. *Asian J. Pharm. Sci.* **2014**, *10*, 255–274. [CrossRef]

53. Ben-Arfa, B.A.E.; Salvado, I.M.M.; Frade, J.R.; Pullar, R.C. Guidelines to adjust particle size distributions by wet comminution of a bioactive glass determined by Taguchi and multivariate analysis. *Ceram. Int.* **2019**, *45*, 3857–3863. [CrossRef]

54. Bentzon, M.D.; Andersen, L.O.; Goul, J.; Bodin, P.; Vase, P. Influence of the Powder Calcination Temperature on the Microstructure in Bi(Pb)-2223 Tapes. *IEEE Trans. Appl. Supercond.* **1997**, *7*, 1411–1414. [CrossRef]

55. Ben-Arfa, B.A.E.; Salvado, I.M.M.; Pullar, R.C.; Ferreira, J.M.F. The influence of processing parameters on morphology and granulometry of a wet-milled sol-gel glass powder. *Ceram. Int.* **2018**, *44*, 12754–12762. [CrossRef]

56. Olhero, S.M.; Fernandes, H.R.; Marques, C.F.; Silva, B.C.G. Additive manufacturing of 3D porous alkali-free bioactive glass scaffolds for healthcare applications. *J. Mater. Sci. Biomater.* **2017**, *52*, 12079–12088. [CrossRef]

57. Eqtesadi, S.; Motealleh, A.; Pajares, A.; Miranda, P. Effect of milling media on processing and performance of 13-93 bioactive glass scaffolds fabricated by robocasting. *Ceram. Int.* **2015**, *41*, 1379–1389. [CrossRef]

58. Cesarano, J., III; King, B.H.; Denham, H.B. Recent Developments in Robocasting of Ceramics and Multimaterial Deposition. In Proceedings of the International Solid Freeform Fabrication Symposium, Austin, TX, USA, 10–12 August 1998; pp. 697–704.

59. Azzolini, A.; Sglavo, V.M.; Downs, J.A. Novel method for the identification of the maximum solid loading suitable for optimal extrusion of ceramic pastes. *J. Adv. Ceram.* **2014**, *3*, 7–16. [CrossRef]

60. Liu, D.M.; Tseng, W.J. Influence of solids loading on the green microstructure and sintering behaviour of ceramic injection mouldings. *J. Mater. Sci.* **1997**, *32*, 6475–6481. [CrossRef]

61. Boccaccini, A.R.; Chen, Q.; Lefebvre, L.; Gremillard, L.; Chevalier, J. Sintering, crystallisation and biodegradation behaviour of Bioglass®-derived glass-ceramics. *Faraday Discuss.* **2007**, *136*, 27–44. [CrossRef]

62. Eqtesadi, S.; Motealleh, A.; Wendelbo, R.; Ortiz, A.L.; Miranda, P. Reinforcement with reduced graphene oxide of bioactive glass scaffolds fabricated by robocasting. *J. Eur. Ceram. Soc.* **2017**, *37*, 3695–3704. [CrossRef]

63. Ben-Arfa, B.A.E.; Salvado, I.M.M.; Ferreira, M.F.; Pullar, R.C. The effects of Cu2+ and La3+ doping on the sintering ability of sol-gel derived high silica glasses bioglass. *Ceram. Int.* **2019**, *45*, 10269–10278. [CrossRef]

64. Ben-Arfa, B.A.E.; Palamá, I.E.; Salvado, I.M.M.; Ferreira, M.; Pullar, R.C. Cytotoxicity and bioactivity assessments for Cu^{2+} and La^{3+} doped high-silica sol-gel derived bioglasses: The complex interplay between additive ions revealed. *J. Biomed. Mater. Res. Part A* **2019**, *107*, 2680–2693. [CrossRef]

 *processes*

Article

Injectable Chitosan Scaffolds with Calcium β-Glycerophosphate as the Only Neutralizing Agent

Piotr Owczarz [1,*], Anna Rył [1], Marek Dziubiński [1] and Jan Sielski [2]

[1] Department of Chemical Engineering, Lodz University of Technology, 90-924 Lodz, Poland;
    anna.ryl@edu.p.lodz.pl (A.R.); marek.dziubinski@p.lodz.pl (M.D.)
[2] Department of Molecular Engineering, Lodz University of Technology, 90-924 Lodz, Poland;
    jan.sielski@p.lodz.pl
*   Correspondence: piotr.owczarz@p.lodz.pl; Tel.: +48-42-631-37-32

Received: 11 March 2019; Accepted: 16 May 2019; Published: 19 May 2019

Abstract: The presented work describes the method of preparation of thermosensitive chitosan hydrogels using calcium β-glycerophosphate salt as the only pH neutralizing agent and supporting the crosslinking process. The presence of calcium ions instead of sodium ions is particularly important in the case of scaffolds in bone tissue engineering. Rheological and physicochemical properties of low concentrated chitosan solutions with the addition of calcium β-glycerophosphate were investigated using rotational rheometry techniques, Zeta potential (by electrophoresis), XPS, and SEM analysis together with an EDS detector. It was found to be possible to prepare colloidal solutions of chitosan containing only calcium β-glycerophosphate (without sodium ions) undergoing a sol-gel phase transition at the physiological temperature of the human body. It has also been shown that it is possible to further enrich the obtained cellular scaffolds with calcium ions. Using the addition of calcium carbonate, hydrogels with a physiological ratio of calcium to phosphorus (1.6–1.8):1 were obtained.

Keywords: sol-gel phase transitions; injectable scaffolds; chitosan; calcium β-glycerophosphate; rheology; bone tissue engineering

1. Introduction

The development of tissue engineering is associated with the search for new therapies for both chronic and urgent diseases. The possibility of supporting pharmacological treatment or reconstruction of damaged tissues can be facilitated by highly porous, three-dimensional structures called scaffolds [1]. Materials used for the production of scaffolds are mostly natural or synthetic polysaccharides, bioceramics and their composites [2]. Among natural polysaccharides, chitosan is still growing in popularity. It is not without reason that it is called the 21st century polymer for biomedical applications [3–6]. Its widely known antibacterial, antifungal and antiviral properties make it a very good biomaterial for medical purposes [7–10]. In addition, this polysaccharide is characterized by bioadhesiveness, bioactivity, non-toxicity, biodegradability, cytocompatibility, and is well absorbed by living organisms [3,11,12]. Chitosan also has anti-inflammatory, anti-cancer, and hypolipemic and antioxidant properties [13–17]. This biopolymer, similar in terms of its chemical structure to hyaluronic acid (often used in engineering of cellular scaffolds), is much cheaper.

Thermosensitive hydrogel undergoing a sol-gel phase transition, is one of the materials used for the preparation of polymer scaffolds applied in tissue engineering [2]. The local formation of chitosan hydrogel can be induced by temperature [18–22], pH [23,24], or ultraviolet radiation [25]. Thermosensitive hydrogels are characterized by relatively low viscosity and predominance of viscous properties in the sol phase as well as much higher viscosity and predominance of elastic properties in the gel phase. The main difference between the two phases is the observed flow phenomenon characteristic of viscous fluid and the lack of flow for the viscoelastic fluid [26]. Due to their smaller invasiveness as compared to implant scaffolds, in recent years they have been the object of research by many authors [27,28]. The use of injection scaffolds makes it also possible to better fill the existing tissue defects [28]. Their additional advantage is the ability to control important properties of the scaffolds such as porosity, size, geometry, and the degree of pore in order to the mimic the topological and microstructural characteristics of the extracellular matrix (ECM) [2]. The interest in hydrogels in bone tissue engineering (BTE) is also due to their similarity to cartilage tissue which is highly hydrated and composed of chondrocytes located in type II collagen and glycosaminoglycans (GAG) [29].

Chitosan is a cationic polysaccharide that is hydrophobic at neutral pH of the medium, it dissolves only in an acidic environment. Chitosan hydrogel qualify as a physical gel [30] obtained by diffusion limited aggregation (DLA) [31,32]. In this case, the non-covalent interactions (van der Waals and hydrogen bonds) between the polymer chains as a result of the conversion of the polymer from hydrophilic to hydrophobic form, associated with a change in the degree of dissociation of the amino groups, are decisive. The formation of the gel network is associated with a change in the degree of dissociation (pKa) of the amino groups. Lack of positive charge causes a change (decrease) in the thickness of the solvation shell around this group and enables the aggregation of chains. The disappearance of the polyion charge leads to a change in the nature of the chain from hydrophilic to hydrophobic, and consequently, the precipitation of the chitosan from the solution and the aggregation of the polymer particles. The phase transition point of such systems is at a temperature above 50 °C. Thus, it prevents their use as injection scaffolds formed in vivo. The addition of disodium glycerophosphate (Na_2GP) to the chitosan solution in acid results in the formation of an unstable colloidal system (low value of Zeta potential) and a reduction in the temperature of the sol-gel phase transition point [19,32–34]. The chitosan colloidal solutions obtained by dissolving the polymer in hydrochloric acid are commonly described in the literature, however, the role of glycerophosphate is discussed [19–22,33,35]. The process of chitosan aggregation from solutions with glycerophosphate occurs at much lower temperatures and its kinetics depend on the current state of protonation of amino groups—the reaction-limited aggregation (RLA) [20,31,32,35]. Based on the solution of the Poisson-Boltzmann equation for chitosan (polyion) solutions containing disodium glycerophosphate (counterion), Filion and Lavertu [20,35] showed that a sufficient condition for the thermal neutralization of polyion is that $dpK_0/dT < dpK_a/dT$, where pK_0 is the intrinsic dissociation constant of chitosan and pK_a is the dissociation constant of the buffer coupled to chitosan, i.e., counterion. Furthermore, the authors stated that it is possible to use alternative compounds for glycerophosphate, which can cause thermal phase transition of chitosan systems. According to Filion and Lavertu, the key role of the buffer added to chitosan is its ability for heat-stimulated absorption of protons released from chitosan. On the other hand, Eeckman et al. [36] found that the phosphate salts containing bivalent and trivalent metal ions exert a greater influence on the lowering of the critical temperature (LCST) compared to monovalent metal salts. Also, the addition of small quantities of inert salts reduces the LCST. In this case, in addition to the concentration, the type of salt added and the valence of the anion are also important. The addition of an electrolyte can disrupt the highly ordered structure of water molecules surrounding the polyion, causing an increase in its hydrophobicity and, as a result, lowering LCST and faster precipitation of molecules from the solution, resulting in their agglomeration—forming the gel structure [36,37].

The methods for preparation of chitosan hydrogels described so far in the literature have been based on the use of disodium β-glycerophosphate [18–21]. The addition of this salt into the acidic chitosan solution causes an increase of pH to the physiological value (about 7) and a simultaneous shift of the sol-gel phase transition temperature to approximately 37 °C. The literature presents few studies on the use of glycerophosphate salts other than sodium salts, e.g., calcium [38,39] and magnesium [40]. However, in all published studies, these salts have always been an addition to the commonly prepared chitosan scaffolds with disodium β-glycerophosphate never being the sole source of glycerophosphate. Moreover, the effect of calcium or magnesium ions is unclear, especially on rheological properties, because the introduction of both salts also causes a change (increase) in the concentration of glycerophosphate.

The use of only calcium β-glycerophosphate for the preparation of chitosan hydrogels follows from the fact that this compound potentially combines the benefits of the buffering properties of glycerophosphate as well as the replacement of sodium ions unfavorable to organisms, in particular for bone tissues, with calcium ions. The addition of glycerophosphate salt (change in pH of the solution) allows to reduce the sol-gel phase transition temperature [30,34,35], which is extremely important when designing scaffolds formed in vivo.

Based on researches conducted with the use of disodium glycerophosphate, it was found that the glycerophosphate residue is an osteogenic compound [41]. On the other hand, the presence of calcium ions is particularly important when thermosensitive hydrogels are used in bone tissue engineering. It results from the fact that calcium and phosphate ions are necessary to form ceramics whose structure is similar to hydroxyapatite—a mineral which together with collagen builds a scaffold responsible for mechanical properties of bone tissue [42]. Studies available in the literature [38,39] discussing osteoblast cultures in chitosan hydrogels indicate that the addition of calcium β-glycerophosphate salts into commonly prepared chitosan hydrogels improves cell proliferation and viability. The quoted study also points to the non-monotonic dependence of the proliferation of osteoblast cells and their viability versus the β-glycerophosphate salt content. In studies using only calcium-phosphate bioceramics, it was found that the best conditions for the viability of osteoblasts are ensured when the calcium to phosphorus ratio was 1.6:1.0 [43]. Due to the mineral composition of bones, in the case of engineering of these tissues, there is a need to ensure an adequate ratio of calcium to phosphorus in the prepared scaffold. For this purpose, admixtures of ceramic materials are used in many scaffolds in BTE. Among the many chitosan–ceramic composites, chitosan–calcium phosphate systems are particularly popular. This is due to the similarity of their chemical composition to inorganic bone components [44]. The need to add bioceramics into the chitosan scaffolds results from their inadequate mechanical properties [45,46] and insignificant osteoconductivity of chitosan [47]. The most commonly used ceramic materials are calcium phosphate [43], β-tricalcium phosphate [48–50] and hydroxyapatite [47,49,51–53]. Among the compounds that can also be used to improve mechanical properties and osteoconductivity is, for example, calcium carbonate [54] which in an acidic environment dissociates to calcium ions, carbon dioxide and water.

The aim of the work was to propose a method for the preparation of chitosan scaffolds containing the addition of calcium β-glycerophosphate as the only substance neutralizing pH and causing a temperature decrease in the sol-gel phase transition.

2. Materials and Methods

Thermosensitive colloidal chitosan solutions were prepared by dissolving 400 mg of crab-derived chitosan (Sigma-Aldrich, Poznan, Poland, product number: 50494) in 20 mL of 0.1 M acetic acid (Sigma-Aldrich, Poznan, Poland, product number: 695092). The thoroughly mixed solution was left for 24 h to allow the polysaccharide to completely dissolve. Next, a suspension of calcium β-glycerophosphate cooled for 2 h was added dropwise to the chitosan acetate solution. The suspension was obtained by mixing from 0.2 g to 2.0 g of salt in such an amount of distilled water (from 1 mL to 5 mL) as to obtain the appropriate molar concentrations of calcium-β-glycerophosphate salt, as shown in Table 1.

The basic physicochemical parameters of chitosan such as the degree of deacetylation and molecular weight were determined using the titration method and gel permeation chromatography. The degree of deacetylation was determined as a difference DD = 1-DA. The degree of acetylation (DA) was determined by titrimetric method. The molecular weight (Mw) of the tested samples was determined by gel permeation chromatography (GPC/SEC) with the use of high-performance liquid chromatography (HPLC) on a Knauer Smartline company apparatus, equipped with an analytical isocratic Pump1000 and DRI detector (S-2300/2400, Knauer, Berlin, Germany). The studies carried out indicated that the degree of deacetylation of chitosan was DD = 81.8%, and the average molecular weight was Mw = 680 kDa.

Table 1. The composition of the solutions used in the studies with their designations.

The Molar Concentration of β-CaGP [mol/dm^3]	Sample Number	
	Without CaCO$_3$	With CaCO$_3$
0.00	1	1CC
0.05	2	2CC
0.10	3	3CC
0.19	4	4CC
0.38	5	5CC

In order to increase the ratio of calcium ions to phosphorus ions in the solution, calcium carbonate was added before introducing calcium β-glycerophosphate into the colloidal solution (Ubichem, Redditch, UK, product number: C044H). The use of an appropriate amount of salt was to ensure the assumed molar ratio of calcium to phosphorus ions is equal to 1.6:1.

2.1. Characteristics of the Mechanical Properties of Thermosensitive Hydrogels

Rheological measurements of the obtained solutions were carried out using the Anton Paar Physica series MCR301 (Anton Paar, Warszawa, Poland) rotational rheometer with a cone-plate measuring system (50 mm cone diameter, 1° cone angle and 48 μm cone truncation). In order to determine viscoelastic properties, measurements were carried out under isothermal conditions in a wide range of angular frequencies ($\omega = 0.005$ s^{-1}–500 s^{-1}) at temperatures of 5 °C, 25 °C, 30 °C, 35 °C, and 40 °C. The conditions of the sol-gel phase transition were also determined by performing non-isothermal oscillatory measurements at a constant heating rate (1 deg/min) in the range 5 °C–60 °C and isothermal measurements at 37 °C. In both cases a constant deformation value was used (angular frequency $\omega = 5$ s^{-1}, strain amplitude $\gamma = 1\%$).

2.2. Stability of Colloidal Solutions (Zeta Potential) and pH Measurements

The Zeta potential changes, indicating the stability of the tested solutions, were determined using Malvern ZetaSizer Nano ZS (Malvern Instruments Ltd., Malvern, UK) with a folded capillary cell DTS1060 type. The measurements were carried out in non-isothermal conditions every 1 K in the temperature range 5–50 °C (heating rate 1 K/min). A series of 100 measurements were carried out for each temperature and the results were averaged.

The pH measurements of the sol phase were conducted at 4 °C using a pH-meter ELMETRON CP-401 (ELMETRON Sp. j., Zabrze, Poland), equipped with a dedicated electrode for viscous liquids ERH-12-6.

2.3. SEM Morphology and Analysis of the Atomic Composition of the Surface Layer with the Energy-Dispersive X-ray Spectroscopy Technique

The morphology of scaffolds was analyzed using the scanning electron microscope SEM (FEI, Quanta 200F, Hillsboro, OR, USA). With the use of an EDS detector (Oxford Instruments, X-Max, Manchester, UK), an analysis of the atomic composition of the surface layer of the scaffolds obtained was also carried out. During the tests, an energy beam of 3.5 eV was used. The measurements were carried out under low vacuum (about 100 Pa) in nitrogen atmosphere.

2.4. Analysis of Surface Composition by the X-ray Photoelectron Spectroscopy Technique

The atomic composition of the tested material was analyzed using the XPS AXIS Ultra spectrometer from Kratos Analytical Ltd. (Manchester, UK). For the analyzed samples, emission spectrum was made in the full binding energy range from 1200 to 0 eV. The measurements were carried out at an absolute pressure of 1×10^{-6} Pa.

3. Results and Discussion

3.1. Calcium Glycerophosphate Versus Disodium Glycerophosphate

Measurements determining the effect of the glycerophosphate salt cation on viscoelastic properties and the stability of colloidal systems were carried out for the same molar concentrations (0.38 mol/dm^3) and were based on the concentration of the sodium glycerophosphate salt most commonly used in the literature.

The effects of the addition of glycerophosphate salt and the type of cation contained in this salt were determined by measurement of Zeta potential. The obtained results are shown in Figure 1. The value of the electrophoretic mobility of the particles in the electric field and the value of ζ-potential determined on the basis of the Henry equation can be used to compare the dispersion stability of the colloid systems studied [55]. This stability is defined as the result of the sum of attractive forces (Van der Waals, hydrogen bonds) and repulsive forces (electrostatic forces). Based on the changes in the Zeta potential obtained for the solution without glycerophosphate, in all of the tested temperatures, the particles in the suspension had a high positive potential, 40–50 mV, which indicates a system with good stability. Moreover, the Zeta potential value increases with increasing temperature. According to the interpretation adopted in the literature [55], this indicates that they do not tend to repel each other or tend to agglomerate. In the case of solutions containing the addition of glycerophosphate salt, the measured ζ-potential values are much lower, 3–12.5 mV for disodium glycerophosphate and 0–13.5 mV for calcium glycerophosphate. The low values of the Zeta potential of the molecules indicates that the solution containing them is an unstable system. The macromolecules are beginning to attract each other and aggregation occurs. The influence of the addition of glycerophosphate on the stability of chitosan (chloride) solutions was confirmed in the study by Owczarz and coworkers [32].

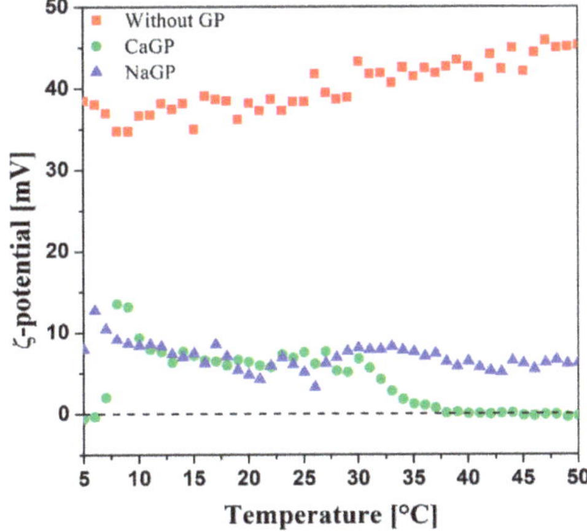

Figure 1. Experimental curves of Zeta potential for samples without glycerophosphate, with disodium glycerophosphate and calcium glycerophosphate under non-isothermal conditions.

The analysis of Zeta potential curves for samples with the addition of glycerophosphate salts shows that at temperatures below 30 °C, the value decreases with increasing temperature, from 12.8 to 6.9 and from 13.6 to 6.69 mV for sodium and calcium glycerophosphate, respectively. Moreover, it does not depend on the type of salt used. Significant differences in the obtained Zeta potential curves are visible in the gelation region (above 30 °C). In the case of solutions containing the commonly used disodium glycerophosphate salt, further heating of the experimental medium did not change the value of the ζ-potential and thus the stability of the system. According to the authors [55], such systems should be classified as highly unstable systems. In contrast, in the case of calcium glycerophosphate salts, with the temperature ranging from 30 °C to 37 °C, a further decrease in the Zeta potential to 0 mV was observed—the system is in the phase of coagulation of colloidal particles [55]. Thus, the conducted research allowed to clearly determine that the type of glycerophosphate salt (the valency of the cation forming it) has an important role in the stability of the colloidal system [56].

Analysis of the results of isothermal measurements made for samples 1–5 in a wide range of angular frequencies ω, indicates that the use of the highest concentration of β-glycerophosphate, corresponding to the commonly used molar sodium salt concentration [19,33], causes a significant change in the dominant properties of the medium—cf. Figure 2a,b. Already at 5 °C, almost in the entire range of angular frequencies, the predominance of elastic properties over viscous ones is observed—the system shows properties characteristic of the gel; shown in Figure 2a.

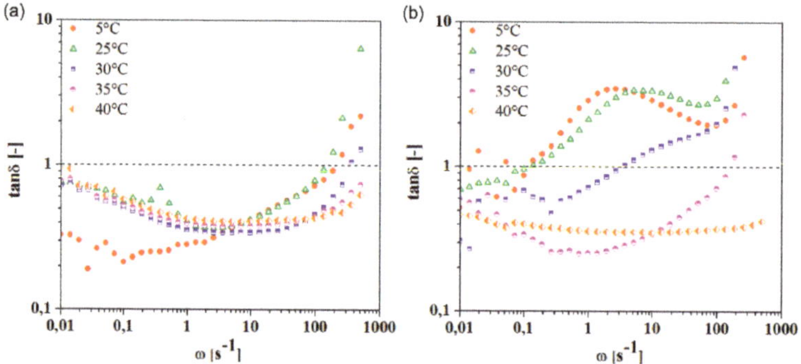

Figure 2. Comparison of mechanical spectra of colloidal chitosan solutions with calcium β-glycerophosphate (**a**) and sodium β-glycerophosphate (**b**).

Moreover, based on the non-isothermal rheological measurements, it was found that the use of calcium glycerophosphate salt concentration similar to the concentration of sodium glycerophosphate commonly used in the literature [33], caused a significant lowering of the sol-gel phase transition point. The use of the same concentration and thus an equal amount of phosphate residues derived from β-glycerophosphate decreased the temperature of the sol-gel transition point from 39 °C to about 18 °C.

3.2. Optimal Concentration of Calcium Glycerophosphate

Due to such a violent change in the viscoelastic properties of colloidal chitosan solutions as well as the gelation conditions, it was necessary to determine the optimal concentration of calcium glycerophosphate salt in order to ensure adequate properties for the injection application of the tested chitosan solutions (G′ < G″), i.e., the predominance of viscous features at low temperatures and the expected sol-gel phase transition at about 37 °C. The results of non-isothermal measurements of rheological properties of colloidal chitosan solutions with various concentrations of calcium β-glycerophosphate are shown in Figure 3.

It was found that the dependence of calcium β-glycerophosphate salt concentration on the temperature of gelation of chitosan hydrogels, are, as to the direction, consistent with the studies on sodium β-glycerophosphate salt presented in the literature [33,57]. It means that the change in the cation in the glycerophosphate salt does not change the quality of the obtained results; clearly, the same effect of the increase in concentration on the reduction of the gelation temperature is observed. However, obtained phase transition temperatures are significantly lower than when sodium glycerophosphate salts are used.

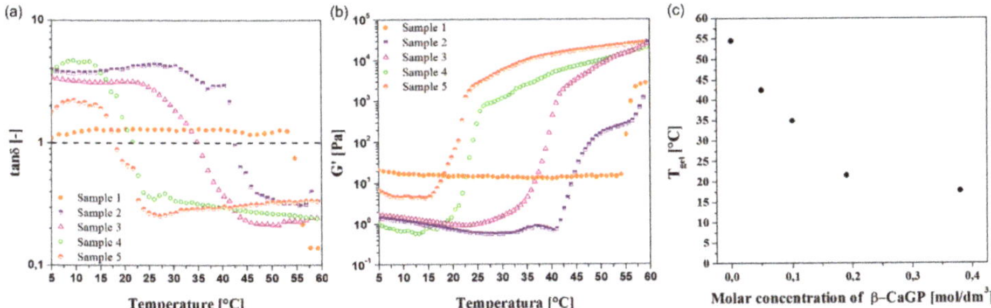

Figure 3. The effect of calcium β-glycerophosphate on (**a**) the curve of damping factor tanδ in non-isothermal conditions, (**b**) the change of storage modulus G′ under non-isothermal conditions, and (**c**) the dependence of sol-gel phase transition temperature.

Analysis of changes in the experimental curve and the obtained values of storage modulus G′ as a function of temperature (Figure 3b) indicate that the addition of calcium glycerophosphate salt reduces the mechanical resistance at low temperatures, but finally leads to its increase in the gel phase.

Reduction of the content of calcium glycerophosphate salt caused a gradual change of the dominant properties of the medium to viscous ones at low temperatures and an increase of phase transition temperature. It was found that the optimal parameters had a colloidal chitosan solution containing 0.1 mol/dm^3 of calcium β-glycerophosphate—sample 3. This solution has thermosensitive properties, as shown in Figure 4f. It is visible that the sol phase was maintained up to a temperature of about 25 °C. Subsequent heating of the solution led to its structural change which is observed in the case of non-isothermal (Figure 3a) and isothermal experimental curves (Figure 4c–e). The determined sol-gel phase transition point as a crossover point of storage (G′) and loss (G″) moduli curves occurred at 35 °C (Figure 3a). A complete formation of the structure occurred at above 40 °C. Furthermore, it was found that with the formation of an unlimited three-dimensional structure, improvement of the mechanical properties of the obtained scaffold was visible. This is seen as a plateau in the curve of changes in the value of the storage modulus G′ at angular frequency ω—cf. Figure 2b. This is also confirmed by the values of G′ modulus which are much higher than the ones obtained in the experiments carried out at lower temperatures. The storage G′ and loss G″ moduli at 5 °C are shown in Figure 4a and represent a highly flexible region in which a network structure is formed and a glass transition region occurs between the highly flexible and glassy state [56,57]. As the temperature of the measurement increases, the obtained experimental curves are shifted towards the strong dominance of the glass transition region. At 35 °C and 40 °C, experimental curves characteristic of the glassy state are observed [58].

Analysis of damping factor—tanδ = G″/G′, (see Figure 4f) allows us to state that for the measurements carried out at temperatures of 5 °C and 25 °C, the analyzed polysaccharide solution is characterized by high molecular weight and the presence of long side chains. The difference in the value of tanδ is due to the decrease in mechanical resistance characterized by storage modulus G′. This biopolymer occurs in both temperatures in an amorphous form. At 30 °C, the curve becomes flattened, which may indicate gradual structural changes occurring in the studied medium—the occurrence of a weakly crosslinked, amorphous biopolymer, a so-called soft rubber. At the highest temperatures of 35 °C and 40 °C, a change in the structural form of the medium is observed. In the whole range of angular frequencies ω, elastic properties predominate over viscous ones (G′ > G″). For measurements carried out at 40 °C, it was found that the mean value of the tangent of loss angle tanδ was about 0.3. This indicates the occurrence of a polymer in a glassy form.

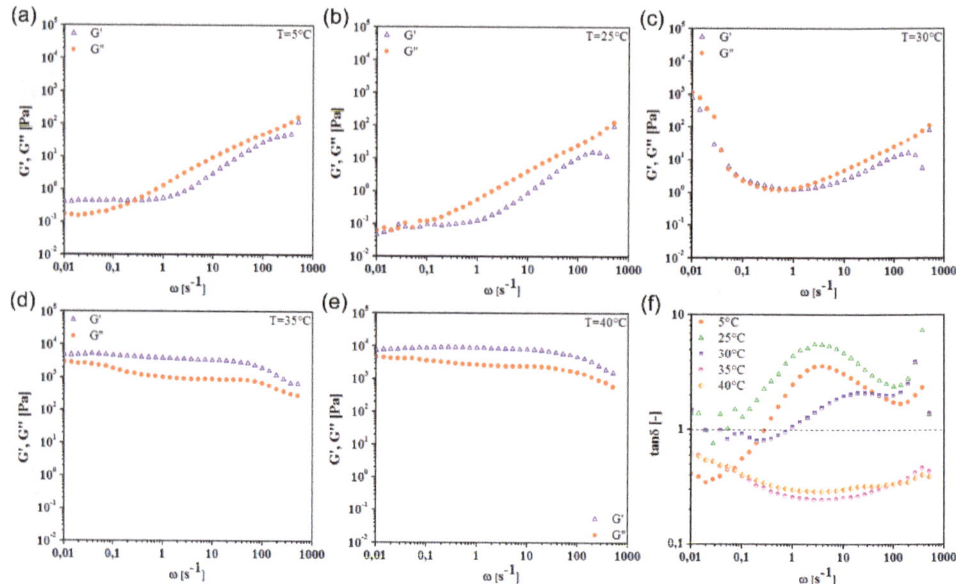

Figure 4. Relations of storage G′ and loss G″ moduli (**a–e**) and damping factor tanδ (**f**) as a function of angular frequency ω for sample 3 at different temperatures.

Table 2 presents the results of measurements of the sol-gel phase transition kinetics determined at 37 °C. It has been found that the use of a higher concentration of calcium-β-glycerophosphate leads to a shorter time needed to form the spatial structure of the polymer network. Simultaneously, it should be noted that for samples 4 and 5, although the time of scaffold formation is short, the phase transition temperature is too low. For these samples, there is a risk of starting the gelation process at room temperature. This is connected with a significant increase in the viscosity of the system, and thus difficult injection application.

Table 2. Results of non-isothermal and isothermal measurements for the tested samples and analysis of the gelation kinetics curve.

Sample	pH	Non-Isothermal Measurements	Isothermal Measurements		Analysis of the Gelation Kinetics Curve	
		Tgel [°C] Based on F-L Method	Tgel [s] Based on tan = 1 Method	Value of G′ [Pa] for tan = 1	G′$_{min}$ [Pa] Value	Reference Time [s]
1	5.03	54.9		Not tested		
2	5.88	43.1	2760	5.52	0.16	2681
3	6.29	35.4	700	3.24	0.11	810
4	6.58	22.0	163	2.18	0.15	121
5	6.78	18.0	145	23.10	1.00	87
1CC	6.17	54.0 *		Not tested		
2CC	6.19	39.2	1873	7.21	0.79	1733
3CC	6.21	29.6	403	5.96	0.61	580
4CC	6.48	17.2 *	*	*	*	*
5CC	6.73	13.0 *	*	*	*	*

Samples 1 and 1CC were not tested in isothermal measurements, due to the too long expected measurement time. * Samples which in the entire measuring range were characterized by predominance of elastic properties over viscous ones.

In order to normalize the obtained data, an analysis was performed in which the reference time of fast gelation was determined; shown in Figure 5a. This value was defined as a difference between the time of reaching the same value of storage modulus G′ for all tested samples at the fast gelation region (determining the same state of material elasticity), and the time of reaching the minimum value of storage modulus G′, which is identified with the end of the viscoelastic flow phase and the start of the fast gelation stage [19]. The set value of storage modulus G′ was chosen so that for all cases it was above the gelation point defined as the intersection point of the curves of G′ and G″ moduli. In the considered cases, it was G′$_{REF}$ = 50 Pa. The analysis shows that the reference time of fast gelation is a non-linear, decreasing function of the calcium β-glycerophosphate concentration—cf. Figure 5b.

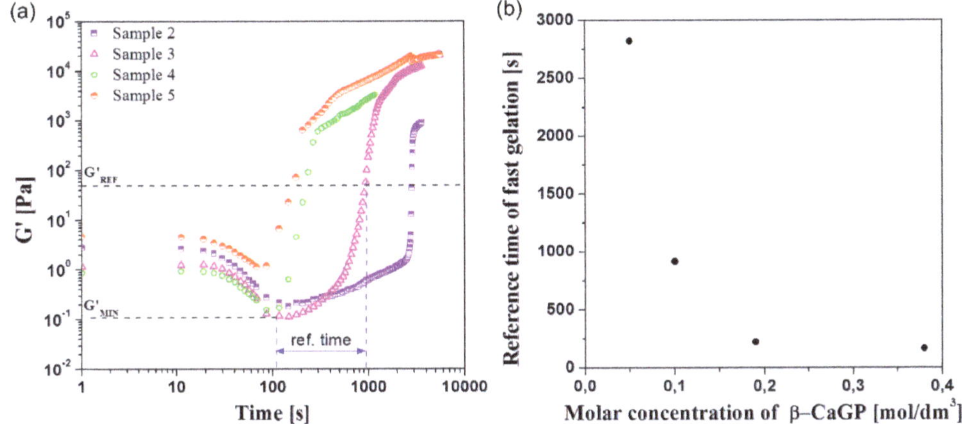

Figure 5. Kinetics of changes in the value of storage modulus G′ along with the presentation of a method for determining the reference time of rapid gelation (**a**) and the dependence of the reference time of gelation on molar concentration of calcium-β-glycerophosphate (**b**).

3.3. Enrichment with Calcium Ions

In order to increase the proportion of calcium in relation to the content of phosphorus to a similar ratio naturally occurring in bone tissue i.e., 1.6/1, the calcium carbonate was added to the colloidal solution of chitosan. The presence of calcium carbonate in the solution was in each case associated with a change in the mechanical properties of the tested solutions; however, this effect depended on the concentration of calcium glycerophosphate salt.

The experimental curves of G′ and G moduli for sample 3 (Figure 6a–c), which underwent phase transition in conditions similar to the physiological temperature of the human body (cf. Figure 3a) and the corresponding sample containing calcium carbonate, sample 3CC (Figure 6d–f), are shown in Figure 6. Analysis of the results of measurements of oscillating spectra obtained at temperatures up to 25 °C allowed us to state that in both solutions, samples 3 (Figure 6a) and samples 3CC (Figure 6d), the predominance of viscous features over elastic properties is visible (G′ < G″). Slight changes in the experimental curves of storage G′ and loss G″ moduli, as compared to sample 3, are observed in the range of low angular frequency ω for measurements carried out at temperatures 5 °C and 25 °C. Further heating of samples 3 and 3CC to 30 °C, 35 °C and 40 °C leads to changes in the curves of storage and loss moduli; cf. Figure 6. The addition of calcium carbonate to sample 3 (which gives the composition of sample 3CC) results in the occurrence of characteristic shapes of the curves of storage G′ and loss G″ moduli, cf. Figure 6b–e, towards a stronger domination of the glass transition and glassy state regions [58–60]. The addition of calcium carbonate results in obtaining for sample 3CC similar experimental curves for storage G′ and loss G″ moduli that were obtained for sample 3 in the case of tests carried out at lower temperatures; cf. Figure 6.

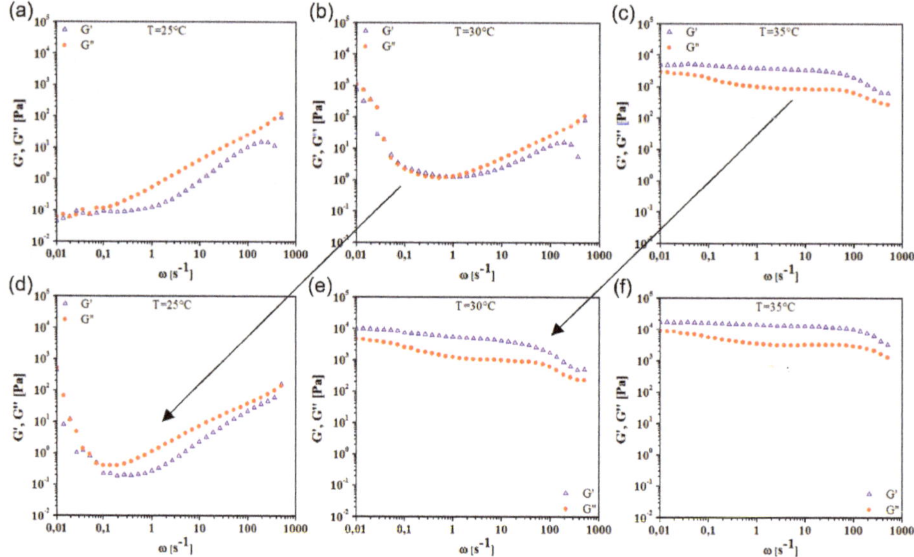

Figure 6. Comparison of experimental curves for storage G′ and loss G″ moduli for sample 3 (**a–c**) and sample 3CC (**d–f**) at 25 °C, 30 °C and 35 °C.

Performing analogous analyses for samples 4 and 5, it was found that in the case of sample 5, the addition of calcium carbonate (sample 5CC) increased the mechanical resistance of the obtained media at all tested temperatures. Simultaneously, there were no distinctly different experimental curves of storage modulus G′ and changes in the rheological properties of the medium. Samples 5 and 5CC were characterized by the predominance of elastic properties over viscous ones already at low measuring temperatures. The addition of calcium carbonate salt to sample 4 (sample 4CC) was related to a rapid change in its viscoelastic properties at low measuring temperatures. Analyzing the results of tests conducted at 5 °C for sample 4CC, it was found that for a solution containing an additional source of calcium, the predominance of elastic properties over viscous properties was observed. At the same time, it was found that the improvement of mechanical properties of sample 4CC caused by the addition of calcium carbonate, was not as significant as in the case of sample 5CC.

The effect of the addition of calcium carbonate on the conditions of sol-gel phase transition was determined based on the analysis of the results of non-isothermal oscillatory measurements. However, only in the case of 2CC and 3CC solutions it was possible to determine the sol-gel phase transition temperature as the intersection point of the dynamic modules (tanδ = 1). The remaining solutions containing the calcium carbonate were characterized by the predominance of elastic properties over viscous ones already at low temperature, hence it was not possible to apply the above method to determine the gelation temperature. For this reason, the method proposed by Fredrickson and Larson [61] was used. This theory is based on the order-disorder of block copolymers considering the aggregation of hydrophobic acetylated blocks of chitosan chains in systems with glycerophosphate [34,62], and it can be successfully applied even in the case of systems with dominance of elastic properties over viscous ones. As a consequence, the dependence of temperature as a function of the glycerophosphate concentration with the addition of calcium carbonate was determined. It is clearly visible that in each case the addition of calcium carbonate reduces the sol-gel phase transition temperature; see Figure 7b.

Taking into account the application aspect, the use of solutions with a predominance of elastic properties seems to be disadvantageous because their injection can be difficult and cause discomfort to the patient. On the other hand, the 1CC sample undergoes a phase transition at too high a temperature which will prevent the formation of the scaffold structure in vivo. Consequently, the best solution seems to be using a 3CC sample, whose phase transition temperature is around 30C. Simultaneously, this sample undergoes a sol-gel phase transition in the typical form (from dominance of viscous properties at low temperature, to dominance of elastic properties at high temperature).

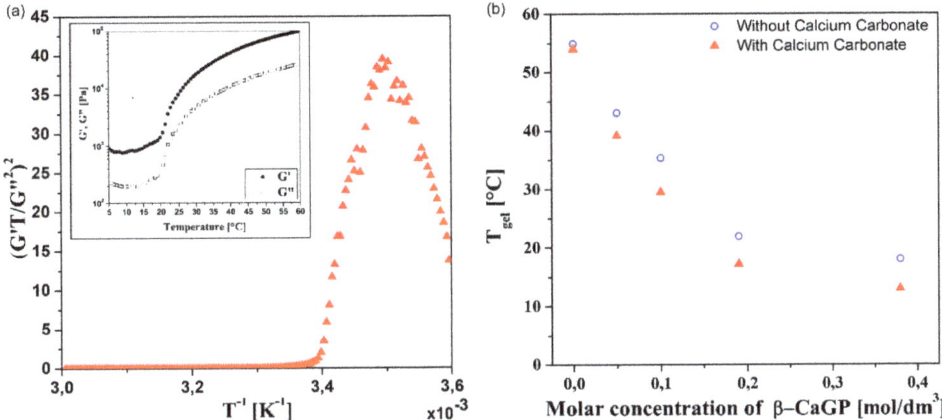

Figure 7. (**a**) Application of the Fredrickson-Larson method to determine the phase transition point for the medium with the predominance of elastic properties—sample 5CC. (**b**) Effect of the addition of calcium carbonate on the gelation temperature.

In the case of sample 3CC, the predominance of viscous properties over elastic properties is observed in measurements carried out at low temperatures. Differences in the experimental curves of mechanical spectra in the temperature range of 25 °C to 35 °C, as shown in Figure 6, are confirmed by non-isothermal measurements of gelation point temperature, cf. Figure 8a. Equalization of the values of storage G' and loss G'' moduli (the gelation point of sample 3CC) takes place at about 30 °C, as shown in Figure 8a. For comparison, in the case of sample 3, without the addition of calcium carbonate, the obtained sol-gel phase transition temperature is about 5 °C higher, as shown in Figure 6a. Differences in the sol-gel phase transition temperatures result from different viscoelastic properties of both media—cf. Figure 6c,e.

The introduction of an additional source of calcium also caused a significant increase in the mechanical resistance of the medium tested; shown in Figure 8b and Table 3. A particularly large difference in the values of modulus G' is observed at 37 °C. This is due to the fact that at this temperature, both samples are in different phases of the gelation process. Sample 3, devoid of calcium carbonate addition, occurs in the area of fast gelation, while sample 3CC is already in the slow gelation region [57]. Comparison of results obtained for the isothermal measurements indicates that the addition of calcium carbonate (sample 3CC) shortens the time after which intersection of the curves of storage and loss moduli was observed, with a simultaneous increase in the value of storage modulus G' (Table 2). The presence of calcium carbonate also shortens the reference time of fast gelation.

The critical role of glycerophosphate ions on the formation of three-dimensional spatial structure was confirmed by rheological tests for a sample containing calcium carbonate and is devoid of calcium β-glycerophosphate—sample 1CC. Analysis of changes in the values of storage modulus G' obtained in non-isothermal measurements (Figure 8d) indicates that despite the dominance of the elastic properties of the medium at low temperatures, formation of a three-dimensional cross-linked structure is possible only at temperatures above 55 °C. The determined temperature (position of the gelation point) is comparable with the phase transition temperature of the colloidal chitosan solution without the calcium β-glycerophosphate and calcium carbonate. This means that the introduction of a different type of salt into the solution without the addition of calcium β-glycerophosphate (pH buffering) does not provide the possibility to form a three-dimensional cross-linked structure at temperatures below 40 °C, and at low temperatures it causes only changes in the properties of the experimental medium from a viscous fluid to the fluid with predominant elastic properties; shown in Figure 8c.

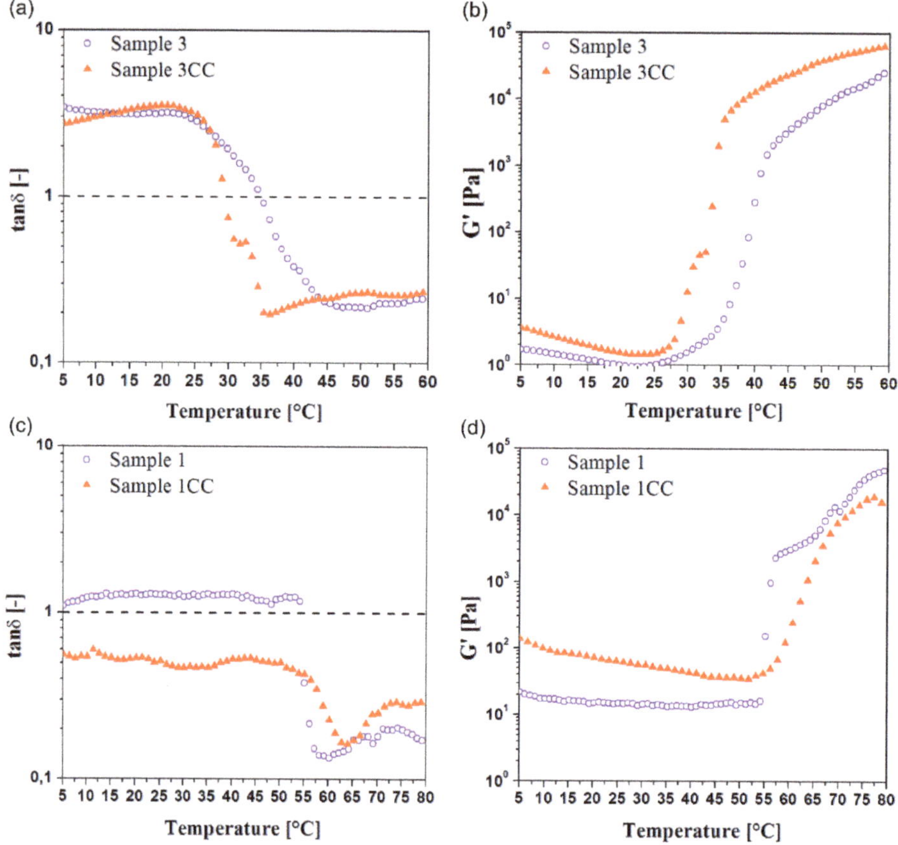

Figure 8. Experimental curves tanδ as a function of temperature for sample 3 and 3CC (**a**), and sample 1 and 1CC (**c**). Experimental curves of storage modulus G′ as a function of temperature for sample 3 and 3CC (**b**), and sample 1 and 1 CC (**d**).

Table 3. The values of the storage modulus G′ and the dynamic viscosity η determined on the basis of non-isothermal measurements at critical temperatures.

	Storage Modulus G′ [Pa]			Dynamic Viscosity [Pa·s]		
	5 °C	20 °C	37 °C	5 °C	20 °C	37 °C
Sample 3	1.72	1.00	14.38	1.18	0.63	1.72
Sample 3CC	3.68	1.53	8020.80	2.00	1.08	1604.20

	Experimentally Determined Gelation Temperature Tgel [°C]	Storage Modulus G′ [Pa] at Tgel	Dynamic Viscosity [Pa·s] at Tgel
Sample 3	35.4	4.25	0.85
Sample 3CC	29.6	7.00	1.41

3.4. Structural Properties

3.4.1. SEM Morphology

SEM images of the prepared chitosan scaffolds are shown in Figure 9. It can be seen that the concentration of calcium β-glycerophosphate salt has a significant influence on the scaffold architecture. An increase in the concentration of calcium β-glycerophosphate causes higher corrugation and cornification of the structure as well as reduction of pore diameters, i.e., significant surface development. In the case of the highest concentrations of calcium β-glycerophosphate (samples 4 and 5), no additional porosities were observed in the scaffold structure. For comparison, scaffolds completely devoid of calcium β-glycerophosphate form vast, flat geometries with numerous smaller pores; shown in Figure 9a.

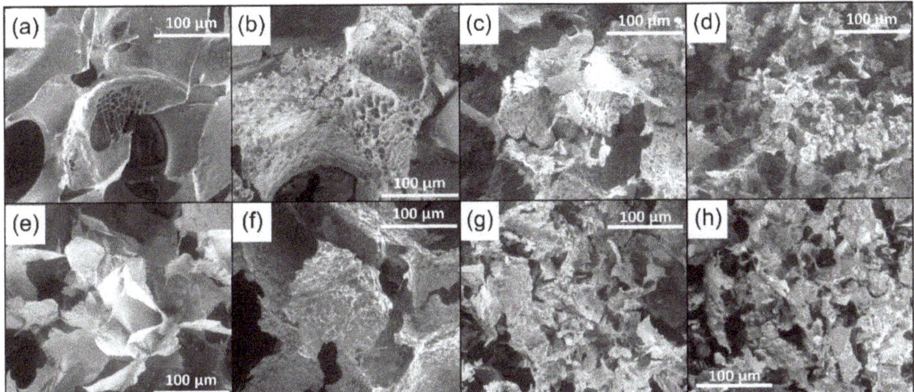

Figure 9. Microscopic images of chitosan scaffolds at 500× magnification: (**a**) sample 1, (**b**) sample 3, (**c**) sample 4, (**d**) sample 5, (**e**) sample 1CC, (**f**) sample 3CC, (**g**) sample 4CC, and (**h**) sample 5CC.

It was found that among scaffolds containing calcium β-glycerophosphate, the best and most developed structure had the scaffold prepared from a solution containing 0.1 mol/dm^3 of this salt, as shown in sample 3 (Figure 9b). The use of this concentration made it possible to obtain pores of two sizes: large ones of approx. 100 μm and numerous smaller pores with dimensions from several to a dozen μm, whose presence should positively affect the growth and proliferation of cells. An additional source of calcium has a slight effect on the obtained scaffold architecture. Regardless of the concentration of calcium glycerophosphate salt used, it was found that calcium carbonate is deposited on the surface of the resulting chitosan scaffolds in the form of flakes (Figure 9e–h), simultaneously not altering the resulting spatial structure (cf. Figure 9a–d), without the addition of calcium carbonate.

3.4.2. EDS Analysis

Table 4 shows the results of the analysis of elemental composition of the surface layer, determined using the EDS technique. On the basis of the obtained results, it was found that the introduction of an additional source of calcium in the form of calcium carbonate made it possible to obtain a ratio of calcium to phosphorus ions equal to (1.6–1.8):1. Such a ratio corresponds to calcium-phosphate ceramics commonly used in tissue engineering [63,64].

Table 4. Results for EDS analysis.

Number of Sample	Atomic Composition % (Standard Deviation)				Calcium/Phosphorus Ratio
	Carbon	Oxygen	Phosphorus	Calcium	
Sample no. 3CC	47.75 (2.67)	42.71 (1.03)	2.63 (0.98)	4.33 (1.86)	1.65
Sample no. 4CC	45.59 (0.72)	44.55 (0.24)	3.54 (0.19)	6.32 (0.40)	1.79
Sample no. 5CC	36.48 (1.52)	48.10 (0.41)	5.94 (0.25)	9.48 (1.54)	1.60

3.4.3. XPS Study

Example of the wide scan XPS spectrum, dependence of intensity against the binding energy, obtained for the scaffold enriched with calcium carbonate (sample 5CC), is shown in Figure 10. While qualitatively analyzing the results, the peaks typical for phosphates, including the hydroxyapatite, was determined [65]. The spectra reveal the formation of chitosan–calcium phosphate composites. From the quantitative analysis (Table 5), it follows that on the tested sample surface, the ratio of calcium (13.3 mass concentration %) to phosphorus (8.26 mass concentration %) is approx. 1.6:1—cf. Table 5. It can therefore be assumed that, in quantitative terms, a structure similar to hydroxyapatite was obtained.

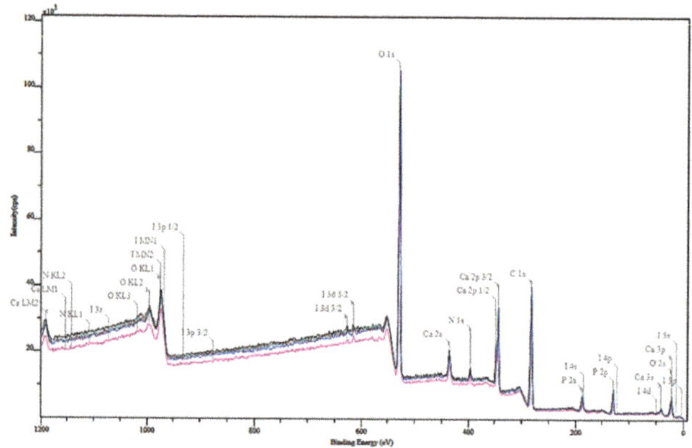

Figure 10. XPS spectrum of a sample 5CC.

Table 5. Detailed data of XPS analysis for considered peaks—sample 5CC.

Peak	Position BE (eV)		FWHM (eV)		Mass Conc %	
	Value	SD *	Value	SD	Value	SD
O 1s	532.65	0.06	3.00	0.01	41.01	1.00
Ca 2p	399.28	0.07	1.44	0.07	13.30	0.17
P 2p	347.25	0.05	1.33	0.01	8.26	0.18
C 1s	286.41	0.07	1.41	0.05	35.57	1.17
N 1s	133.15	0.09	1.65	0.04	1.86	0.05

* SD—Standard Deviation.

This value confirms the results obtained using the EDS technique—cf. Table 4. This proves the importance of introducing calcium carbonate into chitosan solutions with β-glycerophosphate as an additional source of calcium.

4. Conclusions

On the basis of the conducted tests, it has been found that it is possible to prepare thermosensitive chitosan hydrogels completely devoid of disodium β-glycerophosphate salts. The use of only calcium β-glycerophosphate during the preparation of chitosan hydrogels followed from the fact that this compound combined the benefits of buffering properties of glycerophosphate while replacing at the same time the harmful sodium ions with calcium ions. It was also shown that the use of identical concentrations of two glycerophosphate salts with sodium and calcium atoms caused a rapid change in viscoelastic properties and different conditions of phase transition of colloidal chitosan solutions. Analysis of the results of research carried out for various concentrations of calcium β-glycerophosphate has shown that the use of four times lower concentration of calcium β-glycerophosphate, relative to the molar concentration of sodium salt commonly used in studies published in the literature [19,33], makes it possible to obtain a colloidal solution that undergoes phase transition at a temperature of about 35 °C. The dependence of the position of the gelation point on the concentration of calcium β-glycerophosphate is similar to the one presented in the literature for solutions containing disodium β-glycerophosphate [19,33]. However, the obtained values of phase transition temperatures are much lower, as shown in Figure 3b.

The introduction of an additional source of calcium in the form of calcium carbonate to the experimental medium caused a change in its rheological properties, namely a decrease of phase transition temperature, as shown in Figure 7, and an increase of mechanical resistance of the obtained scaffolds, as shown in Figure 8 and Table 3. It is worth mentioning that this change significantly depended on the used concentration of calcium glycerophosphate salt. It turned out that the 3CC sample best meets the requirements for injection scaffolds in tissue engineering applications.

The conducted studies, based on the Zeta potential curves, allowed to clearly determine that the type of glycerophosphate salt (the valency of the cation forming it) has an important role in the stability of the colloidal system. This effect is more visible in the gelation region (above 30 °C), where in the case of solutions containing calcium glycerophosphate, coagulation of particles occurs. This phenomenon was not observed with commonly used sodium glycerophosphate.

The conducted analyses of elemental composition of the surface layer by EDS and XPS techniques confirmed that the obtained chitosan-hydroxyapatite composite had a calcium to phosphorus ratio characteristic for this type of bioceramics which amounted to approx. 1.6:1. Analysis of the morphology of the scaffolds indicates that the scaffolds for tissue culture have a highly interconnected surface and spatial structure characterized by two classes of pore size. The use of calcium β-glycerophosphate instead of sodium β-glycerophosphate salt enables wider use of thermosensitive chitosan hydrogels in bone tissue engineering. It results from the preparation of scaffolds with much better mechanical properties, which simultaneously contain tissue-generating elements in proportions characteristic for chemical compounds which are natural building materials of bone tissue.

The conducted research provides new valuable information both in terms of application as well as for the mechanism of sol-gel phase transitions of colloidal chitosan solutions, which is still discussed in the literature.

5. Patents

The research presented in the manuscript has been submitted for patent protection at Patent Office of the Republic of Poland (P.423822).

Author Contributions: Conceptualization, P.O., A.R. and M.D.; Data curation, P.O. and A.R.; Formal analysis, P.O. and A.R.; Investigation, P.O., A.R. and J.S.; Methodology, P.O. and A.R.; Resources, P.O. and M.D.; Supervision, M.D.; Visualization, A.R.; Writing—original draft, P.O. and A.R.; Writing—review & editing, P.O. and A.R.

Funding: This research received no external funding.

Acknowledgments: The authors would like to thank Jacek Balcerzak for performing XPS measurements and useful discussions.

Conflicts of Interest: The authors declare no conflict of interest.

References

1. Hollister, S.J. Porous scaffold design for tissue engineering. *Nat. Mater.* **2005**, *4*, 518. [CrossRef] [PubMed]
2. Garg, T.; Singh, O.; Arora, S.; Murthy, R. Scaffold: A novel carrier for cell and drug delivery. *Crit. Rev. Ther. Drug Carr. Syst.* **2012**, *29*, 1–63. [CrossRef]
3. Dash, M.; Chiellini, F.; Ottenbrite, R.M.; Chiellini, E. Chitosan—A versatile semi-synthetic polymer in biomedical applications. *Prog. Polym. Sci.* **2011**, *36*, 981–1014. [CrossRef]
4. Di Martino, A.; Sittinger, M.; Risbud, M.V. Chitosan: A versatile biopolymer for orthopaedic tissue-engineering. *Biomaterials* **2005**, *26*, 5983–5990. [CrossRef]
5. Paul, W.; Sharma, C.P. Chitosan, a drug carrier for the 21st century: A review. *STP Pharma Sci.* **2000**, *10*, 5–22.
6. Sahoo, D.; Sahoo, S.; Mohanty, P.; Sasmal, S.; Nayak, P.L. Chitosan: A New Versatile Bio-polymer for Various Applications. *Des. Monomers Polym.* **2009**, *12*, 377–404. [CrossRef]
7. Chirkov, S.N. The Antiviral Activity of Chitosan (Review). *Appl. Biochem. Microbiol.* **2002**, *38*, 1–8. [CrossRef]
8. Ing, L.Y.; Zin, N.M.; Sarwar, A.; Katas, H. Antifungal Activity of Chitosan Nanoparticles and Correlation with Their Physical Properties. Available online: https://www.hindawi.com/journals/ijbm/2012/632698/ (accessed on 16 December 2017).
9. No, H.K.; Young Park, N.; Ho Lee, S.; Meyers, S.P. Antibacterial activity of chitosans and chitosan oligomers with different molecular weights. *Int. J. Food Microbiol.* **2002**, *74*, 65–72. [CrossRef]
10. Rabea, E.I.; Badawy, M.E.-T.; Stevens, C.V.; Smagghe, G.; Steurbaut, W. Chitosan as antimicrobial agent: Applications and mode of action. *Biomacromolecules* **2003**, *4*, 1457–1465. [CrossRef]
11. Cheung, R.C.F.; Ng, T.B.; Wong, J.H.; Chan, W.Y. Chitosan: An Update on Potential Biomedical and Pharmaceutical Applications. *Mar. Drugs* **2015**, *13*, 5156–5186. [CrossRef]
12. Dutta, P.K.; Dutta, J.; Tripathi, V.S. Chitin and chitosan: Chemistry, properties and applications. *J. Sci. Ind. Res.* **2004**, *63*, 20–31.
13. Friedman, A.J.; Phan, J.; Schairer, D.O.; Champer, J.; Qin, M.; Pirouz, A.; Blecher-Paz, K.; Oren, A.; Liu, P.T.; Modlin, R.L.; et al. Antimicrobial and anti-inflammatory activity of chitosan-alginate nanoparticles: A targeted therapy for cutaneous pathogens. *J. Investig. Dermatol.* **2013**, *133*, 1231–1239. [CrossRef]
14. Gibot, L.; Chabaud, S.; Bouhout, S.; Bolduc, S.; Auger, F.A.; Moulin, V.J. Anticancer properties of chitosan on human melanoma are cell line dependent. *Int. J. Biol. Macromol.* **2015**, *72*, 370–379. [CrossRef]
15. Pan, H.; Yang, Q.; Huang, G.; Ding, C.; Cao, P.; Huang, L.; Xiao, T.; Guo, J.; Su, Z. Hypolipidemic effects of chitosan and its derivatives in hyperlipidemic rats induced by a high-fat diet. *Food Nutr. Res.* **2016**, *60*. [CrossRef] [PubMed]
16. Qi, L.; Xu, Z. In vivo antitumor activity of chitosan nanoparticles. *Bioorg. Med. Chem. Lett.* **2006**, *16*, 4243–4245. [CrossRef] [PubMed]
17. Yen, M.-T.; Yang, J.-H.; Mau, J.-L. Antioxidant properties of chitosan from crab shells. *Carbohydr. Polym.* **2008**, *74*, 840–844. [CrossRef]
18. Chenite, A.; Chaput, C.; Wang, D.; Combes, C.; Buschmann, M.D.; Hoemann, C.D.; Leroux, J.C.; Atkinson, B.L.; Binette, F.; Selmani, A. Novel injectable neutral solutions of chitosan form biodegradable gels in situ. *Biomaterials* **2000**, *21*, 2155–2161. [CrossRef]
19. Cho, J.; Heuzey, M.-C.; Bégin, A.; Carreau, P.J. Physical gelation of chitosan in the presence of beta-glycerophosphate: The effect of temperature. *Biomacromolecules* **2005**, *6*, 3267–3275. [CrossRef]
20. Lavertu, M.; Filion, D.; Buschmann, M.D. Heat-induced transfer of protons from chitosan to glycerol phosphate produces chitosan precipitation and gelation. *Biomacromolecules* **2008**, *9*, 640–650. [CrossRef]
21. Qiu, X.; Yang, Y.; Wang, L.; Lu, S.; Shao, Z.; Chen, X. Synergistic interactions during thermosensitive chitosan-β-glycerophosphate hydrogel formation. *RSC Adv.* **2011**, *1*, 282–289. [CrossRef]
22. Supper, S.; Anton, N.; Seidel, N.; Riemenschnitter, M.; Schoch, C.; Vandamme, T. Rheological Study of Chitosan/Polyol-phosphate Systems: Influence of the Polyol Part on the Thermo-Induced Gelation Mechanism. *Langmuir* **2013**, *29*, 10229–10237. [CrossRef]
23. Wang, T.; Turhan, M.; Gunasekaran, S. Selected properties of pH-sensitive, biodegradable chitosan–poly(vinyl alcohol) hydrogel. *Polym. Int.* **2004**, *53*, 911–918. [CrossRef]
24. Zhu, Y.-J.; Chen, F. pH-Responsive Drug-Delivery Systems. *Chem. Asian J.* **2015**, *10*, 284–305. [CrossRef]

25. Kim, S.; Kang, Y.; Mercado-Pagán, Á.E.; Maloney, W.J.; Yang, Y. In vitro evaluation of photo-crosslinkable chitosan-lactide hydrogels for bone tissue engineering. *J. Biomed. Mater. Res. Part B Appl. Biomater.* **2014**, *102*, 1393–1406. [CrossRef]

26. Jeong, B.; Kim, S.W.; Bae, Y.H. Thermosensitive sol–gel reversible hydrogels. *Adv. Drug Deliv. Rev.* **2012**, *64*, 154–162. [CrossRef]

27. Liu, M.; Zeng, X.; Ma, C.; Yi, H.; Ali, Z.; Mou, X.; Li, S.; Deng, Y.; He, N. Injectable hydrogels for cartilage and bone tissue engineering. *Bone Res.* **2017**, *5*, 17014. [CrossRef]

28. Solouk, A.; Mirzadeh, H.; Amanpour, S. Injectable scaffold as minimally invasive technique for cartilage tissue engineering: In vitro and in vivo preliminary study. *Prog. Biomater.* **2014**, *3*, 143–151. [CrossRef]

29. Drury, J.L.; Mooney, D.J. Hydrogels for tissue engineering: Scaffold design variables and applications. *Biomaterials* **2003**, *24*, 4337–4351. [CrossRef]

30. Hennink, W.E.; van Nostrum, C.F. Novel crosslinking methods to design hydrogels. *Adv. Drug Deliv. Rev.* **2002**, *54*, 13–36. [CrossRef]

31. Zaccone, A.; Winter, H.H.; Siebenbürger, M.; Ballauff, M. Linking self-assembly, rheology, and gel transition in attractive colloids. *J. Rheol.* **2014**, *58*, 1219–1244. [CrossRef]

32. Owczarz, P.; Ziółkowski, P.; Modrzejewska, Z.; Kuberski, S.; Dziubiński, M. Rheo-Kinetic Study of Sol-Gel Phase Transition of Chitosan Colloidal Systems. *Polymers* **2018**, *10*, 47. [CrossRef]

33. Chenite, A.; Buschmann, M.; Wang, D.; Chaput, C.; Kandani, N. Rheological characterisation of thermogelling chitosan/glycerol-phosphate solutions. *Carbohydr. Polym.* **2001**, *46*, 39–47. [CrossRef]

34. Owczarz, P.; Ziółkowski, P.; Dziubiński, M. The Application of Small-Angle Light Scattering for Rheo-Optical Characterization of Chitosan Colloidal Solutions. *Polymers* **2018**, *10*, 431. [CrossRef]

35. Filion, D.; Lavertu, M.; Buschmann, M.D. Ionization and Solubility of Chitosan Solutions Related to Thermosensitive Chitosan/Glycerol-Phosphate Systems. *Biomacromolecules* **2007**, *8*, 3224–3234. [CrossRef]

36. Eeckman, F.; Amighi, K.; Moës, A.J. Effect of some physiological and non-physiological compounds on the phase transition temperature of thermoresponsive polymers intended for oral controlled-drug delivery. *Int. J. Pharm.* **2001**, *222*, 259–270. [CrossRef]

37. Dhara, D.; Chatterji, P.R. Phase Transition in Linear and Cross-Linked Poly(N-Isopropylacrylamide) in Water: Effect of Various Types of Additives. *J. Macromol. Sci. Part C* **2000**, *40*, 51–68. [CrossRef]

38. Kamińska, M.; Kuberski, S.; Maniukiewicz, W.; Owczarz, P.; Komorowski, P.; Modrzejewska, Z.; Walkowiak, B. Thermosensitive chitosan gels containing calcium glycerophosphate for bone cell culture. *J. Bioact. Compat. Polym.* **2017**, *32*, 209–222. [CrossRef]

39. Skwarczynska, A.L.; Kuberski, S.; Maniukiewicz, W.; Modrzejewska, Z. Thermosensitive chitosan gels containing calcium glycerophosphate. *Spectrochim. Acta Part A Mol. Biomol. Spectrosc.* **2018**, *201*, 24–33. [CrossRef]

40. Lisková, J.; Bačaková, L.; Skwarczyńska, A.L.; Musial, O.; Bliznuk, V.; De Schamphelaere, K.; Modrzejewska, Z.; Douglas, T.E.L. Development of thermosensitive hydrogels of chitosan, sodium and magnesium glycerophosphate for bone regeneration applications. *J. Funct. Biomater.* **2015**, *6*, 192–203. [CrossRef]

41. Wang, L.; Stegemann, J.P. Thermogelling chitosan and collagen composite hydrogels initiated with β-glycerophosphate for bone tissue engineering. *Biomaterials* **2010**, *31*, 3976–3985. [CrossRef]

42. Florencio-Silva, R.; Sasso, G.R.D.S.; Sasso-Cerri, E.; Simões, M.J.; Cerri, P.S. Biology of Bone Tissue: Structure, Function, and Factors That Influence Bone Cells. Available online: https://www.hindawi.com/journals/bmri/2015/421746/ (accessed on 16 December 2017).

43. Liu, H.; Yazici, H.; Ergun, C.; Webster, T.J.; Bermek, H. An in vitro evaluation of the Ca/P ratio for the cytocompatibility of nano-to-micron particulate calcium phosphates for bone regeneration. *Acta Biomater.* **2008**, *4*, 1472–1479. [CrossRef]

44. Silva, T.S.N.; Primo, B.T.; Silva Júnior, A.N.; Machado, D.C.; Viezzer, C.; Santos, L.A. Use of calcium phosphate cement scaffolds for bone tissue engineering: In vitro study. *Acta Cir. Bras.* **2011**, *26*, 7–11. [CrossRef]

45. Albanna, M.Z.; Bou-Akl, T.H.; Walters, H.L.; Matthew, H.W.T. Improving the mechanical properties of chitosan-based heart valve scaffolds using chitosan fibers. *J. Mech. Behav. Biomed. Mater.* **2012**, *5*, 171–180. [CrossRef]

46. Wu, G.; Yuan, Y.; He, J.; Li, Y.; Dai, X.; Zhao, B. Stable thermosensitive in situ gel-forming systems based on the lyophilizate of chitosan/α,β-glycerophosphate salts. *Int. J. Pharm.* **2016**, *511*, 560–569. [CrossRef]

47. Venkatesan, J.; Kim, S.-K. Chitosan composites for bone tissue engineering–an overview. *Mar. Drugs* **2010**, *8*, 2252–2266. [CrossRef]
48. Kucharska, M.; Walenko, K.; Lewandowska-Szumieł, M.; Brynk, T.; Jaroszewicz, J.; Ciach, T. Chitosan and composite microsphere-based scaffold for bone tissue engineering: Evaluation of tricalcium phosphate content influence on physical and biological properties. *J. Mater. Sci. Mater. Med.* **2015**, *26*, 143. [CrossRef]
49. Shavandi, A.; Bekhit, A.E.-D.A.; Ali, M.A.; Sun, Z.; Gould, M. Development and characterization of hydroxyapatite/β-TCP/chitosan composites for tissue engineering applications. *Mater. Sci. Eng. C Mater. Biol. Appl.* **2015**, *56*, 481–493. [CrossRef]
50. Zhang, Y.; Zhang, M. Calcium phosphate/chitosan composite scaffolds for controlled in vitro antibiotic drug release. *J. Biomed. Mater. Res.* **2002**, *62*, 378–386. [CrossRef]
51. Escobar-Sierra, D.M.; Martins, J.; Ossa-Orozco, C.P. Chitosan/hydroxyapatite scaffolds for tissue engineering manufacturing method effect comparison. *Rev. Fac. De Ing. Univ. De Antioq.* **2015**, 24–35. [CrossRef]
52. Kong, L.; Gao, Y.; Lu, G.; Gong, Y.; Zhao, N.; Zhang, X. A study on the bioactivity of chitosan/nano-hydroxyapatite composite scaffolds for bone tissue engineering. *Eur. Polym. J.* **2006**, *42*, 3171–3179. [CrossRef]
53. Zhang, J.; Nie, J.; Zhang, Q.; Li, Y.; Wang, Z.; Hu, Q. Preparation and characterization of bionic bone structure chitosan/hydroxyapatite scaffold for bone tissue engineering. *J. Biomater. Sci. Polym. Ed.* **2014**, *25*, 61–74. [CrossRef]
54. Maleki Dizaj, S.; Barzegar-Jalali, M.; Hossein Zarrintan, M.; Adibkia, K.; Lotfipour, F. Calcium Carbonate Nanoparticles; Potential in Bone and Tooth Disorders. *Pharm. Sci.* **2015**, *20*, 175–182.
55. Patel, V.R.; Agrawal, Y.K. Nanosuspension: An approach to enhance solubility of drugs. *J. Adv. Pharm. Technol. Res.* **2011**, *2*, 81–87.
56. Aoki, T.; Muramatsu, M.; Torii, T.; Sanui, K.; Ogata, N. Thermosensitive Phase Transition of an Optically Active Polymer in Aqueous Milieu. *Macromolecules* **2001**, *34*, 3118–3119. [CrossRef]
57. Cho, J.; Heuzey, M.-C.; Bégin, A.; Carreau, P.J. Chitosan and glycerophosphate concentration dependence of solution behaviour and gel point using small amplitude oscillatory rheometry. *Food Hydrocoll.* **2006**, *20*, 936–945. [CrossRef]
58. Kasapis, S.; Mitchell, J.; Abeysekera, R.; MacNaughtan, W. Rubber-to-glass transitions in high sugar/biopolymer mixtures. *Trends Food Sci. Technol.* **2004**, *15*, 298–304. [CrossRef]
59. Ferry, J.D. *Viscoelastic Properties of Polymers*; John Wiley & Sons: New York, NY, USA, 1980.
60. Owczarz, P.; Ry, A.; Modrzejewska, Z.; Dziubiński, M. The influence of the addition of collagen on the rheological properties of chitosan chloride solutions. *Prog. Chem. Appl. Chitin Its Deriv.* **2017**, *22*, 176–189. [CrossRef]
61. Fredrickson, G.H.; Larson, R.G. Viscoelasticity of homogeneous polymer melts near a critical point. *J. Chem. Phys.* **1987**, *86*, 1553–1560. [CrossRef]
62. Aliaghaie, M.; Mirzadeh, H.; Dashtimoghadam, E.; Taranejoo, S. Investigation of gelation mechanism of an injectable hydrogel based on chitosan by rheological measurements for a drug delivery application. *Soft Matter* **2012**, *8*, 7128–7137. [CrossRef]
63. Bose, S.; Tarafder, S. Calcium phosphate ceramic systems in growth factor and drug delivery for bone tissue engineering: A review. *Acta Biomater.* **2012**, *8*, 1401–1421. [CrossRef]
64. Pina, S.; Oliveira, J.M.; Reis, R.L. Natural-Based Nanocomposites for Bone Tissue Engineering and Regenerative Medicine: A Review. *Adv. Mater.* **2015**, *27*, 1143–1169. [CrossRef] [PubMed]
65. Arce, J.E.; Arce, A.E.; Aguilar, Y.; Yate, L.; Moya, S.; Rincón, C.; Gutiérrez, O. Calcium phosphate–calcium titanate composite coatings for orthopedic applications. *Ceram. Int.* **2016**, *42*, 10322–10331. [CrossRef]

 *processes* 

Article

Green and Facile Synthesis of Dendritic and Branched Gold Nanoparticles by Gelatin and Investigation of Their Biocompatibility on Fibroblast Cells

Quoc Khuong Vo [1], My Nuong Nguyen Thi [2], Phuong Phong Nguyen Thi [1,*] and Duy Trinh Nguyen [3,4,*]

[1]   Faculty of Chemistry, Ho Chi Minh City University of Science, Vietnam National University-Ho Chi Minh City, Ho Chi Minh City 700000, Vietnam; vqkhuong@hcmus.edu.vn
[2]   Faculty of Biology and Biotechnology, Ho Chi Minh City University of Science, Vietnam National University-Ho Chi Minh City, Ho Chi Minh City 700000, Vietnam; ntmnuong@hcmus.edu.vn
[3]   NTT Hi-Tech Institute, Nguyen Tat Thanh University, Ho Chi Minh City 700000, Vietnam
[4]   Center of Excellence for Green Energy and Environmental Nanomaterials, Nguyen Tat Thanh University, Ho Chi Minh City 700000, Vietnam
*   Correspondence: ntpphong@hcmus.edu.vn (P.P.N.T.); ndtrinh@ntt.edu.vn (D.T.N.)

Received: 3 August 2019; Accepted: 3 September 2019; Published: 18 September 2019

Abstract: In this work, gold nanostar (AuNPs) and gold nanodendrites were synthesized by one-pot and environmentally friendly approach in the presence of gelatin. Influence of gelatin concentrations and reaction conditions on the growth of branched (AuNPs) were investigated further. Interestingly, the conversion of morphology between dendritic and branched nanostructure can be attained by changing the pH value of gelatin solution. The role of gelatin as a protecting agent through the electrostatic and steric interaction was also revealed. Branched nanoparticles were characterized by surface plasmon resonance spectroscopy (SPR), transmission electron microscopy (TEM), XRD, dynamic light scattering (DLS) and zeta-potential. The chemical interaction of gelatin with branched gold nanoparticles was analyzed by Fourier transform infrared spectroscopy (FT-IT) technique. Ultraviolet visible spectroscopy results indicated the formation of branched gold nanoparticles with the maximum surface plasmon resonance peak at 575–702 nm. The structure of both nanodendrites and nanostars were determined by TEM. The crystal sizes of nano-star ranged from 42 to 55 nm and the nanodendrites sizes were about 75–112 nm. Based on the characterizations, a growth mechanism could be proposed to explain morphology evolutions of branched AuNPs. Moreover, the branched AuNPs is high viability at 100 µg/mL concentration when performed by the SRB assay with human foreskin fibroblast cells.

Keywords: nanodendrites; nanostar; fibroblast cells; gelatin; one-pot synthesis

---

1. Introduction

In recent years, multi-branched gold nanoparticles (AuNPs) have gained focus as the most promising nanomaterial for biomedical applications due to their unique properties such as surface functionalization and biocompatibility, which highly depend on structural features of the nanoparticles [1]. The use of diverse gold nanostructures in drugs delivery and site-specific cell binding have been effectively studied. For instance, some bioactive molecules such as amino acid, peptide or protein could be specifically conjugated to the multi-branched AuNPs, or the rod-shaped nanoparticles have been researched to enhance higher cellular uptake [2,3]. Thus, to expand the array of application, many synthetic methods are focused on tuning size [4], shape [3,5], composition, assembly [6] and encapsulation of AuNPs [7]. However, these synthetic methods could rely on organic solvents or toxic ingredients which limit the application of AuNPs in biological and medical fields [3,8]. This calls for the development of green preparation routes for the controlled synthesis of multi-branched AuNPs. Pathways for the synthesis of multi-branched gold nanoparticles consist of seed-mediated and one-pot synthesis. The former method is capable of producing AuNPs with tunable morphology. However, the synthesis of the branched gold nanoparticles commonly requires cetyl trimethyl ammonium bromide (CTAB) to control the growth of crystal face [9]. This surfactant is toxic and thus limiting AuNPs synthesized in biological applications [10]. Therefore, to prepare a biocompatible system containing multi-branched AuNPs with non-toxic reducing and stabilizing agents, another technique that is environmentally friendly is required.

As a natural polymer, gelatin has been recently used as a 'green' stabilizing agent to prepare AuNPs because of its non-toxicity biocompatibility [8,11] and biodegradable properties [8]. This natural polymer can be adsorbed effectively onto the particle surface and prevent the nanoparticles from aggregating. The interaction between gelatin and AuNPs has been proposed through its abundant functional groups.

Recently, though gelatin has been widely incorporated into many modified AuNPs materials such as nanocarriers for drug delivery or engineering tissue [11], the structure-directing effect of gelatin is rarely investigated. Due to the structural transformation in different conditions, gelatin could be an appropriate component for controlling the growth of gold nanoparticles. Furthermore, this hydrophilic coating allows better dispersion of the synthesized AuNPs, which could be suitable for testing in biological experiments. Suarasan et al. have reported the role of gelatin as a stabilizer in the green synthesis of gold nanoparticles. The obtained gelatin coated AuNPs was tested for stability and biocompatibility with Osteoblast cells. Additionally, the interaction between gelatin and AuNPs were also revealed through the steric and electrostatic mechanisms in aqueous media [6]. The distinct stages in the synthesis of AuNPs were investigated by Yi-Chen Wang and co-workers using the mixture of $HAuCl_4$/HEPES and gelatin solution. Based on their proposed mechanism, it was suggested that the shape and size of AuNPs were strongly dependent on the nucleation time and heating temperatures [12]. The stability of AuNPs at high ionic strength solution was greatly improved in the presence of gelatin, as reported by Sorina Suarasan et al. The obtained gelatin@AuNPs was demonstrated as the effective substrate for surface-enhanced Raman scattering (SERS) detection of Rose Bengal fluorophore [13]. By using an in-situ synthesis method with sodium citrate, the size of AuNPs was also shown to be controllable by varying concentration of gelatin solution [14].

The formation of AuNPs is substantially influenced by not only the reducing and stabilizing agents, but also by the experimental conditions. In this work, we propose a direct approach to prepare AuNPs with multi-branched morphology using gelatin as a protecting agent. Unlike previous procedures, one-pot synthetic method is developed to obtain highly branched gold nanocrystals. The advantage of this method is the absence of template or surfactants, most of which are insufficiently environmentally-friendly to be applied in biomedical engineering, medicine and biotechnology. In addition, our method requires no hazardous chemicals and occurs in aqueous solution. The multi-branched structures of AuNPs are controlled by changing the experimental conditions such as temperature, pH value and stabilizing factors. The synthesis reaction proceeded in about 45 min, and then the branched AuNPs was formed without heating. To the best of our knowledge, this preparation process for multi-branched AuNPs requires the lowest temperature. Finally, the stability of gelatin at different pH values was also investigated to propose mechanism growth of branched and dendritic structures.

2. Experimental

2.1. Materials

Gelatin (Type A) from porcine skin was purchased from Sigma-Aldrich (St. Louis, MO, USA). Tetrachloroauric (III) acid trihydrate ($HAuCl_4.3H_2O$, 99.99%), acetic acid and trichloroacetic acid were purchased from Merck (Darmstadt, Germany). Ascorbic acid ($C_6H_8O_6$), trisodium citrate dihydrate ($C_6H_5Na_3O_7.2H_2O$), anti-vimentin/anti-cytokeratin 19 anti-bodies, collagenase and sulforhodamine B (SRB) were purchased from Sigma-Aldrich. Dulbecco's Modified Eagle Medium (DMEM) and Nutrient mixture F-12 were purchased from Gibco (Dún Laoghaire, Ireland). Tris base solution was purchased from Promega (Madison, WI, USA). Camptothecin was purchased from Calbiochem (Darmstadt, Germany) and used as received. All chemical reagents and solvents were of analytical grade, and aqueous solutions were prepared with deionized water (conductivity was below 4.3 µS/cm).

2.2. Preparation of AuNPs

Prior use, glassware was cleaned using aqua regia (1:3, HNO_3: HCl, v/v), followed by rinsing with deionized water. To prepare colloidal gold solution, $HAuCl_4$ underwent reduction by ascorbic acid in the presence of gelatin. Prior to synthesis, the gelatin powder was freshly dissolved in ultrapure water and vigorously stirred for 10 min at 50 °C to avoid degradation of gelatin during storage. In a typical procedure, 100 µL of 25 mM $HAuCl_4$ solution was added into 1.0 mL gelatin (0.1%, *w/v*) solution at 40 °C of temperature, and then the mixture was magnetically stirred at 700 rpm for 45 min. Subsequently, an aqueous solution of ascorbic acid (0.5 mL, 0.1 M) was added into reaction mixture. DI water was finally added to adjust the total volume to 10 mL. During this time, the color of the reaction solution changed from transparent to pink, purple-blue and to deep blue, which is indicative of branched AuNPs. The influences of the gelatin concentrations and different experimental conditions on the synthesis of AuNPs were investigated. Additionally, shapes and sizes of the obtained gold nanoparticles were experimentally altered by varying the pH value. The certain amount of 1.00 M NaOH was added to change the pH value of the solution.

2.3. Characterization

The branched AuNPs colloidal solutions were characterized by UV-Vis adsorption spectroscopy on a UV-Vis-NIR-V670 spectrophotometer (JASCO, Japan), operated between 300 and 800 nm with scanning rate 200 nm per minute using quartz cuvettes (1-cm path length). FT-IR analysis was performed on a Tensor 27-Bruker spectrometer. 3.00 mL of analysis sample was mixed with KBr at ratio 2–5%, scanning wavenumber from 400–4000 cm^{-1}. The XRD pattern was recorded using D8 Advance—Bruker, Germany with Cu-Kα radiation in the 2θ range from 30° to 80°, operated at 40 kV. As-synthesized AuNPs colloids were centrifuged and subsequently dried under vacuum at room temperature to obtain dry powders. These powder samples were placed on glass and subjected to X-Ray, and then characterized at scanning rate of 0.02° per second. Wide-angle XRD showed a face-centered cubic structure, five peaks are assigned to (111), (200), (220), (311), and (222) facets. The morphology of nanoparticles was investigated by Transmission Electron Microscopy (TEM) using JEM-1400, Japan. The samples were dropped small amount of AuNPs colloidal solution onto the carbon-copper grids (300-mesh, Ted Pella, Inc., Redding, CA, USA) and evaporated at room temperature. The average diameter and size distribution of branched AuNPs were determined statistically with the mean value of 200 particles in several chosen areas in microphotographs.

2.4. Primary Fibroblast Culture

Human foreskin fibroblast cells were obtained from a healthy tissue donor at Department of Andrology in Binh Dan Hospital, Ho Chi Minh City, Vietnam. This work was ethically approved by the Vietnam National University, Ho Chi Minh City. Informed consent forms of tissue donors were obtained before the interview, physical examination and sample collection commenced. Tissues were collected and used for research purposes only. Tissues after collection were stored in PBS, and then split into small pieces with dimensions of 2–3 mm × 5 mm. Incubation was carried out with 0.2% (*w/v*) collagenase (Sigma) at 37 °C for 2 h. After being filtered through a cell strainer of about 70 µm diameter, the cell suspension was centrifuged at 200x *g*, 10 min and then resuspended in Dulbecco's Modified Eagle Medium: nutrient mixture F12 (DMEM/F12) (Gibco). Cell culture was performed in the DMEM/F12 media supplemented with 10% (v/v) FBS, 20 mM HEPES, 0.025 µg/mL amphotericin B, 100 IU/mL penicillin G, 100 µg/mL streptomycin at 37 °C, 5% CO_2. To allow identification of primary fibroblasts, anti-vimentin/anti-cytokeratin 19 anti-bodies (Sigma) stain was used and cell morphology was observed.

2.5. SRB Assay

The cytotoxicity of branched AuNPs was estimated by applying an SRB assay. First, 96-well plates were inoculated with cells at a culture density of 10,000 cells per well for 24 h, followed by AuNPs treatment at different concentrations for 48 h. Before washing and staining, a cold 50% (*w/v*) solution of trichloroacetic acid (Merck) was used to maintain the total protein of treated cells constant for 1–3 h. Washing and staining were carried out with 0.2% (*w/v*) Sulforhodamine B (Sigma) for 20 min in which the dye was first washed five times with 1% acetic acid (Merck), followed by dissolution in 10 mM Tris base solution (Promega). A 96-well micro titer plate reader (Synergy HT, Biotek Instruments) was used to measure the absorbance at two wavelengths of 492 nm and 620 nm. Each experiment was conducted in triplicate to produce the mean result and standard deviation. The growth inhibitor value (Inh %) was calculated as follows: Inh % = (1–[ODt/Odc] × 100) % where ODt and ODc denoted the optical density value of the test sample and the control sample, respectively. The positive control sample was camptothecin (Calbiochem).

3. Result and Discussion

3.1. Influence of Gelatin Concentration

The spherical AuNPs could be achieved with synthesis routes involving only either ascorbic acid or gelatin [13–15]. However, when being fabricated with the mixture of ascorbic acid and gelatin in an appropriate condition, produced AuNPs is highly branched. Furthermore, the presence of NaOH solution could play an important role in branched particle formation through enhanced the reduction ability of ascorbic acid [16]. Besides, the influence of gelatin on the evolution of particles shape without template was not clearly discussed. UV-Vis absorption spectra of Au colloidal solutions obtaining at different concentrations of gelatin (0, 0.5, 1.0, 1.5, 2.0 and 2.5%) were examined, as shown in Figure 1 and Table 1. Without using of gelatin (curve a), a red colloid was obtained with maximum absorbance peak located at 532 nm, indicated the optical signature for the spherical AuNPs in solution [17]. As the concentration of gelatin was increased to 2.5% (*w/v*), a significant red shift of surface plasmon resonance (SPR) band from 532 to 576 nm was recorded. The color of colloidal gold solution rapidly changed from wine red, red-purple and dark blue. Additionally, the new emerging band at 576 nm also revealed the transverse plasmon resonance of the tips, which was typically observed in branched structure of AuNPs [18]. From the UV-Vis results, the intensity of absorbance peak gradually increased from 1.091 to 1.188, corresponded to the increase of gelatin concentration (from 0.5 to 2.0% *w/v*) and the increase of the gold nanoparticles in colloidal solution. In contrast, as the gelatin concentration was 2.5%, the decrease of intensity of absorbance peak could be caused by the decrease of the formed AuNPs. Corresponding TEM results in Figure 2 could also confirm the formation of branched AuNPs. The evolution of morphology from quasi-spherical (images 2-a) to branched nanoparticles (images 2-b) was observed when increased the concentration of gelatin. The average diameter of 92.5 nm was measured by TEM (Figure 2b) with the concentration of gelatin at 2.5% (*w/v*), which was consistent with the UV-Vis results. In fact, as the diffusion of particles through the solution significantly decreases due to the increasing of gelatin densification, the possibility of a particle reaching the crystal surface is also restricted. Thus, the branched structure is specially formed under diffusion-limited growth [19]. Otherwise, a higher hydrodynamic diameter of 117.3 nm was observed from dynamic light scattering (DLS) data, providing more information about the protecting layer of gelatin on the surface of nanoparticles. According to above results, the presence of gelatin can play both of protecting and growth controlling roles in the formation of branched particles.

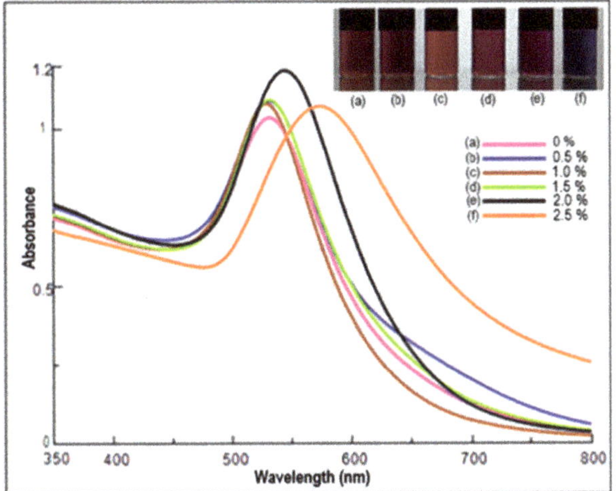

Figure 1. UV-Vis adsorption spectra of branched gold nanoparticles (AuNPs) synthesized by gelatin with different concentrations of (**a**) 0, (**b**) 0.5, (**c**) 1.0, (**d**) 1.5, (**e**) 2.0, (**f**) 2.5% (*w/v*).

Table 1. UV-Vis results of branched AuNPs synthesized by gelatin with different concentrations of gelatin.

Sample	Gelatin C % (*w/v*)	λmax (nm)	Absorbance
G1	0	532	1.037
G2	0.5	532	1.091
G3	1.0	528	1.082
G4	1.5	532	1.094
G5	2.0	546	1.188
G6	2.5	576	1.074

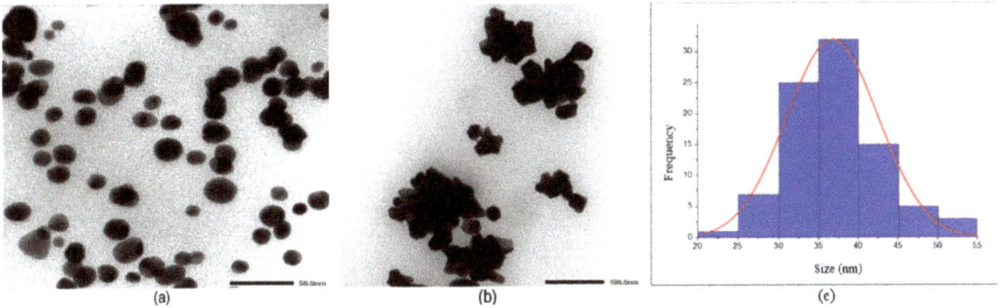

Figure 2. Representative TEM images of synthesized AuNPs with (**a**) low gelatin concentration (1.0% *w/v*) (scale 50 nm), (**b**) high gelatin concentration (2.5% *w/v*) (scale 100 nm) and (**c**) histogram of particle size distribution of synthesized AuNPs with low gelatin concentration (1.0% *w/v*).

3.2. Influence of Reaction Temperature

The growth rate of AuNPs was extremely dependent on the temperature reaction, which also determined the viscosity and stability of gelatin on the colloidal solution [20]. However, if the synthesis of AuNPs was carried out at elevated temperature (80 °C), gelatin would be more degradable in the reaction solution. According to the previous reports [20,21], the conformation of gelatin structure can be changed from triple helices to random coil when elevating the reaction temperature above the gel melting point (~35 °C). Regarding the influence of temperature on AuNPs at different temperature, the spherical AuNPs was synthesized with 1.5% (*w/v*) gelatin concentration at pH 7.0 and temperature of 40, 50, 60, 70, 80 °C. Based on the UV-Vis data (Figure 3 and Table 2), the intensity of maximum wavelength decreased when elevating the temperature from 40 °C to 80 °C and the highest intensity was obtained at 40 °C (1.14, λ_{max} = 529 nm, curve a). The maximum wavelength of SRP of synthesized AuNPs samples were around of 526–529 nm that indicated the formation of monodispered spherical nanoparticles (Figure 3, curve a–e). The results proved that the reaction temperature influenced directly on the production rate of the particles. Therefore, the reaction temperature of 40 °C was chosen for later investigation.

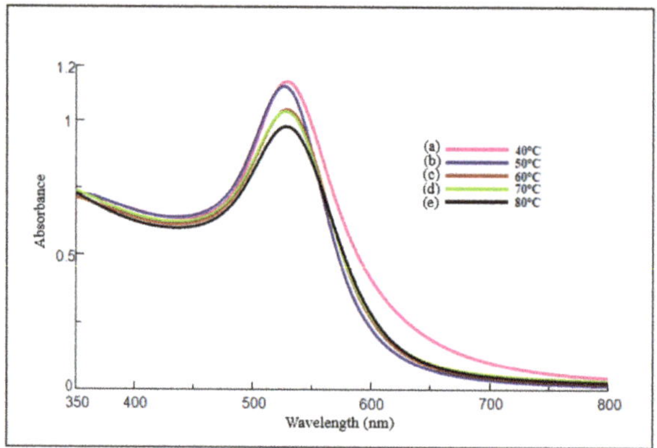

Figure 3. UV-Vis adsorption spectra of AuNPs synthesized by gelatin at different temperature of (**a**) 40, (**b**) 50, (**c**) 60, (**d**) 70 and (**e**) 80 °C.

Table 2. UV-Vis results of branched AuNPs synthesized by gelatin at different temperatures.

Temperature	λmax (nm)	Absorbance
40 °C	529	1.144
50 °C	526	1.127
60 °C	529	1.040
70 °C	528	1.035
80 °C	528	0.977

3.3. Influence of pH Condition

pH value also is a key factor for the growth of branched AuNPs. The rate of redox reaction may change when adjusting the pH value, which can be explained by the reductant forms in various pH conditions. In the previously suggested reduction mechanism, gold (III) chloride ion was reduced to Au° atom by two reductant forms: protonated (H_2Asc) and deprotonated ($HAsc^-$, Asc^{2-}) at different pH conditions [15]. At pH value of 3.0, the shoulder band located at 703 nm in Figure 4 could be the spectral overlap of multiple peaks, which were corresponded to the longitudinal plasmon resonance of the elongational tips [18]. Furthermore, it is interesting to note that the gold nanodendrites obtained at pH 3.0 (Figure 5a) was not previously reported in the literature. This dendritic structure was typically achieved with mixed solution of the cationic surfactant and anionic surfactant [15]. However, to our knowledge, it is the first time gold nanodendrites are obtained with protecting agent of gelatin. At pH 5.0, a remarkable blue shift of absorbance peak from 608 to 553 nm was observed, indicating the change of AuNPs morphology (Table 3).

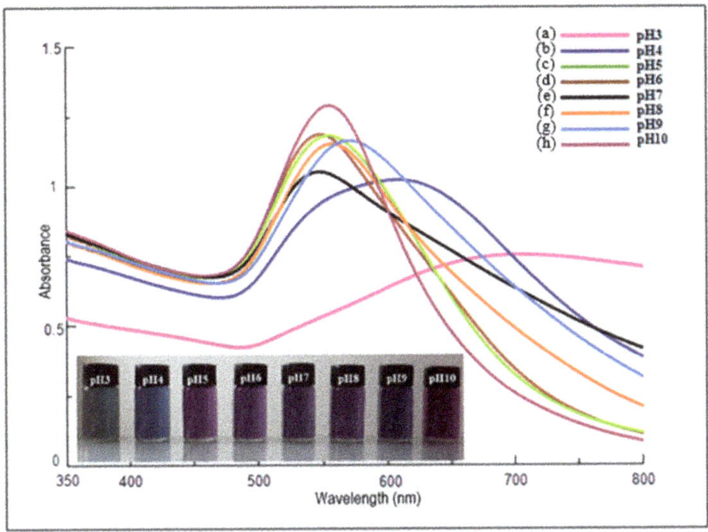

Figure 4. UV-Vis adsorption spectra of branched AuNPs synthesized by gelatin at various pH value: (**a**) pH 3.0; (**b**) pH 4.0; (**c**) pH 5.0; (**d**) pH 6.0; (**e**) pH 7.0; (**f**) pH 8.0; (**g**) pH 9.0; (**h**) pH 10.0.

Table 3. UV-Vis results of branched gold nanoparticles (AuNPs) synthesized by gelatin at various pH value.

pH Value	Wavelength (nm)	Abs
pH3	702	0.751
pH4	608	1.022
pH5	553	1.179
pH6	548	1.183
pH7	550	1.049
pH8	557	1.150
pH9	571	1.162
pH10	555	1.287

This point was also supported by TEM images where quasi-spherical and polyhedral AuNPs with diameter ranging from 21 to 32 nm were observed (Figure 5c,d). As the pH value was increased to 9.0, the absorbance peak shifted to 571 nm that could be featured for the morphology conversion. This red-shift also relates to both of an increase in particle size and the self-organization of gold nanoparticles. The intensity of SRP band in Figure 4 showed that the number of nanoparticles could be higher than forming at low pH values. According to the previous reports [19,22], this might be explained by to the viscosity and structural conformation of gelatin at various pH levels, which determined the electrostatic and steric interaction between gelatin and the AuNPs. Besides, the transfer of electron between gelatin and gold ions are strongly dependent on densification and structural transformation of gelatin. TEM image of branched AuNPs at pH 10.0 (Figure 6) showed that the tips of the branched structure were slightly longer and the average particles size was about 55 nm. The zeta potential of colloidal solution was of −5.2 mV and electrophoretic mobility mean was -0.4×10^{-5} cm^2/Vs, which is consistent with the diffusion limited growth mechanism of branched nanoparticles. The average particle size (74.3 nm) and size distribution of branched AuNPs (gelatin concentration 2.5%, *w/v* at pH

10.0) measured by DLS (Figure 6d) were slightly different from TEM images (Figure 6c). This may be due to the count of specific sizes from TEM or the layer of gelatin on the surface of nanoparticles.

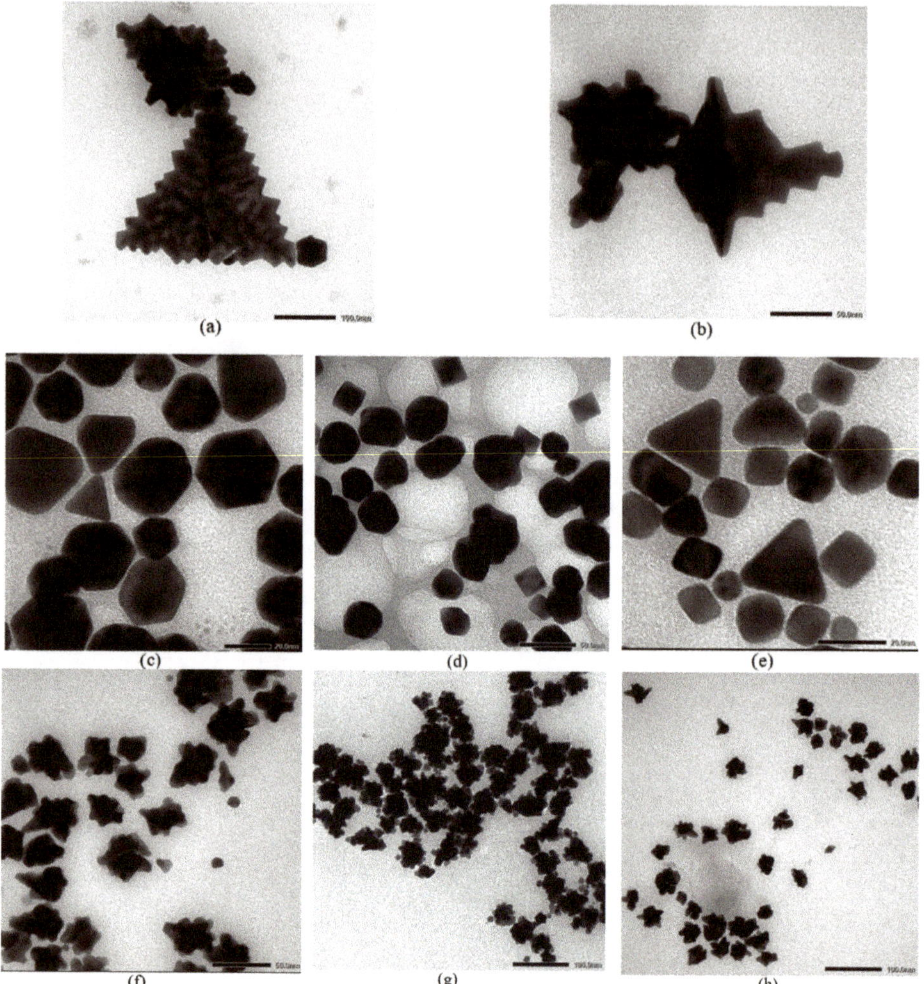

Figure 5. Representative TEM images of synthesized AuNPs at different pH value: (**a**) pH 3.0 (scale 100 nm); (**b**) pH 4.0 (scale 50nm); (**c**) pH 5.0 (scale 20 nm); (**d**) pH 6.0 (scale 50 nm); (**e**) pH 7.0 (scale 20 nm); (**f**) pH 8.0 (scale 50 nm); (**g**) pH 9.0 (scale 100 nm) and (**h**) pH 10.0 (scale 100 nm).

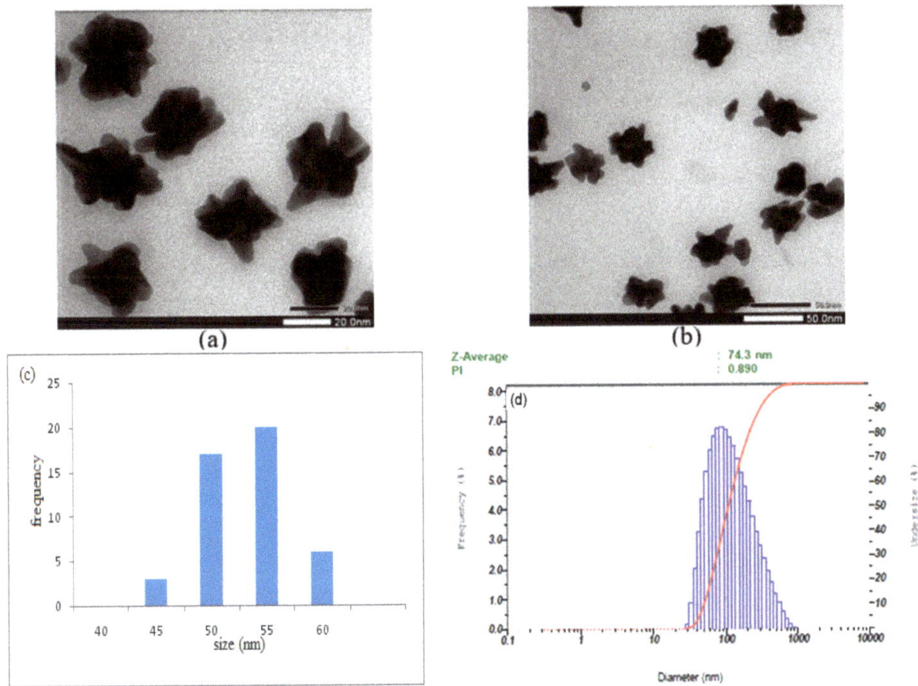

Figure 6. TEM images branched AuNPs (**a**) scale 20 nm; (**b**) scale 50 nm, (**c**) size distribution of branched gold nanoparticles (gelatin concentration 2.5%, *w/v* at pH 10.0) from TEM images and (**d**) size distribution of branched gold nanoparticles (gelatin concentration 2.5%, *w/v* at pH 10.0) from dynamic light scattering (DLS) result.

3.4. Proposed Mechanism of Branched AuNPs

The growth process of gold nanoparticles could be proposed through multi-steps. These steps are: (1) the $AuCl_4^-$ ions are reduced by ascorbic acid to generate $Au°$ atom, (2) diffusion of the pre-formed $Au°$ atom to the growth surface of gold crystal, (3) adsorption of the $Au°$ atom onto the growth surface, (4) and surface growth via the incorporation of $Au°$ atom onto the surface. Due to the limited protection of gelatin on the growth surface of Au crystal, these preformed nanoparticles can extend in other directions for branching. Interestingly, the residue of Au^{3+} ions in solution could be also reduced to form $Au°$ atom by the functional groups of gelatin molecules. At low pH value, the protonated form (H_2Asc) of ascorbic acid could take place, causing the decrease of redox reaction rate and in turn the decrease of nucleation rate. Based on the diffusion-limited growth mechanism [19], when the amount of generated $Au°$ atom in solution is low due to a slow reaction rate, the growth process is regulated by the diffusion of $Au°$ atom to the particle surface. Thus, the diffusion-controlled morphologies including dendritic and branched nanoparticle are predominantly formed in colloidal solution. Besides, the disruption of intermolecular bond led to the transformation of structure from triple helices to random coils at low pH condition. With the coils conformation, many functional groups such as thiol and amine are exposed, facilitating their interaction with the {111} crystalline facets [13], and resulting in the slow aggregation of $Au°$ atom onto growth surface to form nanodendrites. On the contrary, at higher pH value (from 5.0 to 8.0), deprotonated forms of ascorbic acid such as $HAsc^-$ and Asc^- can appear in solution, leading to the increased redox reaction rate [15]. Subsequently, the supply of growth species is also increased, and polyhedrons are formed through the crystal growth near the equilibrium state [19]. Therefore, appropriate variations of parameters such as concentration of growth species, viscosity of solution and reaction temperature could control the morphology conversion of gold nanoparticles. When the pH value of the gelatin solution reaches its isoelectric point (IEP) (PI 9.0), the gelatin dispersed in solution mainly as triple helices conformation. This caused the viscosity of solution to increase and significantly reduced the diffusion rate of growth species. The branched AuNPs was predominantly formed through the diffusion-controlled process [19]. Additionally, an increase in the solution viscosity led to a decrease in the diffusion rate of the ion Au^{3+}, and the electrostatic interaction caused by the carboxyl and amino functional groups in molecular structure could be transformed at high pH condition.

3.5. X-Ray Diffraction Analysis of Gold Nanoparticles

From XRD pattern of the AuNPs prepared in the presence of gelatin, the diffraction peaks are observed in 2-Theta values of 38.0°, 44.0°, 64.5°, 77.3° (Figure 7), corresponding to {111}, {200}, {220}, {311} lattice planes of the face-centered cubic (fcc) of gold crystalline structure, respectively [14]. The strongest intensity peak is assigned to plane {111} and the broad peaks could be due to the small nanoparticle size.

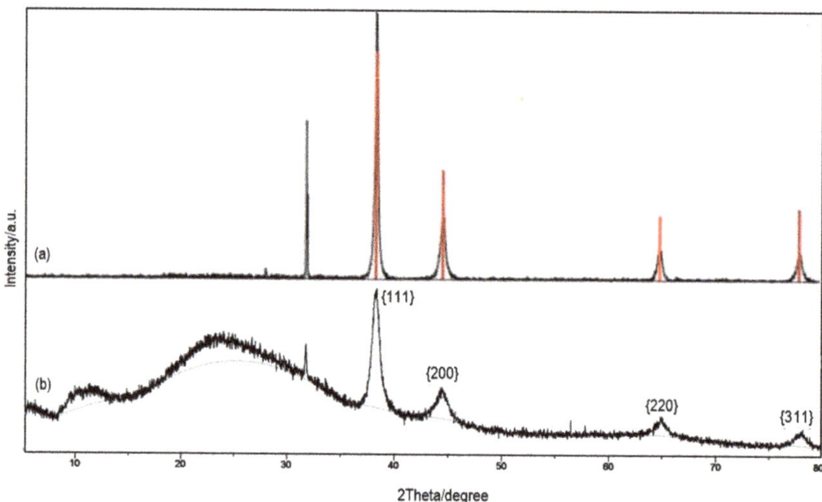

Figure 7. XRD pattern of (**a**) bare gold nanoparticles and (**b**) branched gold nanoparticles stabilized with gelatin concentration of 2.5% (*w/v*).

3.6. FT-IR Analysis of Gelatin-AuNPs Interaction

The polymeric chains of gelatin contain both of amide I, amide II and carboxyl functional groups. The molecular structure and interaction of gelatin with AuNPs were analyzed by FT-IR spectroscopy. FT-IR spectra of pure gelatin was shown Figure 8 and band positions were summarized in Table 4. In pure gelatin spectrum (curve b), the adsorption band at 1652 is attributed to amide I, representing the vibration of C=O stretching. The broadening in band at 3411 cm^{-1} (curve b) is assigned to the O-H vibration of retained water. The decrease of intensity peak at 3080 cm^{-1} (N-H stretching) (curve b) was observed in the gelatin-AuNPs spectrum (curve a), indicated the coordination of amine groups to the gold nanoparticles. N-H bending vibration of amide II groups at 1545 cm^{-1}, in plane N-H bending coupling to C-N stretching vibrations (amide III band) at 1243 cm^{-1} and the COO^{-} symmetrical stretching vibration from 1400 to 1500 cm^{-1} are the characteristic bands of pure gelatin (curve b) [13,23]. The bands from 1660–1650 cm^{-1} and 1640–1620 cm^{-1} are observed in α-helix and β-antiparallel sheets, respectively [24]. When interacted with the gold nanoparticles, the amide I band at 1652 cm^{-1} (curve b-pure gelatin) disappeared in AuNPs-gelatin spectrum (curve a), indicating that the gelatin structure was changed. Additionally, the amide III band located at 1243 cm^{-1} (curve b) is assigned to the β-antiparallel sheets secondary structure of gelatin. Based on previous reports [24,25], α-helix and β-antiparallel sheets secondary structure are characterized with the amide III band at 1330–1295 and 1250–1220 cm^{-1}, respectively. The amide III band at 1243 cm^{-1} (curve b-pure gelatin), which was attributed to C-N stretching vibration, disappears in spectrum of AuNPs-gelatin (curve a). Furthermore, the intensities of the amide I, amide II and carboxylic bands (1500–1400 cm^{-1}) also decrease dramatically. The difference in intensity can be explained due to the blocking of vibration, corresponding to the interaction between functional groups and AuNPs. Consequently, the active sites of gelatin for binding to gold nanoparticles such as amide and carboxyclic can be determined from the FTIR results.

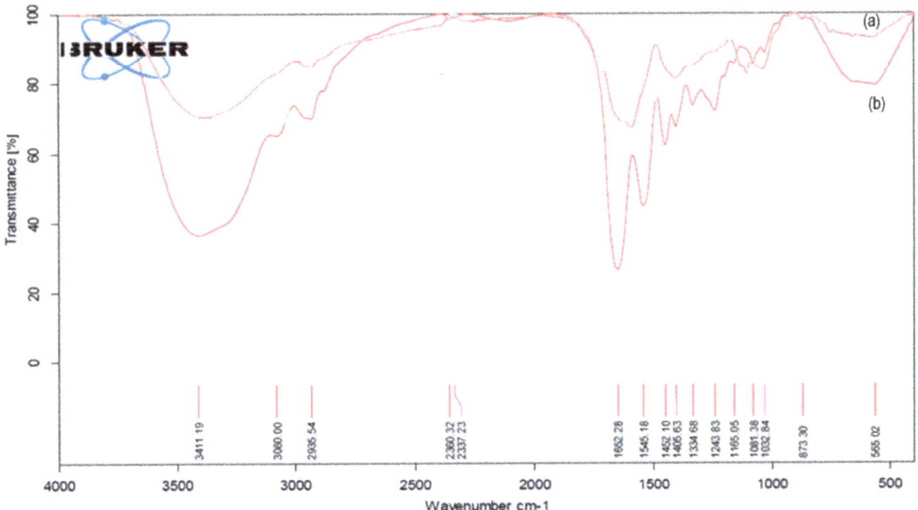

Figure 8. FTIR spectra of gelatin-AuNPs (curve **a**) and gelatin (curve **b**).

Table 4. FTIR bands position of pure gelatin and gelatin-AuNPs presented in Figure 8.

Position (cm^{-1}) Pure Gelatin (Curve b)	Position (cm^{-1}) Gelatin—Gold Nanoparticles (Curve a)	Assignment
3411	3404	Hydrogen bonds of retain water
3080	3075	N-H stretching
1652		Amide I in β-antiparallel sheets secondary structure
1640–1620		Amide I in α-helix structure
1545	1536	Amide II N-H bending vibration
1243		Amide III, in plane N-H bending coupling to C-N stretching vibrations
1330–1295		Amide III in α-helix
1250–1220		Amide III in β-antiparallel sheets
1500–1400		COO$^-$ symmetrical stretching vibration

3.7. Cell Cytotoxicity of AuNPs-Gelatin

The cytotoxic of AuNPs-incubated human foreskin fibroblast cells was evaluated by standard SRB assay and the results were presented in Table 5 and Figure 9. After incubating for 48 h, the growth inhibitor value (Ihn %) of AuNPs-incubated fibroblast cell was 17.86% (Table 5), at a dose of 100 μg/mL AuNPs-gelatin colloidal solution. In the case of treating with gelatin-AuNPs, Fibroblast cells maintained the higher survival rate than incubating with AuNPs treated cells (Table 5) at pH value 6.0~7.0, especially at high concentration (100, 200, 400 ppm). Furthermore, when treated solely with gelatin, the viability of fibroblast cell increased to 119.46%, which could be due to the presence of the natural protein. The results proved that the branched AuNPs with gelatin coating could provide highly biocompatible materials for *in-vitro* applications.

Table 5. Growth inhibitor value of fibroblast cells incubated with AuNPs and AuNPs-gelatin.

Sample	Concentration (µg/mL)	Growth Inhibitor Value (%)			
		1	2	3	Average
Gold nanoparticles	400	47.69	42.29	32.50	40.83 ± 7.70
	200	44.62	35.18	26.07	35.29 ± 9.27
	100	24.62	16.60	15.36	18.86 ± 5.02
	50	10.77	13.44	19.64	14.62 ± 4.55
	20	15.38	14.62	17.86	15.96 ± 1.69
AuNPs-gelatin	400	29.62	29.64	32.14	30.47 ± 1.45
	200	18.85	20.55	25.00	21.47 ± 3.18
	100	21.54	15.81	16.07	17.81 ± 3.23
	50	18.46	20.95	17.86	19.09 ± 1.64
	20	7.31	12.25	13.93	11.16 ± 3.44
Gelatin		−13.54	−19.34	−25.56	−19.46 ± 6.01

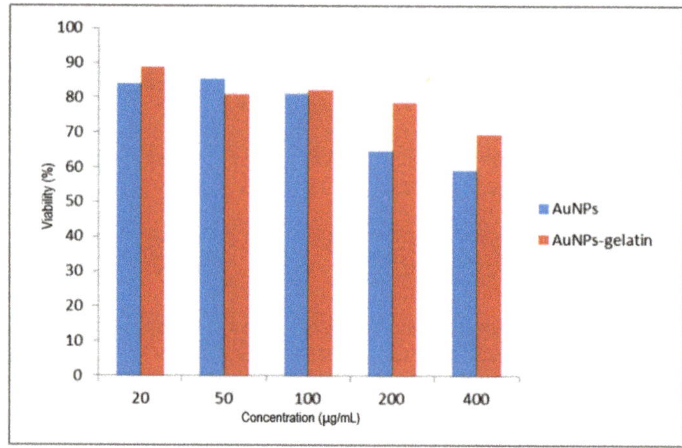

Figure 9. Cell viability of fibroblast cells treated with AuNPs and AuNPs gelatin colloidal solution after 48 h incubation.

4. Conclusions

In the present work, gold nanodendrites and gold nanostars were successfully synthesized with the natural polymer gelatin by using one-pot green synthesis method. Gelatin concentration and pH value play important roles in controlling the shape of gold nanoparticles. It was shown that a low pH value can enables the synthesis of dendrite nanoparticles with appropriate reagent ratio, while at a higher pH value, polygonal structures can be predominantly formed. Especially, when pH value of gelatin is at its IEP, branched AuNPs are synthesized. On the basic of characterization, the morphological conversion was also revealed through the proposed mechanism of branched nanoparticles growth. The interaction between of branched AuNPs and gelatin was also demonstrated, and the good biocompatibility of AuNPs was obtained in-vitro, testing with Human foreskin fibroblast cells. The branched AuNPs-gelatin could be a promising material to develop for future applications in drug delivery and cosmetics.

Author Contributions: Investigation, Q.K.V., M.N.N.T., P.P.N.T. and D.T.N.; Writing—original draft, Q.K.V.

Funding: This research funded by Vietnam National University HoChiMinh City (VNU-HCM) under grant number C2018-18-09.

Conflicts of Interest: The authors declare no conflict of interest.

References

1. Prashant, K.J.; Ivan, H.E.S.; Mostafa, A.E. Au nanoparticles target cancer. *Nano Today* **2007**, *2*, 18–29.
2. Parth, M.; Tapan, K.M. Recent advances in gold and silver nanoparticle based therapies for lung and breast cancers. *Int. J. Pharm.* **2018**, *553*, 483–509.
3. Li, J.J.; Wang, S.C.; Zhao, J.; Weng, G.J.; Zhu, J.; Zhao, J.W. Synthesis and SERS activity of super-multibranched Au-Ag nanostructure via silver coating-induced aggregation of nanostars. *Spectrochim. Acta A* **2018**, *204*, 380–387. [CrossRef] [PubMed]
4. Sun, L.; Pu, S.; Li, J.; Cai, J.; Zhou, B.; Ren, G.; Ma, Q.; Zhong, L. Size controllable one step synthesis of gold nanoparticles using carboxymethyl chitosan. *Int. J. Biol. Macromol.* **2019**, *122*, 770–783. [CrossRef]
5. Canpean, V.; Gabudean, A.M.; Astilean, S. Enhanced thermal stability of gelatin coated gold nanorods in water solution. *Colloids Surf. A Physicochem. Eng. Asp.* **2013**, *433*, 9–13. [CrossRef]
6. Li, X.; Wang, Z.; Li, Y.; Bian, K.; Yin, T.; Gao, D. Self-assembly of bacitracin-gold nanoparticles and their toxicity analysis. *Mater. Sci. Eng. C* **2018**, *82*, 310–316. [CrossRef] [PubMed]
7. Razavi, S.; Seyedebrahimi, R.; Jahromi, M. Biodelivery of nerve growth factor and gold nanoparticles encapsulated in chitosan nanoparticles for schwann-like cells differentiation of human adipose-derived stem cells. *Biochem. Biophys. Res. Commun.* **2019**, *513*, 681–687. [CrossRef]
8. Jurate, V.; Rajender, S.V. Green synthesis of metal nanoparticles: Biodegradable polymers and enzymes in stabilization and surface functionalization. *Chem. Sci.* **2011**, *2*, 837–846.
9. Murphy, C.J.; Sau, T.K.; Gole, A.M.; Orendorff, C.J.; Gao, J.; Gou, L.; Hunyadi, S.E.; Li, T. Anisotropic metal nanoparticles: Synthesis, assembly, and optical application. *J. Phys. Chem. B* **2005**, *109*, 13857–13870. [CrossRef]
10. Alkilany, A.M.; Nagaria, P.K.; Hexel, C.R.; Shaw, T.J.; Murphy, C.J.; Wyatt, M.D. Cellular Uptake and Cytotoxicity of Gold Nanorods: Molecular Origin of Cytotoxicity and Surface Effects. *Small* **2009**, *5*, 701–708. [CrossRef]
11. Ali, N.; Harpinder, S.; Wayne, C.; Ryan, T.S.; Robert, R.; Mehdi, N. Gold nanorod-incorporated Gelatin-based Conductive Hydrogels for Engineering Cardiac Tissue Constructs. *Acta Biomater.* **2016**, *41*, 133–146.
12. Wang, Y.C.; Gunasekaran, S. Spectroscopic and microscopic investigation of gold nanoparticle nucleation and growth mechanism using gelatin as a stabilizer. *J. Nanopart Res.* **2012**, *14*, 1200. [CrossRef]
13. Sorina, S.; Monica, F.; Dana, M.; Simion, A. Gelatin-nanogold bioconjugates as effective plasmonic platforms for SERS detection and tagging. *Colloids Surf. B.* **2013**, *103*, 475–481.
14. Madhav, N.; Lee, S.J.; Park, I.S.; Lee, M.H.; Bae, T.S.; Kuboki, Y.; Uo, M.; Fumino, W. Synthesis of gelatin-capped gold nanoparticles with variable gelatin concentration. *J. Nanopart. Res.* **2011**, *13*, 491–498.
15. Magdalena, L.B.; Krzysztof, P.; Marek, W.; Krzysztof, F. The kinetics of redox reation of gold (III) chloride complex ions with L-ascorbic acid. *Inorg. Chim. Acta* **2013**, *395*, 189–196.
16. Sihai, C.; Zhong, L.W.; John, B.; Stephen, H.F.; David, L.C. Monopod, Bipod, Tripod, and Tetrapod Gold Nanocrystals. *J. Am. Chem. Soc.* **2003**, *125*, 16186–16187.
17. Njoki, P.N.; Lim, I.S.; Mott, D.; Park, H.; Khan, B.; Mishra, S.; Sujakumar, R.; Luo, J.; Zhong, C. Size correlation of optical and spectroscopic properties for gold nanoparticles. *J. Phys. Chem. C* **2007**, *111*, 14664–14669. [CrossRef]
18. Colleen, L.N.; Hongwei, L.; Jason, H.H. Optical Properties of Star-Shaped Gold Nanoparticles. *Nano Lett.* **2006**, *6*, 683–688.
19. Hiroaki, I. Self-Organized Formation of Hierachical Structures. *Top. Curr. Chem.* **2007**, *270*, 43–72.

20. Damjan, P.; Sanjin, M.; Miroslav, P.; Mario, B. Role of microscopic phase separation in gelation of aqueous gelatin solution. *RSC Adv.* **2014**, *10*, 348–356.
21. Segtnan, V.H.; Isaksson, T. Temperature, sample and time dependent structural characteristics of gelatine gels studied by near infrared spectroscopy. *Food Hydrocoll.* **2004**, *18*, 1–11. [CrossRef]
22. Zhang, W.; Huang, Y.; Wang, W.; Huang, C.; Wang, Y.; Yu, Z.; Zhang, H. Influence of pH of Gelatin Solution on Cycle Performance of the Sulfur Cathode. *J. Electrochem. Soc.* **2010**, *157*, A443–A446. [CrossRef]
23. Shihua, T.; Youqun, L. Interaction via in situ binding of CdS nanorods onto gelatin. *J. Colloid Interface Sci.* **2011**, *360*, 71–77.
24. Hashim, D.M.; Man, Y.B.C.; Norakasha, R.; Shuhaimi, M.; Salmah, Y.; Syahariza, Z.A. Potential use of Fourier transform infrared spectroscopy for differentiation of bovine and porcine gelatins. *Food Chem.* **2010**, *118*, 856–860. [CrossRef]
25. Fu, F.-N.; Deoliveira, D.B.; Trumble, W.R.; Sarkar, H.K.; Singh, B.R. Secondary Structure Estimation of Proteins Using the Amide III Region of Fourier Transform Infrared Spectroscopy: Application to Analyze Calcium-Binding-Induced Structural Changes in Calsequestrin. *Appl. Spectrosc.* **1994**, *48*, 1432–1441. [CrossRef]

Article

Analysis of the Degradation Process of Alginate-Based Hydrogels in Artificial Urine for Use as a Bioresorbable Material in the Treatment of Urethral Injuries

Jagoda Kurowiak *, Agnieszka Kaczmarek-Pawelska *, Agnieszka G. Mackiewicz and Romuald Bedzinski

Department of Biomedical Engineering, Institute of Material and Biomedical Engineering, Faculty of Mechanical Engineering, University of Zielona Góra, Licealna 9 Street, 65–417 Zielona Gora, Poland; a.mackiewicz@ibem.uz.zgora.pl (A.G.M.); r.bedzinski@ibem.uz.zgora.pl (R.B.)
* Correspondence: j.kurowiak@ibem.uz.zgora.pl (J.K.); a.kaczmarek@ibem.uz.zgora.pl (A.K.-P.);
  Tel.: +48-683282359 (J.K.); +48-683282576 (A.K.-P.)

Received: 31 January 2020; Accepted: 4 March 2020; Published: 6 March 2020

Abstract: Hydrogels from natural polymers such as sodium alginate have great potential in regenerative medicine because of their biocompatibility, biodegradability, mechanical properties, bioresorption ability, and relatively low cost. Sodium alginate, a polysaccharide derived from brown seaweed, is the most widely investigated and used biomaterial in biomedical applications. Alginate dressings are also useful as a delivery platform in order to provide a controlled release of therapeutic substances (e.g., pain-relieving, antibacterial, and anti-inflammatory agents). In our work, we aimed to analyze process of degradation of alginate hydrogels. We also describe an original hybrid crosslinking process by using not one, as usual, but a mixture of two crosslinking agents (calcium chloride and barium chloride). We proved that different crosslinking agents allow producing hydrogels with a spectrum of mechanical properties, similar to the urethra tissue. Hydrogels were formed using a dip-coating technique, and then examined by mechanical testing, FTIR (Fourier-Transform Infrared Spectroscopy), and resorption on artificial urine. Obtained hydrogels have a different degradation rate in artificial urine, and they can be used as a material for healing of urethra injuries, especially urethra strictures, which significantly affect the quality of life of patients.

Keywords: sodium alginate; hydrogel material; regenerative medicine; urethra

---

1. Introduction

In tissue regeneration, biodegradable materials are desirable, because they can support tissue healing in the time of need. Biodegradable hydrogels are well known, and they have great potential as biomaterials in regenerative medicine. Currently, hydrogel materials are used on the commercial market as wound dressings, in stomatology, as a coating for catheters, and as drug carriers [1]. Hydrogels made from synthetic polymers more often contain polyethylene glycol (PEG), polyvinyl alcohol (PVA), polyvinylpyrrolidone (PVP), or polyamide [2,3]. Hydrogels made from natural polymers have great potential in regenerative medicine because of their biocompatibility, biodegradability, mechanical properties, bioresorption ability, and relatively low cost. In the group of natural polymers, the most popular and desired hydrogels are gelatin, chitosan, and sodium alginate. They are in use for clinical applications since the 1960s, when they were initially used in ocular applications including contact lenses and intraocular lenses due to their favorable oxygen permeability and lack of irritation leading to inflammation and foreign body reaction, which was observed with other plastics. Hydrogels as crosslinked polymeric networks contain hydrophilic groups that promote swelling due to interaction with water. Polymer chain arrangement and the presence of almost 75% water allows building a structure similar to the extracellular matrix, thereby supporting tissue integration [1,4]. The presence of water in hydrogel and its swelling in the bodily fluids support the diffusion of ions, proteins, and other components that are important for tissue healing. The water which accumulates between the hydrogel polymer chains may also be a carrier for drugs. The swelling and degradation ratio of hydrogel drug carriers influence the drug release close to the target, e.g., skin injury. The polymer chain arrangement and high water content are important in the proliferation phase of wound formation, where epithelization occurs and newly formed granulation tissue consisting of endothelial cells, macrophages, and fibroblasts begins to cover and fill the wound area by producing a new extracellular matrix (ECM). The presence of the new extracellular matrix is crucial for proper healing because it provides conditions for sustaining the cells and blood vessels, which provide nutrients needed to restore the tissue integrity and homeostasis [5]. The extracellular matrix (ECM) also serves as a porous and pliable scaffold for supporting the movement of the cells, nutrients, and growth factors through the wound environment. Studies on the chemical composition of the ECM during wound healing indicated that the deposition of a number of matrix components is different in chronic and acute wounds. The native ECM has the ability to coordinate stromal cells for synthesizing new tissue (e.g., if injured) with control over the tissue structure through the regulation of cell phenotype. Biomaterials act as an artificial ECM and, in the context of a scaffold, they must have biological and mechanical properties that match the native body tissue, and they must facilitate the localization and delivery of cells and/or transforming factors to the desired sites in the body. This includes providing a two- or three-dimensional space for the formation of new tissues with appropriate structure and guiding the development of new tissues with appropriate function [6]. Therefore, drug-incorporated scaffolds with a structure similar to ECM are particularly promising for synergistically accelerating the healing process of chronic wounds [7]. A wide range of applications of hydrogels results from their valuable properties, primarily their ability to swell, which allows water to be accumulated in a particular or desired localization, e.g., in a slow-healing wound. Polymeric hydrogels, due to their specific three-dimensional (3D) structure, are highly biocompatible with soft tissues, and they are able to regenerate the damage tissues and restore their functions. These characteristics make them potential candidates for use in biomedical and pharmaceutical applications such as tissue engineering [8–10]. Chen et al. [11] proved that the 3D microenvironment, especially the elastic and relaxation moduli of the cell culture hydrogel matrix, can regulate the metabolic properties of the living cells. The authors showed that the molecular weight and chemical properties of the polymers used for the cell culture matrix influence the three-dimensional cell culture system. Summa et al. [12] assumed that hydrogel films based on sodium alginate and their polysaccharide chains organized in a 3D structure on a micro level enhance wound healing and induce cell proliferation, whereas the presence of antiseptic povidone iodine in the films prevents bacterial infections.

Sodium alginate, a polysaccharide derived from brown seaweed, is the most widely investigated and used biomaterial in biomedical applications. Alginate is highly hydrophilic and able to absorb wound exudate while maintaining a moist microenvironment. Alginate dressings are also useful as a delivery platform in order to provide controlled release of therapeutic substances (e.g., pain-relieving, antibacterial, and anti-inflammatory agents) [12]. Moreover, their high biological affinity supports tissue integration, and they have biomechanical properties similar to tissues [13–18]. Biomechanical compatibility with tissue is a crucial factor that determines successful tissue regeneration [19]. Researchers, like Barros at al., made attempts to assess hydrogel materials based on sodium alginate, used in ureteral stents, to unblock the ureter and restore the urinary transport function from kidneys to the bladder. The motivation to carry out research on stents for the genitourinary system is the increasing number of male patients with strictures in the urethra, caused by tissue hypertrophy, which impairs the urine flow [20–22]. During the development of novel stents made from hydrogels, particular attention should be paid to the material's mechanical characteristics, especially stress and deformation, as well as the stent shape, which should provide a proper interaction in the place of contact between the implant and the tissue [22]. Sodium alginate is a popular and cheap material obtained from brown seaweed. Sodium alginate is anionic and hydrophilic, which enables obtaining a crosslinked polymeric structure with elasticity and strength proper for tissue regeneration. Sodium alginate contains residues of the α-L-guluronic acid (G-blocks) and β-D-mannuronic acids (M-blocks). The G-blocks have a snake structure, while M-blocks have of a more stretched structure than G-blocks. G-blocks and M-blocks are connected by glycosidic bonds. The guluronic and mannuronic acid residues in the hydrogel structure may arrange in three variants/configurations: two β-D-mannuronic acids residues (MM), two α-L-guluronic acid residues (GG), or alternating connections of the M- and G-blocks (MG). The hydrogels obtained from alginate with a higher number of β-D-mannuronic acid residues have higher elasticity than those made from alginate with a higher number of α-L-guluronic acid residues [13,14,16,17,23]. One of the mechanisms leading to the formation of alginate hydrogels is crosslinking by divalent cations, like Ca^{2+}. The G-blocks in the alginate bond with cation in the polymer chains, forming a structure called an "egg-box". The divalent cation bonds covalently with two polymer chains and four G-blocks (two from each chain). The number of divalent ions and their dimensions determine the physicochemical properties of the hydrogel. The affinity of alginates for divalent ions decreases in the following order: Pb > Cu > Cd > Ba > Sr > Ca > Co, Ni, Zn > Mn [24]. A hydrogel structure crosslinked by divalent ions may be spread by single-valent ions like sodium or potassium, which exchange the divalent ions in the "egg-box" structure, leading to a decrease in the attraction force between G-blocks, thereby increasing the distance between them until they become free from the polymeric structure.

The research described in the present work is focused on the alginate crosslinking process and the resorption on artificial urine, which is rich in sodium and potassium. Motivation for the research and analysis was based on actual problems with tissue regeneration in the urinary tract, especially the urethra. In our research, we developed a hydrogel material able to prevent and to heal urethra strictures, caused by wound healing after surgical treatment, inflections, and other reasons. Alginates are natural polymers, which can build hydrogels with properties similar to soft tissues. In this paper, we aimed to analyze the process of alginate hydrogel crosslinking and the degradation of the alginate hydrogels in artificial urine. We also describe an original hybrid crosslinking process using not one, as usual, but a mixture of two crosslinking agents (calcium chloride and barium chloride). It is assumed that different cross-linking agents allow producing hydrogels with different degradation rates in artificial urine, with a spectrum of mechanical properties, similar to the urethra tissue. The aim was to obtain an alginate-based hydrogel that can be used as a biodegradable material for the healing of urethra injuries, especially urethra strictures, which significantly affect the quality of life of patients. In this work, we analyze two processes: crosslinking and degradation of alginate hydrogels.

2. Materials and Methods

2.1. Materials

All chemical components used in the research were as follows: alginic acid sodium salt (Sigma-Aldrich, 180947) with the parameters viscosity 15–25 cP, 1% H_2O, pH 6.5–8.5, molecular weight 120–190 g/mol, and density 1.601 g/cm^3; calcium chloride (Sigma-Aldrich), barium chloride (Sigma-Aldrich), urea (Sigma-Aldrich), sodium chloride (Sigma-Aldrich), ammonium chloride (Chempur), sodium sulfate (Avantor Performance Materials Poland S.A.), and monobasic sodium phosphate (Avantor Performance Materials Poland S.A.). All solutions were prepared using deionized water. The artificial urine solution was prepared according to the recipe described by Mayrovitz and Sims [25], which is presented in Table 1.

Table 1. Components of artificial urine with the amount needed to prepare 1 L of solution.

Source	Chemical Reagent	Quantity in Grams
Sigma-Aldrich	Urea	124.9
Sigma-Aldrich	NaCl	45.0
CHEMPUR	NH_4Cl	12.85
Avantor Performance Materials Poland S.A.	Na_2SO_4	15.0
Avantor Performance Materials Poland S.A.	NaH_2PO_4	13.7

The sample preparation, as well as the determination of the degradation process, mechanical properties, and FTIR (Fourier-Transform Infrared Spectroscopy) research methodology, is described below. The nomenclature of the samples and the examined concentrations of sodium alginate and crosslinking agents are presented in Table 2. The amounts of alginate and crosslinking agents were selected on the basis of the results of other researchers [21,26].

Table 2. Type and name of prepared materials based on sodium alginate.

Type of Crosslinker	Concentration of Crosslinker	30 mg/mL Sodium Alginate	50 mg/mL Sodium Alginate
$CaCl_2$	1.0 M	3.0Ca1.0	5.0Ca1.0
$CaCl_2/BaCl_2$		3.0CaBa1.0	5.0CaBa1.0

2.2. Samples Preparation

To prepare samples with different amounts of sodium alginate and different crosslinking levels, 30 mg/mL and 50 mg/mL sodium alginate was dissolved in deionized water. The solutions were mixed for 16 h at room temperature using an electromagnetic stirrer. After stirring, the alginate solutions were sonicated using a sonic washer to remove the air bubbles. Formation of alginate hydrogel samples was carried out by crosslinking of the sodium alginate solution using three different crosslinking solutions with 1 M concentration: calcium chloride, barium chloride, and a mixture of calcium and barium chloride (volume ratio 1:1). The samples were formed as beads with dimensions of 10 mm and 2 mm thickness (for resorption and FTIR) and as tubes of 5–6 mm width, 6–7 mm free outer radius, and 0.55–0.63 mm thickness for mechanical testing (Figure 1a). The mechanical examination of the tubular hydrogel samples using a static tensile test is related to the proposed application of the obtained hydrogel as a biodegradable material for tubular stents for healing of the urethra injuries, which have a tubular shape. The alginate beads were formed in a 3D-printed mold, and the alginate tubes were formed using the dip-coating method on a polymer matrix (Figure 1b). The material was manufactured at an ambient temperature of 23 °C.

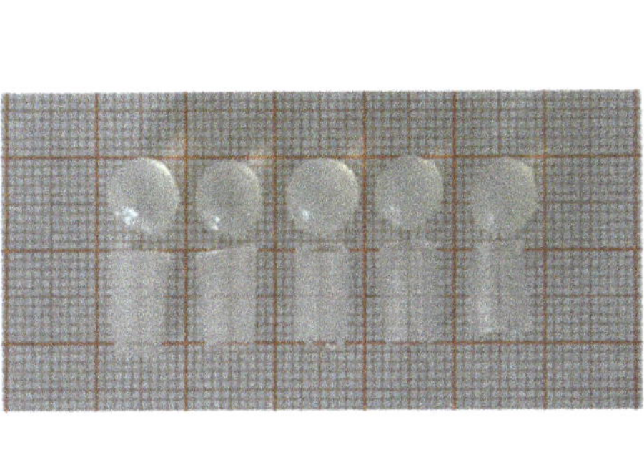

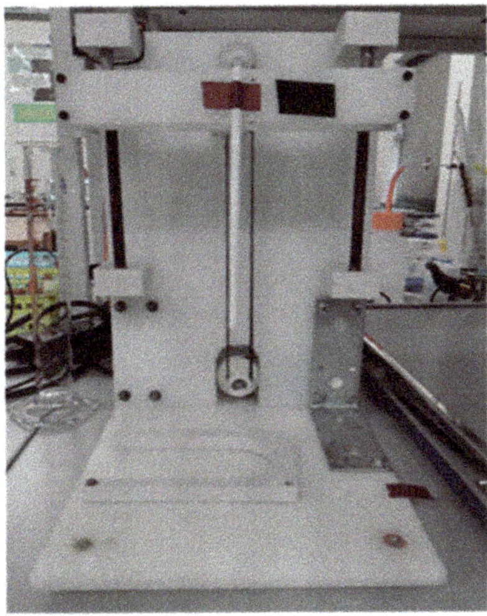

(a) (b)

Figure 1. (**a**) Samples of hydrogel material; (**b**) station for forming hydrogel material in the shape of tubes.

2.3. Degradation, Swelling, and Water Loss

Material degradation studies were performed in an artificial urine solution prepared according to the recipe AU-5 of Mayrovitz and Sims [25]. The resorption was performed for all materials, repeating the procedure three times for each type of sample with specified concentrations of sodium alginate and crosslinker. The resorption studies were carried out for 14 days via continuous immersion of samples in artificial urine. Each sample was immersed in sterile containers with 15 mL of artificial urine. The temperature of the artificial urine solution and test conditions was 23 °C. After the set time of immersion in urine, the hydrogel was drained off with a laboratory filter for 15 s. The material resorption or swelling was calculated based on Equation (1).

$$\text{Resorption or swelling (\%)} = [(M_0 - M_1)/M_0] \times 100, \tag{1}$$

where M_0 is the initial mass of the sample before immersion in the artificial urine solution, and M_1 is the wet mass of the sample after immersion in the artificial urine solution.

The loss of water from the tested samples was determined on the basis of an analysis of change in the sample weight after drying for 2.5 h in a dryer while maintaining a constant temperature of 70 °C. The loss of water was calculated based on Equation (2).

$$\text{Water loss (\%)} = [(M_1 - M_s)/M_1] \times 100, \tag{2}$$

where M_1 is the wet mass of the sample after immersion in the artificial urine solution, and M_s is the sample mass after drying in an oven at 70 °C for 2.5 h.

2.4. Examination of Mechanical Properties

Mechanical testing of the obtained tubular samples (Figure 2) was carried out using a testing machine with an electromechanical actuator Zwick Roel EPZ 005. The alginate hydrogel tubes were examined in a static tensile test in the radial direction with a speed of 5 mm/min, according to Reference [27]. To measure the geometrical parameters of the samples, especially the surface area, the samples were photographed. The photos were analyzed and the surface area was calculated using MS Office 2016. The Young's modulus value was calculated for the linear region of deformation at 0.05 mm/mm. The tests were conducted at a temperature of 23 °C.

Figure 2. Tensile test tubular hydrogel material on the machine.

2.5. Fourier-Transform Infrared Spectroscopy (FTIR)

To determine the composition of the alginate samples and the crosslinking agent bondage, the samples were examined using FTIR-ATR (Fourier-Transform Infrared Spectroscopy-Attenuated Total Reflection). The tests were carried out on a Thermo Scientific™ Nicolet™ iS50 FTIR Spectrometer. The absorption was measured within the range of 500 to 4000 cm^{-1} using an ATR detector with a resolution of 16 scans per spectrum. Before testing, the samples were dried for 12 h to get rid of the water, which could distort the spectrum.

3. Results and Discussion

3.1. Degradation, Swelling, and Water Loss

Resorption of the ball-shaped sodium alginate material was carried out for samples with different concentration combinations of sodium alginate and crosslinker. The results obtained for the pre-drained material are presented graphically in Figure 3. Their analysis shows that the type of crosslinking substance affected the resorption time of the material. It was observed that samples with both 30 mg/mL and 50 mg/mL sodium alginate dissolved in 100 mL of deionized water crosslinked with different combinations of crosslinkers showed different degrees and times of degradation. Samples of the materials crosslinked with Ca^{2+} cations featured slow material degradation, causing point-blank swelling in extreme cases by up to 50%. This was most likely due to the penetration and replacement of free bonds and cation sites by H_2O molecules into the deeper layers of the material, which were exposed after partial degradation in the initial period of immersion in artificial urine. In case of the 5.0Ca1.0 sample during the 14-day test of resorption, the sample mass grew, showing large swelling, while the 3.0Ca1.0 sample did not begin to degrade until 12 days. The hydrogel material crosslinked with the Ca^{2+}/Ba^{2+} cation combination showed degradation throughout the entire 14-day resorption study cycle. Both 5.0CaBa1.0 and 3.0CaBa1.0 samples lost weight. After completion of the degradation studies (14 days), their mass decreased by an average of about 18%. Assuming that the conducted biodegradation tests were accelerated tests, because the sample was immersed continuously in the artificial urine, it can be assumed that the material would be able to stay in the urethra for up to 77 days

(11–12 weeks). The water loss in the material after drying at 70 °C for 2.5 h for all samples was about 75% on average.

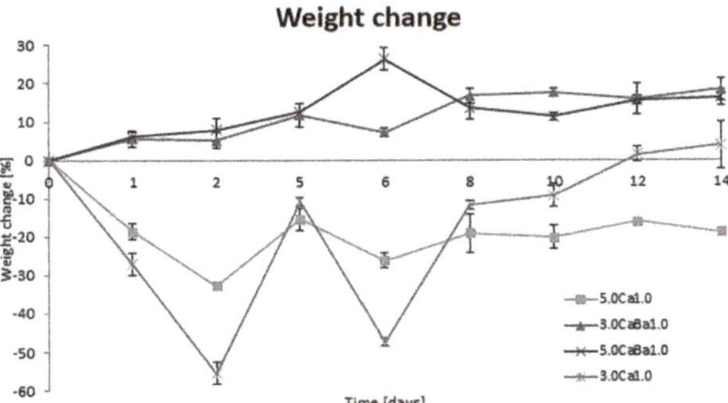

Figure 3. Assessment of wet mass change of hydrogel materials based on sodium alginate crosslinked with Ca^{2+} cations and a Ca^{2+}/Ba^{2+} mixture during continuous immersion in an artificial urine solution after pre-draining in a laboratory filter for 15 s.

The obtained results are comparable to those described in the literature [21,28], where it was also observed that the time and rate of the resorption are affected by the ratio of the concentrations of the polymer used and the crosslinker. An important aspect when analyzing the degradation of the hydrogel materials dedicated to future application as stents of the genitourinary system is the amount of urine excreted throughout the day. According to the research carried out by Barros, it appears that their proposed material is intact for up to three days, while, after 10 days, there is full degradation [21].

3.2. Mechanical Properties of Alginate Hydrogels

Mechanical examinations of the tubular hydrogel samples were carried to determine the strain characteristics and elasticity in terms of the Young's modulus value. The average values of Young's modulus following six repetitions for each type of material are presented in Figure 4. The results show that using the mixture of calcium and barium ions as a crosslinking agent led to the formation of hydrogels with a higher value of longitudinal elasticity in comparison to crosslinking with only calcium ions. Moreover, the Young's modulus value was the highest for samples with 50 mg/mL sodium alginate. The obtained mechanical properties of the alginate hydrogels were similar to those of the urethra, and the potential of using the proposed hydrogels for urethra injury healing is promising. The urethra mechanical properties described in the literature are different. Yao et al. [29] provided evidence of human urethra strength with a Young's modulus of 2.4 MPa. More researchers described the White New Zealand Rabbit urethra properties. Zhang et al. [30] published that the Young's modulus for this tissue is 0.25 MPa, and Feng et al. [31] stated that it is 0.5 MPa. The differences in this value for animal models may be caused by features of individual variation. The tissue properties of animal and human models depend on the age of the model, as well as its health and living conditions. The human urethra strength may also depend on these factors. Mechanical properties of alginate hydrogels were also described by researchers; however, as they used other sodium alginate concentrations and calcium ions as a crosslinking agent, a comparison of obtained results is difficult. In the literature, there are no results related to a mixture of crosslinking ions. A comparison of the obtained results in terms of the mechanical strength for different sodium alginate hydrogels and crosslinking agents described in the literature is presented in Figure 4.

The mechanical characteristics of the urethra show the features of a hyper-elastic material. In order for a stent structure to be used as an implant to unblock fibrosis in the urethra during flow, the stiffness of the material should be close to or higher than the stiffness of the tissue into which it will be inserted. It is best for a hydrogel stent to work in a resilient deformation range. The study showed that, depending on the hydrogel concentration applied and the type of crosslinking agent, values close to or higher than the Young's modulus of a rabbit urethra could be obtained. Further evaluation of the results obtained could be achieved by conducting clinical trials using the proposed materials in the form of a urethra stent.

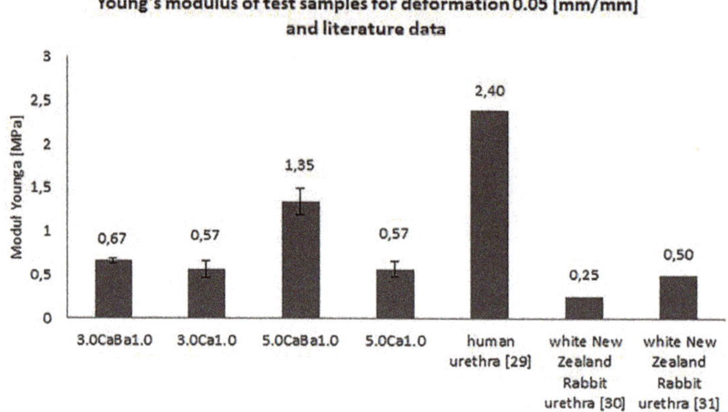

Figure 4. Young's modulus values for tubular alginate samples obtained in mechanical testing.

3.3. Fourier Transform Infrared Spectroscopy (FTIR-ATR)

The FTIR-ATR spectra for the alginate hydrogel samples crosslinked with calcium and a mixture of calcium and barium ions are presented in Figure 5. The analysis revealed differences in the absorbance values for the examined sodium alginate concentrations and crosslinking agents. The absorbance in the range 3000 to 3500 cm^{-1}, characteristic for the stretching of hydroxyl bonds (O–H), was the highest for samples crosslinked with calcium ions. Tensile vibration bands of symmetrical and asymmetrical CO carboxyl bonds and divalent ions were observed at values ranging from 1400 to 1600 cm^{-1}, while bands from 1000 to 1200 cm^{-1} characteristic for C–O bond vibration, at 800 cm^{-1} characteristic for mannuronic acid, and at 700 cm^{-1} characteristic for guluronic acid were also observed upon binding crosslinking ions [32,33].

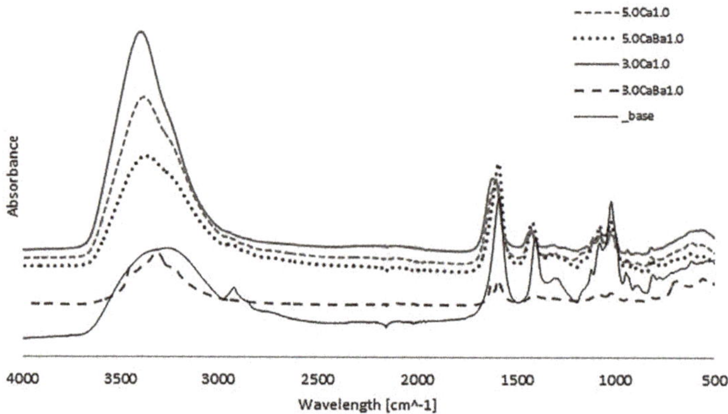

Figure 5. FTIR (Fourier Transform Infrared Spectroscopy) spectrum analysis for alginate samples.

4. Conclusions

Hydrogel materials, in particular those based on natural polymers, have great potential, and they gave promising results in tissue engineering research. Their properties including their 3D structure, high water content, and many others support tissue regeneration. The hydrogels based on natural polymers, especially on sodium alginate, allow increasing the effectivity and development of novel products and knowledge in tissue engineering and regeneration medicine [20,21]. The alginate-based hydrogels developed in the presented research show their potential as a material for genitourinary tissue healing. The obtained alginate hydrogels have mechanical properties similar to the urethra tissue, and they contain 75% water, which is also a value compatible with the water content in tissues. The content of alginate and crosslinking with a mixture of divalent calcium and barium ions increased the tensile strength with an influence on the resorption ratio and swelling process. The results show that further research can be carried out, aimed at even more accurate analysis, leading to the improvement of the material properties of ureteral stents based on sodium alginate. By choosing the right ratio of ingredients, it is possible to obtain a material with the desired physicochemical properties and adjust the resorption process ratio. Hydrogels are materials that can be easily formed in various shapes depending on the needs and place of application in the body. When developing the degradation and strength properties of urethra stents in the future, it is absolutely necessary to take into account the actions of muscles, urine flow simulations, and the technology of introducing the stent.

Author Contributions: Conceptualization, J.K., A.K.-P., and R.B.; methodology, J.K., A.K.-P., A.G.M., and R.B.; software, J.K. and A.G.M.; validation, J.K., A.K.-P., and R.B.; formal analysis, A.K.-P. and R.B.; investigation, J.K., A.K.-P., and A.G.M.; resources, J.K., A.K.-P., A.G.M., and R.B.; data curation, J.K.; writing—original draft preparation, J.K.; writing—review and editing, A.K.-P. and R.B.; visualization, J.K.; supervision, A.K.-P. and R.B.; project administration, A.K.-P. and R.B.; funding acquisition, R.B. All authors have read and agreed to the published version of the manuscript.

Funding: This research was funded by the National Science Center Poland, grant number DEC-2016/21/B/ST8/01972.

Conflicts of Interest: The authors declare no conflicts of interest.

References

1. Saul, J.M.; Williams, D.F. 12-Hydrogels in Regenerative Medicine. In *Handbook of Polymer Applications in Medicine and Medical Devices*, 2nd ed.; Modjarrad, K., Ebnesajjad, S., Eds.; Elsevier Inc. William Andrew: San Diego, CA, USA, 2013; pp. 279–302.
2. Aljohani, W.J.; Wenchao, L.; Ullah, M.W.; Zhang, X.; Yang, G. Application of Sodium Alginate Hydrogel. *J. Biotech. Biochem.* **2017**, *3*, 19–31. [CrossRef]
3. Lee, K.Y.; Mooney, D.J. Alginate: Properties and biomedical applications. *Prog. Polym. Sci.* **2012**, *37*, 106–126. [CrossRef]
4. Budama-Kilinc, Y.; Çakır-Koç, R.; Aslan, B.; Özkan, B.; Mutlu, H.; Üstün, E. Hydrogels in Regenerative Medicine. In *Biomaterials in Regenerative Medicine*; Dobrzanski, L.A., Ed.; IntechOpen Limited: London, UK, 2018; pp. 277–301.
5. Sinno, H.; Prakash, S. Complements and wound healing cascade: An updated review. *Plast. Surg. Int.* **2013**, 1–7. [CrossRef] [PubMed]
6. Naseer, I.; Khan, A.; Asif, A.; Yar, M.; Haycock, J.; Rehman, I. Recent concepts in biodegradable polymers for tissue engineering paradigms: A critical review. *Inter. Mater. Rev.* **2018**, 1–36. [CrossRef]
7. Kim, H.S.; Sun, X.; Lee, J.H.; Kim, H.W.; Fu, X.; Leong, K.W. Advanced drug delivery systems and artificial skin grafts for skin wound healing. *Adv Drug Deliv Rev* **2019**, *146*, 1–31. [CrossRef] [PubMed]
8. Arjmandi, M.; Ramezani, M. Mechanical and tribological assessment of silica nanoparticle-alginate-polyacrylamide nanocomposite hydrogels as a cartilage replacement. *J. Mech. Behav. Biomed. Mater.* **2019**, *95*, 196–204. [CrossRef]
9. Swieszkowski, W.; Ku, D.N.; Bersee, H.E.; Kurzydlowski, K.J. An elastic material for cartilage replacement in an arthritic shoulder joint. *Biomaterials* **2006**, *27*, 1534–1541. [CrossRef]
10. Liu, M.; Zeng, X.; Ma, C.; Yi, H.; Ali, Z.; Mou, X.; Li, S.; Deng, Y.; He, N. Injectable hydrogels for cartilage and bone tissue engineering. *Bone Res.* **2017**, *5*, 1–20. [CrossRef]

11. Chen, J.; Irianto, J.; Inamdar, S.; Pravincumar, P.; Lee, D.A.; Bader, D.L.; Knight, M.M. Cell mechanics, structure, and function are regulated by the stiffness of the three-dimensional microenvironment. *Biophys. J.* **2012**, *103*, 1188–1197. [CrossRef]

12. Summa, M.; Russo, D.; Penna, J.; Marganoli, N.; Bayer, J.S.; Bandlera, T.; Athanassion, A.; Bertorelli, R. A biocompatible sodium alginate/povidine iodine film enhances wound healing. *Eur. J. Pharm Biopharm.* **2018**, *122*, 17–24. [CrossRef]

13. Marković, D.; Zarubica, A.; Stojković, N.; Vasić, M.; Cakić, M.; Nikolić, G. Alginates and similar exopolysaccharides in biomedical application and pharmacy: Controled delivery of drugs. *Adv. Technol.* **2016**, *5*, 39–52. [CrossRef]

14. Sun, J.; Tan, H. Alginate-Based Biomaterials for Regenerative Medicine Applications. *Materials* **2013**, *6*, 1285–1309. [CrossRef] [PubMed]

15. Augst, A.D.; Kong, H.J.; Mooney, D.J. Alginate Hydrogels as Biomaterials. *Macromol. Biosci.* **2006**, *6*, 623–633. [CrossRef] [PubMed]

16. Kaczmarek-Pawelska, A. Alginate-Based Hydrogels in Regenerative Medicine. In *Alginates–Recent Uses of This Natural Polymer*; Pereira, L., Ed.; IntechOpen Limited: London, UK, 2019; pp. 1–12.

17. Bartkowiak-Jowsa, M.; Będziński, R.; Kozłowska, A.; Filipiak, J.; Pezowicz, C. Mechanical, rheological, fatigue, and degradation behavior of PLLA, PGLA and PDGLA as materials for vascular implants. *Meccanica* **2013**, *48*, 721–731. [CrossRef]

18. Gasperini, L.; Mano, J.F.; Reis, R.L. Natural polymers for the microencapsulation of cells. *J. R. Soc. Interface* **2014**, *11*, 1–20. [CrossRef]

19. Gibas, I.; Janik, H. Review: Synthetic polymer hydrogels for biomedical applications. *Chem. Chem. Technol.* **2010**, *4*, 297–304.

20. Barros, A.A.; Rita, A.; Duarte, C.; Pires, R.A.; Sampaio-Marques, B.; Ludovico, P.; Lima, E.; Mano, J.F.; Reis, R.L. Bioresorbable ureteral stents from natural origin polymers. *J. Biomed. Mater. Res. Part B* **2014**, 1–10. [CrossRef]

21. Barros, A.A.; Oliveira, C.; Lima, E.; Rita, A.; Duarte, C.; Reis, R.L. Gelatin-based biodegradable ureteral stents with enhanced mechanical properties. *Appl. Mater. Today* **2016**, *5*, 9–18. [CrossRef]

22. Bartkowiak-Jowsa, M.; Będziński, R.; Chłopek, J.; Filipiak, J.; Szaraniec, B. Comparative analysis of the deformation characteristics of biodegradable polymers considered as a material for vascular stents. *Polymers* **2010**, *56*, 224–231. [CrossRef]

23. Kaczmarek-Pawelska, A.; Winiarczyk, K.; Mazurek, J. Alginate based hydrogel for tissue regeneration: Optimization, antibacterial activity and mechanical properties. *J. Achiev. Mater. Manuf. Eng.* **2017**, *81*, 35–40. [CrossRef]

24. Mørch, Ý.A.; Donati, I.; Strand, B.L.; Skijåk-Break, G. Effect of Ca^{2+}, Ba^{2+}, and Sr^{2+} on Alginate Microbeads. *Biomacromolecules* **2006**, *7*, 1471–1480. [CrossRef]

25. Chutipongtanate, S.; Thongboonkerd, V. Systematic comparisons of artificial urine formulas for in vitro cellular study. *Anal. Biochem.* **2010**, *402*, 110–112. [CrossRef]

26. Straccia, M.C.; Gomez d'Ayala, G.; Romano, I.; Oliva, A.; Laurienzo, P. Alginate Hydrogels Coated with Chitosan for Wound Dressing. *Mar. Drugs* **2015**, *13*, 2890–2908. [CrossRef] [PubMed]

27. Afonso, J.S.; Jorge, R.M.; Martins, P.S.; Soldi Mda, S.; Alves, O.L.; Patricio, B.; Mascarenhas, T.; Sartori, M.G.; Girao, M.J. Structural and thermal properties of polypropylene mesh used in treatment of stress urinary incontinence. *Acta Bioeng. Biomech.* **2009**, *11*, 27–33. [PubMed]

28. Chen, L.; Shen, R.; Komasa, S.; Xue, Y.; Jin, B.; Hou, Y.; Okazaki, J.; Gao, J. Drug-Loadable Calcium Alginate Hydrogel System for Use in Oral Bone Tissue Repair. *Int. J. Mol. Sci.* **2017**, *18*, 989. [CrossRef] [PubMed]

29. Yao, F.; Laudano, M.A.; Seklehner, S.; Chughtai, B.; Lee, R.K. Image-based simulation of urethral distensibility and flow resistance as a function of pelvic floor anatomy. *Neurol. Urodyn.* **2015**, *34*, 664–668. [CrossRef] [PubMed]

30. Zhang, K.; Fu, Q.; Chen, X.; Chandra, P.; Mo, X.; Song, L.; Atala, A.; Zhao, W. 3D bioprinting of urethra with PCL/PLCL blend and dual autologous cells in fibrin hydrogel: An in vitro evaluation of biomimetic mechanical property and cell growth environment. *Acta Biomater.* **2017**, *50*, 154–164. [CrossRef]

31. Feng, C.; Xu, Y.M.; Fu, Q.; Zhu, W.D.; Cui, L.; Chen, J. Evaluation of the biocompatibility and mechanical properties of naturally derived and synthetic scaffolds for urethral reconstruction. *J. Biomed. Mater. Res. A* **2010**, *94A*, 317–325. [CrossRef]

32. Nastaj, J.; Przewlocka, A.; Rajkowska-Mysliwiec, M. Biosorption of Ni(II), Pb(II) and Zn(II) on calcium alginate beads: Equilibrium, kinetic and mechanism studies. *Pol. J. Chem. Tech.* **2016**, *18*, 81–87. [CrossRef]
33. Sadiq, A.; Choubey, A.; Bajpai, A.K. Biosorption of chromium ions by calcium alginate nanoparticles. *J. Chil. Chem. Soc.* **2018**, *63*, 4077–4081. [CrossRef]

Article

Development of Hydrophilic Drug Encapsulation and Controlled Release Using a Modified Nanoprecipitation Method

Jiang Xu [1,2], Yuyan Chen [2,3], Xizhi Jiang [1], Zhongzheng Gui [1] and Lei Zhang [1,*]

[1] College of Biotechnology and Sericultural Research Institute, Jiangsu University of Science and Technology, Zhenjiang 212018, China; xujiangfrank@gmail.com (J.X.); pinzhongsheng@163.com (X.J.); srizzgui@hotmail.com (Z.G.)

[2] Department of Chemical Engineering, Waterloo Institute for Nanotechnology, University of Waterloo, 200 University Avenue West, Waterloo, ON N2L 3G1, Canada; yychensuda@hotmail.com

[3] Institute of Functional Nano & Soft Materials Laboratory, Soochow University, Suzhou 215123, China

* Correspondence: l78zhang@just.edu.cn; Tel.: +86-511-85616716

Received: 28 March 2019; Accepted: 16 May 2019; Published: 1 June 2019

Abstract: The improvement of the loading content of hydrophilic drugs by polymer nanoparticles (NPs) recently has received increased attention from the field of controlled release. We developed a novel, simply modified, drop-wise nanoprecipitation method which separated hydrophilic drugs and polymers into aqueous phase (continuous phase) and organic phase (dispersed phase), both individually and involving a mixing process. Using this method, we produced ciprofloxacin-loaded NPs by Poly (D,L-lactic acid)-Dextran (PLA-DEX) and Poly lactic acid-co-glycolic acid-Polyethylene glycol (PLGA-PEG) successfully, with a considerable drug-loading ability up to 27.2 wt% and an in vitro sustained release for up to six days. Drug content with NPs can be precisely tuned by changing the initial drug feed concentration of ciprofloxacin. These studies suggest that this modified nanoprecipitation method is a rapid, facile, and reproducible technique for making nano-scale drug delivery carriers with high drug-loading abilities

Keywords: diblock copolymers; drug delivery systems; nanoparticles; nanoprecipitation; self-assembly

---

1. Introduction

Since its introduction in the late 1980s by Fessi et al. [1], nanoprecipitation technology has grown into a highly efficient, easily handled, and mature tool to fabricate drug-loaded nanoparticles (NPs) for biomedical researchers. Polymer and hydrophobic drugs were dissolved in the same organic solvents and then mixed with an aqueous solution. By modifying the solution mixing speed, Oil/Water ratio [2], pH [3], polymer/drug ratio [4], block ratio of block polymers [5], and solvent selection [6], the parameters of NPs including size, distribution, and drug encapsulation efficiency can be tuned easily. We previously reported on the encapsulation of the hydrophobic drug ellipticine by using peptide molecules [7,8].

However, the conventional nanoprecipitation procedure has relatively limited potential to encapsulate hydrophilic compounds [9]. This deficiency of encapsulation is mainly due to poor solubility of hydrophilic compounds in most organic solvents [10–14]; it further affects its controllability over drug-loading amount [15]. Efforts have been made to improve the state; however, these involve complex preparation processes, including O/W and W/O/W emulsion methods [16] as well as hydrogel [17], resulting in much larger non-uniform nanoparticles, with sizes from several hundred to several thousand nanometers [18]. Classical nanoparticles methods lead to mass loading of approximately 1 wt% [19].

In this study, we chose ciprofloxacin as our model drug, one of the most commonly used anti-infective agents for ocular treatment because of its low toxicity, broad-spectrum antimicrobial activity, and low resistance to bacteria [20]. Previous research on ciprofloxacin encapsulation have largely been conducted with emulsion methods [16] or hydrogels [17], which usually involved complex preparations or led to much larger and polydisperse NPs. For example, emulsion methods require a homogenization process, which takes a long time to mix; additionally, residue of toxic organic phases can be a potential challenge for biocompatibility. Here, we developed a modified nanoprecipitation method that separates drug and polymer into two different organic and aqueous phases. Monodispersed, ciprofloxacin-loaded NPs were formed with a tunable size via a simple drop-wise mixing process similar to conventional nanoprecipitation's [9]. The drug encapsulation efficiencies of NPs and their in vitro release kinetics were assessed. In addition, a high linear correlation was found between the initial concentration of ciprofloxacin and the mass-loading ability of PLA-DEX and PLGA-PEG NPs. This work provides a new quantitative approach for producing polymeric NPs with high-loading capability of hydrophilic drugs through a simple, one-step nanoprecipitation.

2. Experimental Section

2.1. Materials

Poly (D,L-lactic acid)-Dextran (PLA20kDa-DEX10kDa) was synthesized as previously reported [21]. Briefly, the synthesis can be divided into three stages: (1) reductive amination between aldehyde on the reducing end of dextran and amine group of N-Boc-ethylenediamine cross-linker. (2) Deprotection of the Boc group, which involves cleaving the amide bond between the Boc group and the protected amine moiety as well as deprotonating the $-NH3^+$ end groups which were deprotected. (3) Conjugation of the end-modified dextran with PLA, which was facilitated by adding catalysts N-(3-dimethylaminopropyl)-Nethylcarbodiimide (EDC) and Sulfo-NHS and allowing the reaction to proceed for 4 h at room temperature. PLGA (30~35 kDa, 50:50)-PEG (6 kDa) was purchased from Lakeshore Biomaterials (Birmingham, AL, USA). Both polymers are amphiphilic because PLA and PLGA parts are hydrophobic, and Dextran and PEG parts are hydrophilic [22–26]. Ciprofloxacin, dimethyl sulfoxide (DMSO), and hydrochloric acid (HCl) were purchased from Sigma Aldrich (Oakville, ON, Canada). Simulated tear fluid (STF) was prepared for the in vitro release experiment using a previously described formulation [27].

2.2. Synthesis of PLA-DEX and PLGA-PEG Nanoparticles (NPs)

1 mL PLA-DEX or PLGA-PEG solution (5 mg/mL in DMSO) was added to 10 mL 1 M HCl aqueous solution by a pipette with gentle magnetic stirring for 10 min. Next, the dispersion was filtered by 200 nm filters for further use. The size and polydispersity (PDI) of NPs were determined by dynamic lighting scattering (DLS; 90Plus Particle Size Analyzer, Brookhaven Instruments, NY, USA), $\lambda = 659$ nm at 90°).

2.3. Transmission Electron Microscopy (TEM)

The particle morphology of PLA-DEX and PLGA-PEG NPs were further characterized using transmission electron microscopy (TEM, Philips CM 10, Philips Electronics, Eindhoven, Netherlands) with a lanthanum hexaboride filament (LaB6). The NP solution was prepared according to protocols mentioned above and coated onto a copper grid. A drop of aqueous phosphotungstic acid solution (20 mg/mL) was used to briefly stain the NPs for 10 s and was then removed by absorbent paper. The copper grid was dried at room temperature overnight before TEM imaging.

2.4. Ciprofloxacin Encapsulation by PLA-DEX and PLGA-PEG NPs

PLA-DEX was dissolved in dimethyl sulfoxide (DMSO, 5 mg/mL), and ciprofloxacin was dissolved in 1 M HCl aqueous solution with concentration of 0.5, 1, 2, 4, 5, 6, and 8 mg/mL, respectively. 1 mL

PLA-DEX solution was added to 10 mL ciprofloxacin solution by dropper with gentle magnetic stirring for 10 min. Next, the solution was filtered by 200 nm filters for further use. 1 mL of filtered solution was centrifuged with an Amicon centrifuge tube (MWCO = 3000) for 30 min at 8000 rpm. 1 mL of 1 M HCl aqueous solution was used to resuspend the NPs, and another round of centrifugation (8000 rpm, 30 min) was performed to wash away the un-bonded and loosely associated ciprofloxacin. Subsequently, the NPs were dissolved by a mixed solvent of 1 M HCl aqueous solution and DMSO (*v:v* = 1:1) to release all encapsulated drugs for determination of encapsulation efficiency by UV spectroscopy (see Supplementary Materials). Specifically, a standard curve relationship of concentration of ciprofloxacin in solution and UV absorbance was calibrated first. Second, by comparing experimental UV absorbance to the standard calibration curve, an experimental concentration of ciprofloxacin in the solution could be determined.

The same procedure was applied for encapsulation of ciprofloxacin by PLGA-PEG for comparative analysis. The encapsulation efficiency and mass-loading ability of NPs were calculated by Equations (1) and (2):

$$\text{Encapsulation efficiency} = (\text{Mass of the drug encapsulated}/\text{Mass of the initial drug feed}) \times 100\% \quad (1)$$

$$\text{Mass loading ability} = (\text{Mass of the drug encapsulated}/\text{Mass of the polymer}) \times 100\% \quad (2)$$

2.5. Drug Release Study

After the modified nanoprecipitation (1 mL 5 mg/mL PLA-DEX and 10 mL 5 mg/mL ciprofloxacin), 4 mL filtered solution was centrifuged, washed, and re-centrifuged per the procedure mentioned above. The NPs were resuspended in 10 mL Millipore water, and then transferred into a dialysis membrane (100 kDa, MWCO) against 400 mL simulated tear fluid (STF) in a release bottle with stirring at 37 °C. 1 mL of each sample was collected at 2 h, 4 h, 6 h, 8 h, 10 h, 12 h, 24 h, 48 h, and 144 h to determine the cumulative-released dose by UV spectroscopy. The most commonly adapted formula of STF is: NaCl 0.67 g, NaHCO$_3$, 0.20 g, CaCl$_2\cdot$2H$_2$O 0.008 g in 100 mL DI water (pH = 7.4) [28]. Comparative groups of free drug without NP carriers and of ciprofloxacin-PLGA-PEG NPs were made according to the same protocols.

3. Results and Discussion

3.1. Formation of Ciprofloxacin-Loaded NPs by Block Polymers

Scheme 1 shows a schematic of the formation process of ciprofloxacin-loaded NPs within the modified nanoprecipitation. Nanoprecipitation of drug-loaded NPs occurs because of solvent displacement, in which a "good" solvent condition turns into a "bad" solvent condition [29]. However, in our modified nanoprecipitation, drug (ciprofloxacin) and polymer were dissolved in two different "good" solvents individually, which were then converted into one same "bad" solvent in a one-step mixing. The formation of drug-loaded NPs is due to the interfacial deposition of polymer and drug because of the interfacial solvent displacement between two different unstable liquid phases [1].

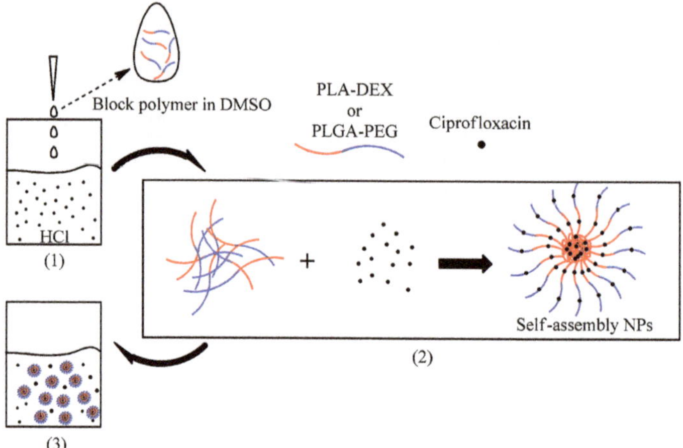

Scheme 1. The formation procedure of ciprofloxacin-loaded Nanoparticles (NPs). (1) Drop-wise nanoprecipitation; (2) Block polymer self-assembled to form core-shell NPs with ciprofloxacin; (3) The NPs and non-encapsulated ciprofloxacin after nanoprecipitation.

3.2. Encapsulation of Ciprofloxacin by NPs

As shown in Table 1, the effective diameters slightly increased from 82 ± 8 nm to 98 ± 11 nm by PLA-DEX NPs, and from 174 ± 1 nm to 205 ± 15 nm by PLGA-PEG NPs, respectively, with the increase of ciprofloxacin concentration from 0.5 mg/mL to 8 mg/mL. Given that other environmental conditions remained the same, this increase of NP size resulted from the increased amount of encapsulated drug content within the NPs (Figure 1), suggesting that we could tune the NP size by controlling the concentration of ciprofloxacin. Moreover, with this modified nanoprecipitation method, the NPs remained a relatively low and similar polydispersity (Table 1) regardless of changes in ciprofloxacin concentration or NP size. This characteristic revealed that our modified nanoprecipitation method has a superior controllability over the morphology of NPs. The TEM images (Figure 2) of the NPs are shown in spherical shape in Figure 2. We analyzed the size distributions of the two NPs in the TEM images. It was 35.83 ± 6.33 nm and 70.15 ± 5.86 nm for PLA-DEX and PLGA-PEG, respectively. Notably, the hydrodynamic diameter of NPs by DLS was much higher than the TEM result, which may be because that hydrophilicity of outer layer of NPs increased intensity-based diameter of NPs detected by DLS [30]. Furthermore, the stain and dry processes might cause the differences.

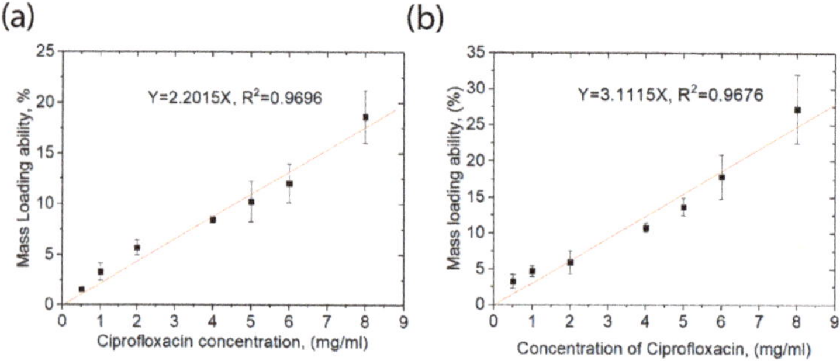

Figure 1. Relationship of NPs' mass-loading ability and initial concentration of ciprofloxacin. (**a**) PLA-DEX NPs, y = 2.2015x, R^2 = 0.9696; (**b**) PLGA-PEG NPs, y = 3.1115x, R^2 = 0.9676. (n = 3; mean ± S.D.).

Table 1. NP parameters of effective diameter, polydispersity, and mass-loading ability for PLA-DEX and PLGA-PEG. (n = 3; mean ± S.D.).

Polymer	C of Cipro (mg/mL)	Size of NPs (nm)	PDI	Encapsulation Efficiency	Mass Loading Ability (%)
PLA-DEX	0.5	82 ± 8	0.137 ± 0.011	1.52 ± 0.23	1.52 ± 0.23
	1.0	89 ± 3	0.129 ± 0.015	1.67 ± 0.42	3.33 ± 0.84
	2.0	89 ± 7	0.124 ± 0.029	1.42 ± 0.15	5.68 ± 0.74
	4.0	93 ± 7	0.141 ± 0.020	1.1 ± 0.04	8.45 ± 0.31
	5.0	95 ± 9	0.138 ± 0.026	1.0 ± 0.20	10.25 ± 1.99
	6.0	94 ± 9	0.133 ± 0.014	1.0 ± 0.32	12.06 ± 1.90
	8.0	98 ± 11	0.133 ± 0.034	1.17 ± 0.16	18.64 ± 2.60
PLGA-PEG	0.5	174 ± 1	0.125 ± 0.014	3.29 ± 1.00	3.29 ± 1.00
	1.0	175 ± 1	0.123 ± 0.002	2.38 ± 0.38	4.76 ± 0.75
	2.0	178 ± 8	0.123 ± 0.007	1.49 ± 0.40	5.97 ± 1.60
	4.0	179 ± 2	0.098 ± 0.005	1.35 ± 0.07	10.79 ± 0.59
	5.0	183 ± 5	0.124 ± 0.007	1.37 ± 0.13	13.71 ± 1.25
	6.0	188 ± 5	0.103 ± 0.015	1.49 ± 0.25	17.84 ± 3.04
	8.0	205 ± 15	0.112 ± 0.022	1.70 ± 0.30	27.24 ± 4.79

(a)   (b)

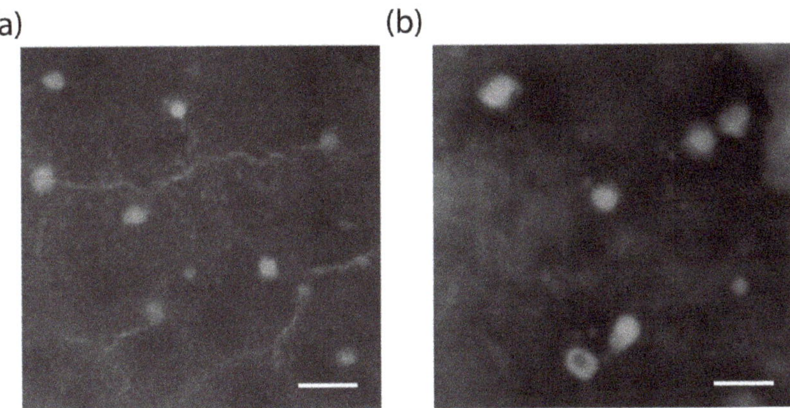

Figure 2. Transmission electron microscopy images demonstrate the spherical shape of (**a**) PLA–DEX NPs and (**b**) PLGA-PEG NPs (Scale bar = 100 nm).

In addition, with this modified nanoprecipitation method, NPs presented a considerable mass loading ability, 18.6 ± 2.6% by PLA-DEX and 27.2 ± 4.8% by PLGA-PEG, respectively. Noticeably, the mass loading ability of NPs, either PLA-DEX or PLGA-PEG, exhibited a liner relationship with the concentration of ciprofloxacin by this modified nanoprecipitation. Both correlation coefficients were very close to 1 (R = 0.9847 by PLA-DEX NPs and R = 0.9837 by PLGA-DEX NPs) according to Figure 1, which indicates a quantitative approach that may control the amount of the encapsulated ciprofloxacin [29] in the NPs by changing the concentration of original drug feed.

Although the encapsulation efficiency (Table 1) of our modified nanoprecipitation method appears relatively lower than those commonly seen in hydrophobic drugs, it is still meaningful because encapsulation of ciprofloxacin by a conventional nanoprecipitation method is difficult to achieve. In addition, the mass-loading capability of our modified nanoprecipitation is even higher than the emulsion method [16]. Moreover, the remaining or non-encapsulated ciprofloxacin in our protocol (after the centrifugation with MWCO = 3000, as described in Section 2.4) can be re-used for another or even multiple runs of modified nanoprecipitation because majority of ciprofloxacin solution remains aqueous and of a high concentration of ciprofloxacin.

3.3. Ciprofloxacin Release Study

Figure 3 illustrates how the ciprofloxacin was released in simulated tear fluid (STF) as a free drug, as encapsulated content by the PLA-DEX NPs, and by PLGA-PEG NPs, respectively. The free drug group had a very quick release up to 83.2 ± 6.9% within the initial 2 h due to the free diffusion of the ciprofloxacin; the free diffusion was almost finished at 6 h. In contrast, the ciprofloxacin encapsulated by both PLA-DEX and PLGA-PEG NPs had a slower burst-release within the initial 2 h, up to 67.5 ± 6.1% and 56.1 ± 9.0%, respectively. These differences indicate that part of the encapsulated drugs was strongly associated to the NPs that were well "covered" or "protected" during the burst-release stage [31] when compared with the free drug group. Both PLA-DEX NPs and PLGA-PEG NPs exhibited a much more controlled release profile than the free drugs. After the first 12 h, a steady-release stage continued within both two groups, culminating at 96.9 ± 5.2% and 95.4 ± 2.5%, respectively, at 144 h. Notably, the PLGA-PEG NPs released ciprofloxacin at a slower rate than the PLA-DEX NPs. This may have been caused by PLA-DEX NPs having more hydrophilic surfaces [32], which are less compact in an aqueous environment, making it easier for ciprofloxacin drugs to detach [4]. The modified nanoprecipitation method is a promising tool for encapsulation and release of hydrophilic drugs and preparation of functional NPs as drug delivery carriers. The cell membrane is a major barrier for drug transport. Most of the small molecules penetrate the membrane by passive diffusion, which is easily affected by concentration as well as the physical and chemical properties of the cell membrane, such as the cholesterol present on the membrane [33,34]. The NPs may help the drug cross the membrane in addition to passive diffusion, providing an optimized combination. Previously, we reported a polypeptide that may help the cancer drug ellipticine penetrate the cell membrane and improve its uptake [7]. In our further work, the uptake mechanism of the PLA-DEX and PLAG-PEG NPs will be identified for further application.

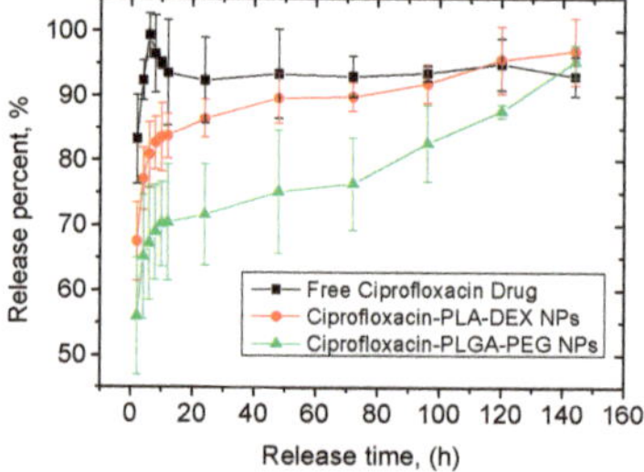

Figure 3. The cumulative release profiles of ciprofloxacin in vitro in simulated tear fluid (STF) at 37 °C as free drugs by PLA-DEX NPs and PLGA-PEG NPs. (n = 3; mean ± S.D.).

4. Conclusions

A novel, one-step, drop-wise, modified nanoprecipitation method was successfully developed through which we could efficiently encapsulate the hydrophilic drug ciprofloxacin by PLA-DEX and PLGA-PEG NPs. NPs of considerable drug-loading ability were synthesized with tunable sizes. We also found that the mass-loading ability of the NPs varied as an excellent linear function of the concentration of ciprofloxacin, which demonstrates a possibility for an accurate control over the drug-loading amount by NPs. Both PLA-DEX and PLGA-PEG NPs exhibited a sustained release in comparison with free drugs. By optimizing the formulation of NPs by changing the drug/polymer ratio and polymer composition, a more effective drug release rate and constant treatment can be achieved in future clinical applications. These characteristics make this modified nanoprecipitation method a promising technique for the encapsulation and release of hydrophilic drugs as well as the preparation of functional NPs as drug delivery tools for further biomedical application.

Supplementary Materials: The following are available online at http://www.mdpi.com/2227-9717/7/6/331/s1.

Author Contributions: J.X. and Y.C. performed experiments. J.X. and L.Z. wrote the paper and analyzed data. X.J. and Z.G. contributed to make comments on the data analysis and manuscript writing. All authors discussed and commented on the manuscript.

Funding: This research and the APC was funded by High-Level Talent Research Fund of Jiangsu University of Science and Technology (Grant number 1732931803).

Acknowledgments: The authors are very grateful to Erin Bedford for TEM characterization. The authors want to thank Sébastien Lecommandoux and Olivier Sandre for their fruitful comments and suggestions. The datasets supporting this article have been uploaded as part of the Supplementary Material.

Conflicts of Interest: The authors declare no conflict of interest.

References

1. Fessi, H.; Puisieux, F.; Devissaguet, J.P.; Ammoury, N.; Benita, S. Nanocapsule formation by interfacial polymer deposition following solvent displacement. *Int. J. Pharm.* **1989**, *55*, R1–R4. [CrossRef]

2. Martín-Banderas, L.; Flores-Mosquera, M.; Riesco-Chueca, P.; Rodríguez-Gil, A.; Cebolla, Á.; Chávez, S.; Gañán-Calvo, A.M. Flow focusing: A versatile technology to produce size-controlled and specific-morphology microparticles. *Small* **2005**, *1*, 688–692. [CrossRef] [PubMed]

3. Bo, Q.; Zhao, Y. Double-hydrophilic block copolymer for encapsulation and two-way pH change-induced release of metalloporphyrins. *J. Polym. Sci. Part A Polym. Chem.* **2006**, *44*, 1734–1744. [CrossRef]

4. Liu, S.; Jones, L.; Gu, F.X. Development of Mucoadhesive Drug Delivery System Using Phenylboronic Acid Functionalized Poly (D,L-lactide)-b-Dextran Nanoparticles. *Macromol. Biosci.* **2012**, *12*, 1622–1626. [CrossRef] [PubMed]

5. Karnik, R.; Gu, F.; Basto, P.; Cannizzaro, C.; Dean, L.; Kyei-Manu, W.; Langer, R.; Farokhzad, O.C. Microfluidic platform for controlled synthesis of polymeric nanoparticles. *Nano Lett.* **2008**, *8*, 2906–2912. [CrossRef] [PubMed]

6. Bilati, U.; Allémann, E.; Doelker, E. Development of a nanoprecipitation method intended for the entrapment of hydrophilic drugs into nanoparticles. *Eur. J. Pharm. Sci.* **2005**, *24*, 67–75. [CrossRef] [PubMed]

7. Zhang, L.; Xu, J.; Wang, F.; Ding, Y.; Wang, T.; Jin, G.; Martz, M.; Gui, Z.; Ouyang, P.; Chen, P. Histidine-Rich Cell-Penetrating Peptide for Cancer Drug Delivery and its Uptake Mechanism. *Langmuir* **2019**, *35*, 3513–3523. [CrossRef] [PubMed]

8. Zhang, L.; Sheng, Y.; Yazdi, A.Z.; Sarikhani, K.; Wang, F.; Jiang, Y.; Liu, J.; Zheng, T.; Wang, W.; Ouyang, P. Surface-Assisted Assembly of Histidine-Rich Lipidated Peptide for Simultaneous Exfoliation of Graphite and Functionalization of Graphene Nanosheets. *Nanoscale* **2019**, *11*, 2999. [CrossRef] [PubMed]

9. Schubert, S.; Delaney, J.T., Jr.; Schubert, U.S. Nanoprecipitation and nanoformulation of polymers: From history to powerful possibilities beyond poly (lactic acid). *Soft Matter* **2011**, *7*, 1581–1588. [CrossRef]

10. Bilati, U.; Allémann, E.; Doelker, E. Nanoprecipitation versus emulsion-based techniques for the encapsulation of proteins into biodegradable nanoparticles and process-related stability issues. *Aaps Pharmscitech* **2005**, *6*, E594–E604. [CrossRef]

11. Nihant, N.; Schugens, C.; Grandfils, C.; Jérôme, R.; Teyssié, P. Polylactide microparticles prepared by double emulsion/evaporation technique. I. Effect of primary emulsion stability. *Pharm. Res.* **1994**, *11*, 1479–1484. [CrossRef] [PubMed]

12. Vila, A.; Sánchez, A.; Pérez, C.; Alonso, M.J. PLA-PEG nanospheres: New carriers for transmucosal delivery of proteins and plasmid DNA. *Polym. Adv. Technol.* **2002**, *13*, 851–858. [CrossRef]

13. Huang, X.; Lowe, T.L. Biodegradable Thermoresponsive Hydrogels for Aqueous Encapsulation and Controlled Release of Hydrophilic Model Drugs. *Biomacromolecules* **2005**, *6*, 2131–2139. [CrossRef] [PubMed]

14. Beck-Broichsitter, M.; Nicolas, J.; Couvreur, P. Solvent selection causes remarkable shifts of the "Ouzo region" for poly (lactide-co-glycolide) nanoparticles prepared by nanoprecipitation. *Nanoscale* **2015**, *7*, 9215–9221. [CrossRef]

15. Pramod, P.S.; Takamura, K.; Chaphekar, S.; Balasubramanian, N.; Jayakannan, M. Dextran Vesicular Carriers for Dual Encapsulation of Hydrophilic and Hydrophobic Molecules and Delivery into Cells. *Biomacromolecules* **2012**, *13*, 3627–3640. [CrossRef]

16. Jeong, Y.-I.; Na, H.-S.; Seo, D.-H.; Kim, D.-G.; Lee, H.-C.; Jang, M.-K.; Na, S.-K.; Roh, S.-H.; Kim, S.-I.; Nah, J.-W. Ciprofloxacin-encapsulated poly (DL-lactide-co-glycolide) nanoparticles and its antibacterial activity. *Int. J. Pharm.* **2008**, *352*, 317–323. [CrossRef] [PubMed]

17. Hosny, K.M. Ciprofloxacin as ocular liposomal hydrogel. *Aaps Pharmscitech* **2010**, *11*, 241–246. [CrossRef]

18. Szoka, F.; Papahadjopoulos, D. Procedure for preparation of liposomes with large internal aqueous space and high capture by reverse-phase evaporation. *Proc. Natl. Acad. Sci. USA* **1978**, *75*, 4194–4198. [CrossRef] [PubMed]

19. Govender, T.; Stolnik, S.; Garnett, M.C.; Illum, L.; Davis, S.S. PLGA nanoparticles prepared by nanoprecipitation: Drug loading and release studies of a water soluble drug. *J. Control. Release* **1999**, *57*, 171–185. [CrossRef]

20. Neumann, W.; Sassone-Corsi, M.; Raffatellu, M.; Nolan, E.M. Esterase-Catalyzed Siderophore Hydrolysis Activates an Enterobactin–Ciprofloxacin Conjugate and Confers Targeted Antibacterial Activity. *J. Am. Chem. Soc.* **2018**, *140*, 5193–5201. [CrossRef]

21. Verma, M.S.; Liu, S.; Chen, Y.Y.; Meerasa, A.; Gu, F.X. Size-tunable nanoparticles composed of dextran-b-poly(D,L-lactide) for drug delivery applications. *Nano Res.* **2012**, *5*, 49–61. [CrossRef]

22. Song, Z.; Feng, R.; Sun, M.; Guo, C.; Gao, Y.; Li, L.; Zhai, G. Curcumin-loaded PLGA-PEG-PLGA triblock copolymeric micelles: Preparation, pharmacokinetics and distribution in vivo. *J. Colloid Interface Sci.* **2011**, *354*, 116–123. [CrossRef] [PubMed]

23. Cai, Q.; Wan, Y.; Bei, J.; Wang, S. Synthesis and characterization of biodegradable polylactide-grafted dextran and its application as compatilizer. *Biomaterials* **2003**, *24*, 3555–3562. [CrossRef]

24. Ma, T.Y.; Hollander, D.; Krugliak, P.; Katz, K. PEG 400, a hydrophilic molecular probe for measuring intestinal permeability. *Gastroenterology* **1990**, *98*, 39–46. [CrossRef]

25. Li, L.; Ding, S.; Zhou, C. Preparation and degradation of PLA/chitosan composite materials. *J. Appl. Polym. Sci.* **2004**, *91*, 274–277. [CrossRef]

26. Lee, S.J.; Han, B.R.; Park, S.Y.; Han, D.K.; Kim, S.C. Sol–gel transition behavior of biodegradable three-arm and four-arm star-shaped PLGA–PEG block copolymer aqueous solution. *J. Polym. Sci. Part A Polym. Chem.* **2006**, *44*, 888–899. [CrossRef]

27. Shen, J.; Deng, Y.; Jin, X.; Ping, Q.; Su, Z.; Li, L. Thiolated nanostructured lipid carriers as a potential ocular drug delivery system for cyclosporine A: Improving in vivo ocular distribution. *Int. J. Pharm.* **2010**, *402*, 248–253. [CrossRef]

28. Cohen, S.; Lobel, E.; Trevgoda, A.; Peled, Y. A novel in situ-forming ophthalmic drug delivery system from alginates undergoing gelation in the eye. *J. Control. Release* **1997**, *44*, 201–208. [CrossRef]

29. Capretto, L.; Cheng, W.; Carugo, D.; Katsamenis, O.L.; Hill, M.; Zhang, X. Mechanism of co-nanoprecipitation of organic actives and block copolymers in a microfluidic environment. *Nanotechnology* **2012**, *23*, 375602. [CrossRef]

30. Salorinne, K.; Man, R.W.; Li, C.H.; Taki, M.; Nambo, M.; Crudden, C.M. Water-Soluble N-Heterocyclic Carbene-Protected Gold Nanoparticles: Size-Controlled Synthesis, Stability, and Optical Properties. *Angew. Chem. Int. Ed.* **2017**, *56*, 6198–6202. [CrossRef]

31. Magenheim, B.; Levy, M.Y.; Benita, S. A new in vitro technique for the evaluation of drug release profile from colloidal carriers—ultrafiltration technique at low pressure. *Int. J. Pharm.* **1993**, *94*, 115–123. [CrossRef]

32. Gref, R.; Domb, A.; Quellec, P.; Blunk, T.; Müller, R.H.; Verbavatz, J.M.; Langer, R. The controlled intravenous delivery of drugs using PEG-coated sterically stabilized nanospheres. *Adv. Drug Deliv. Rev.* **1995**, *16*, 215–233. [CrossRef]

33. Zhang, L.; Zhao, L.; Ouyang, P.-K.; Chen, P. Insight into the role of cholesterol in modulation of morphology and mechanical properties of CHO-K1 cells: An in situ AFM study. *Front. Chem. Sci. Eng.* **2019**, *13*, 98–107. [CrossRef]

34. Zhang, L.; Bennett, W.F.D.; Zheng, T.; Ouyang, P.K.; Ouyang, X.P.; Qiu, X.Q.; Luo, A.Q.; Karttunen, M.; Chen, P. Effect of Cholesterol on Cellular Uptake of Cancer Drugs Pirarubicin and Ellipticine. *J. Phys. Chem. B* **2016**, *120*, 3148–3156. [CrossRef] [PubMed]

Article

The Engineering of Porous Silica and Hollow Silica Nanoparticles to Enhance Drug-loading Capacity

Ngoc-Tram Nguyen-Thi [1,2], Linh Phuong Pham Tran [3], Ngoc Thuy Trang Le [2,3], Minh-Tri Cao [1], The-Nam Tran [1], Ngoc Tung Nguyen [2,4], Cong Hao Nguyen [2,3], Dai-Hai Nguyen [2,3], Van Thai Than [5,6,*], Quang Tri Le [7] and Nguyen Quang Trung [8,*]

[1] Tra Vinh University, No. 126, Nguyen Thien Thanh, Ward 5, Tra Vinh city 940000, Vietnam
[2] Graduate University of Science and Technology, Vietnam Academy of Science and Technology, Hanoi 100000, Vietnam
[3] Institute of Applied Materials Science, Vietnam Academy of Science and Technology, 01 TL29, District 12, Ho Chi Minh City 700000, Vietnam
[4] Center for Research and Technology Transfer, Vietnam Academy of Science and Technology, Hanoi 100000, Vietnam
[5] NTT Hi-Tech Institute, Nguyen Tat Thanh University, Ho Chi Minh City 700000, Vietnam
[6] Center of Excellence for Functional Polymers and NanoEngineering, Nguyen Tat Thanh University, Ho Chi Minh City 700000, Vietnam
[7] Department of Orthopedic, 7A Military Hospital, 466 Nguyen Trai Street, Ward 8, District 5, 72706, Ho Chi Minh City 700000, Vietnam
[8] Nghe An Oncology Hospital, Vinh City 460000, Vietnam
* Correspondence: tvthai@ntt.edu.vn (V.T.T.); nqtrung8910@gmail.com (N.Q.T.)

Received: 23 June 2019; Accepted: 29 August 2019; Published: 4 November 2019

Abstract: As a promising candidate for expanding the capacity of drug loading in silica nanoplatforms, hollow mesoporous silica nanoparticles (HMSNs) are gaining increasing attention. In this study, porous nanosilica (PNS) and HMSNs were prepared by the sol-gel method and template assisted method, then further used for Rhodamine (RhB) loading. To characterize the as-synthesized nanocarriers, a number of techniques, including X-ray diffraction (XRD), transmission electron microscopy (TEM), nitrogen absorption-desorption isotherms, dynamic light scattering (DLS), thermogravimetric analysis (TGA), and Fourier transform infrared spectroscopy (FTIR) were employed. The size of HMSN nanoparticles in aqueous solution averaged 134.0 ± 0.3 nm, which could be adjusted by minor changes during synthesis, whereas that of PNS nanoparticles was 63.4 ± 0.6 nm. In addition, the encapsulation of RhB into HMSN nanoparticles to form RhB-loaded nanocarriers (RhB/HMSN) was successful, achieving high loading efficiency (51.67% ± 0.11%). This was significantly higher than that of RhB-loaded PNS (RhB/PNS) (12.24% ± 0.24%). Similarly, RhB/HMSN also possessed a higher RhB loading content (10.44% ± 0.02%) compared to RhB/PNS (2.90% ± 0.05%). From those results, it is suggested that prepared HMSN nanocarriers may act as high-capacity carriers in drug delivery applications.

Keywords: hollow mesoporous silica; porous silica; high drug loading capacity; nanoparticles; drug delivery system

---

1. Introduction

Various advantages of porous nanosilica (PNS), such as a large surface area, chemical and thermal stability, high biocompatibility and biodegradability, and readiness for surface functionalization, have made it an excellent nanocarrier system for drug delivery applications [1–4]. Due to the unique porous structure of the PNS nanocarriers, anticancer drugs could be effectively encapsulated, and protected from temperature variation and pH-induced enzymatic degradation of the surrounding media [5–7]. However, concerns about toxicity of the carriers have been expressed [8,9]. As a result, the reduction in toxicity of the nanocarrier by improving the drug loading capacity of silica nanoplatforms is of great significance.

A sub class of PNS, hollow mesoporous silica nanoparticles (HMSNs), is a structure of a mesoporous silica shell possessing a large cavity at the core of the particle [10–13], offering a potential solution to the problem of low capacity. It can be said that the HMSN was developed bearing in mind the importance of PNS applications in drug and bio-sensing agents' delivery. The large compartment inside each HMSN vastly increases their drug loading capacity and pore volume compared to conventional PNS [14,15]. Accordingly, it can be theoretically concluded that a lower amount of HMSNs is needed to deliver the same load of drug, and thus to achieve the desired therapeutic effect by comparison with that of PNS, hence minimizing the potential accumulation of foreign materials in the host. Moreover, HMSNs retain the property of a readily conjugatable surface, making them an increasingly recognized platform for fabricating stealth mechanisms and stimuli-responsive functionalization [16–21].

In a recent report, a HMSNs-b- cyclodextrin (CD) nanocarrier was formed by grafting 3-(3,4-dihydroxypheneyl) propionic acid (DHPA)-functionalized beta-cyclodextrin (b-CD) onto the surface of HMSNs, and was further anchored with polyethylene glycol (PEG)-conjugated adamantane (Ada) via host–guest interaction for the delivery of doxorubicin (DOX). It was revealed from thermogravimetric analysis (TGA) results that the DOX encapsulation efficiency of the HMSNs-b-CD/Ada-PEG carrier reached around 10 wt% of DOX. This figure was greater than that of PNS of a similar size, at approximately 5 wt%, which could be explained by the hollow cavity of HMSNs [22]. In another study, positively charged CD_{PEI} nanoparticles were grafted on the pore openings of HMSNs through disulfide bonds. The carrier was later then loaded with DOX and grafted with hyaluronic acid (HA) to yield a more complex structure—DOX/HMSN-SS-CD_{PEI}@HA, resulting a drug loading efficiency of 30.5%. Compared to the DOX-loaded PNS, the HMSN-SS-CD_{PEI}@HA nanoparticles had higher loading capacity and excellent biocompatibility. These advantages are generally due to mesoporous structure, high specific surface area, and particularly, the hollow cavity of the HMNS nanoparticle, suggesting its use as a high capacity nanocarrier for drug delivery [23].

Among the various methodologies to fabricate HMSN, the template assisted method is one of the most controllable methods, regardless of a wide range of template-free routes that have emerged recently [10,11,24]. With regard to template assisted fabrication, the hard template in the form of a solid SiO_2 core ($sSiO_2$) is first synthesized via a modified Stöber method. Afterwards, the sol-gel process of tetrathylorthorsilicates (TEOS) catalyzed by acid or base in the presence of hexadecyltrimethylammnomium bromide (CTAB) was performed to form a controllable PNS layer, which was then used to coat the $sSiO_2$ spheres, prepared as mentioned. The silica cores were then separated through selective etching in a suitable solvent, Na_2CO_3, for etching at a high temperature [25].

In this study, a process for synthesizing PNS and HMSN nanocarriers was reported. The as-synthesized systems were compared for Rhodamine (RhB) loading capacity and efficiency. Both carriers were characterized by a number of techniques, including X-ray diffraction (XRD), transmission electron microscopy (TEM), nitrogen absorption-desorption isotherms, dynamic light scattering (DLS), thermogravimetric analysis (TGA), and Fourier transform infrared spectroscopy (FTIR), respectively to evaluate the crystallinity, morphology, BET surface area, size and size distribution, and composition of the materials.

2. Materials and Methods

2.1. Materials

Reagent grade chemicals, including Tetraethyl orthosilicate (TEOS, 98%), Cetyltrimethylammonium bromide (CTAB, 99%), and Rhodamine B were commercially obtained from Sigma-Aldrich (St. Louis, MO, USA), Merck (USA), and HiMedia (India), respectively. Ammonia solution (NH_3 (aq), 28%) and ethanol were purchased from Scharlau (Barcelona, Spain). All chemicals were used as received.

2.2. Preparation of PNS

The sol gel technique was used to prepare PNS in which a TEOS, CTAB, ethanol, and water, act as the silicone source, structure-directing agent, solvent, and reactant, respectively. The TEOS was hydrolyzed and condensed with ammonia (NH_3) (Figure 1a). Initially, 2.8% NH_3 solution (0.07 M), deionized water (deH_2O), ethanol (2.2 M), and CTAB (2.6 g) were mixed together at 60 °C with a stir-bar for 30 min. Afterwards, the solution was introduced drop-wise with TEOS (0.4 M) for 2 h under stirring at 60 °C, followed by dialysis using MWCO 12–14 kDa membrane (Spectrum Laboratories, Inc., Rancho Dominguez, CA 90220, USA) against an acetic acid (2 M)/ethanol (1:1, *v/v*) solution and then with deH_2O at room temperature. A solid product was finally obtained by calcination at 500 °C for 24 h.

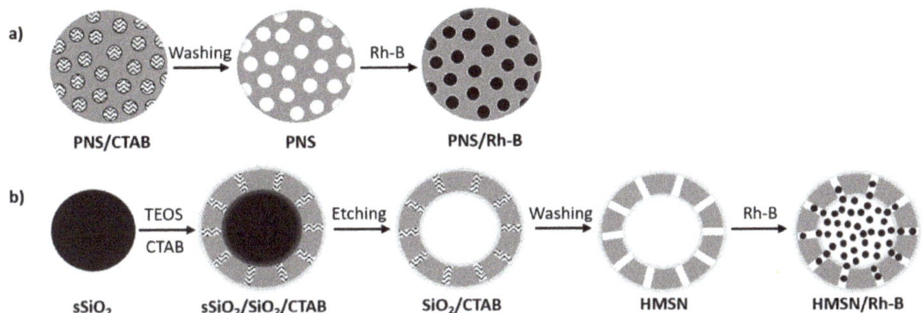

Figure 1. Schematic representation of the formation of porous nanosilica (PNS) (**a**) and hollow mesoporous silica nanoparticles (HMSNs) (**b**).

2.3. Preparation of HMSNs

HMSNs were prepared via a previously reported procedure with minor modification, including three main stages, as shown in Figure 1b [12,26]. The synthesis commenced with the preparation of hard templates of $sSiO_2$ spheres via the Stöber method with modifications [27]. In detail, ethanol (13.5 M), deH_2O (6.0 M), and ammonium hydroxide (0.38 M) were mixed together under constant stirring at 50 °C for 30 min. The resultant solution was then introduced with TEOS solution (0.29 M) at room temperature and subject to another 6h of stirring. Afterwards, dialysis with an MWCO 12–14 kDa membrane and freeze-drying took place to obtain the final solution. In the second step, a structure was formed by coating a thickness-controlled PNS layer on each $sSiO_2$ using CTAB as the organic co-templates [28]. The core@shell structure was denoted as $sSiO_2/SiO_2/CTAB$. Briefly, CTAB (0.05 M) was dissolved in deH_2O under magnetic stirring at room temperature for 15 min. $sSiO_2$ solution (5 mL) was added into the CTAB solution using a magnetic stir bar for 30 min at 50 °C, followed by the addition of a mixture of ethanol/ammonia solution (1.43:0.05, M/M). TEOS solution (0.27) was added into the above mixture under magnetic stirring at 50 °C for 6 h and the obtained solution was dialyzed against deH_2O (MWCO 12–14 kDa), which was changed 5 or 6 times per day, for 4 days. In the third step, the prepared $sSiO_2/SiO_2/CTAB$ and free CTAB were mixed with an aqueous Na_2CO_3 solution (0.2 M) for 9 h under stirring at 50 °C. HMSN spheres, obtained by freeze-drying the above mixture (EyeLA FDU-1200, −38.3 °C, 29.5 Pa), were soaked in acetic acid (2 M)/ethanol (1:1, *v/v*) solution over 24 h and had the CTAB template removed by washing with deH_2O several times.

2.4. Characterization

A Bruker D2 Phaser diffractometer (Germany) with Cu/Kα radiation at a scanning rate of 4°/min (λ = 0.154056 nm, 40 kV, 40 mA) was employed to perform XRD measurements. PNS, $sSiO_2$, $sSiO_2$/SiO_2/CTAB, and HMSNs were morphologically characterized and measured for size via TEM using JEM-1400 (JEOL, Tokyo, Japan) operating at an accelerating voltage of 300 kV. To prepare the sample for TEM analysis, one drop of each solution was added in deH_2O (1 mg/mL) and placed onto a carbon-copper grid (300-mesh, Ted Pella, Inc., Redding, CA, USA), which was then subject to air-drying for 10 min.

A TRISTAR 3000 instrument (Micromeritics, Norcross, GA, USA) operating at 77 K under continuous adsorption conditions was employed to measure N_2 adsorption–desorption isotherms. Before being analyzed, samples were subject to degassing at 103 Torr at 110 °C for 16 h. The size distribution of pores was derived via the Barrett–Joyner–Halenda method from the desorption branch of the nitrogen isotherm. The pore volume was taken at P/P_o-0.97 single point.

To perform TGA, a TGAnalyzer (Perkin Elmer Pyris 1, Perkin Elmer, Hopkinton, MA, USA) operating at a heating rate of 20 °C/min in a nitrogen flow from 100 °C to 800 °C was employed. The measurement of the surface charge of PNS and HMSN was performed at 37 °C using a Zetasizer Nano ZS (ZEN 3600, Malvern Instruments, Malvern, UK), equipped with a 532 nm wavelength. UV-Vis spectroscopy (Shimadzu, Kyoto, Japan) with a wavelength of 544 nm was used to determine the loading content of HMNS with RhB.

2.5. Preparation of RhB/PNS and RhB/HMSN

The equilibrium dialysis method, which was described previously, was used to load RhB into PNS and HMSN [1]. Briefly, solutions with either PNS or HMSN were created by adding 16 mg of the carrier into RhB solution (10 µg/mL). These solutions were stirred for 12 h to allow the encapsulation of RhB into the carriers. Unloaded RhB in the solutions was then removed by dialysis. In the dialysis solution, the content of unloaded RhB ($W_{U.RhB}$) was measured via the UV-Vis absorbance method, in which the standard curve was established by dissolving bare RhB in deH_2O to create multiple solutions with pre-specified concentrations (0–5 µg/mL). The recording of absorbance of the standard and dialysis solutions were performed with a V-750 UV/Vis spectrophotometer (Jasco Co., Tokyo, Japan). Based on the standard curve, which explains the relationship between RhB content and absorbance, and the absorbance results, the contents of RhB in the samples were determined. RhB loading efficiency (LE) and loading capacity (LC) were calculated as follows:

$$LE(\%) = \frac{\frac{(100 - W_{U.RhB})}{100}}{} \times 100$$
$$LC(\%) = \frac{(100 - W_{U.RhB})}{W_{PNS\,or\,HMSN} + 100 - W_{U.RhB}} \times 100$$

In which the 100 (mg) was the initial amount of RhB for loading experiments. $W_{U.RhB}$ and $W_{PNS\,or\,HMSN}$ are the content of unloaded RhB in the dialysis solution and the dried weight of the nanocarrier respectively.

2.6. Statistical Analysis

Microsoft® Excel was used to process the data. The results were expressed as mean ± standard deviation.

3. Result and Discussion

Figure 1 shows the synthesis of PNS and HMSN following a previously described procedure [26]. While Figure 1a represented the procedure of synthesizing PNS, Figure 1b illustrated three main steps for preparation of HMSN, including the preparations of the $sSiO_2$ core, mesoporous shell ($sSiO_2/SiO_2/CTAB$), and hollow core (HMSN). The successful preparation of PNS and HMSN was indicated by FTIR spectra shown in Figure 2. In Figure 2a, a band at 1640 cm^{-1} was attributed to the OH stretching vibration of water molecules present in PNS. Stretching vibrations of Si–O–Si, Si–OH, and Si–O could also be assigned to strong peaks at around 1089, 973, and 836 cm^{-1}, which are typical peaks of silica nanoparticles. The broad band observed at around 3444 cm^{-1} was associated with OH stretching frequency for the silanol group. These aforementioned peaks were also present in the spectrum of HMSN.

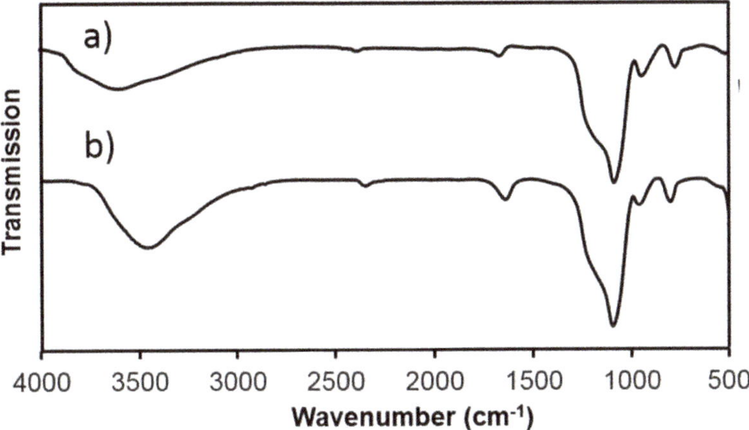

Figure 2. The FTIR spectra of PNS (**a**) and HMSN (**b**).

The TGA curves of PNS (a) and HMSN (b) are shown in Figure 3, in which the temperature was elevated to 800 °C at a constant rate of 10 °C/min. As high temperature leads to decomposition and vaporization of an organic fraction, inorganic remainders are often found at the end of the analysis. Generally, a pattern of mass depreciation could be observed for both PNS and HMSN. First, in the temperature range of lower than 200 °C, the removal of physically adsorbed water produced by hydrolysis of the precursors and the solvent employed during the synthesis took place. In the second range of 200–600 °C, residual organic product (CTAB) from the decomposition of the precursors was lost and the dehydroxylation of silanol (SiOH) groups from the silica surface occurred. In this phase, the weight loss of PNS was approximately 30%, whereas that of HMSN was only 13%, indicating that a small amount of CTAB may have remained in the HMSN nanoparticles.

The prepared nanoparticles were further characterized by TEM for morphological examination and size measurement. Figure 4 revealed the average diameter of PNS to be 63.4 ± 0.6 nm, and sSiO$_2$ to be 104.0 ± 0.7 nm (Figure 4a,b). As reported by Chen el al., the size of sSiO$_2$ was essential in influencing the size of hollow core [11]. Different hollow structures can be obtained in the sizing process of sSiO$_2$ by altering the reagent ratio in the initial stage. In the next step, by coating a mesoporous silica layer, the sSiO$_2$/SiO$_2$/CTAB and HMSN nanoparticles possessed spherical shapes, with significantly increased particle sizes, which were 167 ± 0.5 nm and 134 ± 0.3 nm for sSiO$_2$/SiO$_2$/CTAB and HMSN, respectively (Figure 4c). This could be explained by the binding of the outer PNS layer (Figure 1b), in which case the thickness and pore size of the mesoporous shell with the biodegradable property could be modified by using different organic solvents, adjusting concentration of TEOS, and reaction time [11]. In the last step aiming at selectively etching the solid silica core and protecting the integrity of silica shell, the absorption of positively charged CTAB (CTAB$^+$) onto the surface of sSiO$_2$/SiO$_2$/CTAB took place via electronic attraction to keep the shell intact. After commencing, this mechanism was stimulated by the Na$_2$CO$_3$ existing in the solution, CTAB in the PNS shell, and free CTAB in the media. Following the etching process, the acid acetic acid (2 M)/ethanol (1:1, *v/v*) solution was used to remove CTAB from the mesopores.

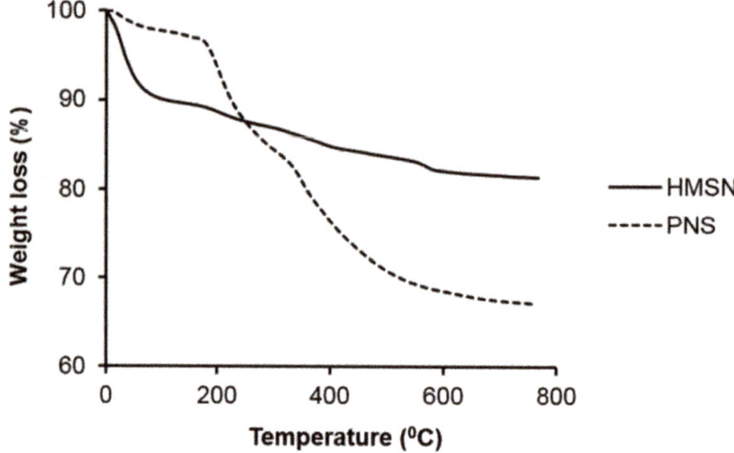

Figure 3. Thermogravimetry analysis for PNS (**a**) and HMSN (**b**).

In the blood circulation, it was suggested that the extravasation and renal clearance of nanoparticles of less than 10 nm in diameter are more expedited than those of larger nanoparticles (<200 nm), which are likely to be retained for a longer period [29,30]. Accordingly, the achieved size of synthesized HMSN products have the potential to allow the particles to bypass early clearance through kidney and vascular leakage.

Figure 5 shows the XRD patterns of PNS (a) and HMSN (b). As shown, the broad XRD reflection peaks of PNS (2θ = 21.1°) and HMSN (2θ = 23.0°) both indicate the amorphous nature of silica nanoparticles [31]. In addition, apart from a small amorphous bump at 37° in the PNS sample that could be attributed to trace CTAB left after calcination, no considerable difference was found between the two XRD patterns of PNS and HMSN, indicating that these nanoparticles likely have similar structures.

In addition, the porosities of the prepared PNS and HMSNs were evaluated by the N_2 adsorption-desorption technique. The results are presented in Figure 6b. It was noticed that the isotherms of both PNS and HMSN belong to the Langmuir type IV, characterized by a hysteresis loop, suggesting that both materials possess a mesoporous structure [32]. The hysteresis loops indicated the capillary condensation associated with mesopores, which occurred much more rapidly at higher pressure in PNS than in HMSN. Another difference between the materials is the type of the hysteresis loop, with PNS being H1, with the adsorption and desorption branches almost vertical and parallel at relative pressures of 0.9 to 1.0. This type of loop is associated with uniform particles and narrow distribution of pore size [33], while the H4 loop of HMSN—in which the adsorption and desorption branches remain nearly horizontal and parallel over a wide range of pressure—indicates irregular but parallel, open-ended, and slit-like pores, with connectivity between intragranular pores, and it indicates the existence of some micropores [33]. The N_2 adsorption percentage of HMSN over 60 min was almost 96%, which was higher than that of PNS (84%). This phenomenon could be explained by the hollow cavity of HMSN. The specific area and pore volume of HMSNs were 983.7 m^2/g and 2.33 cm^3/g, respectively. These figures were starkly contrasted by values found in PNS, which were 155.2 m^2/g and 8.75 cm^3/g, respectively, indicating that the prepared HMSN could be a potential carrier for bioactive molecules, with a high capacity. Having said that, further investigations on strategies to modify the surface of HMSN are still required to improve the controllability of this system, due to its similar nature to silica nanoparticles, whose availability for functionalization is abundant [4,34].

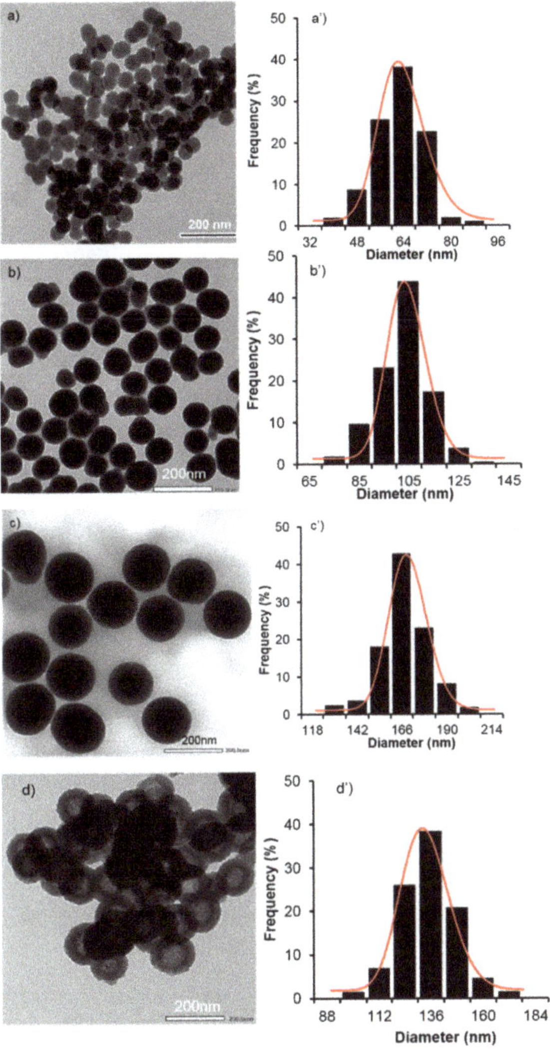

Figure 4. TEM images and the particle size distribution of PNS (**a**,**a'**), sSiO$_2$ (**b**,**b'**), sSiO$_2$/SiO$_2$/CTAB (**c**,**c'**), and HMSN (**d**,**d'**), respectively.

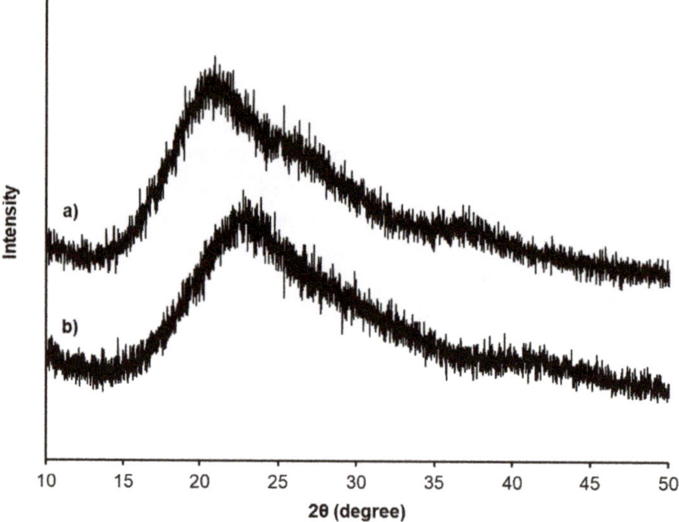

Figure 5. XRD patterns of PNS (**a**) and HMSN (**b**).

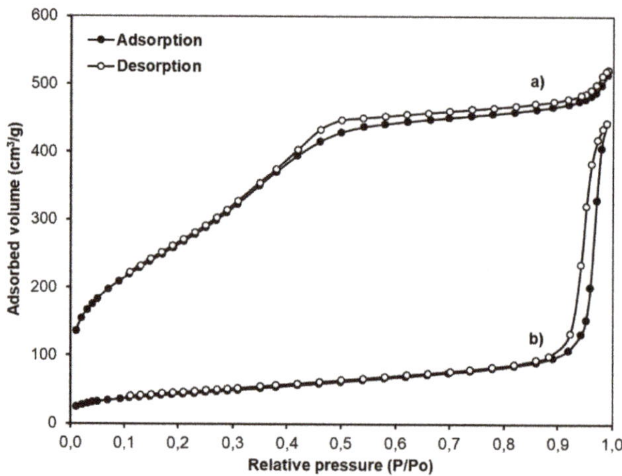

Figure 6. Nitrogen adsorption–desorption isotherms of HMSN (**a**) and PNS (**b**).

The surface charge of the prepared PNS, $sSiO_2$, $sSiO_2$/PNS(CTAB), and HMSN nanoparticles were characterized by zeta potential measurements. Results for the latter three are presented in Figure 7a. Not only does the zeta potential data expresses the stability of $sSiO_2$ nanoparticles' dispersion in water, the technique has also been employed to elucidate the adsorption mechanisms of drugs and biological ligands on the surface of $sSiO_2$ nanoparticles. It is stated that the absolute value of zeta potential is positively associated with the stability of colloidal systems. An absolute value of higher than 30 mV indicates a stable suspension [35]. The surface charges of the $sSiO_2$ and PNS were found to be −43.3 ± 0.87 mV and −44.53 ± 0.23 mV, respectively, in this study. The highly negative charges of $sSiO_2$ and PNS nanoparticles were attributed to the ionization of –OH on the surfaces of silica in the ethanol solutions. However, the zeta potential of $sSiO_2$/PNS(CTAB) increased dramatically to 32.5 ± 0.66 mV, for the adsorption of CTA cations (CTA^+) in CTAB on the $sSiO_2$ surface, via electronic attraction. By contrast, there was then a steep decrease in the zeta potential of HMSN, reaching

−25.45 ± 0.07 mV, different to sSiO$_2$/PNS(CTAB) due to the absence of CTA$^+$ after the etching process. Moreover, an investigation into the stability of the particles in aqueous solution is also presented in Figure 7b, which demonstrated a stable surface charge after 48 h. These results suggest a potential application of PNS and HMSN nanoparticles for drug delivery systems with high stability.

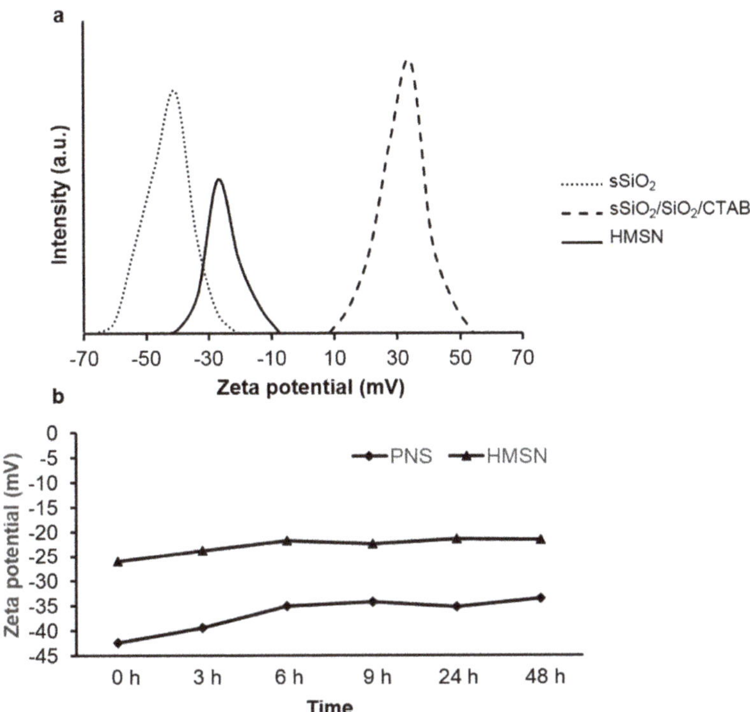

Figure 7. (**a**) Zeta potential of sSiO$_2$, sSiO$_2$/SiO$_2$/CTAB, and HMSN; (**b**) Zeta potential of PNS and HMSN in solutions over the course of 48 h.

Drug loading efficiency and capacity are considered to be essential in any drug delivery system design, since they play a direct role in the therapeutic activity of the system [36]. A high loading capacity is also desirable for our rationale of minimizing the dose-dependent toxicity of the delivery system. In our case, the LE, indirectly derived from the content of non-encapsulated RhB, of PNS and HMSN were 12.24% and 51.67%, respectively. The lower RhB LE in RhB/PNS nanocarriers was attributed to their non-hollow structure. In contrast, RhB/HMSN nanocarriers possessed a significantly higher LE due to the hollow nature of the HMSN's cavity, leading to a significantly improved drug loading capacity. Similarly, RhB/HMSN also showed a higher LC (10.44%) compared to RhB/PNS (2.9%). Therefore, it is suggested that the introduction of a hollow form may improve the LE and LC of HMSN nanoparticles, assisting the tumor-targeted delivery and therapeutic result of a drug that is loaded into HMSN nanocarriers.

4. Conclusions

The present study has reported a successful preparation attempt of PNS and HMSN nanoparticles via sol-gel method. The formed PNS and HMSN existed in spherical shapes with average particle sizes of 63.4 ± 0.6 nm and 134.0 ± 0.3 nm, and were proved to be potential carriers for anticancer drugs. The encapsulation of RhB into the HMSN nanoparticles achieved the LE of 51.67%. The high LE achieved could be a contributing factor to the reduction of dose-dependent toxicity in drug delivery with nanocarriers. However, further studies are required to further improve the HMSN system, especially regarding surface modification for stealth delivery, and to offer more sustainability in cargo release from this system.

Author Contributions: Conceptualization, investigation, and data analysis, N.-T.N.-T.; data curation, writing—original draft preparation, and visualization, L.P.P.T. and N.T.T.L.; writing—reviewing and editing, M.-T.C., T.-N.T., N.T.N., C.H.N.; supervision and project administration, D.-H.N., V.T.T., Q.T.L., and N.Q.T.

Funding: This research was funded by Tra Vinh University (grant number 177/HD.HDKH-DHTV).

Conflicts of Interest: The authors declare no conflict of interest.

References

1. Vo, U.V.; Nguyen, C.K.; Nguyen, V.C.; Tran, T.V.; Thi, B.Y.T.; Nguyen, D.H. Gelatin-poly (ethylene glycol) methyl ether-functionalized porous Nanosilica for efficient doxorubicin delivery. *J. Polym. Res.* **2019**, *26*, 6. [CrossRef]

2. Lu, J.; Liong, M.; Zink, J.; Tamanoi, F. Mesoporous Silica Nanoparticles as a Delivery System for Hydrophobic Anticancer Drugs. *Small* **2007**, *3*, 1341–1346. [CrossRef]

3. Slowing, I.I.; Trewyn, B.; Giri, S.; Lin, V.S.Y. Mesoporous silica nanoparticles for drug delivery and biosensing applications. *Adv. Funct. Mater.* **2007**, *17*, 1225–1236. [CrossRef]

4. Thi, T.T.H.; Cao, V.D.; Nguyen, T.N.Q.; Hoang, D.T.; Ngo, V.C.; Nguyen, D.H. Functionalized mesoporous silica nanoparticles and biomedical applications. *Mater. Sci. Eng. C* **2019**, *99*, 631–656.

5. Bao, B.Q.; Le, N.H.; Nguyen, D.H.T.; Tran, T.V.; Pham, L.P.T.; Bach, L.G.; Ho, H.M.; Nguyen, T.H.; Nguyen, D.H. Evolution and present scenario of multifunctionalized mesoporous nanosilica platform: A mini review. *Mater. Sci. Eng. C* **2018**, *91*, 912–928. [CrossRef] [PubMed]

6. Slowing, I.I.; Vivero-Escoto, J.; Wu, C.W.; Lin, V.S.Y. Mesoporous silica nanoparticles as controlled release drug delivery and gene transfection carriers. *Adv. Drug Deliv. Rev.* **2008**, *60*, 1278–1288. [CrossRef] [PubMed]

7. Tang, F.; Li, L.; Chen, D. Mesoporous silica nanoparticles: Synthesis, biocompatibility and drug delivery. *Adv. Mater.* **2012**, *24*, 1504–1534. [CrossRef] [PubMed]

8. Chauhan, S.; Manivasagam, G.; Kumar, P.; Ambasta, R.K. Cellular Toxicity of Mesoporous Silica Nanoparticle in SHSY5Y and BMMNCs Cell. *Pharm. Nanotechnol.* **2018**, *6*, 245–252. [CrossRef] [PubMed]

9. Murugadoss, S.; Lison, D.; Godderis, L.; Brule, S.V.D.; Mast, J.; Brassinne, F.; Sebaihi, N.; Hoet, P.H. Toxicology of silica nanoparticles: An update. *Arch. Toxicol.* **2017**, *91*, 2967–3010. [CrossRef] [PubMed]

10. Tsou, C.J.; Hung, Y.; Mou, C.Y. Hollow mesoporous silica nanoparticles with tunable shell thickness and pore size distribution for application as broad-ranging pH nanosensor. *Microporous Mesoporous Mater.* **2014**, *190*, 181–188. [CrossRef]

11. Chen, F.; Hong, H.; Shi, S.; Goel, S.; Valdovinos, H.F.; Hernandez, R.; Theuer, C.P.; Barnhart, T.E.; Cai, W. Engineering of Hollow Mesoporous Silica Nanoparticles for Remarkably Enhanced Tumor Active Targeting Efficacy. *Sci. Rep.* **2014**, *4*, 5080. [CrossRef] [PubMed]

12. Farsangi, Z.J.; Rezayat, S.M.; Beitollahi, A.; Sarkar, S.; Jaafari, M.; Amani, A. Hollow Mesoporous Silica Nanoparticles (HMSNs) Synthesis and in vitro Evaluation of Cisplatin Delivery. *J. Nanoanal.* **2016**, *3*, 120–127.

13. Ghasemi, S.; Farsangi, Z.; Beitollahi, A.; Mirkazemi, S.M.; Rezayat, M.; Sarkar, S. Synthesis of Hollow Mesoporous Silica (HMS) nanoparticles as a candidate for sulfasalazine drug loading. *Ceram. Int.* **2017**, *43*, 11225–11232. [CrossRef]

14. Zhu, Y.; Fang, Y.; Borchardt, L.; Kaskel, S. PEGylated hollow mesoporous silica nanoparticles as potential drug delivery vehicles. *Microporous Mesoporous Mater.* **2011**, *141*, 199–206. [CrossRef]

15. Shi, S.; Chen, F.; Cai, W. Biomedical applications of functionalized hollow mesoporous silica nanoparticles: Focusing on molecular imaging. *Nanomedicine* **2013**, *8*, 2027–2039. [CrossRef]

16. Zhu, Y.; Shi, J.; Shen, W.; Dong, X.; Feng, J.; Ruan, M.; Li, Y. Stimuli-Responsive Controlled Drug Release from a Hollow Mesoporous Silica Sphere/Polyelectrolyte Multilayer Core–Shell Structure. *Angew. Chem. Int. Edit.* **2005**, *44*, 5213–5217. [CrossRef]

17. Mei, X.; Yang, S.; Chen, D.; Li, N.; Li, H.; Xu, Q.; Ge, J.; Lu, J. Light-triggered reversible assemblies of azobenzene-containing amphiphilic copolymer with β-cyclodextrin-modified hollow mesoporous silica nanoparticles for controlled drug release. *Chem. Commun.* **2012**, *48*, 10010–10012. [CrossRef]

18. Ma, X.; Zhao, Y.; Ng, K.W.; Zhao, Y. Integrated hollow mesoporous silica nanoparticles for target drug/siRNA co-delivery. *Chem. A Eur. J.* **2013**, *19*, 15593–15603. [CrossRef]

19. Yang, Y.; Guo, X.; Wei, K.; Wang, L.; Yang, D.; Lai, L.; Cheng, M.; Liu, Q. Synthesis and drug-loading properties of folic acid-modified superparamagnetic Fe_3O_4 hollow microsphere core/mesoporous SiO_2 shell composite particles. *J. Nanopart. Res.* **2014**, *16*, 2210. [CrossRef]

20. Patil, P.B.; Karade, V.; Waifalkar, P.P.; Sahoo, S.C.; Kollu, P.; Nimbalkar, M.; Chougale, A.D.; Patil, P.S. Functionalization of magnetic hollow spheres with 3-Aminopropyl) triethoxysilane (APTES) for controlled drug release. *IEEE Trans. Magn.* **2017**, *53*, 1–4. [CrossRef]

21. Vo, U.V.; Tran, T.V.; Nguyen, D.M.T.; Nguyen, C.K.; Thi, N.T.N.; Nguyen, D.H. Effective pH-responsive hydrazine-modified silica for doxorubicin delivery. *Asian J. Med. Health* **2017**, *4*, 1–7. [CrossRef]

22. Liu, J.; Luo, Z.; Zhang, J.; Luo, T.; Zhou, J.; Zhao, X.; Cai, K. Hollow mesoporous silica nanoparticles facilitated drug delivery via cascade pH stimuli in tumor microenvironment for tumor therapy. *Biomaterials* **2016**, *83*, 51–65. [CrossRef] [PubMed]

23. Zhao, Q.; Wang, S.; Yang, Y.; Li, X.; Di, D.; Zhang, C.; Jiang, T.; Wang, S. Hyaluronic acid and carbon dots-gated hollow mesoporous silica for redox and enzyme-triggered targeted drug delivery and bioimaging. *Mater. Sci. Eng. C* **2017**, *78*, 475–484. [CrossRef] [PubMed]

24. Wang, T.; Ma, W.; Shangguan, J.; Jiang, W.; Zhong, Q. Controllable synthesis of hollow mesoporous silica spheres and application as support of nano-gold. *J. Solid State Chem.* **2014**, *215*, 67–73. [CrossRef]

25. Chen, Y.; Xu, P.; Chen, H.; Li, Y.; Bu, W.; Shu, Z.; Li, Y.; Zhang, J.; Zhang, L.; Pan, L.; et al. Colloidal HPMO Nanoparticles: Silica-Etching Chemistry Tailoring, Topological Transformation, and Nano-Biomedical Applications. *Adv. Mater.* **2013**, *25*, 3100–3105. [CrossRef]

26. Chen, Y. *Design, Synthesis, Multifunctionalization and Biomedical Applications of Multifunctional Mesoporous Silica-Based Drug Delivery Nanosystems*; Springer: Berlin/Heidelberg, Germany, 2016.

27. Masalov, V.M.; Sukhinina, N.S.; Kudrenko, E.; Emelchenko, G.A. Mechanism of formation and nanostructure of Stöber silica particles. *Nanotechnology* **2011**, *22*, 275718. [CrossRef]

28. Zhou, X.; Cheng, X.; Feng, W.; Qiu, K.; Chen, L.; Nie, W.; Yin, Z.; Mo, X.; Wang, H.; He, C. Synthesis of hollow mesoporous silica nanoparticles with tunable shell thickness and pore size using amphiphilic block copolymers as core templates. *Dalton Trans.* **2014**, *43*, 11834. [CrossRef]

29. Nguyen, D.H.; Lee, J.S.; Choi, J.H.; Lee, Y.; Son, J.Y.; Bae, J.W.; Lee, K.; Park, K.D. Heparin nanogel-containing liposomes for intracellular RNase delivery. *Macromol. Res.* **2015**, *23*, 765–769. [CrossRef]

30. Tran, D.H.N.; Nguyen, T.H.; Vo, T.N.N.; Pham, L.P.T.; Vo, D.M.H.; Nguyen, C.K.; Bach, L.G.; Nguyen, D.H. Self-assembled poly(ethylene glycol) methyl ether-grafted gelatin nanogels for efficient delivery of curcumin in cancer treatment. *J. Appl. Polym. Sci.* **2019**, *136*, 20. [CrossRef]

31. Nguyen, T.N.; Huynh, T.N.; Hoang, D.; Nguyen, D.H.; Nguyen, Q.H.; Tran, T.H. Functional Nanostructured Oligochitosan–Silica/Carboxymethyl Cellulose Hybrid Materials: Synthesis and Investigation of Their Antifungal Abilities. *Polymers* **2019**, *11*, 628. [CrossRef]

32. Sing, K.S.W.; Everett, D.H.; Haul, R.A.W.; Moscou, L.; Pierotti, R.A.; Rouquerol, J.; Siemieniewska, T. Reporting Physisorption Data for Gas/Solid Systems. *Int. Union Pure Appl. Chem.* **1985**, *57*, 603–619. [CrossRef]

33. Wang, W.; Liu, P.; Zhang, M.; Hu, J.; Xing, F. The Pore Structure of Phosphoaluminate Cement. *Open J. Compos. Mater.* **2012**, *2*, 104–112. [CrossRef]

34. Nguyen, A.K.; Nguyen, T.H.; Bao, B.Q.; Bach, L.G.; Nguyen, D.H. Efficient Self-Assembly of mPEG End-Capped Porous Silica as a Redox-Sensitive Nanocarrier for Controlled Doxorubicin Delivery. *Int. J. Biomater.* **2018**, *2018*, 1575438. [CrossRef] [PubMed]

35. Blanco, E.; Shen, H.; Ferrari, M. Principles of nanoparticle design for overcoming biological barriers to drug delivery. *Nat. Biotechnol.* **2015**, *33*, 941–951. [CrossRef]

36. Shen, S.; Wu, Y.; Liu, Y.; Wu, D. High drug-loading nanomedicines: Progress, current status, and prospects. *Int. J. Nanomed.* **2017**, *12*, 4085–4109. [CrossRef]